Principles and Techniques of Electron Microscopy

Principles and Techniques of Electron Microscopy

Biological Applications

M. A. Hayat
Professor of Biology
Kean College of New Jersey, Union, New Jersey

THIRD EDITION

MACMILLAN
PRESS
Scientific & Medical

Second edition published 1981 by Aspen Publishers Inc.,
Rockville, Maryland
Third edition 1989

Published by
THE MACMILLAN PRESS LTD
Houndmills, Basingstoke, Hampshire RG21 2XS
and London
Companies and representatives
throughout the world

Printed in Hong Kong

ISBN 0–333–45294–1

Contents

Preface

The primary objective of the first and second editions of this book was to provide the reader with a foundation in the biochemical concepts governing preparatory procedures for transmission electron microscopy. The purpose of this revised edition remains the same.

At the outset it must be pointed out that progress in preparatory procedures to obtain optimal preservation of biological specimens has fallen far behind improvement in instrumentation. Unfortunately, the problems of biological specimen preparation have not been attacked with the same converging intensity afforded the improvement of resolving power. So, methodology is a major constraint in obtaining more detailed and accurate information on cell ultrastructure. A case in point is the development of the high-resolution scanning transmission electron microscope (STEM), which has made visualisation of single heavy atoms a reality. Spatial resolution of 0.2 nm is now easily achievable for inorganic materials, whereas spatial information obtained for biological specimens is generally no better than 1.5–2.0 nm

However, the situation is not quite as discouraging as it may appear, for the removal of the 1.5 nm barrier in biological electron microscopy to study the finest details of macromolecular and molecular organisation remains a distinct possibility. Exploration has begun into several avenues to overcome this obstacle. One such avenue is the full utilisation of the STEM, which can provide superior contrast and many types of information simultaneously. This instrument can provide a higher resolution with biological specimens. Another avenue is the use of dark field in the conventional TEM, thus eliminating the need for staining.

Although preparatory methods can be successfully completed from a recipe, real understanding comes from a knowledge of the fundamentals of chemicals. This book discusses in detail the chemistry of the interactions of various reagents with cellular substances. Such information helps in better understanding an electron micrograph of a specimen which has been subjected to fixation, dehydration, embedding, sectioning, staining, vacuum and electron bombardment.

An electron micrograph is an image of a specimen altered compared with its living state. The degree of alteration shown by a specimen apparently varies according to the preparatory procedure used. Since one examines an altered specimen out of necessity, it is important to understand the process that causes the alterations. The range of effects produced by various treatments is significant, and understanding them is essential to a successful extrapolation of what is shown by the electron micrograph to what must have existed *in vivo*. As long as unavoidable artefacts are interpretable, they can be accepted. It is necessary that an electron micrograph be interpreted with respect to the treatments that the specimen has undergone. Such an approach will result in more accurate interpretation of the image shown by an electron micrograph.

During the last decade, while the basic principles have remained unchanged, significant advances have taken place in methodology for both transmission and scanning electron microscopy. The availability of increased information on the chemistry of procedures has necessitated technical modifications. Some of the established procedures have become obsolete. Examples are the undesirability of using $KMnO_4$ as a routine fixative, and the inhibitory effect of uranyl salts, when used *en bloc*, on the glycogen staining. Specimen processing has been simplified, and specific and better contrast-enhancing stains have been introduced. The development of colloidal gold methodology as a label has proven to be of a profound importance in the field of immunocytochemistry. The preservation and staining of lipids with imidazole, *p*-phenylenediamine or malachite green is significant. These advances are included in this new edition. Outdated materials and cumbersome tables and structural formulae have been deleted. Owing to the limited space available, specific preparatory methods could not be included, which, however, have been presented in *Basic Techniques for Transmission Electron Microscopy*, published by Academic Press.

As stated above, some progress has taken place in the understanding of biochemical events that occur during the processing of specimens. For example, the affinity of the PTA reaction with proteins and carbohydrates depends upon the pH used, and the nature of the Schiff bases introduced by glutaraldehyde into the specimen is temporary. An understanding of the interactions involved among the specimens, stain and support film during electron bombardment is also beginning to emerge. The problem of radiation sensitivity of biological specimens, especially in high-resolution work, has been recognised. The limitation presented by support films in the form of background noise in the investigation of the finest details of macromolecular and molecular organisation is now understood. The advantages and disadvantages of at least some preparatory procedures have become known. As a result, it can now be stated that the preparatory procedures stand on firmer ground.

Two new chapters, 'Negative Staining' and 'Low-temperature Methods', are added to this edition. Negative staining is the most important method for visualising the topography and structure of viruses, isolated cell organelles and macromolecules. It is one of the easiest and fastest methods for detecting the viruses. By using immunoelectron microscopy in conjunction with negative staining, viruses of similar ultrastructure can be differentiated with great sensitivity. Immunogold labelling in combination with negative staining has considerable potential as a means of detecting viruses, and viral antigens and antibodies. Colloidal gold marker is especially useful for detecting very small viruses such as picornaviruses and parvoviruses, as well as low levels of viruses.

Cryotechniques, including ultrarapid cooling of cells and tissue, are discussed in detail. This approach is a useful alternative to chemical fixation. The aim of cryotechniques is to stabilise the ultrastructure as it exists *in vivo*. However, this feat is not easy to accomplish. The specimen must undergo alterations to adjust to the changing temperature. Such alterations occur primarily during the interval between the physiological and the immobilisation temperatures. The topics discussed in this chapter were selected for routine use. Esoteric techniques requiring complex and expensive instruments were not included. An example of such techniques is cryoelectron microscopy of vitrified specimens.

The methods have been presented in a self-explanatory form so that the reader can practise them without outside help. Not only are practical instructions on how to process biological specimens provided, but also is included a detailed discussion on the principles underlying the various processes. Alternative procedures and points of disagreement have been presented to help the reader interpret data accurately. The approach taken in this book may contribute to the development and standardisation of methodology. Most of the methods given can be carried out successfully in practical classes. The procedures described have been found to be the best and most reliable. Complete descriptions of the preparation of buffers and fixative solutions are included. It is suggested that before any processing is undertaken, the entire procedure should be read, instruments should be checked, and necessary solutions and media should be prepared. Full author and subject indexes and references with complete titles are provided.

This book is intended not only for students and teachers, but also for technicians and research workers not familiar with the techniques. It is addressed primarily to those interested in electron microscopy, but certain topics are also helpful to those involved in the techniques for light microscopy. The reader is presented with what is known for certain, as well as what gaps exist in our knowledge. Potential research areas have been pointed out wherever possible. It is almost impossible to discuss all the techniques employed in the elucidation of structure in a single volume. Eight other volumes, under the same main title, edited by the author have been published by Van Nostrand Reinhold Company. *Correlative Microscopy*, edited by the author, has just been published by Academic Press. A two-volume series on *Colloidal Gold: Principles, Methods and Applications* is in preparation, and will be published by the same publisher.

Many eminent colleagues have read various sections of this volume, which has resulted in an improved text. Regretfully, it is not practical to acknowledge them individually, because of their large number. I deeply appreciate their help.

Union, New Jersey, 1989 M.A.H.

1 Chemical Fixation

INTRODUCTION

Chemical fixation is the most widely used method for preserving biological specimens for transmission electron microscopy (TEM). Some of the reasons for its universal use are the adequate preservation of many cellular components, including enzymes; the clarity of structural details shown by electron micrographs; and the ease of application to both prokaryotes and eukaryotes. No other fixation method at present available can claim these advantages. Cryofixation is a useful adjunct to chemical fixation.

The goals of fixation are to preserve the structure of cells with minimum alteration from the living state with regard to volume, morphology and spatial relationships of organelles and macromolecules, minimum loss of tissue constituents, and protection of specimens against subsequent treatments, including rinsing, dehydration, staining, vacuum and exposure to the electron beam. Fixed specimens show less shrinkage in the last steps of dehydration as well as during infiltration and embedding. Fixation facilitates transmembranous diffusion of substrates and the trapping agents in enzyme cytochemical studies. For these types of studies, fixation for a short period of time accomplishes a moderate cross-linking of proteins. The role of fixation in enzyme cytochemistry has been presented elsewhere (Hayat, 1981, 1986a). Prefixation with aldehydes prevents loss of proteins from specimens prepared for negative staining. In addition, prior fixation with an aldehyde minimises the possibility of artefacts in specimens treated with cryoprotectants.

Ideally, the aim of a desirable fixation method is satisfactory preservation of the cell as a whole and not the best preservation of only a part of it. In practice, however, fixation is usually of selective rather than of general effectiveness, in the sense that the objective of the study determines the type of fixation method used. For example, in studying lipids, one would select OsO_4 or glutaraldehyde–phenylenediamine instead of glutaraldehyde alone, while for the study of distribution and translocation of water-soluble substances, one would select the freeze-drying method in preference to the use of OsO_4 solution. However, cryofixation is the best method for preserving these substances. Various methods employed for freezing cells and tissues for TEM are presented in Chapter 7.

In the case of certain studies, the best approach would be to eliminate conventional fixation altogether. One such approach utilises the critical point drying method (Anderson, 1951), which is useful for electron microscopy of whole-mounted cells. The advantage of this method is that specimens are preserved in three dimensions with their natural contrast without sectioning. Cell skeleton has been extensively studied by the critical point drying method or the freeze-drying method. Methodology of critical point drying for TEM (Hayat and Zirkin, 1973) and scanning electron microscopy (SEM) (Hayat, 1978) has been presented.

An essential effect of fixation is the separation of the liquid phase from the solid phase of the protoplasm. The degree of smoothness and rapidity in this separation depends upon the type of fixation method used. As a general rule, the freeze-drying method achieves a more rapid and smooth separation than that obtained by chemical fixation. The net physical dislocation of the solids after fixation may or may not be visible. Nevertheless, even in tissues fixed with buffered glutaraldehyde and OsO_4, some movement of solids does take place. Most other changes in the properties of protoplasm are due to the chemical reaction of the fixative with various organic substances, especially active groups of protoplasmic proteins (e.g. amine) and lipids.

Before embedding in water-insoluble resins (e.g. Epon), free water present in the tissue must be replaced by a solvent during dehydration. Since important chemical bonds in living tissues are dependent for their stability upon the presence of water, a fixative should provide more stable bonds which will hold the molecules together during dehydration and subsequent treatments so that they will not be translocated or extracted. This is accomplished primarily by the formation of inter- and intra-molecular cross-links of proteins. In general, fixatives form cross-links not only between their reactive groups and the reactive groups in the tissue, but also between different reactive groups in the tissue. It is recognised that chemical fixation unmasks or frees certain reactive groups in the tissue for cross-linkage which otherwise may not show intermolecular bonding.

As stated above, an important feature of fixation is a change in the appearance and 'nature' of cellular proteins. This change in the physical configuration of protein molecules is called denaturation. Denaturation is followed by loss of many properties (e.g. solubility, specific gravity and crystallisation) of the protein molecule. Some of these changes are responsible, in part, for the stabilisation of proteins. For example, the denatured protein is less soluble in aqueous solutions, since during denaturation, the unfolding of proteins exposes hydrophobic groups which repel water. Denatured proteins are also less susceptible to precipitation than those in the native state.

Although in many cases the denaturation is essentially irreversible, it may also be temporary. The permanency of denaturation depends upon the type and concentration of the fixative used, as well as upon the duration of fixation. In the case of temporary or partial denaturation, protein subsequently either reverts to its original state or remains unstable until completely fixed. The unstable state of the protein is an obstacle in achieving satisfactory preservation. On the other hand, fully denatured proteins become coagulated, and the changes involved are irreversible. Permanently denatured proteins are generally well-stabilised and are less susceptible to extraction during dehydration than those denatured temporarily. For example, the protein of egg white (albumin) is globular and water-soluble in the native state, but becomes fibrous and water-insoluble in the denatured state.

Fixatives differ in their effects on the optically homogeneous ground substance of the living protoplasm.

The reagents that coagulate proteins into an opaque mixture of granular or reticular solids suspended in fluids are called coagulant fixatives (e.g. ethanol). It must not be thought that coagulated fixatives flocculate all proteins; for instance, ethanol does not flocculate nucleoproteins. Essentially, coagulant fixatives permanently flocculate molecules of most of the protoplasmic proteins, and thus cause a considerable change in protein structure. This results in distortion of the fine structure. The coagulant fixatives are, therefore, unfit for use in electron microscopy. Most of the fixatives of this type are also called non-additive, for they fix proteins without becoming a part of them.

In contrast, non-coagulant fixatives transform proteins into a transparent gel. This results in the stabilisation of proteins without much structural distortion of the original state. Non-coagulant fixatives cause very little dissociation of protein from water, and proteins retain at least some of their reactive groups. Another desirable effect of these fixatives is that they can render some proteins non-coagulable by subsequently used coagulant reagents including fixatives. The importance of this change becomes apparent when the fixed tissue is subsequently dehydrated with ethanol. Glutaraldehyde, a non-coagulant, has become the most widely used pre-fixative because of its capacity to stabilise most proteins without coagulation. Other commonly used non-coagulants are OsO_4, acrolein and formaldehyde. On the basis of their chemical affinity for cellular proteins, these fixatives are also called additive fixatives, for they chemically become part of the proteins they fix. These fixatives may also add themselves onto cell constituents other than proteins; for example, OsO_4 stabilises certain lipids by additive fixation. Some of these fixatives (e.g. OsO_4) are heavy metals and as a result impart density to the tissues they fix.

Even a small tissue block (1 mm^3) shows inhomogeneous volume changes during fixation. In other words, it does not show the same volume change throughout its length and width. This can be explained on the basis of uneven volume changes in at least three different regions of the tissue block. Even within a superficially uniform tissue such as liver, various compartments (cellular, interstitial and luminal) exhibit variation in volume changes (Bertram *et al.*, 1986). Furthermore, some tissue blocks may shrink and others may swell during processing. The extent of shrinkage or swelling under a standardised method of fixation and embedding is related to the tissue composition. Such volume variations are a problem in stereological studies.

Fixation tends to bring dimensional changes in the intercellular space. Various tissues respond differently in this respect to the fixative used. The dimensions of the intercellular space in the central nervous tissue are especially sensitive to the method of fixation. Since relatively large intercellular spaces in this tissue are preserved by rapid freezing and since cell structure preserved by such techniques is considered to be a more faithful representation of the *in vivo* structure, fixatives which preserve equally large intercellular spaces are considered reliable for this tissue. The magnitude of the space is also affected by the solvent used for dehydration. The use of ethanol may result in a somewhat larger space than that obtained after dehydration with acetone.

Studies by Van Harreveld and Khattab (1968, 1969), however, indicate that drastic changes in the fluid distribution of central nervous tissue during fixation preclude any direct relationship between the intercellular spaces in the living and fixed tissues. Although the exact nature and dimensions of intercellular spaces present in living tissue remains uncertain, the artefactual effect of fixation on dimensions of the intercellular space is obvious. A primary fixation by perfusion with slightly hypertonic glutaraldehyde followed by osmication seems to produce a more accurate perservation of brain ultrastructure. For additional information on the

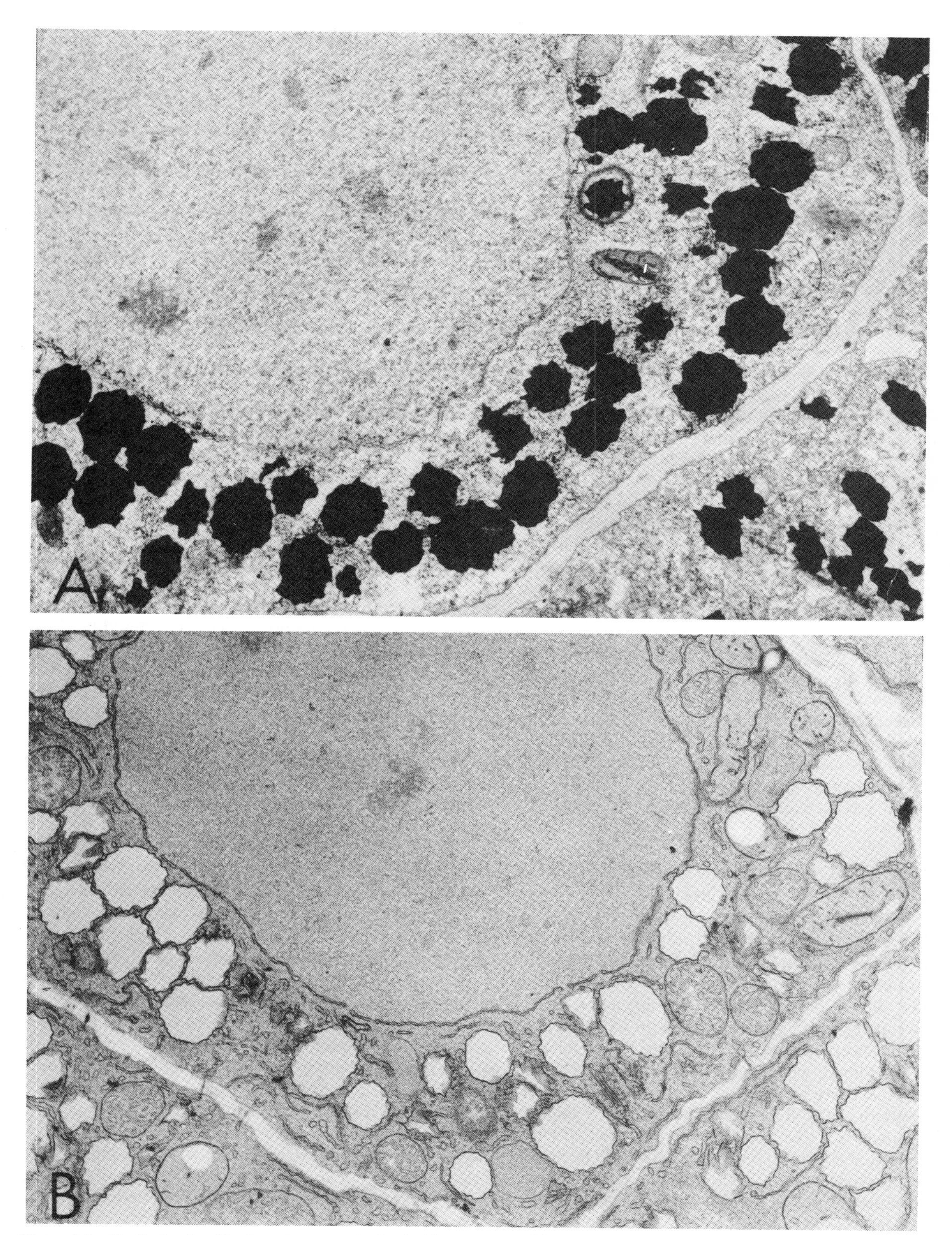

Figure 1.1 Cortical cells of leafy spurge root tip fixed with two types of fixation methods. A: Fixation with glutaraldehyde and OsO_4 results in good preservation of cellular structure including unsaturated lipids (black droplets). B: Fixation with $KMnO_4$ and OsO_4 results in the loss of unsaturated lipids (empty-looking structures). These lipids are extracted during pre-fixation with $KMnO_4$

effect of different fixation methods on the dimension of intercellular spaces in this tissue, the reader is referred to Hayat (1981).

Since the extraction phenomenon greatly influences the contrast and general appearance of an electron micrograph, increasing importance is being given to the problem of extractability or solubility of tissue constituents during and after fixation. Experimental evidence indicates that various amounts of carbohydrates, proteins, lipoproteins, nucleic acids and lipids are lost during fixation. For example, a loss as high as 25% of phospholipids from amoeba has been reported following use of various fixatives and dehydration solutions (Korn and Weisman, 1966). The lipid loss appears to be related to the degree of saturation of fatty acids present in the lipid fraction of the tissue. The best method for preserving phospholipids and triglycerides is ultrarapid cryofixation followed by freeze-drying (Hagler and Buja, 1986).

Among various fixatives currently in use, a double fixation with glutaraldehyde followed by OsO_4 is the most effective in reducing the loss of cell constituents (Fig. 1.1). These two fixatives are employed in order to stabilise the maximum number of different types of molecules. It should be noted, however, that information concerning the interaction between the heavy metal and the organic fixative is insufficient, and the resultant changes in the reactivity and structure of cell components are not fully known. It is apparent that at present the information on the chemical reactions between the fixative solution and specific cellular constituents is inadequate. A correct interpretation of the electron micrographs is dependent upon a better understanding of the chemistry of fixation.

REAGENTS: HAZARDS, PRECAUTIONS AND SAFE HANDLING

Almost all the reagents used in processing specimens for electron microscopy are potentially hazardous to various degrees. Many chemicals employed for fixation, rinsing, dehydration, embedding and staining are potentially capable of causing harm to workers.They can be absorbed either by skin contact or by inhalation. A fume-hood, which is checked periodically to ensure that fumes are sufficiently removed, is essential in the laboratory. The fume-hood should have a capacity of maintaining a face velocity (the velocity of air entering a hood) of about 100 ft/min. Hands should be protected by disposable gloves made of impermeable material. Polyethylene gloves are the most resistant to penetration by reagents such as resins. Latex gloves are less resistant, while vinyl gloves are the least resistant. Eyes should be covered with gas-tight goggles.

One of the most commonly used buffers, sodium cacodylate, contains ~30% arsenic by weight. This arsenic presents hazards of ingestion, contact and inhalation. Chronic exposure to arsenic can cause contact dermatitis, inflammation of mucous membranes, kidney and liver inflammation, nerve paralysis, and cancer of liver, skin and lung. Eyes should be protected and a fume-hood should be used for weighing this reagent and preparing the buffer solution. To avoid the production of arsenic gas, this reagent should not come in contact with acids. Since cacodylate buffer reacts with H_2S, the two should not be used in the same solution. Cacodylate solutions should not be poured down the sink, since they can react with drain cleaners and other chemicals to produce toxic arsenic gas. Sodium barbitone (Veronal) buffer is a poison if taken by mouth, while *s*-collidine buffer is toxic and foul-smelling and therefore should be used in a fume-hood.

OsO_4 volatilises readily at room temperature. It is dangerous because of its toxicity and vapour pressure. Its fumes are injurious to the nose, eyes and throat. Hands or any other part of the body must not be exposed to this reagent. OsO_4 must be handled at all times in a fume-hood, and gas-tight goggles should be worn when handling it (see also p. 55). All aldehydes are hazardous to various degrees and should be handled in a fume-hood; breathing of aldehyde vapours is dangerous. Aldehyde solutions may cause dermatitis if permitted to wet the skin. Spillages of aldehyde can be inactivated by using Fehling's solution, which changes its colour from blue to red. Glutaraldehyde spills can be neutralised by solid glycine. Alternatively, a bottle containing 1 M glycine can be kept on hand to spray on the spill. One volume of this glycine renders an equivalent volume of 50% (or less) glutaraldehyde non-volatile.

A carcinogen can be formed spontaneously in air when formaldehyde and HCl vapours mix. Formaldehyde has been reported to cause nasal and squamous cell carcinoma in animals exposed to 11–15 ppm. About 5 ppm is the limit for any instant exposure, while 2 ppm is the limit for prolonged or repeated exposure and is the level at which formaldehyde can just be smelled. The potential carcinogenic hazard of formaldehyde in the human respiratory tract is currently being debated (Perera and Petito, 1982). A variety of pathological consequences arising from exposure to formaldehyde have been reviewed by Loomis (1979).

Acrolein is a hazardous chemical because of its flammability and extreme reactivity. It is highly toxic through vapour and oral routes of exposure, is irritating to respiratory and ocular mucosa, and induces uncontrolled weeping. Acrolein is moderately toxic through skin absorption and is a strong skin irritant even at low concentrations. Although acrolein is highly toxic by the vapour route, the sensory response to very low vapour concentration gives adequate warning. The physiological perception of the presence of acrolein begins at 1 ppm, at which concentration an irritating effect on the eyes and nasal mucosa is felt. Thus, there is little risk of acute intoxication, because the lachrymatory effect compels one to leave the polluted area. A threshold level between 0.05 and 0.1 ppm of air has been suggested. Above this level, acrolein sensitises the skin and respiratory tract, and causes bronchitis and pneumonia in experimental animals. In human beings, when there is contact with acrolein, cutaneous or mucosal local injury may be observed. Gloves and fume-hood must be used when handling this chemical. In the preparation of solutions, acrolein should be slowly added to water with stirring rather than the reverse. Any acrolein that contacts the skin should be washed off immediately with soap and water. Waste acrolein should be disposed of by pouring it into 10% sodium bisulphite, which acts as a neutraliser.

The components of all the epoxy resins are potentially dangerous. They can be absorbed by skin contact or inhalation. The hazards associated with the use of resins are carcinogenesis, primary irritancy, systemic toxicity, environmental pollution and fire hazards (Causton, 1981). Vinyl cyclohexene dioxide (ERL 2406, Spurr mixture) is carcinogenic and is known to produce tumours in animals. The resin should be used only when necessary—for example, for embedding hard animal tissues and plant tissues. Monomeric resin components (liquid or vapour form) can cause skin, mucosal and eye irritation. For example, corneal injury may be induced by nonenyl succinic anhydride and butyl glycidyl ether, and methyl methacrylate can cause contact dermatitis.

Resins emit vapours, especially during polymerisation, and if the oven is not vented to the outside, vapours will seep into the laboratory. Systemic toxicity can result in heart dysfunction, chest spasms, blood disorders and neurological disorders. Contamination can occur by skin absorption, inhalation or accidental ingestion. These effects may be sudden or gradual. Almost all resins are volatile and form aerosols during polymerisation. These vapours can cause headaches and reduce appetite. Most of the resins are flammable. For example, benzoyl peroxide is potentially explosive when dry and is capable of violent reaction with amines. When propylene oxide is allowed to mix with phosphotungstic acid (PTA) for longer than 5 min, the result is an explosive exothermic reaction. Propylene oxide is potentially carcinogenic. Urea-formaldehyde embedding medium may also be a health hazard.

Unpolymerised resins should be handled in fume-hoods. They should be mixed with a magnetic stirrer or disposable glass rods, and transferred without spillage, by use of disposable pipettes. Before disposal, waste resin should be kept in a fume-hood until polymerised for several days at 60 °C. Resins should not be disposed of in a monomeric form. Items contaminated with resins should be collected in a polyethylene bag in the fume-hood and then sent for burial or burning. Every effort should be made to avoid contaminating handles of doors and refrigerators. All containers used for monomeric resins should be disposable.

Gloves should be worn during infiltration and embedding of the specimens in resins, and hands must be washed with soap and cold water after gloves are removed; most gloves are not impermeable to resins. A major spillage of a resin should be covered with sand or vermiculite. Polymerisation of blocks should be carried out in an oven vented to the outside or in a small oven kept in the fume-hood. No resin block for electron microscopy is completely polymerised. Therefore dust and small chips produced during sawing or filing of 'polymerised blocks' are hazardous. Resin dust should not be allowed to remain on a bench, on the floor or in a wastepaper basket. It should be collected with a damp paper towel or cloth, or a vacuum cleaner. If possible, sawing or filing of the resin blocks should be carried out in a fume-hood. All bottles containing resins should be unbreakable and clearly labelled. The bottles should be tightly capped during storage, and their exterior should be kept free of resins. Resins must never be washed off the skin with a solvent; soap and cold water should be used to wash contaminated skin.

Uranium and its compounds are radioactive and highly toxic. Uranyl acetate is a dual hazard, being both chemically and radiologically toxic. If one holds a portable beta–gamma detector over an open bottle of uranyl acetate, a continuous clicking sound can be heard. The uranium used to make most uranyl acetate is called 'depleted' because its ^{234}U and ^{235}U content has been reduced by ~50%. Hence, although the radioactivity of uranyl acetate is depleted, it remains radioactive. One gram of uranium emits 12 500 decays per second of alpha particles, 25 000 decays per second as beta emission, and some gamma radiation. Beta emission of uranyl acetate is thought to be close to the aforementioned theoretical values. Alpha particles have a range of 10–30 μm in the tissue; they can be stopped by a thick piece of paper.

The chemical toxicity of uranyl acetate is a greater hazard than its radioactivity. The maximum allowable concentration of soluble uranium compounds in the air is 0.05 mg per m^3 of air. The inhalation of powdered uranyl compounds is very dangerous. A daily uptake of 50 mg is considered lethal. Inhalation may cause disorders to the upper respiratory tracts, lungs and liver. Damage can also occur to the kidneys. The threshold toxic dose of uranyl acetate for production of kidney damage in man is estimated to be 7 mg. The site of toxicity is the proximal tubule of the kidneys. Exposure to insoluble compounds of uranium may lead to lung cancer, pulmonary fibrosis and blood disorders.

Uranium compounds, particularly in powder forms, must not be touched by bare skin, inhaled or ingested. Glasswares containing even minute amounts of these salts should not be left exposed. After use of these reagents, the hands and the work area should be washed. Unfortunately, uranium compounds are not labelled radioactive by suppliers. Darley and Ezoe (1976) have discussed potential hazards of uranium and its compounds in electron microscopy.

Lead and mercury salts used for staining are highly toxic. Tannic acid is poisonous, and diaminobenzidine may be carcinogenic. Bismuth, silver nitrate, dimethyl sulphoxide, chromium, potassium ferrocyanide and potassium permanganate are harmful. Lanthanum nitrate is an irritant and readily oxidises.

Many organic solvents used in the laboratory are fire hazards because of their high flammability and volatility, while others are physiological hazards, owing to their toxic fumes. Some of the solvents are hazardous on both counts. Among these reagents are acetone, propylene oxide, ethanol, carbon tetrachloride, benzene, xylene, chloroform and amyl acetate. These reagents should not be used near an open flame. All solutions in the laboratory should be pipetted by pipettes that do not require mouth suction. Propylene oxide may explode if used in the same room with an open flame. This reagent should be used only in a fume-hood. Acetone and xylene should not be used in a sonicator. Benzene should always be kept in a fume-hood, for repeated inhalation of its fumes can result in severe physiological damage, such as reduced circulation of white blood cells, injury to the liver and spleen, and even leukaemia. Such damage may not appear until some months or even years have passed. Ether is equally hazardous. Amyl acetate is an anaesthetic and so it should be used only in a fume-hood. Inhalation of its fumes may cause liver damage.

Lectins and certain enzyme inhibitors are dangerous. Photographic developers may cause dermatitis in susceptible persons. Many dyes (e.g. toluidine blue and methylene blue) are toxic and should not be permitted to come in contact with the skin. Phosphorus pentoxide, a desiccant, reacts violently with water. It should not be discarded in a water sink. Contact of skin or eyes with liquid nitrogen or its gas can cause severe burning. A metal piece cooled with liquid nitrogen is also extremely cold. Liquid nitrogen should always be used in well-ventilated areas so that the build-up of N_2 gas will not deplete the O_2 in the air. Because of the qualitative similarities of basic biological processes among various species of mammals, data on the effects of chemicals on animals can be used to predict human response. Therefore, chemicals proven hazardous for animals should also be considered hazardous for human beings. Positive human data are not needed for a chemical to be considered as likely to be carcinogenic or hazardous; see also pp. 55 and 81. Health and safety hazards in the electron microscope laboratory have also been discussed by Humphreys (1977), Thurston (1978), Weakley (1981), Ringo *et al.* (1982, 1984), Lewis (1983), Smithwick (1985) and Bastacky and Hayes (1985). EMscope Laboratories Ltd (Kingsnorth Industrial Estate, Ashford, Kent) has printed a very useful 'Safety Chart, Chemicals in Electron Microscopy', for free distribution.

FACTORS AFFECTING THE QUALITY OF FIXATION

It is recognised that the pH, total ionic strength, specific ionic composition, dielectric constant, osmolarity, temperature, length of fixation and method of application of the fixative are critical factors in determining the quality of tissue fixation. In certain tissues, similar fixation techniques may show variations in the ultrastructural morphology of an organelle. These variations are related to the functional state of cells at the time of fixation. Circadian rhythm, for example, is known to affect the development of certain organelles. The Golgi complex is one such organelle which shows circadian variations in its morphology. The morphology of the Golgi complex of STH cells in the pars distalis of male mice, for instance, differed depending upon the time of the day at which the mice were killed (Gomez-Dumm and Echave Llanos, 1970). In mice killed at midnight the Golgi complex was small and well defined, whereas mice killed at noon showed a marked hypertrophy of the Golgi complex.

Diurnal variation in the fine structure of endoplasmic reticulum membranes has been established. There are regional differences in the distribution of smooth and rough endoplasmic reticulum within the hepatic lobule (Chedid and Nair, 1972). The amount of smooth endoplasmic reticulum varies with respect to the time of the day. These variations are controlled by numerous factors, including DNA synthesis, mitotic activity, and protein (enzyme) and hormone synthesis. Enzyme rhythm can be correlated with the diurnal rhythm in the endoplasmic reticulum.

Another example is mitochondria, the ultrastructure of which is markedly affected by different physiological states prevalent just prior to fixation. There is evidence indicating that isolated mitochondria (Hackenbrock, 1966) as well as mitochondria *in situ* (Meszler and Gennaro, 1969) show ultrastructural changes with different physiological states. It has been demonstrated, for instance, that in the radiant heat receptors of certain snakes, free nerve endings show mitochondria having a swollen configuration if the organ is unstimulated immediately before fixation, whereas they appear condensed if stimulated by infrared radiation prior to fixation (Meszler and Gennaro, 1969).

TISSUE SPECIMEN SIZE

A simple and reasonable requirement for satisfactory fixation is the uniformity of fixation throughout the tissue specimen. Uniformity in fixation is dependent primarily upon the type and size of the specimen, since most tissue specimens are fixed through successive layers (Fig. 1.2). Even in homogeneous animal tissue such as liver, good preservation is limited to within 2–3 cell layers (90 μm) from the surface of the specimen block fixed by immersion (Fig. 1.3) (Reith *et al.*, 1984). Changes in nuclei are minor compared with the cytoplasm in the successive layers of the tissue block. On the other hand, fixation by vascular perfusion results in uniform fixation. Thus, small size of the specimen is of the utmost importance in achieving uniform fixation by immersion, yet many workers inadvertently fail to take advantage of the small size of the specimen. The origin of most of the failures in achieving satisfactory fixation lies in the large size of the tissue block.

Rapid and uniform fixation with OsO_4 can occur only to a depth of ~0.25 mm in most tissues. Consequently, the ideal size of the tissue block should not exceed 0.5 mm if it is to be uniformly fixed. Although specimens of larger size can be adequately fixed with aldehydes, especially with formaldehyde, it is not encouraged. Moreover, larger blocks have to be cut into smaller pieces for post-osmication. In general, the smaller the size of the specimen, the better and more uniform will be the quality of fixation, irrespective of the type of the tissue and fixative used. If for some reason the tissue cannot be cut into small cubes, it may possibly be cut into thin strips of ~0.5 mm or less in thickness.

Realising the importance of obtaining extremely small tissue blocks with minimum mechanical damage, the Smith and Farquhar' TC-2 tissue sectioner was developed. When using a hand-held razor blade, the cutting should be accomplished by one quick slashing motion; several back and forth movements of the razor blade will result in extensive physical damage, especially to the soft tissues.

In practice, a tissue block fixed by immersion usually produces non-uniform fixation: Fig. 1.2 shows three regions of varying fixation quality. The inward diffusion of the fixative from the surface of the tissue block in direct contact with the fixative produces a gradient of fixative concentration. The surface layers of the block are well fixed and contain the highest concentration of the fixative, but may show cellular changes resulting from mechanical injury and extraction of tissue constituents. The extraction of tissue constituents is greatest in the surface layers of the block. Distortion of cell components in the surface layers can also occur in pathological tissues. Great caution should be exercised, therefore, in the interpretation and evaluation of diseased tissues after fixation by immersion, and it is advisable to trim off a few surface layers of cells. Obviously, no such problem is encountered in the study of whole organisms such as bacteria, algae or fungi.

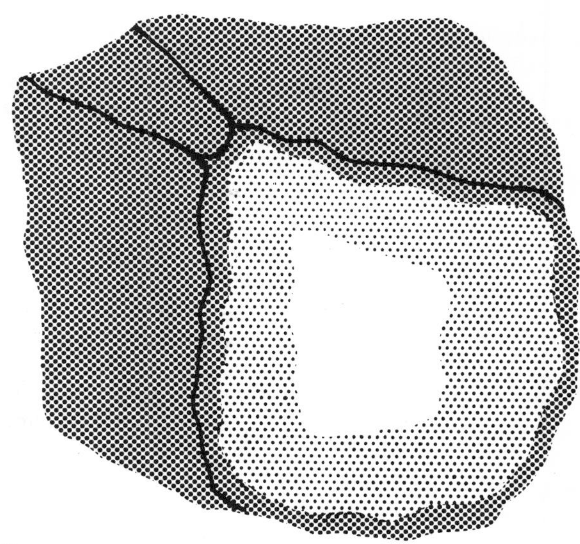

Figure 1.2 Diagrammatic representation of a tissue block (1–2 mm^3) partially penetrated by OsO_4 and showing a peripheral stratification with visible different layers at varying depths from the surface. Note the unfixed core

Immediately beneath the surface layer lies the region of well-fixed cells which are ideal for study. The cells constituting the core of the tissue block are incompletely fixed and may show displacement of cellular materials, though devoid of mechanical injury and excessive extraction of cellular materials. Very nearly uniform fixation, however, can be obtained when possible by the perfusion technique. The non-uniform fixation discussed above is encountered more commonly in tissue blocks which are larger than 0.5–1.0 mm^3 in size.

OSMOLARITY AND OSMOLALITY

It is instructive to define some terms used commonly in discussing the subject of 'osmolarity'. The term 'osmolarity' is used here strictly in terms of the response of cells immersed in a solution. The term 'osmolarity' or

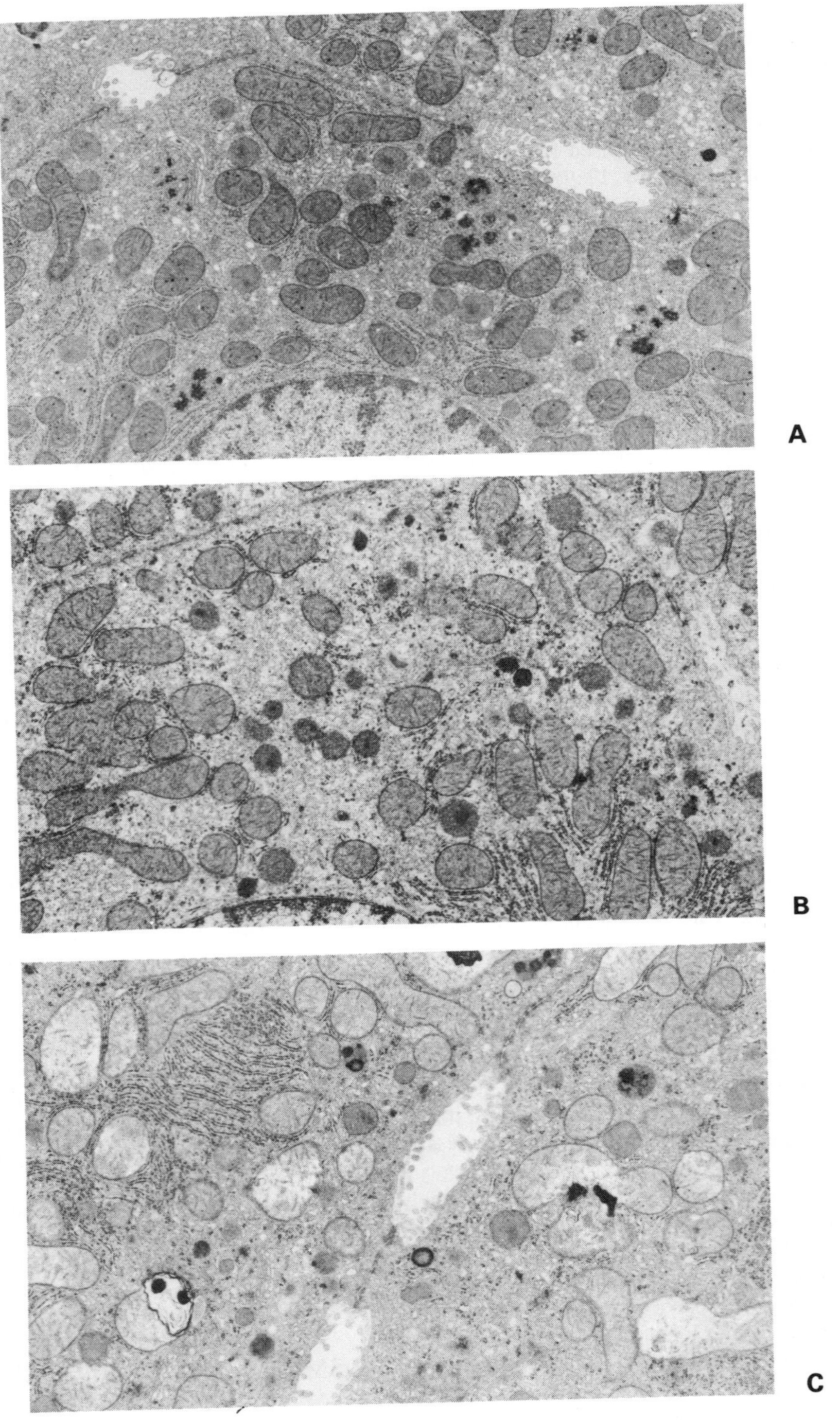

Figure 1.3 Effects of fixation by vascular perfusion and by immersion on the rat liver. A: Fixation by vascular perfusion with 1% glutaraldehyde in 0.1 M cacodylate buffer (pH 7.4) is excellent throughout. B: Fixation by immersion, showing good preservation within ~60 μm from the block surface. C: Fixation by immersion, showing enlarged mitochondria with extracted matrix and distorted cristae and artefactual myelin figures in the deeper cell layers. From Reith *et al.* (1984)

'osmolality' indicates the molarity (mole per litre of solution) or molality (mole per kilogram of solvent) that an ideal solution of non-dissociating substance must possess in order to exert the same osmotic pressure as the test solution. Solutions should be defined in osmolarity measurements, because they are independent of temperature.

Since non-electrolyte solutions do not ionise, the molar and osmolar concentrations are equivalent. On the other hand, electrolyte solutions dissociate into ions and thus exert a greater osmotic pressure than their molar concentration indicates. However, because of ion interaction, the osmotic pressure is lower than would be expected on the basis of number of particles. As a result, two solutions of identical molarities can have different osmolarities.

An isotonic solution exerts an osmotic pressure equal to that exerted by cell cytoplasm. A fixative solution is isotonic with a cell (or tissue specimen) if the cell neither shrinks nor swells when immersed in it. Two solutions are isosmotic with each other if both exert the same osmotic pressure; in other words, both solutions possess the same solute particle concentration. An isosmotic fixative solution may or may not be isotonic.

The osmolarity (tonicity) of a fixative solution has a direct effect on the appearance of the fixed specimen. Cell size and shape may be affected by a change in buffer osmolarity. It appears, however, that various tissue types differ in the degree of their response to the difference in osmolarity between the fixative solution and the cell's normal external environment. The effect produced is probably partly correlated with the degree of 'density' (compactness) of the tissue. The effect of osmolarity on relatively compact or hard tissues appears, in general, to be minimal, whereas osmolarity exerts a marked effect on 'less compact' tissues. This is demonstrated in the examples below.

A classic example of the difference in response to the osmolarity of the fixative solution is exhibited by brain tissue. The brain of the newborn rat shows a profound sensitivity to changes in the osmolarity of glutaraldehyde solutions, whereas the mature brain is affected little when the osmolarity is varied from 280 to 1290 mosm. The sensitivity of the former is due, in part, to its less compact and more hydrated nature. Maser *et al.* (1967) indicated that the morphology of newborn rat epidermis is affected by a difference of as little as 50 mosm. Some of the other cell systems that have shown the effects of changes in the fixative osmolarity are: grasshopper spermatocytes and spermatids (Tahmisian, 1964), rat muscle (Fahimi and Drochmans, 1965), neuromuscular junctions (Clark, 1976) and membranes of *Limnaea* eggs (Elbers, 1966).

Various components of a cell seem to differ in their sensitivity to the changes in fixative osmolarity. Ultrastructure of mitochondria in chick embryo heart is sensitive to as small a change in buffer osmolarity as 20 milliosmoles (mosm), while endocardial lining of the organ is less sensitive to a change in the osmolarity (Pexieder, 1977). (It should be noted that chick embryo heart is isotonic to 250 mosm.) Relatively less sensitivity shown by the endocardium may be due to instantaneous diffusion of the fixative into this lining by vascular perfusion.

Vehicle Osmolarity

The vehicle of the fixative usually consists of a buffer to which may be added electrolytes or non-electrolytes. Although buffers are partly responsible for the vehicle osmolarity, salts or other substances, when added, increase the osmolarity. Since most of the buffers, when used at the physiological range of pH, are hypotonic, the desirability of adding other substances to increase the osmolarity is apparent.

During fixation, specimens are exposed to the osmotic effect of the vehicle of the fixative, the osmotic effect of the fixative and the fixation effect. The last effect is discussed later for each of the fixatives. The osmotic effect of the vehicle plays a dominant role in determining the quality of specimen preservation during primary fixation with glutaraldehyde. A plausible explanation for this phenomenon is in order. The total number of molecules and ions other than water in a fixative solution, and not the nature of these particles, determines the total osmotic pressure. The effective osmotic pressure, on the other hand, refers to the pressure caused by those particles which do not pass freely through the cell membrane. Thus, the effective osmotic pressure is different from the total osmotic pressure as measured by the freezing point depression method. The effective osmotic pressure is primarily a function of the fixative vehicle, because its components do not pass freely through the membrane. Since the total osmotic pressure of a fixative solution is less important than its effective osmotic pressure, the osmolarity of the vehicle becomes more important than that of the glutaraldehyde in the fixative solution.

Glutaraldehyde also exerts some osmotic effect on cells. That isotonic vehicle will cause shrinkage of cells in the presence of glutaraldehyde indicates the osmotic contribution by the dialdehyde. This is confirmed by the finding that a hypertonic solution of glutaraldehyde, in either an isotonic or a hypotonic buffer, causes shrinkage damage to lung tissue, as evidenced by changes in the shape of erythrocytes in capillaries and small blood vessels examined by the TEM or the SEM (Mathieu *et al.*, 1978). Thus, the importance of glutaraldehyde concentration in determining the optimal osmolarity of the fixative solution for tissues in general, and cells in culture or in suspension in particular, cannot be disregarded. The extent of the contribution of osmotic effect by glutaraldehyde is dependent upon its concentration as well as upon the type of specimen. Changes in the glutaraldehyde concentration ranging from 2 to 5% may not show any significant effect on the ultrastructure of intact tissues.

The osmolarity of the buffer used for rinsing and preparing OsO_4 solution is important, because fixation with aldehydes does not destroy the osmotic activity of cellular membranes. At present, most workers use the same buffer for rinsing and preparing OsO_4 solution as that in the glutaraldehyde fixative solution.

Methods for Adjusting the Osmolarity

Several methods are available for adjusting the osmolarity of the fixative solution.

In addition to the osmolarity contributed by the buffer and the fixative, osmolarity can be adjusted by adding appropriate non-electrolytes or electrolytes. Most commonly used non-electrolytes are sucrose, glucose, dextran and PVP. These substances can be used in aldehydes and secondary OsO_4 fixatives, but not in primary OsO_4 fixatives, because they may cause leaching of soluble cellular materials during the fixation. The addition of these substances to aldehyde solutions is safe, for there seems to be no appreciable binding between the two. However, since these substances are usually not available in a pure form, caution is warranted in their use for cytochemical studies. Commercially available PVP, for instance, usually contains small amounts of oxidising substances (e.g. hydrogen peroxide) which may interfere in enzyme reactions.

Alternatively, osmolarity can be adjusted by adding electrolytes such as NaCl and $CaCl_2$. These substances are added to the buffer before adjusting the pH, since they may change the final pH. Care is required in selecting the type and concentration of the substance to be added. If there is any indication that ionising substances will have an adverse effect, non-electrolytes should be used. In some cases, monovalent ions are preferred to divalent ions, for the latter are more likely to produce ion effects. In some cases, Mg^+ is preferred to Ca^{2+}, since the former does not precipitate proteins.

Recommended Osmolality

Each cell type requires a specific optimal osmolality of the fixative solution for its satisfactory preservation. This has to be determined by trial and error, except in the case of a few cell types for which the optimal osmolality is known. It should be noted that any tissue organ having more than one cell type will require different osmolality of the fixative for each cell type. This situation is exemplified by kidney. Indeed, it is difficult to preserve kidney tissues uniformly, because a fourfold range of osmolarities may be present between cells. The tubular elements and the collecting system of the medulla contain fluid markedly hypertonic to plasma, and the interstitial fluid is likewise hypertonic. Renal tubules exposed to fixatives hypertonic to plasma show widening of intercellular spaces, whereas glomeruli are well preserved with such fixatives. The satisfactory fixation of the latter is probably because of the rapid speed of fixation occurring at this site. Satisfactory preservation of renal cortex can be obtained by using fixatives isotonic to the plasma. Interstitial cells in the cortex are well preserved in such fixatives, but interstitial cells in the medulla show signs of swelling and loss of intercellular organelles.

Renal medulla is especially difficult to fix uniformly well. Fixatives which yield satisfactory preservation of other cell types in the kidney cannot be used for the renal medulla, because of the specific osmotic conditions in this tissue. Renal cortical cells of rats and cats are well preserved with fixative solutions containing 1.0–1.5% glutaraldehyde, with a total osmolality of ~300–320 mosm. However, the cells of the inner stripe of the centre zone of the medulla show swelling. The cells of the inner medulla exhibit extreme swelling and empty-looking cytoplasm. Each level of the medulla requires a fixative solution of a specific osmolality.

The problem of determining optimal osmolality of the fixative for kidney is also encountered, to a somewhat lesser extent, in the preservation of not only other complex animal tissues, but also plant tissues. Almost every plant part consists of many cell types; for example, a leaf typically has cell types such as epidermis, spongy, palisade, xylem, phloem, etc. Each of these cell types possesses a specific osmolarity. Even similar cells at different developmental stages have different osmotic pressures. In fact, the osmolarity of a cell changes from moment to moment, and various organelles or compartments within a cell possess different osmolalities. For instance, mitochondria, myofibrils and sarcoplasmic reticulum system in the striated muscle have different osmotic pressures. Average osmolalities for plant and animal specimens are given below.

An osmolality of 800 mosm is recommended for mature, vacuolated plant cells (e.g. mesophyll), while 400 mosm is suggested for meristematic cells (e.g. root tip). As an example, ~3% glutaraldehyde in 0.025 M PIPES buffer (pH 7.2) has a similar osmolality to that of the sap from a 120 mm internode of the flowering stem of a grass; this osmolality is higher than that in a 350 mm internode (Lawton and Harris, 1978).

For most mammalian tissues, the recommended total (fixative and vehicle) osmolality ranges from 500 to 700 mosm, while the vehicle osmolality is ~300 mosm. A somewhat hypertonic vehicle is used for fixing tissue blocks, because total osmolality changes as a result of limitations in the penetration of the aldehyde and the dilution of the fixative during fixation. Somewhat lower total osmolalities (300–350 mosm) are desirable for delicate specimens. Certain sea organisms and viruses are best preserved at higher osmolalities; for example, the core of fowlpox is best fixed with unbuffered glutaraldehyde (0.5%) having a total osmolality of 700 mosm (Hyde and Peters, 1970). Sea organisms such as *Amphioxus* need an osmolality of ~1000 mosm: the osmolality of sea water may be ~1025 mosm. In general, higher osmolalities are needed when fixation is accomplished by vascular perfusion. The classic fixative solution of Karnovsky (1965) has an osmolality of 2010 mosm.

For scanning electron microscopy, the vehicle osmolality should be isotonic to the specimen under study. Such a vehicle should be used to prepare 1–3% solution of glutaraldehyde or a mixture of glutaraldehyde and formaldehyde. For embryonic tissues, the total osmo-

Table 1.1 Approximate osmolalities of vehicles

Vehicle	*Osmolality* (mosm)
0.1 M cacodylate + 0.05% $CaCl_2$ + 0.07 M sucrose	250
0.1 M cacodylate + 0.05% $CaCl_2$ + 0.1 M sucrose	300
0.1 M cacodylate + 0.05% $CaCl_2$ + 0.18 M sucrose	350
0.1 M cacodylate + 0.05% $CaCl_2$ + 0.23 M sucrose	400
0.2 M cacodylate	350
0.15 M Sörensen	300
0.18 M Sörensen	350
0.2 M Sörensen	400
0.2 M collidine + 0.16 M sucrose	400

Table 1.2 Approximate osmolalities (mosm) of buffers and glutaraldehyde fixatives (pH 6.85)

Buffer	*Buffer only*	*Buffer + 1% GA*	*Buffer + 2% GA*	*Buffer + 4% GA*
0.1 M phosphate	212	345	472	710
0.1 M cacodylate	220	340	470	720
Isotonic phosphate	310	441	575	800
Caulfield's	280	390	504	730
Dalton's dichromate	345	503	630	895

Note that 0.1 M phosphate or cacodylate buffers are hypotonic.

Table 1.3 Approximate total osmolalities of glutaraldehyde fixative in various vehicles

Glut. Conc.	*Vehicle*	*Osmolality* (mosm)
1.5%	0.07 M Sörensen + 0.015% $CaCl_2$	300
1.8%	0.05 M Sörensen	250
1.8%	0.05 M Sörensen + 0.05 M sucrose	300
1.8%	0.05 M Sörensen + 0.1 M sucrose	350
1.7%	0.1 M Sörensen	350
1.7%	0.1 M Sörensen + 0.05 M sucrose	400
2.5%	0.05 M Sörensen + 0.08 M sucrose	400
2.5%	0.1 M Sörensen + 0.13 M sucrose	450
2.0%	0.05 M Sörensen	300
1.8%	0.1 M Sörensen	350
2.0%	0.1 M Sörensen	400
1.8%	0.08 M Sörensen + 0.015% $CaCl_2$	350
2.2%	0.08 M Sörensen	400
1.8%	0.1 M cacodylate + 0.05% $CaCl_2$	350
2.0%	0.1 M cacodylate	400
1.5%	0.1 M collidine + 0.05% $CaCl_2$	350
3.0%	0.1 M collidine + 0.05% $CaCl_2$	400
2.5%	0.1 M collidine	350

Table 1.4 Average osmolalities of buffered glutaraldehyde formulations of different concentrations

Formulation	*Osmolality* (mosm)	*pH*
Phosphate buffer (0.1 M)	230	7.4
1.2% glutaraldehyde	370	7.2–7.3
2.3% glutaraldehyde	490	7.2–7.3
4% glutaraldehyde	685	7.1–7.3

Table 1.5 Osmolality of 1% aldehyde solutions in 0.1 M cacodylate buffer with or without 8%d sucrose

Aldehyde	*Without sucrose* (mosm)	*With sucrose* (mosm)
Acetaldehyde	415	739
Crotonaldehyde	413	702
Formaldehyde	693	835
Formalin	1081	1300
Glutaraldehyde	362	573

lality should be 300–400 mosm. It should be noted that although a fairly wide range of total osmolality can be employed for certain tissues without apparent damage to the specimen, an optimal osmolality yields consistently reliable results. The desired osmolality of a vehicle can be selected from Table 1.1, and the total osmolality of various fixative solutions can be chosen from Tables 1.2–1.7.

Measurement of Osmolarity

The osmotic pressure is one of the colligative properties of the solution. The depression of freezing point is another colligative property of the solution, which can be used to determine the osmolarity of a solution. Freezing point depression is related to the number of particles supposedly present in the solution. Since it is difficult to calculate the number of particles, osmolarity is determined by referring to the effects of a perfect non-dissociating solute in an ideal solution.

Several methods have been employed for measuring the osmolarity of fixative solutions. The freezing point depression (Δ) method seems to be the most accurate one, considering that different salts dissociate to different degrees. As mentioned above, a convenient method to determine fixative osmolality is to determine its freezing point depression with the aid of a commercial osmometer. The fixative solution is then adjusted by adding saline or water. A depression equivalent to that of 0.8% NaCl is considered to be isotonic to most mammalian cells. The freezing point depression of mammalian blood plasma is ~−0.6 °C.

The contribution of 1% OsO_4 to the freezing point depression is only −0.5 °C. Glutaraldehyde by itself, on the other hand, is capable of exerting considerable osmotic effect, because of the relatively high concentrations generally used and other reasons. It is apparent, therefore, that in the case of OsO_4 formulations, only the vehicle is responsible for freezing point depression, whereas the concentration of glutaraldehyde contributes significantly to the freezing point depression of the fixative solution.

IONIC COMPOSITION OF FIXATIVE SOLUTION

It is known that the maintenance of a delicate balance among various ions (e.g. Na^+, K^+, Ca^{2+} and Mg^+) in the immediate environment of a living cell is necessary for the normal function and structure of the cell. An upset in this balance is likely to bring about a change in metabolism, even if the osmolarity is kept constant. This change, in turn, could cause changes in the structure of the cells. Sufficient data are available which indicate that many of the fixation vehicles at present in use do not meet the physiological requirements of the tissues. Although the degree to which a fixative solution should be 'physiological' is still in question, it is reasonable to assume that an approximation of the fixation conditions with the normal extracellular environment would be helpful in obtaining a more successful fixation. It must be pointed out that while the effect of osmolarity has been recognised in many studies, less attention has been paid to the composition of the fixative solutions. The type of non-electrolyte or ion (especially the valency of cations) and the dielectric constant of the fixative vehicle play an important role in the preservation of cell fine structure.

As mentioned earlier, changes in the cell structure occur because ionic composition of fixative solutions differs from that of the natural extracellular fluids.

Table 1.6 Approximate total osmolalities of 2% OsO_4 fixative in various vehicles

Vehicle	*Osmolality* (mosm)
Distilled water	75
0.1 M cacodylate + 0.05% $CaCl_2$	250
0.1 M cacodylate + 0.05% $CaCl_2$ + 0.05 M sucrose	300
0.1 M cacodylate + 0.05% $CaCl_2$ + 0.09 M sucrose	350
0.1 M cacodylate + 0.05% $CaCl_2$ + 0.14 M sucrose	400

Table 1.7 Average osmolalities of different concentrations of glutaraldehyde and OsO_4 in various buffers

Buffer	*Glutaraldehyde* (%)	*Osmolalities* (mosm)
Krebs buffer	–	280
Millonig buffer	–	275
Krebs buffer[a]	0.1	320
Krebs buffer	2.0	550
Millonig buffer[a]	0.1	305
Millonig buffer	2.0	595
Collidine buffer + 1% OsO_4	–	170
Millonig buffer + 1% OsO_4	–	305
Millonig buffer + 2% OsO_4	–	360

[a] These fixatives were proven best for preserving mouse red blood cells. Modified from Arnold *et al.* (1971).

These changes may result in the extraction of cellular materials or deposition of fixative components, especially during the initial period of fixation. The extraction of cellular materials can be minimised by stabilising them through maintaining an approximately Ringer type of balance among the various ions surrounding the cells, until the fixing agent has penetrated into the tissue in sufficient concentration to bring about permanent fixation. In practice, this is accomplished by making the fixative solution somewhat similar to the extracellular fluids. This is done by adding non-electrolytes or electrolytes to the vehicle. The advantages of the addition of these substances outweigh the possible disadvantages (which are presented later).

FIXATIVE pH

It is known that proteins are charged both positively and negatively, and their isoelectric pH is dependent upon the relative number of acidic and basic groups present in the molecules. At their isoelectric pH, proteins are electrically neutral—i.e. they show minimal solubility and viscosity, the lowest osmotic pressure and the least swelling. In other words, various properties exhibited by proteins are minimal at the isoelectric pH. Any shift from the isoelectric point (which is defined as the hydrogen ion concentration where the mean charge on the protein is zero) affects the physicochemical properties of a protein. In general, proteins are capable of combining with cations on the alkaline side of the isoelectric point and with anions on the acid side.

Under normal conditions, the pH of cells is on the alkaline side of the isoelectric pH of proteins. For instance, the pH of cytoplasm in various types of animal cells has been found to be in the close neighbourhood of 7.0, while for nuclei it appears to be in the range 7.6–7.8. These values are higher than the isoelectric pH (5.0–6.0) of the majority of the natural tissue proteins. This means that the majority of the cytoplasmic proteins would carry a net negative charge. Since a pH shift in either direction from the isoelectric point results in increased ionisation, the osmotic pressure of protein solutions increases. This increase in osmotic pressure is partly responsible for the osmotic attraction of fixing fluids into the cells. The isoelectric points of a number of proteins have been compiled by Young (1963).

Since the isoelectric point of a protein depends upon the dissociation constants of acid and basic groups, it is readily affected by the nature of the fixing fluid and is altered by fixation. It is known that OsO_4 lowers the isoelectric point of proteins, probably owing to the destruction of basic amino acids. In other words, the net negative charge of the proteins is significantly increased by the reaction with OsO_4. Other fixatives, such as glutaraldehyde, formaldehyde and potassium dichromate, also lower the isoelectric point of proteins (Hopwood *et al.*, 1970). The pH of the fixative must remain close to the pH of the tissue, because a change in the tissue pH, brought about by the fixative, is bound to alter radically the structure and behaviour of tissue proteins in solution.

Buffer solutions are universally present in living cells. Although the inherent buffer system of protoplasm neutralises to some degree the penetrating ions of a fixative, electrolytes in a fixative exert a profound effect not only on the structural relations of protein molecules, but also on the activity of proteins as enzymes. Disaggregation of polyribosomes in L cells grown in cultures at pH 6.0 has been shown (Perlin and Hallum, 1971). Even during the normal growth of cells in a culture, the pH rises first, followed by acidification as a result of cell metabolism. Not only the growth of cells but also other parameters of cellular metabolism may be affected by pH variations. Wrigglesworth and Packer (1969) demonstrated that pH-dependent conformational modifications of protein (the term 'conformational change' is defined as a change in the secondary, tertiary or quaternary structure of the protein molecule) in mitochondrial membranes occur, which lead to reversible alterations in membrane ultrastructure.

Small changes in the pH of a fixative bring about large changes inside the cell, including the consistency of protoplasm, the selectivity of the cell membrane and the activity of enzymes. It is known that the combination of proteins with anions or cations to form insoluble compounds is closely related to the pH of the fixing fluid. Since proteins are responsible for the characteristic structure of a cell and the molecular weight of proteins is strongly pH-dependent, the importance of pH of the fixative is obvious.

As stated above, a knowledge of not only the pH of the proteins but also the actual hydrogen ion concentration of the cell interior is important in order to control the quality of fixation. It is not uncommon to find differences between the pHs of the body fluid and the interior of the cell. This is exemplified by red blood cells; the pH of these cells is slightly higher than the extracellular pH. This difference increases at 4 °C compared with that at room temperature. The intracellular pH shifts to the alkaline side with a decrease in temperature. There are differences in the pH level even among the various parts of a cell. A case in point is the difference in the pH between the vacuoles and cytoplasm of a cell. The difference is apparent especially in plant cells, since they possess more prominent vacuoles. The sap is relatively weakly buffered, so that the penetrating fixative readily shifts its pH. Since the major part of mature cells of plants constitutes a large vacuole or vacuoles surrounded by a thin layer of cytoplasm, these cells are more susceptible to shifts in their pH during fixation than most animal cells.

Since the average pH value of most animal tissues is 7.4, the best preservation of fine structure is obtained by keeping the pH of the fixative within narrow limits (i.e. 7.2–7.4) of this value. However, protein cross-linking with glutaraldehyde is enhanced at pH values higher than 7.4. A more alkaline pH (i.e. 8.0–8.4) is preferred for certain highly hydrated tissues, such as protozoan, invertebrate and embryonic. Helander (1962), for instance, indicated that gastric mucosa is preserved much better at pH 8.5. The improved preservation obtained under alkaline conditions was presumably due to the more efficient inhibition of the proteolytic action of the pepsin. A fixative of relatively high tonicity is also useful for fixing these tissues, because they are characterised by a high plasma osmotic pressure. Plant cells, with their relatively large sap vacuoles, also possess a relatively high internal osmotic pressure.

The average pH of plant cells is lower than the average pH of animal cells. The pH of the vacuolar sap in plant cells ranges from 5.0 to 6.0, while the pH of the cytoplasm of the same cells is usually between 6.8 and 7.1. Although these pH values are relatively low, the recommended pH during pre-fixation with glutaraldehyde (in PIPES buffer) is 8.0, and a pH of 6.8 during post-fixation with OsO_4 (in PIPES buffer) is desirable. The reason for using pH 8.0 is to compensate for the acidity contributed by the vacuolar sap which is released into the cytoplasm immediately after the tissue block comes in contact with glutaraldehyde. The release of the vacuolar sap not only lowers the pH, but also dilutes the fixative solution. These damaging effects can be transmitted through plasmodesmata to outer cells situated deep in the tissue block.

In certain cases, acidic pHs during fixation seem to be desirable. Some examples are cited here. Gastrin cell granules show greater electron density when fixation is carried out at pH 5.0–6.0 (Mortensen and Morris, 1977). Neurophysin–hormone complex, dense core of granules in type II cells of the pars intermedia of the pituitary of the eel (Thornton and Howe, 1974) and *in vitro* insulin granules (Howell *et al.*, 1969) are better preserved at pH 5.0–7.0. Studies of the effect of fixative pH on the appearance of neurosecretory granules indicate that at pH 7.0 the granules are pleomorphic with respect to electron density, whereas at pH 5.0–6.0 all granules remain electron dense (Nordmann, 1977).

Lower pHs (~6.0) have been recommended for stabilising nuclear constituents, including delicate fibrils of mitotic spindle (Roth *et al.*, 1963). Coupland (1965) indicated that the most intense chromaffin reaction occurred when the tissue was fixed at pH 5.8. According to Tooze (1964), the extraction of haemoglobin was reduced when the erythrocytes were fixed with OsO_4 at pH 6.2 instead of at 7.4. Optimum fixation of fowlpox virus core was obtained with unbuffered glutaraldehyde at pH 3.0 (Hyde and Peters, 1970). It is, however, pointed out that the selection of the optimal pH for a certain type of tissue is determined only after experimentation. Quite obviously, no definite statement as to the exact intracellular pH can be made, for pH varies from cell to cell and from moment to moment in the same cell. In addition, possible errors in measurements of intracellular pH cannot be disregarded.

FIXATIVE PENETRATION

An ideal fixative should kill specimens quickly, causing minimum shrinkage or swelling. Since the speed of killing is dependent primarily upon the rate of penetration of a fixative into the tissue, chemicals of low molecular weight such as formaldehyde are usually most effective. However, a direct relationship between molecular weight and rate of penetration of a fixative does not always exist. A case in point is mercuric chloride, which penetrates rapidly, although it has a relatively high molecular weight. The rate of penetration is also affected by the number of reacting radicals in the molecule of a fixative independent of its molecular weight. Some of the intrinsic properties of a fixative that influence the rate of penetration are its solubility in lipids and the polarity of its molecule.

Among the commonly used fixatives, formaldehyde penetrates faster than does either glutaraldehyde or OsO_4, and glutaraldehyde penetrates faster than does OsO_4. The K values for OsO_4, glutaraldehyde and formaldehyde are 0.2, 0.34 and 2.0, respectively. In 24 h, OsO_4 alone penetrates deeper than does a mixture of OsO_4 either with formaldehyde or with glutaraldehyde; the latter mixture penetrates the least. Tissue turns pale yellow in colour, owing to glutaraldehyde penetration, while OsO_4 penetration into the tissue is indicated by a brown to black colour. However, penetration by OsO_4 may not always show up as a change in colour.

Other factors that influence the rate and depth of penetration are tissue type, temperature, duration and mode of fixation, concentration of the fixative, osmolality of the fixative vehicle and the addition of electrolytes or non-electrolytes to the vehicle. All fixatives penetrate faster at room temperature than in the cold. Up to certain limits, an increase in the fixative concentration and duration of the fixation results in enhanced penetration of tissue specimens. Penetration is expedited by vascular perfusion.

Compact or dense tissues permit relatively slow penetration. The presence of air in the tissue (e.g. lung) and/or waterproof substances, such as cutin and waxes on the tissue surface (e.g. leaf), hinders the penetration of a fixative. As a result, these tissues float instead of being submerged during immersion fixation. Flotation of specimens is highly undesirable. The rate of penetration into these tissues, however, can be enhanced by employing vascular perfusion, vacuum or dilute solutions of common detergents prior to fixation. Hard exoskeletons of certain organisms (e.g. mite) prevent penetration of the fixatives. This difficulty can be overcome by immobilising the organism in a stendor dish by carefully melting, with a hot needle, a small amount of paraffin around each leg. The organism is then chilled and flooded with a chilled fixative. Using microscalpels, the exoskeleton is removed and the internal organs are exposed to the fixative.

The rate of penetration can also be enhanced by rupturing or damaging the cell membrane and/or cell wall. A novel approach consists of the use of a laser to achieve instant penetration of impermeable specimens by the fixative. Cole and Schierenberg (1986) used a nitrogen-pumped dye laser coupled to the microscope to puncture nematode embryos at the desired stage of development. The egg is punctured with a single pulse of light using the orange laser dye Rhodamine 6G (Lambda Physics, Göttingen), with a wavelength of 645 nm. This allows the penetration by glutaraldehyde through the shell, and cytoplasmic movement stops immediately and transient structures such as mitotic spindles and growing cell membranes are preserved. Somewhat rapid and uniform penetration results when specimens are stirred during fixation. The factors affecting the rate of penetration of individual fixatives are discussed later.

TEMPERATURE OF FIXATION

In general, higher temperatures enhance the speed of chemical reactions between the fixative and the cell constituents, including enzymes. Higher temperatures increase the rate of penetration (diffusion) of the fixative into the tissue and simultaneously the rate of autolytic changes. Lower temperatures depolarise the membrane and increase its resistance to ion penetration. It is most likely that a longer fixation time at higher temperature is accompanied by excessive extraction of cellular materials. It appears logical to assume, therefore, that effects on the quality of preservation, due to a change in temperature, are controlled by all these factors. It should be noted, however, that various fixatives respond differently to changes in temperature in terms of their net effect on the quality of preservation. Also, various tissues differ in their response to changes in temperature of fixatives.

The use of low temperatures during fixation is justified by decreased extraction and a slower rate of autolysis. It is likely that low temperatures affect less the rate of diffusion but more the rate of autolysis. In certain types of specimens, into which the penetration of the fixative in the cold is a problem, fixation at higher temperatures may be necessary. The fixation of bacterial spores, for example, was facilitated at 40 °C (Thomas, 1962). Cold labile structures, such as cytoplasmic microtubules, are best preserved at room or body temperature. It is well known that swelling is slightly more pronounced when the tissues are fixed in the cold.

In general, the effects of higher temperatures on the quality of preservation with rapidly acting fixatives differ from the effects with slowly acting fixatives. This is explained by the fact that, with the former fixatives, a reasonable increase in temperature results in an extremely rapid fixation, which more than compensates for the increased extraction effect. On the other hand, a similar increase in temperature with the latter fixatives would lead to excessive leaching, because the speed of actual fixation is too slow to compensate for the increased extraction.

The fixatives that penetrate and fix rapidly (e.g. acrolein) commonly do not require higher temperatures or longer durations of fixation; the fixatives that penetrate and fix slowly (e.g. OsO_4) tend to require longer durations of fixation but lower temperatures to minimise extraction of cellular constituents; the fixatives that penetrate slowly but fix rapidly (e.g. glutaraldehyde) may be used at 4 °C or at higher temperatures provided that the fixation time is reduced. The fixatives that penetrate rapidly but fix slowly (e.g. formaldehyde) are not preferable for the preservation of fine structure except when they are employed in combination with other fixatives or for special studies. In any case, longer fixation at higher temperatures should be avoided.

Most likely, various cell organelles differ in their response to changes in temperature with respect to their structural appearance. It is known, for instance, that labile microtubules are not always preserved if fixation in glutaraldehyde is carried out in the cold, although other cytoplasmic details are preserved satisfactorily. Lower temperatures also affect some metabolic processes necessary to maintain the normal structure of certain organelles; these temperatures affect, as far as is known, mainly the appearance of mitochondria.

The effect on various cell organelles of changes in temperature, even during dehydration, may differ. Ito (1961) reported that changes in temperature during dehydration affect the preservation of delicate structures such as smooth endoplasmic reticulum. No satisfactory explanation is available to account for the difference in response of various cell organelles to changes in temperature of dehydration agents, except that dehydration at room temperature causes excessive loss of lipids.

It must be pointed out that the effect of rate of penetration on the quality of tissue preservation has been recognised in many of the studies, but too little attention has been given to the speed at which the actual process of fixation takes place. Similarly, while the role of temperature in the rate of diffusion has been explored, hardly any information is available on the optimal temperature required to expedite actual fixation. The importance of the latter temperature becomes clear when one considers that although most fixatives kill the cell instantaneously at conventional temperatures used, they do not instantly fix it. Thus, an appreciable period of time elapses between exposure of the intracellular to the extracellular environment and complete fixation of cell macromolecules by the fixative. This brief pre-fixation period of interaction between the intracellular and extracellular environments most probably has a profound influence on the final quality of tissue preservation.

It is erroneous to assume that fixation occurs at a macromolecular level because the structural organisation of cell components is fixed. Park *et al.* (1966), for example, showed that chloroplasts isolated from spinach leaves infiltrated with 6% glutaraldehyde possessed substantial Hill activity. These chloroplasts were considered fixed on the basis of their structural stability. It is apparent, therefore, not only that a satisfactory fixative should penetrate and kill rapidly, but also that an optimal temperature should be provided to expedite actual fixation.

DURATION OF FIXATION

The optimal duration of fixation for most tissues is not known. For most purposes, an arbitrary standardised duration of 1–4 h at room temperature or 4 °C is currently in use. The optimal duration of fixation for a specific tissue is controlled by the type of fixative, specimen size, type of specimen, temperature, buffer type, staining method to be used and the objective of the study. Very little is known of the effects of overfixation except that it results in the extraction of tissue constituents. Since the extraction caused by chemical fixation is progressive in time, the need to determine the optimal duration is obvious. A few examples of the effects of overfixation are given below.

Synaptic vesicle flattening is affected by the preparatory procedure used, including duration of fixation, type of fixative, osmolality of fixative and vehicle, temperature and degree of stimulation before fixation. Paula-Barbosa (1975) indicated that prolonged (up to 20 h) immersion in glutaraldehyde during fixation (with or without vascular perfusion) might cause flattening of synaptic vesicles in the rat cerebellum. The appearance of hormone granules in gastric cells is affected by the duration of fixation. The proportion of dense-cored granules decreased when the pre-fixation time exceeded 30 min (Mortensen and Morris, 1977). Although no change in the transfascicular area of rat peroneal nerve occurred when the duration of fixation was increased from 1 h to 12 h, shrinkage (~20%) of axis cylinders took place with prolongation of fixation beyond 1 h (Ohnishi *et al.*, 1974). Motion picture analysis of fixation of amoeba indicates that a short fixation time helps to minimise distortion of cell structures (Griffin, 1963).

Duration of fixation is more critical with OsO_4 than with glutaraldehyde, since the former does not cross-link many proteins, which are probably being extracted by the vehicle, while the tissue is being fixed. The implications of fixation time, with respect to the quality of tissue preservation by each of the fixatives, are discussed in detail later. It is, however, pointed out that the fixation times at present employed are safe but longer than necessary. It seems, therefore, that modifications to the present conventional fixation schedules will have to be evolved to retain and render visible relatively small amounts of cellular materials which are extracted because of longer than necessary exposure to fixatives. It is anticipated that, in general, by reducing the duration of fixation from the present conventional fixation periods the tissue would be capable of retaining an increased range of cellular materials in a demonstrable form.

CONCENTRATION OF FIXATIVE

In general, low concentrations of fixatives require longer durations of fixation. Longer durations of fixation tend to cause diffusion of enzymes, extraction of cellular materials, and shrinkage or swelling of the tissue. Higher concentrations of fixatives are also undesirable, since concentrated solutions destroy enzyme activity and damage the cellular fine structure. Oxidising fixatives such as OsO_4 are effective cross-linking agents only when used in correct concentrations; higher concentrations of OsO_4 can cause the oxidative cleavage of protein molecules. This could result in the loss of peptide fragments.

Different cell components differ in their response to variations in the concentration of a fixative. Baker and McCrae (1966) have indicated that low concentrations (0.25%) of formaldehyde disrupted the endoplasmic reticulum, whereas mitochondria were insensitive to changes in concentration. However, mitochondria are generally more sensitive to changes in the osmolarity of the buffer system.

The effect on the tissue of varying the concentration differs, depending upon the type of fixative employed. Glutaraldehyde has a wide range of effective concentration, provided that optimal osmolarity of the buffer system is maintained. The effective concentration of OsO_4, on the other hand, is of a rather limited range. The optimal concentration of a fixative apparently varies within an appropriate range, depending upon the specimen under study. In general, higher concentrations are needed to preserve satisfactorily hydrated specimens such as embryonic tissue. It should be noted that the concentration of a fixative usually falls during fixation. The most suitable concentration, for general purposes, of each of the fixatives discussed is indicated later.

EFFECTS OF ADDED SUBSTANCES

The effects of the addition of substances to the fixative vehicle can be explained not only on the basis of osmolarity, but also by the type of substance used. There may be interactions between the added substances and the cell constituents, on the one hand, and the fixative agent on the other hand. It is known, for example, that the denaturation of proteins and nucleic acids is affected by the salts added to the fixative solution.

The addition of non-electrolytes usually minimises the extraction of certain cellular constituents. A significant reduction in the extraction of chlorophyll in isolated chloroplasts by adding sucrose to the fixative is well known. The addition of dextran to glutaraldehyde improves the preservation of myelin and node of Ranvier, and prevents swelling of brain slices during incubation *in vitro* and subsequent fixation. Dextran or polyvinyl pyrrolidone (PVP) even offsets the swelling of chloroplasts caused by detergent treatment (Deamer and Crofts, 1967). In perfusion fixation, the addition of dextran to glutaraldehyde prevents the artefactual enlargement of extravascular space, especially in pancreas (Bohman and Maunsbach, 1970).

The effectiveness of non-ionising, osmotically active substances in preventing swelling is related, in part, to the extremely slow or non-penetrating property of these substances. The distribution of water in cells is normally a function of the relative concentrations of internal and external osmotically active species. During fixation, the concentration of these species can be controlled. It is apparent, therefore, that the amount of water in the cell, and thus cell volume, is controlled primarily by the external concentration of non-penetrating, neutral molecules of substances such as sucrose, dextran, etc.

Because of their extremely slow rate of penetration through membranes, non-electrolytes may also act as osmotic stabilisers where the permeability properties of membranes persist, as, for example, after fixation with glutaraldehyde. Furthermore, the addition of non-electrolytes may help to prevent the collapse of certain cell structures by keeping them expanded when their active transport and permeability properties are destroyed by the fixative. This effect is related to the hydration property of non-electrolytes. It has been demonstrated, for instance, that the addition of sucrose to unbuffered OsO_4 leads to the preservation of well-defined vesicles bound by clear unit membrane in the bacterial mesosomes (Burdett and Rogers, 1970). The protective action of PVP or dextran against damage associated with freezing and thawing, in biological systems, is well known.

The addition of electrolytes (especially $CaCl_2$) to the fixative solutions has many beneficial effects, including (1) decrease in the swelling of cell components, (2) maintenance of cell shape, (3) reduction in the extraction of cellular materials, and (4) membrane and cytoskeletal stabilisation. Electrolytes seem to be superior to non-electrolytes in the control of swelling and maintenance of cell shape. The presence of an electrolyte at isotonic concentration is an essential condition for shape preservation during the fixation of red blood cells (Morel *et al.*, 1971).

Electrolytes are added to the fixation solution not only to prevent swelling, but also to lessen the amount of extraction of cellular materials. Divalent cations stabilise polypeptides and phospholipids by binding to them. The effect of electrolytes on the solubility of proteins has long been known, and the capacity of these salts to precipitate or to dissolve various proteins is a familiar phenomenon in protein chemistry. The salts that combine chemically with amino acids are the ones that either increase or decrease the aqueous solubility of polypeptides. Different amino acids differ in their response to the addition of a specific salt to the fixation vehicle. It is known, for example, that the aqueous solubility of leucine, tryptophan and phenylalanine is decreased in the presence of NaCl, while glycine and cysteine show no apparent reaction to the presence of this salt. In general, the type and concentration of electrolyte ions added to the fixative vehicle profoundly affect the solubility of amino acids. Stabilisation of phospholipids occurs presumably through binding of cations to membranes. Hydrolysis of membrane phospholipids by Ca^{2+}-activated phospholipases can be avoided in the presence of $MgCl_2$ (Hauser *et al.*, 1980).

Numerous attempts have been made to lessen the amount of extraction of cellular materials by the addition of electrolytes, especially by divalent cations. A few examples will suffice. It was demonstrated by Tooze (1964) that the addition of Ca^{2+} ions to a final concentration of 0.01 M in the fixative completely suppressed extraction of haemoglobin and stabilised erythrocytes, whereas the addition of monovalent sodium ions had no stabilising effect. The ultrastructure of chromosomes has been shown to be affected by Ca^{2+} (Barnicot, 1967). The concentration of divalent ions influences chromatin condensation. The addition of Ca^{2+} solution prevented the precipitation artefact in the nuclei of frog erythrocytes (Davies and Spencer, 1962), and spindle filaments were stabilised by either using a pH of 6.1 or adding Ca and Mg (Harris, 1962).

Probably the most dramatic effect of the presence or absence of divalent ions (e.g. Ca^{2+}) during fixation is exhibited by bacteria. The morphological pattern of bacterial membranes is influenced by Ca^{2+} deficiency during fixation. Phospholipid losses during dehydration increase in bacteria when Ca^{2+} ions are omitted from the fixative. Mesosome configuration is profoundly affected by the ionic composition of the OsO_4 fixative. It has been demonstrated that fixation of bacteria under conditions of Ca^{2+} deficiency results in symmetrical membranes and lamellar mesosomes, while in the presence of Ca^{2+} the membranes appear asymmetrical and mesosomes become vesicular (Silva, 1971).

The available information on the exact role played by electrolytes in reducing the extraction of cellular materials is insufficient. Stabilisation of membranes, especially by divalent ions, possibly plays some role in decreased loss of cellular materials. The ionic environment of cells during and after fixation with an aldehyde may affect membrane structure. Calcium added to the primary fixative usually binds to membranes, resulting in the stabilisation of membranes in various cell and tissue types.

Effects of the addition of divalent cations to the buffer are exceedingly complex. Calcium and magnesium cations have a stabilising or destabilising effect on microtubules and microfilaments, depending on the ion concentration. Low concentrations of these cations destabilise cytoplasmic components (Wild *et al.*, 1987). This destabilisation can be avoided by adding 250 μmol $CaCl_2$ and 250 μmol $MgCl_2$ to 0.1 M cacodylate buffer used with glutaraldehyde and OsO_4. Addition of 250 μmol $MgCl_2$ to 0.05 M cacodylate for buffering glutaraldehyde destabilises the membranes in parathyroid cells (Wild *et al.*, 1987). This results in the swelling of rough endoplasmic reticulum combined with cell shrinkage, loss of membranes as well as ground cytoplasm, and production of light and dark cells in the parathyroid tissue.

Calcium and magnesium cations reduce cell volume in the presence of a buffer of low osmolarity, but they prevent excessive cell shrinkage in the presence of buffers of high osmolarity. Stabilisation of cell components depends on the concentrations of the added ions as well as the buffer and the fixative. Both Ca^{2+} and Mg^{2+} can be added to the buffer during fixation by immersion or vascular perfusion. Magnesium ions should be added in a concentration twice as high as that of Ca^{2+} to stabilise cell components (Wild *et al.*, 1987).

Possible deleterious side-effects of the presence of substances added to the fixative solutions cannot be overlooked. Available data indicate that the rate of fixative penetration into the tissue may be retarded by the addition of substances to the fixative vehicle. The reasons for this decreased rate are not known, although the effect could be due to greater conformational stability in diffusion barriers when other molecules or ions are present. Increased viscosity and competition with the fixative molecule could also be a partial explanation. The addition of electrolytes, especially divalent ions of Ca and Mg, however, results in only a slight decrease in the rate of penetration of the fixative, in comparison with the decrease caused by the addition of non-electrolytes.

The addition of divalent cations may cause protein precipitation and excessive granularity. For this reason, it is usually preferred to use only low concentrations of these ions to adjust osmolarity. Monovalent ions impart only negligible granularity, although fixative solutions containing only monovalent cations are, in general, less effective in suppressing the extraction of cellular materials. Divalent cations are 10–100 times more effective than are monovalent ions in this respect. A disadvantage of adding Ca^{2+} is considerable movement of chromosomes relative to each other (Skaer and Whytock, 1976). The addition of Ca^{2+} ions may bring about physiological effects such as stimulation of myofibrils or triggering of secretion (Glauert, 1974).

Calcium ions seem to disrupt the regular pattern of microtubules (Schliwa, 1977). The extraction of microtubules in the presence of bicarbonate ions has been shown. Phosphate buffers may cause precipitation. This is exemplified by the appearance of spherical dense granules on the alveolar surfaces of the lung fixed primarily with phosphate-buffered OsO_4. It should be remembered that artefactual precipitation is not easy to detect.

BUFFERS

The capacity of buffers occurring naturally in cells for maintaining a constant pH in the presence of weakly buffered or unbuffered fixatives is rather limited. Also, it is recognised that tissues fixed in unbuffered fixative solutions form artefacts, since acidification of tissues precedes their fixation. Tissues in unbuffered OsO_4 solution, for instance, cause the pH to drop from 6.2 to 4.4 in a 48 h period. It is known that a shift to lower pH is usually associated with cell death. The acidification may be explained, at least partly, on the basis of an irreversible dissociation of protein macromolecules to low molecular weight proteins and the consequent increase of ionisable carboxyl groups. The reduction of OsO_4 to an acid residue can also cause acidification of the tissue. By buffering the fixing solution, the probable acidification can be neutralised, and thus damage to the tissue is minimised.

Buffer solutions contain a weak acid and its salt or a weak base and its salt, and they resist changes in hydrogen ion concentration when small amounts of a strong acid or base are added to it. Thus, during fixation the acidification very slowly changes the pH of the buffer, since the salt suppresses the dissociation of the acid by the common-ion effect. The necessity of maintaining a proper H ion concentration during fixation is, therefore, apparent, especially when using slow-penetrating fixatives such as glutaraldehyde or OsO_4.

The relative tolerance of some specimens to variations in pH does not mean that there is no optimal pH for these specimens. In general, not only various tissue types but also different cells and organelles respond differently to pH variations. Available data are quite sufficient to indicate the importance of pH in specimen preservation. It is indeed possible that inadequate preservation of deep layers of tissue specimen is a result not only of autolysis and slow penetration of the fixative, but also of inadequate buffering. It should be noted that although dilution of a buffer changes the pH slightly, it proportionately decreases its buffering capacity, since the concentration of the buffer determines its capacity to absorb acid or base at any pH.

Temperature has a significant effect on the pK_a values of many buffers; for example, Tris buffer, with a pH of 8.4 prepared in a cold room, shows a pH of 7.8 at room temperature and a pH of 7.4 at 37 °C (Good *et al.*, 1966). This change in the pH is important, since the efficiency of a buffer system varies at different pH levels. For instance, phosphate buffer has poor buffering capacity above pH 7.5, whereas tris(hydroxymethyl)aminomethane has poor buffering capacity below pH 7.5. It is, therefore, recommended that the pH be measured immediately prior to fixation.

Comparatively little attention has been given to the effect of specific ionic composition of buffers on the preservation of fine structure. The quality of fixation is influenced not only by the pH, but also by the type of ions present in the buffer, since buffers are active reagents in the fixation process. In other words, the colligative nature of fixing fluids has a significant effect on the fine structure. Presumably, ions in the buffer vehicle interact with certain chemical groups within the tissue, and thus affect the fixation quality. Variations in the specific constitution of the buffer produce significant differences in the stainability and morphological appearance of cells and organelles. Ionic species in the buffer may mask or uncover certain sites of organic molecules which, without these species, may interact differently with a fixative and a stain. The chemical composition of the buffer, for instance, affects the density of the cytosomal matrix in renal proximal tubular cells of the rat kidney (Ericsson *et al.*, 1965).

If the buffer contains physiologically active ions (e.g. Ca^{2+}, Na^+, K^+ and Cl^-), biochemical alterations may occur in the cells long before their immobilisation by the fixative (Schiff and Gennaro, 1979). Calcium, often added to the fixative vehicle to stabilise membranes, can form several insoluble phosphate salts that can precipitate within the tissue and are not removed during rinsing, post-fixation and dehydration. Such precipitates may appear as fine electron-dense granules scattered throughout the specimen. Certain other components of a buffer can also form insoluble salts within the tissue. It was shown that the amount of proteins extracted during fixation in the presence of Ca^{2+} was different from that in the presence of Na^{+}, even though in both cases the osmolalities were nearly equal (Wood and Luft, 1965).

The use of two different buffers, one during pre-fixation with glutaraldehyde and the other during post-fixation with OsO_4, may result in the formation of artefactual electron-dense granules scattered throughout the specimen, but especially along the membranes. This artefact may occur when cacodylate and Millonig buffers are used during pre-fixation and post-fixation, respectively. It has been suggested that the artefact is a chelate compound composed of the decomposition products of glutaraldehyde, substances dissolved from the tissue being fixed, cacodylate, some component of the Millonig buffer and OsO_4 (Kuthy and Csapó, 1976). Such artefacts, however, can be prevented by using an extremely pure glutaraldehyde, by using the same buffer throughout the fixation procedure or by using veronal as a second buffer during post-fixation.

BUFFER TYPES

Different tissue types differ in their reaction to the same buffer, and similar cellular components respond differently to different buffers. Various buffers extract different proteins from similar tissues fixed with glutaraldehyde; the same is true in the case of unfixed tissues (Kuran and Olszewska, 1977). This is exemplified by nuclear components, the fine structure of which is affected differently by different buffers. Various buffers and their calcium content affect differently the size of the threads in the artefactual network formed in the nuclear sap during fixation with glutaraldehyde (Skaer and Whytock, 1977). In cacodylate the threads range in size from 5 to 20 nm in diameter, whereas in non-chelating HEPES containing 1.25 mmol calcium, the size is mostly 10 nm. In HEPES nuclear sap generally forms a finer network. Studies of the influence of various buffers on rat salivary gland secretory granules indicated that with phosphate buffer, filaments appeared aggregated into large, fibrillar structures; with cacodylate buffer, the filaments tended to be dispersed; and with collidine buffer, the filaments as well as vesicles were present within the granules (Simson, 1977).

Although all the buffers currently used in electron microscopy have fairly desirable dissociation constants, solubilities and reactivation, no single buffer can claim universal superiority over the others. Each buffer system has certain advantages and disadvantages, which are given later for each of the buffers. The most efficient buffering action of various buffers cannot be predicted according to a constant, but must be determined for each buffer for a specific pH range and for a specific type of specimen. Most probably, there is no one ideal buffer for a given pH range. It cannot be assumed that various particulate systems and soluble enzyme systems in cells respond the same way to different buffer systems. The objective of the study best determines the type of buffer to be used. It should be remembered that many experiments have failed only because of the imperfections of the buffer used.

A number of buffers are in use in electron microscopy, several of which are called 'physiological buffer systems', because they are quite effective in the pH range 7.2–7.4. The advantages and limitations of these buffers are presented below.

Cacodylate Buffer

Cacodylate buffer is quite effective in the pH range 6.4–7.4. Cacodylate avoids the presence of extraneous phosphates which may interfere with cytochemical studies. It is incompatible with uranyl salts. Cacodylate is not a very efficient buffer in terms of stabilising the pH of the fixative (Coetzee and van der Merwe, 1987). This buffer causes changes in membrane permeability and a redistribution of cellular materials along osmotic gradients. It is resistant to bacterial contamination during specimen storage. Calcium can be added to the fixative solution in the presence of cacodylate without the formation of precipitates. According to Hulstaert and Blaauw (1986), cacodylate buffer is superior to phosphate buffer for animal tissues. The opposite opinion was given by Reith *et al.* (1984). They suggested deeper fixation by immersion in the presence of phosphate buffer compared with cacodylate buffer. However, this superiority is related to higher osmolality (300 mosm) of the phosphate buffer, in contrast to the lower osmolality (200 mosm) of cacodylate buffer used in this study.

Sodium cacodylate contains arsenic, which is a health hazard. If inhaled or absorbed through the skin, it can cause dermatitis, liver and kidney inflammation, etc. Hands should be protected by gloves and fume-hoods should be used for weighing out the reagent and preparing the buffer solution. The reagent should be prevented from coming in contact with acids, in order to avoid the production of arsenic gas.

Collidine Buffer

Collidine buffer, when half neutralised by HCl, is very efficient in the neighbourhood of pH 7.4. Its buffering capacity is in the biological range of pH (6.0–8.0). It neither reacts nor complexes with OsO_4. When OsO_4 is added to this buffer, the pH does not change any more than when an equivalent amount of distilled water is added. It is biologically stable at room temperature almost indefinitely.

Of six different buffers tested for the lung tissue, collidine buffer proved the best (Gil and Weibel, 1968). Collidine is a pyridine derivative, and since pyridine is a classical phospholipid extracting agent, an excessive extraction of cellular substances in the presence of this buffer is likely. According to Luft and Wood (1963), OsO_4 buffered with collidine extracted more proteins from rat liver during and after fixation than it did with seven other buffers. Collidine is considered superior to phosphate buffers for tissue storage in 10% formaldehyde solution. The former is effective in the fixation of large tissue blocks, because it extracts cellular materials and this facilitates the penetration of the fixative. Membranes may be damaged when formaldehyde is used with collidine (Carson *et al.*, 1972).

Collidine buffer depletes monoamines stored in the synaptic vesicles core. It has been shown that collidine abolishes the core osmiophilia and chromaffin reaction from rat pineal gland and vas deferens nerves (Tomsig and Pellegrino de Iraldi, 1987). This effect appears to be due to a loss of the neurotransmitter itself rather than a chemical blockage of reactive glycol groups in the neurotransmitter molecules (Tomsig and Pellegrino de Iraldi, 1988). This buffer is toxic and has a strong smell. Only pure collidine should be used. Collidine cannot be recommended for routine electron microscopy.

HEPES Buffer

HEPES (*N*-2-hydroxyethylpiperazine-*N*′-2-ethanesulphonic acid) buffer is one of the zwitterionic buffers and has a pK_a of 7.31 at 37 °C. HEPES is a tertiary amine heterocyclic buffer, and thus may interfere with tissue amine–aldehyde reactions. Massie *et al.* (1972) indicate that this buffer is a non-toxic substitute for bicarbonate buffers in cell culture media. It seems to stabilise the lipid component of cell membranes during fixation. Plasma membranes fixed in the presence of this buffer appear to resist chemically initiated extraction of cellular materials. HEPES buffer seems to be compatible with divalent cations and does not bind significant amounts of these ions. The use of this buffer during fixation results in the production of relatively small amounts of acid.

MOPS Buffer

MOPS (3-[*N*-morpholino] propanesulphonic acid) buffer is a member of the zwitterionic buffers and has a pK_a of 7.1 at 37 °C. It is an amine related to group B tertiary amines. MOPS is one of the few buffers that when reacted with aldehydes produces hardly any acid (Johnson, 1985).

Phosphate Buffers

Phosphate buffers are more physiological than any other buffer, because they are found in living systems in the form of inorganic phosphates and phosphate esters. The extracellular fluid of animal tissues contains mainly Na^+. These buffers are non-toxic to cells grown in culture. Phosphate and collidine buffers stabilise the pH of the fixative more efficiently than any other buffer (Coetzee and van der Merwe, 1987). The pH of the buffers seems to change little at different temperatures, and they can be stored for several weeks in the cold, provided that glucose or sucrose has not been added. However, these buffers gradually become contaminated and precipitates may appear during storage.

In spite of the fact that sodium monophosphate and sodium diphosphate are present as effective buffer systems in animal tissues, the use of phosphate buffers is not always desirable. For instance, they produce artefacts in the form of electron-dense, spherical particles of different sizes on the luminal surfaces of the types I and II pulmonary epithelial cells in the lung. Post-incubation of specimens fixed in phosphate-buffered glutaraldehyde in the same buffer may cause a decrease in the nuclear mass (Kuran and Olszewska, 1974). Phosphate buffers decrease nuclear mass more than does veronal buffer. Phosphate buffers extract non-chromosomal proteins in the nucleus, while veronal buffer removes part of the chromatin proteins. There is some evidence indicating that phosphate buffers may cause swelling of intracellular organelles, while cacodylate does not seem to exert this effect. Phosphate buffers tend to precipitate polyvalent cations and uranium salts. The buffers are undesirable when the addition of certain concentrations of calcium to the fixation solution is necessary.

PIPES Buffer

PIPES (piperazine-*N*,*N*′-bis[2-ethanesulphonic acid])

is one of the zwitterionic buffers and has a pK_a of 6.66 at 37 °C. It is an amine related to group A tertiary amines. This property may interfere with tissue amine–aldehyde reactions. Small amounts of acid are produced when this buffer is used with glutaraldehyde or formaldehyde during fixation. A number of studies indicate the superiority of PIPES buffer over other buffers (Baur and Stacey, 1977; Haviernick *et al.*, 1984). It is thought that this buffer allows increased retention of cellular materials and reduces lipid loss, especially from the plasma membranes (Schiff and Gennaro, 1979). PIPES buffer appears to be compatible with divalent cations and does not bind significant amounts of these ions.

However, like other buffers, PIPES buffer is not suitable for every type of study. It has been reported to produce artefactual multivesicular myelin figures in the rat cerebral cortex (Schultz and Wagner, 1986). It has been suggested that the non-toxic nature of PIPES buffer allows the formation of this artefact, whereas phosphate and cacodylate buffers, especially the latter, interfere with cellular activity during fixation. Specimens fixed in the presence of PIPES buffer show relatively low elemental concentration (Baur and Stacey, 1977).

Tris Buffer (tris(hydroxymethyl) aminomethane)

Tris buffer has a poor buffering capacity below pH 7.5 and is a biological inhibitor. Tris buffer, like other primary amines, appears to react with glutaraldehyde, and therefore should be avoided except when no alternative is available. It causes excessive extraction of cellular materials.

Veronal Acetate Buffer

Veronal acetate buffer is most effective between pH 4.2 and 5.2, and thus inoperative at pH 7.2–7.5. It should not be used with aldehydes, since it reacts with these fixatives and the reaction product has no buffering value in the physiologically important range of pH. Veronal acetate cannot be stored in the absence of the fixative, because it easily becomes contaminated by bacteria and mould. It has the advantage of not precipitating uranyl acetate. Membranes appear to be preserved better when OsO_4 is buffered with veronal than with other buffers.

CHOICE OF BUFFER

The role of the buffer in fixation is exceedingly complex. In addition to stabilising the acid–base balance during fixation, the buffer influences the tonicity of the environment in which the fixative acts. The type of buffer used as a vehicle affects the rate of protein cross-linking with glutaraldehyde (Coetzee and van der Merwe, 1985a). Also, the pH of the buffer influences both the cross-linking capacity of glutaraldehyde and the speed of cross-linking. Since each buffer has a pH range at which it is most effective, the type of buffer used becomes relevant in protein cross-linking. Moreover, various buffers have different osmotic effects, even when used at the same molarity. Apparently, the buffer should co-create the microenvironment in which the fixative is most effective.

An ideal buffer used as a vehicle with glutaraldehyde during fixation should have sufficient buffering capacity to minimise lowering of the pH due to the rapid adduct formation between glutaraldehyde and tissue amines. Acid production results from the fact that the pK_a value of tissue amines is usually 2–3 pH units higher than that of the newly formed adducts (Johnson, 1985). To minimise the decrease in pH, the buffer should have a pK_a of 0.2–0.3 units less than the pH at which fixation is to be accomplished. As mentioned earlier, the pK_a changes as a function of temperature.

The concentration of the buffer should be about threefold higher than the concentration of primary amines in the tissue available to glutaraldehyde. The buffer should not have an amino acid component which will react with glutaraldehyde. The buffer should not bind divalent cations. HEPES buffer, with a pK_a of 7.0–7.3 (37 °C), offers most of these desirable characteristics, except that it is an amine heterocyclic buffer. The characteristics of a number of buffers, presented earlier, can be considered in selecting a buffer.

PREPARATION OF BUFFERS

Cacodylate (0.2 M)

Solution A	
Sodium cacodylate ($Na(CH_3)_2AsO_2 \cdot 3H_2O$)	42.8 g
Distilled water to make	1000 ml
Solution B: 0.2 M HCl	
Conc. HCl (36–38%)	10 ml
Distilled water	603 ml

The desired pH can be obtained by adding solution B, as shown below, to 50 ml of solution A and diluting to a total volume of 200 ml.

Solution B (ml)	*pH of buffer*
18.3	6.4
13.3	6.6
9.3	6.8
6.3	7.0
4.2	7.2
2.7	7.4

The buffer should be prepared under a fume-hood.

Collidine

Stock solution

s-Collidine (pure)	2.67 ml
Distilled water to make	50.0 ml (approx.)

Buffer

Stock solution	50.0 ml
1.0 N HCl	9.0 ml (approx.)
Distilled water to make	100 ml

The pH is 7.4, which can be obtained by adjusting with HCl.

HEPES

A 0.2 M solution of *N*-2-hydroxyethylpiperazine-*N'*-2-ethanesulphonic acid is prepared in distilled water. The pH is adjusted with 0.2 M NaOH. The osmolarity of 0.179 M HEPES buffer containing 2% glutaraldehyde is ~300 mosm.

MOPS

A 0.2 M solution of 3-(*N*-morpholino)propane-sulphonic acid is prepared in distilled water. The pH is adjusted with 0.2 M NaOH.

PIPES (0.3 M)

Distilled water	50 ml
Piperazine *N,N*-bis(ethane sulphonic acid)	9 g

Add enough 0.1 N NaOH (0.4%) to adjust the pH; at pH 5.5–6.0 the powder is completely dissolved. After the required pH has been reached, more distilled water is added to make up 100 ml. The stock solution is stable for several weeks at 4 °C.

Phosphate (Millonig, 1964)

$NaH_2PO_4 \cdot H_2O$	1.8 g
$Na_2HPO_4 \cdot 7H_2O$	23.25 g
NaCl	5.0 g
Distilled water to make	1000 ml

The pH is 7.4 and the osmolality is 440 mosm.

Phosphate (Sörensen)

Solution A

Sodium phosphate, dibasic ($Na_2HPO_4 \cdot 2H_2O$)	11.876g
Distilled water to make	1000 ml

Solution B

Potassium phosphate, monobasic (KH_2PO_4)	9.08 g
Distilled water to make	1000 ml

The desired pH can be obtained by adding enough solution B to solution A, as shown below, to make a total volume of 100 ml.

Solution A	*pH of buffer*
0.6	4.9
2.3	5.3
4.9	5.6
12.1	6.0
26.4	6.4
49.2	6.8
61.2	7.0
67.0	7.1
72.6	7.2
77.7	7.3
81.8	7.4
85.2	7.5
88.5	7.6
93.6	7.8
96.9	8.0

Tris(hydroxymethyl)aminomethane maleate (0.2 M)

Solution A

Tris(hydroxymethyl)aminomethane	30.3 g
Maleic acid	29.0 g
Distilled water to make	500 ml
Charcoal	2 g

The solution is shaken, allowed to stand for about 10 min, and filtered.

Solution B: 4% NaOH

NaOH	4 g
Distilled water	96 ml

The desired pH can be obtained by adding solution B to 40 ml of solution A and diluting to a total volume of 100 ml with distilled water.

Solution B (ml)	*pH of buffer*
15.0	6.4
18.0	6.8
19.0	7.0
20.0	7.2
22.5	7.6
24.2	7.8
26.0	8.0

Veronal acetate (Ryter and Kellenberger, 1958)

Stock solution	
Sodium veronal (barbitone sodium)	2.94 g
Sodium acetate (hydrated)	1.94 g
Sodium chloride	3.40 g
Distilled water to make	100 ml
Buffer	
Stock solution	5.0 ml
Distilled water	13.0 ml
1.0 M $CaCl_2$	0.25 ml
0.1 M HCl	~7.0 ml

The HCl is added dropwise until the required pH is obtained. The buffer should be prepared fresh.

ALDEHYDES

The use of aldehydes, with the exception of formaldehyde, as fixatives is a comparatively recent development. In 1959 Luft demonstrated that a monoaldehyde, acrolein, could be usefully applied to solve certain problems in the preservation of fine structure. However, the extensive studies by Sabatini *et al.* (1963) brought the usefulness of aldehydes, especially glutaraldehyde, to the attention of electron microscopists. These studies indicated that a primary fixation with glutaraldehyde, followed by a secondary fixation with OsO_4, yielded satisfactory preservation of fine structure and enzyme activity in a wide variety of specimens. This double fixation has become the standard procedure for preserving both plant and animal specimens, as well as prokaryotes. A mixture of glutaraldehyde and formaldehyde (freshly prepared by depolymerising paraformaldehyde) has the additional advantage of more rapid penetration into the tissue specimen, and is preferred for many specimens. Acrolein in combination with glutaraldehyde has also been employed for certain studies.

Because OsO_4 is a heavy metal oxide, it has the capacity to destroy enzymatic activity by oxidation, and it is not used as a primary fixative in enzyme cytochemistry, whereas many cytochemical reactions can be performed on tissue specimens after an aldehyde fixation. Also, relatively large tissue blocks can be fixed in glutaraldehyde and still larger blocks can be preserved with acrolein or formaldehyde. Prior to further treatment, aldehyde-fixed tissue blocks can be cut into small pieces with minimal mechanical damage, since cross-links introduced by aldehyde impart some degree of consistency to the tissue. Pre-fixation with aldehydes thus lessens the distortions introduced by mincing fresh tissues to be fixed subsequently with OsO_4. It is known that aldehydes form both intramolecular and intermolecular cross-links with protein molecules, which result in the formation of more rigid heteropolymers.

The reaction of aldehydes with proteins proceeds through aldehyde condensation reactions with amino groups to yield α-hydroxyamines, which can condense with additional amino groups to affect cross-linking. Formation of a methylol or substituted methylol derivative is the first step. These reactions are discussed in detail later.

Various aldehydes differ greatly in the rate and extent of their reaction with the reactive groups of proteins. The extent of protein reaction is influenced by many factors, including the duration of fixation, pH, temperature, conformation of the protein molecule, and the type and concentration of aldehyde used. The variations in the interaction between proteins and aldehydes are considered to be related in part to the effects of the latter on enzyme activity.

GLUTARALDEHYDE

$$\mathrm{OHC{-}CH_2{-}CH_2{-}CH_2{-}CHO}$$

Glutaraldehyde is a five-carbon dialdehyde of relatively simple structure. The molecular formula above shows that a straight hydrocarbon chain links two aldehyde moieties. Glutaraldehyde has a molecular weight of 100.12, and is characterised by a relatively low viscosity. It is regarded as slightly to moderately toxic and non-corrosive to stainless steel.

Comparative studies by Sabatini *et al.* (1962, 1963, 1964) and Barrnett *et al.* (1964) demonstrated conclusively that of all the aldehydes tested, glutaraldehyde proved to be the most effective for preserving fine structure. Glutaraldehyde is effective in preserving both prokaryotes and eukaryotes, including fragile specimens such as marine invertebrates, embryos, diseased cells and fungi. The dialdehyde stabilises intracellular systems of labile microtubules, rough and smooth endoplasmic reticulum, mitotic spindles, platelets and pinocytotic vesicles better than does any other known fixative. Computer-generated reconstructions of mitotic spindle microtubules show that fixation with glutaraldehyde or freeze-substitution results are indistinguishable with respect to microtubule numbers, lengths and arrangements (Heath *et al.*, 1984). However, in principle, freeze-substitution is likely to give better preservation of labile mitotic spindles.

Tissue specimens can be left in this fixative for many hours without apparent deterioration; this flexibility of the pretreatment is one of its major assets. At present, glutaraldehyde is the most efficient and reliable fixative for the preservation of biological specimens for routine electron microscopy.

Although glutaraldehyde does not cause significant protein conformational changes, some structural modification of the protein molecule does occur, especially in that of the α-helix. Available data indicate that proteins are not denatured to any marked extent by fixation with glutaraldehyde; protein tertiary structure seems to remain intact after such fixation. Conformation and biological activity of proteins remain mostly unimpaired after their moderate cross-linking by glutaraldehyde. A possible explanation for this phenomenon is that amino groups, which are the primary target of glutaraldehyde, are usually abundant on the surface of proteins.

Nature of Commercial Glutaraldehyde

As a preliminary to any study of the chemical nature of cross-linking reaction, it is necessary to elucidate the nature of commercial glutaraldehyde. Furthermore, determination of at least the approximate extent of impurities in glutaraldehyde is necessary, since these affect pH, osmolarity and aldehyde concentration in the fixative solution, which, in turn, influence the rate of penetration and cross-linking of proteins. Another reason for knowing the nature of glutaraldehyde is that various impurities differ in the extent and nature of their reactions with OsO_4 during post-fixation, which determines in part the final image of the specimen. Certain impurities in glutaraldehyde solution enhance the formation of osmium black precipitates. Some other impurities may cause the glutaraldehyde solution to show a white precipitate when diluted with a buffer or water.

Glutaraldehyde is generally supplied as a 25% solution with a pH of 3–6. In this state it exists as a hydrate and stays relatively stable for long periods of time if stored at sub-freezing temperatures. With time, however, the solution turns yellow and the pH falls, which indicates an excessive polymerisation of glutaraldehyde. As long as the colour stays faintly yellow, there is not much polymerisation or loss of aldehyde groups. Any glutaraldehyde solution with a pH less than 3 is suspect. Stock solutions of glutaraldehyde with a pH above 4 yield the best preservation of ultrastructure. However, pH is a rather poor indicator of the condition of a glutaraldehyde solution.

The rate of glutaraldehyde polymerisation is dependent upon many factors, including its concentration, pH, temperature and age. Distilled glutaraldehyde tends to polymerise unless all traces of water are removed; even small amounts of water catalyse polymerisation. In fact, at room temperature, concentrated solutions of commercial glutaraldehyde polymerise rapidly, whereas dilute solutions (4% or less) keep better. It deteriorates rapidly at room temperature, in the presence of oxygen, acids or bases, but remains unaffected by the presence of light. The polymerisation is slow in diluted, slightly acidic aqueous solutions.

It is difficult to obtain the dialdehyde in a pure state as a monomer, because most commercial lots contain an indeterminate amount of impurities, although some dealers now supply purified glutaraldehyde. Commercially available '25% glutaradehyde solution' may contain 79% water, 3% glutaraldehyde and 18% derivatives (impurities) of high molecular weight, which may or may not be broken down to glutaraldehyde. The main impurities include glutaric acid, acrolein, glutaraldoxime, ethanol, methanol, 3,4-dihydro-2-ethoxy-2H-pyran, and various polymers (e.g. α,β-unsaturated aldehydes) and products of oxidation and photochemical degradation.

High-performance liquid chromatography analysis of unpurified glutaraldehyde solutions shows a new type of trimer (paraglutaraldehyde), 2,4,6-tris(4-oxobutyl)-1,3,5-trioxane, which exhibits absorption maximum at 280 nm (Tashima *et al.*, 1987). Since monomeric glutaraldehyde also absorbs UV light at 280 nm, this trimer is difficult to distinguish from the monomeric form by the conventional UV absorption method. Paraglutaraldehyde seems to be primarily responsible for the precipitation in alkaline glutaraldehyde solutions, although pentamer and heptamer oligomers may also cause some precipitation in weakly alkaline solutions. Oligomers such as paraglutaraldehyde are thought to be produced during the synthesis of glutaraldehyde solution. As stated above, the chemical nature of these impurities is of considerable importance in understanding the phenomenon of cross-linking of proteins by glutaraldehyde. Since agreement on which of the species plays a key role in protein cross-linking is lacking, various points of views are presented in the next section.

Rubbo *et al.* (1967) suggest that the aldehyde exists as a monomer and that an equilibrium is established between the free aldehyde molecule (open chain) and the hydrated ring structure (cyclic hemiacetal of its hydrate). Hardy *et al.* (1969, 1972) suggest that the monomer exists as a mixture of hydrated forms in aqueous solutions. They consider α,β-unsaturated aldehyde to be a minor component of the organic materials present in aqueous solutions of glutaraldehyde, provided that the solutions are prepared from pure glutaraldehyde and analysed at neutral pH.

Boucher (1972, 1974) and Boucher *et al.* (1973) propose that the monomeric glutaraldehyde is the active species, and the ability of polymeric species to revert to the active monomer is dependent upon pH. It is known that pH plays an important role in determining the type of polymers formed in glutaraldehyde solution. According to these authors, polymers formed at an alkaline pH cannot revert to the monomeric form, because room temperature and time tend to encourage the formation of a more irreversible polymer. On the other hand, polymers formed under the neutral or acidic conditions are thought to revert easily to the monomeric form.

According to another point of view, polymeric species are the major component of commercial glutaraldehyde. As stated above, there is no doubt that glutaraldehyde is polymeric and may contain oligomers such as dimer, cyclic dimer, trimer, bicyclic trimer and other high polymeric species. The dimer which is thought to be the principal polymer in commercial glutaraldehyde is α,β-unsaturated dimer, which forms slowly under ordinary storage conditions but forms relatively rapidly when stored in a buffer at an alkaline pH. In other words, glutaraldehyde is to some extent in equilibrium with its cyclic hemiacetal and polymers of the cyclic hemiacetal at an acid pH, but at neutral or even slightly alkaline pH the dialdehyde undergoes an aldol condensation with itself rather rapidly followed by dehydration, which results in the formation of the dimer. At a certain high alkaline pH, the dimer will precipitate from solution.

As stated above, the dimer absorbs in the ultraviolet at 235 nm because of the α,β-unsaturated bond, and a linear relationship exists between glutaraldehyde concentration and absorption at 235 nm (Munton and Russell, 1970). The dimer can be distinguished from other impurities by differences in their absorption in the UV light; glutaric acid, for instance, has been found to absorb maximally at 207 nm (Gillett and Gull, 1972). The polymerisation of the dimer is enhanced by heat, acids and bases.

The scheme below explains the formation of the dimer (Robertson and Schultz, 1970).

$$\underset{\textbf{(I)}}{2CHO\text{—}(CH_2)_3\text{—}CHO} \xrightarrow{OH^-} \underset{\textbf{(II)}}{CHO\text{—}(CH_2)_2\text{—}CH(CHO)\text{—}CH(OH)\text{—}(CH_2)_3\text{—}CHO}$$

$$\xrightarrow[H^+,\Delta H^-]{OH^-} \underset{\textbf{(III)}}{CHO\text{—}(CH_2)_2\text{—}C(CHO){=}CH(CH_2)_3\text{—}CHO + H_2O}$$

In the case of glutaraldehyde (**I**), the condensation product is a secondary alcohol (**II**), which then is able to eliminate H_2O, forming an α,β-unsaturated aldehyde (**III**); the second reaction is endothermic ($-\Delta H$). The initial oxidation of some aldehyde groups results in the formation of acids, and the presence of even small amounts of these acids, in turn, catalyses an aldol condensation, forming a disubstituted α,β-unsaturated aldehyde. It is known that dilute alkaline solutions cause aldehydes, containing at least one hydrogen in position α, to undergo polymerisation to hydroxyaldehydes as a result of aldol condensation (Fieser and Fieser, 1956).

Commercial glutaraldehyde solutions usually have two peaks (at 280 nm and 235 nm) in their UV absorption spectra. The 280 nm carbonyl peak is due to monomeric glutaraldehyde, while the 235 nm peak is attributed to the presence of a dimer. The purity of glutaraldehyde solution is evaluated by employing a purification index defined as A_{235}/A_{280}; a lower purification index value apparently corresponds to a higher purity. However, the size of these two peaks does not indicate the absolute content of monomeric and polymeric glutaraldehyde. The reason is that some glutaraldehyde solutions contain appreciable amounts of impurities in spite of its low purification index value. This is explained on the basis of the presence of impurities that absorb UV light at or near 280 nm.

The optimal ratio between the monomeric form and the dimer in glutaraldehyde solution, to accomplish satisfactory fixation, seems to be 1:1 to 2:1; ratios outside this range may cause poor fixation. Since for different uses of glutaraldehyde different grades of purity are needed, standardised supplies should be available; such supplies are also needed in order to achieve reproducible results. It is desirable that the suppliers give some indication of the impurities present in each batch of glutaraldehyde.

Reaction with Proteins

It is pointed out that detailed discussion of the reactions of glutaraldehyde with proteins has been presented elsewhere (Ford, 1978; Hayat, 1981; Cheung and Nimni, 1982a, b; Cheung *et al.*, 1985). A brief account of the progress in understanding glutaraldehyde interaction with proteins and its applications has been provided by Hayat (1981, 1986a). A summary of the reactions of this dialdehyde with proteins is given below.

Since the reaction products of glutaraldehyde with tissue proteins have been neither isolated nor identified, the chemistry of its reaction with proteins has been the subject of much debate. The presence of numerous polymerised forms in glutaraldehyde discussed above is partly responsible for the difficulty in chemical identification of the reaction products. One established fact is that the reaction of glutaraldehyde with proteins involves ε-amino groups of lysine side-chains and results in a reaction product with an absorption maximum at ~265 nm. Glutaraldehyde also reacts with other amino acids, including tyrosine, tryptophan, histidine, phenylalanine and cysteine. This dialdehyde is very reactive towards α-amino groups of amino acids, N-terminal amino groups of some peptides, and sulphhydryl groups of cysteine. The imidazole group of histidine is thought to be partially reactive with glutaraldehyde, and the presence of a free amino group seems to be necessary for this reaction. The phenolic ring of tyrosine is also partially reactive in *N*-acetyl tyrosine ethyl ester and glycyl tyrosine (Habeeb and Hiramoto, 1968). Nevertheless, overwhelming evidence indicates that lysine is the most important component of protein involved in the reaction with glutaraldehyde.

In contrast to the established view that the 265 nm absorbing material is the final stable reaction product, Cheung and Nimni (1982a) suggest the presence of at least one more product that shows absorption at 325 nm. They indicate that the 265 nm product is unstable and that the 325 nm product is the major stable reaction product. It is further suggested that as the reaction proceeds, the absorption at 265 nm decreases, this decrease being accompanied by an increase in the absorption at 325 nm.

Reactions of α,β-unsaturated aldehydes (e.g. heated glutaraldehyde) with amines readily yield 325 nm absorbing products. In contrast, monomeric glutaraldehyde yields hardly any 325 nm absorbing products. It is possible that the reactions involve intermediates such as conjugated Schiff's base and the Michael adducts of amines to α,β-unsaturated aldehydes proposed by Peters and Richards (1977). Molin *et al.* (1978) have proposed a two-step reaction explaining the reaction of glutaraldehyde with amino groups through the formation of Schiff's bases. The implication is that Schiff's bases are intermediates in the formation of more stable products that absorb at 325 nm. The information presented above indicates the complexity of the reactions between glutaraldehyde and proteins and the likelihood that cross-linking of proteins proceeds through more than one step or through several routes.

An essential feature of protein cross-linking with glutaraldehyde is oxygen uptake at a high rate (Johnson, 1986). The reason is that oxygen-consuming pathways are the major reaction routes in glutaraldehyde–primary amine reactions. This oxygen uptake competes with the oxygen used in respiration. Available oxygen may be increased by continuously aerating the fixative with oxygen during fixation. Minnasian and Huang (1979) inhibited respiration by adding sodium azide to glutaraldehyde, and reported improvement in tissue preservation deep (up to several millimetres) within the immersion-fixed blocks. Sodium azide inhibits the transfer of electrons to oxygen in the mitochondrial electron transport chain, while it does not affect the oxygen uptake in amine–glutaraldehyde reactions (Johnson, 1986).

MECHANISM OF PROTEIN CROSS-LINKING

Richards and Knowles (1968) and Hardy *et al.* (1976a,b; 1977) have presented mechanisms of cross-linking and proposed the structures of reaction products resulting from glutaraldehyde interaction with proteins. These and other reactions are summarised in Fig. 1.4.

Chemical studies indicate that pyridine derivatives are the major reaction products of amine–glutaraldehyde reactions. Hardy *et al.* (1976a,b; 1977) have isolated pyridine from the products of model amino acid–glutaraldehyde and protein–glutaraldehyde reactions. Studies by Johnson (1986) also indicate that pyridines and pyridine polymers are produced from amine–glutaraldehyde precursors. It is thought that these pyridine polymers provide cross-links that bridge randomly spaced primary amino groups in cells. Johnson (1986) proposes that during fixation glutaraldehyde and soluble amino acids provide the precursors for the pyridine polymer bridges between insolubilised amines (proteins). The extent and rate of cross-linking depend on the availability of polymer precursors. In the absence of sufficient amine precursors, slower cross-linking may occur via aldol condensation between the glutaraldehyde side-chains. The extent of cross-linking in cells varies, depending on the concentration of soluble amines or the glutaraldehyde:amine ratio. It should be noted that free glutaraldehyde becomes depleted at the early stages of fixation; the cross-linking

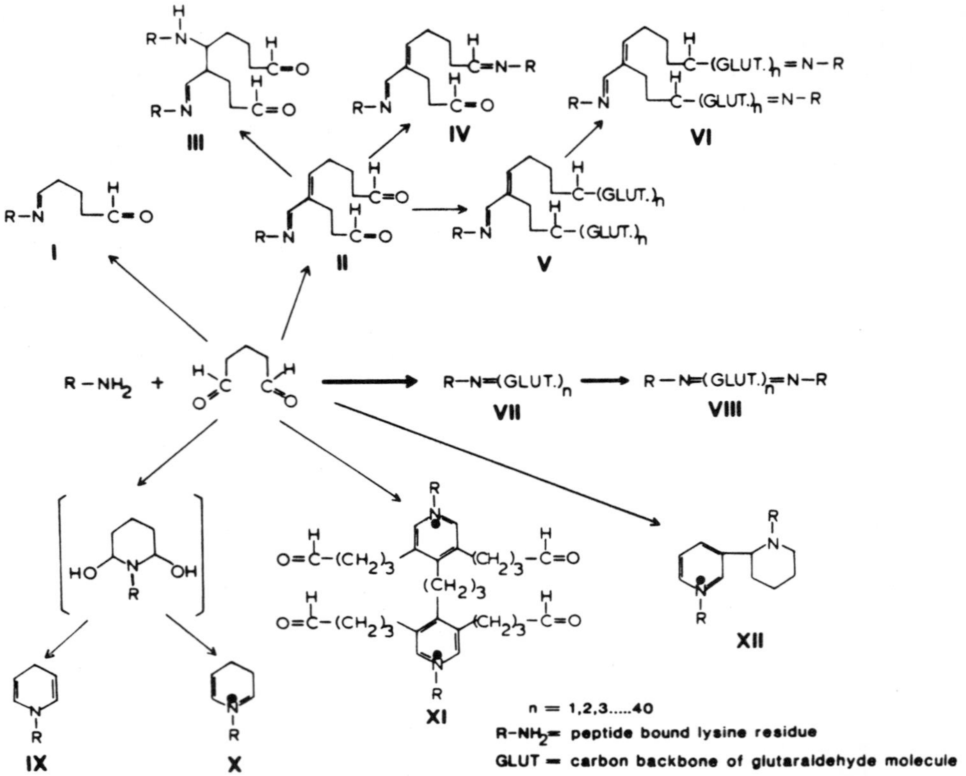

n = 1,2,3.....40
$R-NH_2$ = peptide bound lysine residue
GLUT = carbon backbone of glutaraldehyde molecule

Figure 1.4 Schematic representation of possible reactions of the ε-NH_2 groups of peptide bound lysyl residues with glutaraldehyde. Compound I is a Schiff's base formed between glutaraldehyde and the amine. Compound II is the α,β-unsaturated (conjugated) Schiff's base. Compound III is a subsequent Michael addition product of compound II. Compound IV is derived from the reaction between compound II and another amine. Compounds V and VI represent the glutaraldehyde polymer adducts or cross-links which are also derived from compound II via Schiff's base-catalysed polymerisation. Compounds VII and VIII may be structures similar to V and VI except through some other unknown reaction mechanisms. The exact chemical structures of these compounds are not yet understood. Only structures such as III, IV, VI and VIII represent the actual cross-links formed between peptide chains. Compounds IX and X are possible dead-end dihydropyridine or dihydropyridinium products which follow the ring closure of an intermediate. Compound XI is an earlier model of the cross-link proposed by Hardy *et al.* (1976a). Compound XII is their most recent suggested cross-link—anabilysine. These latter structures, however, do not reflect the larger-molecular-weight products observed by Cheung and Nimni (1982a,b). From Cheung and Nimni (1982a)

reaction can be re-initiated by adding more glutaraldehyde while fixation is in progress.

Significant contributions towards understanding the mechanism of protein cross-linking with glutaraldehyde have also been made by Hopwood (1967a,b; 1969a, 1972), Hopwood *et al.* (1970), Korn *et al.* (1972), Monsan *et al.* (1975), Blass *et al.* (1976) and Tashima *et al.* (1987).

Reaction with Lipids

The reactions between glutaraldehyde and lipids have not been extensively explored. Some studies indicate that glutaraldehyde does not prevent the potential loss in phospholipid bilayer. However, many studies indicate that glutaraldehyde probably reacts with those phospholipids containing free amino groups. The preservation of lipids such as phosphatidylserine and phosphatidylethanolamine with glutaraldehyde is mediated by their amino groups (Roozemond, 1969). Studies of the effect of glutaraldehyde fixation on the extraction of lipids from the mouse brain, kidney and liver also indicate that phosphatidylethanolamine is cross-linked with protein and retained in the tissue. According to Doggenweiler and Zambrano (1981), the retention of phosphatidylserine is mainly due to the presence of strong hydrophobic bonds between it and proteins in the membrane. Fixation of rod and cone membranes with glutaraldehyde rendered 38% of the phospholipids unextractable with chloroform–methanol (Nir and

Hall, 1974). Phosphatidylcholine and phosphatidylinositol are not cross-linked with glutaraldehyde, because they lack primary amines. Preliminary studies by the author indicate that glutaraldehyde renders some of the phospholipids unextractable during dehydration, and very little phospholipids are extracted by this fixative. Glutaraldehyde could conjugate a phospholipid molecule to a protein.

With the exception of certain phospholipids, the reaction of most lipids in the tissue with glutaraldehyde is slight. Even if lipids were retained completely during fixation in glutaraldehyde, most of them would be completely extracted during dehydration and embedding procedures. When glutaraldehyde-fixed cells are not osmicated prior to dehydration, lipoprotein membranes have negative images; osmication after dehydration does not help. Glutaraldehyde may cause some phospholipids to pass into solution, which form myelin-like figures when OsO_4 reacts with these lipoproteins during post-fixation. These artefactual myelin figures are more abundant in large tissue blocks. Formation of these figures can be almost eliminated by adding $CaCl_2$ (1–3 mmol) to the glutaraldehyde fixative. Addition of $CaCl_2$ also minimises the loss of lipids during dehydration and improves the preservation of mitochondria.

Reaction with Nucleic Acids

Only scanty information is available on the possible interactions between glutaraldehyde and nucleic acids. Electron micrographs of specimens fixed with glutaraldehyde show that some nucleic acids are retained in the specimen. Because dialdehyde-treated DNA shows considerable resistance to deoxyribonuclease, it is assumed that dialdehydes react with the amino groups of cystidine and guanine (Brooks and Klamerth, 1968). Since, after treatment with glutaraldehyde, purified DNA can reduce the silver nitrate–methenamine solution, it has been suggested that this dialdehyde may react with DNA (Thiéry; see Millonig and Marinozzi, 1968). However, the association of proteins with nucleic acids cannot be disregarded in explaining the 'retention' of nucleic acids. It is known that various polyamines and histones are associated with nucleic acids.

Glutaraldehyde seems to cross-link only the protein moiety of chromatin, whereas formaldehyde cross-links both protein and DNA. In the absence of cross-links between DNA and protein components of chromatin, it can be inferred that these two components may move relative to each other during subsequent processing for electron microscopy (Sewell *et al.*, 1984). However, Sewell *et al.* indicate that cross-links are formed between HI and the chromatin core histones by glutaraldehyde, which may pin the DNA to the core and give rise to the zigzag appearance of such fibres. This is expected, since glutaraldehyde is the most efficient protein–protein cross-linker. *In vitro* studies by Hopwood (1975) indicate that at temperatures up to 64 °C no reaction occurs between native DNA and glutaraldehyde. At temperatures above 75 °C, the reaction follows pseudo-first-order kinetics, proceeding more rapidly at higher temperatures. Similar reactions occur between the dialdehyde and RNA, except that they may start above 45 °C. The probable reason for the lack of reaction of glutaraldehyde with nucleic acids at low temperatures and the reaction at higher temperatures is that the nucleic acid tertiary structure is maintained by hydrogen bonds which are weakened or broken only at higher temperatures.

To what extent the information derived from *in vitro* studies can be applied to *in situ* nucleic acids is uncertain, for the situation is infinitely more complex in tissue specimens than that in models. Although cross-linking of proteins with glutaraldehyde is well established, what role these glutarated compounds play in the stabilisation of nucleic acids is not known. Thus, a definitive statement on the reaction of glutaraldehyde with RNA and DNA in tissue specimen during fixation cannot be made.

Reaction with Carbohydrates

Some carbohydrates, especially glycogen, are retained in the aldehyde-fixed tissue. Approximately 40–65% of the total glycogen is retained in tissues (e.g. liver and heart) after fixation with glutaraldehyde. Glycogen in relatively thick sections of glutaraldehyde-fixed tissue can be stained with ammoniacal silver solution, without pretreatment with periodic acid (Millonig and Marinozzi, 1968). In this reaction, the silver solution brings out glycogen, presumably by reacting with the free aldehyde groups of glycogen-bound glutaraldehyde. Glutaraldehyde most probably reacts with polyhydroxyl compounds to form polymers. Additional information is needed to explain the reaction of glutaraldehyde with carbohydrates.

Osmolarity of Glutaraldehyde

As stated earlier, it is generally recognised that the osmolarity of the fixative exerts a significant influence on the quality of preservation of tissue fine structure. In the total osmolarity of the fixative solution, the concentration of the fixative vehicle (buffer salts and additives such as sucrose) seems to be more important than that of glutaraldehyde, although this aldehyde does exert considerable osmotic effect on certain tissues, such as kidney and lung and plant cells. It has already been emphasised that glutaraldehyde concentration can be changed over a wide range with relatively little change in the morphology of fine structure.

An agreement on the degree of contribution of glutaraldehyde to the effective osmotic pressure of the fixative solution is lacking. Conflicting reports have been published. Recent studies indicate that the contribution of glutaraldehyde to the effective osmotic pressure is affected by the type and molarity of the buffer, the concentration of the glutaraldehyde and the water potential of the specimen (Coetzee and van der Merwe, 1985a). These interactions are exceedingly complex and their effects are non-linear. Addition of glutaraldehyde to buffers increases their dissociation coefficient, with a corresponding increase in their osmolarity (Coetzee and van der Merwe, 1985b). This increase is more than can be expected after the simple arithmetic addition of the osmolar values of the buffer and the aldehyde. The increase in buffer osmolarity may be as much as 107 mosm. Na–Na phosphate buffer (0.05 M, pH 7.2) has an osmolarity of 188 mosm. When 5% glutaraldehyde is added to this buffer, the osmolarity is increased to 225 mosm. Of the four buffers tested, Na–Na phosphate showed the lowest increase in osmolarity (Coetzee and van der Merwe, 1985b). The total osmolarity of the fixative solution changes from the time it begins to be used to the completion of fixation. The total osmolarity of the fixative solution tends to increase on standing, and this increase is contributed by the glutaraldehyde.

Strong evidence exists for the importance of the concentration of substances in the fixative vehicle. As an example, liver and muscle tissue are best fixed with moderately hypertonic fixatives (400–500 mosm) containing 2% glutaraldehyde, whereas strongly hypertonic fixatives (570 mosm) and isotonic fixatives, both containing 2% glutaraldehyde, give poor fixation in that the extraction of cellular constituents is considerable. Hampton (1965) reported that the concentration of sucrose in glutaraldehyde bears importantly on the morphology of fixed mouse Paneth cell granules. He demonstrated that these cells were preserved much better with glutaraldehyde containing 1% sucrose than with glutaraldehyde without the non-electrolyte.

Evidence has been obtained that after fixation with glutaraldehyde and other aldehydes the osmotic properties of membranes and other metabolic activities remain partly preserved, although significant changes in the mechanical behaviour of fixed cells do occur. Cells fixed with glutaraldehyde maintain their impermeability to solutes such as sucrose, whereas OsO_4-fixed cells become permeable to sucrose and other solutes of similar size. That Hill activity of chloroplasts isolated from glutaraldehyde-fixed leaves is retained has been shown by Park *et al.* (1966). It has been established that fixation with glutaraldehyde prevents conformation and volume changes in isolated mitochondria and chloroplasts, and allows electron transport to continue. Dielectric techniques have shown that even though membrane resistances of glutaraldehyde- or acrolein-fixed cells are, in general, much lower than those of normal cells, a major portion of the fixed membrane still acts as an insulating barrier. It should be remembered that membrane resistance is considered to be an accurate criterion for determining membrane condition.

Studies on the shape of agranular synaptic vesicles indicate that even brief storage of aldehyde-perfused nervous tissue pieces in cacodylate buffer, prior to osmication, has a severe flattening effect on the vesicles of peripheral cholinergic axon endings (Bodian, 1970). This and other cytological and biochemical studies lead one to question the completion of fixation at the molecular level by glutaraldehyde in intact tissues. Membranes remain sensitive to a change in the osmotic pressure of the rinsing and dehydration solutions. Precautions must, therefore, be taken to minimise the differences between the osmolarity of rinsing solutions and that of the fixative. One way to increase the osmolarity of rinsing solutions is by addition of either electrolytes or non-electrolytes. In some cases, glutaraldehyde may destroy the selective permeability property of the cell membrane. This has been demonstrated in the isolated myocardium (Harper, 1986).

Osmolality of Glutaraldehyde

The presence of osmotically active impurities in glutaraldehyde solutions results in osmolalities generally much higher than the theoretical values. For example, osmolalities as high as 570 mosm of a 3% solution of commercially supplied glutaraldehyde have been recorded (Maser *et al.*, 1967), whereas the osmolality of a 3% solution of pure glutaraldehyde is only 300 mosm. Since even the latter solution, when buffered, is hypertonic to the majority of biological fluids bathing in the tissue, it is obvious that the buffered fixatives containing the former solution would be very hypertonic. The deleterious effect of excessively hypertonic fixatives has already been discussed. Moreover, the impurities may themselves affect the quality of fixation. The recovery of enzymatic activity, for instance, is inversely related to impurity content. Thus, to obtain optimum fixation and maximum retention of enzymes, glutaraldehyde should be 'pure' and its osmolality known. In other words, the quality of fixation can be controlled more effectively when the amount of the active component in glutaraldehyde is known.

Polymerisation of glutaraldehyde, on the other hand, results in decreased osmolalities. When monomeric glutaraldehyde is polymerised to dimers or trimers, there is a decrease in the osmolality. This is to be expected, because the osmolality of a solution is a measure of the total number of solute particles in solution per unit of solvent. When two or more individual monomeric molecules form a single larger dimer or trimer, the number of separate monomeric molecules is reduced and the osmolality decreases. Osmolalities of aqueous 2% commercial glutaraldehyde solution from different batches vary from 190 to 280 mosm.

Average osmolalities of phosphate-buffered glutaraldehyde formulations of different concentration are given below. The commercially available glutaraldehyde was purified by charcoal treatment, and the osmolalities were measured with a commercial osmometer (Chambers *et al.*, 1968).

Formulation	*Osmolality* (mosm)	*pH*
Phosphate buffer (0.1 M)	230	7.4
1.2% glutaraldehyde	370	7.2–7.3
2.3% glutaraldehyde	490	7.2–7.3
4.0% glutaraldehyde	685	7.1–7.3

A slightly hypertonic solution (400–450 mosm) of buffered glutaraldehyde is recommended for general fixation. For mammalian cells and tissues the recommended formulation is: 2% glutaraldehyde in 0.1 M cacodylate buffer containing 0.1 mol sucrose (pH 7.2) with total and vehicle osmolalities of 510 and 300 mosm, respectively.

Temperature

Since the reaction between glutaraldehyde and cellular reactive groups is increased at high temperatures, the rate of penetration and fixation by glutaraldehyde is enhanced at room temperature. However, in order to minimise the extraction of cellular constituents due to autolysis, low temperatures are preferred. Fixation with glutaraldehyde at low temperatures also reduces the shrinkage of mitochondria, the granularity of cytoplasm and volume changes. For the preservation of very labile enzymes, fixation by immersion or vascular perfusion should be carried out in the cold. Low temperatures minimise the formation of artefacts, especially in enzyme studies, due to excessive polymerisation of glutaraldehyde when it has been stored at sub-freezing temperatures. Treatment with glutaraldehyde for 2 h at 0–4 °C is generally preferred for routine fixation of plant and animal tissues.

However, glutaraldehyde can be used in many tissues at temperatures ranging from 0 to 25 °C, with little apparent difference in the appearance of the fine structure. Sometimes the temperature of fixation can be critical. For example, labile structures, such as certain microtubules, may be lost when glutaraldehyde is employed in the cold, owing to rearrangement. Certain types of dense specimens may not allow adequate penetration by glutaraldehyde at cold temperatures. In such cases, it is desirable to carry out fixation at room temperature. Alternatively, fixation can be accomplished at room temperature following a brief preliminary fixation in the cold. In vascular perfusion, glutaraldehyde should not be used below body temperature, otherwise it may cause vasoconstriction.

Concentration

The process of protein cross-linking is affected by the concentration of the glutaraldehyde solution. Although changes in the concentration of the dialdehyde are always accompanied by changes in its osmolarity, this change within a reasonable limit may not affect the quality of tissue preservation. However, the quality of fixation of isolated cells is readily influenced by changes in the concentration of the fixative. Slow penetration by glutaraldehyde might prompt the investigator to use higher concentrations. This practice, however, is not desirable, because the use of excessively high concentrations results in cell damage. The most desirable concentrations of glutaraldehyde solutions, including the vehicle, are hypertonic, but in these solutions the concentration of glutaraldehyde is usually hypotonic. In general, high concentrations (e.g. 37%) cause an excessive shrinkage, while low concentrations (e.g. 0.5%) result in marked extraction of cell constituents from the tissue.

Although glutaraldehyde can be used safely in a relatively wide range of concentrations, 1.5–4.0% solutions are recommended for fixing plant and animal tissues. Higher concentrations of glutaraldehyde have proved useful for obtaining superior fixation of certain tissues. A few examples are given. Lung tissue can be fixed with 8% aged, polymerised glutaraldehyde in 0.1 M phosphate buffer (pH 7.2) for 2 h. This fixative solution has an osmolality of ~1000 mosm. Nervous tissues, including CNS, can be fixed with 6% glutaraldehyde in 0.1 M cacodylate buffer with an osmolality of ~900 mosm. Glutaraldehyde in a concentration as high as 19% has been used as an initial perfusate for preserving CNS prior to a conventional perfusate of low concentration (Schultz and Case, 1970). This approach seems to minimise considerably the production of artefacts such as myelin figures and vacuolated mitochondria. Relatively high concentrations of glutaraldehyde (6%) seem to be desirable for fixing plant tissues. For example, *Avena* coleoptile is preserved better with 6% glutaraldehyde than with solutions of low concentrations.

An interesting approach was utilised by Thornthwaite *et al.* (1978) to determine optimum glutaraldehyde concentration. They used electronic cell volume analysis to determine ideal concentration of glutaraldehyde for fixing nucleated mammalian cells. In this study, high-resolution electronic cell volume spectra of glutaraldehyde-fixed cells were employed to determine the glutaraldehyde concentration which produced spectra most closely resembling those of the unfixed cells. The optimum concentrations of glutaraldehyde in 0.05 M cacodylate buffer (pH 7.4) for fixing mastocytoma tumour cells, human lymphocytes, and human granulocytes and monocytes were 3.8%, 4.9% and 4%, respectively.

On the other hand, certain specimens are fixed better by using low concentrations of glutaraldehyde. Superior fixation of single cells is obtained with low concentrations of glutaraldehyde (0.2–1.0%). For example, preliminary fixation with a low concentration of glutaraldehyde (0.1%) followed by a second fixation with 3% glutaraldehyde seems to preserve platelets in suspension better; platelet fixation is thought to occur without coagulation of plasma proteins by using this approach. Relatively low concentrations of glutaraldehyde (1–2%) are useful for enzyme localisation. One should be aware that too low a concentration of glutaraldehyde may increase the number of artefacts, including the number of myelin figures in tissue specimen. It is interesting to note that similar artefacts may be introduced by purified glutaraldehyde.

pH

The importance of pH in fixation has been emphasised earlier. Among all the factors that influence the interaction of glutaraldehyde with proteins, pH is considered to be the most important in obtaining the maximum binding of aldehyde groups with proteins. An increase in pH usually results in an increase in the binding capacity of glutaraldehyde. For example, the maximum uptake of glutaraldehyde by collagen and cross-linking of collagen by glutaraldehyde occurs at pH 8.0. The guanidino groups of arginine may react only at pHs higher than 9.0. It has been shown that glutaraldehyde tans more rapidly with decreasing hydrogen ion concentration (Habeeb and Hiramoto, 1968).

A higher pH (8.0) seems to be more effective during pre-fixation of plant tissues with glutaraldehyde. Improved preservation of animal tissues by raising the pH from 7.0 to 8.0 has also been reported (Peracchia and Mittler, 1972). However, at pHs higher than 7.5, glutaraldehyde tends to polymerise rapidly. This excessive polymerisation may introduce artefacts. Therefore, fixation at higher pHs for routine electron microscopy cannot be recommended until more information on the mechanism of protein cross-linking by glutaraldehyde and the role of pH in it under fixation conditions becomes available.

METHOD FOR USING GLUTARALDEHYDE AT HIGHER pHS

Specimens are fixed by immersion in 3–6% glutaraldehyde in a buffer (pH 7.0) for 1 h at room temperature. The pH of the glutaraldehyde solution is then raised to 8.0 in three steps of ~0.3 pH unit each, at 30–60 min intervals by adding drops of 1 N NaOH while stirring. While approaching close to pH 8.0, 0.1 N NaOH is used to avoid too rapid a change in the pH. The specimens are allowed to remain in glutaraldehyde at pH 8.0 for 30–60 min. For further details, see Peracchia and Mittler (1972).

CHANGES IN pH DURING FIXATION WITH GLUTARALDEHYDE

When glutaraldehyde is added to distilled water (pH 7.4) in the absence of a buffer, the pH is lowered to about 4.3 (Johnson, 1985).The pH of buffers is also lowered when glutaraldehyde is added to them. The pH usually drops during fixation. Any drop in pH will adversely affect the protein cross-linking, because the optimal pH at which glutaraldehyde is most efficient cross-linker of proteins is somewhat higher (7.5–8.0) than the generally used physiological pH values of 7.2–7.4. Purified glutaraldehyde contains about 0.002 equivalents of hydrogen ions per mole of glutaraldehyde.

Even in the presence of a buffer with glutaraldehyde, the pH may drop during fixation, depending in part on the type of buffer used. The presence of a buffer is supposed to resist changes in hydrogen ion concentration, and buffer does minimise decreases in pH during fixation with glutaraldehyde. Some buffers are more effective than others in resisting a change in hydrogen ion concentration or pH, and this effectiveness depends primarily on the pK_a value of a buffer. The pK_a values of buffers commonly used in electron microscopy are given below.

Buffer	pK_a (37 °C)
Cacodylate	6.19
PIPES	6.66
MOPS	7.10
Phosphate	7.21
HEPES	7.31

Since pH stands in an inverse relationship to the hydrogen ion concentration, an increase in hydrogen ion concentration means acidification resulting in lowering of the pH. Although several mechanisms exist within living cells to consume hydrogen ions (Roos and Baron, 1981), the large amount of hydrogen ions released during glutaraldehyde fixation is thought to overwhelm these mechanisms. The acidification of solutions containing amino acids and proteins in the presence of glutaraldehyde or formaldehyde is not uncommon. Significant but transient lowering of the pH occurs within cells being exposed to glutaraldehyde (Johnson, 1985). By reacting with free amino groups, glutaraldehyde causes the following equilibrium to shift to the right:

$$R{-}NH_3^+ \rightleftharpoons R{-}NH_2 + H^+$$

This release of hydrogen ions contributes to the pH changes (acidification) in the tissue during fixation. The total amount of hydrogen ions released is proportional to the concentration of primary amines (Johnson, 1985).

Changes occurring in the intracellular and/or extracellular pH during fixation are expected to influence the nature and degree of protein cross-linking. During fixation with glutaraldehyde, for instance, there is a decrease in the overall positive charge of the amino groups. This decrease is related to the amount of acid released, which, in turn, depends on the availability of active hydrogen atoms from the functional groups (e.g. primary amines) in the tissue. However, the positive charge is restored within minutes, because of the formation of positively charged pyridinium end-products of amine–glutaraldehyde reactions (Hardy *et al.*, 1976a,b; 1977, 1979).

Rate of Penetration

The rate of glutaraldehyde penetration depends primarily on both its concentration and the type of the specimen, and secondarily on the pH and concentration of the fixative and ambient temperature. The rate of penetration is also influenced by the type of buffer used, although the ionic strength of the buffer has much less effect on the penetration. Using a bovine serum albumin model system, it has been shown that the rate of glutaraldehyde penetration is only slightly affected by the ionic strength of the buffer and then only if it rises above 0.2 M (Coetzee and van der Merwe, 1985a).

Penetration at room temperature is definitely faster than in the cold. Glutaraldehyde (2%) in a buffer (200 mosm) penetrates into soft animal tissues (e.g. liver) ~0.7 mm in 3 h at room temperature, while good fixation reaches to a depth of only 0.5 mm in the same period of time. After 24 h, glutaraldehyde penetrates to a depth of ~1.5 mm, while good fixation reaches to a depth of 1.0 mm. According to Chambers *et al.* (1968), however, the maximum penetration of human liver by 4% glutaraldehyde in 24 h at room temperature and in the cold was 4.5 mm and 2.5 mm, respectively. A mixture of glutaraldehyde (2%) and formaldehyde (2%) in the buffer (200 mosm) penetrates human liver to depths of 2.0, 2.5 and 5.0 mm in 4, 12 and 24 h, respectively (McDowell and Trump, 1976).

Even similar tissue types obtained from different species react differently to glutaraldehyde diffusion. Chambers *et al.* (1968) showed that 4% glutaraldehyde, for example, penetrated rabbit liver and human liver 1.5 mm and 3.0 mm, respectively, in 9 h at room temperature. The rapid penetration into the human liver is presumably related to more efficient conduction of the fixative by its vessels.

It is apparent from the above data that in order to obtain uniform fixation of all cells in a tissue specimen within a period of 2 h, the dimensions of the specimen should not exceed 0.5 mm on any side. Prolonged fixation would not necessarily improve the quality of fixation, because penetration and fixation should be completed prior to the onset of autolytic changes.

The depth of fixation is considered to be the same as the depth of the penetration by glutaraldehyde. The penetration of the dialdehyde into the tissue is indicated by a pale-yellow colour (which is primarily due to the formation of Schiff's bases when the dialdehyde reacts with basic amino acids) and firm appearance. Since glutaraldehyde introduces Schiff positivity, at least initially, to the tissue fixed in it, the presence of aldehyde groups in the region showing pale-yellow colour can be confirmed by its pink staining with Schiff's reagent.

The slow penetration by glutaraldehyde becomes a serious problem in the fixation of dense specimens (e.g. seed) and specimens with relatively impervious walls (e.g. yeast). However, this problem can be solved by using glutaraldehyde in combination with rapid penetrants such as formaldehyde or acrolein. A mixture of glutaraldehyde (2%) and formaldehyde (2%) was employed to preserve grooves (cleft-like invaginations) on the surface of plasma membrane of yeast cells (Ghosh, 1971). These grooves were not preserved when the cells were fixed with glutaraldehyde alone.

The rate of penetration can also be increased by adding dimethyl sulphoxide (DMSO) to the fixative, although its effects on enzymes and ultrastructure are not fully known. Glutaraldehyde solutions containing 2.0–10% DMSO are recommended. It has been demonstrated that DMSO enhances the cell permeability of animal tissues, plant tissues and micro-organisms. Excellent preservation of embryonic tissues has been obtained by employing a mixture of glutaraldehyde (3%), formaldehyde (2%), acrolein (1%) and DMSO (2.5%) (Kalt and Tandler, 1971). The addition of DMSO to the fixative is known to improve preservation of frozen tissues (Etherton and Botham, 1970). The advantages as well as the disadvantages of using DMSO for preserving both ultrastructure and enzymatic activity have been discussed by Hayat (1973, 1981).

Specimen Shrinkage

Aldehyde fixative solutions used routinely for electron microscopy cause shrinkage of the cell. The degree of shrinkage in cytoplasm may differ from that in the nucleus. Glutaraldehyde causes more shrinkage in nuclei than does OsO_4 or formaldehyde, but just the opposite is true in the case of cytoplasm. Thus, when comparing subcellular dimensions in fixed cells, the possibility of different degrees of change of dimensions induced in different parts of the same cell by the fixative cannot be ignored.

It should be noted that fixatives used routinely are hypertonic. Fixatives with a higher osmolarity usually cause greater shrinkage than that caused by fixatives with lower osmolarity. A change in the concentration of an aldehyde, even within reasonable limits, may have an effect on the volume of single cells or tissues.

Various types of tissues show different degrees of shrinkage after fixation with glutaraldehyde. The wide range of shrinkage shown by the following examples is also due to differences in the preparatory procedures employed in different laboratories. Specimens may shrink ~5% in linear dimensions, as compared with their size before fixation (Weibel and Knight, 1964). Rat brain tissue fixed with 4% glutaraldehyde and embedded in Epon showed a shrinkage of 9% (Hillman and Deutsch, 1978), while mouse ova shrank 8%, compared with live cells (Konwiński *et al.*, 1974). Glutaraldehyde (4%) caused a shrinkage of ~6% of rat liver in 18 h at 4 °C (Hopwood, 1967a), while 2% glutaraldehyde caused 5–10% shrinkage of calf erythrocytes (Carstensen *et al.*, 1971). Glutaraldehyde reduced the surface area of lymphocyte nuclei by 4–6% (Maul *et al.*, 1972). Although fixation with glutaraldehyde causes shrinkage of chloroplasts, subsequent treatments may neutralise this shrinkage (Diers and Schieren, 1972). Isolated cells usually shrink more as a result of fixation than do intact tissues. The extent of shrinkage is affected by the duration of post-fixation with OsO_4 and its concentration, vehicle osmolarity, aldehyde concentration (either too low or too high) and other post-fixation conditions. Age-dependent dimensional changes should be kept in mind. It is known, for example, that young brains shrink more than old ones.

To reduce dimensional changes in the tissue, it has been suggested that osmolarity should be adjusted at all stages of fixation, including post-glutaraldehyde rinse and post-osmication (Louw *et al.*, 1986). Osmolarity of the buffer used for rinsing should be equivalent to that of the buffer vehicle used for glutaraldehyde fixative (Lee *et al.*, 1982).

Limitations of Glutaraldehyde

Being an organic reagent, glutaraldehyde is unable to impart contrast or electron opacity to tissue to result in electron staining. Treatment of cells with glutaraldehyde monitored by light microscopy indicates that structural changes occur during fixation (Skaer and Whytock, 1976; Arborgh *et al.*, 1976). A highly pleomorphic lysosomal system in chick cells is vesiculated by standard glutaraldehyde fixation (Buckley, 1973a,b). However, this system can be stabilised in the presence of a relatively high concentration of calcium ions, but the best results are obtained with prior equilibration and fixation in the presence of $CaCl_2$ (0.1 mol) and $MgCl_2$ (0.02 mol) plus sucrose (0.3 mol). A pleomorphic network which forms a dominant component of the living cytoplasm of plant hair cells is completely transformed into vesicles by glutaraldehyde (O'Brien *et al.*, 1973; Mersey and McCully, 1978). Glutaraldehyde causes clumping of chromatin and transforms nuclear sap into a coarse network.

Glutaraldehyde is incapable of rendering most lipids insoluble in organic solvents used during dehydration. Accordingly, glutaraldehyde-fixed tissues show cellular membranes as negative images. Longer durations of fixation with the dialdehyde may increase the production of artefactual myelin figures. The production of these figures in the CNS is probably related to delayed fixation by solutions of low concentrations of glutaraldehyde. It is assumed that this artefact results from mobilisation of complex lipids by glutaraldehyde, followed by their molecular reorganisation and staining with OsO_4.

As stated above, glutaraldehyde is a poor fixative for phospholipids, and so it causes a number of artefacts in membranes, especially those undergoing rapid structural changes. A few examples suffice. The clearing of intramembrane particles from the plasma membrane overlying the secretory granule during exocytosis commonly seen in glutaraldehyde-fixed cells is an artefact (Tanaka *et al.*, 1980). A second example is the formation of blebs, usually at the site of membrane fusion, in glutaraldehyde-fixed cells. These blebs appear to pinch off, giving rise to many vesicles within exocytotic pockets (Chandler, 1984). Artefactual blebs may also be seen in cells that are not involved in fusion activity (Hasty and Hay, 1978). All of these and other similar artefacts can be explained on the basis of the inability of glutaraldehyde to fix lipids. Glutaraldehyde is unable to prevent movement of phospholipids and intramembrane particles. Membrane fluidity is mostly unaffected by glutaraldehyde. On the other hand, OsO_4 reduces membrane fluidity to zero (Jost *et al.*, 1973).

The preparation of specimens for electron microscope autoradiography is accompanied by significant translocation and intercellular redistribution of radiolabelled lipids, causing spurious labelling patterns. Thus, the problem of redistribution and intercellular transfer of natural phospholipids in the glutaraldehyde-fixed cells and tissues cannot be ignored, and the results of autoradiographic methods used for ultrastructural localisation of lipids should be interpreted with caution.

Glutaraldehyde is not an effective fixative at cold temperatures (−20 °C). Acrolein is most effective, even at −80 °C. Although this aldehyde provides adequate preservation of the cytoplasm, fixation of membranes is less than satisfactory. Osmium tetroxide can also be used at temperatures of −40 °C to −50 °C, for example, for freeze-substitution. Glutaraldehyde fixation in the cold may result in the loss or rearrangement of labile microtubules and a more dispersed pattern of ribosomes. Glutaraldehyde is unable to satisfactorily preserve viruses (e.g. wheat streak and tobacco mosaic viruses) in plant tissues.

The contraction of extracellular space in the nervous tissue during glutaraldehyde fixation is a well-known phenomenon. Van Harreveld and Khattab (1968) have demonstrated that perfusion of cerebral cortex with glutaraldehyde causes an increase in the impedance of the tissue, an accumulation of chloride into cellular elements and a contraction of extracellular space. This transport of extracellular material is thought to be a consequence of an increase in the sodium permeability of the plasma membrane of cells which take up chloride and water during fixation with glutaraldehyde. Studies by Van Harreveld and Fifkova (1972) indicated that glutamate was released from the intracellular into the extracellular compartment in thick retina during fixation with glutaraldehyde, and it was suggested that the action of glutamate on the plasma membrane was responsible, in part, for the contraction of extracellular space.

Caution is warranted in the use of glutaraldehyde for enzyme studies. In general, glutaraldehyde is a powerful inhibitor of the activity of most enzymes and antigens. Another type of artefact is due to the ability of glutaraldehyde to immobilise metabolites such as amino acids. Glutaraldehyde is thus capable of binding free amino acids to tissue constituents. This becomes significant in autoradiographic studies where amino acids labelled with radioisotopes are employed. Peters and Ashley (1967) have described a possible artefact caused by glutaraldehyde in the study of protein formation in the presence of labelled amino acids. On the other hand, this binding capability of glutaraldehyde can be utilised advantageously to immobilise diffusible compounds, including enzymes containing amino acid groups. In this connection, it is known that this aldehyde penetrates into the tissue before labelled amino acids are lost from the cell by diffusion.

Studies of the effect of glutaraldehyde on the levels of amino acids in rat brain indicated that, following perfusion fixation, significant increases in the levels of brain glutamic acid, alanine, valine, isoleucine, leucine and tyrosine occurred compared with unfixed controls (Davis and Himwich, 1971). This is expected since, through cross-linking, glutaraldehyde binds amino acids to proteins initially and then larger amounts can be extracted for the analyses. Some of the increase in the amino acid level is also attributed to the shrinkage of the tissue, because the level of amino acids is based on wet tissue weight.

Fixation with glutaraldehyde does not completely protect the specimen from the extraction of cellular materials (e.g. some proteins) during subsequent processing, such as rinsing in a buffer. This loss is less at low temperatures. Some nuclear proteins, for instance, in glutaraldehyde-fixed (as well as unfixed) specimens undergo extraction by phosphate or veronal buffer. Different buffers extract different types and amounts of proteins.

Some of the problems mentioned above can be overcome by post-fixation with OsO_4. Post-fixation stabilises the fine structure already maintained by glutaraldehyde so that it can withstand embedding in resins. It is recognised that DNA of chromatin and general cytoplasmic details are preserved much better and more completely by double fixation than by either OsO_4 or glutaraldehyde. Post-fixation with OsO_4 is absolutely essential for electron cytochemical methods that are based on the 'osmiophilic principle'. Fixation with glutaraldehyde without subsequent treatment with OsO_4 is considered unsatisfactory for routine electron microscopy, except for some cytochemical studies.

Purification of Glutaraldehyde

The impurities present in glutaraldehyde can be removed by vacuum distillation, by filtering through activated charcoal or by ion exchange. The simplest method is treatment with charcoal. However, this method yields a less pure glutaraldehyde, for charcoal is unable to remove certain impurities such as inorganic materials. When very high purity is desired, the method of choice is vacuum distillation, either at atmospheric pressure or at reduced pressure. The concentration of glutaraldehyde in the distillate is determined with a recording spectrophotometer. Purified glutaraldehyde should show an absorption maximum in the UV at a wavelength of 280 nm; absorption at any other wavelength is caused by impurities. Since a direct linear relationship is found between concentration or osmolarity and the optical density at 280 nm, this parameter can serve to determine both the concentration and the osmolarity of the distilled glutaraldehyde.

A very minor and slow, but possibly significant, polymerisation of purified glutaraldehyde occurs even under mildly alkaline conditions of fixation, and both animal and plant specimens catalyse the formation of polymers which absorb at 235 nm (Jones, 1974; Goff and Oster, 1974). Significant amounts of these polymers can be found in inadequately purified glutaraldehyde. Molin *et al.* (1978) found an index of 1.0–1.2 (235/280 nm) in glutaraldehyde purified by absorption with activated charcoal. Purified glutaraldehyde polymerises faster at an alkaline pH and at room temperature than at an acidic pH and in the cold. Although purified glutaraldehyde is commercially available, scientists in some countries lack access to it. Several methods for purifying glutaraldehyde are given below.

THE CHARCOAL METHOD

Approximately 200 ml of commercial glutaraldehyde solution (50%) is added to 30 g of activated charcoal (Merck & Co., American Norit Co. or Fisher Sci.) in a large flask or beaker. The mixture is thoroughly shaken for ~1 h at 4 °C and then vacuum filtered through Whatman No. 42 filter paper mounted in a Buchner funnel. The filtrate is remixed with 20% (w/v) fresh, activated charcoal and refiltered. The process is repeated at least twice; the number of washings depends upon the amount of absorption at 235 nm, which shows the extent of impurities. The final yield of purified glutaraldehyde is ~20–30 ml.

It should be noted that each charcoal wash reduces both the volume of solution and its glutaraldehyde concentration. Starting with 200 ml of 25% glutaraldehyde and using 30 g of Norit Ex charcoal for each wash resulted in ~150 ml in a concentration of ~22% after the first wash (Garrett *et al.*, 1972). After the second wash, the volume was reduced to ~95 ml in a concentration of ~18%. Repeated washings with alkaline charcoal result in the elevation of pH of the glutaraldehyde solution. Starting with 25% glutaraldehyde with pH 3.1, after one, two, three and four charcoal washes the pH was raised to 5.8, 6.9, 7.7 and 8.0, respectively (Trelstad, 1969). The resultant alkaline pH of the purified glutaraldehyde solution encourages its rapid polymerisation. However, the rate of polymerisation can be reduced by slightly acidifying the solution by the addition of a few drops of HCl.

THE DISTILLATION METHOD

According to the simple method (Smith and Farquhar, 1966), the distillation is carried out at atmospheric pressure. The distillate is collected at ~100 °C in 50 ml aliquots which are monitored by measuring the pH. Any sample showing a pH lower than 3.4 is discarded. A pure glutaraldehyde solution in a concentration of 8–12% is obtained.

Alternatively, a single-stage distillation under moderate vacuum (Anderson, 1967) yields a glutaraldehyde of equivalent purity. Approximately 250 ml of commercial glutaraldehyde is charged into a 500 ml Vigreaux distilling flask heated by an electrical heating mantle and connected to a Liebig condenser. The distillation is performed under vacuum at 15 mmHg. The temperature is raised to ~65 °C and distillate is collected, which is a viscous clear liquid. Upon interrupting the vacuum, the distillate is immediately diluted with an equal volume of freshly boiled demineralised distilled water by slow addition of the latter (75 °C) to the magnetically stirred distillate under a stream of nitrogen. It may be useful to remember that the boiling point of 25% aqueous solution of glutaraldehyde is ~101 °C.

Fahimi and Drochmans (1965) developed a method to obtain monomeric glutaraldehyde by vacuum fractionation distillation. According to this method, the glutaraldehyde solution is distilled twice to obtain a high concentration of the monomer. Dijk *et al.* (1985) have developed a method which yields a high concentration of the monomer after one distillation, which has been reported to have a purification index less than 0.2.

Storage of Glutaraldehyde

As stated earlier, impurities increase spontaneously when glutaraldehyde is kept under ordinary conditions of storage. It is thought that under these conditions, glutaraldehyde is oxidised to glutaric acid, which is the final oxidation product of this dialdehyde. Since barium glutarate is insoluble, some workers store glutaraldehyde over solid barium carbonate in order to neutralise the glutaric acid as it forms. This practice is undesirable, since neutralisation of the acid promotes further conversion of glutaraldehyde to an acid. More importantly, however, impurities that increase under ordinary conditions of storage show absorbance at 235 nm instead of at 207 nm, which is the characteristic of glutaric acid. The increase in absorbance at 235 nm is due to the formation of a polymer of glutaraldehyde rather than a product of oxidation. This increase is temperature-dependent, only slightly affected by the presence of oxygen and unaffected by the presence of light (Gillett and Gull, 1972).

The two most important factors in the storage of glutaraldehyde are temperature and pH. The degree of polymerisation is highest when solutions are stored at room or higher temperatures. In fact, the process of polymerisation increases exponentially with the temperature (Rasmussen and Albrechtsen, 1974). However, polymerisation is almost independent of temperature in the range 1–25 °C, and a linear relationship does not exist between temperature and the presence of polymeric glutaraldehyde above 0 °C.

Glutaraldehyde polymerises rapidly at high pHs. For instance, at pH 8.5 or over, polymerisation occurs so rapidly, even at 4 °C, that it is advisable not to mix the buffer and fixative until immediately prior to use (Rasmussen and Albrechtsen, 1974). Approximately 50% polymerisation of purified glutaraldehyde occurs at 4 °C in 7 weeks at pH 6.5, in 3 weeks at pH 7.5 and in 6 days at pH 8.5. The desired pH can be obtained by the addition of a few drops of dilute solution of HCl. It should be noted that the rate of glutaraldehyde polymerisation differs in various buffers at the same pH. Less polymerisation occurs when glutaraldehyde is stored in cacodylate buffer compared with that in phosphate buffer at the same pH (7.4).

Purified glutaraldehyde remains relatively stable for several months if stored at 4 °C or below, provided that the pH is lowered to ~5.0 (Trelstad, 1969). Purified glutaraldehyde can be stored for ~6 months at −14 °C and for ~1 month at 4 °C without significant polymerisation. The most effective way to minimise the deterioration of purified glutaraldehyde is by storing it as an unbuffered, 10–25% solution at sub-freezing temperatures (~−20 °C). According to Rasmussen and Albrechtsen (1974), however, there is a sharp rise in polymerisation on either side of neutrality. Probably, somewhat different storage conditions are required for purified and unpurified glutaraldehyde. There is no great advantage in storing glutaraldehyde in the dark or under inert gas, since commercial lots already contain sufficient acid to catalyse polymerisation. However, the purified glutaraldehyde may be stored under oxygen-free conditions.

Glutaraldehyde-containing Fixatives

A large number of fixative mixtures containing glutaraldehyde and other reagents are available. These mixtures possess the advantages of glutaraldehyde, and the limitations of the latter are compensated by other appropriate reagents. For example, since the reaction of glutaraldehyde with lipids is slight, the retention of certain lipids can be achieved by using this dialdehyde in combination with digitonin, malachite green or filipin. Glutaraldehyde can be used in combination with acrolein, alcian blue, caffeine, digitonin, dimethyl sulphoxide, formaldehyde, haematoxylin, hydrogen peroxide, lead acetate, malachite green, OsO_4, phosphotungstic acid, picric acid, potassium dichromate, potassium ferricyanide (or ferrocyanide), potassium permanganate, ruthenium red, spermidine phosphate, tannic acid, trinitro compounds or uranyl acetate. The preparation, applications, and advantages and limitations of these mixtures have been presented elsewhere (Hayat, 1981, 1986b).

FORMALDEHYDE

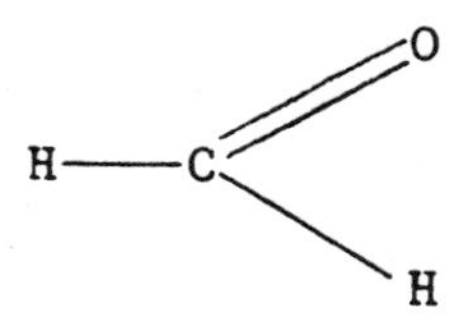

Formaldehyde is a colourless gas. It is easily soluble in water, and is available commercially as formalin (73–40%). It contains a small amount of formic acid (<0.05%) and a considerable amount of methanol (6–15%). The methanol in formalin hinders polymerisation by breaking down high molecular weight oligomers of polymethylene glycols, forming hemiacetals. Because the latter are more soluble than the former, precipitate formation is prevented.

$$HOCH_2OH + CH_3OH \rightleftharpoons HOCH_2OCH_3 + H_2O$$

$$HOCH_2OCH_2OH + CH_3OH \rightleftharpoons HOCH_2OCH_2OCH_3 + H_2O$$

Another mechanism may play an important role in preventing paraformaldehyde precipitation. According to this mechanism, methanol stabilises formalin solution by (1) inducing depolymerisation, which causes a decrease in the concentration of the higher and less soluble homologues, and (2) formation of more soluble products (hemiacetals) (Dankelman and Daemen, 1976).

Formaldehyde, a monoaldehyde, is the simplest

member of the aldehyde fixatives. It reacts in an aqueous solution as methylene glycol (**A**), and in acid

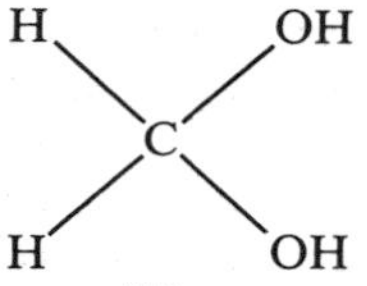

(A)

conditions it contains higher concentrations of the more reactive electrophile (**B**).

Formalin consists of free formaldehyde, methylene glycol and polyoxymethylene glycols, and, in very small concentrations, methylal, methylformate, trioxane and acetals of the polyoxymethylene glycols (Walker, 1964; Gruber and Plainer, 1970). Monomeric formaldehyde probably exists as $HOCH_2OH$ in solution. It has a strong tendency to polymerise into a dimer, trimer, etc., with the general formula $HO(CH_2O)_nH$. Only a small part (11%) of the formaldehyde is monomeric in formalin; when formalin is diluted to 2%, the monomer predominates.

Since formalin contains various impurities, formaldehyde produced by the dissociation of paraformaldehyde powder is more efficient as a fixative; details of its preparation are given on page 40. Formaldehyde is, in general, not recommended for preserving ultrastructure except in special cases or in combination with glutaraldehyde or with other fixatives. For example, this monoaldehyde has proved useful for fixing very dense tissues, such as seeds, which are not penetrated easily by glutaraldehyde. In this connection, it has been used alone or mixed with glutaraldehyde. Since chromic acid reacts with both nucleic acids and proteins, this reagent in combination with formaldehyde has proved effective in stabilising nucleic acids of phages and viruses, and their integrity, for the most part, is preserved during dehydration and embedding (Langenberg and Sharpee, 1978). Formaldehyde (4%) is effective in pre-fixing large slices of surgically removed tissues before they are cut into smaller pieces and fixed with glutaraldehyde; a pre-fixation for 30 min at 4 °C is adequate. Fox *et al.* (1985) have discussed in detail the use of formaldehyde for light microscopy. Formaldehyde fixatives are described elsewhere (Hayat, 1986b).

Reaction with Proteins

The reactions of formaldehyde with proteins are numerous and well understood. Formaldehyde is thought to cause cross-linking of peptide chain, and the reactive functional groups so far identified include amino, amido, guanidino, thiol, phenolic, imidazolyl and indolyl. The participation of lysine in the cross-

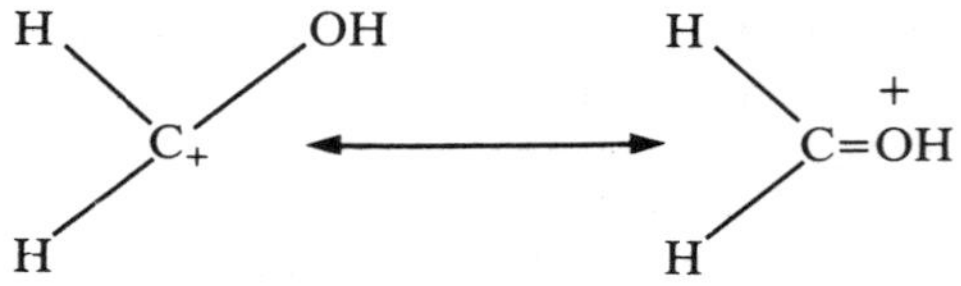

(B)

linking reaction has been confirmed (Caldwell and Milligan, 1972).

In the reaction of formaldehyde with proteins, the first step involves the free amino groups with the formation of amino methylol groups, which then condense with other functional groups such as phenol, imidazole and indole to form methylene bridges ($—CH_2—$). The reactions of formaldehyde with proteins are surveyed below.

Formaldehyde reacts readily with compounds containing an active hydrogen atom, and forms additive compounds such as hydroxymethyl. The addition compounds are formed freely with amino, imino and peptide groups. The reactions given on p. 39 show, for example, that with hydroxyl and sulphhydryl groups, the addition compounds formed are hemiacetal and hemithioacetal, respectively. The addition compounds, in turn, react (condense) with other compounds containing an active hydrogen atom, which results in the formation of methylene bridges ($—CH_2—$). The occurrence of these bridges is considered to be responsible for the fixation of proteins by formaldehyde under conditions appropriate for electron microscopy. Studies by Russo *et al.* (1981) indicate that the major cross-linked species in the histones is heterologous dimers and not homologous dimers. In this study, cross-linking was carried out with 1% formaldehyde for 1 h at 4 °C.

It should be noted that the majority of the reactions are reversible, and formaldehyde for the most part is removable by washing with water. The reactions of formaldehyde with proteins are influenced by several factors, including the concentration of the fixative solution, temperature, pH and the duration of fixation. In general, higher values of these parameters result in an increased binding of formaldehyde. The maximum binding seems to occur at pH 7.5–8.0. By using interference microscopy, it was calculated that after 2 h fixation with formalin the bound formaldehyde constituted 3.6% of the dry mass of isolated nuclei (Abramczuk, 1972). At higher pHs, formaldehyde transforms collagen fibrils into a gel, which resists degradation *in vivo* as well as *in vitro* by collagenase. The presence of 3–8 cross-links per molecule of collagen is sufficient to retard collagenolysis.

1. Addition:

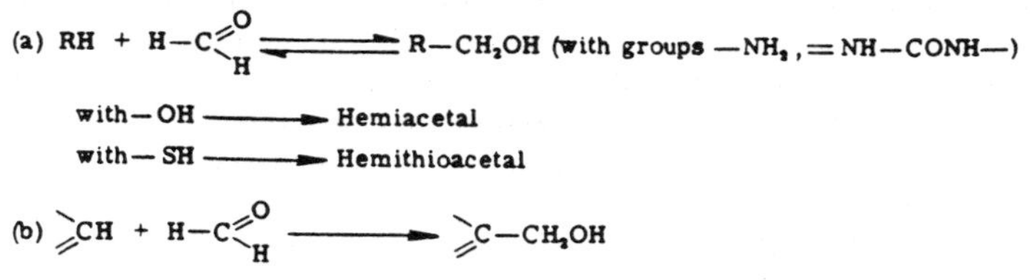

(a) $RH + H{-}C(=O){-}H \rightleftharpoons R{-}CH_2OH$ (with groups $-NH_2$, $=NH-CONH-$)

with $-OH \longrightarrow$ Hemiacetal

with $-SH \longrightarrow$ Hemithioacetal

(b) $>CH + H{-}C(=O){-}H \longrightarrow >C{-}CH_2OH$

2. Condensation:

(a) $-NH_2 + H{-}C(=O){-}H \rightleftharpoons -N{=}CH_2 + H_2O$

(b) $C_6H_5{-}OH + CH_2(OH)_2 + HO{-}C_6H_5 \longrightarrow C_6H_5{-}O{-}CH_2{-}O{-}C_6H_5 + 2H_2O$

(Phenoplasts)

(c) $R{-}CH_2OH + RH \rightleftharpoons R'{-}CH_2{-}R' + H_2O$

(Formation of methylene bridges)

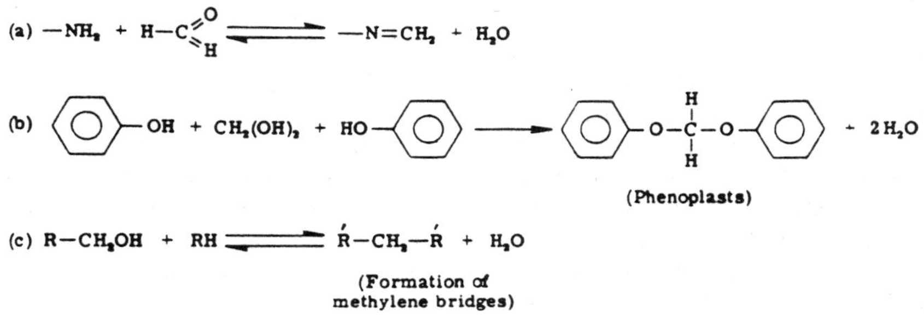

Reaction with Lipids

Although formaldehyde-fixed tissues fail to show lipids after dehydration and embedding, formaldehyde is capable of changing, to some extent, the physical and chemical properties of lipids. Available evidence indicates that formaldehyde reacts at least with unsaturated fatty acids in tissues during fixation and that the site of the reaction is double bonds. Jones (1969a) isolated and characterised the new products formed in the reaction of formaldehyde with pure unsaturated fatty acids under conditions identical with those of fixation. It is also known that carbonyl groups introduced by formaldehyde during fixation are demonstrable by Schiff's reagent. However, since formaldehyde-fixed specimens fail to show lipids after dehydration, it is thought that this monoaldehyde is the fixative of choice when lipid extraction is desired. After fixation with formaldehyde, the lipid-depleted membranes consist largely of protein.

Reaction with Nucleic Acids

Formaldehyde is used extensively in structural and functional studies of nucleic acids and nucleoprotein as an agent for causing denaturation as well as for preventing renaturation of these biomolecules. The most important features of formaldehyde interaction with nucleic acids and nucleoproteins are: (1) it reacts with both the proteins and nucleic acids without destroying polypeptide or polynucleotide chains; (2) it modifies the bases and forms cross-links in nucleic acids; (3) it preserves the conformation of nucleoproteins; and (4) its small molecule penetrates through the protein shell into nucleic acids. The use of formaldehyde as a probe for nucleic acid structure has provided significant information on the mechanism of reaction of the monoaldehyde with DNA and RNA. Formaldehyde has been used extensively as a probe for determining the secondary structure of DNA as well as understanding the mechanism of DNA unwinding (Stevens *et al.*, 1977).

It has been shown that formaldehyde reacts with the amino groups of DNA nucleotides, and that the reaction with formaldehyde proceeds much more rapidly with free nucleotides or denatured DNA than with native DNA (Stollar and Grossman, 1962). The studies of native calf thymus DNA demonstrated that the overall reaction can be formulated as an equilibrium conformational 'opening' step, followed by a slow chemical reaction of formaldehyde with nucleotide amino groups normally involved in interchain hydrogen bonding (von Hippel and Wong, 1971).

It is known that histones are fixed on DNA in nucleohistones if the latter are treated with formaldehyde (1%) at 0 °C for 24 h; probably, covalent bonds are involved in the fixation of histones on DNA. As a result of this reaction, methylene bridges occur on the two neighbouring amino groups of basic residues as well as on the three aminated nucleotide bases. After formaldehyde fixation, histones are no longer acid- or salt-dissociable from DNA. Formaldehyde also cross-links DNA and other proteins.

Formaldehyde is widely used for the stabilisation of viruses. This monoaldehyde has been used to produce RNA–protein complexes in rod-like plant viruses. Such complexes are resistant to the action of high temperatures, detergents and mercaptoethanol (Mazhul *et al.*, 1978).

In general, it is thought that the major reaction of formaldehyde with nucleic acids is largely reversible. It is well known that the binding of formaldehyde with amino groups of the bases is reversible. Eyring and Ofengand (1967) indicated, for instance, that immediately after the reaction of formaldehyde with nucleotides, hydroxymethylation occurred which was reversible. Similarly, the reaction of formaldehyde with polynucleotides was completely reversible (Haselkorn and Doty, 1961). The reaction occurs in two steps for helical polynucleotides: the first is the denaturation of the helix and the second is the formaldehyde addition to the amino groups freed by the denaturation. On the other hand, Lewin (1966) found that formaldehyde reacts not only with basic amino groups of adenine, cytosine and guanine, but also with acidic imino groups of thymine and guanine. Imidazole ring nitrogens of histidine residues also react with formaldehyde to form *N*-hydroxymethyl derivatives (Martin *et al.*, 1975).

It should be noted that, under certain conditions, the reaction between formaldehyde and nucleic acids can be irreversible. Collins and Guild (1969) showed that at least one reaction of formaldehyde with DNA occurs at 100 °C at pH 8.0, which is irreversible at 20–37 °C. Other studies have also suggested that formaldehyde may form stable methylene bridges (—NH—CH_2—NH—) between nucleotides (Alderson, 1964). Finally, it may be said that the binding of formaldehyde with nucleic acids ranges from easily reversible to only partially reversible, even after prolonged dialysis.

Reaction with Carbohydrates

Formaldehyde does not preserve soluble polysaccharides, but prevents the extraction of glycogen provided that the duration of fixation is not too long. Quantitative studies indicate that glycogen is preserved in the rat brain perfused with formaldehyde (Guth and Watson, 1968). During initial ischaemia, glycogen is reduced to ~60%, but after the arrival of the aldehyde fixative, glycogen remains more or less constant. Acid mucopolysaccharides are not preserved by formaldehyde unless they are bound to proteins. Formaldehyde, on the other hand, is very effective in fixing mucoproteins.

Preparation of Formaldehyde Solution

Formaldehyde generated from paraformaldehyde powder is the most effective monoaldehyde for preserving enzyme activity, and is preferred to formalin (as it is commercially available), because the latter contains undetermined amounts of impurities such as methanol and formic acid. Paraformaldehyde polymer is slowly and partially soluble in water at 60 °C. The addition of 1 N NaOH accelerates and completes dissolution of the powder. When paraformaldehyde is dissolved in the alkaline form of the buffer containing disodium phosphate or sodium cacodylate, it may be cleared without the addition of NaOH. Water produces partial hydrolysis of the polymer, which is completed by the presence of a small amount of alkali. Solutions should be prepared immediately prior to use.

ACROLEIN

Acrolein (CH_2=CH—CHO) is a highly reactive, volatile liquid, which owes its common name to its acid odour when formed by the scorching of oils and fats. It is an olefinic aldehyde with three carbon atoms and conjugated double bonds:

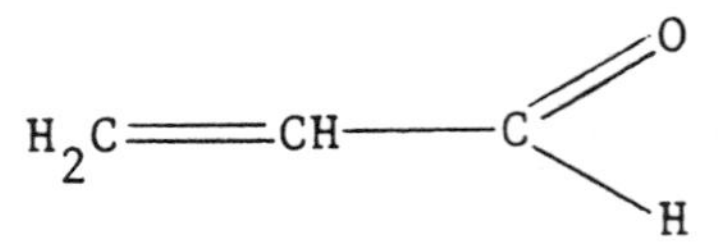

Essentially, it is an α,β-unsaturated carbonyl derivative, and is the simplest in the series. It has a molecular weight of 56.062 and a viscosity of 0.393 centistokes at 20 °C. It is freely soluble in water, and absorbs intensely in the neighbourhood of 211 nm in aqueous solutions. Acrolein undergoes most reactions of ethylenic compounds, including addition of halogens and hydrogen halides at the ethylenic bond. It has a propensity to react with substances that bear the sulphhydryl group or thiols.

Luft (1959) introduced acrolein as a primary fixative for electron microscopy. Because it penetrates and reacts faster than do most other fixatives and causes little shrinkage, it was originally recommended as an alternative to OsO_4 as a primary fixative. Since the rate of glutaraldehyde penetration is relatively slow (~0.4 mm in rat liver in 1 h), the core of the tissue block (1 mm^3) is not well fixed. Although glutaraldehyde does accomplish uniform fixation when used via vascular perfusion, many types of specimens (e.g. surgical biopsy samples and human tissues, plant tissues, yeast and fungi) cannot be fixed by perfusion. Therefore, acrolein, which penetrates much faster (~1 mm in rat liver in 1 h) than does glutaraldehyde or formaldehyde, is useful where fixative penetration is a problem.

The problem of fixative penetration is encountered with specimens which are large, dense or covered by impermeable substances such as waxes and chitin.

Thus, it is effective in the fixation of large tissue blocks which, for practical reasons, cannot be dissected into smaller pieces. Acrolein can also be utilised in studies where only surface layers need to be examined, which is true of most of the studies carried out by SEM. The surface layers of large tissue blocks can be fixed by short exposure (~20 min) to acrolein. Homogeneous tissues such as muscle often do not require overall fixation.

Acrolein fixatives are usually prepared without purifying commercially available lots. The fixative solutions are usually prepared with 10% acrolein and are hypotonic. Osmolarity can be raised by adding calcium or magnesium chloride, but not sucrose. Specimens fixed with acrolein are usually post-fixed in OsO_4.

Acrolein-containing Fixatives

Acrolein is too reactive to be used alone. It can be used in combination with other aldehydes such as glutaraldehyde and formaldehyde for fixing hard (very dense) specimens. Details of the preparation of these fixative mixtures are presented elsewhere (Hayat, 1981, 1986b).

Reaction with Proteins

Of all the monoaldehydes tested, acrolein is probably the most reactive. It shows a high specificity for proteins; is thought to form linkages with amino groups; and yields initially

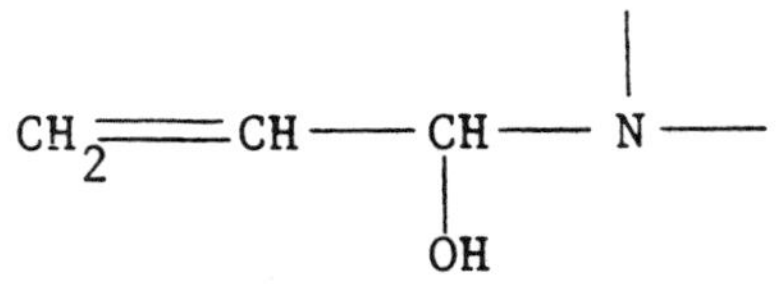

$$CH_2{=}CH{-}\underset{\displaystyle OH}{\underset{|}{CH}}{-}\overset{|}{N}{-}$$

which then is polymerised (van Winkle, 1962). Acrolein reacts rapidly at room temperature with sulphhydryl groups through its ethylene linkage. Reactions of this type lead to tissue-bound aldehyde groups. The reaction leading to a protein-bound aldehyde group, however, is due to a reaction not only of sulphhydryl groups, but also of certain other groups present in proteins. It has been suggested, for instance, that the double bond of acrolein reacts with SH, aliphatic NH_2, NH and imidazole groups (Jones, 1969b). The carbonyl groups introduced by acrolein can be demonstrated with Schiff's reagent. Fixation with acrolein has been employed to bring out protein by staining the free aldehyde group of the protein-bound acrolein with Schiff's reagent or silver nitrate methenamine (van Duijn, 1961; Marinozzi, 1963a).

Reaction with Lipids

Some information is available regarding the effect of acrolein on the extraction of tissue lipids. Published data on the effects of acrolein and formaldehyde on the changes of lipid constituents of rat brain indicate that the former is much more rapid in action and causes less extraction of lipids than the latter (Norton *et al.*, 1962). Dog erythrocytes fixed with pure acrolein, with a concentration of up to 3%, show less than 5% loss in cell phospholipids, whereas 10% acrolein and acrolein containing hydroquinone stabiliser cause much higher loss of phospholipids (Carstensen *et al.*, 1971). The above and other studies imply that acrolein reacts readily with lipids, especially with membrane phospholipids. The resulting new, active site on the phospholipid may achieve cross-linking with nearby protein. Acrolein is thought to react with fatty acids.

It is likely that at lower concentrations of acrolein, after the initial reaction, the resulting active sites on phospholipids undergo a reaction with an adjoining site of protein, which results in the cross-linking. At higher concentrations, the new reactive sites react with a second molecule of acrolein instead of protein, which results in the formation of derivatives that are extractable (Carstensen *et al.*, 1971; Jones, 1969b). This assumption could be a possible explanation for the increased extraction of phospholipids after fixation with concentrated solutions of acrolein.

Precaution in the Handling of Acrolein

Acrolein has a strong tendency to polymerise on exposure to light, air, and certain chemicals with generation of heat. The containers for storing this reagent must be cleaned thoroughly before use, and special care should be taken to prevent contamination, since even traces of contamination can initiate polymerisation. Commercially available acrolein contains an oxidation inhibitor (usually 0.1% hydroquinone), and thus treated it can be stored in a cool place for several months without significant polymerisation. The inhibitor can be removed by distillation (Carstensen *et al.*, 1971); however, purification of acrolein is not necessary. For routine use, the reagent should be stored in small-size brown bottles in a cool place and must be kept tightly stoppered. Acrolein is considered to be contaminated if its solution is turbid or if a 10% solution in tap-water shows pH below 6.4. Such acrolein must be redistilled before use. The method of distillation has been explained in detail by Albin (1962).

Acrolein is a hazardous chemical because of its flammability and extreme reactivity. It is highly toxic through vapour and oral routes of exposure, is irritating to respiratory and ocular mucosa, and induces uncontrolled weeping. Acrolein is moderately toxic through skin absorption, and is a strong skin irritant even at low concentrations. Although acrolein is highly toxic by the vapour route, the sensory response to very low vapour concentration gives adequate warning. The physiological perception of the presence of acrolein begins at 1 ppm, at which concentration an irritating effect on the eyes and nasal mucosa is felt. Thus, there is little risk of acute intoxication, because, as stated above, the lachrymatory effect compels one to leave the polluted area. It does not have toxicological effects, even from repeated exposure to low tolerated concentrations. However, laboratory safety manuals quote a threshold level between 0.05 and 0.1 ppm of air. Above this level, acrolein sensitises the skin and respiratory tract, and causes bronchitis and pneumonia in experimental animals. In the case of human beings, when there is contact with acrolein, cutaneous or mucosal local injury may be observed. Human and animal toxicity caused by exposure to acrolein has been discussed by Izard and Libermann (1978).

It is strongly recommended that gloves and a fumehood be used when handling this chemical. In the preparation of solutions, acrolein should be slowly added rather than the reverse. Any acrolein that contacts the skin should be removed immediately by washing with soap and water. Waste acrolein should be disposed of by pouring into 10% bisulphite solution, which acts as a neutraliser.

ALDEHYDE MIXTURES

The rationale justifying the use of glutaraldehyde and formaldehyde mixture is that rapidly penetrating monoaldehyde temporarily fixes the specimen until the slower-penetrating dialdehyde irreversibly cross-links the proteins. However, formaldehyde alters the primary amine–glutaraldehyde chemistry by significantly decreasing the oxygen uptake and altering the yield of UV-absorbing glutaraldehyde–amine reaction products (Johnson, 1986). Formaldehyde seems to participate in the condensation reactions and thus may change the final reaction products. In addition, formaldehyde tends to reduce the adverse effect of hypoxia on the tissue. Another contribution by formaldehyde is its ability to cross-link protein to DNA. Such cross-linking is not accomplished by glutaraldehyde. According to Kirkeby and Moe (1986), glutaraldehyde and formaldehyde in some unknown way react with each other, catalysing protein cross-linking. The fixation effect is stronger with the mixture than that obtained with the two aldehydes used separately. Further studies are needed to elucidate biochemical reactions responsible for the improved tissue fixation with a mixture of glutaraldehyde and formaldehyde. The use of glutaraldehyde in combination with formaldehyde and/or acrolein has been discussed elsewhere (Hayat, 1981).

OSMIUM TETROXIDE

Infrared absorption spectra and X-ray diffraction studies indicate that the OsO_4 molecule is tetrahedral, perfectly symmetrical and therefore, as a whole, nonpolar. The non-polarity of OsO_4 facilitates its penetration into charged surfaces of tissues, cells and organelles, which is in part responsible for its effectiveness as a fixative and a stain. Specimens can be immersed in an aqueous solution of OsO_4 or exposed to its vapour.

Metallic osmium occurs naturally in close association with iridium and is innocuous, but OsO_4 slowly formed on exposure of the spongy metal to air is responsible for its toxicity. Osmium tetroxide has a molecular weight of 254.2, melts at 41 °C and boils at 131 °C. It exists as faintly yellow monoclinic crystals with an acrid chlorine-like odour. The reagent dissolves in water rather slowly; solubility in water is ~7.24% at 25 °C. It dissolves in neutral distilled water without change of pH. It also dissolves in benzene, paraffin oil, carbon tetrachloride (CCl_4) and saturated lipids. Osmium tetroxide is 518 times more soluble in CCl_4 than in water, has an absorption peak close to 250 nm and volatilises readily at room temperature. It is convenient to remember that 1% OsO_4 is 0.04 M. Since aqueous solutions of OsO_4 are hypotonic, it is necessary to increase the osmolarity of the fixing solution by adding electrolytes or non-electrolytes.

Osmium tetroxide acts not only as a fixative, but also as an electron stain, and this is its major advantage over most other known fixatives. Reduced osmium imparts high contrast to the osmiophilic structures in the specimen. It also acts as a mordant: OsO_4-dependent enhancement of lead staining is well documented.

Osmium tetroxide is a non-coagulant type of fixative—i.e. it is able to stabilise some proteins by transforming them into clear gels without destroying many of the structural features. Tissue proteins that are stabilised by this reagent are not coagulated by alcohols during dehydration. The most important application of OsO_4 is its use as a post-fixative, for it preserves many lipids and imparts electron density to cell components.

The quality of tissue preservation obtained through fixation with glutaraldehyde followed by OsO_4 has not been surpassed by any other fixation combination. The fixative has also proved very effective in freeze-substituted tissues. The main disadvantages of OsO_4

are its slow rate of penetration into most tissues and its inability to cross-link most proteins and preserve carbohydrates. As a result, the fine structure may be changed considerably prior to completion of fixation. Even when tissue blocks range in size from 0.5 to 1 mm^3, the core of the block may remain incompletely fixed. Consequently, OsO_4 is not used as a primary fixative in routine electron microscopy. On the other hand, the slow rate of penetration by OsO_4 during post-fixation is not detrimental, since cell structure has already been partially stabilised by an aldehyde.

Reaction with lipids

There is ample evidence that indicates that fixation with OsO_4 results in at least partial retention of lipids. It has been demonstrated, for instance, that fixation with OsO_4 is indispensable for subsequent demonstration of lipid droplets with Sudan dyes in blood leukocytes (Coimbra and Lopes-Vaz, 1971). Since sudanophilia is not altered by removing the reduced osmium, the staining results are due to immobilisation of finely dispersed free lipids rather than to an interaction between Sudan dyes and osmium.

Since saturated fatty acids are not altered chemically by OsO_4, it is most likely that the unsaturated fatty acids are preferentially involved in the fixation process. Moreover, the facts that both oleic acids and olein (which contain double bonds) reduce OsO_4, while palmitic and stearic acids and their triglycerides (which do not contain double bonds) do not reduce OsO_4, lead to the conclusion that OsO_4 oxidises olefinic double bonds. It is also known that hydrogenation and bromination of the double bonds prevent the fixation of phospholipids. The presence of a double bond in the lipid molecule is a prerequisite to the reaction between lipids and OsO_4 under the conditions of preparative electron microscopy.

The following explanation of the reaction of OsO_4 with unsaturated lipids is based on the discussion presented by Behrman (1984). Addition to olefins of osmium(VIII) results in the production of an osmate ester along with osmium(VI) (Schröder, 1980). The rate of this reaction depends on the chemical nature of the olefin. This reaction is shown in Equation (1), in which the reaction product is a dimeric monoester. Single crystal X-ray studies by Collin *et al.* (1974) demonstrate that the reaction product has the structure shown in Equation (1). The rate of this reaction also depends on the nature and concentration of ligands, if present. This reaction product in solution exists as a mixture of the isomers shown in Equation (2) (Marzilli *et al.*, 1976).

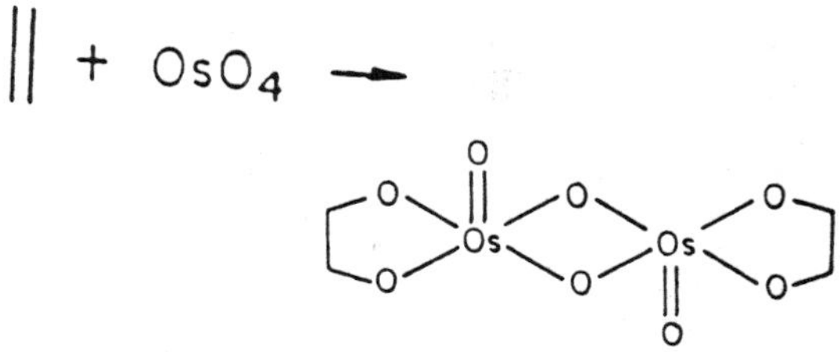

$$\| + OsO_4 \longrightarrow \quad (1)$$

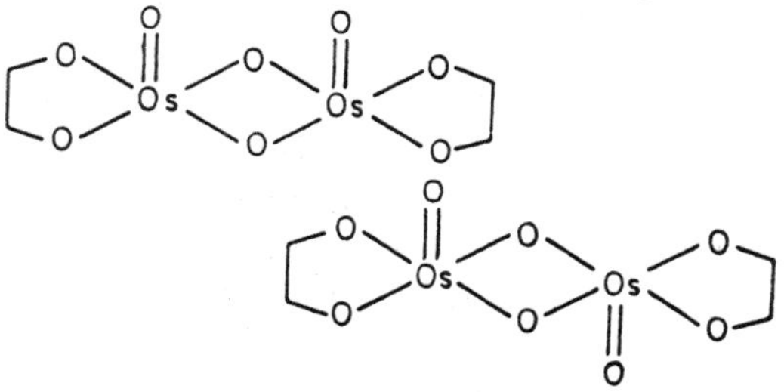

$$(2)$$

In the presence of suitable ligands such as pyridine, the reaction takes a different course, leading to the production of hexacoordinate ester structure which is much more stable both to hydrolysis and to various exchange reactions than is the structure shown in Equation (1). The structures of the osmium(VI) esters are different when formed in either the presence or the absence of ligands. The rate of ester formation is significantly increased in the presence of ligands. Reaction of the simplest oxo-osmium(VI) species (the osmate ion) with a diol in the presence of ligands leads to the same osmate(VI) ester as is formed by reaction of OsO_4, ligands and the olefin (Criegee *et al.*, 1942) shown in Equation (3).

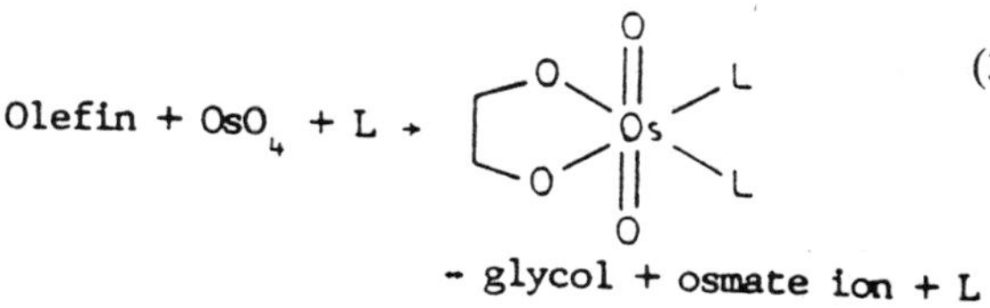

$$\text{Olefin} + OsO_4 + L \rightarrow \quad (3)$$

$$\rightleftharpoons \text{glycol} + \text{osmate ion} + L$$

The ester shown in Equation (3) undergoes three important reactions. The first is a ligand exchange reaction (Daniel and Behrman, 1975), shown in Equation (4). The second is the reverse of Equation (3)—hydrolysis to the osmate ion and the glycol. The third is a transesterification reaction with another glycol. As stated earlier, the rates and equilibria of these reactions are significantly affected by the nature and concentration of the ligands involved. These reactions are summarised in Scheme 1. Two generalisations are made: (1) bidentate ligands such as 1,10-phenanthroline stabilise these esters far more than can be achieved with monodentate ligands such as pyridine; (2) the esters can also be stabilised by increasing the concentration of externally supplied ligands.

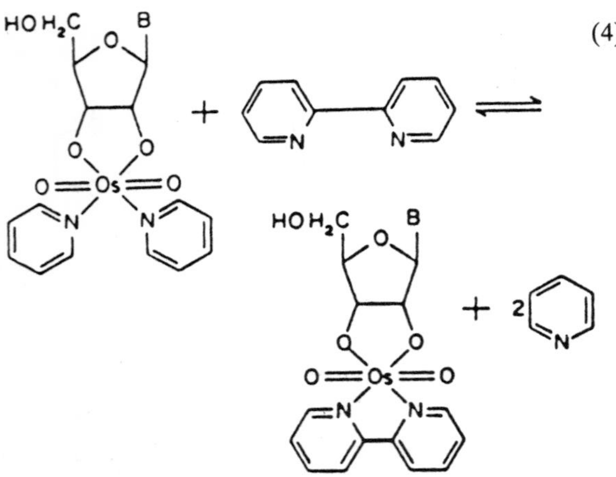

(4)

Scheme 1

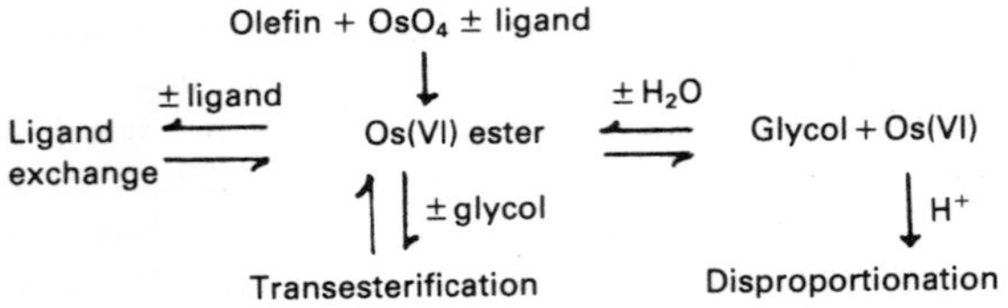

Olefin + OsO_4 ± ligand

Ligand exchange ± ligand

Os(VI) ester

± H_2O

Glycol + Os(VI)

± glycol

H^+

Transesterification

Disproportionation

Scheme 2

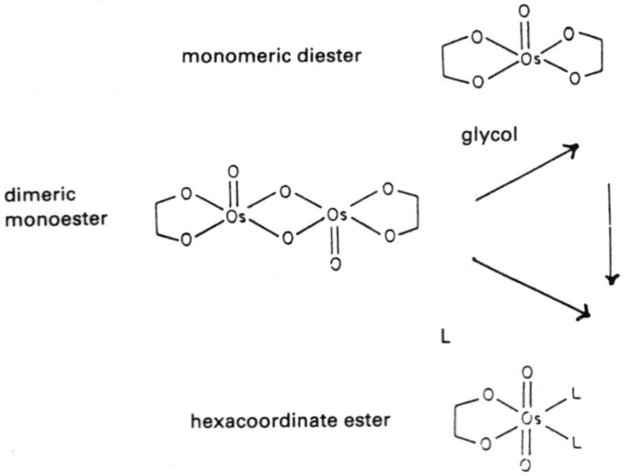

monomeric diester

glycol

dimeric monoester

L

L

hexacoordinate ester

The reactions of dimeric monoester in the absence of a ligand as shown in Equation (1) are very rapid with water, glycols and ligands (Subbaraman *et al.*, 1972). Reactions of the monomeric diester shown at the top of Scheme 2 depend strongly on the nature of the glycol component (Marzilli *et al.*, 1976).

The osmate(VI) ion can undergo reactions with ligands other than glycols to yield oxo-osmium(VI) dimers or monomers of the types shown in (A) (Galas *et al.*, 1981).

X = L (A)

Staining of Unsaturated Lipids

Overwhelming evidence indicates the deposition of reduced osmium $(OsO_2 \cdot nH_2O)^-$ at the polar (hydrophilic) end of the lipid molecule. This does not mean that the polar end is the primary site of reaction. Indeed, the C=C double bonds are the primary site of reaction with OsO_4, while secondary reactions may involve the polar spectra of phospholipid. Phospholipids show a complete disappearance of the C=C absorption band after reaction with OsO_4. One molecule of OsO_4 reacts with one double bond. For every diester bond formed, one osmium atom in the form of a lower oxide is produced.

At least three schemes have been proposed to explain the trilaminar appearance of the membrane.

(1) Osmium oxides migrate from the hydrophobic region and are subsequently deposited at the hydrophilic interface. The preponderance of the Os(IV) and Os(III) in fixed membranes (White *et al.*, 1976) favours the migration scheme. Since stained membranes contain considerable amounts of lower-valency, polar osmium adducts, it is quite likely that the final osmium products are deposited at the polar ends.

Indeed, if there are C=C present in the membranes, these are almost certainly the initial site of reaction. The reaction of OsO_4 with olefins is extremely rapid and highly exothermic, which indicates a low energy of activation and a considerable free energy decrease. White *et al.* (1976) have observed the solid-phase reaction of OsO_4 with cholesterol at −80 °C. It is likely that at least some of those initially formed products would hydrolyse and that the osmium-containing hydrolysis product (e.g. $OsO_2(OH)_4^{-2}$) could then diffuse away. It then might be bound by co-ordinating ligands (N, O, S) occurring in polar head groups and surface membrane proteins. Such bound osmium could serve as initiation sites for the formation of oxo-bridged osmium dimers and polymers. Thus, some osmium would remain in the hydrophobic interior of the bilayer, but much more would be found at the hydrophilic interface, producing the observed contrast.

(2) Osmium oxides are moved to the hydrophilic interphase by bending of the lipid osmate esters (Riemersma, 1963; Stoeckenius and Mahr, 1965). However, marked bending of the hydrocarbon chain of a lipid osmate ester would seem, *a priori*, to be unlikely on thermodynamic grounds. Furthermore, this mechanism would require such bending to take place far more often than cross-linking via Os(VI) diesters or dimeric monoesters to account for the observed con-

trast in the stained membrane.

(3) Unsaturated lipids may be cross-linked via covalent bonds that do not directly involve osmium (White, 1978, personal communication). For example, the osmate ester, initially formed by the reaction of OsO_4 with a C=C, could undergo homolytic C—O or O—Os bond scission to yield a free radical intermediate. This radical could then attack the double bond on an adjacent hydrocarbon chain. The osmium would act as a free radical chain initiator. Anionic or cationic mechanisms can be similarly postulated. Such mechanisms could explain the lack of electron density in the centre of OsO_4-stained membranes.

It has been proposed that reduced osmium behaves as an anionic dye, and migrates from its site of formation at double (ethylene) bonds by interacting with cationic groups of phosphatide molecules towards the polar groups at the lipid–protein interface of membranes (Riemersma, 1970). An osmate anion bound to a lipid alkyl chain—i.e. an osmic acid monoester group—is shifted into the lipid polar group region by virtue of electrostatic interactions with the quaternary ammonium ions present. This proposed migration and ultimate deposition of osmium at the lipid–protein interface explain the trilaminar appearance of membranes. Further reduction of Os(VI) esters to osmium dioxide hydrate is brought about by dehydration solvents (ethanol) and by reducing materials present in the tissue. Osmium dioxide is expected to be one of the final osmium-containing products, and the fact that it does not diffuse is probably a major factor in obtaining reproducible morphology.

Some evidence has been presented showing that osmates or osmium dioxide are likely to migrate towards the cations of phospholipid 'head' groups (Adams *et al.*, 1967). It was demonstrated that the black reaction product of OsO_4 with the olefinic bonds in propylene bonds in polythene was not removed by other quaternary cationic compounds such as 20% aqueous cetyltrimethylammonium bromide or 2% alcian blue.

In summary, in the absence of ligands, the predominant reaction product is the monomeric diester formed by initial reaction with a single olefinic site, partial hydrolysis to a diol and formation of the bridged monomeric diester. This diester is easily hydrolysed, and any free osmate ion so formed can disproportionate, leading to osmium(IV) dioxide and osmium(VIII) tetroxide. The trilaminar appearance of the membranes treated with OsO_4 seems to be due to the preferential deposition of OsO_2 at the two hydrophilic faces of the membrane, with some residual osmium at the original olefinic site.

Reaction with proteins

Much of the information on the reaction of OsO_4 with proteins has been obtained through studies of the addition of OsO_4 to protein solutions or films. The results of these types of studies can be applied, with some reservations, to the actual situation existing in cells, owing to the fact that the protein concentrations forming gels are not vastly different from the overall concentrations of proteins in cells. These studies have demonstrated that weak solutions of OsO_4 can form gels with proteins such as albumin, globulin and fibrinogen (e.g. Millonig and Marinozzi, 1968); however, these proteins differ in their reactivity with OsO_4. For example, fibrinogen and globulin react with OsO_4 at lower concentrations and at much more rapid rates than does albumin. This difference in reactivity is claimed to be related to the amount of tryptophan present. Since OsO_4 is an additive fixative, it probably reacts at the double bonds of tryptophan.

That OsO_4 reacts with various amino acids, peptides and proteins under the conditions of preparative electron microscopy has been demonstrated (e.g. Hake, 1965). Osmium tetroxide seems to react at alkaline pH readily with cysteine and methionine; moderately with tryptophan, histidine, proline and arginine; and only slightly with lysine, asparagine and glutamine. According to Deetz and Behrman (1981), histidine reacts rapidly with OsO_4, and only sulphur-containing residues react faster than histidine. Sulphur-containing amino acids appear to be oxidised to their sulphone derivatives. The synthesis and characterisation of stable and well-defined osmium(VI)–amino acid complexes have been reported (Roth and Hinckley, 1981). Six amino acids (glycine, alanine, valine, leucine, isoleucine and phenylalanine) tested produced similar complexes. The rate of precipitate formation decreased with an increase in their molecular weight.

Hake (1965) showed the formation of carboxylic acids through oxidation of α-amino acids. Studies of the reaction between OsO_4 and proteins of red blood cell membranes indicated that the amino acids of these proteins differed in their ability to reduce OsO_4 (Eddy and Johns, 1965). Available evidence indicates that oxidative deamination of tissue proteins by OsO_4 is not uncommon. Oxidative deamination is accompanied by the evolution of ammonia, which probably originates from side-chain amino groups.

By employing the OTAN (OsO_4-α-naphthylamine) reaction, which detects bound osmium in the tissue, Elleder and Lojda (1968a,b) suggested that certain protein-rich tissue components are able to bind osmium. Furthermore, since neurosecretory substances do not contain a large amount of lipid but are rich in protein-bound sulphhydryl groups, a conspicuous staining of neurosecretory cells with OsO_4 and ethyl gallate may be due to the presence of reactive sulphhydryl (—SH) groups (Wigglesworth, 1964). It has been suggested that an increase in the osmiophilia of axonal membranes of crayfish as a result of electrical stimulation is due to the unmasking of —SH groups in membrane proteins and their reaction with OsO_4 (Peracchia and Robertson, 1971). When the —SH groups are blocked by maleimide or *N*-ethylmaleimide before fixation with OsO_4, the increase in osmiophilia does not appear. The reduction potential of —SH groups is sufficient to reduce OsO_4 very rapidly.

From the foregoing discussion it is evident that OsO_4 does interact with certain proteins, and a small amount of the fixative certainly reacts and blocks sulphhydryl, disulphide, phenolic, hydroxyl, carboxyl, amino and certain heterocyclic groups. These groups differ in the extent of their reactivity with OsO_4; for instance, —SH groups are far more reactive than are disulphide bonds.

Both soluble and membrane proteins undergo alterations in their secondary structure when treated with OsO_4 (Lenard and Singer, 1968). Approximately 40–60% of the α-helical content of proteins in red blood cell membranes was lost after treatment with 2% OsO_4 in phosphate buffer (pH 7.5) for 30 min at 4 °C; such loss was increased to ~70% when the specimen was pretreated with glutaraldehyde. Mitochondrial proteins are also partially aggregated after treatment with OsO_4 (Wood, 1973). Osmium tetroxide probably causes alterations in both the primary and secondary structures of proteins, resulting in their extraction. The isoelectric point of proteins is lowered following treatment with OsO_4, which indicates the disappearance of basic groups.

Although the appearance of a black or brown colour is an unequivocal indication of the reactivity of OsO_4 and the accumulation of lower oxides of osmium, the absence of colour does not necessarily mean that the cellular structure is non-reactive with OsO_4. The absence of a detectable change in colour may be due to the fact that only small amounts of lower oxides form, or osmium is bound in a higher valency state, or after reaction lower oxides migrate to other sites. Thus, a change in colour alone should not be used as the only criterion for the chemical reaction between OsO_4 and proteins. It is conceivable that certain amino acids react with OsO_4 without reducing it to lower oxides of osmium. The cross-linking of certain proteins would thus be accomplished without an apparent increase in their electron density. Osmium tetroxide may stain proteins weakly or not at all, but it still may preserve them. It is emphasised that even if OsO_4 is responsible for introducing relatively few cross-links in proteins, it may still contribute significantly to the preservation of fine structure.

In summary, the fixing capability of OsO_4 includes not only the preservation of unsaturated lipids, but also the stabilisation of certain proteins. Osmium tetroxide functions as an intermolecular or intramolecular protein cross-linking agent as well as introducing cross-links between protein and unsaturated lipids. The initial reaction of OsO_4 with amino acids may increase the rate of reaction with double bonds in lipids. The chemical nature of the reaction products arising from OsO_4 reactions with amino acids has been elucidated (Nielson and Griffith, 1979).

Reaction with lipoproteins

In order to understand the mechanism of fixation, it is imperative to study the nature and extent of reaction of OsO_4, not only with lipids and proteins, but also with lipoprotein complexes, which make up most of the membranes. Studies of lung lipoprotein myelinics treated with OsO_4 showed that the metal reacted readily with the lipoprotein and that the primary site of osmium deposition was the aqueous phase (Dreher *et al.*, 1967). Although protein reacted readily with OsO_4, the reduction did not destroy the lipoprotein monolayers, even though the structural properties apparently were altered. After OsO_4 treatment, the hydrophobic layer was doubled in thickness, whereas the hydrophilic layer remained unchanged in thickness. The variations observed in the thickness of cell membranes appear to be related to the number of double bonds near the aqueous phase, the extent of penetration of hydrophilic protein into the hydrophobic lipid layer, and the relative amounts of lipid and protein present. This is a significant step towards explaining the variations in thickness of various membranes.

Osmium tetroxide may fix the relative positions of amphipathic proteins in membranes and immobilise molecular motion in the lipid bilayer (Jost and Griffith, 1973). It may build bridges between the aliphatic chains of lipids and the peptide bonds of certain membrane proteins (Litman and Barrnett, 1972; Nermut and Ward, 1974). OsO_4 is known to interfere with membrane cleavage and membrane fusion (Poste and Papahadjopoulos, 1976). The exact role played by protein in the binding of osmium in lipoprotein complexes is rather difficult to assess; however, the heavy metal is capable of introducing cross-links between lipids and proteins. The functional groups in proteins and lipids that react with OsO_4 have been discussed earlier.

Reaction with nucleic acids

The information on the interaction of OsO_4 with nucleic acids has important biochemical implications. Such information may be obtained by various methods, including X-ray crystallography (Rosa and Sigler, 1974), direct visualisation of base sequences with the TEM (Whiting and Ottensmeyer, 1972) and correct interpretation of cell structures treated with this metal. Osmium tetroxide in the presence of pyridine reacts with the pyrimidine moieties (thymine, uracil and cytosine) in polynucleotides (Chang *et al.*, 1977), whereas adenosine and guanosine under similar conditions are not oxidised (Burton, 1967). Preliminary data indicate that thymine is attacked approximately ten times more rapidly than is uracil. The overall reactivity order for the common pyrimidine residues is thymine > uracil > cytosine.

The usual reactive sites in nucleic acids are the 2,3-glycol in a terminal ribose group (Daniel and Behrman, 1975) for the Os(VI) reaction and the 5–6 double bond of uracil and thymine residues for the Os(VIII) system. Osmium(VIII) reagents also oxidise thio bases such as 4-thiouridine, and react rapidly with the isopentenyladenine group. Similar reactions occur at 0 °C with denatured DNA, but not with double-stranded DNA (Beer *et al.*, 1966). Osmium tetroxide is known to be very reactive towards single-stranded regions but not double-stranded regions in DNA in a wide range of pH values (Lilley and Paleček, 1984). The low reactivity of OsO_4 to double-helical DNA seems to be due to steric effects.

Using higher concentrations of OsO_4 and higher temperatures (23 °C), Beer *et al.* (1966) demonstrated that OsO_4 in the buffer solution that did not contain a ligand reacted predominantly with the thymine base of denatured DNA. They proposed that the base is converted to 4,5-dihydroxythymine, and that 1 mole of osmium reacted with 1 mole of the nucleotide. In the presence of ligands such as pyridine (which contains tertiary nitrogen), stable osmate ester derivatives are formed (Daniel and Behrman, 1976). Exposure of yeast tRNA crystals to a mixture of Os(VI) and pyridine produced a derivative containing approximately one atom of osmium and two molecules of tRNA (Rosa and Sigler, 1974). The adduct is stable and cannot be reversed by treatments known to disrupt the secondary and tertiary structure of the tRNA molecule.

Similarly, DNA can be converted to a new product which remains largely unbroken and linear, but in which the majority of the thymine residues are converted to an addition product containing one osmium atom and two cyanide atoms (Di Giamberardino *et al.*, 1969). A single atom of osmium coupled to thymidine residues does not provide adequate contrast to clearly visualise the base. Cyanide may provide additional electron-scattering groups for the ester by binding additional negative charges. One can conclude from the above and other biochemical studies that although nucleic acids in the tissue seem to be inert to OsO_4, the above-mentioned preferential reactions are important in the development of methods (base-specific markers) for electron microscopy of the base sequence in DNA and RNA.

It is useful to present a brief discussion of the effects of OsO_4 on the nucleic acids in the tissue. After fixation in OsO_4, coalescence of DNA fibres into coarse aggregates during alcohol dehydration is a common phenomenon. This type of clumping of intramitochondrial DNA fibres has been demonstrated in a wide variety of species. The clumping of DNA can be prevented under certain conditions of fixation. Bacterial nuclei can be stabilised by employing OsO_4 in the presence of Ca^{2+} and amino acids (Kellenberger *et al.*, 1958). The gelation of a solution of DNA or nucleohistones by OsO_4 in the presence of Ca^{2+} and tryptophane has been reported (Schreil, 1964). Calcium tends to increase the stability of the double helix of DNA.

The clumping of DNA can also be prevented by post-fixation with uranyl acetate prior to dehydration or by fixation with $KMnO_4$. The stabilising effect of these methods becomes quite clear when one considers that intramitochondrial DNA in a clumped state shows a thickness of up to 25 nm, while in a stabilised state the thickness ranges from 1.5 to 5.0 nm. The gelation of DNA by the above procedures prevents the damaging effects of alcohol dehydration, and results in little shrinkage or formation of coarse aggregates. Apparently, the formation of gel prior to dehydration is necessary in order to preserve the fine fibrillar structure of DNA.

Ribonucleoprotein and/or RNA have been considered to be responsible for the staining of nucleoli with OsO_4 at the light microscope level (Battaglia and Maggini, 1968; Stockert and Colman, 1974). However, it is possible that the preservation and staining of nuclei do not depend solely upon the interaction between OsO_4 and nucleic acids. It is known that nucleic acids in most eukaryotic organisms are chemically linked with basic proteins, and it has been suggested that a high content of arginine and lysine in the nuclear histones may account in part for staining of the nucleus (Wigglesworth, 1964). Furthermore, the presence of unsaturated lipids in the nucleus cannot be ruled out. It is known that chromatin contains lipids of various kinds. Thus, the staining of nuclei by OsO_4 may also be due to the interaction of unsaturated lipids with this fixative. Staining methods using OsO_4 have been presented later and elsewhere (Hayat, 1986b).

Reaction with carbohydrates

Osmium tetroxide, in general, does not react with carbohydrates at a rate significant enough to be useful in electron microscopy. However, there are some exceptions. The preparation and characterisation of osmium–carbohydrate polymers derived from glucose in aqueous or dilute buffer solutions have been reported (Hinckley *et al.*, 1982). The compounds produced are soluble and characterisable as macromolecules. Osmium(VI) species seem to react with diols in the presence of ligands. The reaction of osmium(VI)–ligand complexes with *cis*-glycols is known to produce osmate esters identical with those formed by the oxidation of olefins by osmium(VIII)–ligand reagents (Subbaraman *et al.*, 1973). Resch *et al.* (1980) have also shown that methyl glycosides react with osmium(VI) species to give stable products containing one osmium atom per sugar.

The possibility of a slow oxidation of sucrose to oxalic acid by OsO_4 was suggested by Bahr (1954). Glycogen solutions are known to blacken after a prolonged treatment with OsO_4 at 50 °C. Glycogen in the sections of OsO_4-fixed tissue has been shown to reduce silver nitrate–methenamine solution even in the absence of a preliminary oxidation with periodic acid (Millonig and Marinozzi, 1968). This reaction indicates that OsO_4 oxidises glycogen with the release of aldehyde groups.

Osmium tetroxide probably produces vicinal hydroxyl groups by oxidation, which are, in turn, split to form aldehydes by, for instance, periodate oxidation. Also, OsO_4 is reduced by thiosemicarbazones when condensed with the aldehydes that are formed by periodate oxidation of polysaccharide 1,2-glycol groups (Hanker *et al.*, 1964; Seligman *et al.*, 1965). An attempt has been made to explain the chemistry of linking of free hydroxyl groups in glucose by OsO_4 (Luzardo-Baptista, 1972). It has also been shown that OsO_4 reacts with solutions of pure amino sugars, producing black droplets (Wolman, 1957).

The above-mentioned data indicate the difficulty in accurately assessing the degree of interaction of OsO_4 with carbohydrates. It must be mentioned, however, that most carbohydrates in the OsO_4-fixed tissue are extracted during washing and dehydration. Neutral polysaccharides such as glycogen are relatively less readily water-soluble and thus show less leaching. Since glycogen shows little increase in electron density after fixation with OsO_4, it is apparent that the interaction of this fixative with glycogen does not involve binding of the lower oxides of osmium to the latter. However, the electron density of glycogen can be enhanced selectively by adding 0.05 M $K_3Fe^{III}(CN)_6$ to 1% OsO_4 during post-fixation (de Bruijn, 1968). Electron density of lipid droplets and membranes is also increased by this method. If the objective of the study is to differentiate between ribosomes and glycogen, sections should not be post-stained.

The mechanism involved in the glycogen staining reaction with OsO_4–$K_3Fe(CN)_6$ or $K_2OsO_2(OH)_4$–$K_4Fe(CN)_6$ mixture has been partly elucidated (de Bruijn and den Breejen, 1976). It is thought that the Os(VI)–Fe(II) complex reacts selectively with unchanged diols in tissue glycogen, and that this complex is potentially able to stain tissue aldehydes and carboxyl groups when present in tissue or introduced by oxidising agents. Hydrogen peroxide and OsO_4 are potentially able to introduce carboxyl groups in glycogen.

The possibility that OsO_4 oxidises certain carbohydrates without the formation of a black precipitate cannot be ruled out. In a case where OsO_4 interacted with carbohydrates, the fixation capacity of the reagent would be altered. Such interaction may reduce the effective concentration of OsO_4 in a given volume of the fixation mixture. A reduction in the concentration of OsO_4 would result in a slower rate of penetration and fixation.

Reaction with phenolic compounds

The first suggestion that OsO_4 reacts with phenol-rich regions in plant cells was made by Schultze and Rudneff as early as 1865. In recent years, the presence of phenol-containing regions in cells of higher (e.g. Baur and Walkinshaw, 1974; Mueller and Beckman, 1976) and lower plants (e.g. Fulcher and McCully, 1971; Evans and Holligan, 1972) and animals (e.g. Tranzer *et al.*, 1972) has been reported. Sufficient evidence is now available indicating that phenol-containing regions of cells are osmiophilic. These regions show electron density with OsO_4 alone, without post-staining with uranyl acetate and lead citrate.

When phenolic-containing cells are fixed with glutaraldehyde followed by OsO_4, phenolics leach from the vacuoles into the cytoplasm, where they subsequently react with OsO_4. This results in a dense, osmiophilic cytoplasm, the details of which are obscured. Leaching of the phenolics from the vacuoles can be minimised by pre-fixation with low concentrations of glutaraldehyde (Mueller and Beckman, 1974). The most effective approach to prevent leaching of phenolics is to add caffeine, nicotine or cinchonine to glutaraldehyde during pre-fixation (Mueller and Rodehorst, 1977). These alkaloids react with phenolics in the vacuoles and prevent their leaching. Caffeine causes the phenolic material to condense into globules, while the other two alkaloids precipitate the phenolics into amorphous masses against the tonoplast. Caffeine is most effective at concentrations of 0.1–1.0%, while nicotine and cinchonine should be used at a concentration between 0.05 and 0.1%.

Method Root specimens are fixed in 2.5% glutaraldehyde containing 0.5% caffeine in 0.05 M phosphate buffer (pH 6.8) for 2 h at room temperature. The specimens are washed in the buffer containing caffeine for 1 h, followed by post-fixation in 1% OsO_4 without caffeine for 1 h. Thin sections are stained with aqueous uranyl acetate (2%) for 5 min, followed by lead citrate for 5 min.

Reaction with alkaloids

Osmium tetroxide reacts with alkaloids, such as isoquinoline, pyridine and quinuclidine, when present in plant tissues. These and other structurally related alkaloids are potentially osmiophilic. The structure of the reaction products is described below (Wright *et al.*, 1981). Reaction of OsO_4 with quinuclidine initially yields osmium(VIII) adduct, $OsO_4 \cdot C_7H_{13}N$, a highly reactive, five-coordinated trigonal bipyramidal species (**I**) shown below. The reaction of this adduct with

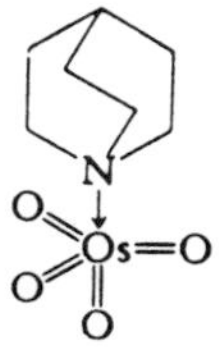

(**I**)

quinine yields osmium(VI) complex, a dimeric species (**II**) with a highly unusual asymmetric oxygen bridge shown below. Further reaction of this species (**II**) with

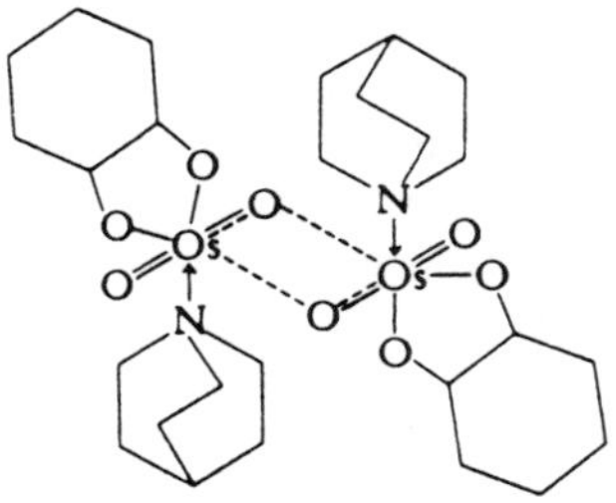

(**II**)

tertiary amines (imidazole) gives osmium(VI) complex (**III**), where L = isoquinoline and L′ = quinuclidine, imidazole, etc.

In the standard fixation and staining environment, the adduct (**I**), on reduction with ethanol, would yield an osmium(VI) complex (**IV**).

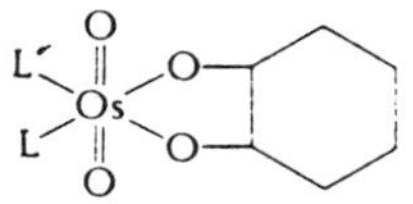

(**III**)

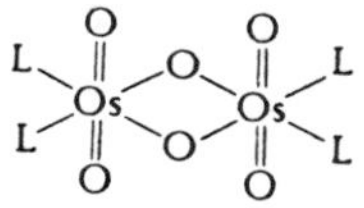

(**IV**)

Loss of lipids

It has been assumed in the past that most lipids are preserved during fixation by OsO_4.

One of the reasons for this misconception is that the total amount of masked lipids present in the living cell has been underestimated. It has been shown that the insoluble fragment of protoplasm in living cells contains as much as 45% lipids (Smith *et al.*, 1957). It is recognised that even chromatin contains a high proportion of lipids. The fact that fixatives penetrate many times more slowly into tissues than into protein gels containing equal amounts of protein also indicates the presence of large amounts of lipid in the tissues. It is accepted that one of the major barriers to the fixative penetration is the lipid component of cytoplasmic membranes.

Another reason for the misunderstanding is that the morphology of some organelles changes little, even though a considerable amount of lipid component is lost. For example, although lipids are necessary for electron transfer in mitochondria, the characteristic morphology of mitochondria can persist even after extraction of phospholipids with aqueous acetone. A detailed study of the fine structure of bovine heart mitochondria after lipid extraction showed that the outer membrane was lost only when more than 80% of the lipid was removed, and in the remaining membranes the characteristic triple-layered appearance of the membrane was preserved even when more than 95% of the lipid was extracted (Fleischer *et al.*, 1967). Similar results were obtained with myelin figures (Napolitano *et al.*, 1967). These results may be explained on the basis of OsO_4 reaction with residual lipids and/or with proteins or even with some carbohydrates.

Studies of fixed chloroplasts showed that after 20% of the glycolipids and 40% of the chlorophyll were extracted, the characteristic structural features were retained, but when most of the lipids were extracted with chloroform–methanol, structural features could not be demonstrated (Ongun *et al.*, 1968). It is apparent that in chloroplasts some lipid fixation is necessary for retention of ultrastructure. It follows from these considerations that the absence of any apparent change in the morphology of cell components cannot be indicative of the preservation of lipids, and that various organelles differ in their requirement of the amount of fixed lipids necessary to maintain the ultrastructure.

The visualisation of electron-opaque masses in sections of tissues fixed with OsO_4 probably does not represent accurately lipids present in the living cells. It is conceivable that electron-opaque masses conceal the 'fixed' lipids. In fact, it has been reported that empty spaces are left behind after the removal of osmium by oxidation from the electron-opaque osmium reaction products (Casley-Smith, 1967). It is also worth considering that dehydration and even embedding may affect the final relative concentration of reduced osmium in the tissue. That reduced osmium deposits are soluble in xylene was indicated by Marinozzi (1963b), although no experimental data are available on the effect of ethanol or acetone on reduced osmium.

X-ray diffraction studies of frog sciatic nerves indicate that fixation by OsO_4 is not enough to prevent extraction of cholesterol during dehydration (Moretz *et al.*, 1969). Bahr (1955) and Dallam (1957) also pointed out a considerable loss of unsaturated lipids by ethanol extraction from tissues fixed with OsO_4. Although fixation with glutaraldehyde followed by OsO_4 seems to minimise lipid loss, especially loss of neutral lipids, a considerable amount of lipid is lost during dehydration. It should be noted that only minute amounts of lipids are extracted during fixation; the major loss occurs during dehydration.

Some lipids are extracted by OsO_4 itself. Proteins diffuse out in OsO_4-fixed cells unless pre-fixed with an aldehyde. These two events cause cell membranes to become freely permeable to small ions and molecules. This is the reason that the osmolarity of OsO_4 vehicle is not very important.

The varying results discussed above suggest that the degree of extraction is dependent not only upon the type of intracellular lipids, but also on the type of tissue involved. Moreover, the degree of extraction of indigenous tissue lipid and that of absorbed lipid are not similar. It also appears that lipid loss is not related entirely to the degree of saturation of fatty acids present in the lipid fraction of a cell. The details of the loss of free lipids during dehydration and embedding are discussed in the chapter on 'Embedding'.

Dehydration at 4 °C or a partial dehydration procedure reduces lipid loss. In the partial dehydration procedure, the steps of 100% ethanol and propylene oxide are eliminated, and the final dehydration is completed by using Epon monomer (see Idelman, 1964, 1965). Retention of free fatty acids can be improved by adding $CaCl_2$ to the fixative (Mitchell, 1969; Strauss and Arabian, 1969); $CaCl_2$ forms a highly insoluble salt with free fatty acids. Cholesterol can be preserved in the tissue by adding digitonin in aqueous solution to glutaraldehyde and OsO_4 fixatives. By using the partial dehydration procedure, a loss of radioactivity of only 11.2% was detected when intracellular triglycerides had been labelled by injection of palmitic acid to rats (Stein and Stein, 1967).

Loss of proteins

The biphasic effect (i.e. the initial gelation and then extraction) of OsO_4 upon tissue constituents is well known. It is recognised that tissues fixed with OsO_4 lose proteins during both fixation and dehydration. For example, mitochondria in rat liver tissue lost ~22% of their proteins during fixation in OsO_4 and a further 12% during dehydration (Dallam, 1957). Studies of the effect of OsO_4 on the conformation of protein molecules, by means of circular dichroism measurements in the spectral region of the peptide absorption band, indicated that OsO_4 fixation either alone or preceded by pre-fixation with glutaraldehyde caused the loss of helical structure of the proteins (Lenard and Singer, 1968). It was also indicated that from one-quarter to one-third of the membrane protein is in a helical conformation and the remainder is in the random-coil form. The importance of these studies cannot be overemphasised in view of the role played by the three-dimensional folding (the conformation) of the individual protein chains in determining the structure of cell components.

Extraction of proteins occurs not only when OsO_4 is used as a primary fixative, but also when it follows pre-fixation with glutaraldehyde. This has been demonstrated in erythrocyte membranes (McMillan and Luftig, 1973). According to Lenard and Singer (1968), conformational changes of proteins are more severe after glutaraldehyde fixation followed by OsO_4 than with either fixative alone. It seems that pre-fixation with glutaraldehyde does not prevent leaching of some proteins, especially when specimens are post-fixed with OsO_4.

The degree of extraction is primarily dependent upon the duration of fixation and dehydration, the type of buffer employed and the type of proteins involved. The difference in the extraction of different proteins was indicated by Bahr (1955), who demonstrated that in rat liver the amount of extracted protein exceeded 50% of its dry weight after a fixation of 4 h with OsO_4, whereas muscle showed an extraction of less than 50% of its dry weight after a similar treatment. He also showed that liver, muscle and tendon of rat exhibited an increase in the extractability of proteins with time, whereas skin did not show such an increase. The extraction effect of OsO_4 has been utilised advantageously for studying the ultrastructure of aortic elastica (Cliff, 1971). Rat aorta fixed for periods up to 72 h with OsO_4 revealed an underlying fibrillar structure due to the differential extraction of the amorphous matrix material.

A prolonged fixation by OsO_4 usually results in progressive destruction of cellular proteins, which leads to an increased extractability during dehydration. Although the exact mechanism of extraction of proteinaceous substances as a result of prolonged fixation with OsO_4 is not known, a few possible mechanisms are discussed below. It has been suggested that the initial, partial denaturation of proteins is followed by further oxidation of certain proteins, leading to the production of soluble end-products which are capable of being washed out of the cells after long fixation with OsO_4 (Wolman, 1955). Another explanation of the loss of proteins is that OsO_4, like other oxidising agents, causes cleavage of certain protein molecules. Oxidising agents can cause cleavage of proteins through disulphide bridges in histidine, tryptophan and tyrosine, or through other groups (Joly, 1965). The cleavage of protein molecules could result in the loss of peptide fragments especially with OsO_4, since it possesses low cross-linking ability. Studies by Hopwood (1969b) utilising Sephadex G-50 separation demonstrated that OsO_4 causes oxidative cleavage of proteins.

An explanation of the loss of protein in zymogen granules after exposure to OsO_4 was suggested by Amsterdam and Schramm (1966). Zymogen granules isolated from rat parotid and pancreas lost a large part of their protein content within a few minutes after treatment with OsO_4. Since these granules are quite stable without OsO_4 treatment, it was assumed that the fixative reacted with the granule membrane and increased its permeability without cross-linking the proteins inside the granules. As a result, the soluble proteins diffused out of the granules.

As stated earlier, initially OsO_4 stabilises proteins against dissociation by intermolecular cross-linking, but subsequently it cleaves peptides. Such cleavage is difficult to envisage, because it may not result in an apparent altered morphology. On the other hand, OsO_4 may change morphology without protein cleavage. Protein cleavage by OsO_4 is significantly reduced in the presence of tertiary amines (Emerman and Behrman, 1982). Baschong *et al.* (1984) have shown that protein cleavage in polyheads by OsO_4 is substantially reduced when tetraethylenediamine (2 mmol) is added to the specimen prior to treatment with OsO_4. However, tertiary amines may cause structural distortions. Protein cleavage can also be reduced by shortening the time of exposure to OsO_4.

Osmium tetroxide may also dissociate some protein assemblies such as actin filaments (Maupin-Szamier and Pollard, 1978). However, F-actin can be protected against such dissociation by adding virotoxins (Gicquaud *et al.*, 1983). Actin filaments can also be protected from the damaging effect of OsO_4 by adding tannic acid to the glutaraldehyde fixative before post-osmication (Maupin and Pollard, 1983). Since tannic acid tends to increase the diameter of actin filaments, a better approach to improve the preservation and staining of these filaments is the use of amines (especially diamines) with glutaraldehyde, followed by osmication (Boyles *et al.*, 1985).

Changes in specimen volume

Osmium tetroxide hardens the tissue slightly and usually causes some gross swelling during fixation. The degree of swelling or shrinkage is influenced both by the osmolarity and the type of ions present in the fixation vehicle, and by the type of specimen under study. The final condition of the specimen is also affected by the methods used for dehydration and embedding. Post-osmicated tissues tend to show fewer dimensional changes than do non-osmicated specimens.

Only a limited amount of data on the changes in volume and weight of the tissue during fixation with OsO_4 is available. Isolated cells usually show swelling even in iso-osmotic solutions of OsO_4. According to Bahr *et al.* (1957), animal tissues exhibit a marked and rapid swelling when fixed with 1% OsO_4. In the case of liver tissue, for instance, they found as much as 30% swelling after 4 h fixation, and at least half of this value was reached after only 15 min in the fixative. Comparative studies by these workers indicated that animal tissues such as brain, spleen, kidney, muscle and liver do not differ significantly in their fundamental pattern of swelling during fixation. This swelling is nearly neutralised by the shrinking action of the dehydration solvents. Further shrinkage of the specimen occurs during infiltration and polymerisation of the embedding resin. Although the above data were obtained by using methacrylate, which has been largely replaced by other better embedding resins, the information is valuable as an aid to understanding the general pattern of swelling and shrinkage.

Kushida (1962a) investigated the swelling effect of 1% OsO_4 in sea-water on sea urchin eggs, and found

3.2% increase in the cell volume. However, this swelling was temporary because the cells underwent 12.2% shrinkage in ethanol and additional 1.7–21.1% shrinkage in standard embedding media, including epoxy and polyester resins. *Limnaea* eggs fixed in 1% OsO_4 solution in distilled water at pH 6 showed a swelling of ~25%, while when fixed in a 2% isotonic solution of OsO_4, a swelling volume of ~10% was still found (Elbers, 1966).

Fixation temperature seems to exert little influence on the extent of volume changes, and swelling is a consequence of the chemical action of OsO_4. An increase in the volume of the tissue during fixation is usually closely followed by an increase in its weight. The increase in the specific weight of the tissue appears to be primarily an expression of the binding of osmium in the tissue during fixation. This increase, however, reverses during dehydration and finally increases during infiltration by embedding resins.

Specimen swelling during primary fixation in OsO_4 can be prevented by adding $CaCl_2$ or NaCl to the fixative vehicle, while during secondary fixation the swelling can be avoided by adding either an electrolyte (NaCl) or a non-electrolyte (glucose or sucrose). The addition of $CaCl_2$ causes the cross-linking of negatively charged proteins, which results in a reduced osmotic pressure within the cell. Divalent cations tend to maintain cell volume by affecting the gelled state of cytoplasm. This is accomplished by achieving an equilibrium between the swelling pressure of the cytoplasmic gel and the osmotic pressure of the fixative solution (Wild *et al.*, 1987). The addition of non-electrolytes to OsO_4 solution during primary fixation results in less than satisfactory fixation of albumin in model experiments (Millonig and Marinozzi, 1968) and loss of cellular materials. Non-electrolytes should not be added to prevent swelling of isolated cells, for these materials do not prevent the flow of water into the cells. Sea urchin eggs show more swelling in OsO_4 containing sucrose than in OsO_4 without sucrose (Millonig and Marinozzi, 1968). To prevent swelling, a final concentration of 1–3 mmol of $CaCl_2$ in OsO_4 solution is recommended. Caution is advised, for $CaCl_2$ may cause a granular precipitation of cell proteins and phosphates present in the buffer, and Ca^{2+} and Mg^{2+} may cause excessive shrinkage.

PARAMETERS OF FIXATION

Concentration of Osmium Tetroxide

Osmium tetroxide is most effective when used in optimal concentration; higher concentrations can cause oxidative cleavage of protein molecules, which would result in the loss of peptide fragments. At present, the most commonly used concentration of OsO_4 ranges from 1 to 2% in a buffer. Osmium tetroxide solutions less than 1% may be desirable for cytochemical and certain morphological studies. Concentrations ranging from 0.2 to 0.5% are desirable for particulate specimens. Some ciliates have been post-fixed satisfactorily with 0.1% OsO_4 (Shigenaka *et al.*, 1973).

Temperature of Fixation

It has already been pointed out that various cell components differ with regard to their appearance as a result of changes in fixation temperature, which can cause both qualitative and quantitative alterations in tissue. Certain types of microtubules are lost when cells are fixed in cold OsO_4, presumably because of its slow rate of penetration. Nevertheless, fixation should usually be carried out at ~4 °C, since OsO_4 is a slow penetrant and autolytic activity is reduced at low temperatures. Furthermore, low temperatures reduce leaching or extraction of cell constituents during fixation. Also, if a relatively long exposure of the tissue to OsO_4 is necessary, it can be accomplished with less damage at low than higher temperatures. Fixation in OsO_4 at low temperatures has proven valuable in improving the uniformity of the quality of preservation. Fixation at higher temperatures may cause shrinkage of mitochondria and increased granularity of the cytosol. Alternatively, a preliminary brief fixation by OsO_4 can be carried out in the cold followed by main fixation at room temperature. Perhaps the quality of ultrastructure preservation, with a few exceptions, is not significantly affected by varying the temperature, provided that the specimen size is very small and duration of fixation is not very long.

Rate of Penetration

A knowledge of the rate of penetration and the amount of osmium uptake by the tissue is important in order to interpret electron micrographs correctly. This information may indicate to what extent an externally introduced heavy metal participates in the morphological appearance of a micrograph. In other words, the amount of osmium uptake by the tissue in a given period of time greatly influences the contrast and general appearance of the electron micrograph. Measurements of the rate of penetration are also important for determining the optimal duration of tissue fixation. The rate of penetration is controlled primarily by the diffusion gradient at the front of penetration, and the diffusion gradient, in turn, is influenced by several factors, some of which are discussed below.

The speed of OsO_4 penetration is partly dependent upon the tissue density when solutions of equal concentrations are used. In general, the higher the tissue density, the slower will be the speed of penetration. The concentration of OsO_4 in the fixative solution is another factor which influences the rate of diffusion. The rate of diffusion usually increases with an increase in the concentration of OsO_4; however, this relationship is not linear. Higher temperatures and fixation by perfusion also accelerate penetration by OsO_4.

The addition of fixation vehicles to balance the OsO_4 solution osmotically with the cell interior is another important factor that significantly influences the rate of penetration. The addition of vehicles results in a slower rate of penetration, since OsO_4 in water alone penetrates most rapidly. The net decrease in the rate of penetration obviously depends upon the types and quantities of salts employed. In general, the more osmotically balanced the fixation mixture, the less the swelling of the tissue, and also the slower the rate of penetration. Usually the addition of electrolytes such as NaCl results in a decreased rate of penetration, but this decrease is less than that caused by the addition of non-electrolytes such as sucrose. It is advisable, therefore, to increase the duration of fixation and the concentration of OsO_4 in the presence of salts in the fixation mixture.

On the other hand, since the diffusion gradient decreases with the increase of diffusion time, the rate of OsO_4 penetration is expected to decrease with the continuance of fixation, irrespective of the density or physiological condition of the tissue. The diffusion constant of OsO_4 into gelatin–albumin gel falls off with time, which suggests the formation of a barrier. This barrier may also hinder the penetration by solvents and embedding media. However, the rate of penetration throughout the duration of fixation is controlled by many factors. In most tissues, the rate of penetration varies during the total duration of fixation. For example, the uptake of OsO_4 by rat tail tendons is slower in the beginning than in the later stages of fixation. This initial slow rate is probably due to the tightly packed collagen fibrils, diffusion barriers and the slow rate of chemical reaction between the cellular materials and the fixative. In contrast, some tissues such as fat exhibit a quicker uptake in the beginning. In this case only a thin outer layer reacts quickly with OsO_4, which then prevents a deeper penetration by the fixative.

The size of the tissue specimen is apparently also responsible for variation in the rate of penetration, which would be different at different depths from the surface of the specimen. It is obvious, therefore, that the smaller the size of the specimen, the more uniform will be the rate of penetration. Since deeper layers of relatively large specimens will be penetrated rather slowly, the cells of these layers will be exposed to low concentrations of OsO_4. Thus, only the outer layers of a tissue block are well fixed by the slowly penetrating OsO_4 (Fig. 1.2). It is generally recommended, therefore, to use only the peripheral layers for examination unless these have been injured mechanically.

Unlike most polar oxidising agents, OsO_4 is able to penetrate both hydrophilic and hydrophobic lipids. The reagent penetrates tissues very slowly but reacts rapidly. It is excessively superficial in its action, so that specimens larger than 0.5 mm or so in diameter are often not fixed uniformly. For this reason the tissue specimen should be cut into small pieces (less than 1 mm^3) or, alternatively, sectioned with an automatic sectioner.

In most types of cells and tissues the maximum speed of OsO_4 penetration is reached within 1 h. The rate of penetration is calculated to be ~800 μm deep during the first hour of fixation. After this period the rate is progressively slowed, because the fixed outer layers of cells resist deeper penetration of the fixative. This slow penetration is also due to a progressive decrease in the concentration of OsO_4 fixative. It is apparent, therefore, that the fixation of deeper layers of a specimen 0.5 mm across requires a long time. In conventional fixation, the core of the specimen is not as well fixed as the outer layers. Osmium tetroxide penetrates slightly more slowly than does glutaraldehyde.

Slow penetration of OsO_4 might tempt the investigator to use higher concentrations of the fixative. This is not practical because, as stated earlier, OsO_4 is poorly soluble in water. Rate of penetration, however, can be increased by using OsO_4 in combination with potassium dichromate or ferricyanide.

The total uptake of osmium per unit weight of tissue from the fixing fluid differs, depending primarily upon the tissue type. For example, pancreas, liver and kidney tissues contain about three times as much osmium as muscle and skin tissues after an equivalent time of fixation (Bahr, 1955). This higher uptake by the former tissues is probably due to the fact that these tissues are richer in lipoprotein membranes. The uptake of osmium per unit membrane is probably the same in the two groups of tissues. The accumulated reduced osmium may account for as much as 46% of the tissue dry weight after rat heart has been fixed by perfusion with OsO_4 (Krames and Page, 1968).

Duration of Fixation

The duration of fixation is intimately linked to the

fixation temperature in all respects. As a general rule, the most desirable duration of fixation is a compromise between the two simultaneous effects of the fixative: (1) fixation and (2) extraction of cellular materials. In other words, length of fixation should be determined on the basis of achieving the best possible fixation and the least possible extraction and alteration of tissue components. Leaching or extraction of cellular materials is undesirable, except in some cases where an increased image contrast of unextracted elements is the primary objective. A prolonged treatment with OsO_4 is known to increase contrast of cystine-rich proteins such as keratin. Since OsO_4 is unable to make all cellular constituents insoluble in water, prolonged fixation causes extraction of cell constituents, especially of proteinaceous substances. Therefore, if possible, a short fixation time should be employed.

It is difficult to recommend a definite duration of fixation, because the rate of OsO_4 uptake varies in different types of tissues. Thus, each type of tissue has its own specific requirement in terms of an ideal duration of fixation. The size of the tissue specimen and the type of buffer employed also influence the optimal duration of fixation. Other factors which influence the fixation time include the concentration of OsO_4 and the concentration of organic matter in the cell. It is known that cells containing very low concentrations of organic matter are difficult to fix with OsO_4. It has to be admitted that the optimal duration of fixation for most tissues is yet to be determined.

For isolated cells and particulate specimens, a few minutes of fixation may be satisfactory. For dense tissue blocks, 30 min–2 h may be needed. Ordinarily, a 15 min–2 h fixation time is ample. Osmium tetroxide has been employed as a post-fixative for as long as 12 h at room temperature to obtain preservation and staining of early embryonic tissues (Kalt and Tandler, 1971). Fixation at a cold temperature requires longer fixation times. For most purposes, a 1% OsO_4 solution having a pH of 7.2–7.4 with an osmolality of 300 mosm is recommended. A better preservation of nuclear structures and spindle fibres is obtained when OsO_4 is employed in a slightly acidic medium.

REMOVAL OF BOUND OSMIUM FROM SECTIONS

The specificity of certain staining techniques is interfered with by the bound osmium in the tissue which has been treated with OsO_4. Methods are available for removing this osmium from thin sections. Unmounted, thin sections are treated (by floating) with 10% aqueous solution of periodic acid for 20–30 min at room temperature. Alternatively, sections are exposed either to a saturated solution of potassium periodate for 30–60 min or to a 1.5% solution of hydrogen peroxide for 10–15 min at room temperature. It should be noted that these treatments may themselves selectively extract cellular substances. An example of such an extraction is sulphur-rich keratohyalin granules from sections of keratinocyte.

OSMIUM BLACKS

Osmium blacks can be defined as $OsO_2 \cdot nH_2O$ (Riemersma, 1970) or coordination of polymers of Os(IV) (Hanker *et al.*, 1967). When OsO_4 is reduced by the unsaturated lipid components of tissue, osmium blacks, along with some osmium oxide as a by-product, are formed. Since OsO_4 is soluble in lipids, its reduction by alcohol results in further blackening during dehydration. The excessive blackening, however, can be minimised by fixing the tissue in darkness. The amount of osmium blacks formed is dependent primarily upon the concentration of OsO_4 present in the fixative.

Osmium blacks are amorphous and generally insoluble in tissue constituents. Because of their electron-scattering properties, the atoms of coordination compounds of osmium remaining in the tissue contribute to the formation of contrast in electron micrographs. Because these polymers of osmium are insoluble in water, in the organic solvents used for dehydration, and in the epoxy and acrylic monomers used in the preparation of thin sections, they are extremely useful for the ultrastructural demonstration of enzyme activity. The demonstration of oxidoreductase by utilising diaminobenzidene (DAB) and Oso_4 and producing osmium blacks is well known. Hanker (1975) has explained how the oxidation of DAB and interaction with OsO_4 result in the formation of osmium blacks. The disubstituted indigo dyes are amenable to osmication, resulting in the production of osmium blacks for localising 5′-nucleotide phosphodiesterase activity. In this reaction probably olefinic linkage of the indigo molecule is involved.

Besides their use in localising enzyme activity, osmium blacks have been used in a periodic acid–*p*-fluorophenylhydrazine reaction for localising mucosubstances (Bradbury and Stoward, 1967) and for enhancing the contrast of osmicated lipid-containing membranes following their treatment with osmiophilic thiocarbohydrazide which are post-treated with OsO_4. This reaction results in the bridging of osmium to osmium through thiocarbohydrazide (Hanker *et al.*, 1966).

PREPARATION AND PRECAUTION IN THE HANDLING OF OSMIUM TETROXIDE

Osmium tetroxide is dangerous to handle because of its toxicity and vapour pressure; the solid has a vapour pressure of 11 mm at 25 °C (Griffith, 1965). Its fumes are injurious to the nose, eyes and throat. Hands or any other part of the body must not be exposed to this reagent. It is commonly used at concentrations of 1–2% in water or buffer. This reagent is a strong oxidising agent, and is readily reduced by the presence of organic matter and exposure to light. It should be noted that even the smallest amount of organic matter will reduce it to hydrated dioxide, which is worthless as a fixative. However, the reduction can be avoided by complete exclusion of dust and organic matter and by use of a brown glass bottle. Extreme care should be taken to avoid contamination by dust particles and exposure to light.

Osmium tetroxide must be handled in a fume-hood, and because the reagent is rather expensive, unstable and volatile, solutions should be prepared with utmost care. The first step in the preparation of its aqueous solution is to remove the label (after reading it!) from the glass ampoule containing the OsO_4 crystals. (Osmium tetroxide is also supplied as an aqueous solution in glass ampoules.) The glass ampoule, a glass-stoppered bottle and a heavy glass rod are carefully cleaned with concentrated nitric acid (to remove all the organic matter), and then thoroughly washed with distilled water to eliminate all traces of the acid.

The bottle and the rod should be dried in an oven; they should never be wiped with a paper or cloth towel, because these materials invariably leave behind some lint which would reduce the solution to hydrated dioxide. A measured amount of distilled water, buffer or other vehicle is added to the bottle. The glass ampoule is frozen in the freezer of a refrigerator for several hours, and after opening it by scoring its neck with a file, the osmium crystals are poured into the bottle. Ampoules with pre-scored necks are also available. The purpose of freezing is to dislodge the crystals from the inner walls of the ampoule. If the ampoule is not frozen, the scored ampoule is gently placed into the bottle. After breaking the ampoule with a heavy glass rod, the bottle is quickly stoppered and is shaken vigorously. Several hours are required to dissolve OsO_4 crystals completely in the vehicle (the solution can be prepared in a few minutes by using a sonicator). After the crystals have dissolved, the bottle is completely wrapped in aluminium foil and stored in a refrigerator. This solution should be protected from exposure to light and contamination caused by organic matter and laboratory dust.

It is emphasised that since OsO_4 is extremely volatile and its solutions rapidly decrease in concentration, solutions should be prepared in small quantities and stored in a tapered flask fitted with a glass stopper and Teflon sleeve. It is also emphasised that the use of ground-glass stoppers is not recommended, for they do not prevent decrease in the concentration of OsO_4, even when maintained at 4 °C. The only effective way of keeping OsO_4 solutions is to use Teflon liners on glass stoppers. The flask must be tightly stoppered, wrapped in aluminum foil and stored in a refrigerator. The solution is thought to be stable for several months under the above conditions of storage. Alternatively, it may be stored in a ground-glass-stoppered bottle, but it is less stable under these conditions.

Studies on the volatility of OsO_4 solutions indicate that the strength of a 2.22% aqueous solution of OsO_4 falls to 2.15% in 24 h when stored in an ordinary glass-stoppered bottle at 0 °C; more dilute solutions tend to deteriorate more rapidly (Frigerio and Nebel, 1962). Variations in the concentration of OsO_4 solutions lead to erroneous interpretation of electron micrographs, particularly in quantitative electron microscopy. It is obvious that in order to know the concentration of OsO_4 within the tissue specimen, the exact concentration of the OsO_4 solution must be known.

It should be noted that when OsO_4 solutions are stored in a refrigerator, all the internal surfaces will be discoloured by the leaking fumes, which may also affect other items in the refrigerator. Osmium tetroxide fumes can penetrate plastics. If the OsO_4 solution needs to be disposed of down a sink, large amounts of running water should be used. However, burial in a guarded area designated for toxic waste is preferable.

OSMIUM TETROXIDE AND GLUTARALDEHYDE MIXTURE

There are some indications that sequential double fixation, glutaraldehyde followed by OsO_4, has undesirable effects on certain specimens, especially on isolated cells and membranes. One possible explanation for these effects is that during pre-fixation the specimen is subjected to the detrimental effects of glutaraldehyde, such as shrinkage and lipid extraction, prior to the application of OsO_4. Moreover, OsO_4 induces marked conformational changes in proteins when the specimen has been pre-fixed with glutaraldehyde. It is known that cells fixed with glutaraldehyde followed by OsO_4 are less stable against mechanical stress (sonication) than those fixed with glutaraldehyde only.

Evidence is accumulating which indicates that a prolonged wash in a buffer between the pre-fixation with glutaraldehyde and post-fixation with OsO_4 produces undesirable effects such as uneven fixation, cell shrinkage, widened extracellular spaces, extraction or swelling of mitochondrial matrices and clumping of chromatin. It has also been shown that the storage of glutaraldehyde-perfused tissue for as little as 30 min in cacodylate buffer, prior to hardening by OsO_4, has a profound flattening effect on granular synaptic vesicles of cholinergic nerve endings (Bodian, 1970). Other reported disadvantages of sequential double fixation are: the loss of arrangement of microtubules when fixed in the cold, myelinisation of lipids and a more dispersed pattern of ribosomes.

As stated above, lipids undergo profound changes during and after fixation with glutaraldehyde. Glutaraldehyde cross-links proteins without reducing the fluidity of the lipid bilayer (Jost and Griffith, 1973). Lipids in the cytoplasm as well as in the membranes, which have not been cross-linked by glutaraldehyde, are free to form not only single and multilayered vesicles, but also multivesicular mounds. Free blebs and vesicles or intramembrane particle-free membrane blebs (blisters) present in the aldehyde-fixed tissues are considered to be artefacts of aldehyde fixation (Shelton and Mowczko, 1977); post-fixation with OsO_4 does not prevent the formation of such artefacts. Such artefacts are virtually absent in freeze–fracture replicas of corneal fibroblasts, which are frozen without aldehyde fixation and in sections of the tissue fixed with a mixture of glutaraldehyde and OsO_4 (Hasty and Hay, 1978). The mixture prevents post-fixation movement of membrane lipids, especially the negatively charged fluid lipids, which are capable of considerable mobility after aldehyde fixation.

Fixation by the mixture results in sharp membrane definition and especially good preservation of nucleo- protein-, lipid- and polysaccharide-containing structures. Granules and vesicles stain distinctly, but glycogen is defined poorly. This procedure seems to preserve polyribosomal structure especially well in plant tissues, and cytoplasmic microtubules appear intact even after fixation at low temperatures. Fixation with the mixture yields definitely better preservation of cellular materials in certain specimens such as single cells than that obtained by double fixation with commercial glutaraldehyde.

The mixture has been used successfully for fixing white blood cells, which are not well fixed by sequential double fixation. The quality of preservation and contrast can be further improved by post-fixing the cells with 2% uranyl acetate prior to dehydration. Several other types of cells in suspension or in monolayer cultures have been successfully fixed with this mixture (Hirsch and Fedorko, 1968). The mixture has been recommended for preserving ciliates (Shigenaka *et al.*, 1973), fragile structure in pathologically altered cells (Laiho *et al.*, 1971), *Tetrahymena* and chicken myoblasts (Kolb-Bachofen, 1977) and *Actinophrys* (Ockleford and Tucker, 1973). Fixation with the mixture, followed by osmication in the cold, has been employed for processing algae; fungi; lichens; roots, stems and leaves of higher plants; various animal tissues, including insect midgut (Brunings and Priester, 1971); and isolated subcellular fractions (Franke *et al.*, 1969). A similar procedure has proved effective in the preservation of the interfacial zone which separates the intracellular structures of vesicular-arbuscular mycorrhizal fungi from the cytoplasm of the root cells of the host plant (Carling *et al.*, 1977). Contrary to previous reports, the interfacial zone is completely filled with a granular matrix. Excellent preservation of polyribosomes attached to the thylakoidal membranes in the chloroplast of bean leaf has been obtained by this method (Falk, 1969).

Although aldehydes reduce OsO_4 and usually it is desirable that excess aldehyde be removed from the tissue prior to post-fixation with OsO_4, these two fixatives can be used as a mixture, because the reaction between glutaraldehyde and OsO_4 has a temperature-dependent lag period; at 18 °C the lag is longer than 30 min (Hopwood, 1970). Thus, if fixation with the mixture is completed within ~30 min at 4 °C, osmium precipitates may not develop. The mixture should be prepared immediately prior to use.

As a result of the reaction between OsO_4 and aldehydes, the net concentration of fixatives available to react with tissue is decreased. Furthermore, both OsO_4 and glutaraldehyde may compete for the same amino acid residues. Also, glutaraldehyde alone is more effective in cross-linking proteins than when used in combination with OsO_4. Other limitations in the use of the mixture include the presence of electron-dense coating on the plasma membrane, availability of a limited duration of fixation, and the necessity of lower temperatures of fixation, resulting in the loss of some microtubules. The above-mentioned problems have prompted a reappraisal of the usefulness of the mixture. It has been indicated that by using vacuum-distilled glutaraldehyde followed by OsO_4, cells in suspension are well fixed and the preservation is reproducible (Gillett *et al.*, 1975).

Glutaraldehyde – OsO_4 mixture

2% glutaraldehyde	1 part
2% OsO_4	1 part

Both glutaraldehyde and OsO_4 solutions are prepared in a buffer, and the two solutions are mixed immediately prior to use. Specimens are fixed in the mixture at 4 °C, washed thoroughly in the buffer and post-fixed in 2% OsO_4.

PERMANGANATES

Potassium permanganate was introduced as an alternative fixative to OsO_4 by Luft (1956) prior to the introduction of glutaraldehyde fixation. The primary reasons for the popularity of the permanganate fixatives were the clarity of membranes and the ease of fixation. Soon it became known that membranes stand out primarily because of an extensive extraction of background cellular substances. This extraction takes place during fixation and/or subsequent dehydration. Major alterations observed in $KMnO_4$-fixed tissues include swelling of mitochondria and plastids; loss of ribosomes, lipid droplets, microtubules (cytoplasmic and neurotubules), neurofilaments, myofilaments, nuclear annuli, interchromatic and perichromatic granules and soluble cytoplasmic proteins; and enlargement of nuclear pores. Although internal membrane systems of mitochondria and plastids are sharply defined, the matrix of these organelles, which shows considerable density when fixed with glutaraldehyde and OsO_4, appears to be wholly removed by $KMnO_4$.

Permanganates penetrate into the tissue faster than do more commonly used fixatives, although the penetration of the former into deeper cells of the tissue block is a problem. The rate of penetration by $KMnO_4$ into gelatin–albumin gel model of protoplasm is ~1 mm in 1 h (Bradbury and Meek, 1960). In tissues, membrane systems are the first to be 'fixed' (within 15–30 min) by $KMnO_4$; longer durations of fixation (1–4 h) lead to an increase in the electron density of some other structures such as nucleolus, chromatin and matrix of mitochondria. In fact, the appearance of the tissue fixed with $KMnO_4$ depends primarily upon the duration of fixation. Since $KMnO_4$ is not in common use, it will not be further discussed in this volume; for additional details, the reader is referred to Hayat (1981).

METHODS OF FIXATION

There are four major modes of fixation: (1) vascular perfusion; (2) immersion; (3) dripping on the surface of the organ; and (4) injection into the organ. Although each of these modes has certain advantages and disadvantages, fixation by vascular perfusion is decidedly superior in most cases to other modes of fixation. Fixation by vascular perfusion also yields better preservation of the fine structure (e.g. mitochondria) in the ischaemic tissues such as myocardium (Sanan *et al.* 1985). The superiority of vascular perfusion to immersion fixation is shown in Fig. 1.3. This is true for both transmission and scanning electron microscopy. In fact, for certain types of tissues (e.g. lung, brain and kidney), fixation by vascular perfusion is indispensable.

VASCULAR PERFUSION

Briefly, the advantages of vascular perfusion are as follows.

(1) Fixation begins immediately after the arrest of systemic circulation, which results in minimum alterations in cell structure. It is always desirable to shorten the interval between death and fixation.

(2) A rapid and uniform penetration of the fixative into all parts of the tissue is accomplished. The increased rate of penetration is probably related to a rapid and extensive flow of the fixative via the vascular bed into the tissue.

(3) Rapid fixation of heterogeneous tissue components is achieved with a minimum of traumatisation.

(4) Tissue is stabilised against excessive dissolution and translocation of cellular substances.

(5) An accurate estimation of enzymatic activity is facilitated, for perfusion removes blood, enzymatic activity of the serum is eliminated and natural inhibitors are suppressed.

(6) Tissue is sufficiently fixed and hardened prior to direct handling. Manipulation of tissues after arrest of systemic circulation, but prior to fixation, results in the introduction of artefacts.

Vascular perfusion, however, has certain limitations. This method obviously cannot be used with human subjects. In fact, the majority of the tissue types cannot be fixed by perfusion. Appropriate anaesthesia and analgesia are necessary. A sufficient amount of a naesthetic should be injected intravenously to induce immediate deep anaesthesia, without a stage of excitation or heart arrest. However, too deep an anaesthesia should be avoided. Animals must not be stressed during the operation.

The success of the perfusion method depends, in part, upon the complete exclusion of blood from the vascular system and prevention of vasoconstriction (blocking or narrowing of blood vessels). An isotonic buffered saline or Ringer solution can be used to flush out the blood from the vascular bed. It is necessary that the optimal amount of saline be used. A delay in starting the flow of the fixative perfusate may be caused by the use of large amounts of saline or by the slow flow of the saline. The amount of saline and the duration needed to remove the blood depend upon the extent of vasculature to be perfused. The amount can be calculated in millilitres equivalent to body weight in grams. In general, 15% body weight is desirable for perfusion of the whole body, 5% for perfusion of the head, with or without forelimbs, and 1–2% for perfusion of the CNS exclusively. However, the amount of saline introduced is best controlled by timing the flow rather than introducing a certain volume of saline. A duration of 10–30 s is recommended.

Drugs such as papaverine, sodium nitrite, procaine hydrochloride and lidocaine chloride can be employed as vasodilators with perfusates. Approximately 1% sodium nitrite in 0.9% sodium chloride can be added to the saline perfusate. An anticoagulant, such as heparin (150 000 USP) in a dose of 0.5–1.0 ml/kg body weight, can be used either in the perfusate or mixed with the injected anaesthetic. Finally, the perfusion method is rather elaborate, and a considerable amount of skill and experience is needed to perform dissection and cannulation in the minimum possible time.

Because vapours arising from the aqueous solutions of fixatives are damaging to the eyes, nose and mouth, great care should be taken to avoid the fumes during the lengthy procedures of dissection, cannulation and perfusion. Some workers prefer to carry out the procedures under a fume-hood, while others wear goggles to protect the eyes.

In order to obtain consistently satisfactory fixation, it is essential that optimal conditions prevail during vascular perfusion. The important parameters include the pH, temperature, duration of washing and fixation, composition and concentration of the perfusate, osmotic pressure of the perfusate, hydrostatic pressure employed during perfusion, the actual rate of flow of the perfusate and the route of perfusate instillation. The best results are obtained when the pH is maintained within the physiological range (7.2–7.4). The temperature of the perfusate should not be below body temperature of the animal, otherwise vasoconstriction may occur, which would impair the effectiveness of the perfusion.

The osmotic pressure of the perfusate should be identical with that of the blood of the animal under study. To accomplish this, however, is not easy, because the osmotic pressure of blood varies among species. Optimal osmolalities for some tissue types are listed in Table 1.8. In order to raise the colloid osmotic pressure of the fixative, dextran or polyvinyl pyrrolidone (PVP) (2%) or gum acacia (1%) can be added to the fixative. Such increased pressure minimises the expansion of extravascular spaces. The addition of PVP to the fixative is especially helpful in improving the quality of preservation of brain tissue.

Table 1.8 Recommended osmolality of perfusate for selected tissues

Tissue type	*Osmolality* (mosm)
Rat heart	300
Rat brain	330
Rat lung	330
Rat kidney	420
Rat renal medulla[a]	
inner stripe of outer zone	700
outer level of inner zone	1000
middle level of inner zone	1300
papillary tip	1800
Rat developing renal cortex[b]	828
Rat (or cat and fowl) liver	450
Rat (or chicken) embryo liver	420
Rat skeletal muscle	475
Frog liver	300
Rabbit brain	820

[a] The addition of 3% dextran to the fixative helps to preserve interstitial structures and intercellular relationships (Bohman, 1974).
[b] With 6% glutaraldehyde (Larsson, 1975).

Agreement on the optimal concentration of the fixative during the initial fixation is lacking. According to some workers (Brightman and Reese, 1969), aldehydes of low concentrations (1% paraformaldehyde and 1.25% glutaraldehyde) are desirable. Schultz and Case (1970), on the other hand, advocate the use of 1% glutaraldehyde preceding the main perfusion. The latter approach will result in rapid initial fixation. We use aldehyde concentrations of 2–4%.

As stated above, in addition to the perfusate pressure, perfusion (hydrostatic) pressure (which primarily controls the rate of flow of the perfusate) affects the quality of tissue preservation. The ideal perfusion pressure, apparently, is equivalent to the pressure experienced by a given tissue *in vivo*. Table 1.9 gives desirable perfusion pressures for some tissue types. It should be noted, however, that even when perfusion pressure equal to the normal blood pressure of the animal is employed, certain cell types of a complex tissue (e.g. kidney) may not be well fixed. A case in

Table 1.9 Recommended perfusion pressure for selected tissues

Tissue type	*Perfusion pressure* (mmHg)
Rabbit spleen	110–120
Rabbit (or rat) brain	150
Monkey brain	110
Baboon lung	40
Rat or mouse heart	103–120
Rabbit arteries	100–120
Rat kidney	100–120
Rat (or cat and fowl) liver	100
Rat (or chicken) embryo liver	50
Rat skeletal muscle	100

point is the inner zone of medulla, which is fixed satisfactorily when perfusion pressure (200 mmHg) higher than the normal blood pressure of the animal is employed.

In general, fixation occurs more rapidly when higher perfusion pressures are used together with an ample flow of fixative solution. This practice will effectively force the blood out of the capillaries and improve the diffusion of the fixative into the tissue. The perfusion pressure should be at least equal to the systolic pressure. However, too high a perfusion pressure may cause swelling of the tissue.

For obtaining optimal results, the maintenance of constant perfusion pressure, within the physiological range, during the entire procedure of vascular perfusion is essential; higher pressures may increase the size of the organ, resulting in the introduction of artefacts. A transition in the perfusion pressure may also result in the introduction of artefacts. The transition occurs when two perfusates are used, which necessitates a switching from one perfusate to the other. Such a transition may be accompanied by a temporary decrease in the perfusion pressure. Kidney tissue is especially prone to this type of artefact because it is a functionally distended organ.

It should be noted that the resistance of the cannula–blood vessel–organ system differs from experiment to experiment. Thus, the perfusion pressure required may not be the same for similar organs of the same animal species. The rate of flow of the perfusate is controlled not only by the perfusion pressure, but also by several other factors, including the size of the needle used as a cannula and the diameter of the blood vessels.

The most effective route for vascular perfusion of a given tissue organ is available. However, if several organs must be perfused simultaneously in a single animal, a major part of the body can be perfused through the left ventricle and arch of the aorta. When this route is used carefully, fixation of liver, kidney, brain and peripheral nervous system, gastro-intestinal tract, spleen and gonads can be accomplished. This route is also ideal for studying proliferative liver lesions, because the method produces arterial perfusion of the liver, and the lesions have almost exclusively an arterial blood supply (Jones *et al.*, 1977).

It is pointed out that even when vascular perfusion is carried out under optimal conditions, relatively large organs, such as brain and kidney, show uneven fixation. A case in point is the presence of solitary dark neurons observed with the light microscope, which are attributed to post mortem or pressure on the poorly fixed parts of the brain (Cammermeyer, 1978). The grey matter of the brain is fixed better than the white matter, because the former is vascularised more extensively than is the latter. As a result, the grey matter is infiltrated rapidly and with larger amounts of the fixative. This artefact may be formed when the flow of the fixative is delayed or the brain is autopsied after termination of the perfusion. Similarly, various parts of the kidney exhibit different qualities of fixation.

Methods of Vascular Perfusion

GENERAL METHOD

A relatively simple and inexpensive apparatus (Fig. 1.5A,B) was constructed by Rossi (1975) to carry out perfusion fixation in small animals such as mice, rats, hamsters, rabbits and young dogs. All organs except lungs and those caudal to the catheter (needle) can be perfused by using this apparatus. The apparatus and procedure suggested by Rossi (1975) are described below.

The apparatus consists of pressurised vessels for the solutions, devices for pressurisation and for measuring pressure, vinyl tubing, needles or catheters to deliver solutions into the vascular bed. Two glass containers (coffee jars) of 1 litre capacity with screw-on lids made airtight by suitable O-rings are needed. Two holes (3–4 mm each) are pierced into each lid, through which vinyl tubings are inserted and cemented into place with an epoxy resin. One tube provides connection between the air spaces in the two jars through vinyl tubing in which is inserted a four-way joint for connection with a manometer and with a rubber-bulb syringe (a in Fig. 1.5A), fitted with a one-way valve for pressurisation.

One end of the second tubing through each lid extends to the bottom of the jar and the other ends are joined by a three-way plastics valve. The third arm of this valve carries the catheter for entering the abdominal aorta. One jar is filled with Ringer and the second with the fixative, and lids are screwed on tightly. The

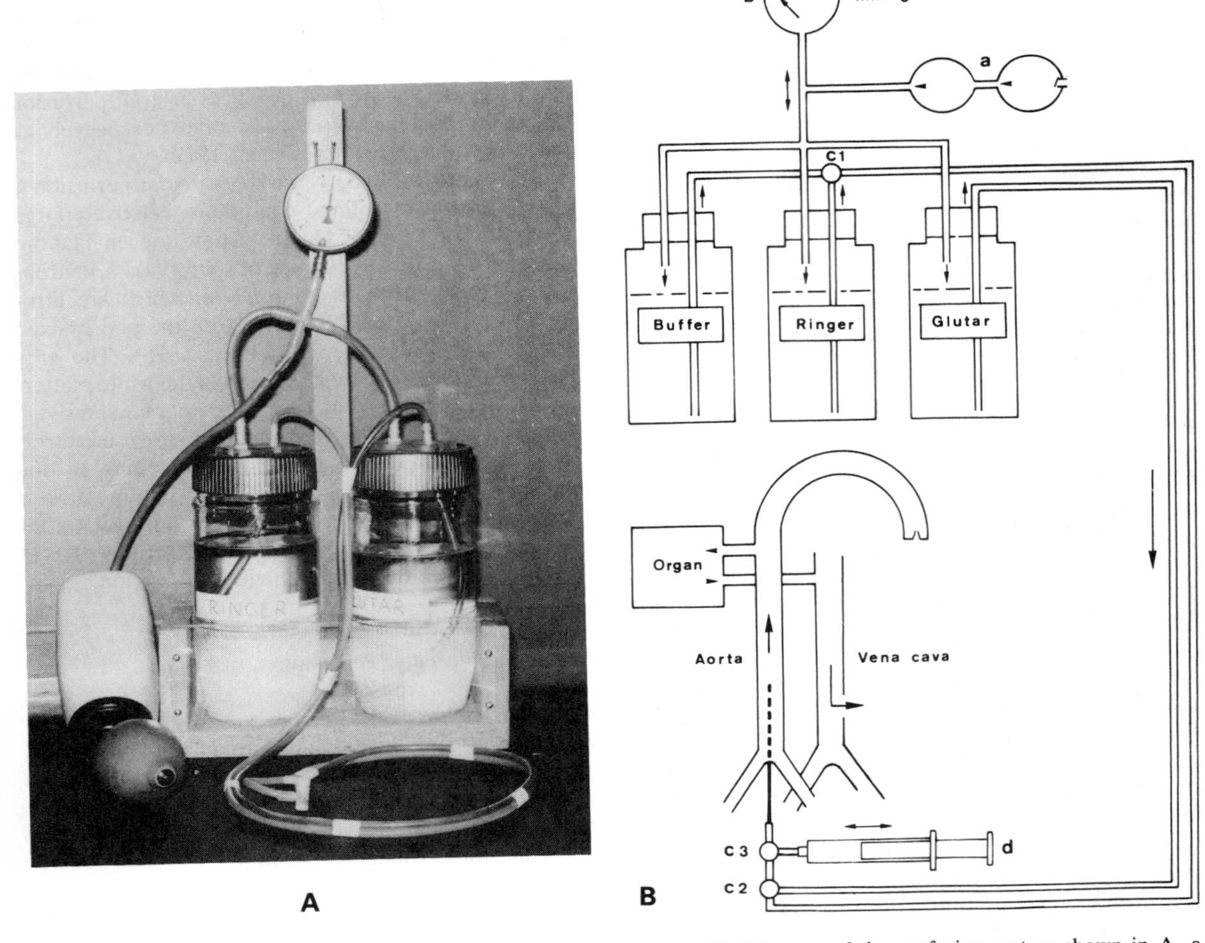

Figure 1.5 A: A perfusion apparatus with two bottles for perfusates. B: Diagram of the perfusion system shown in A. a, Rubber ball syringe; b, manometer; c, 3-way valve. From Rossi (1975)

system is pressurised and freed of air bubbles by manipulating the three-way valve. The valve is turned to an intermediate position to prevent further loss of solutions and the pressure is raised to 120–130 mmHg.

The abdomen of the anaesthetised animal is opened, and viscera displaced until the aorta and vena cava are visible from the renal arteries distally. The aorta bifurcation is freed with round-tipped forceps and the aorta clamped with round-edged surgical forceps just distal to the renal arteries. The aorta is opened with fine scissors at the bifurcation and a catheter or needle is inserted into it at the level of the surgical clamp, which is removed after the catheter has been locked in place with a grooved clamp.

Ringer is admitted into the aorta through the three-way valve at ~120 mmHg to ensure flow against aortic pressure. Immediately the vena cava is cut and within ~2 min blood has been removed by Ringer and the valve turned to admit the fixative solution. Fixation is completed within 10–15 min.

AORTA

The animal (e.g. rabbit) is killed by intravenous injection of 10% urethane anaesthesia. The dorsal aorta from the arch to approximately the first intercostal artery is removed and placed in 0.9% saline. Each arterial section is tied to a hydrostatic column pre-set to a height calculated to produce a pressure head equivalent to either 80 or 125 mmHg. Containers for heparinised saline and for fixative are connected (Fig. 1.6).

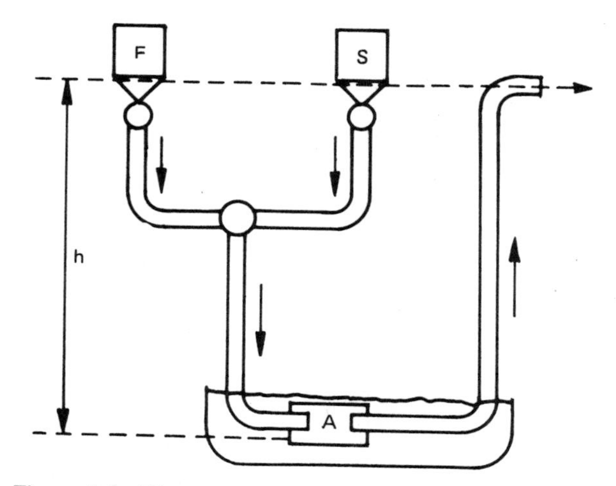

Figure 1.6 Diagram of the apparatus used for vascular perfusion of aorta. Arrows indicate the direction of perfusate flow. From Swinehart *et al.* (1976)

The tied arterial section is immersed in the saline bath and the T-connection opened to allow heparinised 0.9% saline to flow through. Any leaks or opened side vessels are clamped.

After the leaks have been closed, the perfusion with the saline is terminated and 3% glutaraldehyde in 0.1 M phosphate buffer (pH 7.4) is allowed to flow through the vessel. The fixative can be dyed with methylene blue in order to monitor the proper flow of the fixative. The fixative can be recovered as it flows out of the system and poured back into the container for recirculation to maintain the desired height of the column (h in Fig. 1.6) of the fixative. Small pieces of the tissue are immersed in fresh glutaraldehyde solution and post-fixed in OsO_4. For further details, see Swinehart *et al.* (1976).

ARTERIES

The artefacts caused by conventional fixation of arteries include wavy appearance of elastic laminae, narrowing of the lumen by a decrease in the inner circumference, partial detachment of endothelial cells, and disintegration of leucocytes and their adhering to subendothelial structures after removal of endothelium. The elastic laminae of the aorta are considered to be stretched and straight *in vivo*. These fixation artefacts can be minimised by employing the following method of perfusion.

Animals are anaesthetised and ~1 cm of the abdominal aorta near the renal arteries and 1 cm at the bifurcation is dissected, leaving ~7 cm of the vessel untouched. The vessel is perfused with 10 ml of Krebs–Ringer solution through a silastic catheter which has been introduced proximally. During this washing, a second catheter is introduced distally. The outflow of the second catheter is placed 60 cm above the aorta (~82 mmHg pressure) (Haudenschild *et al.*, 1972).

The washing is followed by perfusion with 1.5% glutaraldehyde in 0.1 M phosphate buffer (pH 7.4) having an osmolality of 500 mosm for ~20 min at room temperature. The containers with washing and fixation solutions are placed 100 cm above the aorta (~136 mmHg pressure). Small blocks from the perfused vessel are re-fixed by immersion with the same fixative for 1 h and then post-fixed with 2% OsO_4 in the buffer for 1 h at room temperature.

CENTRAL NERVOUS SYSTEM

The following method is recommended for simultaneously localising the fluorescence with the light microscope and fixing the tissue for electron microscopy (Furness *et al.*, 1977, 1978). A mixture of glutaraldehyde and formaldehyde produces a fluorophore with catecholamines in the peripheral and central nervous systems. The fluorescence reaction is produced at room temperature and is stable in aqueous solutions.

Perfusate

Formaldehyde	4%
Glutaraldehyde	1%
Phosphate or cacodylate buffer (0.1 M, pH 7.0)	

Procedure

The animal (e.g. rat) is anaesthetised with sodium pentobarbital (40 mg/kg body weight) and injected with heparin (4000 μg/kg body weight). The chest is opened and the heart is exposed. The tip of an 18 gauge cannula connected to a perfusate apparatus (Fig. 1.7) is introduced into the aorta via the left ventricle and held in position with a clamp across the ventricles. The blood is flushed out with a 1% solution of sodium nitrite in 0.01 M phosphate buffer (pH 7.0). This step is accomplished in 10–30 s of perfusion at 120 mmHg and is followed immediately by perfusion with the fixative solution for 10 min at the same pressure. About 200 ml of the fixative solution is perfused. The fixed brain is placed in the fresh fixative at room temperature. Small pieces of the fixed brain are further fixed for 30 min and then post-fixed with 1% OsO_4.

EMBRYO

The pregnant animal (e.g. mouse) is anaesthetised by an intraperitoneal injection of sodium pentobarbital (30 mg/kg body weight). The animal is laparotomised and

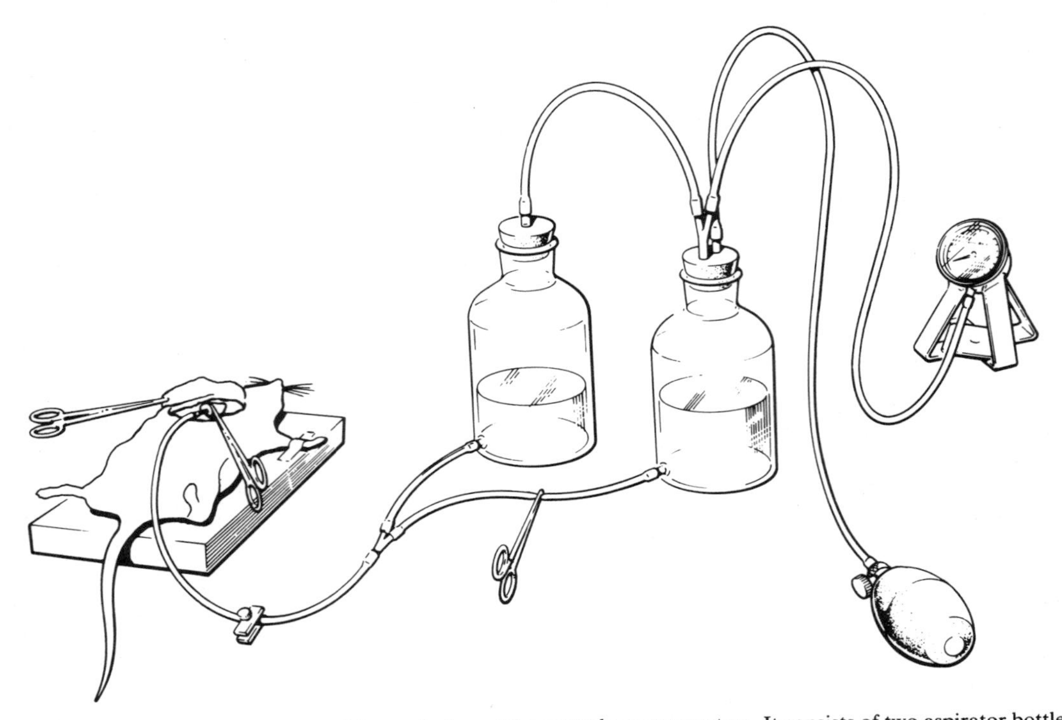

Figure 1.7 Diagram of the apparatus used for perfusion of the central nervous system. It consists of two aspirator bottles, each holding 1 litre, which are connected to a sphygmomanometer bulb and gauge to maintain and monitor the perfusion pressure. One bottle contains buffered sodium nitrite solution, and the other carries the fixative solution. Clamps are used to direct the perfusate selectively from one or the other of the bottles. All tubing used to connect the apparatus is kept as short as practicable and is of wide bore (10 mm inside diameter) so that any pressure difference between the gauge and the cannula is kept to a minimum. From Furness *et al.* (1978)

the uterus is exposed. An opening showing the yolk sac is dissected in the uterus wall under a stereomicroscope. Care should be taken not to disturb the circulations of the conceptus. The beating heart of the embryo is located and a micropipette with a tip diameter of 25–50 μm is inserted through enveloping membranes and precardiac wall into the atrium (Fig. 1.8). The micropipette is immediately retracted, thus leaving an opening in the wall of the atrium through which the blood could flow out. The tip of the pipette is placed in the lumen of the ventricle and the microperfusion by the fixative is started. The start of the perfusion is accompanied by the escape of blood through the opening in the atrium.

The embryo is bleached and becomes yellowish immediately. The speed of fixation can be observed by adding alcian blue to the fixative; all the capillaries show fixative penetration within the first few seconds. The perfusion is continued for 30 s–2 min. Small pieces of the embryo are fixed further for 2 h in the same fixative. For further details, see Abrunhosa (1972).

HEART

Perfusates

(1) Ringer solution containing 0.1% procaine. The pH is adjusted to 7.2–7.4 with 1 N HCl, and the osmolality to 300 mosm with NaCl. Perfusion takes ~3–5 min.

(2) 2% glutaraldehyde in 0.045 M cacodylate buffer. The pH is adjusted to 7.2–7.4 and osmolality to 300 mosm. Perfusion takes ~5 min.

Procedure

In Fig. 1.9, containers (B) filled with various perfusates are placed in a thermoregulated bath (A). The height of the containers from the working table should be adjustable, to regulate the rate of perfusion. The perfusates, maintained at 0–4 °C, flow through a siphon (C) made of soft plastics tubing. The cannula (F), made of Teflon, is attached to the tubing (D) by a middle piece which has a cone-shaped distal opening in which slides the cannula. This system allows rapid change from one solution to another, and the same cannula can

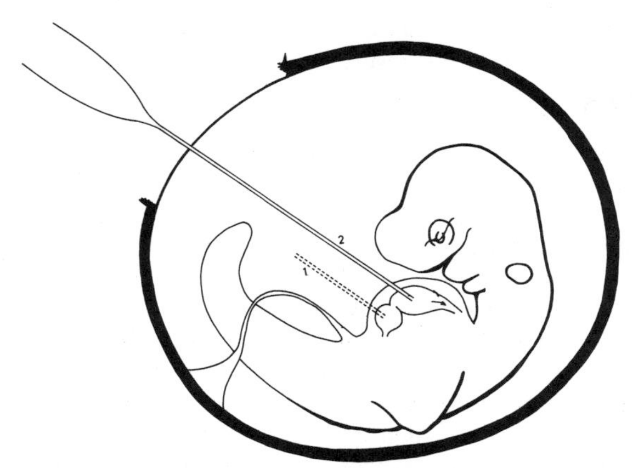

Figure 1.8 Diagram of microperfusion fixation through the embryonic heart. The pipette is first introduced through enveloping membranes and the precardiac wall into the atrium (1). It is then immediately retracted, thus leaving an outflow opening. In the second step, the tip of the pipette is forwarded into the lumen of the ventricle (2), and the perfusion is started. From Abrunhosa (1972)

be used for all perfusates. The distal end of the cannula is obliquely pointed.

Since the tubing (D) is not heat-insulated, the temperature of the perfusates will rise to ~6 °C. With adequate flow and pressure, the temperature of the perfusate coming out of the animal does not exceed 10 °C. The containers are placed ~140 cm above the working table, in view of the average arterial pressure of 103 mmHg in rats. During perfusion, a flow rate of ~40 ml/min is necessary.

The rats are anaesthetised by intraperitoneal injection of 125 mg urethane/100 g of body weight. The animal, on its back, is tied to the working table and the ventral abdominal wall is opened wide with an incision along the white line. The intestine is pushed aside gently and the abdominal aorta is exposed. Two silk threads are slipped under the aorta, the first under the renal arteries and the second above the bifurcation of iliac arteries. The two threads, separated by a distance of ~5 mm, delineate the segment of the aorta where a slit is made for the insertion of the cannula.

The cannula filled with Ringer solution containing procaine is inserted into the lumen of the aorta. The cannula is then tied in place to the aortic wall with the first thread. The second thread is used to tie the aorta; this prevents haemorrhage. After the cannula is in place, perfusion is started. The proximal end of the cannula is slipped into the conical opening of the middle piece (D), and the solution flows into the circulatory system of the animal. Simultaneously, the inferior vena cava is cut, to allow the evacuation of blood and perfusate. For further details, see Forssmann *et al.* (1967).

KIDNEY

The animal (e.g. cat) is anaesthetised with Nembutal (40 mg/kg body weight), a tracheal tube is inserted and one of the femoral arteries is catheterised with PE 100 polyethylene tubing. The catheter is advanced until the tip lies at the junction of the left renal artery and the aorta. A loose ligature is placed around the aorta above the left renal artery, and another one below the right renal artery. The placing of these ligatures minimises an accidental stimulation of nerves at the hilus. The renal arterial system can thus be isolated with the tying of these two ligatures. Another PE 100 polyethylene catheter is inserted into a femoral vein and pushed forward until the tip lies in the vena cava between the two renal veins.

A loose ligature is placed below the junction of the vena cava and the right renal vein. Another ligature is placed around the vena cava above the left renal vein. When these two ligatures are tied during perfusion, they can substantially isolate the renal venous drainage from the remainder of the venous circulation. Approximately 45 min is needed to complete the surgical procedures.

Heparinised saline is perfused by means of a Sigma peristaltic pump into the animal via the catheter inserted into the femoral artery. During saline perfusion, the four ligatures described above are tied as rapidly as possible; this can be accomplished in less than 1 min. This is followed by perfusion with the fixative solution. The fixative is 1% glutaraldehyde in 0.1 M phosphate buffer (pH 7.2) with an osmolality of ~320 mosm. The rate of perfusion is set at 40 ml/min, which is approximately equal to the renal blood flow in cats of this size. The effluent is usually free of blood within 3–4 min. After perfusing for 10 min, the kidneys are removed and cut into small pieces, which are fixed for a further 2 h in the fixative at room temperature. This is followed by washing in the buffer and post-fixation with 2% OsO_4 in the buffer for 1–2 h. For further details, see Yun and Kenney (1976).

LIVER

Perfusate

25% glutaraldehyde	40 ml
0.15 M phosphate buffer (pH 7.2) to make	500 ml

The solution should have an osmolality of ~450 mosm, which can be adjusted with sucrose.

Procedure

The anaesthetised animal (e.g. rat) is tied to the operating board with its back down. The abdominal cavity is opened by a midline incision with lateral

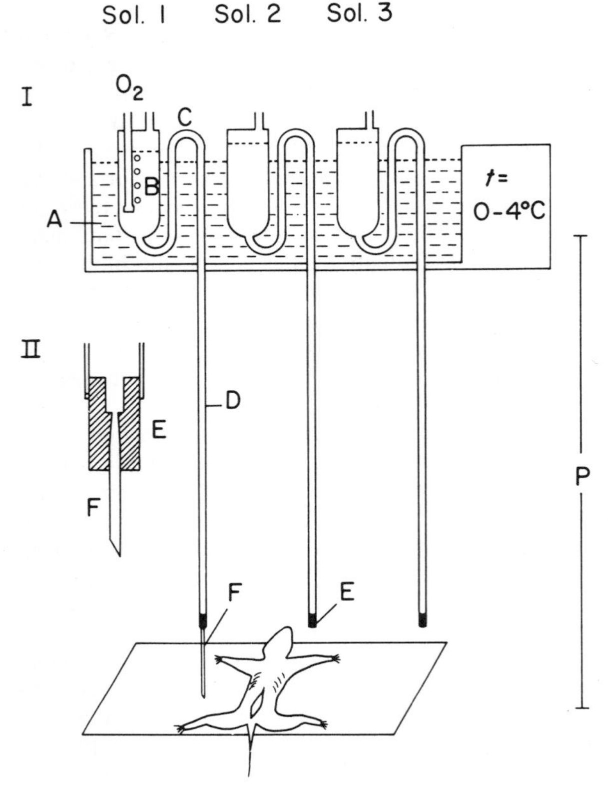

Figure 1.9 Diagram of the apparatus used for perfusion of the heart. (I): A, thermoregulated bath; B, perfusate containers; C, siphon; D, plastics tubing; E, middle piece; F, cannula. (II): Details of the middle piece. From Forssmann *et al.* (1967)

extensions, and the intestine is gently moved to the left side. The portal vein is exposed and two ligatures are passed behind it. The distal ligature is tied to block the flow of venous blood from the portal vein and the flow of hepatic arterial blood to the liver. At the site where the portal vein is branched to different liver lobes, a 1 cm, 20 gauge syringe needle is inserted and secured by the second ligature. Before inserting the needle, the perfusion pressure is adjusted to 20 mmHg, which subsequently falls to ~10 mmHg during perfusion. The needle has been previously connected, via an ordinary clinical intravenous infusion set, to a flask containing the perfusate.

The inferior vena cava below the diaphragm is cut open to relieve the pressure in the right heart. The flask containing the fixative is hung 25–30 cm above the animal. The rate of flow should be 5–10 ml/min. Care should be taken to prevent air bubbles from entering the portal vein or the perfusion system.

Within 3 min after the start of the flow of fixative through the portal vein, the thoracic cavity of the animal is opened and the right thoracic wall is removed. A syringe needle is inserted into the main stem of the hepatic vein and secured with a ligature. The needle has been connected previously to a second flask of fixative, which is hung 15 cm above the animal. The retrograde perfusion should begin ~5 min after the start of flow of the fixative through the portal vein. The flow rate of the retrograde should be ~5 ml/min.

The perfusion is usually completed in 10–15 min, after which time only clear fixative solution comes out of the portal vein. Within 30–40 s after the start of perfusion, the colour of the liver changes from dark reddish-brown to light brown. The completion of fixation can be detected grossly on the surface of the liver. The consistency changes from soft to rather stiff, resembling that of a boiled egg. Uniform overall fixation can be checked by immersing the liver slices in distilled water; poorly fixed areas will show white discoloration. Fixation uniformity can also be checked by examining 0.2–0.5 μm thick sections stained with 1% toluidine blue in 1% borax with a light microscope. Immediately after the completion of perfusion, the liver is removed and cut into small segments (1 mm^3) which are immersed in glutaraldehyde for ~2 h.

In the well-fixed liver, Disse and sinusoidal spaces are clearly delineated by extensions of endothelial and Kupffer cells, and the sinusoids are patent and practically free of blood cells and floccular material. In addition to hepatocytes, endothelial, Kupffer and fat-storing cells are well preserved and easily identified. Endothelial cells are smoothly apposed to the sinusoidal wall. Kupffer cells are attached to broad gaps in the endothelial lining. Fat-storing cells are intercalated between hepatocytes and cells lining the sinusoid. Glycogen is stained dark and is present in clumps.

LUNG

The following method uses a simple perfusion apparatus for pulmonary perfusion of lungs (Coalson, 1983). Expensive equipment such as peristaltic pumps, pressure transducers and recorders is not needed. The pressure is provided by a constant inflow of air from the air inflow valve present in any routine laboratory. Isosmotic–isoncotic electrolyte solutions and fixatives are employed to prevent oedema in the lung. The experimental set-up uses a Harvard respiratory pump, which should be set for a tidal volume delivery appropriate to the species under study.

The perfusates are transferred to quart-size, wide-mouthed canning jars placed in a water-bath at 37.5°C. The jars are fitted with size 12 rubber stoppers. One of the stoppers is bored with two holes. One piece of the 5 mm glass tubing should extend about 3 cm below the stopper. The other piece of tubing should extend to

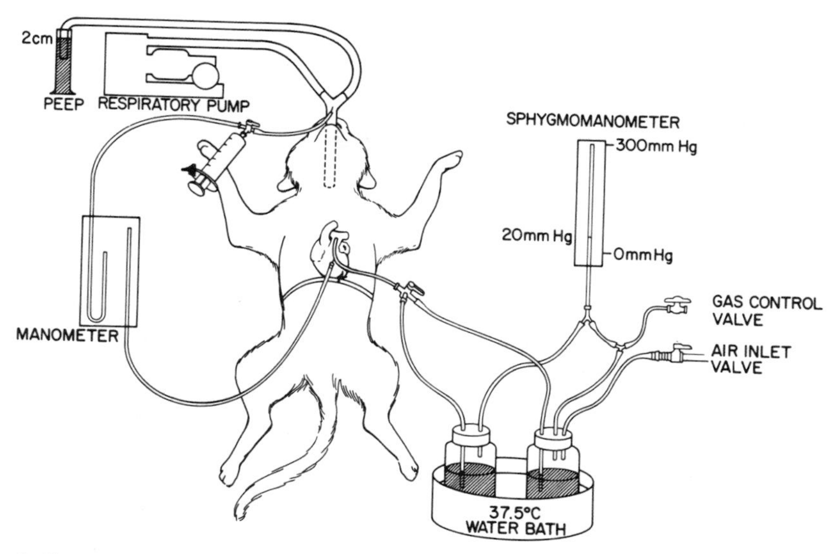

Figure 1.10 Schematic diagram of the apparatus that utilises air from a laboratory air inlet valve to provide the pressure for the system. From Coalson (1983)

within 2 cm of the bottom of the jar. Both pieces of glass tubing should extend 5–8 cm above the top of the rubber stopper. The pieces of glass are sealed with silicone grease to prevent air leaks. The jar that receives the air flow from an air inlet valve is constructed in a similar manner, except that there is a third bored hole to accommodate one additional piece of glass tubing. The short glass tubes are designed to deliver the air and build enough pressure to drive the perfusates into the longer pieces of glass tubing.

Tygon tubing is used to interconnect the system. One piece of Tygon tubing should be run from a compressed air source through the air inlet valve to one of the short pieces of glass tubing in the bottle with the three bored holes. Two additional pieces of Tygon tubing are connected to the two short pieces of glass tubing within the two canning jars. These are joined by a Y-connector to a gas control valve with a needle-point stem (0.25 in) and to a mercury sphygmomanometer gauge (Tycos, Taylor Instruments, Arden, SC). This gauge allows for direct monitoring of the pressure, which is controlled by regulating the air egress from the system with the gas control valve.

The two pieces of Tygon tubing are attached to the longer pieces of glass tubing that extend below the level of the perfusates. These two pieces of Tygon tubing are connected to a three-way stopcock that allows for an immediate sequential switch from one prefusate to another during the perfusion. A single piece of tubing is then constructed which is adapted with appropriate intramedic polyethylene tubing for a suitable cannula to fit the main pulmonary artery. Owing to the diameter changes in the tubing system as it enters the heart and its possible change in resistance, a needle should be inserted into the line at the level of the inflow into the main pulmonary tract. This allows monitoring of infusion pressure with a linear water manometer. Infusion pressure can be regulated by altering the perfusate flow by adjusting the gas control valve. The lines should be cleared of air bubbles and filled totally with the perfusate prior to perfusion.

The animal (cat) is anaesthetised and restrained in a supine position. The trachea is cannulated and lungs are ventilated on air with a Harvard respiratory pump (Fig. 1.10) with a tidal volume of 30 ml and 2 cmH_2O pressure PEEP. A midline incision extending from the upper neck to the lower abdomen is made. The thorax is opened and the sternum is reflected back. The pericardium is opened and the apex of the heart is secured by suture to the diaphragm. A ligature is loosely placed around the main pulmonary artery. After a small incision is made in the right ventricle, the perfusate cannula is quickly inserted into the main pulmonary artery and secured by the ligature.

An isoncotic–isosmotic Krebs–Henseleit solution containing 1% sucrose, 3.5% dextran (mol. wt. 70 000) and heparin (5000 U-USP/L) is perfused under 20 mmHg pressure for 1 min to flush the blood vessels. Krebs–Ringer solution can be used with equal success. The pressure is created by forcing air from an air inlet valve into a closed reservoir system. The pressure is monitored by a sphygmomanometer gauge and adjusted with an air control valve. The left auricle is cut to allow the perfusate to drain from the lung. The abdominal aorta and inferior vena cava are transected to decrease the blood flow into the right heart. This transection is followed by perfusion with 1% glutaraldehyde in 0.1 M cacodylate buffer (pH 7.4) containing 2% sucrose and 6% dextran (mol. wt. 70 000), with total osmolality of 348–363 mosm for 15 min at 37 °C. Simultaneously, the ventilator is stopped and the airway pressure is adjusted to 10 cmH_2O (or any desired pressure). The perfusates are aspirated from the posterior thoracic cavity. Both the airway and perfusion lines are clamped, and the lungs are removed intact. Care should be taken that the lungs are not compressed or deformed. The lung is floated in the fixative for 24 h, cut into specimens for light and/or electron microscopy and further processed accordingly.

MUSCLE (SKELETAL MUSCLE OF RAT HIND LIMB)

Perfusate

A. Locke's solution containing 0.1% $NaNO_2$

NaCl	9.2 g
KCl	0.42 g
$CaCl_2 \cdot 2H_2O$	0.24 g
$NaHCO_3$	0.15 g
sucrose	1.0 g
$NaNO_2$	10 g
distilled water to make	1000 ml

The final solution contains 2 mmol $CaCl_2$.

B. 12% glutaraldehyde

25% glutaraldehyde	60 ml
0.1 M cacodylate buffer (pH 7.2) to make	500 ml

The final solution contains 2 mmol $CaCl_2$ and has an osmolality of 474 mosm.

Procedure

The animal (e.g. rat) is anaesthetised intraperitoneally with 40 mg of Nembutal/kg of body weight. The abdominal aorta is cannulated just above the bifurcation to the hind limbs, and the vena cava is cut to allow free flow of solutions. The temperature of the perfusate is maintained at 40 °C in a water-bath. The perfusate is pumped with the aid of a peristaltic pump (Hughes Hi-Lo) at a constant pressure of 100 mmHg into the hind limb. Perfusion with solution A is carried out for 2–3 min to dilate the vascular bed and remove most of the blood. This is followed by perfusion with solution B for ~10 min. The tissue is cut into small pieces and fixed in solution B for an additional period of 1 h, rinsed in the buffer and post-fixed in 1% OsO_4 for 1–2 h. For further details, see Bowes *et al.* (1970).

OVARY

Perfusate

glutaraldehyde	1%
paraformaldehyde	1%
2,4,6-trinitrocresol	0.01%
cacodylate buffer	0.1 M

Procedure

The animal (e.g. guinea-pig) is anaesthetised by an intraperitoneal injection of sodium pentobarbital (30–35 mg/kg of body weight). Each of the two ovaries is perfused independently of the other. A Holter peristaltic pump (Extracorporeal Medical Specialties Inc., King of Prussia, Pa; model 911, fitted with size D pump chamber) can be used to deliver the perfusates. The perfusion can be started at a pressure of 35–50 mmHg. The abdomen is opened with a U-shaped incision extending from the pubis to the ribs. A ligature (4-0 surgical silk) is placed around both the ovarian artery and vein, and tied loosely in an overhand knot. A length of the uterine artery is carefully separated from its companion vein, and a second ligature is tied loosely around the artery. Ligatures are placed in similar locations around the blood vessels of the other ovary.

The uterine artery is grasped somewhat caudal to the ligature with fine forceps, lifted and pulled slightly taut. The cannula (26 gauge needle), with Krebs–Ringer–bicarbonate buffer (containing 2 mg glucose/ml, pH 7.4, 300 mosm) flowing from it at a rate of 2 ml/min, is inserted into the uterine artery, guided past the ligature and secured into place by tightening the ligature. Immediately a nick is made in the uterine vein to provide an outflow for the perfusate, and the ligature around the ovarian artery and vein is tightened, isolating the ovary from the circulatory system.

After a 2–3 min buffer wash, fixation is begun and continued for 10–30 min. Shortly after the fixative reaches the ovary, it begins to harden, indicating a favourable perfusion. After the perfusion is under way on one side, perfusion of the other ovary is started by using the same sequence. The ovary is cut into small pieces, which are placed directly in 0.1 M cacodylate buffer containing 5% sucrose for a wash, and post-fixed with 1% OsO_4 in the same buffer for 1 h in the cold. For further details, see Paavola (1977).

SPLEEN

Perfusates

I. Modified Ringer solution

KCl	0.3 g
NaCl	8.78 g
distilled water to make	1000 ml

Procaine hydrochloride is added to a final concentration of 0.1%. The pH is adjusted to 7.4. Procaine is added to prevent arteriolar spasm and thus allow free flow of the perfusates.

II. Fixation fluid consists of glutaraldehyde (1.7%) in Sörensen's buffer (0.08 M) with an osmolality of ~310 mosm. Procaine hydrochloride (5%) in water is added to a final concentration of 0.1%.

Procedure

The animal (e.g. rabbit) is anaesthetised by intravenous pentobarbitone sodium (~30 mg/kg body weight). A nylon catheter with a three-way stopcock is inserted in the right femoral artery, with the tip between the left renal and coeliac arteries. The aorta is dissected free in order to control the position of the tip of the catheter, and to prepare the clamping of the aorta above the origin of the coeliac artery, just below the diaphragm.

Modified Ringer solution containing 0.1% procaine is perfused at a perfusion pressure of 110–120 mmHg. This pressure is equal to the systolic blood pressure. The aorta is clamped above the coeliac artery and a small cannula is inserted into the inferior vena cava, or one or two mesenteric veins are severed to allow free blood flow. After ~20 min of perfusion (300–400 ml of Ringer solution), the spleenic vessels appear pale, and the perfusate is changed to the fixation fluid at the same perfusion pressure. The perfusion fixation is maintained for 15–20 min. After fixation, the spleen is removed and small tissue blocks are immersed in 2.5% glutaraldehyde for 1–2 h, followed by post-fixation in 1–2% OsO_4 for 1–2 h. For further details, see Elgjo (1976).

IMMERSION FIXATION

The disadvantages of fixation by immersion are many; some are discussed here. Varied durations elapse between the separation of tissue from the body and its actual fixation. Upon dissection, the immediate alterations in blood pressure induce volume changes in the tissue. Moreover, anoxia and spontaneous post-mortem changes may cause alterations in the tissue morphology. The fixed outer surface of the tissue block may offer resistance to further penetration by the fixative. The diffusion rates of various components of the buffered fixative may differ. Thus, the rate and depth of penetration by fixatives into the tissue block are less than satisfactory. In addition, actual fixation is rather slow and uneven. It is known that fixatives such as glutaraldehyde and OsO_4 penetrate the tissue slowly, and thus the core of the tissue block is rarely well fixed. Both internal and external surfaces of the tissue may show artefacts. Tissue specimens fixed by vascular perfusion show more mitotic figures than those fixed by immersion. The preservation of certain intracellular macromolecules is adversely affected by immersion fixation. For example, glycogen may be completely depleted in the cerebral tissue of mice 1 min after decapitation (Lowry *et al.*, 1964). On the other hand, glycogen can be preserved in the brain by vascular perfusion. Some of the examples of artefacts caused by fixation by immersion are presented below.

Platelet aggregates are known to be abundant in the immersion-fixed tissue sinuses, which probably are formed because of blood stagnation. When similar tissues are fixed by vascular perfusion, the platelet aggregates are smaller and fewer. Wrinkling of normal longitudinal folds, crenation of red blood cells and the presence of bridges in the vascular endothelium fixed by immersion have been reported. In the rat myocardium, fixed in diastole state by vascular perfusion, sarcomeres usually appear relaxed, whereas fixation of a similar tissue by immersion may cause the sarcomeres to contract.

The mode of fixation affects the endothelial lining in the kidney. Both a completely closed endothelium and a discontinuous endothelium having gaps of variable diameter are artefacts caused by immersion fixation. On the other hand, a fenestrated endothelium differentiated into sieve plates and endothelial cell processes obtained with vascular perfusion is considered to be closer to the *in vivo* condition. The latter morphology of the endothelial lining is also obtained after freeze-etching. The collapse of the proximal tubule in kidney tissue specimens fixed by immersion is well documented. This results in considerable change in the size and shape of cells.

As stated above, in spite of the drawbacks of fixation by immersion, many tissue types have to be fixed by this method. It is absolutely imperative that immersion fixation be carried out under optimal conditions in order not to compound the problem. Certain tissue types, such as skin and bone, which can tolerate a brief interruption to their blood supply and still retain their structure, can be fixed satisfactorily by immersion. Fixation by immersion is apparently easy.

DRIPPING METHOD

The advantage of the dripping method is that the tissue remains attached to the body of the animal, at least during pre-fixation. Thus, normal blood supply and metabolism of the cells are maintained until the instant of fixation. Because fixation is initiated *in situ*, autolytic changes are minimised. Furthermore, maintenance of *in situ* conditions during fixation appears to maximise the depth of satisfactory fixation. Also, there is evidence which indicates that the adverse effects of hypotonic or hypertonic fixing solutions are minimised by an intact blood circulation during fixation. *In situ* fixation is especially important in the fixation of muscles which remain attached and extended during fixation.

The disadvantage of the dripping method is that it allows the study of only the surface layers of an organ, since the fixative does not penetrate deeper layers. In the case of kidney, for instance, only the outer 200–400 μm of the cortex, which comprises two to four layers of tubules and the outer row of glomeruli, is well fixed. Obviously, the method is more useful for SEM than for TEM.

For fixation by dripping, the animal is anaesthetised and the tissue exposed by dissection. In some cases, it is necessary to expose the selected tissue completely by removing the overlying tissue and membranous covering in order to facilitate direct flooding of the tissue. The enclosing connective tissue membranes can be removed by peeling them off with a fine pair of forceps. However, utmost care is necessary during the removal of the tissue covering, otherwise this operation may damage the structure of the organ. Also, injury to blood vessels and nerves should be kept to a minimum during dissection and subsequent operation.

Immediately after the tissue is exposed, it is flooded with an ample supply of a suitable chilled (~2 °C) fixative. In general, buffered (pH 7.2–7.4) glutaraldehyde (3%) or a mixture (1:1) of glutaraldehyde and paraformaldehyde is dripped immediately on the exposed tissue. In order to maintain the tissue freshly bathed, drops of the fixative are added every few seconds. After 2–3 min, a thin pad of cotton is placed over the tissue surface and the fixative is added less frequently (a few drops after every 2–3 min). After fixation and hardening have continued for ~20 min, the organ is removed and immersed in a fresh cold fixative. While immersed in the fixative, the organ is cut into small pieces which are fixed for an additional period of 1–3 h to complete the primary fixation.

In order to minimise the contamination and dilution of the fixative, it is necessary that oozing blood and tissue fluids be flushed away with excess fixative. The animal should be kept slightly tilted during the whole procedure so that the used fixative can drain away slowly from the tissue being fixed. In view of the harmful nature of the fixative fumes, one must avoid inhaling them.

INJECTION METHOD

Since the injection method is also carried out *in situ*, it offers the same advantages as those of the dripping method. This method is used for preserving the internal structure of organs. A suitable, chilled fixative is injected slowly and directly into the selected living organ for ~2–20 min after exposing it from the overlying tissues. The fixative is injected by using a micropipette or hypodermic needle. The organ is removed and cut into small pieces, which are further fixed by immersion for 1–3 h. Free-hand injection has the disadvantages of subjecting the tissue to variations of pressure and/or of flow of the fixative solution.

It has been reported that blood and tissue fluid can be washed out from small tissue pieces (~3 × 5 mm) by puncture perfusion (Murakami, 1976). After the tissue block is rolled in a wet Cellophane sheet, it is punctured by a syringe needle (diameter ~0.3 mm), and a Ringer and glutaraldehyde solution is gently perfused. The specimen is further fixed for ~24 h by immersion with glutaraldehyde.

Fixation of cat autonomic ganglia was obtained by a modified injection method (Koelle *et al.*, 1974). The abdominal aorta of the anaesthetised animal is tied just below the diaphragm, the heart is exposed rapidly, the right superior vena cava is nicked, and ~100 ml of cold fixative solution is injected via the left ventricle. Ganglia are removed, cut into small pieces and fixed for an additional period of 2 h.

Human lung tissue can be preserved immediately post mortem by transthoracic injection of stained glutaraldehyde. The method described below was used by Bachofen *et al.* (1975). The fixative solution consists of equal parts of 5% glutaraldehyde and 0.025 g of indocyanine green dissolved in 19 ml of distilled water and 1 ml of pasteurised plasma protein solution for stabilisation. The dye acts as a label to facilitate collection of the fixed parts of the lung.

The fixative solution is injected into peripheral air spaces of the intact lung, without opening the thorax, within 30 min after death. Injection is accomplished with a long, thin needle (20 gauge, 10 cm) introduced into the lung through an intercostal space. After passing the rib, which serves as a landmark for the proximity of the pleura, the needle is introduced into the lung to a depth of 2–3 cm. While 5–10 ml of the fixative solution is gently injected, the needle is slowly withdrawn, to achieve adequate distribution of the fixative. The success of the procedure depends partly upon the exact location of the injection. Axillary, lateral or dorsal injection sites are preferred; the latter should not be far below the scapular angle. Ventral injection sites have the disadvantage of the possibility of injecting the solution into the pericardium, which may result in a rapid draining of the solution through the bronchial tree, and thus inadequate spreading of the fixative.

Tissue samples are removed at autopsy and fixed with 2.5% glutaraldehyde in potassium phosphate buffer (pH 7.4) for an additional period of 1 h. The specimens are post-fixed with 1% OsO_4 buffered at pH 7.4 and adjusted to an osmolality of 345 mosm with sodium cacodylate for 2 h.

Since the tissue is fixed *in situ*, the geometric arrangement of air spaces and capillaries is preserved in the well-inflated lung. However, caution is required while injecting the fixative solution; excess force should be avoided, in order to ensure that the fixative spreads in the air spaces with as little pressure as possible.

ANAESTHESIA

Several equally effective anaesthetics are available for animals of different body weights. Small animals (e.g. rat and guinea-pig) can be anaesthetised with ether or halothane. For large animals, an intravenous injection with Inactin, urethane, pentobarbital (Nembutal), Pentazocine or chloral hydrate is recommended. Alternatively, an intraperitoneal injection can be given, but it is not recommended. Pentobarbital anaesthesia usually results in better fixation than that obtained after ether anaesthesia. Animals can be anaesthetised by injecting 50–60 mg of pentobarbital per kg of body weight or by injecting 129 mg of Inactin per kg of body weight. Anaesthesia and analgesia can also be accomplished by injecting 30 mg of Pentazocine followed by 30 mg of pentobarbital per kg of body weight. If one is not certain about the correct dose, it is desirable to use an overdose. For finding a correct dose of an anaesthetic for various animals, the reader is referred to Lumb (1963) or Soma (1971).

SIMULTANEOUS FIXATION FOR LIGHT AND ELECTRON MICROSCOPY

In histopathological studies it is sometimes desirable to examine the tissue with the light microscope followed by the TEM. Such combined studies may necessitate a prolonged storage of tissues of relatively large size. Thus, there is a need for a fixative which is relatively stable and in which large-size tissue blocks can be fixed and stored for routine automated histological processing as well as for electron microscopy.

The usefulness of electron microscopy in examining surgical and biopsy specimens for tumour diagnosis has been recognised. Electron microscopy is used extensively in the examination of renal biopsy specimens (Seigel *et al.*, 1973). Diagnostic electron microscopy is finding increased application in liver, skin, muscle, nerve, bone marrow, and other biopsy and autopsy procedures. The importance and use of diagnostic electron microscopy are expected to increase rapidly in the near future.

Several fixatives have been developed for specimens to be examined with both light and electron microscope (Zimmerman *et al.*, 1972; Carson *et al.*, 1973; Yanoff, 1973). The best fixative at present available seems to be a mixture of 4% paraformaldehyde and 1% glutaraldehyde in a buffer having an osmolality of 176 mosm (McDowell and Trump, 1976). It is recommended that tissue blocks should not exceed 3 mm in width, and that specimens for electron microscopy be taken from the outer parts of the tissue block. The gross penetration of this fixative into the soft tissues (liver) is 2.5 mm in 4 h, while the depth of good fixation is 1.75 mm in the same period of time. The appearance of tissue blocks stored in the fixative for as long as 12 months remains almost unchanged. The fixative is stable for at least 3 months when stored at 4 °C.

Although the above-mentioned formulation is most satisfactory, an inexpensive and easily prepared fixative for light and electron microscopy is prepared by using a commercial solution of formaldehyde and Millonig buffer (Carson *et al.*, 1973).

Technical grade formaldehyde (37–40%)	10 ml
Tap-water	90 ml
Sodium phosphate (monobasic)	1.86 g
NaOH	0.42 g

Very large (2–3 cm^3) tissue blocks require fixatives which have the maximum penetration power. Such fixatives can be developed by using acrolein or paraformaldehyde, to which may be added trinitro compounds. One such formulation is given below.

2% paraformaldehyde
2% glutaraldehyde
0.1 M phosphate buffer (pH 7.3)
0.2% trinitrocresol
1.5 mmol calcium chloride

TISSUE STORAGE

The most desirable practice in the preparation of specimens for electron microscopy is to fix and embed them immediately after their collection. Loss of cellular materials during processing may increase when specimens have been stored prior to fixation. However, in some cases, owing to unavoidable circumstances, it is not possible to process specimens immediately after they have been collected or excised from the source, so they must be stored for a period of time. The necessity of storage may arise, for instance, in the case of surgical or biopsy specimens, where the facilities needed to process the specimen for electron microscopy are not at hand. In some cases, electron microscopy of such specimens may have to await the results obtained by other means.

The size of the tissue specimen and the objective of the study primarily determine the type of the fixative solution to be used. Specimens of very small size can be stored, especially for relatively short periods of time, in the aldehyde fixative solutions used routinely for electron microscopy. Medium-size (~3 mm) tissue specimens can be stored for up to 12 months in a mixture of formaldehyde (4%) and glutaraldehyde (1%) in a buffer (pH 7.4) having an osmolality of ~200 mosm. Large-size (2–3 cm) tissue blocks require fixatives which have the maximum penetration power. Such fixatives can be developed by using acrolein or formaldehyde, to which may be added trinitro compounds. One such formulation is given on page 69. Another fixative for large specimens consists of 10% paraformaldehyde in 0.2 M collidine buffer, which is thought to extract tissue components and thus facilitate the penetration (Winborn and Seelig, 1970).

Storage of aldehyde-fixed specimens in a buffer is not recommended. Adverse effects (morphological damage and diffusion of cellular materials) of storing aldehyde-fixed specimens in a buffer have not been adequately explored.

POST-MORTEM CHANGES

After death or removal of tissue from the body, anoxic spontaneous post-mortem effects cause alterations in ultrastructure and enzyme activity. Autolytic effects involve alterations in the shape, size, electron density and location of cell components. An increase (initial) (e.g. ornithine transcarbamylase and acid ribonuclease) or decrease in enzyme activity also occurs. Certain cellular materials may appear or disappear, owing to autolysis; examples of such materials are lipids, glycogen and lysosomal enzymes. Since cellular death and ultrastructural changes frequently accompany pathological processes, an understanding of the normal post-mortem changes is necessary in order to distinguish between cellular alterations caused by disease and those resulting from normal post mortem. For example, in assessing lesions in muscle at autopsy, post-mortem alterations should be taken into account.

When organs are initially fixed *in situ* by vascular perfusion, it is necessary to complete the fixation by immersion of small pieces of tissue in the same fixative for an additional period of time. Even with this procedure, the core of the tissue block may show autolytic (post-mortem) alterations, because, in comparison with vehicles, fixatives are slow to penetrate the core. Thus, the vehicle may cause damage before the fixation has effectively begun. It is apparent that the nature and extent of autolytic changes need to be understood.

Factors implicated in post-mortem changes include ischaemia, post-mortem glycolysis and proteolysis. Autophagocytosis may be accompanied by activation or increased amounts of digestive enzymes. The number and volume of lysosomes increased in rat hepatocytes up to 1 h during autolysis (Riede *et al.*, 1976). Similarly, increased autophagocytosis was observed in rat pancreatic acinar cells during autolysis *in vitro* (Jones and Trump, 1975). On the other hand, recent studies by Nevelainen and Anttinen (1977) indicated no increased autophagic (lysosomal) activity in the similar cells during autolysis. It has been reported that autophagocytosis requires some energy and production of ATP (Arstila *et al.*, 1974). The autophagic phenomenon was not increased in the latter study, probably because there was rapid decline of cellular energy in the tissue specimens kept at room temperature. The above-mentioned discrepancy indicates that additional studies are needed to understand the role of various internal and external factors in the onset of autolytic changes.

It is likely that undetectable structural alterations in cell components following death may render organelles physiologically ineffective long before any significant changes in certain biochemical constituents are detected. The non-appearance of any structural damage after death does not necessarily mean the absence of physiological or biochemical damage. In other words, morphological integrity of cell organelles may not have a direct bearing on their functional activity. For example, loss of enzyme activity is not directly related to morphological disruption of cell organelles. It should be noted, however, that metabolic efficiency of an organelle may be impaired immediately after death, not because of a loss of any particular cellular material but owing to disorientation of enzymes. This is exemplified by mitochondria, which lose their efficiency in respiratory units rapidly after death, although the amount of their enzymes remains almost constant.

Although autolysis usually causes a gradual loss of the highly ordered structural organisation, various tissue types differ in their response to autolysis (Table 1.10). Striated muscle and liver tissues are examples of

Table 1.10 Conditions of subcellular structures of different tissue types kept at 6 °C for a period of time and fixed in glutaraldehyde

A. *Brain:*	*Neuron*	*Astrocyte*	*Oligodendrocyte*
Time delay (h)	24	24	24
Chromatin material	+	−	+
Golgi apparatus	−	−	−
Microtubule	++	++	++
Neurofilament	++	++	++
Nuclear envelope	+	+	+
Nuclear pore	+	−	+
Plasma membrane	+	+	+
Rough endoplasmic reticulum	+	+	+
Ribosomes	++	++	++
Smooth endoplasmic reticulum	+	+	+

B. *Neuropil*	
Myelin sheath	+
Synaptosome	++
Spine apparatus	−
Synaptic vesicles	+

C. *Other tissues:*	*Cardiac muscle*	*Liver*	*Glomerulus*	*Proximal tubule*	*Distal tubule*	*Skeletal muscle*	
Time delay (h)	24	24	24	24	24	24	48
Basement membrane			+	+	+	+	+
Chromatin material	+	+	+	++	+		
Endoplasmic reticulum	−	+		+	++	+	−
Foot process			−				
Golgi apparatus		−					
Microvilli in lumen				++			
Mitochondria	++	+		++	++	+	−
Myofibril	++					++	++
Nuclear membrane	+	+	+	+	+	+	+
Nuclear pore	−	−	−				
Ribosomes	++	++		++	++	++	++
Sarcolemma/plasma membrane	+	+	+	+	++	+	+
Sarcotubule	+					+	−

Blank indicates subcellular structure not observed.
Very well preserved, ++.
Fairly well preserved, +.
Poorly preserved, −.
Courtesy of Lim and Solomon (1975).

tissue types which resist post-mortem changes. Heart muscle is more susceptible than is skeletal muscle to post-mortem effects.

Not only different tissue types but also different cell types in a tissue or an organ may show different degrees of susceptibility to autolysis. A well-known example is kidney tissue. Various cell organelles also differ in their response to autolysis (Table 1.10), and this response seems to be related to their chemical composition. Proteins are most resistant, and lipoproteins (membranes) are least resistant. The presence of greater enzyme activities on the membrane surfaces may be one of the reasons for this vulnerability.

In the striated muscle stored at 40 °C, the Z and I bands degenerate in less than 4 h, whereas the A band is resistant to change up to 24 h. In general, actin shows greater susceptibility to proteolytic digestion during post mortem than does myosin. Probably, bonds between the Z band and the I band are weakened during post mortem. Degradation of Z bands in the muscle during post mortem is thought to be caused by endogenous proteolytic enzyme. Catheptic enzyme has also been implicated in the degradation of myofibrils. In the sarcoplasmic reticulum, transverse tubules are relatively resistant to post-mortem effects. In this tissue, glycogen either disappears or decreases rapidly, depending upon the speed of fixation.

It is known that autolysis is accompanied by degenerative nuclear changes. The earliest and most prominent nuclear change is clumping and condensation of chromatin along the nuclear membrane and around the nucleolus, followed by its complete disappearance. A uniform distribution of chromatin within the nucleus is closer to *in vivo* conditions. In the initial stage, nuclear membrane may show crenation; and during the advance stage of autolysis, this membrane may be disrupted.

Mitochondrial changes comprise rounded shape, condensation, swelling, rupturing of membranes, formation of myelin figures, and appearance of flocculent densities or loss of matrix density. The swollen mitochondria, while retaining their outer double membrane, exhibit a progressive disruption of the cristae. For example, in the liver, initial post-mortem changes comprise loss of cristae and swelling of mitochondria in the parenchymal cells. Mitochondria in different cell types exhibit different degrees of alterations during the same time interval after death. This organelle in cardiac muscle is relatively resistant to the effects of autolysis both at room temperature and at 4 °C.

Smooth endoplasmic reticulum is less resistant to post-mortem changes than is rough endoplasmic reticulum; the former begins to form chains of vesicles within 30 min after the liver sample has been collected (Trump *et al.*, 1962). Autolytic changes in rough endoplasmic reticulum comprise degranulation, swelling, and vesiculation and dilation of cisternae. Changes in the Golgi complex are not very conspicuous, although its cisternae become dilated.

The plasma membrane may become indistinct and ruptured, which may allow cell organelles to escape to the extracellular space. Microvilli appear swollen and myelin figures may form at their tips. Microvilli of the bile canaliculi may disappear within 1 h after death, but before their disappearance, dense cytoplasmic bodies containing amorphous material may become apparent in them. Microtubules and microfilaments may break down completely. Peripheral zones of zymogen granules may become less electron dense than is the core. Sinusoidal spaces may dilate as early as 1 h after death.

Although the available information on the length of time cells of different tissue types remain unchanged morphologically after death of the animal is insufficient and in certain cases contradictory, this duration for most cells is probably in the range of less than 1 min to 30 min in a moist atmosphere. The degree of resistance of various tissues and cell organelles to autolysis is given in Table 1.10.

Not all organelles of the same type in the same cell may show damage in the early stages of autolysis. In a cell, for instance, some mitochondria may show damage while others may appear normal. When the tissue remains in the cadaver, it is exposed to fewer physicochemical stresses compared with those imposed upon excised tissues. Autolytic changes set in sooner in diseased organs compared with those in normal organs. Low temperatures retard the onset of autolysis. In spite of the above-mentioned relative resistance of certain cells and organelles to autolysis, the selected tissue should be placed in the fixative as rapidly as possible (see p. 73 for information on plant cells).

PLANT SPECIMENS

Fixative formulations used for the preservation of plant specimens have, in general, been those developed for fixing animal tissues. Although such formulations give adequate preservation of a variety of plant tissues, fixatives specifically suited for the best preservation of plant specimens have yet to be developed. Differences in the protein contents of plant and animal tissues suggest that ideal fixatives for the latter may not be best suited for the former. In general, animal tissues are richer in protein than are plant tissues and viruses.

The relatively low protein content of plant cells must be taken into account when devising fixative formulations, because studies of model protein systems indicate that glutaraldehyde is unable to cross-link low concentrations of proteins. It has been shown that glutaraldehyde is unable to fix albumin in concentrations less than 3–5% (Millonig and Marinozzi, 1968). Similarly, this aldehyde fails to fix gelatin concentrations of 2% or less (Langenberg, 1979). It can be inferred from these studies that a minimum protein concentration in a cell is necessary for it to be fixed by glutaraldehyde.

Other major differences between typical plant and animal tissues, which are important in the fixation process, include a non-living outermost boundary (cell wall) and (a) large central vacuole(s) in the former. These two characteristics create a problem in obtaining a reliable and satisfactory fixation of plant cells.

Because of the presence of a large vacuole and cell wall, plant cells have considerably higher osmotic pressure than that of the surrounding media. If the plant cell is killed before fixation, the compartmentalisation of acids is lost because of the damaged vacuolar membranes. It is known that vacuolar sap is acidic (pH ~5.0–6.0). Thus, in order to minimise the adverse effect of the released sap, fixative solutions with a relatively high pH (~8.0) give superior preservation. Since the average pH of plant cytoplasm is 6.8–7.1, it is desirable to neutralise the acidic pH of the vacuolar sap by using fixative solution at higher pH. The best approach seems to be to use a higher pH during pre-fixation with glutaraldehyde, followed by a lower pH (~7.0) during post-fixation. Protozoa are also fixed better at higher pH values.

A second effect of the vacuole involves a considerable dilution of the fixative by the large volume of vacuolar sap. This problem is almost unique to cells of higher plants and apparently to highly hydrated specimens. In order to counter the dilution effect, it is desirable to use fixative solutions of relatively high concentrations. Concentrations of 5% and 2% of glutaraldehyde and OsO_4, respectively, are desirable. An osmolality of 800 mosm and 400 mosm is recommended for mature, vacuolated cells and meristematic cells, respectively (Salema and Brandão, 1973). On the other hand, since cells of terrestrial plants possess a large vacuole, they are particularly sensitive to plasmolysis during fixation with hypertonic fixatives. Such damage can be avoided by immersing the cells first in a slightly hypotonic fixative solution with sufficient time allowed for equilibrium to occur before gradually making the fixative osmotically stronger (Taylor, 1988).

NaOH–PIPES buffer is preferred to cacodylate and phosphate buffers as a vehicle for many plant tissues. Phosphate, cacodylate and veronal buffers have been reported to be impermeant in *Nitella flexilis* cells for up to 2 h (Taylor, 1988). On the other hand, collidine buffer (>50 mM) rapidly penetrated these cells. Cacodylate buffer was not toxic to these cells in the short term.

Conductive tissue, such as xylem and phloem, can induce a very rapid reaction in the specimen by the release of their pressure. Such reactions can cause passive displacement of cell contents. The controversy about the contents of sieve plate pores may be primarily due to the rupture of internal pressure. The solution to this problem may be to practise *in situ* fixation.

Many plant tissues contain large amounts of air, including gases, especially in the intercellular spaces (e.g. spongy layers in the leaf). The presence of air in the tissue is a hindrance to the penetration of the fixative solution. Therefore, it is helpful to accomplish initial pre-fixation under a gentle vacuum (1 atm for roots). Within a minute or so after the vacuum is applied, one can see a stream of tiny bubbles being released from the tissue.

Surfaces of most plant organs (leaves, stems and flowers) are usually covered by substances (waxes) which resist the penetration of aqueous solutions of the fixative. In certain cases these substances are hydrophobic. The presence of tiny appendages (e.g. hair) on the surface of certain plant organs may trap air and thus impede the penetration of the fixative solution. These problems can be somewhat alleviated by dissolving the impervious substances by treating the specimen very briefly with a dilute solution of a household detergent. Sometimes gentle rubbing of the specimen surface, while in the fixative solution, with a brush helps to dislodge the air bubbles and even some surface substances. Even vigorous shaking of the specimens in the fixative solution in a vial may help. Sinking of the tissue blocks in the fixative solution within a few minutes is almost a prerequisite for a satisfactory fixation.

Although plant tissues do not undergo post-mortem changes as such, desiccation occurs rapidly, which causes changes at the ultrastructural level. Therefore, like animal tissues, plant specimens should be fixed by immersion immediately after their removal from the plant. Attempts have been made to initiate fixation *in vivo*—i.e. to start the fixation prior to removing the specimen completely from the plant. Fixative solutions can be introduced into the intact tissue with a syringe prior to its excision. During excision of differentiating xylem in willow, the stem was irrigated with cold fixative solution (Robards, 1968). Another example is fixation of a relatively thin leaf by sealing a metal ring with lanolin into its upper surface and then filling the ring with the fixative solution. Other procedures can be devised to begin fixation before separating the tissue from the plant organ.

Plasma and vacuolar membranes are particularly sensitive to fixation conditions; the appearance of the latter membrane can be used as a criterion for evaluating the quality of fixation. The smooth contours of vacuolar membranes are thought to be a more life-like condition than the irregular appearance. In a well-fixed cell, the tonoplast is clearly visible as an intact single membrane which forms a distinct boundary with the cytoplasm. Fixation quality is substandard when tonoplast has pulled away from the cytoplasm or plasma membrane has shrunk away from the cell wall. Several studies indicate that vacuolar membrane is easily damaged during fixation; the tonoplast of cells with large central vacuoles seems to be especially sensitive.

Since vacuoles contain hydrolytic enzymes, such as nuclease and protease, their release, due to rupturing of the tonoplast during fixation, may cause damage to the cytoplasm. This damage, which is an artefact, should be distinguished from the changes occurring due to natural senescence preceding the differentiation of cells such as vessel elements, sieve tubes and fibres. It should be noted that tonoplast breakdown is an almost universal phenomenon during cell differentiation, disease, nutrient deficiency, and chemical and physical damage. Fixation of plant specimens has also been discussed by Roland (1978) and Hayat (1981).

CRITERIA OF SATISFACTORY SPECIMEN PRESERVATION

It is difficult to establish a criterion for determining satisfactory preservation of biological specimens, for two main reasons: (1) components of a cell in the living state have not been viewed in the electron microscope; and (2) incomplete information on the molecular basis of fixation. However, there are several reliable procedures and criteria which can be used to test the quality of preservation. The process of fixation in the living cells can be observed under a phase contrast or Nomarski interference contrast microscope, and can be recorded by cinematography. Plant cells (e.g. hair cells) are well suited for this purpose. Thick sections (0.5–2.0 μm) of tissue blocks processed for electron microscopy can be examined with the light microscope or with the high-voltage electron microscope. Ordered structures such as myelin sheath or myofibril can be used to follow dimensional changes within a cell. Mitochondria, being extremely sensitive to changes in the osmotic pressure, can be used as a test for shrinkage or swelling. Microvilli also act as osmometers, for they are more osmotically sensitive than is the remainder of the cell surface. Also, comparison may be made between the dimensions and orientation of the ultrastructure of fixed cells and data collected with the aid of other techniques such as polarisation microscopy, X-ray diffraction or freeze-etching. The X-ray diffraction technique has been effectively used in the case of highly ordered structures such as myelin sheath and collagen fibres (McPherson, 1976), while the freeze–fracture technique has provided extremely useful data on the structure of cell components, especially membranes (Frank *et al.*, 1987).

One of the criteria for satisfactory fixation is the preservation of the continuity of membranes, without distortion or breakage, and in some cases their trilaminar configuration. While such continuity may not be proof that other structures have also been preserved well, the presence of breaks is an obvious indication of defective preservation. It is generally recognised that in satisfactory preservation, the spaces between the membranes are filled with granular material, although some workers consider such material to be the dissolution of substances from various cell components. While some dissolution may occur, it is apparent that the spaces between the membranes in living cells are not empty. No empty spaces are found in the ground substance or within organelles in living cells. Although the absence of enlarged intercellular spaces is a sign of satisfactory fixation, the presence of relatively large intercellular (extracellular) spaces is not uncommon in the well-fixed CNS. It should be noted that the presence of widened and clear spaces is common in pathological tissues.

In the case of the kidney, the criteria of satisfactory fixation include open tubular lumina, uniform tubular cytoplasmic density without any swelling, absence of prominent extracellular compartments, presence of an interstitial space between tubules and uncompressed capillaries. In the living kidney, proximal tubules have large open lumina which close within 5–20 s after clamping the renal artery or interrupting the blood supply.

Subjectively, a satisfactory image is distinct, orderly and overall greyish in electron density, especially within mitochondria, plastids and lumina of blood and lymph vessels. The characteristic ultrastructural appearance of cellular components in well-preserved specimens is summarised in Table 1.11.

ARTEFACTS

All fixation procedures introduce artefacts into the specimen, and it is impossible to eliminate artefacts in a fixed specimen. The best one can accomplish is to drastically minimise their introduction. By using utmost care, the production of artefacts can be substantially reduced. It should be remembered that once an artefact has been introduced, it cannot be removed. Avoidable artefacts must be prevented. However, most importantly, the process which causes artefacts should be understood and the range of effects produced by various fixation procedures be recognised. It is necessary that the appearance of the specimen structure be interpreted with respect to the fixation and other treatments which it has undergone. This understanding is essential in order to relate what is shown by electron microscopy to what must have existed *in vivo*. As long as unavoidable artefacts are interpretable, they can be accepted. In fact, the major goal of improved fixation methodology is the elimination of uninterpretable artefacts.

The number of artefacts in a specimen depends upon manifold fixation parameters such as mode of fixation (immersion or vascular perfusion); type of the fixative and its concentration, pH and temperature; type and osmolality of the vehicle; addition of electrolytes or non-electrolytes; duration of fixation; specimen type, and pre- and post-fixation treatments. Less than optimal conditions of fixation cause known and unknown artefacts. Below are presented some of the known fixation and related artefacts observed in the specimens.

Fixation can bring about diffusion and/or precipitation of cellular materials. A case in point is the production of a coarse artefactual network in the

Table 1.11 Appearance of well-fixed cellular components

Component	*Appearance*
Cell wall	Dense and essentially layered; no breaks
Cytoplasmic ground substance	Finely granular precipitate showing no empty spaces
Endoplasmic reticulum (rough)	Flattened cisternae uniformly arranged in long profiles with attached ribosomes
Endoplasmic reticulum (smooth)	Branching tubules with intact membranes not associated with ribosomes; tubular appearance instead of vesicular
Glycogen	Dark clumps
Golgi apparatus	Intact membranes
Lipid	Uniformly dark
Microbodies	Dense matrix bound by an unbroken single membrane
Mitochondria	Neither swollen nor shrunk; outer double membrane and cristae intact; dense matrix
Nuclear envelope	Double membrane intact and essentially parallel to each other; two membranes different in width and may show pores
Nuclear contents	Uniformly dense with masses of chromatin scattered adjacent to nuclear membrane
Plasma membrane	Single and intact
Plastids	Outer double membrane and lamellae intact; dense stroma
Vacuoles	Bound by unbroken single membrane; absence of intravacuolar membranes

nuclear sap during fixation with glutaraldehyde; the formation of this artefact can be observed as living cells are subjected to fixation (Skaer and Whytock, 1976, 1977). This network could be mistaken for chromatin fibres, since the former stains in the same manner as the latter. Moreover, the threads of the network might add on to chromatin fibres, resulting in the artefactual thickening of the latter. Fixation with formaldehyde seems to cause the nuclear sap to appear as granules or beads (~10 nm in diameter), which are approximately the same size as histone nucleosomes; this is another possible source of confusion in the understanding of the fine structure of intranuclear chromosomes and chromatin.

Not every osmiophilic droplet observed in the fixed cell is necessarily present in the living cell. The appearance of certain types of osmiophilic droplets could be artefactual because of inadequate fixation with glutaraldehyde. Artefactual osmiophilic droplets have been observed on membranes of erythrocytes, endothelial and epithelial cells, within pinocytotic vesicles, mitochondria and plastids.

In certain cases, following either sequential or simultaneous fixation with glutaraldehyde and OsO_4, electron-dense granules of varying sizes occur in thin sections (Fig. 1.11). These granules may occur in stained or unstained sections, as well as with or without *en bloc* staining with uranyl acetate. Apparently, these granules are not staining artefacts, but are caused by some defect in the fixation procedure. Although the exact cause of such artefacts is not known, the presence of reduced osmium in the specimen seems to be a prerequisite to their formation.

In order to prevent the formation of this artefact, Kuthy and Csapó (1976) suggested the use of the same buffer for both pre- and post-fixation, or the use of veronal buffer during post-fixation. However, some of these artefacts are independent of the buffers, and their production is affected by the duration of rinsing between primary fixation and post-fixation. Longer times of rinsing minimise the occurrence of this artefact. The inclusion of glucose (3%) in the buffer used for preparing OsO_4 solution also minimises the occurrence of these artefacts. Mollenhauer and Morré (1978) recommended OsO_4 post-fixation at 0 °C and the addition of sucrose (0.05 M) to the fixative and buffer solutions. When this artefact is already present, the most effective way for its removal from sections is by using oxidising agents such as periodic acid and hydrogen peroxide. These oxidising agents are routinely used for removing bound osmium to obtain specific staining of polysaccharides (Hayat, 1975).

Thin sections on nickel grids or Marinozzi rings (Marinozzi, 1961) are treated with either freshly prepared 1% periodic acid or 3% hydrogen peroxide for 5–10 min at room temperature (Ellis and Anthony, 1979). After thoroughly washing with distilled water, sections are post-stained with 2.5% uranyl acetate in 50% methanol, followed by lead citrate (Fig. 1.11). Copper grids should not be used, for they are not resistant to these oxidising agents. Periodic acid is preferred to hydrogen peroxide.

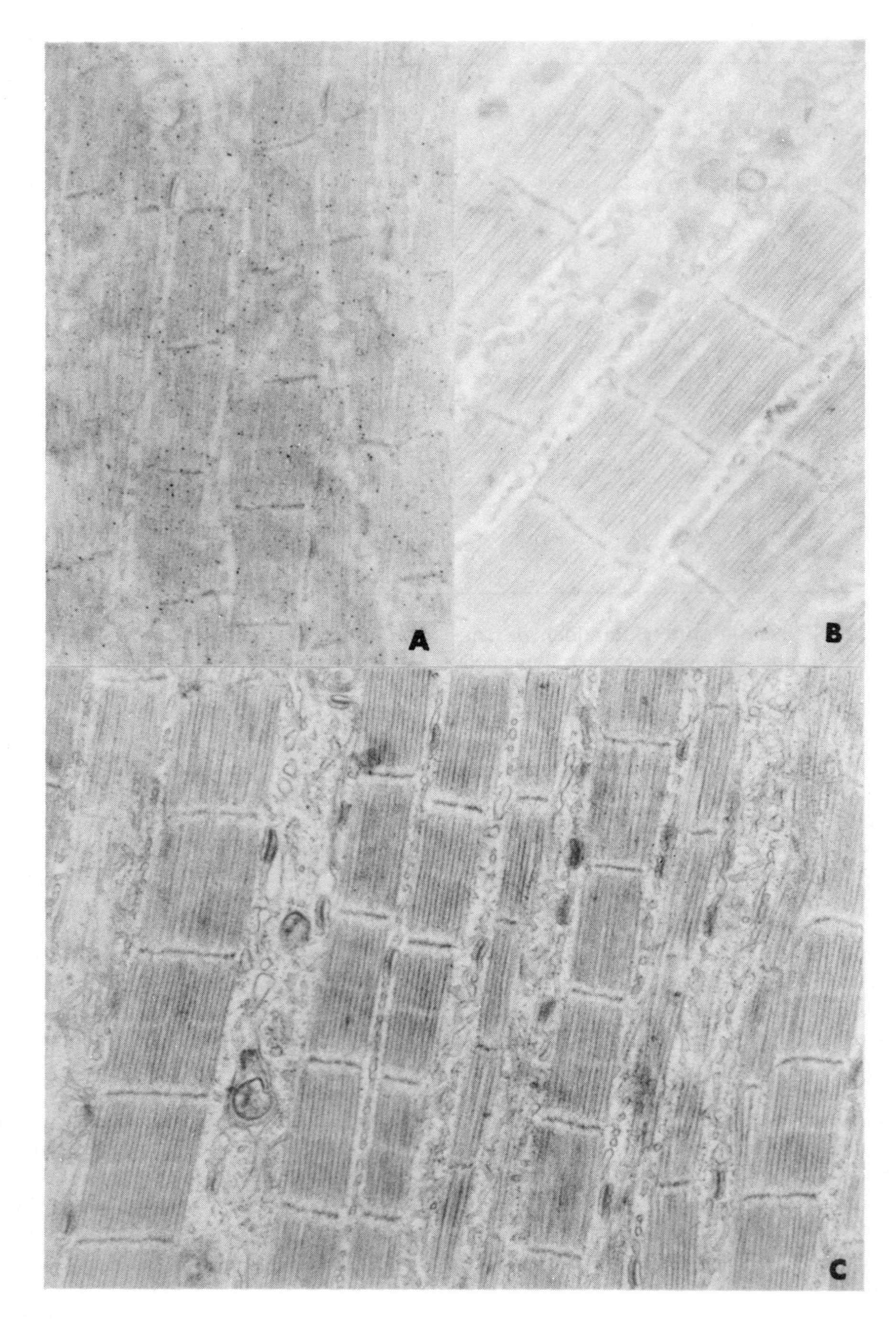

Figure 1.11 Muscle sections (17 000×) before and after oxidation and post-staining. All micrographs are from the same nickel grid. A: Fine granular precipitate is present; no post-staining. B: After 5 min of periodic acid oxidation, the precipitate is removed; no post-staining. C: After oxidation and post-staining with uranyl acetate and lead citrate. From Ellis and Anthony (1979)

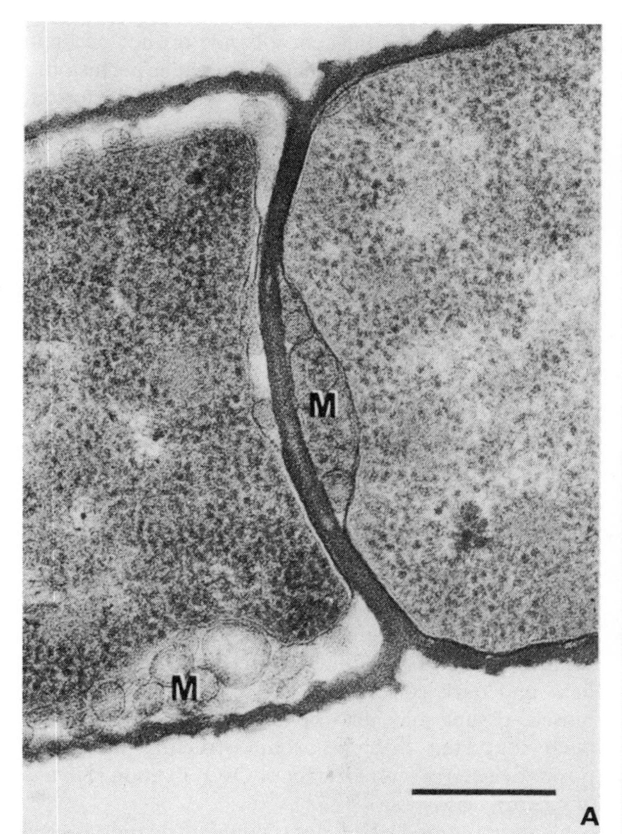

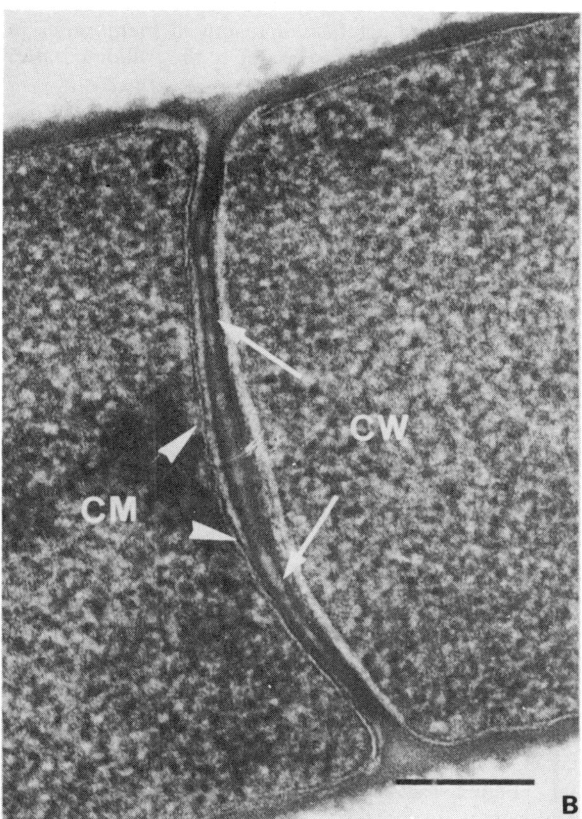

Figure 1.12 Vegetative cells of *Bacillus cereus* in the long phase and at the early stage of sporulation. A: Fixed with 3% glutaraldehyde and 1% OsO_4 in the presence of 0.15% ruthenium red to improve the contrast of carbohydrate components of the cell wall. Artefactual mesosomes are present as invaginations of the plasma membrane, randomly distributed along the developing septum during cell division. The hypertonic environment of the cells during fixation induces a reduction of the invagination size. The mesosomal vesicles are present between the cell wall and the plasma membrane. B: Cryofixed with the propane jet method and freeze-substituted with methanol containing 1% OsO_4, 0.5% uranyl acetate and 3% glutaraldehyde. Mesosomes are absent. CW, cell wall; CM, plasma membrane; M, mesosome. The bar equals 300 nm. Courtesy of H. R. Ebersold

Fixed specimens, however well preserved, cannot reveal any spatial displacement of cell constituents which might take place during the process of fixation. It has been suggested that during fixation, a contraction or molecular rearrangement within the cytonet material results in the deformation of the surface membrane and consequent formation of pinocytotic vesicles and condensation of cytonet around the plain synaptic vesicles. The implication is that the coats of coated vesicles could be artefactual condensations of the proteinaceous material. An example of this type of fixation artefact is the presence of a greater number of pinocytotic vesicles and coated pinocytotic vesicles in the central nervous system (Paula-Barbosa *et al.*, 1977).

The above-mentioned and certain other types of vesicles can be formed in a wide variety of cells. Certain vesicles can be formed as a result of unnatural fusion of membranes. Although membrane fusion in cells is not an uncommon natural phenomenon occurring during certain kinds of metabolic activity, not every membrane fusion observed in the electron microscope is necessarily an expression of an event taking place in the living cell. The presence of small, closed vesicles in the fractions of outer and inner mitochondrial membranes isolated by homogenisation and centrifugation (Sottocasa, 1967) is a well-known example of fusion of membranes during the preparatory procedure.

Other examples of unphysiological membrane fusions include demembranated ciliary axonemes within the pellicular regions of the ciliate *Pseudomicrothorax* (Hausmann, 1977). The membranes of the cilia are thought to fuse with the plasma membrane and are in part incorporated in it. The presence of bridge-like connections or even the fusion of the outer leaflets of adjacent ciliary membranes in the ciliate *Paramecium* seems to be a fixation artefact (Hausmann, 1977). The above-mentioned artefactual fusion of membranes in the two ciliates appears to be due to the collidine buffer used as a vehicle. Even the membranes of two different organisms may fuse during the preparatory procedure. The fusion of plasma membranes in *Aphelidium* (the parasite) with those in *Scenedesmus* cells (the host) has been reported (Schnepf, 1972). These plasma membranes not only fuse, but also occasionally form vesicles. Membranous structures inside a plant cell vacuole are fixation artefacts.

The appearance of internal membranes (mesosomes) in most bacteria is a result of chemical fixation. Such membranes are formed in response to glutaraldehyde fixation and/or centrifugation. Low concentrations of the fixative fail to stabilise membranes and intracellular pools of membrane precursors, which may result in the formation of artefactual intracellular membranes. Such membranes may form by invaginations of the plasma membrane. It is known that artefactual membranes can also form in eukaryotic cells during fixation with glutaraldehyde.

Fixation with a mixture of glutaraldehyde and OsO_4 reduces the production of this artefact. Phosphate and Tris buffers and sucrose should be avoided during fixation (Aldrich *et al.*, 1987). Bacterial cells should not be stored overnight in a buffer or a fixative. In many cases, cryofixation prevents the formation of the artefact. Figure 1.12 shows the presence and absence of mesosomes in bacteria after chemical fixation and cryofixation, respectively. Although intracellular membranes are present in chemolithotrophs and phototrophs (Van Iterson, 1984), caution is needed in accepting their presence in most other bacteria processed with weak fixatives.

The production of artefactual contraction bands and hypercontraction of sarcomeres in fresh cardiac tissue obtained by the biopsy procedure has been reported (Adomian *et al.*, 1977). Probably, the biopsy procedure activates the contractile mechanism in the muscle. Such artefacts should be distinguished from the contraction bands associated with various cardiac pathological states. Artefactual contraction bands do not occur in specimens obtained after death or vascular perfusion.

Fixation of the kidney, even by vascular perfusion, may not result in a uniformly good fixation of the entire organ. Glomeruli from poorly perfused and poorly fixed areas of the kidney show endothelial, epithelial and mesengial swelling similar to that seen in pathological (glomerulonephritis) or immersion-fixed specimens, or during early autolysis. It should be noted that glomeruli are thought to be relatively resistant to the effects of poor fixation, in contrast to the more labile proximal tubular epithelial cells. Mesengial cell swelling is an important index of poor fixation of renal glomeruli, corresponding to the rapid swelling of proximal tubular epithelial cells (Johnson *et al.*, 1973). Swelling of mesengial cells is absent in well-fixed specimens. Endothelial and epithelial cells may show swelling, even with good fixation of other parts of the organ.

Light and dark cells of the adrenal cortex are thought to be fixation artefacts rather than different functional states of individual cells. Osmotic phenomena may, in some way, be responsible for the occurrence of these light and dark cells. Light and dark cells in smooth muscle tissues may also be fixation artefacts. It has been suggested that mesosome structure in Gram-positive bacteria is an artefact of OsO_4 fixation (Silva *et al.*, 1976).

Caution is urged when glutaraldehyde fixation is used to avoid certain artefacts (e.g. dissociation or conformational changes) during sedimentation analysis of mixtures containing different kinds of ribosomal particles, for dimerisation or aggregation of ribosomal particles may be induced by this aldehyde. The dimerisation is especially noticeable with 30*S* subunits (Garcia-Patrone and Algranati, 1976). The addition of albumin to the mixture decreases the aggregation, but does not prevent it. The fixation does not seem to induce association between 30*S* and 50*S* subunits. Artefacts caused by immersion fixation are also discussed on p. 67.

2 Rinsing, Dehydration and Embedding

INTRODUCTION

Since specimens are extremely small, they must be embedded in a suitable material for safe handling. Furthermore, specimens are very fragile, porous and brittle, so they must be permeated with an appropriate material for thin sectioning. Attempts have been made to section unembedded but glutaraldehyde-fixed tissues (Pease, 1982). The resinless section technique (reversible embedding), in which tissues are embedded in a wax or resin that is removed after sectioning, has also been used (see p. 129). However, these techniques are of limited usefulness. The most suitable materials for embedding are resins. Both water-immiscible and water-miscible resins are in use, although the former are most commonly employed. Water-immiscible resins are usually polymerised by heat, while water-miscible resins are cured either by heat or by UV irradiation at low temperatures (see p. 111).

Resins that are miscible with water can be employed for dehydration, thereby avoiding the use of ethanol or acetone and preserving certain cellular components that would otherwise be extracted. These resins also save the specimens from denaturation by heat during polymerisation. Their role in preserving immunoreactivity is well known. However, these resins also extract cellular substances. Recently developed water-miscible embedding media, Lowicryls, LR White and melamine resins, used at low temperatures are useful, especially for immunocytochemical studies. Commonly used formulations of both water-immiscible and water-miscible resins are given later. Reliable embedding methods for specific types of specimens are presented elsewhere by Hayat (1986b).

RINSING

After the specimens have been fixed, they should be rinsed before dehydration. In order to minimise the possible reaction between the fixative and the dehydration agent, most of the excess fixative should be washed off. In double fixation the excess aldehyde should be removed prior to post-fixation with OsO_4. If traces of glutaraldehyde are left in the specimen, they may react with OsO_4 and produce a fine, dense precipitate of reduced osmium. This could ruin the specimen. The time required to remove all the aldehyde varies for different specimens. If the specimens fixed in buffered solutions are washed with water, dissolution and progressive disintegration of some unfixed cellular material will take place as soon as the electrolytes or non-electrolytes of its fixative vehicle are removed by washing with water. The washing, therefore, is carried out in the same buffer (vehicle) as that used in the fixative mixture. Another advantage of using the same buffer and vehicle is that a sudden drastic change in the environment of the tissue specimen is avoided. This is important, since the membranes in the aldehyde-fixed tissue partially maintain their property of differential permeability.

Either non-electrolytes or electrolytes may be added to the rinsing solution to maintain the same osmolarity as that of the fixative mixture. It is thought that the imbibition of water by the specimen during rinsing is reduced when the pH is at or below neutrality and the electrolytes (e.g. 0.5% NaCl) are present in the rinse solution. The presence of sucrose, on the other hand, may facilitate dissolution and may not inhibit imbibition.

In general, two or three quick rinses in the buffer for a total period of 10–20 min is adequate for most specimens. However, some tissues require a longer time; for example, sugar-beet leaves fixed in glutaraldehyde must be washed with a buffer for 3–4 h prior to fixing them with OsO_4, otherwise the penetration and reaction of the latter is extremely slow. Some tissues require as much as 12–24 h of washing with buffers. Relatively long durations are required for removing the fixative before incubation for enzyme cytochemistry. As a matter of principle, the washing should not be longer than necessary. It should be noted here that some of the cross-links introduced by glutaraldehyde are potentially reversible, and for this reason long washings may affect the final appearance of the fine structure. Preliminary experiments may be necessary to establish the optimal duration for a specific type of specimen. It is advisable to carry out washing at 0–4 °C, provided, of course, that the fixation has been completed in the cold.

Caution is advised in the use of rinses in the study of certain cell components. Synaptic vesicles, for instance, are known to be adversely affected by rinsing the specimen. Chloroplasts swell by ~6% of their initial volume during the two rinsing stages. According to Ockleford (1975), certain specimens can be processed satisfactorily without resorting to rinsing either before dehydration or between different fixatives. Admittedly, there are few published data on the adverse effects of a reasonably brief rinse or no rinse at all. Systematic experimentation is needed on this aspect of the preparatory procedure.

DEHYDRATION

When water-immiscible resins are used for embedding, all the free water from the fixed and rinsed specimens must be replaced with a suitable organic solvent before infiltration and embedding. The removal of water is unnecessary when the specimens are to be embedded in water-miscible resins. The water is removed by passing the specimens through a series of solutions of ascending concentrations of an organic solvent. Since epoxy resins are soluble in ethanol and acetone, dehydration is commonly carried out with one of these solvents. Because epoxy resins are more readily soluble in propylene oxide (1,2-epoxy propane; CH_3CHCH_2O), this solvent is frequently used immediately prior to infiltration with the resin. Propylene oxide is a simple homogeneous compound possessing an epoxy radical completely miscible with water. This solvent has the additional advantage of not separating from the epoxy resins during polymerisation. Therefore, no harm is done when small amounts of propylene oxide are left in the embedding resin.

The use of propylene oxide at the last stage of dehydration is a necessity when ethanol is used as the main dehydration agent. The reason is that most resins are not readily miscible with ethanol. In fact, polyester resins are completely immiscible with ethanol. When acetone is employed, the use of propylene oxide as a transitional solvent is unnecessary.

It should be noted, however, that propylene oxide is very reactive, even at low temperatures, and thus may combine with reactive groups in the cells, which may affect certain histochemical and staining reactions. As stated above, the residual traces of this solvent become a part of the polymerised tissue block. Even vacuum evacuation prior to cure does not materially affect this retention. Propylene oxide may react with the epoxy groups of the epoxy monomer molecule and thus may inhibit polymerisation of epoxy monomers; this adversely affects the hardness and cutting properties of the cured block.

Some evidence indicates that propylene oxide easily extracts lipids, especially those not fixed by OsO_4. Owing to its highly reactive properties, propylene oxide might extract even the fixed lipids. Studies of extraction of absorbed lipid (linoleic acid-1-[^{14}C]) from rat intestinal epithelium prepared for electron microscopy indicated a loss of ~26% (Bushman and Taylor, 1968). The major portion (17%) of this loss occurred in propylene oxide at room temperature; the loss in cold OsO_4, cold ethanol and water amounted to ~3.6%, 4.6% and 1.0%, respectively. It is obvious that propylene oxide is primarily responsible at least for the extraction of absorbed lipids. However, extraction of lipids by this solvent can be reduced by lowering the temperature and duration of dehydration. Another undesirable effect of propylene oxide is its ability to dissolve phosphotungstic acid (PTA) after *en bloc* staining.

There is some evidence that acetone, a relatively non-reactive solvent, causes less specimen shrinkage and extraction of lipids than does ethanol. Dehydration with acetone causes little change in the dimensions of striated muscle filaments, whereas ethanol causes a shrinkage of ~10% (Page and Huxley, 1963). Acetone dehydration does not destroy the X-ray diffraction pattern of actin-containing filaments, whereas ethanol dehydration destroys such a pattern. Brain tissue seems to be preserved better when it is dehydrated with acetone instead of ethanol. Acetone also has the advantage of not being reactive with residual OsO_4 in the tissue. Furthermore, no evidence has yet been found indicating chemical interaction between epoxy monomers and acetone. Since acetone has a keto rather than an epoxy group, it would not be expected to inhibit polymerisation of epoxy monomers; by contrast, as stated above, a reactive monoepoxide (such as propylene oxide) could inhibit polymerisation by combining with reactive groups of epoxy monomers.

In spite of its avidity for atmospheric water vapour, acetone has proved to be a very effective dehydration agent when rapidity of action is desired. Dehydration is faster in acetone than in ethanol. Dehydration is also faster when specimen vials are shaken than when they are stationary. After acetone treatment, the amount and nature of phospholipids remain unaltered; even from unfixed tissues, phospholipids are not extracted by acetone (Adams and Bayliss, 1968). On the other hand, according to Ashworth *et al.* (1966), acetone does extract phospholipid from aldehyde-fixed hepatic tissue but far less than does ethanol. The available evidence indicates that acetone extracts hydrophobic lipids (cholesterol, cholesterol esters, triglycerides and free fatty acids), while the extraction of phospholipids is affected by the fixation method, the temperature and duration of dehydration, and the tissue type. In the freeze substitution technique, the leaching of water-soluble and exchangeable Ca^{2+} in histoautoradiography of plant tissues can be prevented by using acetone as the dehydration agent (Gielink *et al.*, 1966). However, sufficient evidence is not available to indicate the superiority of acetone over ethanol in all cases. When acetone or ethanol is mixed with water, the temperature rises from 19 °C (292 K) to 26 °C (299 K) (Helander, 1987). This interaction may affect certain cell constituents.

The use of methanol (a more highly polar organic solvent) instead of ethanol as a dehydrating agent seems to result in better preservation of DNA that has relatively little associated protein (Carlemalm *et al.*, 1985). Ultrastructure of DNA in mature HSV-1 virions is also better preserved with methanol as a dehydrating agent (Puvion-Dutilleul *et al.*, 1987).

Acidified 2,2-dimethoxypropane (DMP) has also been used as a dehydration agent (Müller and Jacks, 1975). According to Thorpe and Harvey (1979), DMP seems to reduce the migration of water-soluble cell components during dehydration, compared with ethanol and acetone. Mitochondrial size tends to be larger after DMP dehydration, as compared with that obtained after ethanol dehydration (Buchanan, 1982). However, the overall superiority of this reagent to acetone and ethanol has yet to be shown. That dehydration with DMP extracts active lipids has been demonstrated (Bechmann and Dierichs, 1982). Such extraction appears to be caused by a reoxidation of lower osmium oxides, with subsequent attack on the unfixed lipids by the DMP. Therefore, DMP is not recommended for routine electron microscopy. Vinylcyclohexane (VCD), a polar dehydrant, was used by Stratton (1976). This reagent, too, cannot be recommended, owing to its carcinogenic effects and lack of superiority to other solvents. Another polar solvent, 1,4-dioxane (Shearer and Hunsicker, 1980), shows promise and deserves further experimentation. To minimise extraction of cellular constituents, the duration of dehydration should be as short as possible without subjecting the tissue to osmotic shock. However, it should be noted that incomplete dehydration is a common cause of incomplete infiltration by resin. Ideally, dehydration and infiltration with resin should be gradual and continuous. An apparatus has been devised for continuous dehydration and infiltration of specimens (Rostgaard and Tranum-Jensen, 1980). Two typical procedures are given below.

Procedure I

(1) 5% ethanol	5 min
(2) 10% ethanol	5 min
(3) 20% ethanol	5 min
(4) 30% ethanol	5 min
(5) 40% ethanol	5 min
(6) 50% ethanol	5 min
(7) 60% ethanol	5 min
(8) 70% ethanol	5 min
(9) 80% ethanol	5 min
(10) 90% ethanol	5 min
(11) 100% ethanol	5 min
(12) 100% propylene oxide (two changes)	5 min each

Procedure II

(1) 5% acetone	5 min
(2) 10% acetone	5 min
(3) 20% acetone	5 min
(4) 30% acetone	5 min
(5) 40% acetone	5 min
(6) 50% acetone	5 min
(7) 60% acetone	5 min
(8) 70% acetone	5 min
(9) 80% acetone	5 min
(10) 90% acetone	5 min
(11) 100% acetone (two changes)	5 min each

All steps are completed with continuous stirring. Longer periods are required to dehydrate hard animal tissues (e.g. connective tissue), highly vacuolated cells, senescent cells and most plant cells. On the other hand, if immediate results are needed, small tissue blocks can be dehydrated in a few minutes, rapidly infiltrated in a vacuum and polymerised in less than 1 h at high temperatures (Hayat and Giaquinta, 1970). If the fixation and washing were carried out in the cold, then the dehydration should also be completed at 4 °C. The tissue specimens are brought to room temperature during the last change in absolute acetone. If trimming of the tissue specimen is required prior to infiltration, it can be safely done in 90% acetone or ethanol.

It is necessary that the acetone used be of reagent type. To reduce water contamination, absolute acetone should be kept in small tightly stoppered bottles. For *en bloc* staining, it should be noted that although uranyl acetate and PTA are not soluble in absolute acetone, they are soluble in dilute solutions of acetone.

UNDESIRABLE EFFECTS OF DEHYDRATION

Since dehydrating agents are strong organic solvents, they inevitably cause shrinkage and extraction of cell constituents. Dehydration with ethanol caused a linear shrinkage of ~9.3% in the liver; most of the shrinkage occurred in 96% ethanol (Hanstede and Gerrits, 1983). Considerable amounts of lipids, carbohydrates and proteins are lost during dehydration. Experiments with amoebas show that, after fixation with glutaraldehyde, almost all neutral lipids are retained but up to 90% of these lipids are lost during dehydration (Korn and Weisman, 1966). After fixation with OsO_4 most of the neutral lipids and some of the phospholipids are extracted by ethanol and Maraglas. In certain specimens, dehydration may cause polarisation of cellular materials such as glycogen. The extent and type of cellular materials lost are determined by the size and type of the specimen and the method of dehydration, especially its duration in non-aqueous solvents. Fig. 2.1 shows excessive extraction of cellular materials caused by prolonged dehydration. In addition, the method of fixation employed has a profound effect on the degree of alterations in the ultrastructure occurring during dehydration. The role of various fixatives in the preservation of different cellular materials has been discussed in Chapter 1.

To minimise the above-mentioned undesirable effects, the duration of dehydration should be as short as possible, provided that the duration of each of the steps is long enough to accomplish a gradual replacement of water with the solvent. A very rapid dehydration tends to cause violent osmotic changes in tissue components, and surface tension forces cause major distortion of the structure. The optimum duration of dehydration is dependent primarily upon the size and type of the tissue specimen. However, since in most studies the tissue specimens are extremely small, a rapid processing should not cause distortion due to osmotic changes. The likelihood of the presence of some soluble proteins after fixation may require that the concentration of the dehydration agent at the start of dehydration should be high enough (e.g. 70%) to insolubilise these proteins by denaturation.

Lipid loss can be minimised by employing partial dehydration. With this procedure, the use of ethanol concentrations higher than 70% as well as propylene oxide is avoided. This is possible because Epon is soluble in 70% ethanol. The schedule is given below (the first four steps are carried out at room temperature and the remaining steps at 37 °C).

(1) 30% ethanol	15 min
(2) 70% ethanol	30 min
(3) 70% ethanol	30 min
(4) 1:1 mixture of 70% ethanol and Epon embedding mixture	30 min
(5) Same as above	30 min
(6) Epon	1 h

The loss of certain lipids is reduced even further by lowering the temperature to 0 °C by employing the incomplete dehydration method given below.

Incomplete Dehydration for Lipid Preservation

70% ethanol in water	5 min at 0 °C
70% ethanol in water	5 min at 0 °C
95% ethanol in water	5 min at 0 °C
95% ethanol in water	5 min at 0 °C
Resin	1 h at 0 °C
Resin	1 h at 0 °C
Resin	1 h at 0 °C
Resin mixture	overnight at 4 °C
Embed in resin mixture	

Partial dehydration and incomplete dehydration are not recommended for routine processing.

INFILTRATION AND POLYMERISATION

INFILTRATION

A complete and uniform penetration of tissue specimens by a suitable embedding medium is a prerequisite for satisfactory sectioning. This is accomplished through infiltration and embedding. Infiltration essentially involves a gradual and continuous replacement of the dehydration agent with an embedding medium, while embedding consists of a complete impregnation of the interstices of a tissue specimen with the medium. Embedding must precede microtoming, because very few tissues are sufficiently rigid to be cut into thin sections without an additional support. An embedding medium not only impregnates and supports the tissue internally, but also attaches the tissue to a block sufficiently strong to be handled safely during sectioning. It is, however, possible to obtain ultrathin sections of unembedded tissue specimens such as muscles (Gilëv and Melnikova, 1968; Pease, 1980). The fixed specimen is dried on a filter paper at room temperature for about 3 days. The specimen is then glued to a plastics rod for sectioning. This procedure cannot be recommended.

As stated above, infiltration is accomplished by gradually decreasing the concentration of the solvent and proportionately increasing the concentration of the embedding medium. If the dehydration agent is miscible with the embedding medium, then the tissue can be placed directly in a mixture of the agent and the medium. In the event that the dehydration agent is immiscible with the embedding medium, a transitional solvent that is miscible with the embedding medium must replace the dehydration agent prior to infiltration.

Infiltration should be gradual and continuous but

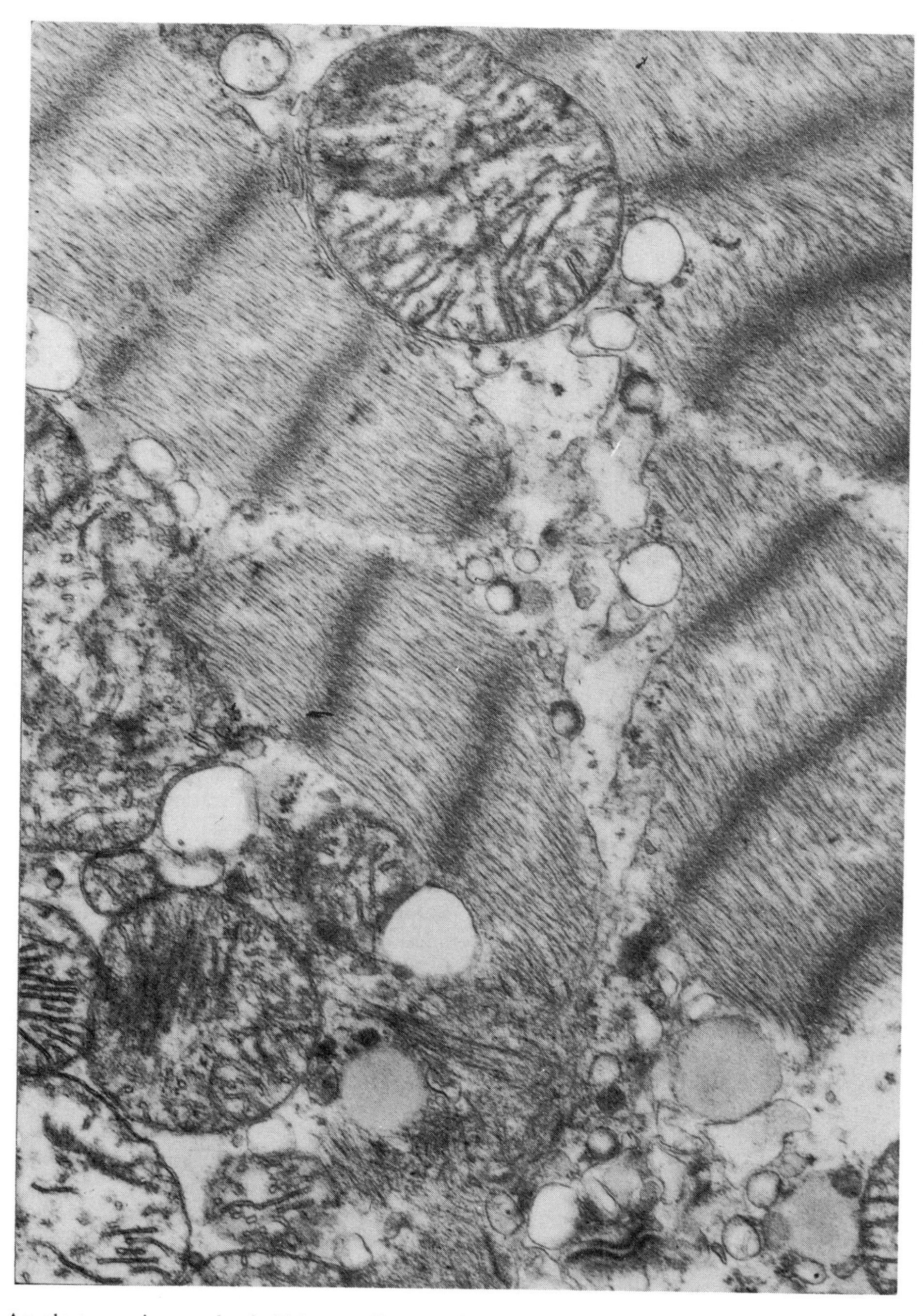

Figure 2.1 An electron micrograph of chicken cardiac muscle showing the effect of prolonged dehydration with ethanol. Excessive extraction of cellular materials is apparent, especially in mitochondria, myofibrils and ground cytoplasm

should not be prolonged unnecessarily, for extraction and other changes in the cellular constituents continue during filtration. The duration of infiltration is dependent on the type of the tissue and resin used. A routine procedure is given below.

(1) Last change in 100% acetone	
(2) 100% acetone and resin (3:1)	20 min
(3) 100% acetone and resin (2:1)	20 min
(4) 100% acetone and resin (1:1)	20 min
(5) 100% acetone and resin (1:2)	20 min
(6) 100% acetone and resin (1:3)	20 min
(7) Resin mixture	20 min
(8) Resin mixture to polymerise	

All the steps prior to polymerisation are completed with continuous stirring and at room temperature. Hard specimens require longer times. Final embedding is most conveniently done in pre-dried gelatin or polyethylene capsules or flat embedding moulds. The capsules or moulds containing the embedding mixture and the specimens are polymerised in an oven for 12 h at 45 °C followed by a final treatment at 60 °C until polymerisation is completed. The time required to complete polymerisation varies for different resins. As an average, a period of 48 h at 60 °C is sufficient to complete polymerisation. Blocks can be sectioned immediately after they are cooled. For a rapid procedure of infiltration and embedding, see p. 124. For infiltration at low temperatures, see p. 110.

It is emphasised that in an embedding mixture all the ingredients must be mixed very thoroughly. Failure to achieve a complete mixing results in an uneven impregnation of the tissue specimen and a block of uneven hardness. This can be expected, because various ingredients of a mixture possess different viscosities and rates of polymerisation, and thus differ in their rates of penetration. For example, the hardener DDSA is less viscous than Araldite; therefore, the former penetrates into the tissue faster than does the latter. It is likely that some of the earlier reports of incomplete impregnation of tissues were due to incomplete mixing of the infiltrating and embedding ingredients rather than to an inherent defect in the resin employed. Accumulation of heavy metals, used as *en bloc* stains prior to embedding, in the tissue may also hinder uniform penetration by the embedding resin. For example, a prolonged exposure of the tissue specimen to uranyl acetate tends to result in an uneven impregnation of the tissue.

The presence of high environmental relative humidity (60% or higher) during dehydration and infiltration may produce poorly embedded specimens. Thin sections of such specimens may show intercellular and intracellular holes. This problem is almost eliminated by keeping the relative humidity below 50%. The final epoxy resin mixture may be kept in the evacuated desiccator to prevent possible hydration. Embedding capsules or flat moulds can be placed in an evacuated desiccator over a 2 cm layer of Drierite for 24 h at room temperature, followed by 60 °C for 3 days to complete the polymerisation of Epon–Araldite mixture (Dellmann and Pearson, 1977).

Viscosity and Agitation

The rate of penetration of the infiltration mixture into the tissue specimen is important, for the final appearance of the tissue is influenced by the duration of the infiltration process. The appearance of the tissue is also affected by the temperature at which the infiltration was carried out. For this reason, the viscosity of the embedding medium used in the infiltration mixture is important. It is known that the viscosity of fluids is caused by the internal friction resulting from their molecular cohesion. In general, the rate of penetration varies inversely with the size of the molecule and the viscosity, and an increase in temperature lowers the viscosity of fluids. This is the reason that infiltration with resins is commonly carried out at room temperature. If it is important to infiltrate a specimen at a low temperature, the duration of infiltration should be increased. The exact duration for any given type and batch of resin has to be worked out by trial and error. The range of viscosity of each of the commonly used resins is given in Table 2.1. High viscosity is an obstacle in achieving rapid infiltration.

A uniform interchange of internal and external fluids should be an essential feature of tissue infiltration. Although the penetration of an almost stationary infiltration mixture into a static tissue specimen is at its minimum, unfortunately it has become the usual practice to place in a vial the specimens for infiltration until the appropriate time has elapsed. This is an oversight which brings the penalties of uneven and slow infiltration. In addition, it is possible that some of the ingredients of an infiltration mixture may separate if the mixture is stationary. This will also result in an uneven infiltration. The advantages of keeping the specimen and the infiltration mixture in continuous motion are therefore obvious.

The fact that shorter infiltration time results in less extraction of cellular materials, without subjecting the tissue to osmotic shock, has prompted workers recently to transfer the specimens after dehydration directly to a 1:1 or even 1:3 mixture of the solvent and the embedding medium. Since not only the chemical properties of the infiltration mixture but also its viscosity have a direct bearing on the penetration rate, more viscous

Table 2.1 Approximate viscosity at 25 °C of various monomers used in electron microscopy

Monomer	*Viscosity* (cP)
Araldite 502	3000
DDSA	290
DER 736	30–60
DER 332	40–55
DER 732	55–100
DER 334	500–700
DMAE (dimethyl aminoethanol)	60
DMP-30	20.5
Durcupan	100
EMbed	4.1
Epon 812	15–21
ERL 4206 (VCD)	7.8
Glycol methacrylate	50.8
HHPA (hexahydrophthalic anhydride)	solid
JB4	5.4
LX-112	7.2
LR White	8
LEMIX	70.8
Maraglas 655	500
Nanoplast	48
NMA (MNA)	275
NSA (nonenyl succinic anhydride)	117
Pelco Medcast	5.4
Poly/Bed 812	9.1
Quetol 651	15
Quetol 523	23
Quetol 653	40
Quetol 812	140
Styrene	0.7

mixtures currently preferred are quite slow to infiltrate. Thus, by use of continued mechanical agitation, the time of infiltration can be reduced. The duration of infiltration can also be reduced by increasing the temperature to ~40 °C. This temperature lowers the viscosity of resins without causing significant polymerisation, within a reasonable time.

Uncommonly hard tissues such as bones and seeds, and delicate specimens such as marine bryozoa, pose special problems in infiltration. In the case of hard tissues, slow diffusion of the embedding medium is a problem, while delicate specimens tend to collapse on coming in contact with the embedding medium. Hard tissues that are difficult to infiltrate should be placed in a vacuum oven for 1–4 h at 45 °C and then subjected to standard polymerisation treatment. Some workers employ a brief period (~10 min) of vacuum treatment routinely for plant tissues. This treatment also eliminates the air bubbles trapped around the tissue specimen. Vacuum treatment is best carried out at 0.5 atm; lower pressures tend to cause bubble formation and disruption of the cells. Various additional treatments are used to facilitate infiltration of the hard tissues, and the reader is referred to the methods given in the original work on a specific type of tissue. Delicate tissues can be infiltrated without much distortion if the process is carried out in an apparatus that allows a slow but continuous rotation of the tissue container in an oblique plane.

Various types of mechanical shaker are in use to achieve a uniform and rapid infiltration. In general, rotational shaking is more efficient than back-and-forth shaking; the latter may cause damage to delicate specimens. The rotary movement imparts a continuous swirling motion to the medium in the vials; the specimens float in the centre of the vial while the medium moves around it. In this way mechanical damage is reduced to a minimum. It is essential that the vials be rotated at an angle, for even very viscous liquids exhibit streaming movement if the container is rotated in an oblique plane. The speed of rotation should be slow enough for gravity to be able to act effectively against the forces of adhesion, but fast enough to achieve maximum streaming movement. Care should be taken that the medium covers the bottom of the vial containing the specimens when it is tilted to an angle. Vials should not be filled to more than one-third of their capacity, because very little mixing occurs at the bottom if they are filled nearly to capacity.

A versatile rotary shaker (Fig. 2.2) is commercially available. The shaker can hold 20 vials on a metal pan ~5 cm deep. The pan is designed to contain ice for cooling the specimen vials during fixation or to maintain the desired temperature by warming during embedding. The angle of inclination of the plate holding the vials can be adjusted over the range 4–15° with two speeds: 65 rev/min and 130 rev/min. By appropriately combining motor speed and angle of inclination, various rates of swirling suitable for fixation, dehydration and embedding can be achieved. Kushida (1969) has indicated desirable rotational speeds, angles of inclination and temperatures for facilitating penetration of various dehydration solvents and embedding media.

POLYMERISATION

Some resins (e.g. LR White) can be polymerised either by heat or by long-wavelength ultraviolet (UV) irradiation, although the former method is more common. However, the most widely used embedding resin, Epon, and its substitutes, are usually polymerised by heat. The former method consists of heating at high temperatures (35–60 °C), while polymerisation by the

Figure 2.2 A general view of the rotary shaker, which facilitates the infiltration of specimens with viscous embedding media. The shaker can also be used to shorten fixation and dehydration schedules

latter method is accomplished at room temperature or in the cold.

Polymerisation by UV Irradiation

The advantage of the UV irradiation method is that it avoids introducing the heat effects, including specimen shrinkage and bubble formation. Volume shrinkage of Epon, Araldite, Vestopal W, butyl methacrylate and methyl methacrylate by heat polymerisation is 3%, 3.6%, 6.5%, 15% and 20%, respectively. However, the actual shrinkage of the specimen is less, because most of the shrinkage occurs in the liquid phase of the resin before gelation.

Polymerisation by UV irradiation is desirable for cytochemical studies where high temperatures must be avoided. As stated above, UV irradiation is commonly employed to polymerise water-miscible embedding media such as GMA and Lowicryls. Other resins, such as Vestopal (Rostgaard and Buchmann, 1974), have also been polymerised by UV irradiation. Satisfactory polymerisation by UV irradiation is achieved only when the ambient temperature, the temperature in the capsule, and the wavelength and intensity of the irradiation are controlled. The least ultrastructural distortion occurs when polymerisation is accomplished at controlled, low temperatures. It should be noted that the temperature inside the capsule is higher than the ambient temperature at which polymerisation is carried out, for polymerisation is a strongly exothermic process.

Various types of apparatus have been used to obtain UV-induced polymerisation. Bartl (1964) achieved a very rapid polymerisation by using a device consisting of a high-pressure mercury lamp and a liquid-cooled polymerisation chamber. In contrast to the expensive device described by Bartl (1964), the apparatus introduced by Cole (1968) is less expensive and easier to construct. The apparatus used by Rostgaard and Buchmann (1974) allows unilateral or circumferential UV irradiation of capsules. Various types of UV radiation sources can be used: Westinghouse Fluorescent Sun Lamp FS 20, 20 W, 24 in tube; Philips CSI 250 W lamp; General Electric black light lamp. Methacrylate and Vestopal can be polymerised in 15 and 35 h, respectively. The lamp can be placed ~25 cm from the capsule.

Polymerisation of Lowicryls and LR White by UV irradiation is discussed elsewhere in this volume (p. 111).

Polymerisation by UV irradiation is rather tedious and may not be very effective in exceedingly complex cross-linking of epoxy resins. Furthermore, the depth of polymerisation into the OsO_4-fixed and Vestopal-embedded specimen is only 0.06–0.08 mm (Rostgaard and Buchmann, 1974). Consequently, the specimen size should be exceedingly small. When UV irradiation is applied unilaterally, the resin opposite to the source may be poorly polymerised. Therefore, the capsule should be irradiated from all sides. The discussion mentioned above indicates that polymerisation by UV irradiation is a complex and poorly understood process. Also, there is some evidence which indicates that prolonged UV polymerisation causes more inhibition of enzyme activity than that caused by heat polymerisation. Nevertheless, polymerisation by UV irradiation at low temperatures is recommended for water-miscible embedding media, with aldehyde-fixed specimens without post-osmication for immunocytochemical studies.

Polymerisation by Heat

Available data are insufficient to discourage polymerisation by heat. Polymerisation by heat does not require elaborate apparatus; any standard oven is adequate. Polymerisation can be carried out directly at 60 °C. In general, the cutting quality of resins is not significantly affected by rapid or slow heat polymerisation, provided that the blocks are adequately cured and the tissue is infiltrated completely.

Mechanism of Cross-linking

The following brief discussion of the mechanism of cross-linking applies primarily to epoxy resins. Marshall (1980) has also summarised the general mechanism of epoxy hardening. Like polyesters, epoxy resins can be transformed from a thermoplastic state to hard thermoset solids by the addition of curing agents. Curing agents are basically of two types: (1) cross-linking and (2) catalytic.

Cross-linking agents participate directly in the reaction and are absorbed into the resin chain. Numerous chemicals function as cross-linking agents (hardeners), but in practice only organic anhydrides (e.g. hexahydrophthalic, nadic methyl and dodecenyl succinic) have proved to be usable. These anhydrides react primarily with intermediate hydroxyl groups and form cross-bridges between the resin molecules.

Catalysts, on the other hand, promote primarily epoxy-to-epoxy or epoxy-to-hydroxyl reactions forming long-chain polymers. This becomes clear when one considers that chemically reactive epoxide and end-groups of epoxy resins are highly strained three-membered rings which open readily and attach themselves to groups containing reactive hydrogen atoms, especially amines. This is why tertiary amines such as benzyldimethylamine and tridimethylaminoethyl phenol are most commonly employed as catalysts (accelerators).

Catalysts, in contrast to hardeners, do not themselves serve as direct cross-linking agents. Since a reaction between the hardener and the resin is slow and normally requires a high temperature, the addition of a catalyst serves to increase the reactivity rate and enables the cure temperature to be lowered to a level desirable for the embedding process.

The physical properties of the thermose epoxy compound are determined by the nature of its chemical linkages. Three types of major reactions are responsible for the curing of epoxy resins: (1) epoxy-to-epoxy linkage, (2) epoxy-to-hydroxyl linkage and (3) epoxy-to-curing agent linkage. The reaction of anhydrides with epoxy groups essentially involves opening of the anhydride ring with an alcoholic hydroxyl to form the monoester, reaction of the nascent carboxylic group with the epoxide to yield an ester linkage, and reaction of the epoxide groups with hydroxyl groups to produce an ether linkage. The role of a catalyst, for example, in epoxy-to-epoxy polymerisation involves opening of an epoxy group; the ions of the opened group, in turn, open another epoxy group. This process continues until the stable ether linkages form a three-dimensional cured system.

Long epoxy-to-epoxy linkages occur because, as stated above, the epoxy group is opened by ions, hydrogens and tertiary amines used as curing agents. During the early stages, the three-dimensional linkages are scattered, but as the polymerisation continues, the linkages become uniform throughout, which results in an exceedingly dense, cross-linked and tough cured system. Uncatalysed epoxy resins are very slow to react with hydroxyl groups, but once catalysed they react rapidly with these groups of various sources and continue reacting until a high degree of cross-linkage is achieved. It is apparent, therefore, that if a mixture of epoxy resin, amine and anhydride is heated, three-dimensional, irreversible polymerisation will take place, resulting in a stable and inert substance very resistant to heat and solvents.

Since most epoxy resins cured with anhydrides and amines are hard and rigid, appropriate modifiers (plasticisers and flexibilisers) are used to increase extensibility and impact strength of the blocks. It should be noted that modifiers tend to cause a reduction in the tensile strength of the cured system. The modifiers may also lower the viscosity, improve the pot life and lower the exothermicity of resin mixtures. Although both plasticisers and flexibilisers impart higher flexibility to the cured block, the former do not react chemically with epoxy resins, whereas the latter do react with epoxy resins and become an integral part of the cured system.

In order to obtain a cured block with the desired characteristics, it is imperative to employ the exact amount of the hardener and the accelerator, and to control the duration and temperature of cure. It is emphasised here that gravimetric methods are more precise than volumetric methods for measuring viscous plastics. The optimum amount of a hardener is dependent upon the available reactive groups in the epoxy resin, while the optimum amount of an accelerator is dependent upon the anhydride and resin employed and cure schedules. Since part of the reaction in cross-linkage can take place independently in the acid medium, the ratio between anhydride and epoxy is less critical than the ratio between anhydride and an amine. It is almost impossible to achieve a desirable cure if the percentage of the amine catalyst in the embedding mixture is not exact. In other words, the time of setting is critically dependent upon the amount of catalyst employed. Any deviation from the optimum amount can alter the characteristics of the final block. For example, even slightly excess amine tends to block chain synthesis at a low molecular weight, which results in a block of dark colour too brittle for satisfactory sectioning. Also, the rate of polymerisation in the presence of excess amine is too fast for adequate infiltration of the resin into the tissue specimen. On the other hand, too little amine fails to effect adequate polymerisation.

Considering the inconsistency in the epoxy equivalent of different batches of the same epoxy resin, it is difficult to calculate a constant optimum amount of the hardener and the catalyst. The optimum amount of a hardener and a catalyst needed to achieve desired properties is therefore determined by trial and error. In general, the lower the amount of the catalyst the slower the rate of cross-linking, and consequently the longer time and higher temperature are necessary to accomplish adequate curing.

Stock mixtures of resins and anhydrides should not be prepared in advance of their use. Such mixtures slowly polymerise, even at room temperature, and consequently the structure of the embedding resin will vary from day to day. Hardeners and accelerators should never be mixed together alone, because of the possibility of explosition. The gelatin or polyethylene capsules should be clean and dry, and the final embedding mixture poured into the capsules should be free of bubbles. Aliphatic amines and anhydrides are capable of causing serious irritation and rash. Therefore, all human contact with these materials should be avoided.

UNDESIRABLE EFFECTS OF INFILTRATION AND POLYMERISATION

The loss of cellular materials and denaturing of proteins continue during infiltration and polymerisation, although they are on a much smaller scale than during fixation and dehydration. The extent and type of cellular materials lost in these steps of the preparatory procedure are determined by the method and duration of infiltration, as well as by the preceding fixation and dehydration. There is little change in the volume of the tissue specimen during infiltration, although isolated cells tend to show considerable shrinkage in this step. There is also little shrinkage (~2%) of tissue specimens during polymerisation of epoxy, polyester and glycol methacrylate resins. Preservation of antigenicity is much better when infiltration and polymerisation are carried out at low temperatures. The same is true when polymerisation is accomplished by UV irradiation instead of heat.

Not only dehydration solvents but also embedding resins are lipid solvents. A considerable body of information exists which suggests that the majority of commonly used embedding resins act as powerful lipid solvents. Even if the lipids are not extracted during dehydration following conventional fixation, they are lost during embedding. Extensive loss of cholesterol oleate from macrophages during embedding in Araldite has been reported (Casley-Smith and Day, 1966). Cope and Williams (1968), using macrophages, studied the loss of incorporated tritiated glycerol oleates during ten different embedding schedules, and found that none of the schedules allowed complete retention of neutral glycerides, losses of which ranged from 50% to 100%. The amount of radioactive cholesterol loss in mouse lung during Epon infiltration was almost as much as that during the dehydration (Darrah *et al.*, 1971).

However, it is possible to minimise lipid loss by omitting higher concentrations of ethanol or acetone and propylene oxide during infiltration. Further improvement in the preservation of lipids can be obtained by fixing the tissue in the presence of 0.2% digitonin and avoiding dehydration in 100% ethanol, acetone or propylene oxide. The role of digitonin in fixation is discussed elsewhere (Hayat, 1981, 1989a).

It is also possible to reduce lipid loss with varying degrees of success by subjecting the tissue to post-chromation during fixation (Casley-Smith, 1967).

Attempts to improve lipid preservation by carrying out infiltration and polymerisation of epoxy resins at low temperatures (Shinagawa *et al.*, 1962) have had only moderate success.

Because conventional resins are efficient lipid solvents, efforts have been made to develop water-miscible embedding media suitable for lipid preservation. However, lipid extraction is not altogether prevented, even when tissues are embedded in these media. For example, embedding in glycol methacrylate at 30 °C and curing by UV irradiation do not reduce lipid extraction to any significant degree. Water-miscible embedding media are also lipid solvents. An objectionable feature of the older water-miscible embedding media is that they are difficult to section, and in most cases the quality of sections as well as ultrastructural preservation are relatively unsatisfactory. For instance, only the peripheral area of the urea-formaldehyde-embedded tissue specimen is suitable for sectioning. Inert dehydration (Pease, 1966a) is also ineffective in preserving lipids.

In the absence of any better technique available at the present time, the application of water-miscible embedding media at low temperatures is the best approach to minimise the extraction of lipids. The properties and embedding procedures of commonly used water-miscible embedding media have been discussed later in this chapter. Improved preservation of lipids may be achieved by using freezing methods in conjunction with water-miscible embedding media; this approach is discussed elsewhere in this volume. Bartl (1962) has demonstrated that frozen tissue specimens can be infiltrated by precooled glycol methacrylate at temperatures ranging from −40 °C to −60 °C. Other resins that can be used during infiltration at low temperatures include Lowicryls and melamine. It is suggested that better preservation of lipid can be obtained by developing embedding media containing a high proportion of water, which may also necessitate cryosectioning.

Morphometric changes occur not only during fixation and dehydration but also during infiltration, polymerisation, sectioning and section stretching and transfer onto grids or glass slides. The final dimensions of the sections and their microscopic images depend on the response to surface tension at the water surface and the temperature at which semithin sections are mounted on a glass slide (Gerrits *et al.*, 1987). An understanding of the physical changes that occur in the cells during resin embedding is important.

STANDARD PROCEDURE FOR FIXATION, RINSING, DEHYDRATION AND EMBEDDING

Materials needed are as follows: buffered fixatives in glass-stoppered bottles (appropriately labelled); complete buffer (recently checked for pH); two plastics Petri dishes (absolutely clean); tweezers; dissecting needles; new single-edge razor blade (or scalpel) that has been cleaned with acetone; crushed ice; small vials; pipettes; saturated aqueous solution of uranyl acetate; container for discarding fluids; complete embedding mixture; embedding capsules or containers for flat embedding; holder for capsules; labels; plastics syringe without the needle; acetone series of ascending concentrations. If available, self-supported capsules are preferred (Fig. 2.3).

After a relatively large piece of tissue has been removed from the organism, it is immediately placed directly in the aldehyde fixative in a small vial in a fume-hood for ~15 min; enough fixative is used to completely cover the tissue. The tissue is transferred to a Petri dish whose bottom contains dental wax, and then cut into small pieces (2–3 mm^3) with a razor blade and transferred (using a fine brush) to a second dish containing several drops of the fixative. If dental wax is not available, plastics Petri dishes can be used. The pieces are further cut cleanly until they measure less than 1 mm^3. A tissue chopper is preferred for obtaining tissue slices of 0.5 mm^3 or less. These tissue pieces are picked up with a fine brush and dropped into prelabelled vials containing the fixative; short, wide-mouth vials are satisfactory. The fixative is at least 12 times greater in volume than the specimens. The vial is kept standing in crushed ice or at room temperature. The primary fixation with aldehydes is completed in 1–3 h at 4 °C or 0.5–2 h at room temperature.

The aldehyde fixative is removed from the vial with a fine pipette and discarded into a suitable container in the fume-hood. The buffer is immediately poured into the vial, and the specimens are rinsed twice for a total period of 10–20 min. The buffer is replaced with 1% OsO_4 in 0.1 M cacodylate or phosphate buffer and post-fixed for 1–2 h at 4 °C or at room temperature. After a thorough rinse in the buffer, the specimens can be treated with 0.5% aqueous solution of uranyl acetate for 15–30 min. (Glycogen staining is adversely affected by this treatment.) Throughout these procedures, specimens must remain wet. Specimens in the same vial are gradually dehydrated while being stirred in a series of ascending concentrations of acetone or ethanol, as given on p. 81.

Since tissue specimens are very small, they should not be transferred from one vial to another. The

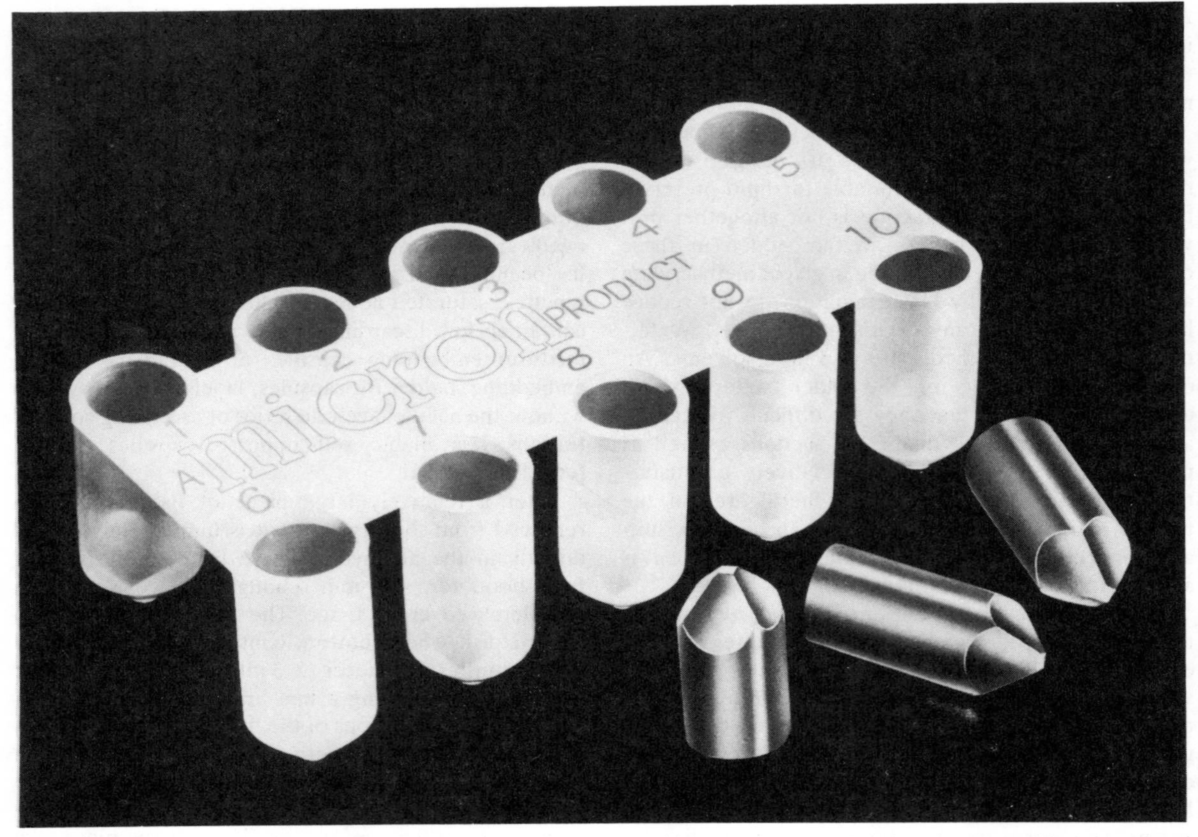

Figure 2.3 Micromould. Ten self-supported polyethylene capsules (8 mm internal diameter) are connected by a membrane numbered 1 to 10. The capsules produce blocks equivalent to No. 00 BEEM capsules, with 1 mm^2 tip which is pyramidal or square in shape. Reusable self-supported moulds are also available (Polysciences Inc.)

solution should be either withdrawn with a micropipette or poured out of the vial and immediately replaced with the other. Great care should be taken that the tip of the pipette does not damage the specimen. When removing solutions, specimens are often sucked into the pipette tip by surface tension, which results in damage to them or loss. The problem can be minimised by employing a non-wettable pipette with an orifice sufficiently small to prevent loss of the tissue pieces. Such pipettes are readily available as disposable pipette tips (C20 tip for a Model P200 or P20 Gilson-Ranin Pipetman). A disposable tip connected to a water aspirator by a flexible hose can be used for withdrawing solutions from the vials (Mollenhauer, 1978).

All the solution changes during fixation, washing, dehydration and infiltration should be carried out with speed so that the specimens do not become dry. Specimens should stay in the original vial used for fixation until they are ready to be transferred to gelatin capsules for embedding.

Specimens are gradually infiltrated (while being continuously stirred) at room temperature with the following mixtures:

(1) Acetone plus embedding mixture (3:1) — 30–60 min
(2) Acetone plus embedding mixture (1:1) — 30–60 min
(3) Acetone plus embedding mixture (1:3) — 30–60 min
(4) Embedding mixture — 30–60 min

Most epoxy resins are polymerised at 45 °C in an oven for 18–24 h and then at 60 °C for an additional 24 h. Alternatively, polymerisation can be accomplished directly at 60 °C for 48 h. Vinylcyclohexene dioxide (Spurr mixture) is polymerised at 70 °C in an oven for 8 h. Gravimetric measurements of resins are preferred to volumetric measurements; an accuracy of ± 0.1 g is acceptable.

The use of Reichert EM Tissue Processor capsules (Fig. 2.4) is recommended, to avoid mechanical dam-

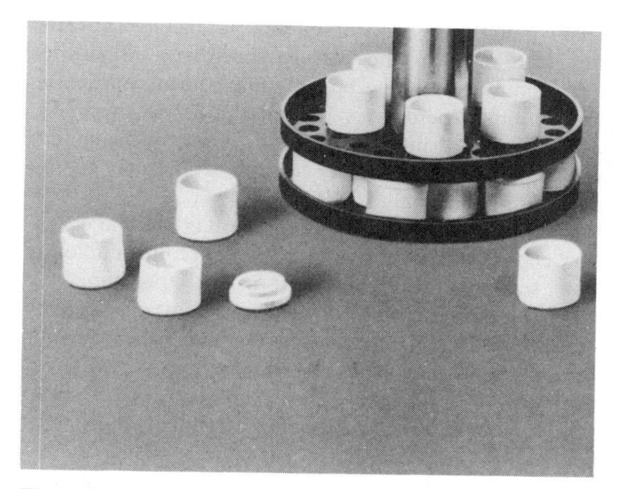

Figure 2.4 Reichert EM Tissue Processor capsules are shown with their holder used in the unit

age to and loss of specimens during processing, including rinsing, dehydration, staining *en bloc* and infiltration. These capsules are especially useful for processing thick (~50 μm) slices of fixed tissue encapsulated with 7% agar and cut with a tissue sectioner for enzyme cytochemical studies. These flat-bottomed, porous capsules (12 mm in diameter and 10 mm high) are made of sintered high-density polyethylene and have tightly fitting caps. They are freely permeable to aqueous solutions, organic solvents, incubation media and resins. Specimens can be processed in these capsules through the final step of infiltration with 100% resin. Low-viscosity resins pass through the pores of these capsules more rapidly than do high-viscosity resins. Since reagents are carried over to the next step involving the walls of the porous capsules, extra changes of 100% dehydration agent and 100% resin are recommended. Although these capsules are permanently blackened by OsO_4, they can be cleaned with acetone for re-use. They are commercially available (Flo-Thru: Normco Inc., Silver Spring, MD).

PROCEDURE FOR MINIMISING CHEMICAL HAZARDS DURING SPECIMEN PREPARATION

Exposure to hazardous chemicals during routine preparation of specimens can be drastically reduced by using a simple device designed by Redmond and Bob (1984). All the materials needed to construct the device are available in the laboratory. Specimens can be processed in this device from the time of initial fixation up to final embedding. The spout is removed from the cap of a plastics squeeze bottle (2 oz) having a sealable spout. The two minute air-holes located on either side (2–3 mm from the end) of the base of the spout are sealed with epoxy resin; care should be taken not to obstruct the opening in the flared end. The BEEM capsule (size 00) is prepared by cutting off the cap and then puncturing a hole (slightly smaller in diameter than that of the bottle spout) in the apex of the capsule. The puncture is made by using a heated metal needle and pushing it through the plastics, producing a hole of the desired diameter. The pointed end of the spout is inserted through the open end of the capsule until it passes through the hole in the apex. The spout is pushed firmly to make a tight seal between the capsule and the flared end (base) of the spout. This puncture is sealed from the outside with epoxy resin to make it airtight.

The centre of the BEEM capsule cap is removed with a cork borer, leaving the circular band. A small square of metal mesh is placed over the open end of the capsule. The mesh is held in place by using the circular band obtained earlier from the cap. The capsule assembly is placed into the squeeze bottle, and allowed to rest on the bottom. The spout is cut 3–4 mm below the rim of the bottle. The capsule assembly is inserted into the hole in the cap, and the height of the assembly is adjusted so that it hangs about 1 mm above the bottom of the bottle. The level of the band holding the mesh is marked with a grease pencil. The device is now ready for use (see Hayat, 1986b).

The fixative solution is poured into the squeeze bottle to the level of the grease pencil line. The specimen is placed in the capsule, the mesh is secured in place, the assembly is placed into the bottle and the top is firmly tightened. The syringe plunger is withdrawn to the 2–3 mm mark, and then its tip is inserted into the hole of the bottle top. To cover the specimen, the fluid from the bottle is drawn into the capsule by pulling up on the syringe plunger. Care should be taken that the fluid just fills the capsule; if the fluid is drawn into the spout and syringe, it may be contaminated. The fluid is removed from the capsule by pushing the plunger down. The air column within the syringe prevents backflow of the fluid. The bottle is gently swirled to disperse the diluted fluid, and the fluid is drawn again over the specimen.

A separate squeeze bottle may be used for each of the steps: fixation, rinsing, post-fixation, dehydration and infiltration. The specimen is transferred easily by unscrewing the cap (with syringe in position) and then lifting the entire capsule assembly out and screwing it in place in the next bottle. Little fluid is transferred, because of the air column in the assembly. While one bottle is in use, the other can be cleaned. The entire procedure should be carried out in a fume-hood.

EMBEDDING MEDIA

Since most tissues are insufficiently firm to be cut into thin sections, they are embedded in a suitable material of sufficient strength to allow a thin section to be cut. Moreover, specimen size is too small to be handled with ease without a support. Not many substances are suitable for use as embedding media for electron microscopy. Some of the major factors that determine the suitability are: (1) satisfactory preservation of the fine structure; (2) little extraction of cellular constituents; (3) ready solubility in monomer form in standard solvents; (4) monomer viscosity neither so low as to induce specimen shrinkage during polymerisation nor so high as to impede rapid and uniform penetration of the specimen by the embedding mixture; (5) uniform polymerisation; (6) little change in volume during polymerisation and hardening; (7) thermostability under electron bombardment; (8) satisfactory sectioning quality of the embedded specimen, which includes homogeneity, hardness, plasticity and elasticity to absorb the force of the cutting impact; (9) sufficient cohesion and compactness; (10) density of the polymerised embedding medium sufficiently low so that distinct imaging is not adversely affected; (11) resistance to heat generated by sectioning; (12) adequate specimen stainability; (13) miscibility with other embedding media; (14) consistency in the quality of different batches of the same embedding medium; (15) easy availability.

An ideal embedding medium should not react with cellular constituents in such a manner as to cause changes in the already fixed ultrastructure. If the medium is sublimed through exposure to the electron beam, the tensional forces will produce deformation of the fine structure. It is known that the appearance of the membranes is influenced by the type of embedding medium used. The capability of a polymerised resin to withstand irradiation depends on its ability to stabilise the radicals that are produced when an electron strikes the resin backbone. Aliphatic chains and saturated rings are less efficient at accommodating the extra electron than are the conjugated species such as aromatic rings and certain unsaturated chains and rings (Causton, 1980).

A good embedding medium should not act as a barrier to the penetration of solutions of heavy metal stains during post-staining of thin sections. The medium should possess sufficient elasticity to absorb the strain of the cutting impact and recover its original form when the strain is removed. Miscibility between embedding media is important for developing better embedding mixtures having desirable qualities of more than one medium. Low viscosity (i.e. mean molecular chain length) is desirable, for it facilitates rapid penetration into the tissue specimen. Significant changes in the volume of a medium during hardening are bound to deform or disrupt the tissue specimen. The stability of ultrathin sections is partly dependent upon the denseness and amorphous nature of the embedding material.

Since granular or crystalline materials lack the necessary cohesion and compactness, they are unsuitable for thin sectioning. It is apparent that paraffin, the most common medium for optical microscopy, cannot be used for electron microscopy, although paraffin-embedded tissues can be deparaffinised and re-embedded in resin for electron microscopy (see p. 134). The injury caused by paraffin treatment limits the application of this method to structures, such as secretory granules, that are relatively resistant to paraffin embedding.

In general, it is difficult to obtain several batches of the same medium having a uniform quality. This lack of uniform quality is the primary reason for inconsistent results obtained by different workers using the same embedding medium. Different batches of Epon 812 or its substitutes, for example, vary in their epoxide content.

Modern resins, more than any other substances, fulfil most of the above requirements. The choice of a resin is determined by the type of the specimen and the objective of the study. The materials most commonly used as embedments are epoxy resins, polyester resins and acrylic methacrylates. X-ray diffraction studies (Balyuzi and Burge, 1970) indicate that the approximate sizes of ordered structures within the commonly used embedding media are: Araldite, 0.7 nm; Vestopal, 1 nm; Epon, 1.2 nm; polymethacrylate, 1.8 nm. Although these data indicate that, for high-resolution electron microscopy, Araldite with minimal rigid structure is the most ideal embedding medium, similar data on recently introduced water-miscible embedding media (e.g. Lowicryls and melamine resins) are not available.

Although epoxy resins and Lowicryls are quite satisfactory, and are used most widely, they are far from being ideal for electron microscopy. In fact, none of the embedding media currently in use possess all of the above-mentioned desirable qualities. Certain desirable properties may be incompatible with others. For instance, media having a relatively high viscosity usually show least shrinkage on polymerisation.

WATER-IMMISCIBLE EMBEDDING MEDIA

Epoxy Resins

Epoxy resins are transparent yellowish resins and are far superior to other embedding media in many respects. They range from viscous liquids to fusible solids, depending upon the molecular weight. An epoxide is a triangular configuration of an oxygen atom bridging two carbons of an organic molecule. Two or more of these epoxide groups on the same molecule yield a product having a potential for forming resins. Chemically, epoxy resins are polyaryl ethers of glycerol having terminal epoxy groups and hydroxyl groups spaced along the length of the chain. They are made by condensing epichlorohydrin with polyhydroxy compounds. Undesirable characteristics of epoxy resins include degradation on exposure to light and relatively high viscosity (e.g. Araldite).

Epoxy resins require the addition of curing (cross-linking) agents to convert them to a tough, extremely adhesive and highly inert solid. This can be accomplished by various bifunctional setting agents which add across the terminal epoxy groups of the molecule and produce a three-dimensional structure. Hardening is accomplished without significant shrinkage (<2%) or uneven polymerisation. The relatively low shrinkage is due to the fact that cross-linking is an addition process and that the resin is highly associated in the unpolymerised state. The addition process is not dependent upon an initiator, and as a result the process is not affected by the presence of impurities. In contrast, even specimen constituents can initiate polymerisation in the methacrylates.

Epoxy resins probably introduce intermolecular cross-linkages, and this may enable them to act as fixatives for some proteins and nucleic acids. It has been reported that specimens fixed with glutaraldehyde followed by OsO_4 and embedded in epoxy resins can be used for enzymatic hydrolysis of specific proteinaceous components, provided that the sections are oxidised by periodic acid (PA) prior to incubation in enzyme solutions (Anderson and André, 1968). These resins are relatively stable with respect to light, heat and oxygen. Sections are relatively stable under electron bombardment, for only ~25% of the mass of the section is lost under normal operating conditions (Cosslett, 1960). However, the loss can be minimised by exposing the sections first to a low-intensity beam and then gradually to a more intense beam. This procedure allows the sections to become stable as a result of carbon deposition on and in the sections before being exposed to an intense beam.

As stated above, although epoxy resin sections show some loss under electron bombardment, this loss is not followed by any flow of the remaining embedding resin. For this reason the three-dimensional structure of the tissue specimen is maintained. The remarkable stability of epoxy resin sections under electron bombardment is the primary reason for the excellent clarity of the picture obtained. A minor disadvantage of the resistance to degradation is a reduction in the contrast between tissue and background. However, the advantages of stability under electron bombardment far outweigh this minor disadvantage. Furthermore, the contrast can easily be enhanced by heavy metals and by mounting the sections, which are sufficiently strong by themselves, on unsupported grids. Supporting films or membranes should be avoided unless they are absolutely necessary.

The problem of poor embedding, even in epoxy resins, is not uncommon. Poorly embedded specimens are difficult to section, and even when sections are obtained, they may disintegrate upon contact with water or under the electron beam. The presence of holes in the section makes it unstable. Several factors cause poor embedding, including the presence of water in absolute dehydration solvents and in the ingredients of the embedding mixture and inadequate dehydration of the specimen. High environmental relative humidity also produces poor embedding. Embedding should be carried out at relative humidities below 50%. Dellmann and Pearson (1977) recommend the evacuation of flat embedding moulds containing the resin in a vacuum desiccator for 24 h at room temperature prior to polymerisation. Epoxy resins are useful for embedding a variety of tissues, because a wide hardness range of this resin can be achieved to suit a specific tissue type by using two different anhydride curing agents (DDSA and NMA).

Since epoxy resin is hygroscopic, extreme care should be taken to prevent contamination by water. For this reason the tissues must be dehydrated completely with reagent grade acetone. If the epoxy mixture is stored in a refrigerator, the container should be tightly stoppered and wrapped in aluminium foil. Before opening, the container should be kept at room temperature for at least 1 h to prevent condensation. The stock mixtures should be kept carefully covered when not in use.

EPON 812 (LX-112, POLY/BED 812, PELCO MEDCAST, EMBED 812)

Epon 812 (known as Epikote in Europe) has been the most widely used embedding medium for electron microscopy. It can be easily sectioned and the specimens can be stained without difficulty. Thin sections of Epon can tolerate the intense heat and strong vacuum in the electron microscope. Although the production of Epon was discontinued in the middle 1970s, it is still available from some suppliers. The existing stocks will soon be exhausted. Substitutes for Epon are EmBed 812, LX-112, Pelco Medcast and Poly/Bed 812.

Epon 812 and its replacements are light-coloured, glycerol-based aliphatic epoxy resins of relatively low viscosity (Table 2.1). They penetrate into the tissue specimens faster than Araldite, because the latter has a much higher viscosity. Sections of Epon and its substitutes show greater contrast in the electron microscope than that shown by comparable Araldite sections. However, the inherent granularity of Epon and its substitutes may limit high-magnification and high-resolution studies. They can be easily hardened uniformly at relatively low temperatures with the addition of acid anhydrides and an amine accelerator. Commonly used acid anhydrides are DDSA and NMA, and the amine accelerators are 2,4,6-tri(dimethylaminomethyl) phenol (DMP-30) and benzyldimethylamine (BDMA). DMP-30 produces blocks of better cutting quality than does BDMA. Slight shrinkage occurs during curing.

Epon and its substitutes have been studied for their physical characteristics, such as rate of flow, density, viscosity and hardness (Mascorro and Kirby, 1986). The replacements are similar in their density and mass weights, whereas they differ in terms of viscosity and rate of flow. The viscosities of Epon and its replacements are given in Table 2.1.

Since LX-112 has the lowest viscosity among epoxy resins, it is recommended for dehydration when conventional dehydrating solvents are not used in order to minimise extraction of tissue constituents. Poly/Bed 812 is recommended as an embedding medium when acetone or ethanol is used for dehydration, for this medium causes ~10% less loss of phospholipids than that resulting from embedding in LX-112 (Stratton *et al.*, 1982).

Embedding Formulation

The physical properties of polymerised epoxy resins depend on the molar ratio between anhydride and epoxide. The proportions of these two components affect intermolecular cross-linking during polymerisation. A low proportion of anhydride leads to a soft block. To avoid brittleness, an anhydride:epoxide ratio of 0.63:1.0 is recommended.

Since Epon substitutes are a generic replacement for Epon 812, the proportions of DDSA, NMA and DMP-30 in the substitutes are the same as in the Epon 812 mixture. Since the W.P.E. (weight per epoxide equivalent) values of the epoxy resins vary from lot to lot, the only reliable way to ensure reproducible hardness of the block is by using the W.P.E. for the embedding formulations. The W.P.E. value can be utilised for embedding on a weight basis by using the following equation:

weight of anhydride (DDSA and NMA) required

$$= \frac{100}{\text{W.P.E.}} \times \begin{matrix}\text{anhydride}\\ \text{molecular}\\ \text{weight}\end{matrix} \times \begin{matrix}\text{ratio between anhydride}\\ \text{and epoxy resin}\\ \text{equivalents}\end{matrix}$$

100 = grams of epoxy resin (the value is arbitrary, being used only as an example). W.P.E. = weight of epoxy resin containing one equivalent weight of epoxide. Molecular weights of NMA and DDSA are 178 and 266, respectively. The ratio between anhydride equivalent and epoxy resin equivalent is 0.63:1.0.

Table 2.2 indicates the W.P.E. values and the required proportions of DDSA and NMA. The higher the W.P.E. value, the lower the proportions of DDSA and NMA.

For accuracy of measurement, the gravimetric rather than the volumetric measurement is recommended for the preparation of embedding formulations. If, for example, the W.P.E. of LX-112 resin is 166, the following formulation is recommended:

Mixture A	
LX-112	80 g
DDSA	90 g
Mixture B	
LX-112	100 g
NMA	75 g

The components of mixture A are put in an 8 oz bottle, which is capped and shaken vigorously until the two components are completely mixed. Solution B is prepared in a similar way. Mixtures A and B are mixed in the ratio of 3:2 to produce a block of medium hardness. Then 0.14 g (ml) of DMP-30 is added per 10 g of this mixture. Even a slightly incomplete mixing may produce blocks of poor sectioning properties. To enhance infiltration of the specimen, polymerisation is carried out at 45 °C for 20 h and then at 60 °C for an additional 24 h. If necessary, polymerisation can be accomplished directly at 60 °C.

Mixtures A and B can be stored for as long as 6 months under inert gas (Freon or argon) and continuous refrigeration. They should be warmed to room temperature before the container is opened; otherwise water will condense in the resin, causing poor sectioning. Also, the presence of water may produce slow hydrolysis of the anhydrides to free acids, resulting in

Table 2.2 Proportions of DDSA and NMA determined by the weight per epoxide equivalent (W.P.E.) of epoxy resins (Epon 812 or its substitutes)

Resin W.P.E.	*Weight of DDSA, Mixture A*	*Weight of NMA, Mixture B*	*Resin W.P.E.*	*Weight of DDSA, Mixture A*	*Weight of NMA, Mixture B*
140	106	89	158	94	79
141	106	88	159	94	78
142	105	88	160	93	78
143	104	87	161	93	77
144	103	86	162	92	77
145	103	86	163	91	76
146	102	85	164	91	76
147	101	85	165	90	76
148	101	84	166	90	75
149	100	84	167	89	75
150	99	83	168	89	74
151	99	83	169	88	74
152	98	82	170	88	73
153	97	81	171	87	73
154	97	81	172	87	72
155	96	80	173	86	72
156	95	80	174	86	72
157	95	79	175	85	71

Note: The amount of the resin is 100 g in each mixture, and the weights of DDSA and NMA are in grams.

defective blocks. Freshly prepared embedding mixtures are preferred.

Since the molecular weights and viscosities of various ingredients differ, it is imperative to mix all the ingredients very thoroughly. Most failures in epoxy resin embedding reported in the past were probably caused by inadequate mixing. Complete mixing, without beating air into the mixture, can be easily obtained with a Teflon stirrer (Zacks, 1963); other special stirring rods can also be used (Minick, 1963). If these stirrers are not available, the ingredients can be mixed with a glass rod (blunted and made smooth on a bunsen burner), on average for at least 20 min.

Mixing can be expedited by warming the resin, the hardener and the container to 60 °C prior to mixing. Immediately before use, the two mixtures are blended, and the accelerator DMP-30 is added in the proportion of 1.5–2%. The amount of accelerator added is crucial in obtaining satisfactory polymerisation (setting). Therefore, the accelerator should be measured carefully; otherwise the block becomes dark in colour and too brittle for satisfactory sectioning. Gravimetric measurements of very small quantities of this accelerator are difficult to obtain. Since the specific gravity of DMP-30 is 0.973 g/ml, both the weight and volume measurements are basically the same—i.e. 0.14 g $\approx$ 0.14 ml. Thus, in this case either value can be used.

The hardness of the final embedment depends upon the ratio between mixture A and mixture B in the final embedding mixture. An increase in the proportion of mixture B will harden the block. In other words, the higher the proportion of NMA the harder the block (Table 2.3). Thus, the hardness can be varied to suit various sectioning conditions. In the author's laboratory a mixture of 1:1 has proved most successful for general use. These mixtures are moderately viscous and can be stored under refrigeration for several months, although freshly prepared mixtures are preferred.

Table 2.3 Proportions of DDSA and NMA and the resulting quality of the block (Epon 812 or its substitutes)

Mixture A (g)	*Mixture B* (g)	*Total weight* (g)	*Accelerator* DMP-30 (g) (ml)	*Relative hardness*
10	0	10	0.14	Softest
7	3	10	0.14	Soft
5	5	10	0.14	Medium
3	7	10	0.14	Hard
0	10	10	0.14	Hardest

If the block softens at normal room temperature, the addition of triallyl cyanurate (TAC) to the embedding mixture tends to alleviate the problem. This addition imparts a high softening temperature to the polymer. The hardness of the block decreases as the concentration of TAC increases.

If the reader encounters a problem of incomplete infiltration by the Epon mixture containing DMP-30 throughout the tissue block, the following formulation is recommended:

Epon 812 (or a substitute)	26 g
DDSA	22 g
NMA	9 g
Benzyldimethylamine	2–3%

ARALDITE

Araldite was introduced as an embedding medium for electron microscopy by Glauert and Glauert (1958). It is a glycerol-based aromatic epoxy resin that shows very little volume shrinkage after polymerisation. It is very viscous (Araldite CY 212:1300–1650 cP at 20 °C), and has a rather low softening temperature. Araldite is a transparent yellowish resin that can be varied in hardness by suitable plasticisers such as dibutyl phthalate. The tissues to be embedded in this resin can be dehydrated with any of the commonly used organic solvents; however, when ethanol or methanol is used, the application of a transitional solvent such as propylene oxide is necessary because residual traces of non-reactive alcohols adversely affect the sectioning quality. Araldite resin is mixed with suitable proportions of a liquid anhydride hardener and an amine accelerator, and polymerised usually by heating. The blocks are heated in an oven at 60 °C for 24 h or longer.

Araldite sections can be mounted on unsupported grids and easily stained with heavy metals. Since Araldite sections show relatively less contrast, it is customary to apply post-staining. It has, however, the disadvantage of possessing the greatest electron density, owing, in part, to its comparatively great thermal stability. The loss of thickness from the sections of this resin under normal operating conditions amounts to ~20% (Cosslett, 1960). Araldite seems to be less grainy at very high resolution than any other known embedding resin. The resin is stable and can be stored in glass-stoppered bottles for many months at room temperature.

Embedding Formulations

Araldite 502	10 g
DDSA	7.8 g
DMP-30	1.5%

Araldite 502	68 g
DDSA	19 g
DMP-30	3 g
TAC	10 g

Araldite CY 212	10 g
DDSA	10 g
BDMA	0.4 g

The final hardness of the block can be varied by changing the proportion of hardener or plasticiser. It should be noted that a slight change in the proportion of the accelerator will drastically affect the colour and brittleness of the block. Elimination of the plasticiser will result in a brittle block. The addition of TAC results in a high softening temperature. A 10% concentration of TAC produces blocks of medium hardness; a decrease in the proportion of TAC results in harder blocks. Table 2.4 shows the W.P.E. values of Araldite 502 and the required proportions of DDSA; similar information is given for Araldite 6005 in Table 2.5.

Table 2.4 Proportions of Araldite 502 and DDSA determined by the weight per epoxide equivalent (W.P.E.) of the resin

Araldite 502 W.P.E.	*Araldite 502* (g)	*DDSA* (g)
232–234	100	80
235–237	100	79
238–240	100	78
241–243	100	77
244–246	100	76
247–249	100	75
250	100	74

Note: Prior to use, 0.136 g (0.14 ml) of DMP-30/10 g of Araldite 502–DDSA mixture is added.

Table 2.5 Proportions of Araldite 6005 and DDSA determined by the weight per epoxide equivalent (W.P.E.) of the resin

Araldite 6005 W.P.E.	*Araldite 6005* (g)	*DDSA* (g)
182	100	91
183–184	100	90
185–186	100	89
187–188	100	88
189–190	100	87
191–192	100	86

Note: Prior to use, 0.411 g (0.46 ml) of benzyl dimethylamine/20 g of Araldite 6005–DDSA mixture is added.

VINYLCYCLOHEXENE DIOXIDE (ERL 4206, SPURR MIXTURE)

ERL 4206 is a cycloaliphatic diepoxide (Trigaux, 1960)

with an epoxide equivalent of 74–78. It possesses one of the lowest viscosities (7.8 cP) of all the known resins used in electron microscopy. The very low viscosity facilitates its rapid penetration into tissues, and it is especially suitable for penetration of plant tissues with hard lignified cell walls and bone tissue. This resin is miscible with ethanol (or acetone) and, thus, does not require a transitional solvent such as propylene oxide following ethanol dehydration. However, the resin is not completely miscible with ethanol. The compact diepoxide structure of this resin yields linear polymers which are highly cross-linked and the sections are unusually resistant to the electron beam. Spurr (1969) introduced ERL 4206 as an embedding medium for electron microscopy.

Embedding formulation

ERL 4206 (epoxide)	10 g
DER 736 (epoxide, flexibiliser)	6 g
NSA (hardener)	26 g
DMAE (accelerator)	0.4 g

Variations in these compositions can be obtained by reference to Table 2.6. The exact weight in grams of each of the components should be used for best performance. If bubbles form, they can be drawn off with a gentle vacuum applied to the mixing container. NSA and DMAE should never be mixed together alone, owing to the possibility of a rapid exothermic reaction. (Benzyldimethylamine or DMP-30 may also be used if a faster cure is needed.) Minimum exposure to atmospheric moisture is recommended. Continuous mild agitation is desirable during infiltration. If specimens are difficult to infiltrate, vacuum treatment may help. Curing should be accomplished in a BEEM capsule with the cap sealed to prevent moisture contamination. The standard cure is 8 h or more at 70 °C in an oven.

The desired hardness of the block can be obtained by altering the proportion of the flexibiliser DER 736; an increase in the proportion of DER 736 will result in a softer block (Table 2.6, mixture C). An increase in the proportion of accelerator DMAE will result in shorter pot life, increased viscosity and faster rate of polymerisation (mixture D). It should be noted that any modification of the proportions of ingredients favouring one characteristic usually results in the sacrifice of other qualities. For example, mixture D has a rapid rate of polymerisation, but has a short pot life. The average viscosity of the above mixtures is only ~60 cP at 25 °C, and 24 h after preparation it increases to 140 cP at 25 °C, which is still markedly lower than the viscosity of many other embedding mixtures. For the preparation of Spurr mixture for embedding undecalcified bone, see Table 2.7. Special care must be observed in using Spurr mixture, because of its carcinogenic properties.

Table 2.6 Suggested variations of vinylcyclohexene dioxide embedding formulations

	Mixtures				
	A	*B*	*C*	*D*	*E*
Ingredients (g)					
ERL 4206	10	10	10	10	10
DER 736	6	4	8	6	6
NSA	26	26	26	26	26
DMAE	0.4	0.4	0.4	1	0.2
Polymerisation time (h) at 70 °C	8	8	8	3	16
Hardness	firm	hard	soft	—	—
Pot life (days)	3–4	3–4	3–4	2	7

From Spurr (1969) and Glauert (1974).

Spurr resin and possibly other resins are relatively easy to section with glass knives when the embedding medium contains a surfactant such as lecithin (0.5–1.0%) (Mollenhauer, 1986a). Lecithin can be obtained in the form of tablets from health food stores; these tablets contain soybean lecithin and glycerine, with total phosphatide content >61%. Since lecithin is not easily soluble in the resin, it is first dissolved (1:1) in peanut oil. Solubilisation of lecithin in the oil requires several days at room temperature or several hours at 60 °C (333 K). This method is useful only for conventional studies when one faces difficulty in obtaining satisfactory sections with glass knives. The following formulation is recommended (Mollenhauer, 1986a).

Vinylcyclohexene dioxide	5 g
DER 736	3 g
NSA	13 g
Lecithin	0.1–0.4 g
Peanut oil	0.1–0.4 g
DMAE	0.3–0.4 g

MARAGLAS

Maraglas 655 is a clear epoxy resin with a viscosity of 500 cP at 25 °C, which cures to a solid state in 72 h at 60 °C. Since Maraglas has a heat distortion point of 87.8 °C after 72 h cure at 60 °C, it is stable under the electron beam. It is readily miscible with acetone, propylene oxide, styrene, methyl methacrylate and DER 732, but not with alcohols. The use of a transitional solvent is therefore necessary if the tissues are dehydrated in ethanol.

Acetone is preferred to propylene oxide as the solvent in dehydration and infiltration, because more propylene oxide than acetone is retained after cure of Maraglas. The retention of propylene oxide appears to

Table 2.7 Spurr medium for embedding undecalcified bone

Ingredients	*Epoxide equivalent*	*Quantity* (g)	*Molar ratio*
ERL 4206 (epoxide)	75	10	2.6
DER 736 (epoxide)	175	5	0.6
NSA (anhydride, hardener)	224	27.5	2
DMAE (accelerator)	98	0.25	—

From Schulz (1977).

be due to a chemical interaction with the monomer rather than a physical retention, and this interaction is catalysed by benzyl dimethylamine (BDMA). The retention of acetone, on the other hand, does not seem to result from a chemical interaction. There is some evidence indicating that acetone dehydration may result in harder Maraglas blocks than does ethanol dehydration. The retention of either solvent results in softer cured Maraglas, which leads to substandard cutting properties. Maraglas can be stored at room temperature for several months without any apparent partial polymerisation. It is relatively difficult to obtain ribbons from a Maraglas block.

Embedding formulations

I

Maraglas 655	48 g
Cardolite NC-513	40 g
TAC	10 g
DMP-30	2 g

The above mixture has a relatively low viscosity, and is polymerised at 65–70 °C in only 8 h. The block is of medium hardness.

II

Maraglas 655	36 g
DER 732	8 g
DP	5 g
BDMA	1 g

Tissue specimens should be allowed to remain in the above mixture for at least 4 h before the initiation of polymerisation by heat. The hardness of the blocks can be adjusted by varying the proportions of DER 732 and Maraglas; higher concentrations of DER 732 will produce softer blocks. To ensure the exclusion of air bubbles, the capsules containing the unpolymerised resin and tissue should be degassed in a vacuum desiccator for ~30 min. In order to achieve uniform polymerisation, embedding should be carried out in polyethylene capsules instead of gelatin capsules.

QUETOL 651

Quetol 651 (EM Zairyosha. Tokyo), an epoxy resin with a viscosity of 15 cP at 25 °C, is an ethylene glycol diglycidyl ether. It is readily miscible with water, ethanol and acetone, and acts as a dehydration agent. Also, it can combine chemically with other epoxy resins.

Quetol 651	33 g
NSA	67 g
DMP-30	1.5–2.0 g

Dehydrated specimens are infiltrated according to the following schedule:

Ethanol+n-BGE (n-butylglycidyl ether) (1:1)	15–30 min
n-BGE	15–30 min
n-BGE + Quetol mixture (1:1)	1 h
Quetol mixture	two changes, 1 h each
Polymerisation for 24 h at 60 °C	

DOW EPOXY RESINS

Dow epoxy resins (DER) are derived from diglycidyl ether or bisphenol A and aliphatic diepoxides. Their primary use is as flexibilisers. The cutting properties of a resin block can be modified by adding plasticisers (dibutyl phthalate) or flexibilisers (DER or Cardolite NC-513). The latter are preferred because as epoxides they become part of the cross-linked structure. They decrease the hardness of the block by reducing the anhydride:epoxide ratio (Ellis, 1986). Many flexibilisers also reduce the viscosity of the resin mixture.

DER 732 was introduced by Erlandson (1964) as a flexibiliser in place of Cardolite NC-513 (a dark-coloured monofunctional flexibiliser) in a Maraglas 655 epoxy formulation. It is a polar diepoxide flexible epoxy resin based on a polypropylene glycol–epichlorohydrin condensation product. DER 732 has a relatively low viscosity (55–100 cP at 25 °C) and light colour, and reduces the viscosity of Maraglas. Its low viscosity and polar character facilitate easy penetration into tissues. Being a diepoxide rather than a monoepoxide, DER 732 is able to attain a high degree of cross-linking with Maraglas. It was selected by Spurr (1969) as a flexibiliser for the Spurr mixture, because of its shorter molecule and low viscosity. The resultant blocks yield sections which are quite stable under the electron beam. Although it is a true epoxy resin and becomes an integral part of a polymerised system, the epoxide groups are separated by hydrocarbon chains that contain no reactive sites. DER 732 can be stored at room temperature almost indefinitely without any apparent change in properties.

DER 332 is a pure epoxide and was introduced as an embedding medium in combination with a polyglycol diepoxide (DER 732) by Lockwood (1964). DER 332 is heated to 60 °C to reduce its viscosity prior to mixing with other components. As the proportion of DER 732 is reduced, the hardness of the block is increased.

DER 334 was introduced by Winborn (1964). It belongs to the bisphenol-A/epichlorohydrin group of epoxy resins. DER 334 has an epoxide equivalent of 178–186 and a viscosity of 500–700 cP at 25 °C. It is readily miscible with propylene oxide, acetone, styrene and xylene, but does not mix with ethyl or methyl alcohols. Tissues dehydrated with alcohols, therefore, must be treated with a transitional solvent such as propylene oxide prior to infiltration with DER 334. The liquid resin can be stored for months in a refrigerator without any apparent damage.

Sectioning Properties

The degree and nature of cross-linking during polymerisation, as well as the hardness of the final block, determine the sectioning properties. A linear, shorter polymer containing a few cross-links is easy to section. The sectioning quality is dependent partly upon the ratio between anhydride chemical equivalent and epoxy chemical equivalent in the embedding mixture. The optimum ratio between the two can be obtained only if the exact content of epoxide in the epoxy batch is known. The epoxide equivalent (weight of Epon in grams containing 1 gram-equivalent of epoxy) of commercial Epon 812 varies from 140 to 160. Realising the complications which may result from the use of epoxy resins having variable epoxide equivalents, some commercial firms have standardised the epoxy equivalent and have such information printed on each container of the epoxy.

Sectioning is easier when the anhydride:epoxy ratio in the resin is low. Available data indicate that the cutting quality of a block of Epon or its substitutes can be improved by lowering the anhydride:epoxy ratio from 1.0 to 0.6. A reduction in the anhydride results in a shorter polymer with fewer cross-links. Since commercial batches of Epon display variations in anhydride:epoxy ratios, Burke and Geiselman (1971) calculated the proportions of anhydrides and Epon (with known epoxide equivalent), to obtain the mixtures with optimal cutting quality, and have presented them in the form of a table. This table provides accurate proportions of epoxy resin, dodecenyl succinic anhydride (DDSA) and nadic methyl anhydride (NMA) to use with any particular batch of resin with a known epoxide equivalent or weight per unit of epoxy value.

The cutting quality of epoxy resins can be improved by adding Dow Corning 200 fluid silicone plastics additive (0.65 cS) to the resin mixture before the addition of the accelerator (Langenberg, 1982). About 1% silicone plastics is added while stirring. With this procedure, glass knives last longer, since the blocks are relatively soft. One disadvantage of using soft blocks is additional compression of the sections.

Polyester Resins

Polyester resins, in their monomer form, exhibit a wide range of miscibility with other substances. They are characterised by low shrinkage, absence of bubble formation and freedom from specimen damage during polymerisation. These resins are degraded very little under electron bombardment, because of their three-dimensional cross-linked structures. Polymers of polyester resins are insoluble in most organic solvents. On the other hand, although methacrylate resins exhibit a wide range of miscibility, they are neither thermostable nor insoluble in organic solvents. Unlike epoxy resins, polyester resins are polymerised by light, heat or oxygen, and therefore should be kept in a refrigerator and well protected from light. Frozen mixtures of these resins can be kept very well for several months. Polymerisation of these resins is uneven in the presence of air. Vestopal W and Rigolac are the two most commonly used polyester resins. The other polyester resin embedding media are Selectron (Low and Clevenger, 1962), Rhodester (Argagnon and Enjalbert, 1964) and Beetle (Rampley and Morris, 1972). They are not in common use.

VESTOPAL W

Ryter and Kellenberger (1958) first demonstrated that Vestopal W is superior to methacrylate resins and the polyester resins previously used. Vestopal W is produced by the esterification of malic anhydride with glycerol or some other polyhydric alcohol. This resin has a fine grain and the sections are easy to stain. Like methacrylate, Vestopal is characterised by rapid penetration and polymerisation. The blocks are harder than most epoxy resin blocks. Vestopal differs from methacrylates in three important aspects: (1) it hardens without apparent shrinkage, (2) it does not exhibit uneven polymerisation, and (3) its sections are stable under electron bombardment, although ~20% of the mass of the section is easily lost under normal operating conditions. Although Vestopal does decompose to some extent under electron bombardment, it sublimes in such a fashion that no liquid phase is developed. The fine structure of the tissue specimen is not distorted, because there is little chance for surface tension to act. The slight decomposition reduces the density, which results in an improved image contrast. Vestopal has a low intrinsic background structure and provides high contrast. These characteristics make this embedding medium useful for high-resolution studies.

Vestopal, benzoyl peroxide and cobalt naphthenate should be kept in a cold and dark place in order to prevent their decomposition. Vestopal can be stored for many months at 4 °C without any apparent deterioration, whereas the initiator and the activator do not last longer that 1 or 2 months, even when stored at 4 °C. It is, therefore, recommended that the initiator and the activator be replaced every few months. Catalysed Vestopal mixture keeps well for several months in a refrigerator, but keeps only a few hours at room temperature. It is preferable to make up the final mixture before each embedding. The following embedding mixture is recommended:

Vestopal W	100 ml
Benzoyl peroxide	1 ml
Cobalt naphthenate	0.5 ml

The initiator should be well mixed with the Vestopal before the addition of the activator. It takes ~40–60 min to achieve a complete homogenisation of the mixture. The initiator and the activator must not be mixed with each other alone, for they may explode. Vestopal is miscible with acetone and styrene but not with ethanol. The tissues, therefore, are dehydrated, usually with acetone. The procedure for dehydration and embedding is the same as for Epon.

Vestopal with added benzoin as a catalyst can be polymerised at room temperature by ultraviolet irradiation. Polymerisation can also be accomplished by heat with benzoyl peroxide instead of cobalt naphthenate. Sectioning quality of the block can be improved by adding 10% triallyl cyanurate.

The following is a revised formulation presented by Kellenberger *et al.* (1980):

Vestopal 120 (formerly H)	17.5 parts
Vestopal 310 (formerly W)	82.5 parts
Benzoin methyl ether (initiator)	0.6%

The resin mixture is polymerised by UV irradiation (360 nm) (2×Philips TLAD 15W/05 or equivalent) at a distance of 20–30 cm at −20 °C to 25 °C for 1–2 days. A higher proportion of Vestopal 120 produces harder blocks. The embedding mixture is available from Martin Jaeger, Chemin de Nancy 16, CH-1222 Vésenaz, Geneva.

RIGOLAC

Kushida (1960) introduced another polyester embedding medium by combining Rigolac 2004 and Rigolac 70F. This mixture is supposed to be better than Vestopal W in that it has a relatively low viscosity, and thus penetrates readily into tissue; also, the hardness of the polymerised blocks can easily be adjusted by altering the ratio between the two components to suit a specific type of tissue. A relatively long polymerisation time (pot life) is recommended to ensure complete penetration of the Rigolac mixture. In order to prolong the polymerisation time, no accelerator is employed. Polymerisation is accomplished by the addition of benzoyl peroxide paste (50% benzoyl peroxide in tricresyl phosphate), which acts as a catalyst. Polymerisation time is dependent primarily upon the concentration of catalyst used and the temperature of polymerisation. Polymerisation can be accomplished either by heat or by UV irradiation.

Rigolac mixture is insoluble in ethyl alcohol but is readily soluble in acetone, which is, therefore, used as the dehydrating agent. The following mixture is recommended:

Rigolac 2004	75 ml
Rigolac 70F	25 ml
Benzoyl peroxide paste	1 g

The final hardness of the block has been found suitable for sectioning over the range of ratios of Rigolac 2004:Rigolac 70F from 4:1 to 3:2. The procedure for dehydration and embedding is the same as for Epon. Polymerisation is accomplished in 24 h at 55 °C.

STYRENE

Monomeric styrene has an extremely low viscosity (0.7 cP) and thus penetrates rapidly into the tissue. In fact, infiltration and embedding can be carried out without diluting it with solvents. It is soluble in ethanol or acetone. Styrene can be used as a transitional solvent before embedding in epoxy or polyester resins. It can also be used in combination with Rigolac (Shinagawa and Uchida, 1961) or methacrylates (Kushida, 1961a) as an embedding mixture. Styrene is polymerised by light, heat and oxygen. It is supplied with an inhibitor, which is not removed. Polymerisation is accomplished either by UV irradiation (wavelength 340–400 nm) or heat at 60 °C. Polymerisation ceases on removal from the heat or light, and thus it is controlled by the duration of the exposure. Because styrene is volatile, polymerisation should be carried out in closed capsules. The addition of butyl methacrylate, a copolymer, results in decreased volatility and improved plasticity; 5–10% butyl methacrylate in the final mixture is recommended. Monomeric styrene has the disadvantage that the polymerised block suffers a shrinkage of ~17%; however, tertiary butyl styrene suffers only 7% shrinkage.

Embedding Formulations

Embedding mixture 1 (DeLamater *et al.*, 1971)

Monomeric styrene	10 g
Methyl ethyl ketone (activator)	0.4 g
Dibutyl phthalate (plasticiser)	0.15 g

Polymerisation can be accomplished by UV irradiation or exposure to 60 °C for 2–3 days.

Embedding mixture 2 (Kushida, 1962b)

Monomeric styrene	3 parts
n-Butyl methacrylate	7 parts
Benzoyl peroxide (catalyst)	1%

Polymerisation can be accomplished by UV irradiation or exposure to 55 °C for 24 h.

Methacrylates

Methacrylates are colourless and transparent and possess the compactness necessary for embedment of biological materials. Methacrylate monomers are completely miscible with ethanol and acetone and with most other organic solvents, and do not penetrate the tissue specimens until the latter have been completely dehydrated. The resins are relatively less viscous and thus penetrate easily into the dehydrated specimens. The polymerisation of methacrylates is linear, in contrast to the three-dimensional characteristic of epoxy resins. However, the structure of methyl methacrylate can be altered to three-dimensional by adding divinylbenzene as a cross-linking agent (Kushida, 1961).

Methacrylic resins polymerise spontaneously if left for a sufficient length of time; impurities or the specimen can initiate polymerisation. Polymerisation is accelerated in bright light and warm surroundings. Therefore, these resins come mixed with a hydroquinone inhibitor which prevents polymerisation during shipping and storage. Lower temperatures and absence of light appear to retard spontaneous polymerisation. Since the inhibitor does not have any adverse effect on the final properties of the block, it is not absolutely necessary to remove it before polymerisation. The inhibitor influences only the rate of polymerisation; this can be counteracted by employing a larger quantity of the catalyst, such as benzoyl peroxide or 1,2-dichlorobenzoyl peroxide. Usually either catalyst is added to constitute 2% of the total embedding mixture. If needed, the inhibitor can be removed by washing the monomer resin with 20% solution of NaOH in distilled water.

Methacrylates have serious flaws as embedding media. They are characterised by uneven polymerisation accompanied by too much shrinkage (~20%) and gas bubbles. The hardened resin also lacks stability under electron bombardment, and as much as 50% of its mass may be lost upon electron irradiation under normal operating conditions (Reimer, 1959; Cosslett, 1960). This loss is followed by a flow of the remaining resin, which results in the distortion of the macromolecular structure of the tissue. Thus, specimen degradation occurs both during polymerisation and under the electron beam.

More uniform polymerisation of methacrylates can be obtained by adding 0.01% uranyl nitrate (Ward, 1958), which acts as a local catalyst. Probably a similar effect is derived from osmium deposited in the tissue during fixation with OsO_4. Polymerisation with UV irradiation seems to produce better results than those yielded by heat. Kushida (1962b) has devised a procedure for polymerisation by UV irradiation at room temperature using 0.3% uranyl nitrate as a catalyst. The use of tertiary butyl alcohol as a plasticiser and benzoyl peroxide as a catalyst in the proportion of 10–12% and 1%, respectively, of the total embedding mixture has been reported to result in the elimination of gas bubble formation, more even polymerisation and improved sectioning quality (Kalucheva *et al.*, 1966).

The shrinkage damage can be reduced by using partially pre-polymerised resin. Elimination of oxygen from the gelatin capsule during the final polymerisation results in a block with little shrinkage. This is accomplished by embedding the infiltrated specimen in a few drops of pre-polymerised resin between two solid resin cylinders in a gelatin capsule (Kölbel, 1970). Pre-polymerisation involves heating the monomer and thus inducing much of the shrinkage before embedding the tissue. A drawback of heating is increased viscosity, resulting in slower penetration. Although the above procedures help to reduce the damage to tissue embedded with methacrylates, it is not possible to eliminate distortion of the fine structure. In conventional embedding, methacrylates do not yield satisfactory results and thus have largely been replaced with epoxy resins. Water-miscible methacrylates are discussed later.

METHYL METHACRYLATE

Methyl methacrylate is a bifunctional monomer and does not react readily with unsaturated polyester resins, but seems to copolymerise easily with styrene (Rampley, 1967). Methyl methacrylate has some advantages in embedding for phase contrast microscopy. Sections of relatively large size can be easily cut with glass or diamond knives. Sections of this resin stain readily with ordinary dyes. Sections produce good initial contrast, but further electron bombardment causes the membranes to collapse.

The desired hardness of the block can be obtained by adjusting the relative proportion of n-butyl and methyl methacrylates; a higher proportion of methyl methacrylate will produce harder blocks. Relatively soft blocks are best suited for soft tissues such as pancreas, while tough tissues such as tendon and muscle require hard blocks. An embedding mixture of adequate hardness for many tissues can be obtained by mixing n-butyl and methyl methacrylate in the ratio of 4:1.

Methyl methacrylate is commonly used for embedding mineralised or unmineralised tissues for light microscopy. It is usually polymerised at room temperature or higher temperatures. Under these conditions, heat evolution may reach 92 °C. However, it can be polymerised at low temperatures. Chappard *et al.* (1983) polymerised a 3:2 mixture of methyl methacrylate and glycol methacrylate (GMA) at 4 °C without UV irradiation. This method is time-consuming, because it requires the removal of methacrylic acid and hydroquinone from the GMA and of hydroquinone from the methyl methacrylate. Liu (1987) has succeeded in polymerising methyl methacrylate at 5 °C without purifying the two resins. Polymerisation is completed even when methyl methacrylate comprises as much as 80% of the embedding mixture. In this method, polymerisation at a low temperature is achieved by changing the concentration of the catalyst in the embedding mixture and by adding the initiator.

Embedding Formulation

Methyl methacrylate (25 ppm hydroquinone)	85 g
JB-4 solution A*	5 g
Dibutyl phthalate	10 g
Polyethylene glycol 1540 distearate	2 g
Benzoyl peroxide	600 mg

Fixed and dehydrated bone specimens are infiltrated for 7–10 days in three changes of the above mixture; the air from the bone is removed under vacuum to facilitate infiltration (Liu, 1987). To initiate polymerisation, 10 g of the embedding mixture is mixed with 0.25 g of JB-4 solution B* in a 20 ml glass scintillation vial. The infiltrated bone specimen is placed in the vial, the air inside the vial is flushed out with a gentle stream of nitrogen for 10–20 s and the vial is sealed immediately. The vial is placed in a chilled water-bath, and polymerisation is completed in 24–32 h at 5 °C.

Poly-*N*-vinylcarbazole

Poly-*N*-vinylcarbazole (PVK) is an embedding resin for studying thin sections (50 nm) specifically with the photoemission electron microscope (Grund *et al.*, 1982). The advantages of this resin are an extended usable viewing period in the microscope and imaging of biological structures without staining with heavy metals. The resin has a carbazole ring structure and a glass temperature of >200 °C, and therefore a high stability in the microscope. Another advantage is rapid emission without pre-irradiation, which allows observation immediately after the onset of irradiation. The viscosity of PVK is lower than that of Epon. The former has not, as yet, been used for electron microscopy.

Fixed specimens are dehydrated in a 70:30 mixture of Cellosolve and ethylene glycol for 30 min and then in a 90:10 mixture of Cellosolve and ethylene glycol for 30 min. The specimens are treated four times with Cellosolve for a total period of 2 h, followed by two treatments with propylene oxide for a total of 30 min. The resin is melted at 80 °C, and 6% dibutyl phthalate is added as plasticiser and 1% benzoyl peroxide as catalyst. Complete mixing of these ingredients is achieved with a heated magnetic stirrer. The specimens are infiltrated with a 1:1 mixture of PVK and propylene oxide at 60 °C for 1 h, and then transferred to the embedding mixture for 1 h. Polymerisation is carried out at 60 °C for 6 days or at 70 °C for 2 days.

WATER-MISCIBLE EMBEDDING MEDIA

The ideal preparation of specimens for cytochemical studies requires a minimum loss of biochemical activity. Unfortunately, most of the embedding media commonly used to prepare thin sections have a limitation in that they are not miscible with water. Consequently, the specimen must be exposed to strong dehydration agents prior to infiltration and embedding. A long and rather severe dehydration treatment destroys, alters or relo-

*JB-4 solution A contains 80 g of GMA, 16 g of 2-butoxyethanol and 0.27 g of benzoyl peroxide.

* JB-4 solution B contains 15 g of polyethylene glycol 400 and 1 g of *N*,*N*-dimethylaniline.

cates the enzyme activity and antigenicity in the specimens, resulting in misleading final interpretations. The extent of the biochemical changes in specimens caused by the dehydration agents is at present largely unknown. Another serious disadvantage is that these embedding media are efficient lipid solvents. In fact, at present there is no satisfactory embedding medium available for retaining lipids. Still another limitation of using water-immiscible embedding media is that the sections cannot be subjected to treatments which contain water.

The problem of the leaching of cellular constituents due to dehydration with organic solvents becomes a serious obstacle in investigations in the area of ultrastructural cytochemistry, including localisation of enzymatic activity and antigenicity. However, some of the limitations imposed upon this area of study can be avoided by the use of water-miscible embedding media (e.g. Lowicryl K4M) of low solvent power. Specimens can be dehydrated and embedded in these media without employing organic solvents such as ethanol or acetone. These resins can be polymerised in the presence of a high percentage of water, and polymerisation can be accomplished by UV irradiation at low temperatures. Thus, specimens are not exposed to high temperatures, which destroy many cellular activities. Proteins are denatured at temperatures above 60 °C (333 K), and the denaturation is accelerated in the presence of organic solvents. Another advantage of water-miscible embedding media is that polymerisation can be accomplished over a wide range of temperatures.

Since most of the water-miscible resins are not cross-linked, they are easily permeated by cytochemical reagents (e.g. stains and digestive enzymes), which, in turn, find easy access to the embedded specimen. The relatively low viscosity of these resins permits easy infiltration into the specimen at low temperatures. In addition, an easy correlation between light microscopy and electron microscopy is possible, owing to the ease in cutting and examining of thick sections of these resins. Nevertheless, it should be noted that, despite their water miscibility, the monomers are effective organic solvents; at least, some nucleoproteins and lipids are definitely extracted during infiltration and embedding in water-miscible monomers.

An ideal embedding medium for ultrastructural cytochemical studies should have at least these characteristics: (1) it should be completely miscible with water in all proportions, to avoid enzyme inactivation by organic solvents; (2) it should have a low melting point, to avoid enzyme inactivation due to high temperature; (3) it should be chemically inert so as not to inactivate enzymes or antigenicity, or dissolve cellular materials; (4) it should allow the sections to remain specifically reactive to certain cytochemical stains; (5) it should be suitable for treatment with enzymes without being removed; (6) it should differ negligibly in volume as between fluid and solid state, to avoid cellular shrinkage or swelling; (7) it should produce both high quality and consistency in the preservation of ultrastructure; (8) it should polymerise at low temperatures; and (9) it should possess desirable cutting properties.

During the past four decades, efforts have been made to develop suitable water-miscible embedding media. Although in recent years progress has been made, a completely reliable and satisfactory medium is yet to be found. Better-known water-miscible embedding media recommended in recent years are discussed below.

Acrylamide–Gelatin–Jung Resin

A mixture of polymerised acrylamide, gelatin and Jung resin (Jung, Heidelberg) was introduced by Hartmann (1984) for cryosectioning of eggs (snails', fishes' and insects') without rupture of the yolk- and lipid-containing areas. This procedure improves preservation of the ultrastructure and sectioning of large eggs for both light and electron microscopy.

A stock solution is prepared as follows:

Acrylamide (3.938 M) containing 0.048 M bis(*N*,*N*′-methylenebisacrylamide)	20 ml
Sörensen phosphate buffer (0.2 M, pH 7.1)	10 ml
Gelatin (16%)	35 ml
Jung embedding resin	20 ml
N,*N*,*N*′,*N*′-Tetramethylenediamine (TMED)	0.2 ml

This solution is continuously mixed with a magnetic stirrer in a water-bath at 30 °C before use to prevent phase separation. The solution can be stored in the refrigerator for at least 8 weeks. Double-distilled water should be used to prepare all aqueous solutions.

The specimen is placed in 8.5 ml of the well-mixed stock solution, which is overlayered in sequence with 0.5 ml of 2% ammonium peroxodisulphate and 0.1 ml of fixative (a mixture of 10% formaldehyde and 6.25% glutaraldehyde in 0.1 M phosphate buffer at pH 7.1). All components are rapidly mixed and then poured into a paraffin oil-coated plastics box. The specimen is oriented and kept in position until the mixture solidifies within 30–60 s. The block with embedded specimen has a rubber-like consistency. It is rapidly frozen in isopentane cooled by liquid nitrogen. Using a cryotome at −18 °C to −22 °C, 5–20 μm thick sections are cut, which are mounted on a glue-coated glass slide. The glue contains 0.8% agarose, 1.8% glycerol and 0.5% glutaraldehyde (or 0.75% formaldehyde). The slides are coated by vertical dipping in the glue at 80 °C and stored in a dust-free box until use.

Post-fixation is accomplished by exposing the sections mounted on glue-coated slides to vapours of an aldehyde in a moist chamber at 30 °C for 5 min, followed by immersion in 4% formaldehyde in phosphate-buffered solution (PBS) or 2% glutaraldehyde in PBS for 30 min. Before staining, the aldehyde is removed by extensive washings in PBS for 2 days. Staining is carried out either with haematoxylin–eosin or with 2–4% OsO_4 in 0.1 M phosphate buffer (pH 7.2), followed by counterstaining with Feulgen.

Durcupan

Durcupan is a water-miscible epoxy resin which can also be used as a dehydrating agent. It is a colourless resin of relatively low viscosity. It probably has the capacity to act as a fixing agent. Since Durcupan is miscible with other epoxy resins and polyester resins, dehydration with this resin can be followed by embedding in other resins. Contact with skin should be avoided, because Durcupan is a strong irritant.

Because of the availability of superior water-miscible embedding media, Durcupan is not in common use. The only use of this resin is in the demonstration of lipids in normal or diseased tissues (Maxwell, 1975). The size of the specimen should be very small, because Durcupan penetrates rather slowly. Specimens should be agitated throughout their processing. Thin sections should be floated on 10% ethanol. Durcupan sections show a fine granularity at high magnifications, resulting in a loss of resolution. This disadvantage is inherent in many water-miscible embedding media. The quality of overall cellular preservation is lower than that obtained with Lowicryl K4M and LR White.

Although Durcupan is completely miscible with water, it must be completely dehydrated before the hardener and accelerator are added in order for polymerisation to take place. In embedding, therefore, tissues are progressively infiltrated by increasing concentrations of the resin until all moisture has been replaced. Thus, tissues are dehydrated without being exposed to organic solvents. In this connection, it should be noted that even this type of dehydration not only causes some alterations in the tissue, but also extracts some soluble materials.

EMBEDDING FORMULATIONS

Durcupan	5 g
DDSA (hardener)	11.7 g
Dibutyl phthalate (plasticiser)	0.3 g
DMP-30 (accelerator)	1.1 g

The ingredients should be mixed thoroughly for at least 1 h.

Durcupan frequently fails to harden to a sufficient degree after embedding, even without the use of a plasticiser. As a result, thin sections of such blocks tend to develop fine-order chatter. The use of Durcupan as a sole resin is not recommended. To produce a block with satisfactory cutting qualities, Kushida (1964) recommended the following embedding formulation.

Mixture A	
Durcupan	100 g
DDSA	234 g
Mixture B	
Durcupan	100 g
NMA	124 g

Mixture A gives a soft block, whereas mixture B alone is very hard. Thus, the hardness of the block depends upon the ratio between mixture A and mixture B. The most suitable mixture ranges in ratio from 8:2 to 7:3 at 20 °C. The following dehydration, infiltration and embedding procedures carried out at room temperature are recommended. All steps are carried out on a shaker.

(1) 50% Durcupan in distilled water	15–30 min
(2) 70% Durcupan in distilled water	15–30 min
(3) 90% Durcupan in distilled water	15–30 min
(4) 100% Durcupan in distilled water	30–60 min
(5) 100% Durcupan in distilled water	30–60 min
(6) 100% Durcupan with Durcupan mixture (3:1)	1 h
(7) 100% Durcupan with Durcupan mixture (1:1)	1 h
(8) 100% Durcupan with Durcupan mixture (1:3)	1 h
(9) Durcupan mixture with DMP-30	1–2 h
(10) Durcupan mixture with DMP-30	1–2 h

The impregnated tissue is then transferred to fresh Durcupan mixture with accelerator in gelatin capsules and polymerised at 50 °C for ~50 h. Sections are strong enough under electron bombardment so that grids without supporting films can be used.

Glycol Methacrylate

Glycol methacrylate (GMA; 2-hydroxyethyl methacrylate; $Ch_2{=}C(CH_3)COOH_2CH_2OH$) is a colourless liquid with a viscosity of 50.8 cP at 25 °C. It is polar and completely miscible in its monomeric form with water, ether and alcohols, while its polymer is practically insoluble in all common organic solvents and water. Complete miscibility with water is a reliable test to determine whether this resin has begun to polymerise spontaneously. It slowly polymerises on storage.

Although GMA was introduced by Rosenberg *et al.* (1960) for electron microscopy, at present it is widely used for light microscopy. Satisfactory morphological detail and reproducible sections in the range 0.5–4.0 μm are easily obtainable with steel, glass, tungsten, carbide, sapphire or diamond knives. Glycol methacrylate in combination with methyl or butyl methacrylate has proved ideal for embedding undecalcified bone for light microscopy (Cole, 1982; Gruber *et al.*, 1985; Hott and Marie, 1987). It has proved particularly useful for localising enzymatic activity and labile tissue antigens with the light microscope. This resin has also been used for embedding soft tissues for immunocytochemistry with the electron microscope. This is possible because organic solvents such as acetone and ethanol which denature proteins need not be used; GMA can be used to dehydrate tissues. However, GMA is a lipid solvent. In some studies acetone can be used for dehydration with excellent results.

Polymerisation of GMA can be accomplished at a low temperature with UV irradiation. Lipid extraction is diminished when embedding is carried out at low temperatures. Polymerisation by UV irradiation may adversely affect the activity of certain enzymes. Since polymerisation is achieved without cross-linking, the embedding resin and the specimen remain permeable to antibodies. Because enzyme substrates have molecular weights <500 dalton, they easily penetrate the resin matrix, even when it is cross-linked. Since GMA contains water, complete dehydration is unnecessary. It allows easy permeation by water-soluble dyes.

Unlike paraffin, as stated above, GMA can be used as its own dehydrating agent, even at low temperatures. Paraffin methods involve heating the tissue to the melting point of Paraplast (60–63 °C, 333–336 K) for a number of hours to accomplish infiltration. Glycol methacrylate sections can be stained without removing the embedding matrix, thus avoiding clearing agents such as xylene which are used for deparaffinising the sections. Superior preservation of tissue and cell detail is guaranteed with GMA.

Commercial GMA contains hydroquinone and its derivatives varying in concentration from 200 to 2000 ppm; concentrations of hydroquinone higher than 400 ppm are not recommended. Hydroquinone stabilises commercial GMA against spontaneous polymerisation. The inhibitor combines with atmospheric oxygen to yield quinone, which, in turn, can trap free radicals and inhibit autopolymerisation of the monomer.

The undesirable effect of hydroquinone is that it blocks the production of free radicals from benzoyl peroxide (BPO) when the BPO tertiary amine (initiator–accelerator) system is used (Gerrits and van Leeuwen, 1985). Thus, BPO radicals lose their ability to initiate the polymerisation. If the hydroquinone has not been removed from the monomer, BPO will have to be used in excess to initiate the polymerisation. High proportions of BPO may lead to evolution of exothermic heat higher than 100 °C (373 K) within the polymerising block. Such heat accelerates the rate of polymerisation, which results in uncontrolled hardening. Such high temperatures can also denature proteins.

During polymerisation with BPO, hydroquinone reacts to give brown phenolates that are trapped in the polymer, producing brown blocks (Chappard *et al.*, 1986). Such blocks are difficult to section. Apparently, for the above reasons, the inhibitor hydroquinone needs to be removed from GMA monomer before the polymerisation is intiated. However, the blocks of Historesin and Technovit 7100 keep clear.

Another impurity that needs to be removed from the monomer is methacrylic acid (MA). Glycol methacrylate is prepared by esterification of MA, but ~1% of free, unreacted MA remains in the monomer. Methacrylic acid can copolymerise with GMA, producing a polymer with free carboxylic groups which are ionised above pH 4 and interact with cationic dyes to yield background staining in the sections (Gerrits and van Leeuwen, 1987). The pH of 10% aqueous solution of GMA can be measured as described by Cole (1984). Although low-acid GMA is commercially available (SPI Inc., West Chester, PA), Frater (1981) and Chappard *et al.* (1986) have presented simple methods for removing MA.

Purification of diluted GMA with NaOH washes and subsequent rotative evaporation of the solvent removes all the MA and most of the hydroquinone (Chappard *et al.*, 1986). Exactly 1000 ml of crude GMA is diluted with 1000 ml of $CHCl_3$ in a separatory funnel. To this solution is added 250 ml of 0.5 N NaOH and the mixture is shaken vigorously. The mixture is allowed to separate into two phases, and the brownish aqueous phase is discarded. The above addition of NaOH and removal of the brownish phase are repeated five times. The $CHCl_3$ is removed with a rotary evaporator. Purified GMA having a light tint is stored at −20 °C (253 K) to prevent autolysis of the monomer as well as progressive increase in the MA content.

In addition to hydroquinone and MA, other impurities present in commercial GMA include ethylene glycol and ethylene dimethacrylate. By using appropriate proportions of the initiator (BPO), the tertiary amine accelerator (*N,N*-dimethylaniline) and GMA, an adequate polymerisation can be achieved without removing the impurities, including hydroquinone.

Excessive concentrations of *N,N*-dimethylaniline (>35 mmol) accelerate the decomposition of BPO undesirably, resulting in extremely rapid polymerisation and formation of gas bubbles in the resin block (Gerrits and van Leeuwen, 1985). Too low concentrations of *N,N*-dimethylaniline may cause incomplete polymerisation and tacky resin blocks. These effects may be due to the formation of excess benzoic acid. In chemically induced polymerisation, benzoic acid is formed as the reaction proceeds and may suppress the reacting lone electron pair of the nitrogen atom in *N,N*-dimethylaniline (Imoto and Choe, 1955). Ruddell (1971) has summarised the mechanism of acceleration of the BPO-initiated polymerisation of GMA by means of *N,N*-dimethylaniline at room temperature.

As stated earlier, the rate of polymerisation also depends on the efficiency of BPO. Higher concentrations of BPO are required during tissue embedding than the minimum concentration of BPO needed to polymerise pure GMA (Gerrits and van Leeuwen, 1985). Too high a concentration of BPO may lead to an evolution of exothermic heat higher than 100 °C (373 K).

Since polymerisation of GMA is exothermic (50 kJ/ml), excessive heat may accumulate in the centre of the polymerising resin. The evolution of exothermic heat at room temperature (20 °C, 193 K) may exceed 100 °C (373 K). Such increase in temperature can be controlled by using heat-conducting aluminium block holders (Gerrits and van Leeuwen, 1985). To prevent inhibition of polymerisation by atmospheric oxygen, embedding moulds are placed in an atmosphere of nitrogen or a vacuum chamber. Alternatively, a paraffin film or paraffin oil can be used to prevent air contact (Chappard *et al.*, 1987). Both the maximum temperature (T_{max}) during polymerisation and the time of the maximum temperature ($t_{T_{max}}$) are influenced by the concentration of hydroquinone in the GMA monomer.

LIMITATIONS OF GLYCOL METHACRYLATE

Glycol methacrylate is plagued with several inherent problems that become unacceptable at the electron microscopy level. The resin has the tendency to swell in many solvents such as ethanol and glycol, and particularly in water or aqueous solutions of electrolytes. The degree of swelling can be either increased or decreased by controlling the type and concentration of electrolytes. Polymerisation damage introduces swelling artefacts in the specimen. This problem is greatly reduced by using a partially prepolymerised embedding mixture (Cole and Sykes, 1974). A second problem is that the resin is not stable under the electron beam, and specimen sections break when exposed to the beam. The use of support films on the grids somewhat diminishes this problem. A third problem is that the block is often excessively brittle and thus difficult to trim and section. This problem can be solved for light microscopy by adding a plasticiser such as polyethylene glycol 400 (PEG 400) (Spaur and Moriarty, 1977). External plasticisers such as PEG, however, evaporate under the electron beam. Acrylates are recommended for electron microscopy, for they are internal plasticisers, which copolymerise. Depolymerisation of GMA under the electron beam can be minimised by adding a cross-linking agent (ethylene glycol dimethacrylate). The addition of cross-linkers (ethylene glycol dimethacrylate or triethylene glycol dimethacrylate) to the embedding mixture in a final concentration of 2% also eliminates the problem of wrinkles or minifolds occasionally present in the GMA sections (Gerrits and van Leeuwen, 1987).

Sections of GMA tend to stretch on the water surface in the cutting trough. Depending on the composition of the embedding mixture, sections show a linear stretch of 10–15% after stretching and mounting. These and other effects are responsible for the formation of certain artefacts in the tissue embedded in GMA. For example, cells may pull apart, nuclear outline may be distorted and glycogen may not be preserved. During processing, tissues such as liver undergo linear shrinkage of ~9% during dehydration (mostly in 96% ethanol) (Hanstede and Gerrits, 1983). Infiltration with GMA causes a linear swelling of 2–5%, and polymerisation results in a shrinkage of 1–2%. Nevertheless, the occurrence of these artefacts is not a very serious hindrance in using GMA for light microscopy. A definitive study of all aspects of embedding with GMA by Gerrits (1987) indicates that the best results are obtained when GMA mixtures of precise composition are prepared in one's own laboratory.

SERIAL SECTIONS

A drawback of some of the GMA formulations is their inability to produce ribbons. Kulzer & Co. CmbH (P.O. Box 1320, D-6382 Friedrichsdorf) introduced a GMA-based embedding mixture that can form ribbons. This embedding kit is sold in North America under the trade name LKB Historesin and in the UK under the trade name Reichert 2040. It uses a non-toxic catalyst system, avoiding the toxicity of the tertiary amine accelerator (*N,N*-dimethylaniline) (Gerrits and Smid, 1983). The addition of the plasticiser PEG 400 facilitates the formation of ribbons. The presence of barbituric acid in the polymerisation system also helps ribbon formation (Gerrits, 1987).

When using Ralph knives and LKB Historange, the optimal PEG concentration is 0.2 ml per 5 ml of the LKB embedding mixture (Yeung and Law, 1987). The PEG concentration is increased to 0.25–0.35 ml per 5 ml of the embedding mixture when the AO rotary microtome is used with a steel knife. The latter results in a slightly softer block which is less damaging to the knife.

As stated above, although a conventional AO microtome can be used to produce ribbons, they form more easily on the LKB Historange (or Reichert 2040), which has a retractable specimen arm. As the specimen holder retracts during the return stroke, it prevents the block face from dragging across the knife edge and lifting the sections away from the knife (Yeung and Law, 1987). Ribbon formation is helped by adding a thin layer of dental wax to the upper and lower surfaces of the block. The upper and lower edges of the block face should be parallel to each other. Since sections do not adhere to each other when their edges are not entirely resin, trimming into the tissue should be avoided. The knife angle should be 7–12°. A tungsten carbide knife is superior to a glass knife, but the latter is better than a steel knife.

EMBEDDING FORMULATIONS

Embedding mixture 1 (Gerrits *et al.*, 1987)

Solution A	
Glycol methacrylate (200 ppm hydroquinone monomethyl ether) (monomer)	90 g
2-Butoxyethanol (plasticiser)	10 g
Benzoyl peroxide (moistened with 20% water) (initiator)	0.5 g
Solution B	
Polyethylene glycol 400 or 200	15 parts
N,N-Dimethylaniline (accelerator)	1 part

On being added, the accelerator is stirred for 1 min with a magnetic stirrer. Polyethylene glycol functions as a carrier for the accelerator and softens the resin block. Polyethylene glycol 200 is easier to handle at low temperatures (<4 °C) compared with PEG 400; the latter solidifies at 0–4 °C.

Tissue specimens fixed with formaldehyde and dehydrated with ethanol are infiltrated with 1:1 mixture of 100% ethanol and solution A for 2 h and then again with solution A for 18 h. The specimens are transferred into polyethylene-based moulds (Sorvall type, 12 × 16 × 15 mm, Du Pont Instruments, Newton, CT) fitted with aluminium block holders and containing 2.5 ml of a 30:1 mixture of solutions A and B. The moulds are placed in an atmosphere of nitrogen at 20 °C (293 K) to prevent inhibition of the polymerisation by atmospheric oxygen (Gerrits and van Leeuwen, 1985). Polymerisation is complete in 2 h.

Sections (2 μm) are cut under a constant humidity of 60% within a week after polymerisation at room temperature (Gerrits *et al.*, 1987). Stretching of the sections is carried out on water at room temperature, and then they are mounted on glass slides and dried at 20 °C (293 K) or 60 °C (333 K) on a hot plate; the former temperature is preferred. Too hard a block can be softened in a humid chamber, while too soft a block can be hardened by placing it in a desiccator or under UV light. Glycol methacrylate can cause allergic skin reactions, *N,N*-dimethylaniline is toxic, and benzoyl peroxide in dry form is highly explosive and by skin contact causes dermal irritation.

Embedding mixture 2 (Ruddell, 1983)

Ruddell (1983) has recommended the use of cyclic diketone carbon acids for catalysing the polymerisation of GMA. Such mixtures seem to polymerise without apparent exothermicity at temperatures ranging from 20 °C (293 K) to −5 °C (268 K) in the dark without using UV light. According to Ruddell (1983), this mixture is free from artefacts (e.g. air bubbles) associated with GMA–benzoyl peroxide resins. The following formulation is recommended.

Glycol methacrylate	5 ml
1-Pentanol (plasticiser)	0.8 ml
Imidazole	16 ml
Monophenylbutazone	30 mg

Tissues can be embedded in covered gelatin or BEEM capsules.

Embedding mixture 3 (Spaur and Moriarty, 1977)

Glycol methacrylate	66.5 ml
Distilled water	3.5 ml
n-Butyl methacrylate	28.5 ml
Ethylene dimethacrylate	5 ml
Benzoyl peroxide	1.5 g

Embedding mixture 4

Glycol methacrylate	66.5 ml
n-Butyl methacrylate	28.5 ml
Ethylene dimethacrylate	5 ml
Benzoyl peroxide	1.5 g
Polyethylene glycol 400	1 ml

PREPOLYMERISATION PROCEDURE

Prepolymerisation increases the speed of final polymerisation and reduces the formation of swelling artefacts. The embedding mixture is heated to 40–45 °C on a magnetic stirrer hot-plate with continuous stirring until the benzoyl peroxide is dissolved. After removal from the heat, the mixture is transferred to an Erlenmeyer flask with a Teflon-coated magnet. The flask is stoppered with a two-hole stopper and a thermometer is inserted into the stopper at an angle, to permit the movement of the magnet. The thermometer almost reaches to the bottom of the flask. The other hole of the stopper permits ventilation.

The flask is heated at the rate of 7 °C/min with continuous stirring to 98 °C on the hot-plate. The flask is removed and plunged into a dry ice–ethanol bath with rapid swirling to cool the mixture rapidly to ~2 °C. There is a critical temperature above which spontaneous polymerisation occurs, and below which prepolymer is formed. The whole process normally is completed in ~5 min. The prepolymer has the viscosity of a thick syrup at 0–4 °C, and can be stored almost indefinitely in a freezer.

DEHYDRATION AND POLYMERISATION

Glycol methacrylate (80%)	15 min
Glycol methacrylate (100%)	four changes, 15 min each
Embedding mixture	two changes, 1 h each
Prepolymerised mixture	48 h

Specimens are embedded in gelatin (not polyethylene) capsules and filled to the top with the fresh prepolymer mixture. The capsules should be left uncapped for ~30 min to eliminate the air bubbles. They must then be capped with as little air as possible trapped within the capsules; otherwise polymerisation will not occur. Alternatively, before the capsules are capped, they are placed in a vacuum chamber and air bubbles removed for 10 min. Polymerisation is affected by long-wavelength UV light (>315 nm), and is completed in 16–20 h at 4 °C, depending on the viscosity of the prepolymer mixture, the amount of catalyst added and the source of UV radiation. The distance between the bottom of the capsules and the lamp is 10–20 mm.

Hydroxypropyl Methacrylate

Owing to the difficulties encountered with GMA, during the early period of its use, another water-miscible methacrylate, hydroxypropyl methacrylate (HPMA; 2-hydroxypropyl methacrylate), which is derived from a single water-miscible monomer, was introduced for electron microscopy by Leduc and Holt (1965). At present, it is used for light microscopy. In essence, HPMA was selected because, being the next homologue of GMA, it possesses physical properties lying somewhere between those of GMA and butyl methacrylate. Thus, HPMA is adequately miscible with water and possesses better cutting properties than do GMA mixtures. The former is less hydrophilic than the latter, and is incompletely miscible with water (21% water in HPMA). Hydroxypropyl methacrylate penetrates tissues more slowly than does GMA, and thus the techniques involved are comparatively tedious and time-consuming (viscosity: GMA = 5 mPa s, HPMA = 6.2 mPa s at 30 °C).

Commercial HPMA contains 200 ppm hydroquinone monomethyl ether (HQMME), which is an inhibitor of spontaneous polymerisation. The inhibitor can be neutralised by the addition of α,α-azobisisobutyronitrile (azonitrile), which acts as an inhibitor for photopolymerisation or thermal polymerisation. Because HPMA is photosensitive, it should be stored in the dark.

The quality of cellular preservation in HPMA resembles that in the GMA-embedded tissues. Both of these embedding media have been used for embedding soft or hard plant and animal tissues for light microscopy. Like GMA, HPMA is useful for embedding undecalcified bone tissue (Franklin and Martin, 1980). Because bone contains alternating hard and soft tissues, uniform embedding is difficult. Decalcification of the bone usually results in the loss of certain cellular constituents, including enzymatic and antigenic activities. Thus, HPMA serves well for studying the bone tissue.

REMOVAL OF METHACRYLIC ACID

Both HQMME (stabiliser) and methacrylic acid (MA) are present in the HPMA monomer. Hydroquinone can be removed by adsorption onto activated charcoal (Bennett *et al.*, 1976). Even a few per cent of MA present in HPMA can cause background staining when basic dyes are used. Franklin *et al.* (1981) have introduced a method for removing MA from both HPMA and 2-hydroxyethyl methacrylate (HEMA). The monomer is dissolved in ether and then MA is extracted by a weakly basic aqueous solution which forms a second phase. The method is detailed below.

One volume of HPMA monomer is diluted with nine volumes of diethyl ether. Ten volumes of this solution are mixed with one volume of a 5% aqueous solution of sodium bicarbonate ($NaHCO_3$) at 4 °C to extract the acid. This step is repeated twice more; a separatory funnel can be used for this purpose. Residual water is removed by shaking briefly the solution with anhydrous sodium sulphate. The ether is removed by flash evaporation at 25–30 °C, and the final ether residue is removed by briefly raising the temperature to 45 °C.

Hydroxypropyl methacrylate monomer and sodium bicarbonate must be ice-cold. The extraction procedure must be carried out in a cold room, and safety goggles should be worn. The separatory funnel is vented repeatedly during extraction, keeping in mind the explosive nature of the vapours. The flash evaporation must be accomplished in a fume-hood, and the ether is collected with an ice-bath.

EMBEDDING FORMULATIONS

Embedding mixture 1 (Gerrits *et al.*, 1987)

Solution A	
Hydroxypropyl methacrylate (200 ppm HQMME)	90 ml
2-Butoxyethanol (plasticiser)	10 ml
Dibenzoyl peroxide (moistened with 20% water (initiator)	0.5 g
Solution B	
Polyethylene glycol 400 (softener)	15 parts
N,*N*-Dimethylaniline (accelerator)	1 part

The amount of dibenzoyl peroxide depends on the concentration of HQMME.

Embedding mixture 2 (Franklin and Martin, 1980)

Solution A	
Hydroxypropyl methacrylate	90 ml
2-Butoxyethanol	10 ml
Triethyleneglycol dimethacrylate	5 ml
Benzoyl peroxide	0.21 g
Solution B	
Polyethylene glycol 400	30 ml
N,*N*-Dimethylaniline	1 ml

Triethyleneglycol dimethacrylate is a cross-linker and imparts the polymer flexibility combined with structural stability. This reagent also improves the spreading properties of the sections. Dibenzoyl peroxide and *N*,*N*-dimethylaniline accelerate polymerisation. The absence of butoxyethanol results in brittle blocks that shatter on sectioning, while the absence of polyethylene glycol results in very hard blocks.

Fixed and dehydrated specimens are brought into 100% ethanol, and transferred to solution A for infiltration at 4 °C overnight in the dark. The specimens are transferred to a 42:1 mixture of solutions A and B in gelatin capsules overnight at 4 °C in a vacuum desiccator, and then at room temperature in a vacuum for 3–5 h to complete the embedding.

Lemix

Lemix is a water-soluble epoxy resin and is claimed to be stable under the lectron beam; it is of sufficiently low viscosity (70.8 cP) to infiltrate large tissue blocks (5 mm) (EMscope Lab. Ltd). It allows the penetration of a number of aqueous histological stains (e.g. methylene blue–basic fuchsin and toluidine blue). Lemix and JB4 resin (Polysciences, Inc.) are ideal embedding media to take advantage of a new combined light and electron microscope, the LEM 2000 (ISI Instruments). The embedding medium used for this instrument must allow the production of thin as well as thick sections. Furthermore, the resin must allow the penetration of a wide range of stains for both light and electron microscopy.

Lowicryls

Lowicryls are acrylate–methacrylate mixtures that polymerise to form a saturated carbon chain structure (Kellenberger *et al.*, 1980; Armbruster *et al.*, 1982; Carlemalm *et al.*, 1982a). These resins were developed primarily as embedding media of low viscosity that can be used at low temperatures for improved preservation of molecular structure of tissue, including the antigenicity. Low viscosity facilitates rapid and uniform penetration by the resin into the tissues, especially at low temperatures. Low temperatures immobilise molecules, owing to the decrease in thermal vibration during solvent exchange (Frauenfelder *et al.*, 1979). At low temperatures the rearrangement of protein macromolecular systems in non-aqueous liquids is significantly reduced. Lowicryls were also designed in the hope that they would preserve the polarity of tissue proteins and other structures.

Although the quality of ultrastructure preservation with Lowicryls is inferior to that obtained with epoxy resins, loss of tertiary structure in proteins is markedly reduced in the former. Therefore, Lowicryls have become a promising alternative to epoxy resins in immunocytochemistry. Although embedding in a Lowicryl does not increase the antigenic yield, the efficiency and accuracy of immunolocalisation seem to improve. Furthermore, non-specific background immunostaining is reduced.

Lowicryls are polymerised by long-wave (360 nm) indirect UV light at temperatures ranging from −50 °C to 0 °C (223 to 273 K) with the initiator benzoyl methyl ether and from 0 °C to 30 °C (273 to 303 K) with benzoyl ethyl ether. Photopolymerisation is accomplished in the absence of UV-absorbing pigments or osmium deposits in the tissue. The addition of the cross-linker triethyleneglycerol dimethacrylate improves the sectioning properties of these resins. During embedding, oxygen is excluded, since free radical polymerisation is strongly inhibited by oxygen. Lowicryls also provide sections of high ratio-contrast in the STEM.

Salient details of Lowicryl methodology are as follows. Since oxidative, inorganic reagents such as OsO_4 tend to cause cleavage of proteins, fixation is carried out with organic cross-linking agents such as glutaraldehyde. To minimise extraction and displacement of cellular components, dehydration is carried out with a graded series of ethanol (or a more polar organic liquid) in water. Dehydration is started at 0 °C and then the temperature is lowered in conjunction with increasing concentration of ethanol until it has reached −35 °C to −50 °C (238–223 K); freezing the specimens is avoided. Infiltration is accomplished by the basically non-polar resin HM20 or the more polar K4M at the same temperature (Carlemalm *et al.*, 1982a,b). Cross-linking of the resin is accomplished by a UV-initiated free radical reaction with the unsaturation in the monomers at the same low temperature.

Properties of various Lowicryls are given below. It should be noted that Lowicryls are not completely miscible with water.

(1) *K4M* Lowicryl K4M is a polar, hydrophilic resin which is soluble in most polar dehydrating agents. A polar resin is assumed to maintain intramolecular and intermolecular hydrogen bonding and general protein conformation as the water is replaced by resin–solvent mixtures. Lowicryl K4M can be used for infiltration and embedding at −38 °C (235 K) and −48 °C (225 K), respectively. The formulation is as follows:

Hydroxypropyl methacrylate	48.4 g
Hydroxyethyl acrylate	23.7 g
n-Hexyl methacrylate	9.0 g
Triethyleneglycol dimethacrylate	13.9 g

Kits containing Lowicryl resins are commercially available.

(2) *K11M* Lowicryl K11M is a polar, hydrophilic resin which can be used for infiltration and embedding at −63 °C (210 K). It is designed primarily for cryosubstitution into a solvent without the addition of a fixative in the hope of increasing the efficiency of immunolabelling (Carlemalm *et al.*, 1986).

(3) *HM20* Lowicryl HM20 is a non-polar, hydrophobic resin which is miscible with most polar dehydrating agents but not with ethylene glycol. It can be used for infiltration and embedding at −48 °C (225 K).

(4) *HM23* Lowicryl HM23 is a non-polar, hydrophobic resin which can be used for infiltration and embedding at −83 °C (190 K).

DEHYDRATION, INFILTRATION AND EMBEDDING

Ethanol (30%)	0 °C	30 min
Ethanol (50%)	−20 °C	60 min
Ethanol (70%)	−20 °C	60 min
Ethanol (100%)	−35 °C	120 min
Ethanol (100%) + resin (1:1)	−35 °C	60 min
Ethanol (100%) + resin (1:2)	−35 °C	60 min
Pure resin	−35 °C	60 min

Specimens are placed in fresh, pure resin at −35 °C and infiltrated overnight. Before the specimens are immersed in the resin for infiltration, the resin should be degassed with a vacuum pump (Fryer *et al.*, 1983). Embedding capsules should be completely filled with the resin and capped before polymerisation. Polymerisation is carried out by indirect UV irradiation (360 nm) for 24 h, followed by further polymerisation at room temperature for days or weeks to improve sectioning quality (Armbruster *et al.*, 1982). Since polymerised resin is hygroscopic, blocks should be stored under partial vacuum and over desiccant. Diamond knives should be used for sectioning at a speed of 2–5 mm/s.

LIMITATIONS OF LOWICRYLS

Although Lowicryls are sufficiently versatile to allow embedding at different temperatures and under different conditions of resin polarity and water content, they have some limitations. These include specimen shrinkage during polymerisation, uneven polymerisation, bubble formation in the block, evolution of heat during polymerisation, non-reproducible results, and loss of mass in sections under the electron beam.

Evolution of heat during polymerisation of Lowicryls even at low temperature has not been sufficiently recognised. Control of resin temperature during polymerisation is difficult. Polymerisation occurs concurrently with the temperature rise. Thus, actual embedding is accomplished at temperatures higher than those specified. Such rise in temperature may influence the survival of antigenicity and enzymatic activity. Failure to control the resin temperature during polymerisation may explain some of the problems mentioned above. Roth *et al.* (1981) have indicated that the survival of some antigens after embedding in Lowicryl K4M was not as good as expected on theoretical grounds.

The rise in temperature during polymerisation of Lowicryls K4M, K11M, HM20 and HM23 with UV light (350 nm wavelength) at room temperature or at 5 °C is substantial when polyethylene embedding capsules are kept in air (Weibull, 1986). At an ambient temperature of 21 °C in air the rise in temperature in K4M, K11M, HM23 and HM20 is 31 °C, 31 °C, 23 °C,

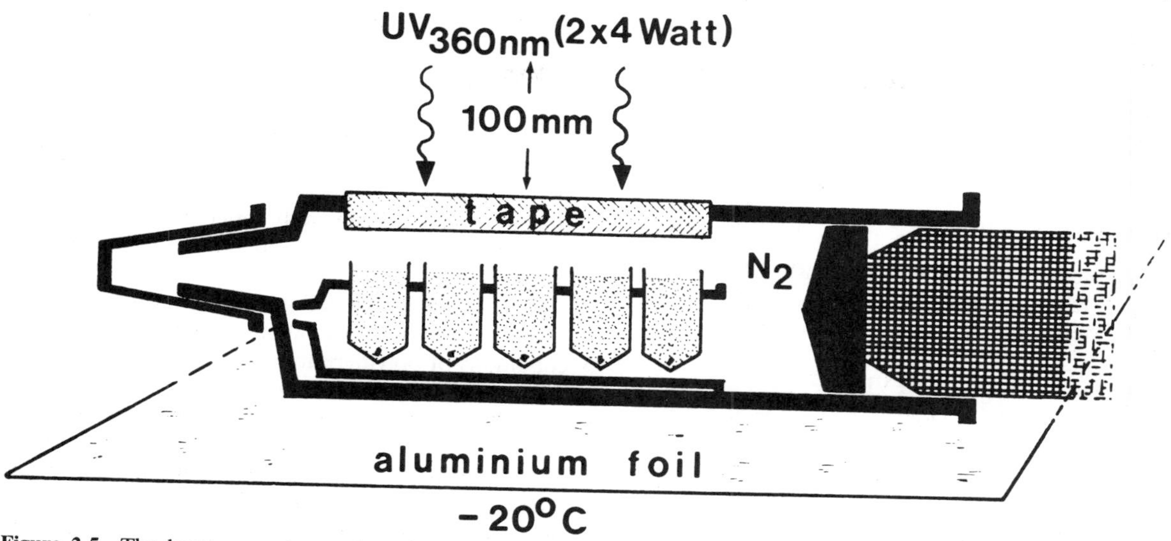

Figure 2.5 The low-temperature polymerisation chamber, made of two plastic syringes. Five BEEM capsules containing Lowicryl K4M and the specimens are surrounded by a protective atmosphere of nitrogen gas (see the text). Sufficient and homogeneous polymerisation occurs only when the chamber is air-tight, eliminating oxygen. From Völker *et al.* (1985)

and 17 °C, respectively. On the other hand, the rise in temperature during polymerisation at −35 °C to −45 °C in air is <2 °C. Similarly, the rise in temperature at 21 °C in the polyethylene capsules immersed in a water-bath is <2 °C. The advantage of low temperatures and the presence of cooling devices surrounding the embedding capsules is apparent. The smaller the volume of the resin in the embedding capsule the less will be the rise in temperature.

For adequate control of temperature, Ashford *et al.* (1986) recommend the use of very small resin volumes, as low a polymerisation temperature as possible, liquid-cooled instead of air-cooled systems, and embedding containers with good heat transfer properties. An apparatus for embedding at Lowicryl temperatures down to −45 °C (288 K) has been developed by Ashford *et al.* (1986). This system can be improved by using a cooled metal block instead of a liquid bath and by increasing the distance between the UV lamp and the resin (Ashford, personal communication).

The problems of oxygen exclusion and condensation of air bubbles at the tissue block encountered during polymerisation of Lowicryl K4M can be avoided by using a simple polymerisation chamber made of two plastic syringes (Fig. 2.5) (Völker *et al.*, 1985). A 20 ml plastic hypodermic syringe serves as the polymerisation chamber. Five uncovered BEEM capsules, containing the resin and the specimen, are fitted into holes, which have been cut into the wall of a smaller syringe (5 ml). This syringe is then inserted into the larger syringe. The air inside the polymerisation chamber is displaced by nitrogen gas, and the larger syringe is completely sealed by its piston at one end and at the other end with an air-tight cap. The whole device is placed on a sheet of aluminium foil in a deep-freezer at −20 °C (253 K). A UV lamp (360 nm, 2×4 W) is positioned at a distance of 100 mm above the polymerisation chamber. To prevent premature polymerisation on the surface of the resin, the top part of the chamber, which would be directly exposed to the light, is covered with a tape. This cover and the aluminium foil ensure that the capsules are irradiated only with diffusely reflected UV light, which results in homogeneous polymerisation of the resin. Gas bubbles escape from the open capsules.

The prolonged processing time required for embedding in Lowicryls may adversely affect sensitive antigens. To minimise such an effect, rapid fixation and embedding with Lowicryl K4M for immunoelectron microscopy and diagnostic work have been carried out (Altman *et al.*, 1984; Simon *et al.*, 1987). However, this approach yields only adequate preservation of the fine structure (see p. 125 for this method).

LR Gold

LR Gold is an acrylic resin (Bio-Rad Lab., Watford, UK) used for embedding large tissue specimens (3×3×3 mm) for enzyme histochemistry or immunocytochemistry (Bendayan *et al.*, 1987). Polyvinyl pyrollidone (PVP; mol.wt. 44 000) is used in combination with methanol for dehydration and infiltration to protect the specimens from osmotic changes. Oxygen inhibits polymerisation of this resin. Gelatin capsules (00 size) are filled completely and lids tightly fitted during polymerisation. Paper labels can be inserted into the capsules. Polymerisation can be accomplished either at room temperature or at low temperatures. Sections can be cut with glass or diamond knives.

PROCEDURE

(1) Methanol (50%) + PVP (20%)	4 °C	15 min
(2) Methanol (70%) + PVP (20%)	−20 °C	45 min
(3) Methanol (90%) + PVP (20%) (2:1)	−20 °C	45 min
(4) Methanol (90%) + PVP (20%) + LR Gold (5:1:5)	−20 °C	30 min
(5) Methanol (90%) + PVP (20%) + LR Gold (3:1:7)	−20 °C	60 min
(6) LR Gold (90%)	−20 °C	60 min
(7) LR Gold + 5% benzoin methyl ether	−20 °C	60 min
(8) LR Gold + 5% benzoin methyl ether		overnight

The tissues are embedded in gelatin capsules, covered and polymerised under UV light at −20 °C for 1 week using the device introduced by Bendayan (1984a). This device consists of two pairs of Philips fluorescent tubes (TL 6/05); one pair is placed 30 cm above the capsules and the other is set at the same distance below the capsules. This sandwich arrangement facilitates uniform polymerisation throughout the capsule. The complete device is placed in a deep-freeze at −20 °C.

LR White

LR White (London Resin Co., Basingstoke, UK) is a polar, monomer polyhydroxylated aromatic acrylic resin capable of thermal or cold polymerisation when an accelerator is used. The resin requires an anaerobic polymerisation. Monomeric LR White is insoluble in water; the resin has a limited miscibility with water to the extent of ~12% by volume. However, sections of polymerised LR White are hydrophilic. This characteristic has the advantage that etching of sections for immunocytochemistry is not necessary. Aqueous solutions of cytochemical reagents (e.g. immunoglobulins, colloidal gold and lectins) easily penetrate such sections. In contrast, etching of sections of heavily cross-linked, hydrophobic resins (epoxy and polyester) is necessary to allow access of antibody to the antigen. Bonds between the tissue and certain embedding resins (e.g. glycol methacrylate) need to be broken before antigenicity can be demonstrated. Etching adversely affects delicate tissue antigens.

Another advantage of LR White is that it shows minimal non-specific staining. Thin sections of tissue embedded in LR White generally demonstrate a positive immunocytochemical reaction after very short incubation times in the primary antiserum, in contrast to equivalent sections of the tissue embedded in Epon (Newman *et al.*, 1983). LR White rapidly infiltrates the tissue because of its low viscosity (8 cP). It has very low toxicity, for it contains only monomers used in medicine and dentistry. A fume-hood is not required during its use.

Like other acrylic resins, the type of cross-linking and the density of LR White can be varied by the conditions of polymerisation. Theoretically, the lower the cross-link density of the resin, the easier it will be for aqueous solutions including immunoreagents to penetrate. For example, gradual heat polymerisation at 50 °C prevents LR White from becoming completely cross-linked (Newman, 1987). The gradual polymerisation also induces a linearity in its molecular arrangement, favouring penetration by aqueous solutions. Gradual polymerisation combined with partial dehydration (in 70% alcohol) facilitates in-depth staining with the immunoperoxidase method (Newman *et al.*, 1983). LR White seems to be superior to epoxy resins for post-embedding immunogold labelling. However, as is the case with other resins, immunogold staining is confined to the surface of the section. LR White is commercially available in pre-mixed formulations.

In view of the advantages mentioned earlier, LR White can be used in cytochemistry and immunocytochemistry. It is useful for both light and electron microscopy. The resin can be polymerised by heat or UV light. It can be polymerised with UV light at either room temperature or low temperatures (it remains liquid at −20 °C). Polymerisation by UV light is not used when embedding certain pigmented or post-osmicated tissues. The necessity of exclusion of oxygen during polymerisation of LR White (or Lowicryl K4M) makes flat embedding problematic. Erdos (1987) has introduced a procedure for flat embedding without the construction of anaerobic chambers.

Although a shelf life of at least 1 year at 4 °C has been claimed for LR White, it should only be used when fresh (definitely less than 6 months old). The reason is that after storage for about 3 months, benzoyl peroxide is incorporated into the monomer and so a significant degree of polymerisation will take place, without any apparent change in the resin. The tolerance of the resin for residual water in the tissue is reduced with age and also infiltration becomes quite poor. Furthermore, tissues that have been partially cross-linked by fixation are often subjected to severe polymerisation artefacts, particularly when the heat-cure method is used (also see Newman, 1989).

Figure 2.6 A cross-section of grasshopper leg muscle embedded in melamine. The specimen was fixed with a mixture of 2% glutaraldehyde and 2% tannic acid in 0.1 M phosphate buffer (pH 7.2) for 24 h. The thin section (36 nm) was cut with a diamond knife at a clearance angle of 11° and a cutting speed of 2 mm/s. To minimise radiation damage, the electron micrograph was taken from a previously unexposed area. The section was not stained with a heavy metal, and was recorded at 1.1 µm defocus. It shows myofilaments with great detail and membranes as trilaminated structures when they are sectioned and photographed exactly perpendicularly (arrow). The inset shows the same muscle at a thickness of 10 nm. From Frösch and Westphal (1985)

Melamine Resins

Three melamine resins are available for electron microscopy. They are highly water-soluble, two-component (melamine–formaldehyde) resin systems (Bachhuber and Frösch, 1983; Frösch *et al.*, 1987). Hexamethylol-melamine-ether is synthesised by the addition of 6 mol of formaldehyde (37%) to 1 mol of melamine, which is subjected to slight etherification with methanol at pH 5–6. After one wash in chloroform, a clear solution of pure methylol-melamine-ether (MME) is obtained. It is a 70% aqueous solution of melamine with a viscosity of 48 mPa s. Unlike Epon, which polymerises by the addition reaction, MME polymerises by polycondensation.

The most promising advantage of MME is that it allows electron phase contrast imaging of unstained specimens (Fig. 2.6). This polar resin is completely miscible with water and remains stable for many months at room temperature. Specimens can be embedded without organic solvents, and sections less than 10 nm thick can be obtained (Frösch *et al.*, 1985). However, a minimal section thickness of at least 30 nm is required for the generation of a substantial amount of phase contrast (Frösch and Westphal, 1985).

Since Nanoplast is compatible with both polar and non-polar solvents, it can be used for freeze-substitution and embedding of unfixed, frozen cells at low temperatures (−82 °C, 191 K). Some fixation may occur because of the presence of residual formaldehyde in the melamine. Nanoplast AME 01 (20% w/v MME in acetone) has been used for this purpose (Frösch *et al.*, 1987).

Another formulation, Nanoplast MUV 116, has been introduced, which is used in a similar way to AME 01, but can be hardened in the cold by UV light (Bachhuber *et al.*, 1987. Unlike Nanoplast AME 01, which is used at ambient temperature, Nanoplast MUV 116 can be used for infiltration and embedding of frozen micro-organisms at −82 °C (191 K), but the final hardening requires a post-curing at 60 °C (333 K) for 24 h. The infiltration and embedding temperatures of the three melamine resins are compared in Table 2.8.

Table 2.8 Embedding temperatures used for various melamine resins

Melamine resin	*Infiltration*	*Polymerisation*
FB 101	40 °C (313 K)	60 °C (333 K)
AME 01	−82 °C (191 K)	60 °C (333 K)
MUV 116	−82 °C (191 K)	−82 °C (191 K)

Limitations of Nanoplast include a lengthy drying and polymerisation, and its blocks are more brittle than those of epoxy resins. Since Nanoplast may release formaldehyde during polymerisation, it must be handled under a fume-hood. Components of the resin are irritating to the skin, eyes and respiratory tract. Protective gloves and goggles should be used. The catalyst is hygroscopic, so it should be kept in a tightly closed container.

Embedding Formulation

Melamine resin (70% in water)	250 ml
Acid catalyst B 52 (*p*-toluene sulphonic acid)	10 g

The catalyst should be added shortly before use and thoroughly stirred. The mixture is allowed to stand for 30 min. Aldehyde-fixed specimens are placed in BEEM capsules which are filled with the resin to a maximum height of 5 mm. The capsules are placed in a desiccator (containing silica gel), which, in turn, is placed in an oven at 40 °C for 2 days. The capsules are removed from the desiccator and polymerised for 2 days at 60 °C (333 K).

Silicon embedding capsules should not be used, for air bubbles form during drying. Thin sections should be cut with a diamond knife at a speed of 2 mm/s, or at a speed of 5–10 mm/s if very thin sections are required. Distilled water should be used in the knife trough. Thin sections are viewed without staining. Low specimen contrast can be improved by adding up to 0.5% uranyl acetate to the embedding medium. If the specimen block is too soft, it can be further polymerised for 5–10 h at 80 °C. Table 2.9 gives the proportions of MME 7002 and the catalyst required in order to obtain sections of various thicknesses. Melamine resins are manufactured by Rolf Bachhuber, P.O. Box 1243, 7900 Ulm, W. Germany, and are also distributed by Agar Aids Ltd; Balzers Union; JB EM Chemicals; Pointe Claire; Dorval; Canada; Plano GmbH; and Polysciences, Inc.

Table 2.9 Proportions of MME 7002 and B 52 and the resulting quality of the block

Type of section	*MME 7002* (g)	*B 52* (g)	*Relative hardness*
Semithin	10	0.15	Soft
Thin	10	0.2	Medium
Very thin	10	0.25	Hard

Polyacrylamide

Polyacrylamide (PAA) was introduced as an embedding medium for immunohistochemical studies using conventional ultramicrotomy (Yamamoto *et al.*, 1980). It is more suitable for immunocytochemistry on thin cryosections (Slot and Geuze, 1982). The stock embedding medium consists of 50% acrylamide and 1.5% bisacrylamide in distilled water. The preparation of final embedding medium is as follows:

Stock embedding medium	4 ml
Formaldehyde (16%)	0.6 ml
Sodium phosphate buffer (1 M, pH 5.7)	0.4 ml
N,N,N′,N′-tetramethylethylenediamine (TEMED)	10 μl
Riboflavin (100 mg/ml)	10 μl

The acid buffer is added to neutralise TEMED. The above recipe indicates how to prepare an embedding mixture of maximal acrylamide concentration (40%); if desired, this can be diluted to a lower concentration. The addition of formaldehyde to the embedding mixture is optional and possibly useful only when very mild fixation is required.

Fixed tissue specimens are transferred to the final embedding medium and agitated overnight at 4 °C in the dark. They are then transferred in drops of the embedding medium to a 9 cm Petri dish. A 3 cm dish is placed rapidly on top of these drops so that each drop is squeezed to a slab. The thickness of this slab is determined by the small rim (0.5 mm) that sticks out from the bottom of the 3 cm dish. The preparation is placed on a light box with fluorescent bulb for polymerisation while being cooled by ice. A PAA gel slab forms within a few minutes, but exposure to the light source is continued for at least 1 h. The presence of formaldehyde in the embedding medium accelerates the polymerisation, which is an exothermal reaction and therefore needs proper cooling.

Polyampholyte

The polyampholyte is prepared by distilling monomeric anionic methacrylate acid (MA) at 10 mmHg pressure; the fraction that boils at 60 °C is collected and stored at 4 °C. After being washed with buffer, fixed tissue specimens are gradually dehydrated and infiltrated with a 1:1 mixture of MA and cationic dimethylaminoethyl methacrylate (DMA) in the following manner. The specimens are immersed in 10%, 20%, 40% and 80% of the mixtures for 30 min at 4 °C at each step. They are then immersed in three changes of 100% monomer mixture for a period of 2 h, followed by three changes in the embedding mixure of 12 h at 4 °C. The specimens are transferred to freshly prepared embedding mixture in gelatin capsules. The embedding mixture is prepared as follows:

Methacrylic acid	20 ml
Dimethylaminoethyl methacrylate	10 ml
Tetramethylene dimethacrylate	3.3 ml
Azodiisobutyronitrile	0.25 ml

MA should be added drop by drop to DMA to avoid thermal polymerisation. Polymerisation is completed by UV irradiation within 48 h at 4 °C.

Polyethylene Glycol (Carbowax)

Carbowaxes are a group of wax-like polymers having different molecular weights (1000–4000) and melting points, depending upon their degree of polymerisation. They have long been in use as embedding material for light microscope histochemistry of lipids, glycogen and enzymes. Polyethylene glycol (PEG) with a molecular weight of ~1000 melts at 32–38 °C, and is miscible with water, paraffin and most organic solvents. It appears to be chemically inert and does not dissolve glycogen and most lipids. The preservation of ultrastructure in the PEG-embedded tissue is adequate. This embedding medium is also useful for producing resin-free sections of animal and plant cells for conventional and high-voltage electron microscopy (p. 131). The embedding procedure is described below.

Fixed tissue specimens are immersed directly into 4 g vials filled with warm (58 °C) 50% PEG (mol. wt. 4000 or 6000) in distilled water. After being placed in a temperature-regulated shaker bath maintained at 58 °C, the vials are agitated during infiltration for 1–6 h. The specimens are removed and, after excess medium is drained on warm filter paper, are transferred to vials containing 70% PEG in distilled water. The specimens are infiltrated on the shaker bath for 1–8 h. This process is repeated twice more with 100% molten PEG. The specimens are transferred to a small paper cup, together with liquid PEG mixture, and solidifed in the refrigerator. Sections about 40 μm thick are cut with a sliding microtome and embedded in Epon 812 or one of its substitutes.

For immunoelectron microscopy, after fixation with a mixture of picric acid and formaldehyde, specimens are incubated at 45 °C in PEG (mol. wt. 1500) with 3–5% (by volume) distilled water (two changes of 15 min each), with occasional stirring (Bosman and Go, 1981). The tissue blocks are left to harden at room temperature. After at least 1 h, sections 6 μm thick are cut at room temperature on a rotary microtome and mounted on glass slides coated with gelatin-chromealum. A single section is gently pressed onto each slide with a thin paint brush until it adheres slightly. The slide is heated at 45 °C on a hot-plate, and after the PEG has melted, a small drop of distilled water is applied around the tissue section with a paint brush, to stretch the section. After careful removal of excess fluid with filter paper, all the slides are dried overnight in an incubator at 37 °C.

Following a wash in phosphate-buffered saline (PBS) for 15 min to remove the remaining PEG, the sections are immunostained. They are then washed for 30 min in PBS and refixed with 2% glutaraldehyde for 20 min. After incubation for 30 min in 0.005% diaminobenzidine (DAB) in 0.15 M Tris–HCl buffer (pH 7.6), the sections are treated for 2 min with the same solution but containing 0.01% hydrogen peroxide. The sections are post-fixed with 2% OsO_4 for 30 min, dehydrated and embedded by using the inverted capsule method (p. 136). The resin is polymerised at 37°C for 1 week. The blocks are separated from the slides by briefly heating the slides on a gas flame.

Polyvinyl Alcohol

Polyvinyl alcohol (PVA) is another water-miscible embedding medium used for electron microscopy (Muñoz-Guerra and Subirana, 1982). This water-soluble polymer can be made insoluble in water by cross-linking with an aldehyde. Specimens are infiltrated with an aqueous solution of PVA, which is then concentrated by dialysis to form a hard gel. The gel is cross-linked with a 10% solution of glutaraldehyde. Water is present throughout the processing, and the final block contains ~10% water. The polymer retains substantially its polarity, since only a small number of hydroxyl groups are blocked during cross-linking. Organic solvents are avoided and the temperature is always maintained below 40 °C. Among the limitations of PVA are that a long time is required to complete the embedding, thin sections may develop holes and certain tissues (e.g. nervous tissue) exhibit poor embedding.

The procedure of embedding consists of infiltrating the specimens with an aqueous solution of PVA, which is then concentrated to a hard gel (Muñoz-Guerra and Subirana, 1982). The gel is cross-linked with glutaraldehyde. Fixed tissue specimens are infiltrated with 1%, 5%, 10% and 20% aqueous solutions of PVA (mol. wt. 14 000; degree of hydrolysis is 99%) at room temperature for 30 min in each step. Two or three specimens are transferred to the bottom of a glass tube (6 mm in diameter; length 50 mm) with one end sealed with a dialysis membrane. About 1 ml of 25% PVA solution is added to the tube, care being taken to ensure that the specimens rest on the membrane surface.

The tube is partially buried in Aquacide and placed in an oven at 40 °C for 2–3 days. The hard gel formed is cut into small cubes, each containing one specimen. These cubes are suspended above a 10% aqueous solution of glutaraldehyde in a small sealed vial for 2 days at 40 °C. The cubes are immersed into the glutaraldehyde solution for 2 days, washed with distilled water, dried on silica gel and mounted onto a resin or Lucite block for sectioning. Sectioning is easy; the highly hydrophilic character of PVA requires a fast speed of sectioning and a careful adjustment of the water meniscus in the trough. Carbon-coated grids should be used. Post-staining is accomplished in a relatively short time.

MIXED RESIN EMBEDDING

Since the majority of the methacrylates, polyesters and epoxy resins are known to be miscible with one another, in recent years attempts have been made to use mixed embedding medium made up of more than one resin. This approach is justified, since a mixed medium possesses the best qualities of more than one resin. An ideal mixed embedding medium should possess, for example, the thermal stability of Araldite, the curing quality of Maraglas and the picture contrast of Epon. The development of such a medium is, of course, dependent upon knowledge of the correct proportions of the various resins involved. It is, therefore, obvious that a considerable degree of skill, patience and experimentation is required to achieve this goal. There is no doubt that further improvement of these methods can be expected.

Mollenhauer (1964) demonstrated convincingly that embedding mixtures section easily as well as show excellent tissue preservation. The composition and properties of some of the embedding mixtures recommended for both plant and animal tissues are given in Table 2.10. DMP-30 should not be used as an accelerator in mixture 3 in this table, since a precipitate may form when mixed with propylene oxide; DMP-30 can be substituted by BDMA in mixture 3. In general, DMP-30 is preferred to BDMA, for sectioning is easier when the former is used. Mixture 2 is easier to section than mixture 1, while mixture 3 is somewhat more difficult to section than either mixture 1 or mixture 2. Mixture 3 is recommended for embedding acetolysed pollen and spore exines. If a relatively hard block is needed, the plasticiser, dibutyl phthalate, can be eliminated. However, the plasticiser improves the sectioning properties of these mixtures. Araldite CY 212 can be substituted by Araldite 502 or Araldite 6005 in the same proportion (Glauert, 1974). These embedding mixtures can be modified to suit a given tissue.

Since Epon and Araldite blocks do not always section easily with glass knives, mixtures given in Table 2.11 can be sectioned well with such knives (Poolsawat, 1973). Soft mixture in this table is especially suitable for cutting relatively large sections.

Table 2.10 Composition and properties of Epon–Araldite mixtures

Composition/properties	*Mixture No.*		
	1	*2*	*3*
Ingredients			
Epon (g)	25	62	—
DDSA (g)	55	100	—
Araldite CY 212 (M)	15	—	—
Araldite 506 (g)	—	81	50
Cardolite NC 513 (g)	—	—	25
Dibutyl phthalate (g)	2–4	4–7	1–2
DMP-30 (fresh)	1.5%	1.5%	1.5%
or			
BDMA	3%	3%	3%
Relative hardness	medium	soft–medium	soft–medium
Image contrast	high	medium	low
Tissue preservation	good	excellent	excellent

From Mollenhauer (1964).

Table 2.11 Properties of cured epoxy resin mixtures

Ingredients (g)	*Hard*	*Rigid*	*Soft*
Epon 812 (or its substitute)	42.74	31.75	31.75
Araldite 502	5.6	17.92	17.92
DDSA	57.6	52.68	52.68
Dibutyl phthalate	—	2	4
Total	105.94	104.35	106.35

To accomplish polymerisation, 10 drops of DMP-30 are added to 10 g of the mixture. From Poolsawat (1973).

The following formulations have proved successful for a wide variety of tissues.

Epon 812 (or its Substitute)–Araldite 502

Stock mixture

Epon 812 (or its substitute)	31 g
Araldite 502	25 g
Dibutyl phthalate	4 g

Final embedding mixture

Stock mixture	8 g
DDSA	20 g
DMP-30	28 drops
or	
BDMA	56 drops

All the components of the above mixtures must be thoroughly mixed; otherwise, uneven infiltration and embedding will result, which will cause difficulty in sectioning. Mixing by shaking is better than by stirring. Furthermore, since, while standing, the ingredients of the mixture may separate, a thorough mixing just before use is advisable. The accelerator should be added just before use.

Epon 812 (or its Substitute)–DER 736 (Kushida, 1967)

Stock mixture

DER 736	40 g
Epon 812 (or its substitute)	10 g
NMA	45 g
DMP-30	1.4–2.0 g

The polymerisation is completed in ~48 h at 50 °C. This mixture is bound to be less viscous than the final embedding mixture of Epon 812. Consequently, DER 736–Epon 812 mixture penetrates the tissue specimens more readily and completely than does Epon mixture alone. The final hardness of the block can be adjusted by changing the ratio between DER 736 and Epon 812; a ratio of 4:1 is desirable.

Epon 812 (or its Substitute)–Thiokol LP-8

Mixture A

Epon 812 (or its substitute)	100 g
NMA	85 g
Thiokol LP-8	15–30 g

Final embedding mixture

Mixture A	100 g

DMP-30	(20–30 °C)	3 g
		or
	(50 °C)	1.5 g

The polymerisation is completed in ~7 days at 20 °C, in ~4 days at 30 °C and in ~24 h at 50 °C. The final hardness and cutting characteristics of the resin may be easily modified by varying the amount of Thiokol LP-8. The greater the amount of Thiokol LP-8, the softer the final block. Blocks polymerised at room temperature are easier to section than those cured at 50 °C; also, the sections of the former show better inherent contrast. The above embedding formulation has a lower viscosity than the usual mixture of epoxy resins, and thus penetrates more readily and completely into the tissue specimen, producing a more homogeneous block that is relatively easy to section.

Thiokol LP-8 is a liquid polythiodithiol polymer terminated by mercaptan (—SH) groups of low viscosity. It acts as both a flexibiliser and hardener, and combines chemically in a polymerised system of epoxy resins. Since polymerisation of the tissue blocks by heat is unsatisfactory, especially for histochemical studies, Thiokol LP-8 is employed because it reacts with epoxy resins at room temperature in the presence of a hardener (NMA) and an accelerator (DMP-30).

Epon 812 (or its Substitute)–Maraglas (Tandler and Walter, 1977)

Epon mixture	
Epon 812 (or its substitute)	21.88 g
DDSA	12.60 g
NMA	11.90 g
DMP-30	0.7–0.8 g
Maraglas mixture	
Maraglas 655	36 g
DER 732	8 g
DP	5 g
BDMA	1 g

A 1:1 mixture of Epon and Maraglas is recommended; a higher proportion of Epon will result in a harder block. This mixture is less viscous than Epon–Araldite mixture.

DER 332–DER 732 (Lockwood, 1964)

	A	*B*	*C*
DER 332	7	7	6
DER 732	3	2	3
DDSA	5	5	10
DMP-30	0.30	0.28	0.38

Before mixing, DER 332 is heated to 60 °C to reduce viscosity. Mixture A is relatively soft and is recommended for embedding soft tissues. Mixture B yields a harder block and is recommended for collagenous tissues. Mixture C is of approximately the same hardness as mixture A but is tougher, owing to the greater content of DDSA, and thus more difficult to section. The hardness of these formulations is controlled by the amount of DER 732 used; the larger the amount of this resin, the softer the final block.

If prolonged infiltration is desired, DMP-30 is replaced by the less reactive DMP-10. Slower polymerisation is useful for some pathological tissues, such as inflamed cells. In the recommended procedure the tissue specimens are dehydrated in ethanol, followed by two changes of 15 min each in propylene oxide. The tissues are then placed in a 1:1 mixture of the resin and propylene, and incubated for 1 h at 45 °C in tightly capped vials. This mixture is decanted and replaced with fresh pure resin mixture. The tissues are left in this mixture for 1 h at 37 °C in open vials, to allow the remaining solvent to evaporate. The tissues are transferred to fresh resin in polyethylene capsules and polymerised at 60 °C for up to 48 h.

Methacrylate–Styrene (Mohr and Cocking, 1968)

N-Butyl methacrylate (stabiliser removed)	70 parts
Styrene (stabiliser not removed)	30 parts
Benzoyl peroxide	1 part

The final embedding mixture is dehydrated by treating with anhydrous calcium sulphate and centrifuged prior to using. Final embedding is completed in pre-dried capsules, using freshly prepared mixture. The blocks are polymerised for 44 h at 55 °C. Methacrylate–styrene is especially suitable for embedding highly vacuolated, senescent or damaged plant tissues. Sections of large size (1 mm) can be cut with relative ease.

Poly/Bed 812–Araldite 6005

Poly/Bed 812	10 ml
Araldite 6005	10 ml
DDSA	24 ml
DMP-30	0.9 ml

Quetol 651–ERL 4206 (Kushida *et al.*, 1986)

Quetol 651	28 g
ERL 4206	6 g
NSA	66 g
DMP-30	1.5–2.0 g

The above mixture is polymerised in 24 h at 60 °C. The mixture has been recommended for three-dimensional observation of thick sections (0.5–5.0 μm) at an accelerating potential of 400 kV.

Quetol 523–2-Hydroxypropyl Methacrylate–Methyl Methacrylate (Kushida *et al.*, 1985)

Quetol 523	10 g
2-Hydroxypropyl methacrylate	65 g
Methyl methacrylate	25 g
2,2′-Azobisisobutyronitrile (catalyst)	1 g

The above water-miscible mixture is polymerised in 12 h at 60 °C. It is recommended for observing identical sites in semi-thin sections (0.3–0.4 μm) with the light microscope and a conventional TEM with an acceleration voltage of 200 kV.

Rigolac–Styrene (Shinagawa and Uchida, 1961)

Rigolac 70F	3 parts
Styrene	7 parts
BP (Benzoyl peroxide)	1%

The styrene:Rigolac ratio determines the hardness of the block.

Silicone (Rhodorsil 6349)–Araldite CY 212 (Scala *et al.*, 1977)

Araldite CY 212	20 ml
Rhodorsil 6349	10 ml
Araldite HY 964	30 ml
Araldite DY 964	1.2 ml

The first two components are mixed thoroughly, with mechanical stirring, for 20 min. The third component is added, with further mechanical stirring for 15 min, and similar stirring is carried out when adding the fourth component. Polymerisation is completed overnight at 100 °C. Rhodorsil 6349 (Rhone Poulenc Soc. Us. Chim., Paris 8e, France) is a colourless liquid with a viscosity of 40 cP at 20 °C. It is inflammable at 54 °C. Rhodorsil requires further testing.

Vinylcyclohexene Dioxide–n-Hexenyl Succinic Anhydride

Vinylcyclohexene dioxide-n-hexenyl succinic anhydride (VCD–HXSA) mixture of low viscosity (20 cP) was introduced by Mascorro *et al.* (1976). According to Oliveira *et al.* (1983), this embedding mixture is especially useful for preserving vacuolar system in plant cells and reducing plasmolysis. It seems that the quality of overall preservation of the fine structure is somewhat less satisfactory than that obtained with epoxy resins. The following embedding mixture is recommended.

Vinylcyclohexene dioxide	10 g
n-Hexenyl succinic anhydride (hardener)	20 g
Araldite RD-2 (modifier)	0.3–0.5 g
Dimethylaminoethanol (catalyst)	0.3 g

Mixing of the components is assured when a magnetic stirrer is used for 2 min. An increase in the proportion of the modifier results in soft blocks.

Specimens are dehydrated in a methanol series. Since VCD–HXSA mixture is completely miscible with methanol, propylene oxide as a transitional solvent is not used. Infiltration is carried out for 1 h in each of the following mixtures of methanol and the embedding medium: 3:1, 1:1 and 1:3. Polymerisation is accomplished in 24 h at 55 °C.

PROPERTIES OF THE FINAL BLOCK

When a resin is polymerised without additives, the final block is unfit for sectioning. Moreover, a resin by itself may take a very long time to polymerise or may never completely cure. Resins such as Epon and vinylcyclohexene dioxide do not possess inherent toughness or electron beam stability. When Epon (or its substitutes) is polymerised by itself, it forms a very brittle solid unsuitable for sectioning. Such Epon blocks also have poor electron beam stability. An ideal block has optimal hardness, elasticity, plasticity and electron beam stability. These characteristics are obtained by adding appropriate hardeners, flexibilisers and catalysts to the resin. The cutting quality of the final resin block depends primarily on the nature and proportion of the ingredients in the embedding mixture. The cutting quality is also influenced by many other factors, such as the quality of dehydration, infiltration and polymerisation. Commonly used ingredients to improve the cutting properties of the blocks are discussed below.

Various components of an embedding mixture do not diffuse into the tissue with the same speed, since they possess varied viscosities. This may result in nonuniform infiltration of the specimen by the embedding mixture and uneven polymerisation. Such blocks may not be easy to section. In routine embedding procedures, resins unfortunately are incompletely polymerised. Since polymerisation does not reach to completion, the sectioning properties of the block are affected by the temperature of polymerisation. Optimal temperature is more important than duration of polymerisation.

HARDENERS

Anhydrides are the ideal hardeners, because they combine long working times with good heat and radiation stability in the polymerised resin (Causton, 1980).

(1) *Dodecenyl succinic anhydride (DDSA)* is a light-

yellow viscous liquid with a viscosity of 300 cP at 25 °C. This compound is important for its long 12-carbon side-chain, which acts to internally plasticise the cured resin. The use of DDSA alone results in the softest block, whereas harder blocks are obtained by replacing some or all of it with HHPA or NMA. Curing of epoxy resins with DDSA alone proceeds at a slow rate and thus requires the addition of a catalytic agent.

(2) *Nadic methyl anhydride (NMA)* is a light-yellow liquid with a viscosity of 175–275 cP at 25 °C. The hardener has a long pot life at room temperature. NMA is essentially a methylated maleic adduct of phthalic anhydride, and possesses excellent elevated temperature properties. Post-cure improves the sectioning quality of the blocks. NMA combines the anhydride group with the electron beam to achieve the stability of the unsaturated nadic ring structure. Hence, if increased electron beam stability is required, it is recommended.

(3) *Hexahydrophthalic anhydride (HHPA)* is a low-melting-point (35 °C) solid and is soluble in liquid resins at room temperature. This compound imparts low mix viscosity and a rapid diffusion, and gives long pot life. The hardness of the final block increases as the proportion of HHPA in the embedding mixture increases. Reduced electron beam stability of the saturated ring may be a disadvantage of this compound.

(4) *Nonenyl succinic anhydride (NSA)* is a branched alkenyl succinic anhydride based on tripropylene, and is water-white in colour. It has a lower viscosity (117 cP at 25 °C) than that of DDSA and NMA.

MODIFIERS

The hardness of the final block can be reduced by adding a plasticiser or a flexibiliser. The latter is preferred, since it reacts with the epoxy resin and becomes part of the cross-linked structure. As a result, it is less likely to be lost under the electron beam. However, plasticisers prevent excessive brittleness and improve the sectioning properties.

(1) *Dibutyl phthalate (DP)* is a non-reactive external plasticiser used to yield a softer block. Hardness of the block can be controlled, without increasing its brittleness, by altering the amount of this compound. If it is used in excess, blocks will not be sufficiently hard for successful sectioning, and may undergo dimensional changes with time.

(2) *Cardolite NC-513* is a long-chain monofunctional epoxide of amber colour. This compound readily co-reacts with resins and becomes an internal, non-extractable flexibiliser. Since it has a viscosity of 25 cP at 25 °C, it also serves as a viscosity reducer. It has the additional advantage of being shelf-stable when mixed with resins.

(3) *Thiokol LP-8* is a clear, amber colour, mobile liquid with a viscosity of 250–350 cP at 27 °C. This flexibiliser is a polysulphide polymer, and is reactive as a curing agent. It reacts slowly with resins when used alone; an amine catalyst such as DMP-30 is employed to accelerate cure.

(4) *DER 736* has been employed as a flexibiliser for vinylcyclohexene dioxide and Epon. Since DER 736 reduces electron beam stability, when it is added, some balancing of the aromatic content of the resin mixture may be necessary.

CATALYSTS

Commonly used catalysts (accelerators) are tertiary amines.

(1) *Benzyldimethylamine (BDMA)* is a tertiary amine, and is used in very small amounts as a catalyst to obtain rapid polymerisation of resins. Both BDMA and DMP-30 are equally efficient during polymerisation of epoxy resins. BDMA has the advantage over DMP-30 in that the former is less viscous than the latter. As a result, the embedding mixture containing BDMA permeates rapidly throughout the tissue block. The author agrees with Glauert (1987) that DMP-30 should be replaced by BDMA in the epoxy mixtures.

(2) *Tri(dimethylaminomethyl) phenol (DMP-30)* is a tertiary amine, and is used as an accelerator for anhydride cures. This compound is one of the most effective and commonly used accelerators, but does not itself serve as a direct cross-linking agent. The accelerating action of the phenolic hydroxyl is probably responsible for the increased reactivity of this amine. DMP-30 is more active than BDMA; hence, the former gives shorter working times. Tertiary amines tend to form adducts with epoxy resins at room temperature. Because DMP-30 has three groups capable of adduct formation, as opposed to the single group of BDMA, some dimers and trimers may form when the former is used (Causton, 1980). As a result, the diffusion of the resin mixture containing DMP-30 is relatively slow, especially into the core of the tissue block. DMP-30 is inactivated by water, and deteriorates with time. Therefore, it should be stored in a vacuum desiccator in the dark, and even then it needs to be replaced regularly.

(3) *Dimethylaminomethyl phenol (DMP-10)* If prolonged infiltration is desired, as when embedding inflamed tissues, the use of this compound rather than DMP-30 is desirable. DMP-10 slows the speed of

polymerisation without changing the qualities of the cured block.

(4) *Azodiisobutyronitrile* is a non-oxidant catalyst used for both heat- and light-initiated polymerisation. This compound is one of the most efficient catalysts for photopolymerisation.

(5) *Dimethylaminoethanol (DMAE or S-1)* is an alkyl alkanol amine with a viscosity of 3.32 cP at 25 °C. The accelerator provides typical tertiary amine cures and imparts a longer useful pot life (3–7 days) to the embedding mixture. It is used in small amounts with vinylcyclohexene dioxide to induce rather rapid cures. Because of the small amount used, DMAE facilitates better penetration of hard specimens by the Spurr mixture.

SPECIMEN ORIENTATION

Since only a very small portion of a tissue specimen is included in a section and only a few cells can be viewed in the electron microscope, it is imperative that sections be cut in the desired plane from an optically selected specific region of the tissue specimen. Such an approach would save a considerable amount of time that would otherwise be spent in cutting and examining a large number of sections. Furthermore, knowledge of the plane of section is essential for the correct interpretation of an electron micrograph. Therefore, the tissue specimen should be oriented properly during embedding so that sections can be cut in a known plane. Before presenting the details of how to accomplish specimen orientation, a brief description of the standard embedding procedure without regard to orientation is in order.

All ultramicrotomes are supplied with chucks to hold cylindrical blocks obtained from gelatin or polyethylene capsules. Such a block has the ideal geometry to be least affected by vibrations arising during sectioning. In general, it is convenient to place the surface of the specimen from which sections are desired next to the tip of the capsule. This is done by pushing the tissue specimen down to the bottom of the capsule with the aid of a pointed wooden stick. Care should be taken not to press the tissue too hard against the tip of the capsule, but to let it lie loosely so that a thin layer of the embedding medium between the tissue and the capsule surface is present.

Since tissue specimens tend to shift in the capsule during polymerisation, owing to the initial decrease in viscosity, orientation is difficult. However, orientation even in the capsule can be attempted after the capsule containing the embedding medium has been in the oven for 4–8 h at 45 °C. Such a heat treatment increases the viscosity sufficiently to minimise displacement of the specimen. Another approach is to place a tiny drop of the embedding medium carrying the tissue, with the blunted tip of a wooden toothpick, at the bottom of the empty capsule. The tissue is oriented and the drop is allowed to attain a higher viscosity before the capsule is filled to the desired level with the embedding medium. Although this approach does not always work, it is easy and can achieve an approximate orientation of certain types of tissues.

As stated above, in general, if tissues are placed in the embedding capsule, the desired orientation is either not accomplished originally or is changed because the tissue shifts during polymerisation. Many different methods have been introduced to achieve desired orientation, with various degrees of success. They are discussed below so that the reader may choose or modify.

Specimens can be pre-embedded in 2% agar blocks which are subsequently placed in the desired plane in capsules and embedded in the resin. A two-stage method also involves agar. Tissue specimens are fixed with glutaraldehyde prior to orientation in liquid agar, which is then allowed to solidify. The solidified agar block containing the specimen is post-fixed with OsO_4, which is followed by dehydration and embedding in a standard resin. The specimens are supported best by 2% agar which is maintained at 45 °C in a water-bath. This method has the added advantage that glutaraldehyde fixed delicate tissues are protected during subsequent processing. Three additional methods are explained here, and the remaining methods have been reviewed by Hayat (1986b). The method introduced by Hwang (1970) is explained below (Fig. 2.7).

Tissue specimens are embedded in a flat mould and the polymerised, rectangular flat block is removed from the mould (A). The cap of the capsule is cut off, and a hole as large as the flat block is made in the cap (B). A part of the capsule may have to be trimmed if it is longer than the desired length. The flat block is inserted into the hole by pushing it from the inside of the cap (C). Some of the inaccurate orientation can be corrected by changing the block-cap angle (D). The capsule is filled with the embedding medium and the cap with the flat block is fitted over the capsule (E). At this stage the orientation can be further corrected by changing the angle of the cap and/or the position of the flat block (F). The capsule is transferred to an oven for polymerisation and the final block is obtained (G,H).

The following method is recommended for oriented embedding of flat or elongated specimens (Prentø, 1985). A polyethylene BEEM capsule (size 00) with lid closed is placed (lid down) on a table (Fig. 2.8a), and

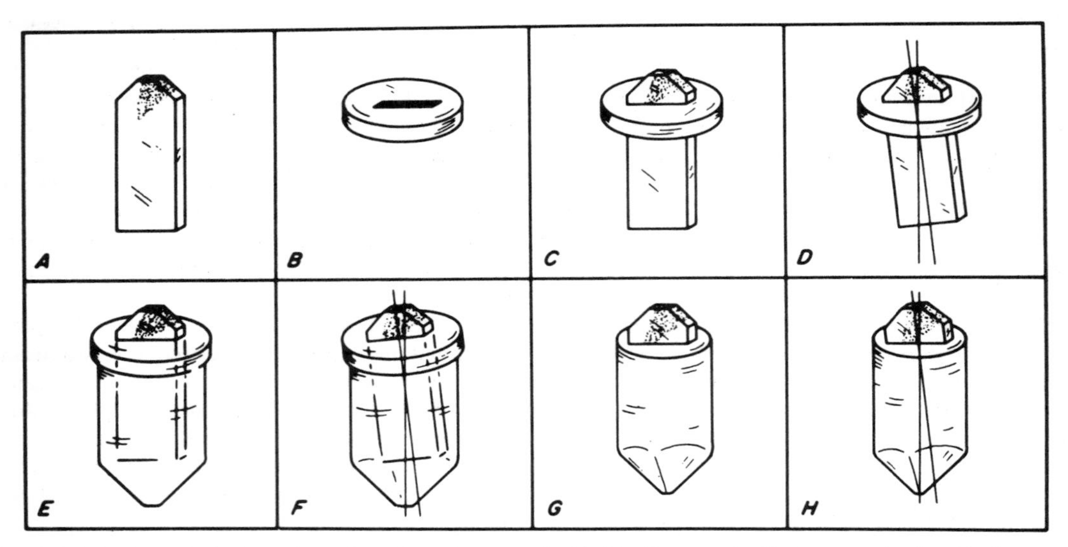

Figure 2.7 Method for obtaining orientation of specimens embedded in flat moulds. Explanation is given in the text. From Hwang (1970)

with a pointed scalpel two parallel cuts are made 3–4 mm apart from the base of the cone to the edge of the lid. Two transverse cuts are made to remove a strip ~3–4 × 12 mm (Fig. 2.8b). The capsule is held horizontally and the resin is injected through the opening created until it is nearly three-quarters filled (~525 μl). Then it is placed horizontally in an oven at 40–50 °C for 16–20 h, to partially polymerise the resin.

The specimen is introduced through the slit, placed on the flat surface, 2–3 mm behind the inner tip of the capsule, and oriented with a needle (Fig. 2.8c). The capsule is filled to the inner rim with prewarmed (40 °C) resin (Fig. 2.8d,e) and placed horizontally in the oven at 60 °C for 2 days, to complete the polymerisation. During embedding and polymerisation, the horizontal capsule may be supported by double-sided adhesive tape or glue.

The following method is useful for maintaining precise orientation of the specimen throughout embedding in water-miscible acrylic resins (Ridgway and Chestnut, 1984). This approach also provides protection for delicate specimens against mechanical damage during processing. Gelatin embedding capsules are preferred to polyethylene ones for these resins; the latter tend to produce blocks of non-uniform hardness. This method combines the already stated procedure of agar pre-embedding with a polyester support platform. The disc-shaped platform serves as a specimen carrier during embedding and, together with the agar support, helps in maintaining specimen orientation during polymerisation.

Polyester support platforms are prepared by first developing a sheet of unexposed photographic film to remove its emulsion, and then punching discs from this material with a standard hole punch. The disc is finished by making two short parallel cuts into one edge with a razor blade, and folding to form a tab (Fig. 2.9A). The tab serves as both a reference point for orientation and a handle for specimen transfer. The disc (7 mm in diameter) fits into size 00 gelatin capsule. A 2% agar solution is kept liquid in a stock bottle maintained in a 45 °C water-bath.

Each fixed specimen is blotted carefully with filter

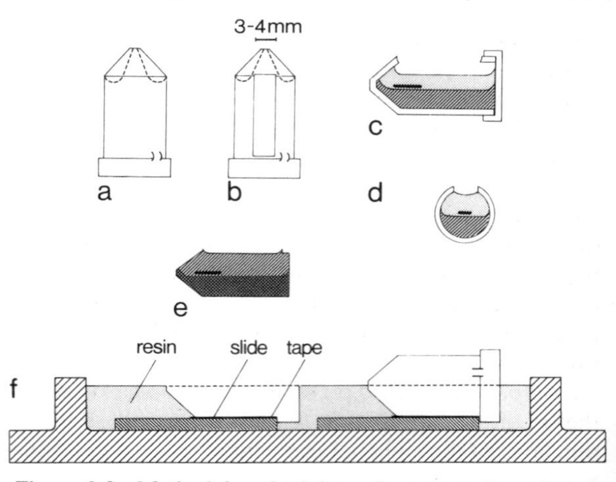

Figure 2.8 Method for obtaining orientation of specimens, especially flat or elongated, during embedding. a: Empty polyethylene BEEM capsule. b: Capsule with slit. c: Longitudinal section through the capsule with oriented specimen. d: Transverse section through resin-filled capsule ready for polymerisation. e: Final block removed from the capsule. f: Section through BEEM capsule support made from the lid of a Ziehl–Neelsen staining jar. From Prentø (1985)

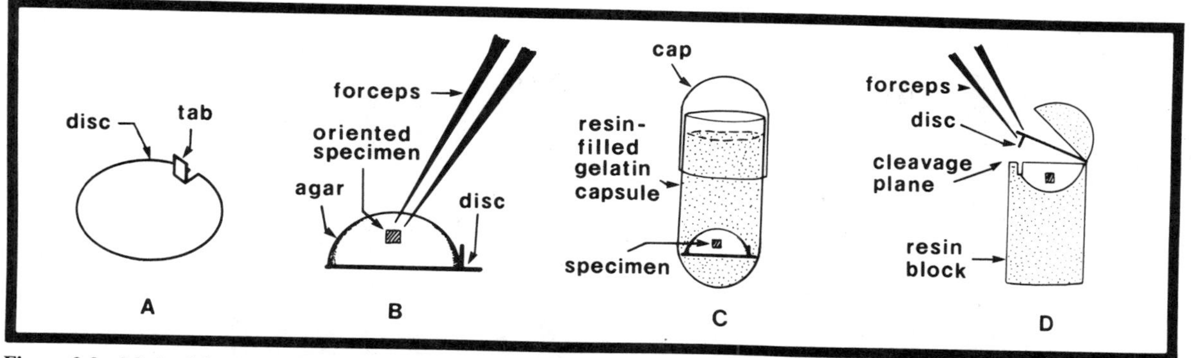

Figure 2.9 Method for improved handling and orientation of small tissue specimens. Explanation is given in the text. From Ridgway and Chestnut (1984)

paper and then positioned on a support disc within a freshly pipetted droplet of agar (Fig. 2.9B). After the desired orientation is achieved relative to the disc, the droplet is allowed to gel for 2–5 min. The specimen–agar droplet–disc is placed into a vial containing the buffer, while other specimens are brought to this stage. The specimens can be post-fixed at this stage or prior to agar pre-embedding. The specimens are dehydrated and infiltrated according to standard procedures.

After two changes of pure resin mixture, each specimen is positioned horizontally in a resin-filled gelatin capsule (Fig. 2.9C), which is capped and heat-cured. After polymerisation, capsules are dissolved in warm water, and each resin block is trimmed down to the level of the disc. The disc is removed with forceps, to reveal the oriented specimen just below the cleavage plane (Fig. 2.9D). The block face containing the area of interest can be trimmed prior to sectioning.

FLAT EMBEDDING

The role of flat embedding in achieving specimen orientation has been discussed earlier. Simultaneous orientation and embedding of thin, flat specimens (e.g. cell monolayers and fungal mycelia) can be accomplished by using flat-bottomed containers modified from standard gelatin capsules (Pépin, 1977). Such containers are obtained by exposing the lower part of the gelatin capsules to water for ~5 min. They are removed from water and, after ~3 min, are pressed into a flat end. Specimens embedded in such flat-bottomed containers can be cut parallel to the flat surface as well as longitudinal to the long axis of the specimen. Alternatively, the conical tip of the polyethylene capsule is cut off, and a cap from another capsule forms the flat bottom. However, embedding in flat moulds is preferred when one needs sections perpendicular to the flat surface of a thin specimen. The use of flat moulds is discussed below.

In the case of certain specimens, a specific plane of orientation can be achieved rather easily by embedding in the silicon rubber flat moulds (Fig. 2.10) (Ladd Research Lab.). Ultramicrotomes equipped with vice-type chucks are available to accept flat blocks. Alternatively, specimens can be embedded in relatively shallow, flat containers such as polyethylene weighing trays, mini-cube ice trays, small disposable beakers cut off ~1 cm from the bottom, small flat-bottomed vials, vinyl cups and caps of vials. It is essential that the monomer resin does not dissolve the container and that the polymerised resin can be easily separated from it. Suitable materials for embedding are aluminium foil, silicone rubber, Cellophane and polyethylene (except for polyester resins). If needed, the container can be protected from the monomer resin by lining it with foil.

The usual method of flat embedding is to embed several specimens in the flat container. Usually a jeweller's saw or mini-hacksaw is used to separate individual specimens from a group of embedded specimens. Alternatively, a tungsten wire (0.2 mm in diameter; used by bacteriologists for making loops) mounted in a jeweller's saw can be used. This wire cuts a very thin channel almost as wide as the wire itself, without damaging the specimen. Both flat and cylindrical blocks can be trimmed with this wire. A segment of the polymerised resin containing the specimen is cut out and glued in the desired orientation to the end of a short Lucite rod with Eastman 910 adhesive. Lucite or Plexiglas rods having the appropriate diameter, which fit in the standard collet-type specimen chuck, are commercially available. During gluing, the exact desired orientation can be achieved. The surfaces to be glued together should be made smooth by rubbing them with sandpaper. This step almost eliminates the possibility of the formation of air bubbles at the interface. The elimination of air bubbles is important, since their presence makes visualisation of the speci-

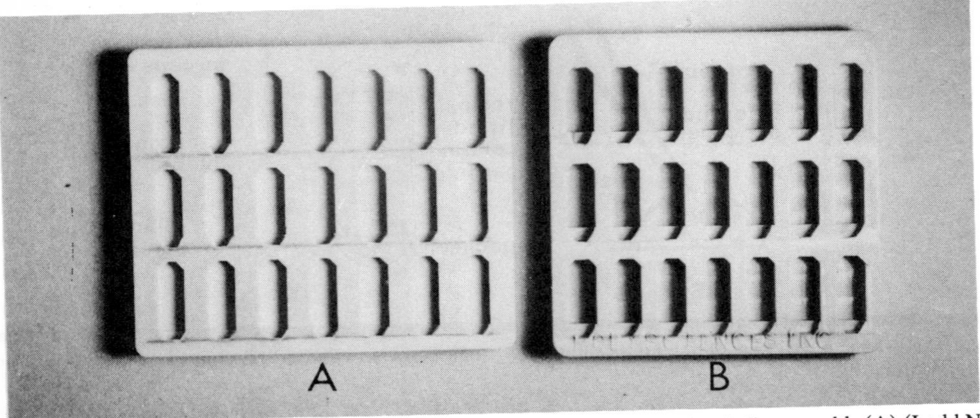

Figure 2.10 Modified flat embedding moulds demonstrating cavity modifications for shallow moulds (A) (Ladd No. 2365), and deeper moulds (B) (Polysciences No. 2615). Note that bottoms of modified tips of troughs in deeper mould are only half as deep as the original cavities. From Shannon (1974)

men difficult and weakens the stability of the bond. A tiny drop of the glue is sufficient to achieve a stable binding. Although this procedure for orientation requires additional time, it is unsurpassed by any other known method in achieving precise orientation for sectioning with an ultramicrotome fitted with a collet-type holder.

LABELLING

Permanent labelling of the embedded tissue is more important than is generally thought. The necessity of permanent labelling becomes quite evident when one has to be absolutely certain regarding the identity of a specimen block which was prepared some time ago and does not show any label. Writing labels on the surface of the specimen block with ink is not recommended, because such labelling is not dependable.

Gelatin or polyethylene capsules are most simply and permanently labelled as shown in Fig. 2.11. A written piece of paper (an index card is preferred; 2.5 cm long) is rolled around the wooden end of a dissecting needle (which acts as a carrier) and lowered in the capsule. By pressing the carrier downward and laterally, the label is made to fit to the curvature of the side walls of the capsule. The label is placed in the capsule before its other contents. After polymerisation is completed, the label becomes permanently embedded in the plastics. For flat embedding, a written piece of paper is placed at the bottom of the embedding dish prior to filling it with the embedding mixture.

RAPID EMBEDDING

A primary limitation in the use of electron microscopy for diagnostic pathology is the long duration needed to process the specimens. Light microscopy is not an entirely reliable means of diagnosing certain diseases such as renal lesions. Electron microscopy can provide an accurate morphological evaluation of biopsy or surgical tissue specimens. Rapid methods are now available which facilitate the completion of fixation through polymerisation of resin blocks in only a few hours.

Rapid methods have been used successfully in the preparation of normal and diseased animal tissues (Hayat and Giaquinta, 1970; Johannessen, 1973; Rowden and Lewis, 1974) as well as plant tissues (Bain and Gove, 1971). In the author's laboratory, other types of specimens, such as tissue culture cells, bacteria, algae and fungi, have also been processed by the rapid schedule, with equal success. Rapid dehydration and infiltration and elevated temperatures during heat polymerisation do not seem to cause distortion of the fine structure. On comparison, little visible difference is found between the tissue processed by the rapid method and that processed by the standard method in terms of the appearance of their fine structure (Fig. 2.12).

Most of the commonly used resins, including Lowicryls, can be used for rapid embedding. The whole process, including fixation, rinsing, dehydration, infiltration and polymerisation, is completed in 3–4 h. Thin sections can be cut when the block is cooled to room temperature. Cooling of the blocks can be expedited at −20 °C. Fixation is carried out at room temperature. The size of the specimen should be as small as possible (<1 mm^3). Longer strips with a diameter of not more than 0.5 mm can also be used. To minimise mechanical damage, biopsy specimens may be immersed in a mixture of glutaraldehyde and formaldehyde or glutaraldehyde for 5–10 min prior to cutting them into very small pieces. The use of formaldehyde alone should be avoided. Stirring of the specimens and the reagents during processing on a rotary shaker is desirable. Embedding capsules should be pre-dried in an oven.

Rapid embedding can be expedited if polymerisation is achieved by microwave heating (McLay *et al.*, 1987). Using a microwave of 400 W, output rating at 2450 MHz, cumulative short exposures with seven 15 s pulses and two 30 s pulses, separated by 30 s cooling intervals, give acceptable polymerisation. In fact, both glutaraldehyde fixation and epoxy resin polymerisation can be carried out in a microwave oven (McLay *et al.*, 1987).

The method developed by Hayat and Giaquinta (1970) for processing normal or diseased animal tissues is given below. Epon can be replaced with any one of its substitutes.

(1) Fixation in glutaraldehyde (4%) — 30 min
(2) Rinse in buffer — two changes, 1 min each
(3) Post-fixation in OsO_4 — 30 min
(4) Acetone (50%) — 4 min
(5) Acetone (70%) — 4 min
(6) Acetone (95%) — 4 min
(7) Acetone (100%) — two changes, 4 min each
(8) Epon mixture:acetone (1:1) — 15 min
(9) Epon mixture — two changes, 10 min each
(10) Embed in fresh Epon for 1 h at 99 °C

The procedure used by Bain and Gove (1971) for processing plant tissues is based on the method described above, except that 6% glutaraldehyde was used and the specimens (0.5 mm) were dehydrated with ethanol and infiltrated with epoxypropane prior to embedding in Epon. Similar methods have been used to process diseased tissues (Johannessen, 1973). Vestopal (Estes and Apicella, 1969) and Quetol 651 (Fujita *et al.*, 1977) have also been used for rapid embedding. Tissue fractions have also been prepared by this method (Takagi *et al.*, 1979).

Rapid embedding methods using Lowcryl K4M have been introduced. The preservation of antigenicity and ultrastructural detail is satisfactory, although membranes are less distinct. Altman *et al.* (1984) introduced rapid embedding in Lowicryl K4M for immunoelectron microscopy; the whole procedure is completed within 4 h. All steps except polymerisation are carried out at room temperature. The infiltration time is shortened and the polymerisation time is also reduced. The reduced distance between the resin specimen and the UV lamp results in a net reduction in irradiation compared with the exposure at a distance of 40 cm for a few days, according to the conventional procedure. Rapid polymerisation is achieved in 45 min or less at 4 °C. The procedure of dehydration and embedding is given below.

(1) Dimethylformamide (50%) — 10 min
(2) Dimethylformamide (75%) — 10 min
(3) Dimethylformamide (90%) — 10 min
(4) Dimethylformamide:Lowicryl K4M (2:1) — 10 min

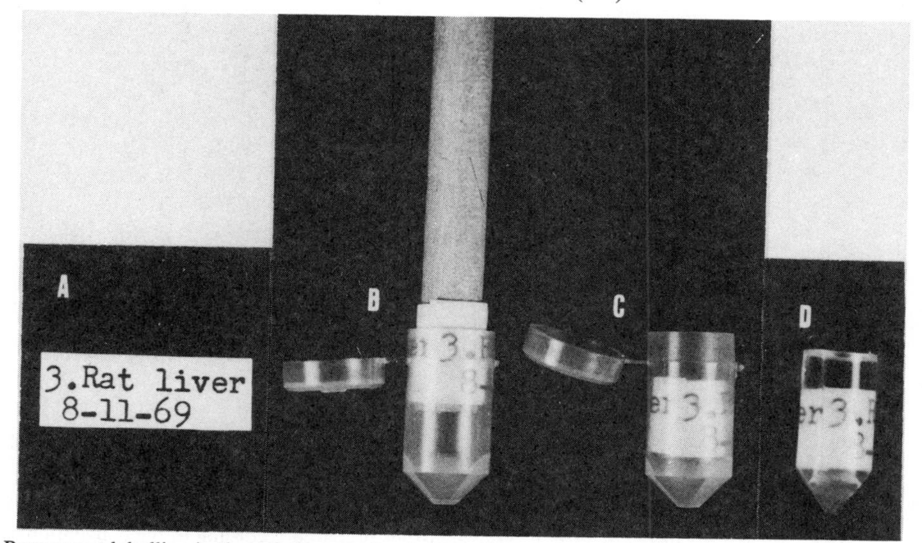

Figure 2.11 Permanent labelling in the gelatin capsule. A: Typed label. B: Label rolled around the wooden end of a dissecting needle being lowered into the capsule. C: Label fitted to the curvature of the side walls of the capsule. D: Polymerised specimen block with embedded label

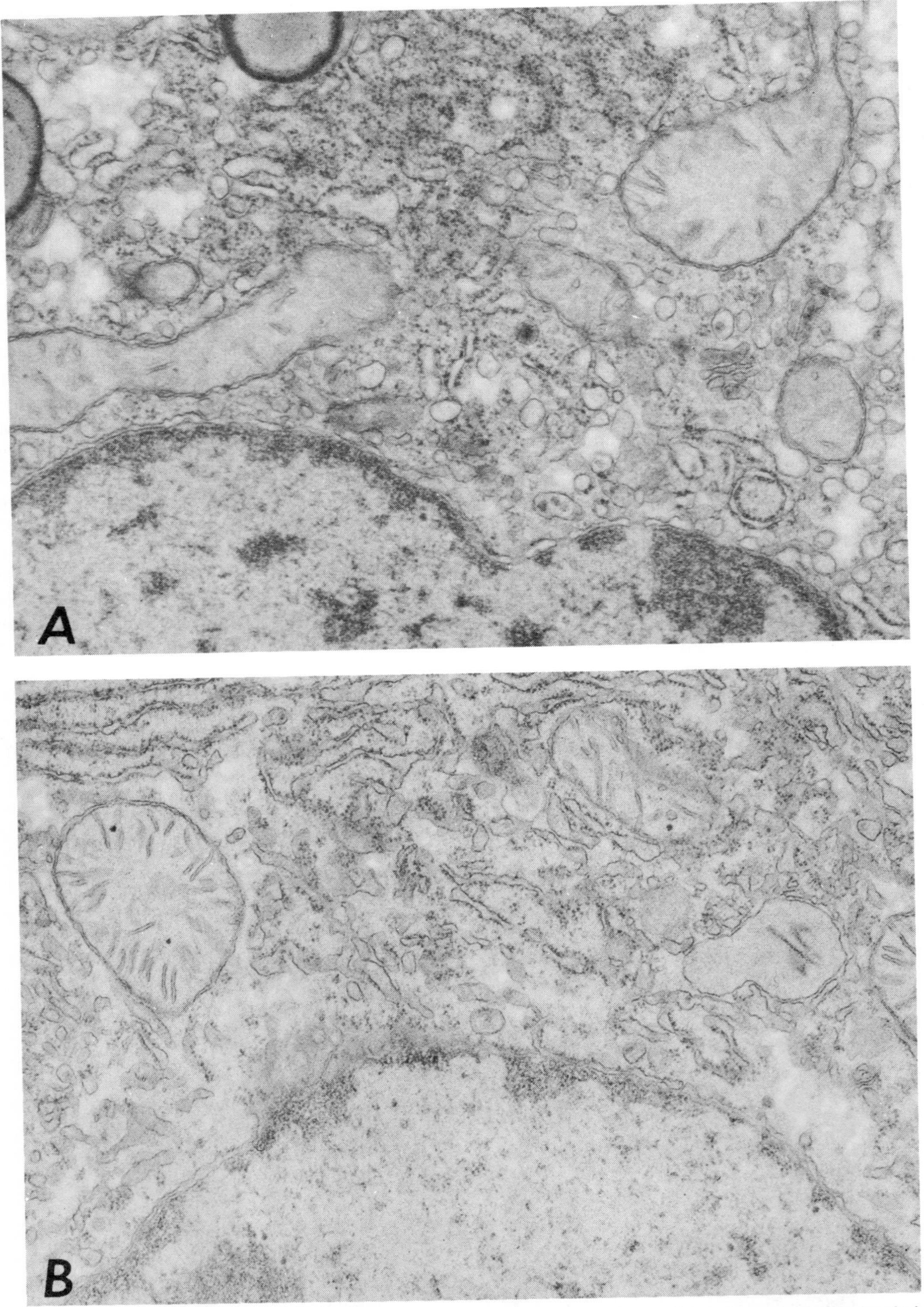

Figure 2.12 Comparison between the ultrastructure of rat liver tissues prepared according to the conventional technique (A) and those prepared by the rapid method (B). There is no visible difference in the appearance of the ultrastructure of the two sections

(5) Dimethylformamide:Lowicryl K4M (1:1) 15 min
(6) Lowicryl K4M 20 min
(7) Lowicryl K4M 25 min

Agitation of specimens in these steps is recommended. Polymerisation is carried out in the following mixture of Lowicryl:

Monomer A	2 g
Monomer B	13 g
Initiator C	75 mg

Tissues are transferred to fresh mixture in BEEM capsules and exposed to UV irradiation (GE 15 W, Black-Lite) at a distance of 10 cm from the lamp for 45 min or less at 4 °C. Altman *et al.* (1984) have introduced a device to accomplish polymerisation.

Another rapid embedding method using Lowicryl K4M and suitable for diagnostic work was introduced by Simon *et al.* (1987). The initiator is replaced with benzoin ethyl ether, and the temperature of polymerisation under indirect UV light is increased from −35 °C (238 K) to −10 °C (263 K) for 18 h, followed by 6 h at room temperature. Although this method takes much longer than that introduced by Altman *et al.* (1984), polymerisation for 24 h is still shorter than earlier protocols. The initiator C is replaced with benzoin ethyl ether in the Lowicryl mixture given on p. 110. During dehydration, the temperature is progressively lowered from 4 °C to −5 °C. The dehydration and infiltration procedure is given below.

(1) Ethanol (30%) 30 min
(2) Ethanol (50%) 30 min
(3) Ethanol (70%) 30 min
(4) Ethanol (95%) 30 min
(5) Ethanol (100%) 30 min
(6) Ethanol plus Lowicryl K4M (1:1) 2 h at −10 °C
(7) Ethanol plus Lowicryl K4M (1:2) 12 h at −10 °C
(8) Lowicryl K4M 8 h at −10 °C

Polymerisation is accomplished in 04 gelatin capsules under indirect UV light for 18 h at −10 °C and then at room temperature under UV light for 6 h.

GRADUAL, PROGRESSIVE DEHYDRATION AND EMBEDDING

There is some evidence indicating that standard methods of dehydration with a solvent and infiltration with a resin introduce artefacts, including considerable specimen shrinkage (~30–40% for soft tissues). At least part of the volume shrinkage through the procedures of embedding is caused by osmotic gradients set up in the tissue (Rostgaard and Tranum-Jensen, 1980). Maximum shrinkage during hydration occurs in the steps between 70% and 100% of the solvent. The only aspect of routine specimen processing that will be discussed here is the adverse effect of incremental increases in the concentration of solvents and resins; advantages and disadvantages of all other aspects of specimen processing have been presented elsewhere in this volume. This adverse effect is especially serious in regard to fragile specimens for TEM and soft specimens processed for SEM. The specimen surface is particularly vulnerable to osmotic changes.

Superior preservation of cellular components, especially specimen surface morphology, can be obtained by employing progressive, gradient-free protocols of dehydration and infiltration. Since very slow and continuous changes of concentration of the preparative media are accomplished, introduction of steep osmotic gradients into the tissue is avoided. Faithful preservation of diffusible cellular substances is not the primary objective in this approach. It has also been shown that the shape and size of erythrocytes are preserved better when glutaraldehyde concentration is slowly increased during fixation (Eskelinen and Saukko, 1982).

Various types of apparatus and procedures have been introduced for processing soft specimens without subjecting them to steep osmotic gradients (Peters, 1980; Rostgaard and Tranum-Jensen, 1980; Jensen *et al.*, 1981; Marchese-Ragona and Johnson, 1982; Brown, 1983a). Another advantage of some of these systems is the saving in manpower, because the procedure can take place automatically. Gradient-free dehydration can be accomplished by using a simple apparatus (Fig. 2.13) introduced by Jensen *et al.* (1981). In this procedure, fixed tissue blocks are pinned to a rubber ring in a beaker filled with buffer which is gradually replaced by acetone over 24 h; the tissue blocks are pinned to a rubber ring by a 26 gauge needle. Magnetic stirring ensures complete mixing of buffer and acetone. This system is useful for the dehydration of relatively large tissue blocks for SEM.

Progressive infiltration (in contrast to incremental increases) by the embedding medium can be accomplished by placing a 4–6 mm layer of 100% acetone (or ethanol) on top of a 15–20 mm layer of pure resin in a glass vial (Marchese-Ragona and Johnson, 1982). The vial is shaken very gently, to cause the boundary between the acetone and resin to become diffuse. The result is an acetone–resin gradient 1–2 mm deep. The dehydrated specimen is transferred from 100% acetone to the acetone layer in the vial. The specimen will sink until it reaches the acetone–resin gradient. As the acetone–resin exchange proceeds, the specimen sinks

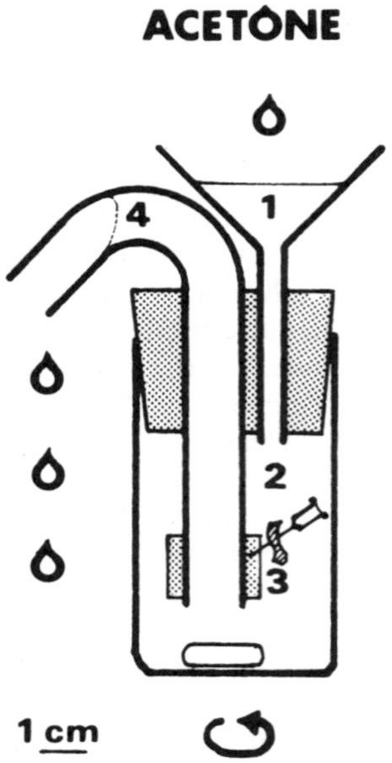

Figure 2.13 Apparatus for gradient-free dehydration for scanning electron microscopy; it can be modified for the dehydration of small specimens for transmission electron microscopy. Apparatus: 1, funnel-intake for acetone; 2, beaker; 3, specimen pinned to rubber ring; 4, outlet of mixture. From Jensen *et al.* (1981)

into progressively increasing concentrations of resin, until it reaches the 100% resin. The tissue specimen (1 mm^3) usually sinks in the pure resin in 30–60 min at room temperature. The acetone–resin gradient and resin are decanted and discarded, and the specimen is immersed in 2–3 changes of fresh resin and then polymerised.

The semiautomatic fluid exchange apparatus (Fig. 2.14) introduced by Brown (1983a) allows the completion of all specimen processing from fixation to embedding in the same exchange chamber, with minimum osmotic stresses. The following description should help in the construction of such apparatus. It consists of a glass fluid exchange chamber with a built-in perforated platform to support the specimen during processing. A captive stirrer located under the platform gently rotates, mixing the contents of the chamber. The stirrer is operated from the outside by a rotating magnet. A reservoir is clamped over the exchange chamber. The reservoir is a 50 ml dropping funnel provided with a Rotaflo GP stopcock, underneath which a glass sidearm is formed for the attachment of a vacuum pump. A 10 ml syringe is used to supply the necessary vacuum.

The fluid exchange chamber is constructed by closing and flattening one end of a Pyrex glass tube, and perforations are formed with a hot needle. The stirring chamber is fused onto the perforated platform, and the glass is drawn out to contain and seal the stirrer within. A length of iron wire sealed inside a thin glass tube forms the stirrer. The outlets of the exchange chamber and reservoir should have the same internal diameter. The magnetic stirrer consists of a horseshoe magnet attached to a wheel 7 cm in diameter which is mounted horizontally onto a hollow shaft and belt-driven by a small variable-speed motor. Inserting the outlet of the exchange chamber through the shaft places the captive stirrer within the rotating magnetic field.

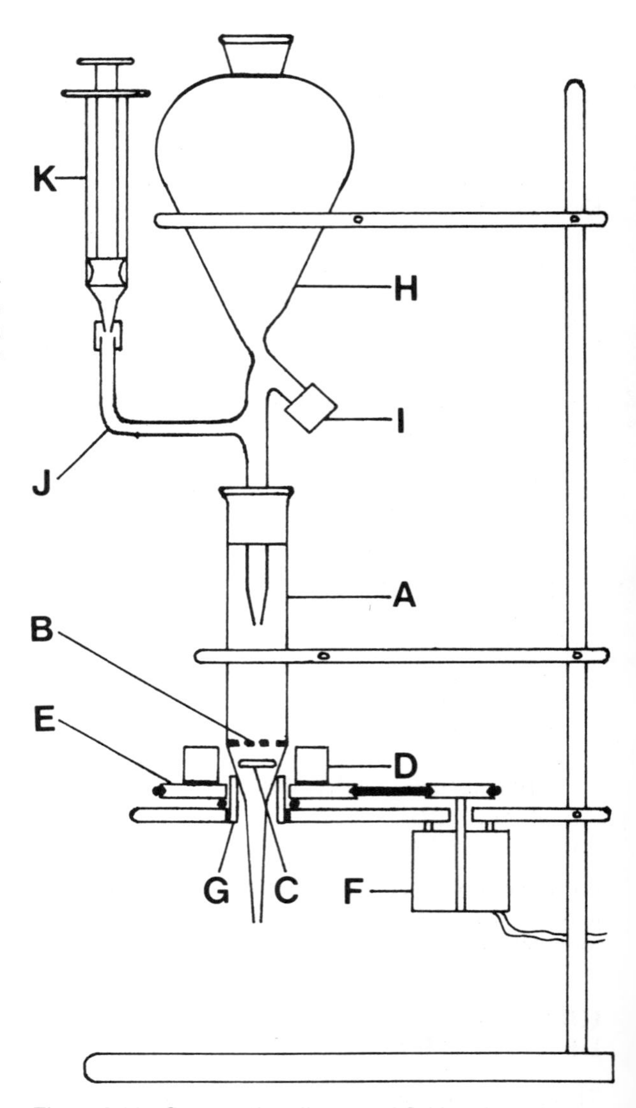

Figure 2.14 Cross-section diagram of fluid exchange apparatus: A, exchange chamber; B, perforated platform; C, captive stirrer; D, magnet; E, rotating wheel; F, motor; G, hollow spindle; H, reservoir; I, Rotaflo GP stopcock; J, sidearm; K, syringe. From Brown (1983a)

The buffer-rinsed (or fixed) specimen, mounted in or on a suitable carrier, is quickly placed into the exchange chamber. The reservoir is clamped over, and the apparatus is attached to the mixer. Using the syringe as a vacuum pump, sufficient fixative is drawn into the exchange chamber to completely cover the specimen. The mixer is switched on, and the speed of the rotating magnet is adjusted to obtain gentle stirring of the contents without causing turbulence. After fixation, the fixative is gently expelled and replaced with buffer solution. Subsequent steps of post-fixation with OsO_4, rinsing and uranyl acetate staining *en bloc* are carried out in a similar way.

During the last buffer (or distilled water) rinse, the reservoir is filled with acetone (about ten times the volume of the contents in the exchange chamber). The stopcock is opened, and the contents of the reservoir are allowed to drop slowly into the exchange chamber over 30 min, mixing with and diluting the buffer. Since the apparatus is airtight, the contents of the exchange chamber remain constant. As the acetone enters the chamber in drops, an equal amount at the same rate will be expelled from the chamber. Thus, there will be a progressive increase in the acetone concentration, the end result being pure acetone. Three further changes of pure acetone are drawn into the chamber at 20 min intervals, completing the dehydration.

To infiltrate, ten volumes of Spurr resin are placed into the reservoir, and the progressive exchange process is repeated. The specimen is removed and transferred into embedding capsules for polymerisation. Cleaning is accomplished by flushing the apparatus with pure acetone, soaking overnight in Taab resin solvent (or any other appropriate solvent) and rinsing with distilled water.

LOW DENATURATION EMBEDDING

Significant protein denaturation and lipid extraction occur during dehydration, infiltration and polymerisation when conventional methods are used. This damage is caused mainly by fixation with OsO_4, dehydration with potent organic solvents and heat polymerisation. Attempts have been made to lower the degree of protein denaturation during embedding by using alternative reagents (Sjöstrand and Barajas, 1968). Other variations of low denaturation embedding have been used (Cohen and Kretzer, 1982; Hoff and MacInnis, 1983). These variations, although claimed to be of significant importance, have not been tested by other workers. Other protocols for lowering protein denaturation include low-temperature dehydration and embedding in Lowicryls (see p. 110) (Carlemalm *et al.*, 1982a), Nanoplast (see p. 113) (Bachhuber and Frösch, 1983) and LR White (see p. 112) (Newman *et al.*, 1983). The goal of these methods is to minimise biochemical perturbations during specimen processing. By maintaining a partially polar environment, antigenic specificity seems to be maintained. The disadvantage of these methods is the long time required to complete it.

The following low denaturation embedding protocol is based on the work of Sjöstrand and Barajas (1968) and Marenus (1985). Very small tissue specimens (<0.1 mm) are cross-linked with 1% glutaraldehyde in Tyrode's buffer (pH 6.8) at 4 °C for 20 min and then rinsed in buffer at 4 °C for 30 min. Specimens are pelleted in a Microfuge (Beckman, Model 11) at 12 000 rev/min for 30 s. The bottom of the Microfuge tube is sliced off, and the sample is removed, blotted on filter paper and resuspended in another Microfuge tube containing 100% ethylene glycol at 4 °C for dehydration. After 150 min, the sample is pelleted, removed from the tube and carefully blotted to remove any excess ethylene glycol. Infiltration is initiated by suspending the sample in a 4:1 mixture of Vestopals 310 and 120L at 4 °C and repelleting. The sample is transferred into a 15 ml tube containing the Vestopal mixture at room temperature, and tumbled at 1 rev/min for 5–8 days to achieve complete infiltration. Then 0.2% benzoin methyl ether (catalyst) is added. After an additional 24 h of tumbling, the samples are transferred to dried gelatin capsules and exposed to long-wave UV radiation (366 nm, 750 μW at 1 m) for 2–3 days at room temperature. Thin sections can be floated on a 0.1 N NaCl solution (pH 7.0) in the trough, to allow for maintenance of antigenicity on the section surface and provide buoyancy for the sections.

REVERSIBLE EMBEDDING

Reversible embedding involves removal of the medium in which the specimen is embedded. This is accomplished after sectioning. Two goals are achieved by using resinless section: (1) improved revelation of certain cellular ultrastructures and (2) easy access to intracellular components by probes coupled to markers visible by light or electron microscopy. Sections divested of their embedding medium, in addition, allow increased depth of penetration by the electron beam, which, in turn, facilitates stereoscopic viewing of relatively thick sections. In other words, in the absence of embedding resin, thicker sections can be examined, since the incident electron beam is not absorbed and scattered by the resin. These sections, of up to 1 μm thick (for detergent-extracted specimens), can be examined in a conventional electron microscope, providing additional spatial cues. When coupled with stereomicroscopy, the specimen provides increase spatial information (Capco and McGaughey, 1966). On the other hand, embedding drastically reduces the general contrast of cellular structures, especially cytoskeletal framework. Epoxy resins also impede interaction between the stain and many cellular constituents. As a result, certain structural entities of the cell are not readily visible by electron microscopy in thin sections of specimens embedded in resins. For example, protein fibres are imaged with high contrast even without staining with heavy metals in resinless sections, since background electron scattering by the resin is absent. Alternatively, if a high-voltage electron microscope is available, cytoskeleton in whole mount preparations of cells without embedding can be observed with this instrument (Porter and Anderson, 1982).

Although resinless sections have been used almost exclusively for investigating the cell cytoskeleton, the method is also useful for revealing certain nuclear components (Lachapelle and Lafontaine, 1987) and the association of virus with the cytoskeleton (Weed *et al.*, 1985). The reversible embedding method can be used in conjunction with conventional or high-voltage electron microscopy, negative staining, critical point drying and rotary shadowing. Resinless sections can also be used for stereo-imaging.

Embedding resins scatter electrons sufficiently to obscure any cellular structure that possesses the same scattering properties as the resin, or closely similar ones. Hence, to increase the contrast among cell structures of similar scattering properties, staining with heavy atoms is in common use. Such staining also differentiates cell structures from the background resin.

A number of embedding media have been used for reversible embedding: polyethylene glycol (PEG) (Wolosewick, 1980), diethylene glycol distearate (DGD) (Capco *et al.*, 1984), polyvinylalcohol (Small, 1984) and polymethyl methacrylate (Plexiglas) (Gorbsky and Borisy, 1985). Steedman's polyester has also been employed as a reversible embedment for combined light and scanning electron microscopy (Norenburg and Barrett, 1987), while PEG has been used for correlative light, scanning and transmission electron microscopy (Kondo and Ushiki, 1985). To determine the quality of ultrastructure preservation in PEG, after removing the PEG embedment, the same specimen is re-embedded in a resin. Images of thin sections thus obtained are comparable to those seen in conventionally fixed, resin-embedded specimens (Nagele *et al.*, 1983). Both DGD and polymethyl methacrylate can be used for obtaining thin or thick embedment-free sections. These two media have several advantages over PEG: brittleness is reduced; sections can float on the water-filled knife trough; ribbons of sections can be produced; thin sections produce interference colours, allowing determination of their thickness; and sections can be easily cut with glass or diamond knives.

Polymethyl methacrylate is a resin that sections more easily than epoxy resins. According to Lachapelle and Lafontaine (1987), polymethyl methacrylate yields better preservation of nuclear ultrastructure than that furnished by DGD and PEG. It is assumed that polymethyl methacrylate undergoes less change in volume during hardening, compared with that shown by the other two embedding media. An excessive change in volume may cause stretching within cells.

The removal of embedding media, especially epoxy resins, from sections is difficult. Waxes used as embedding media are relatively easy to dissolve. In any case, since potent solvents are used for this purpose, damage to cell ultrastructure and dissolution of cellular constituents are inevitable. Transfer of the cytoskeleton from one solution to the next may cause breakage of the cytoskeletal framework. This may result in the loss of some filaments during subsequent sectioning and de-embedding (Capco and McGaughey, 1986). A slight flattening of the portion of a cytoskeleton which is in direct contact with the grid may occur. Shrinkage of the cytoskeleton occurs in de-embedded sections during critical point drying. Fine structural details in such sections are easily damaged by exposure to humid air, resulting in rehydration and subsequent air drying (Kondo, 1984). To avoid this artefact, de-embedded sections should be stored over silica gel until ready for viewing. Considering the possibility of varied artefacts, the de-embedding procedures have a limited goal—i.e. to reveal specific cell structures that are obscured by the embedding medium.

POLYMETHYL METHACRYLATE

Polymethyl methacrylate is composed of linear polymers held together not by covalent cross-linking but by hydrogen bonding. The resin can be extracted from the sections by immersing them in a suitable organic solvent such as acetone. It is useful for reversible embedding. Thin or thick sections are easy to cut.

A piece of polymethyl methacrylate (Rohm & Haas Co.) weighing 120 g is broken into small pieces and placed in a bottle (Gorbsky and Borisy, 1986). About 600 ml of dichloromethane is dried over a molecular sieve for at least 1 week and then transferred to the bottle. The resin is completely dissolved within 2 days with intermittent stirring. About 6 ml of dibutyl phthalate is added as a softening agent. Increased proportion of this reagent results in softer resin. The embedding compound is stable for at least 1 year at room temperature. Since polymethyl methacrylate is a potential carcinogen and dichloromethane is a narcotic, they must be used in a fume-hood.

Fixed specimens are dehydrated in acetone and then carried through two changes of sieve-dried dichloromethane. Infiltration is carried out with a 3:1 mixture of dichloromethane and the resin for 1 h with gentle agitation. The specimens are infiltrated with a 1:1 mixture of dichloromethane and the resin for 2–4 h with agitation, and then transferred to the resin and left to be infiltrated overnight with agitation. Tissue specimens are transferred to wide caps (at least 25 mm in diameter and 14 mm deep) to permit evaporation of the solvent. Prior to evaporation, specimens are covered to at least five times their height in the resin.

To avoid the formation of a surface layer of dried resin during evaporation, a high solvent vapour concentration is maintained above the specimens for keeping the resin surface liquid. This can be accomplished by placing the evaporation caps in a large covered glass Petri dish in a fume-hood. It takes 8–12 h for the solvent to completely escape from the loosely fitting lid of the dish. The slightly soft resin containing the specimen is pried from the cap and placed in an open dish in the fume-hood; the resin is hardened within a few hours.

For critical point drying, sections 0.25–0.5 μm thick are collected on grids coated with Formvar-carbon and immersed in dry 100% acetone for 10 min to remove the resin. After critical point drying, the sections can be coated with a thin layer of carbon or rotary shadowed at an angle of 27° with platinum-carbon, either a very light coat being applied to stabilise structure or a heavy coat to enhance contrast. For negative staining, thin sections are mounted on glow-discharged, Formvar-carbon-coated grids and allowed to dry. The grids are immersed in 100% acetone for 5 min to dissolve the resin. They are rinsed in acetone and then in water, and then immersed in 1% uranyl acetate for 1 min. Excess stain is withdrawn with a torn piece of filter paper. Figure 2.15 shows the results of de-embedding of a tissue culture cell, followed by immunogold labelling of microtubules and critical point drying.

DIETHYLENE GLYCOL DISTEARATE (Capco and McGaughey, 1986)

Cells can be fixed for embedding in diethylene glycol distearate (DGD) or they can first be extracted to produce the cytoskeleton in a medium containing 0.5% Triton X-100, 100 mM NaCl, 300 mM sucrose, 1.2 mM phenylmethylsulphonylfluoride, 3 mM $MgCl_2$, 10 mM PIPES buffer (pH 6.8) and 5 mM EGTA for 5 min at 23 °C. The cells are fixed with glutaraldehyde and OsO_4. They are dehydrated in ethanol according to standard procedures, and then gradually transferred to 100% n-butyl alcohol.

Diethylene glycol distearate (DGD) (Polysciences Inc.) is filtered through Whatman No. 1 filter paper to remove particulates. The DGD is brought to 0.5% (v/v) with DMSO. Because DGD is solid at room temperature, it is maintained at 65 °C to keep it in a molten state. The cells are transferred gradually from 100% n-butyl alcohol to DGD at 65 °C. They are passed through a minimum of three changes of DGD, for 2 h in each change, placed in flat embedding moulds 4–5 mm deep, and then removed from the oven and allowed to harden. During hardening, two drops of molten DGD are added to each well. The depth of the flat mould is important; if the well is too deep, the DGD will fracture upon removal. For this reason, BEEM capsules cannot be used for embedding. These blocks can be placed in a vice-type block holder for trimming and sectioning. These sections are collected on Formvar-carbon-coated grids pretreated with polylysine, and DGD is removed from sections by immersing the grids in 100% n-butyl alcohol for ~2 h, then in a 1:1 mixture of n-butyl alcohol and 100% ethanol for 15 min, and finally in 100% ethanol for 75 min. Grids containing the sections are critical point dried.

POLYETHYLENE GLYCOL (Wolosewick, 1980; Kondo, 1984)

Tissue specimens are fixed and dehydrated according to standard procedures. Vials containing specimens in 100% ethanol are placed in an oven at 60 °C to warm the ethanol. Specimens are immersed overnight in a 1:1 mixture of a 50% solution of PEG 4000 (mol. wt. 3000–3700) in 100% ethanol and 100% ethanol in an oven at 60 °C; caps of the vials are removed to allow ethanol to evaporate. The specimens are removed from the PEG solution, drained of excess PEG on a warm filter paper and transferred into 100% PEG in well-dried gelatin capsules in an oven. After the specimen has sunk to the bottom of the capsule, the capsule is removed from the oven with a long forceps and quickly immersed in swirling liquid nitrogen for 30 s, which results in hardening of the PEG.

Thin sections (180 nm) are cut with a glass or diamond knife and collected on Formvar-carbon-coated grids that have been treated with polylysine. The grids are immersed in 25% ethanol and then in a

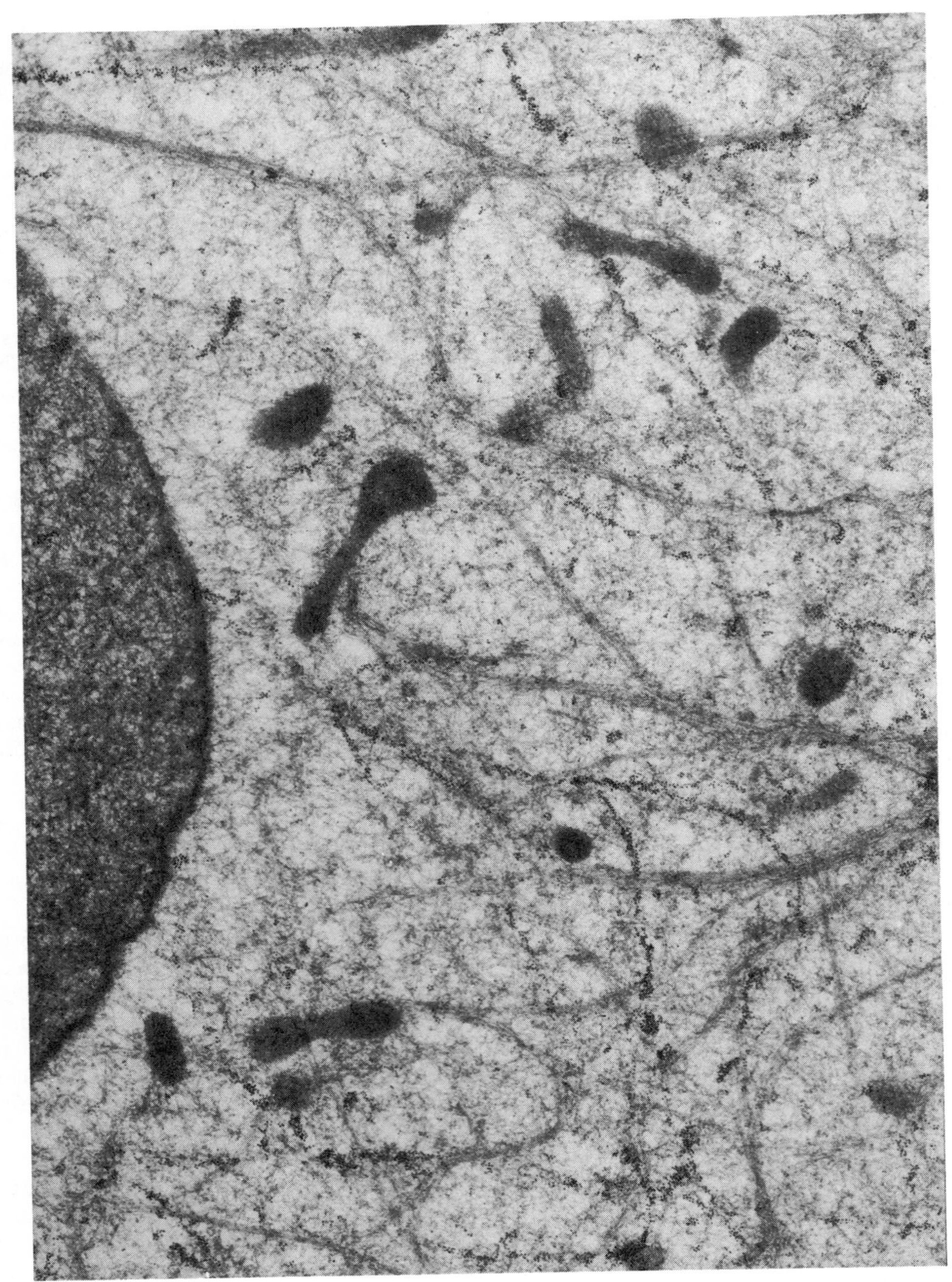

Figure 2.15 Immunogold labelling of microtubules in 0.5 μm section of PtK1 tissue culture cell. Microtubules labelled with 20 nm gold particles course through the cytoplasm among a meshwork of fine cytoplasmic filaments and bundles of intermediate filaments. Resin was removed from the section by immersion in acetone followed by rehydration, antibody labelling, dehydration and critical-point drying. Viewed at 1000 kV. 20 000×. Courtesy G. Gorbsky

graded series of ascending concentrations of ethanol. From 100% ethanol, the grids are critical point dried and stored over silica gel until examined.

Since a certain percentage of sections may become detached from the support film during dehydration and critical point drying, a ribbon of sections is stretched out over one side of a 100-mesh folding grid previously coated with Formvar-carbon and 0.1% aqueous polylysine; the other side of the folding grid is left uncoated (Nagele *et al.*, 1983). The ribbon is sandwiched by closing the grids, and is then submerged in a small dish containing 95% ethanol to dissolve the PEG. After dehydration and critical point drying, the sections can be stabilised by rotary coating with a thin layer of carbon. Embedding in PEG is also described on p. 115.

RE-EMBEDDING

It is not uncommon for a tissue specimen to be found to be poorly embedded in a resin. The resin block may be too soft or too hard, or the tissue may be poorly infiltrated. In each case the production of satisfactory sections is difficult. This problem becomes particularly distressing when no other tissue is available for examination. Examples are biopsy and surgical specimens. The only resort is retrieval of the embedded tissue, which is then reinfiltrated and re-embedded in a resin. Paraffin-embedded tissue blocks or thick sections of these blocks can also be re-embedded in a resin for ultrastructural study. This approach is becoming important because of increasing use of electron microscopy in pathology. Re-embedding of paraffin-embedded tissues in a resin allows electron microscopy in cases in which light microscopy has failed to provide a diagnosis. Additionally, electron microscopy can confirm light microscopic findings. The use of paraffin-embedded tissues allows better control of sample selection, based on a light microscopic survey section. It is easy to randomly exclude a specific area in a very small tissue block used in electron microscopy without prior survey of a large tissue block embedded in paraffin. However, deparaffinisation always results in some damage, especially loss in membrane clarity, to the tissue.

RE-EMBEDDING OF TISSUE POORLY EMBEDDED IN RESIN

If the resin block is excessively soft, the excess resin is trimmed with a razor-blade and then immersed in 100% ethanol saturated with KOH for several hours (Ogura and Oda, 1973). The solution is allowed to stand overnight prior to use; it becomes dark brown in colour. After solubilisation of the resin, the block is rinsed thoroughly with 100% ethanol followed by propylene oxide, and re-embedded in a resin.

According to another method, the tissue block is trimmed free of most of the excess resin with a razor-blade under the dissecting microscope. The tissue is placed in a vial containing propylene oxide for 1 h and then placed in an ultrasonic cleaner (Bauman and Mendell, 1974). This treatment removes much of the remaining resin surrounding the tissue and that which remains becomes softened. Further excess resin is removed with a razor-blade. This tissue is infiltrated with a 1:1 mixture of propylene oxide and Spurr resin with gentle agitation for 18 h at room temperature. This is followed by infiltration with Spurr resin for 48 h with gentle agitation. The tissue is re-embedded for 8 h at 70 °C.

RE-EMBEDDING OF THICK RESIN SECTIONS

Method 1

Araldite 502 can be removed from 10-μm-thick sections of poorly embedded tissue by a relatively simple procedure (McNelly and Hinds, 1975). Thick sections are placed in disposable aluminium weighing dishes and covered with distilled water. The dishes are heated on a hot-plate until the water evaporates, leaving the sections adhered to the bottom. The dishes are filled with the embedding mixture, placed in a vacuum desiccator overnight and then polymerised in an oven at 60 °C for 48 h.

Method 2

Johnson (1976) introduced a rapid and simple method for re-embedding thick sections under microscope control by using the rapidly setting cement cyanoacrylate. A thick section (1 μm) is placed on a drop of water on a No. 1 coverslip and dried on a hot-plate. The section is stained with toluidine blue in 1% borax or with any stain suitable for light microscopy. The coverslip is placed (section side down) on a glass slide without a mounting medium and viewed with the light microscope (a coverslip-corrected objective lens is used).

After a section or a structure for ultrastructural study has been selected, the coverslip with the section is gently removed from the slide and placed on a flat surface with the section side up. A drop of cyanoacrylate is applied to a blank epoxy block, which is quickly pressed against the section and hand-held for several seconds while it sets. After incubation of the section for 1 h in an oven at 60 °C, ice is applied to the coverslip. The coverslip separates from both the section and the block, leaving a smooth surface that may be thin-sectioned without further facing.

Method 3

The following method is useful for re-embedding thick resin sections (mounted on glass slides) that have been examined with the light microscope (King *et al.*, 1982). An embedding capsule filled with a resin and containing a label is inverted over a selected thick resin section mounted on a glass slide. This slide, with the capsule, is placed in an oven at 60 °C long enough for the resin to become firm but not yet completely polymerised; usually this takes 12–24 h. The slide is transferred to an oven at 80–90 °C for a few minutes and then slipped into a special slide holder (Fig. 2.16) (Polysciences Inc.). The slide, held in place by the rim of the slot that fits comfortably in the hand, is grasped firmly and uniformly around the edge, and the capsule is separated from the slide with gentle lateral force (Fig. 2.16). The capsule should easily snap away from the slide. This step should be carried out rapidly, without allowing the slide to cool significantly. The capsule containing the re-embedded section is returned to the oven and allowed to complete polymerisation for another 24 h at 60 °C. This block with the re-embedded section can be sectioned for electron microscopy. Sturrock (1984) has designed a simpler and less expensive holder to help re-embed thick resin sections.

RE-EMBEDDING OF PARAFFIN-EMBEDDED TISSUE IN RESIN

On receipt of the biopsy or surgical specimens, the need for electron microscopy is not always anticipated. When further study with electron microscopy is needed, it may happen that the only specimen available was fixed with formaldehyde and embedded in paraffin. The retrieval of tissue embedded in paraffin will be required. Formaldehyde-fixed specimens, in certain cases, show adequate preservation of the ultrastructure for diagnosis purposes. Various methods are available for re-embedding the tissue from the paraffin block (Lehner *et al.*, 1966; Clark and Rochlani, 1970; Johannessen, 1977; Chien *et al.*, 1982; Bergh Weerman and Dingemans, 1984) or from the paraffin sections (Takeda, 1969; Burns, 1970; Rossi *et al.*, 1970; Bretschneid-

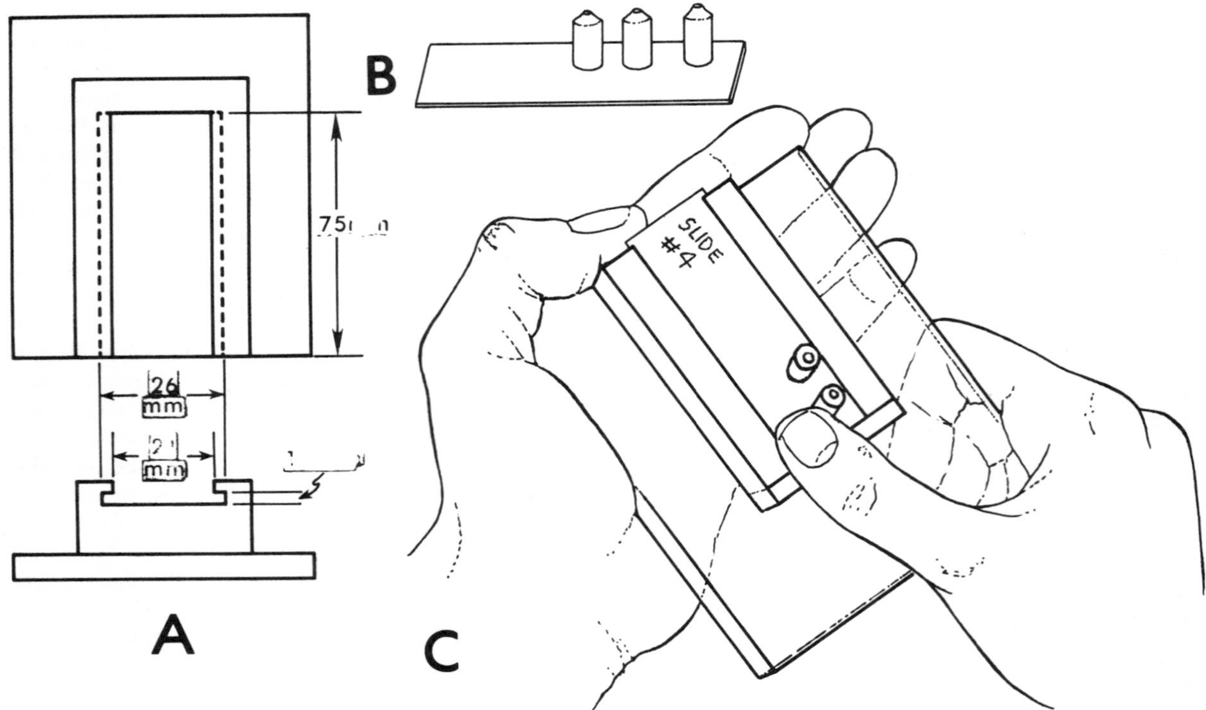

Figure 2.16 Re-embedding thick resin sections. A: Diagram of the slide-holder for assisting with the removal of epoxy capsules containing re-embedded sections from glass slides. The dimensions indicated are appropriate for holding standard slides (24 × 75 × 1 mm). B: The placement of inverted BEEM capsules over selected sections on a slide. C: The slide-holder in use, illustrating a typical grip for applying force to separate capsules from the slide. From King *et al.* (1982)

er *et al.*, 1981; Chien *et al.*, 1982; Lisbeth *et al.*, 1983; Kraft *et al.*, 1983; Yau *et al.*, 1985).

Two-step Method

Aldehyde-fixed and paraffin-embedded tissue blocks are cut into pieces no larger than 3 mm^3 and deparaffinised by placing them in a test-tube containing 100% xylene and rotating them for 1 h. The specimens are rehydrated in an ethanol series of descending concentrations (100%, 95%, 75%, 50% and 15%), for 15 min at each stage, after which there is a final rehydration in buffer overnight. The specimens are retrimmed to pieces no larger than 1 mm^3 and post-fixed with OsO_4. These specimens are then further processed according to standard procedures.

One-step Methods

The following method eliminates rehydration and dehydration, and combines deparaffinisation and osmication in one step (Chien *et al.*, 1982). This simple procedure allows adequate diagnostic interpretation of tissue lesions. An area of interest is removed from the paraffin block and trimmed into cubes 1–2 mm in size. These cubes are immersed in a 2% solution of OsO_4 in xylene for 30–60 min, with gentle stirring. They are then treated with xylene, infiltrated with mixtures of a resin and xylene, and finally embedded in a resin. When the tissues under study are embedded in paraffin-containing plastics polymers (Paraplast), toluene is preferred to xylene.

The following method presented by Bergh Weerman and Dingemans (1984) is faster than that introduced by Chien *et al.* (1982). Paraffin-embedded tissue blocks (1–2 mm^3) are placed in xylene (or toluene) containing 1% (w/v) crystalline OsO_4 and the paraffin is dissolved at 40 °C in an oven for 10 min; gentle agitation of the blocks expedites the dissolution. The specimens are infiltrated (using agitation) with a 1:1 mixture of xylene and epoxy resin at 40 °C for 10 min, and then with pure resin at 40 °C for 10 min. Embedding is accomplished either overnight or for 1–2 h at 100 °C. Workers must be protected from the OsO_4 vapours emitted, especially in the first step.

RE-EMBEDDING OF PARAFFIN-EMBEDDED TISSUE SECTIONS IN RESIN

Method 1

Thick sections of aldehyde-fixed and paraffin-embedded tissue are mounted on glass slides and then deparaffinised by placing the slides in a couplan jar containing 100% xylene. This procedure is followed by rehydration in an acetone series of descending concentrations (100%, 95%, 75% and 15%), 10 min being allowed for each stage, and final rehydration in buffer. The sections on slides are post-fixed with OsO_4 and dehydrated. One or two drops of a 1:1 mixture of acetone and resin are placed over the section, and after 20 min this mixture is replaced by a drop of the pure resin. A gelatin embedding capsule filled with resin is placed face down over the resin-covered thick section on the glass slide and polymerised in an oven at 60 °C. The section adheres to the resin block, which is separated from the slide by immersion in liquid nitrogen or dry ice. Similar procedures can be used for embedding cells grown on coverslips. Before cells are grown on them, the coverslips should be lightly sprayed with Teflon and heated at 250 °C for 20 min.

Method 2 (Kraft *et al.*, 1983)

In the following method, heat is used to separate the section from the glass slide. Tissues are fixed with aldehydes, embedded in paraffin and sectioned at 6–15 μm. The section is placed on a glass slide and stained for light microscopy. Localisation of the region of interest on the slide is accomplished by using a dissecting microscope to make a drawing of the selected areas. After removal of the coverslip with an appropriate solvent, the section is rehydrated rapidly through graded alcohols into 0.12 M PBS (pH 7.2). The section is post-stained with 0.5% OsO_4 in 0.12 M phosphate buffer for 10 min and then rinsed in buffer.

The section is stained with either 1% aqueous methylene blue or 2% toluidine blue and then rapidly dehydrated through graded alcohols to 100% alcohol. Next, n-butyl glycidyl ether (BGE) is applied drop by drop to the slide after it has been drained of excess alcohol. After 5 min, a 1:1 mixture of BGE and Quetol 651 mixture is added to the slide. Five minutes later the slide is drained, and the Quetol mixture is applied twice, 5 min each time. BEEM capsules (size 3), with cap and conical portion removed, are used for embedding. The top rim of the capsule is placed over the selected area on the slide. One hand holds the cap upside down with tweezers while the other rapidly pipettes the embedding mixture into the inverted capsule. When the meniscus is seen above the capsule rim, the cap is placed on it, thus excluding gross air bubbles.

Following 3–4 evacuation cycles in the vacuum oven, the slide (with capsule attached) is left in the oven under vacuum at 60 °C for at least 24 h. After the curing period, the slide is removed from the oven and maintained at 60 °C, because separation becomes difficult, incomplete or even impossible at lower temperatures. Placing the slide on a hot-plate at 62 °C ensures maintenance of the proper temperature. The capsule is grasped near its base with forceps, and the slide is held down with another instrument. Separation is accomplished by snapping off the capsule. The block is now ready for trimming and sectioning.

Epon–Araldite resin gives equally good results and allows the use of BEEM capsules, because the viscosity of the Epon–Araldite resin is greater than that of the Quetol 651 resin. Although the quality of ultrastructural details observed in the sections will not be the same as that in tissues specifically processed for electron microscopy, good results are obtained with this method when specimens have been fixed with glutaraldehyde.

Method 3

Sometimes a paraffin section is the only tissue available for re-embedding. The slide is oriented (section side down) onto a specially designed silicone rubber embedding mould (Fig. 2.17) (Chien *et al.*, 1982) (Polysciences Inc.), allowing the area of interest to be captured in an epoxy block. Any area of a section can be selected

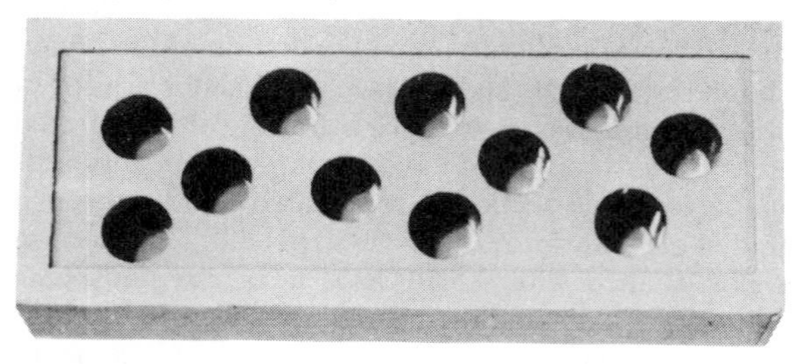

Figure 2.17 A silicone rubber mould for re-embedding paraffin section from a glass slide. From Chien *et al.* (1982)

because the wells in this mould are asymmetrically distributed. After polymerisation, the block is easily removed from the slide by heating it on a hot-plate (100 °C) for 15 s.

Pop-off Method for Re-embedding (Bretschneider *et al.*, 1981)

The pop-off method allows a precise area in a paraffin-embedded section to be re-embedded in a resin. The method is especially useful for cell smears and monolayers, which may be the only specimens available for study. Although ultrastructural detail is often poor, an identical section can be studied under both light and electron microscopes. The area of interest in the stained paraffin-embedded section mounted on a glass slide is lightly circled (the same size as a BEEM capsule) with a diamond pencil on the undersurface of the slide before removing the coverslip. The slide is immersed in xylene to remove the coverslip and then rinsed with fresh xylene to ensure removal of all the mounting medium. The slide is dipped in equal parts of xylene and propylene oxide and then in propylene oxide.

The slide is placed in a 2:1 mixture of propylene oxide and resin for 2 min. After being transferred to a 1:1 mixture of propylene oxide and resin for 2 min, it is placed in a 1:2 mixture of propylene oxide and resin for 10 min. A labelled BEEM capsule is placed in a BEEM capsule holder (Fig. 2.18A). The holder is modified by cutting away a conical area at its base, thus allowing a capsule to be pushed down into the holder so that its top is level with the surface (Fig. 2.18B). The BEEM capsule is filled to overflowing with resin and placed in the holder. The section adhering to the glass slide is inverted gradually over the top of the resin-filled capsule (Fig. 2.18C). If a bubble forms, the slide is removed, 1–2 drops of resin are added and the slide is inverted again.

Polymerisation is carried out overnight in an oven at 76 °C. After cooling to room temperature, the glass slide–BEEM capsule unit is removed from the holder by pushing the capsule up from the underside of the holder. The slide is placed on a hot-plate (100 °C) for 15 s. After the unit has been removed from the hot-plate, the capsule is rocked slightly until it pops the section off the slide. The pop-off section is at the wider surface of the BEEM capsule. The tip of the capsule is cut off as evenly as possible so that the re-embedded section can be viewed under the light microscope by placing the block on top of the condenser lens. The area of interest can be located and marked for ultrastructural studies. The cutting face of the capsule is trimmed into a mesa (a platform with steeply sloping sides). Extreme caution is necessary while sectioning, because the 5–6 μm layer of cells (from the paraffin block) is located immediately at the surface of the block. Particular attention must be paid to ensure

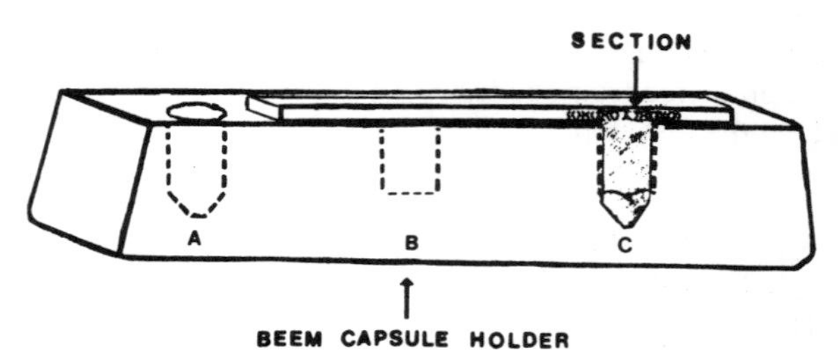

Figure 2.18 BEEM capsule-holder for paraffin section re-embedding. A, Standard capsule well; B, bottom of well cutaway; C, resin-filled capsule in cutaway well. The section is inverted over the capsule. From Bretschneider *et al.* (1981)

accurate orientation of the block face parallel to the knife edge prior to sectioning.

RE-EMBEDDING OF TISSUE CULTURE CELLS (Connelly, 1977)

The following method is used to re-embed tissue culture cells. The cells are cultured in Falcon plastics Petri dishes (35 × 10 mm) in an appropriate medium to confluency or near confluency, depending on the experimental design. They are fixed with 3% glutaraldehyde in 0.1 M phosphate buffer (pH 7.3), rinsed in buffer and post-fixed with 1% OsO_4. Dehydration is carried out in 30%, 50%, 70% and 90% ethanol (propylene oxide should not be used) and then in 90% (two changes), 95% and 100% (two changes) HPMA. This process is followed by infiltration with 1:1, 1:2 and 1:3 mixtures of HPMA and Epon, and then embedment in Epon. The polymerisation is accomplished by placing the uncovered Petri dish overnight in an oven at 60 °C. The embedded cells are removed from the dish by completely breaking off the sides of the dish. With the resin side down, the resin disc is flexed several times and the two layers are separated.

Small rectangular blocks are cut out of the resin layer with either a jeweller's saw or a straight-edge razorblade. These pieces are re-embedded in a small amount of resin reserved from the batch used originally to embed the cells. The re-embedment is accomplished by first half-filling a 4 mm well in a flat embedding mould with the resin. One block of embedded cells is then placed into the well (cell side up), and after the resin has flowed over the surface of the cells, a second block is placed (cell side down) directly over the block already in the well. Care must be taken to avoid air bubbles between the two blocks. Finally, the well is filled to the top with more resin, and the mould is placed in an oven at 60 °C for 24–36 h for polymerisation to be completed. The block is trimmed (Fig. 2.19) and sectioned, with the lines of cells perpendicular to the knife edge, to avoid the breakout of cells due to the possible different states of polymerisation.

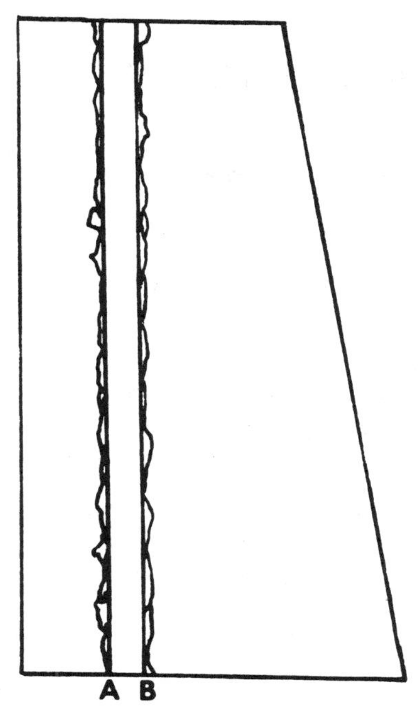

Figure 2.19 Drawing showing typical section shape, which is asymmetrical in order to identify orientation of cell monolayers (A and B) when different populations are embedded together. From Connelly (1977)

3 Sectioning

INTRODUCTION

Electron microscopy of thin sections of tissues and cells provides *in situ* information on the morphology and structure of cellular components and their interrelationship. This approach also provides cytochemical information. No other method of investigation can provide such detailed information. Thin sectioning can be complemented by examining specimen surfaces through freeze-fracturing. Like thick sectioning, thin sectioning has become standardised and reproducible, and is now used routinely for general biological research as well as diagnostic pathology.

The majority of biological specimens are too thick in their natural state to be penetrated by the electron beam. An electron microscope operating at accelerating voltages up to 100 kV normally requires sections ranging in thickness from 10 nm to 100 nm. Thus, the need for thin sections is apparent. In addition, the study of whole sections is necessary in order to understand the structural interrelationship between various parts of a cell or tissue. As the resolution of conventional electron microscopes increases, the demands on the techniques used in the preparation of sections become more exacting. These techniques are now the limiting factor in obtaining the maximum benefit from the use of an electron microscope. Consequently, better-prepared sections are required to take advantage of the increased resolving power of an electron microscope.

To obtain a high-quality electron micrograph, the section must be thin and able to withstand both the high vacuum and electron bombardment. Sections should also be of a known and uniform thickness and free from chatter, wrinkles, breaks or folds. A considerable degree of patience, concentration and finesse is required in order to master the art of thin sectioning. Obtaining good thin sections is difficult; however, the beginning technician should derive comfort and hope from the thought that thousands of high-quality electron micrographs have proved that it can be done.

The major factors that determine the quality of sections are the ultramicrotome, the cutting edge of the knife, the knife angle, the cutting speed, the embedding material, the face of the specimen block and the fluid in the collecting trough. Even the addition of electrolytes or non-electrolytes to a buffer may affect the cutting properties of an embedding medium. Cope (1968) indicated that buffer systems containing sugar or calcium additives had a deleterious effect on the cutting properties of water-miscible methacrylates. The knife edge needed to cut thin sections must be adequately sharp and hard. These special knives require ultramicrotomes of sufficient precision and reliability to cut sections as thin as 5 nm. Several versatile ultramicrotomes are available which can cut sections in the range of 5 nm to several micrometres of thickness: Sorvall Porter-Blum MT-1, MT-2 and MT-5000; LKB Ultrotomes I and III; Reichert OMU2 and OMU3; Cambridge Huxley Mark 2; and Tesla BS 490. These ultramicrotomes operate on the principle of either thermal or mechanical feed.

Sectioning should be carried out in a room free of noise, dust, draught and vibration. Under optimal operating conditions, sections in long unbroken ribbons can be easily obtained. Cutting speed and section thickness can be varied while cutting is in progress. In order to be readily collectable, sections should float on a liquid, which must be in direct contact with the cutting edge of the knife. To accomplish this, the knife must be kept stationary and the specimen mobile. After the section is cut, the specimen moves away, to avoid contacting the knife on its way to the starting position. Either applied force or gravity is employed as the cutting force. Movement of the specimen can be accomplished manually or automatically.

KNIVES

GLASS KNIVES

Latta and Hartmann (1950) first introduced knives made from plate glass in ultramicrotomy. The reasons for using glass are that it is homogeneous, amorphous and hard without being excessively brittle, and has no well-defined slip planes in its structure. It is very strong and does not fail in compression; its theoretical strength is 1–2 million psi. When tested for sharpness by the light-scattering method, a glass edge scatters approximately the same intensity of light as a comparably sharp steel edge. In general, a glass edge is smoother and sharper than a steel edge. However, the edge of a glass knife is not exceedingly sharp but has a wedge-like edge.

The reason for using plate glass is that it can be fractured by using simple breaking methods. Since plate glass has a rather non-uniform consistency, a clean break is not always achieved. However, if controlled and equal forces are applied on both sides of the score line, the fracture runs continuously along the path of stress caused by these forces. Theoretically, the fracture should not deviate from this path of stress. On the other hand, in a material with crystalline structure the fracture may deviate from the path of stress because the forces follow the easiest path. Furthermore, plate glass is characterised by adequate cutting strength.

Most thin sectioning today is being done with glass knives, for they are more easily available, safer to handle and far less expensive than diamond knives. Furthermore, the clearance angle and the cutting speed are of secondary importance. Glass knives have an advantage over diamond knives in that the former can produce semithin sections of large size (1–5 μm). Semithin sections larger than 5 μm can be obtained with the Ralph knife. Knifemakers that produce glass knives with a cutting edge of up to 4 cm are commercially available. However, even the best glass knives have the disadvantage of having to be replaced too often. Each part of the usable edge of even the best knives cuts no more than a few ribbons of good sections. The number of high-quality sections obtained from a given part of the knife edge depends upon several factors, including the composition of the specimen, the hardness of the embedding medium, the size of the section and the section thickness. The softer the specimen block and the smaller and thinner the section, the longer the knife will last. Not all glass knives prepared prove satisfactory, and some will have to be discarded after testing, although ready-made knives are commercially available.

In general, the quality of sections cut with glass knives is somewhat inferior to that of sections obtained with diamond knives. The two surfaces of a section differ in texture—the bottom surface which touches the trough fluid is smoother than the upper surface—and this difference in texture is more pronounced in sections cut with a glass knife than in those cut with a diamond knife. Minimum thickness of sections cut with glass knives is ~30 nm, whereas sections of less than 10 nm thickness can be obtained by using diamond knives. Furthermore, glass knives are unable to yield good sections from many types of tissues being studied today. Hard tissues, such as dense connective tissues containing heavy collagenous fibres or tissues containing calcium or metallic particles, cannot be cut satisfactorily with glass knives. Sections cut with glass knives show less clear images of relatively delicate structures. Another disadvantage of using glass knives is that a considerable amount of time is needed to prepare them, whereas diamond knives come ready to be used. Because of the nature of the fracturing process, glass knife angles cannot be controlled with precision and reproducibility.

However, with the availability of the LKB Knifemaker and the Messer Knifemaker (Japan) the amount of time previously required to prepare glass knives with a pair of pliers can be substantially reduced. These semiautomatic knifemakers produce regular 45° triangular glass knives of 25 mm or longer base, which are used in most ultramicrotomes. A triangular knife of 25 mm base can also be used in the Cambridge Huxley ultramicrotome, which normally requires a trapezoid knife, with the aid of an auxiliary knife holder (Marshall and Sheen, 1970). A larger triangular knife of 38 mm base can also be used in the Cambridge Huxley ultramicrotome if its knife holder is modified. This is accomplished by removing the stop screw and milling a recess in the extreme left-hand side of the back of the holder (Biddlecombe *et al.*, 1971). The Knifemaker II accepts glass strips of three thicknesses (6.4, 8.0, and 10.0 mm), and can produce knives with 40°, 45°, 50°, 55° or 60° knife angle. The usable knife edge per knife is increased when glass knives are made at an included angle of 55° rather than the customary 45° (Ward, 1977). About 42% of knives with an included angle of 55° made with a knifemaker may have more than 33% of their cutting edge free of visible defects.

Glass is subject to flow at room temperature, since it is a supercooled liquid; the thin cutting edge of a glass knife is especially susceptible to this flow. Very little is known regarding the extent of the loss of knife sharpness relative to time, although it has been suggested that knife sharpness is lost in 1 week after preparation. According to Sitte (1984), glass knives can be stored for months without any damage to their sharpness. However, dust particles can damage a stored knife by causing fine nicks in its cutting edge, which may or may not be visible under an optical microscope. It is advisable, therefore, to prepare knives just prior to sectioning.

If glass knives must be stored, they can be kept in sturdy, empty Kodak Electron Microscope Film boxes, which have internal dimensions of 13.5 × 10.6 × 4.0 cm (Newcomb *et al.*, 1984). Two strips of double-sided sticky tape are placed equidistant from each other parallel to the long axis and inside of the box bottom. About 20–25 knives can be stored on the tapes.

Although Schoenwolf (1982) has suggested the use of modified microscope slide boxes to store glass knives, film boxes are easier to use.

The performance of glass knives can be improved by coating the cutting edge with a film of evaporated tungsten metal (see p. 145). Sectioning with glass knives can also be made easy by adding a surfactant such as lecithin to the resin mixture (see p. 98) (Mollenhauer, 1986a). Specimen blocks containing the surfactant are very easy to trim and section. Another approach to facilitate sectioning with glass knives involves the addition of Dow Corning 200 fluid silicone plastic additive to the resin mixture before the addition of the accelerator (Langenberg, 1982 (see p. 94)).

Selection of Plate Glass

Although it is claimed that the quality of a knife is not affected by the type of plate glass used, in the writer's experience different types of glass yield varying qualities of knives, prepared with similar techniques. Specifically, different types of glass vary in their breaking and cutting qualities. The toughness or resistance of a knife edge to the impact of the specimen block face varies, depending, in part, upon the type of glass. Knives made from some grades of glass are more durable than those made from glass of other grades. Differences between the manufacturing processes for different types of glass are probably responsible for variations in their quality, which, in turn, result in differences in the quality of knives.

Because so many different types of plate glass have been successfully used, it is difficult to suggest any one source as being better than any other. However, some workers prefer glass with a high silica content, while others favour white glass (hard glass). Unpolished sheets of glass are thought to be superior to those made smooth by a polishing process, since polishing introduces stresses into the glass.

Probably the most important factor in selecting a glass is the thickness of its sheets. Sheets of plate glass having a thickness of 5–6.5 mm produce knives with a sizable cutting edge. Also, glass sheets of this thickness have proved to be fairly easy to break. In this writer's laboratory, strips (2.5×40 cm) of plate glass (~6 mm thick) sold by Ivan Sorvall Inc., Norwalk, Connecticut and by LKB-Produkter AB, Stockholm, Sweden, have consistently produced knives of good quality. However, it should be remembered that each consignment of glass may vary slightly in its cutting and breaking qualities. These strips apparently cannot be used to obtain a knife by the 'free break' method described later. The use of a 1-in-high knife for the MT-1, MT-2 and LKB microtomes has been proved most satisfactory for sectioning biological materials for general purposes.

Preparation of Glass Knives

The critical element in obtaining good thin sections is the quality of the cutting edge of the knife. One has a better chance of obtaining a 'perfect' cutting edge if reliable techniques are employed to prepare a knife, and several minor but important details are taken into

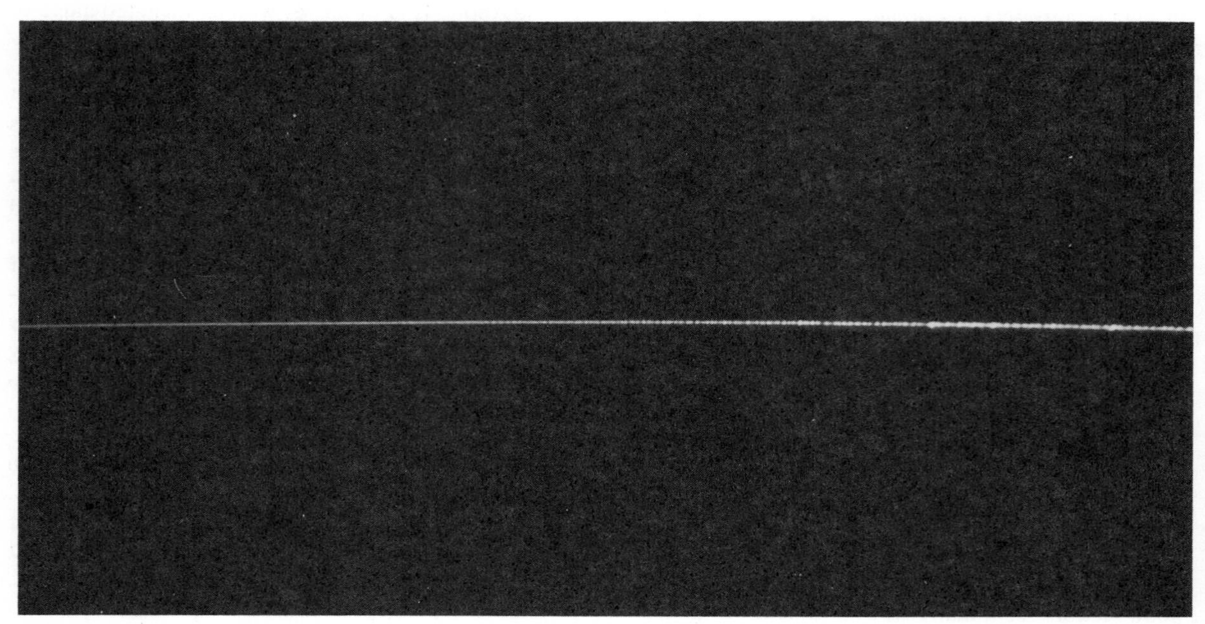

Figure 3.1 Image of a glass knife edge seen with light scattering. The imperfections are seen to the right (nearer the spike) as irregular spots of light. The seemingly flawless part of the edge appears to the left as a thin, even, light line. (From Ward, 1977)

account. A 'perfect' cutting edge may be defined as one which is as straight as the cutting edge of a diamond knife, and in which ~90% of its length is free from minor imperfections. Such a high-quality cutting edge should show only parallel interference fringes and no check marks (whiskers, feathers or striations) under an optical microscope (Fig. 3.1). It is pointed out that, contrary to widely accepted belief, the entire cutting edge possesses extremely fine striations, regardless of whether or not they are visible under an optical microscope. Those striations which are visible only with the aid of a scanning electron microscope are termed 'steps' (Fig. 3.2). The height of these 'steps' ranges from less than 0.1 μm to 0.3 μm (Black, 1971a). The presence of the 'steps' is one of the major reasons why sections cut with a glass knife are inferior to those cut with a diamond knife. These 'steps' are a natural consequence of the glass fracture process.

The novice is assured that with practice and a little patience cutting edges of high quality are obtained. It should be remembered that an *exceptionally sharp and strong* knife edge is capable of cutting even tissues that are not optimally embedded. On the other hand, a *fairly good* knife edge is capable of cutting only optimally fixed and embedded tissues. Only a *perfect* knife can produce flawless sections.

At present, most glass knives are made with a knifemaker (e.g. LKB KnifeMaker II). By applying equal weight and pressure to each side of the score, a straight, controlled break in a strip of glass in any of the three thicknesses (6.4, 8.0 and 10 mm) produces a set of 16 squares (25 × 25 mm square). Each square is broken into equal halves, producing two 45° glass knives; either one or both halves would be good knives. The knifemaker can produce knives with 40°, 45°, 50°, 55° and 60° knife angle. Since the knifemaker accommodates glass strips of up to 10 mm thickness, histology knives with a long cutting edge for cutting semithin sections can also be obtained.

If a knifemaker is not available, only two tools are required to make glass knives: a pair of glazier's pliers (Fig. 3.3) with wide parallel flat jaws and a wheel-type glass cutter (Fig. 3.4). Ordinarily, a 6-mm-thick sheet of plate glass of any convenient size (e.g. 102 × 204 mm) is satisfactory. It is first washed in a detergent (soap) solution and then thoroughly rinsed with hot water. The sheet is then dried by being wiped softly with a piece of clean cloth. Some workers avoid wiping the sheet dry, since rubbing produces electrostatic charges which may attract dust particles. These workers prefer to dry the sheet by letting it stand in a dust-free place. Utmost care should be taken not to touch the clean sheet with bare hands, even if the hands have just been cleaned. It is advisable to cover the working space of the table with thick paper to keep this space clean, since glass chips and dust would contaminate the glass sheet.

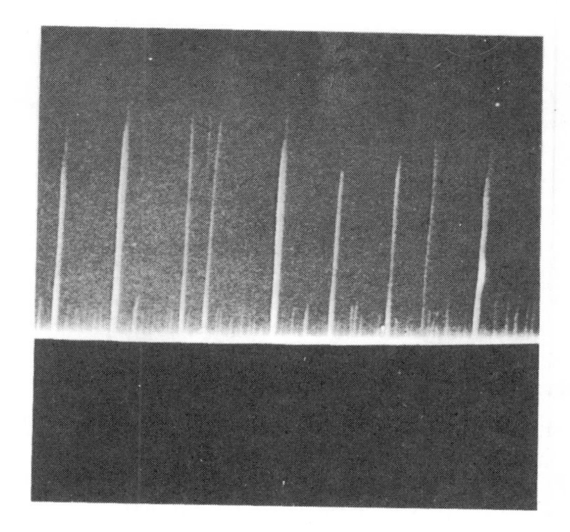

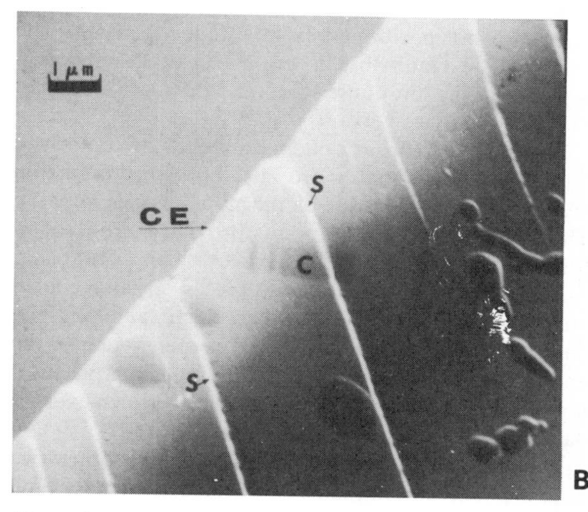

Figure 3.2 Glass knife edge geometry as seen with the SEM. Both major and minor striations start at, and are perpendicular to, the cutting edge. Striations (S) or ridges in the cutting edge (CE) are produced by branching of the crack tip as it approaches the edge. Some contamination (C) and lumps of the deposited film (lower right in B) are visible. A: A low magnification (700×) image. (From Helander, 1984.) B: A high magnification image. (Courtesy J. T. Black)

An optimal pressure and, hence, an optimal time (usually less than 1 s) are needed to obtain a straighter and cleaner break. Sufficient pressure should be applied so that the break should not move too slowly or too fast across the glass. A pressure greater than the minimum needed to break the glass will result in an unsatisfactory cutting edge. A rapid break due to excess pressure is undesirable.

Knives of better quality can be obtained by attaching a narrow strip (~3 mm wide) of adhesive tape on the

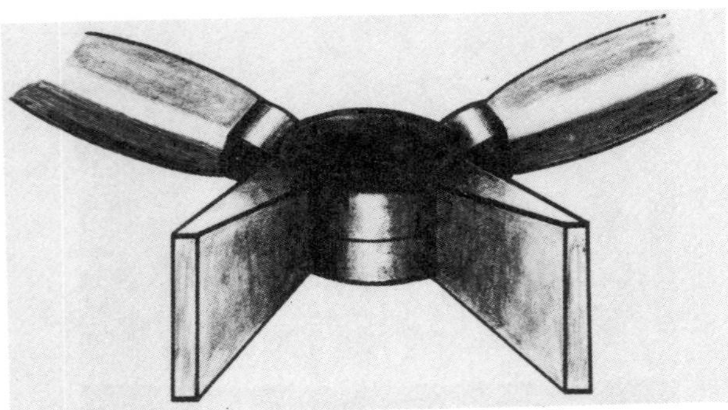

Figure 3.3 A pair of straight-edge glazier's pliers

centre of the lower jaw of the pliers. This central strip creates the fulcrum over which the glass is broken. To break the glass sheet, this strip (fulcrum) is positioned parallel to and below the score line so that the fracture should run in a straight line with the score. If the fulcrum is not positioned exactly under the score, the result will be a faulty break. For instance, if the fulcrum is placed to the left of the score, the free break veers to the right, and vice versa.

The modified glazier's pliers can be further improved by the addition of two more raised pressure points on either side of the centre of the upper jaw (Fig. 3.5). Thus, the three pressure points provide the necessary forces to fracture the glass sheet. This writer has successfully used ~13-mm-long pieces of paper match affixed to the jaws with plastics electrical tape. The paper match should be positioned perpendicular to the ends of the jaws. The tape not only holds the pieces of match in place, but also acts as a cushion which exerts a damping effect on the fracture.

It is necessary that the fulcrum and the other two pressure points be established at the exact position on the lower and upper jaws, respectively. The other two pressure points must be equidistant from the central fulcrum. A slight error in positioning the paper match and the tape may cause an uneven contact of the pressure points with both sides of the glass, and thus result in an uneven distribution of forces when the pliers are squeezed. This slight irregularity is a potential cause of irregular breaks. If the taped jaws pick up chips of glass, the tape should be replaced, as otherwise the pressure will be applied unevenly on the glass. The jaws should be retaped as soon as the pressure points begin to show wear. This is necessary because the height of the fulcrum as well as that of the other two pressure points is critical. The optimal height of these points for fracturing a specific type of plate glass is determined through experimentation.

Ready-made modified pliers are available commercially. The upper jaw of these pliers is ground to a slightly concave shape and the lower jaw to a convex shape (Fig. 3.6). When the concave curvature is pressed against the convex, three points of pressure are created. In this writer's laboratory these pliers have not proved as satisfactory as those with flat jaws. The reason for this probably is that the former possess more curvature than do the latter after modification. These ready-made pliers also lack the advantage of dampening, and the height of their pressure points cannot be reduced. Knives having a smaller knife angle cut better sections and produce less compression than those having a larger knife angle, for the former knives possess a sharper cutting edge. However, a reduction in the knife angle is accompanied by a reduction in the length of the straight part of the cutting edge. The technique described below produces knives which satisfy both requirements of a longer satisfactory cutting edge and a

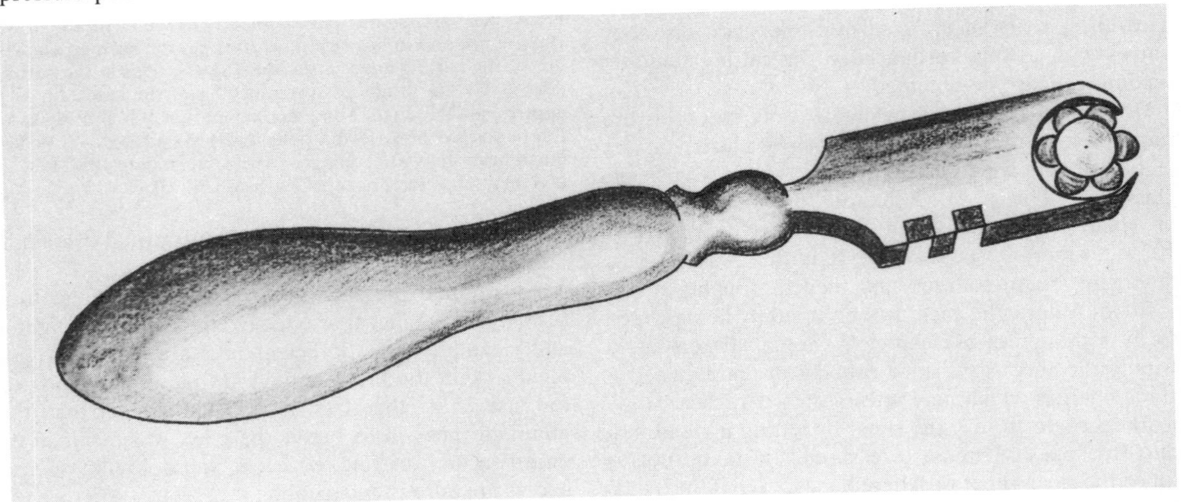

Figure 3.4 Wheel-type glass cutter

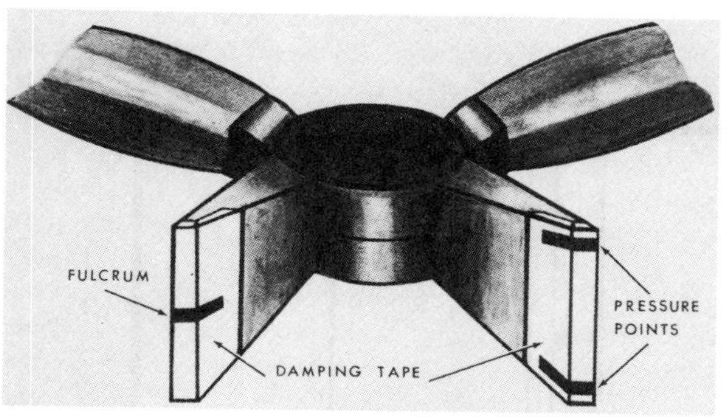

Figure 3.5 A pair of straight-edge glazier's pliers modified with strips of tape

Figure 3.6 A pair of ready-made modified curved-edge glazier's pliers

smaller knife angle. The following procedure is used to prepare knives from a 2×4 in cleaned sheet of plate glass.

Place the sheet on a table, and with a wheel-type glass cutter make a score mark ~½ in long exactly across the centre of the long edge (base) of the sheet (Fig. 3.7). The score must be made perpendicular to the base; to ensure this, some workers use a metal ruler. After scoring the sheet is placed so that its base extends ~½ in from the edge of the table. The base of the sheet should be parallel to the edge of the table. After positioning the fulcrum under the score, an even and gradually increasing pressure is exerted with the pliers until the break is completed. This pressure, equally distributed on either side of the score, should produce an even, slow and almost straight break. The two new surfaces thus produced will be free of major stress marks, except the portion where the score was made, and are potential knife faces. After the completion of the above break, two 2 in squares are obtained.

Break each 2×2 in piece, using the above technique, to obtain a total of four 1×2 in pieces. Using the same technique, break each piece again, to obtain a total of eight 1×1 in pieces. Each of these squares possesses at least two adjacent free break sides (Fig. 3.7). The final diagonal break is directed towards the corner where the two free break sides meet. The square fractured this way usually produces only one knife; the other triangular piece is discarded.

The shape and quality of a knife are primarily controlled by the direction of the diagonal score and its positional relationship with the index of the pliers during the break. The best knives are obtained when the score mark is made very close (<0.2 mm) to the centre (Fig. 3.8). More than half the length of the knife edge produced may be straight and without major flaws. Production of an almost perfect cutting edge is not uncommon when the pliers' jaws are properly positioned over the score mark. To obtain the best results, either the index of the pliers is aligned in a perfect straight line with the score mark or the pliers' jaws are very slightly tipped away from the eventual knife. The preparation of glass knives for thin cryosectioning is explained by Griffiths *et al.* (1983) and van Bergen en Henegouwen (1989).

Examination of Glass Knives

It is advisable to evaluate the cutting edge before it is used for sectioning. This is done under a binocular dissecting microscope (magnification ~30×) using incident light illumination from above. Alternatively knives can be examined while mounted on an ultramicrotome fitted with a focused light source. A beam of light from an appropriate source is focused onto the cutting edge at such an angle that the edge appears as a narrow band of bright light against a dark background. Against such a background the presence of striations (imperfections), usually on one part of the edge, is easy to observe, because they scatter the light. The part free from striations is used for sectioning. This part begins close to the corner where the break started and the speed of the fracture was at its lowest. The sharpest part of the knife does not start from the extreme left corner in Fig. 3.9D but from a few millimetres to the right of this corner, and extends to the point where check marks are present.

The dimension of striations increases in the part where the fracture speed was higher. The part of the edge showing striations under a microscope is unfit for sectioning. It should be noted that the entire edge has striations, but the striations ('steps') on the useful part of the edge are too small to be visualised under an optical microscope (Fig. 3.1, left).

The length of the useful edge is dependent upon the edge of the knife and the fracturing method used. Figure 3.9 illustrates the three configurations of knife edges and the relative length of the useful part of each of

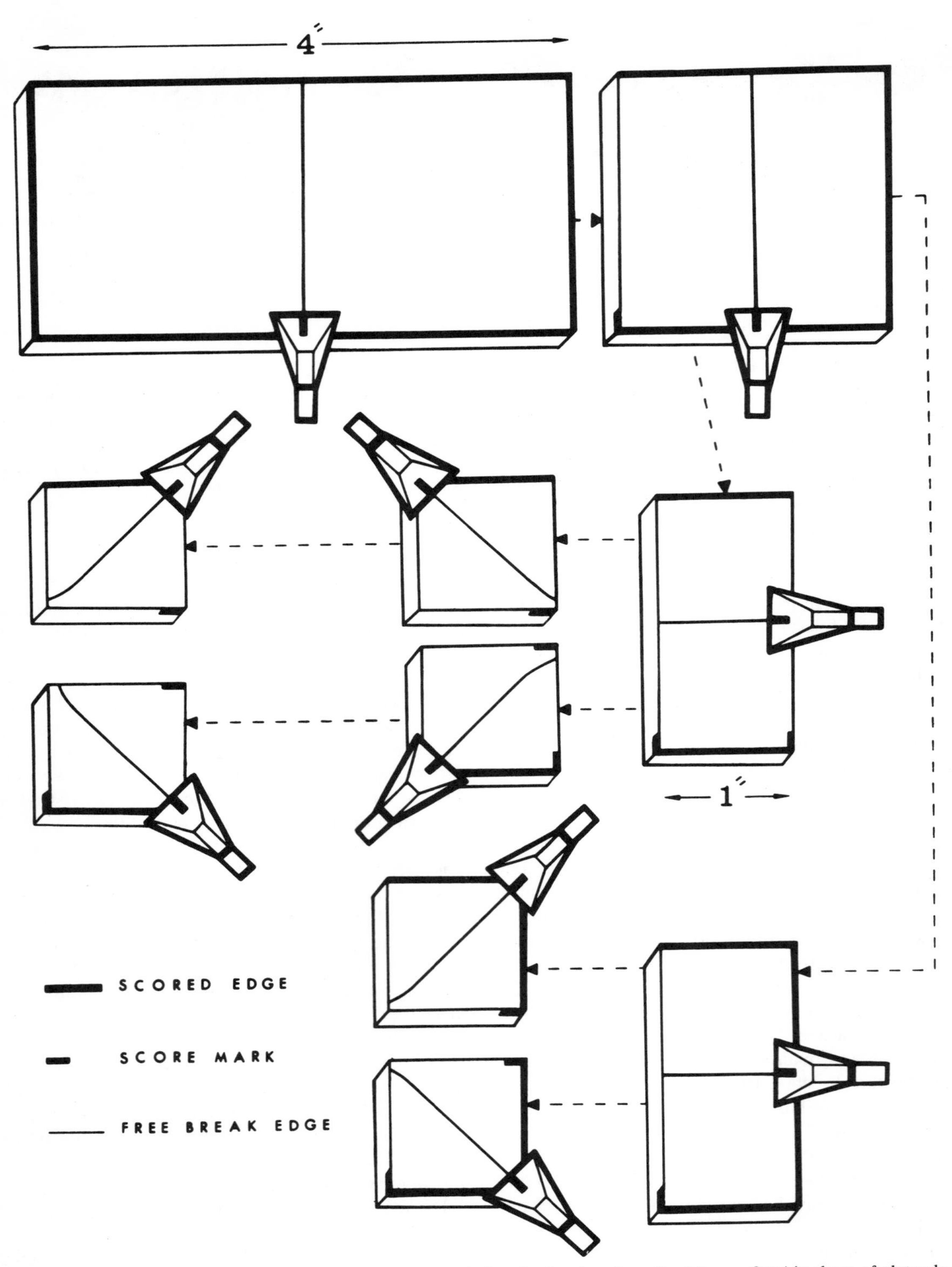

Figure 3.7 Diagram illustrating stages in making 1 in high knives by free break method from a 2 × 4 in sheet of plate glass

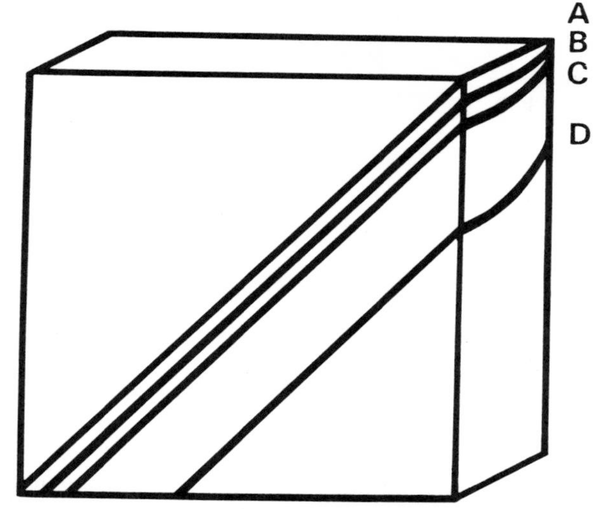

Figure 3.8 Diagram showing the shape of the cutting edge in relation to the distance of the diagonal score from the centre of the square

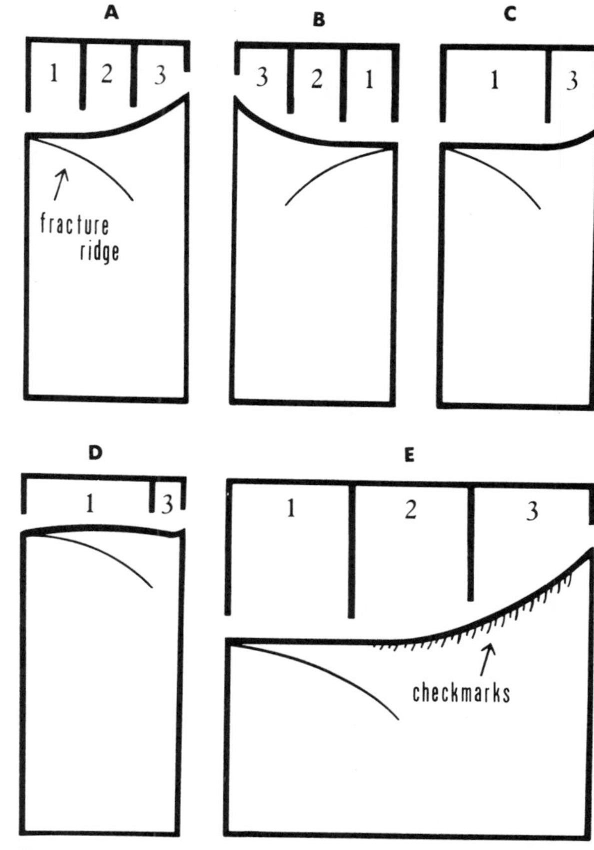

Figure 3.9 Diagram showing the relationship between various configurations of the cutting edge of glass knives and the length of the usable part in each type

them. A knife edge with a slightly convex shape (Fig. 3.9D) is considered superior, in that it often is the sharpest and has the longest useful part. If one is even slightly doubtful regarding the quality of a glass knife, it should be discarded. One must never risk using a knife of doubtful quality.

It should be remembered that a careful examination of the knife under an optical microscope reveals imperfections, but not the degree of sharpness. The sharpness of the cutting edge is determined by the quality of sections. However, the part of the edge without striations is considered sufficiently sharp to yield satisfactory sections of soft plant and animal specimens. It seems appropriate to define the term 'sharpness'. Sharpness relates to the radius of curvature where the two facets of the knife meet at the cutting edge, and is dependent upon the molecular arrangement in the glass (Reid, 1974).

Defects on Glass Knives

Glass knives suffer from two main defects.

(1) Striations (ridges or hackle marks; Fig. 3.2) on glass knives are probably caused by vibrations during the breakage and formation of the knife cutting edge. Both major and minor striations are present perpendicular to the undesirable part of the cutting edge. Some minor striations may also be present on the useful part of the cutting edge. Both types of striations are seen clearly with the SEM (Fig. 3.2). They can also be seen less clearly with light scattering under the light microscope (Fig. 3.1). Each major striation is separated from the next by ~10 μm (Sitte, 1984). The cutting edge also contains steps (rib marks). The presence of these imperfections prevents the production of thin sections with smooth surfaces.

(2) Knife defects caused by dust particles, hard inclusions in the specimen and touching the cutting edge result in isolated marks at irregular distances and in different intensities on the section (Fig. 3.10). These knife marks occur perpendicular to the cutting edge.

Tungsten Coating of Glass Knives

Coating of the cutting edge of glass knives with a thin film of evaporated tungsten metal improves their sectioning quality (Roberts, 1975). Coated knives are more durable and section hard specimens better than standard glass knives. Such knives also improve cryo-ultramicrotomy. The tungsten can be evaporated in an Edwards 12E6 coating unit, using a V-shaped tungsten

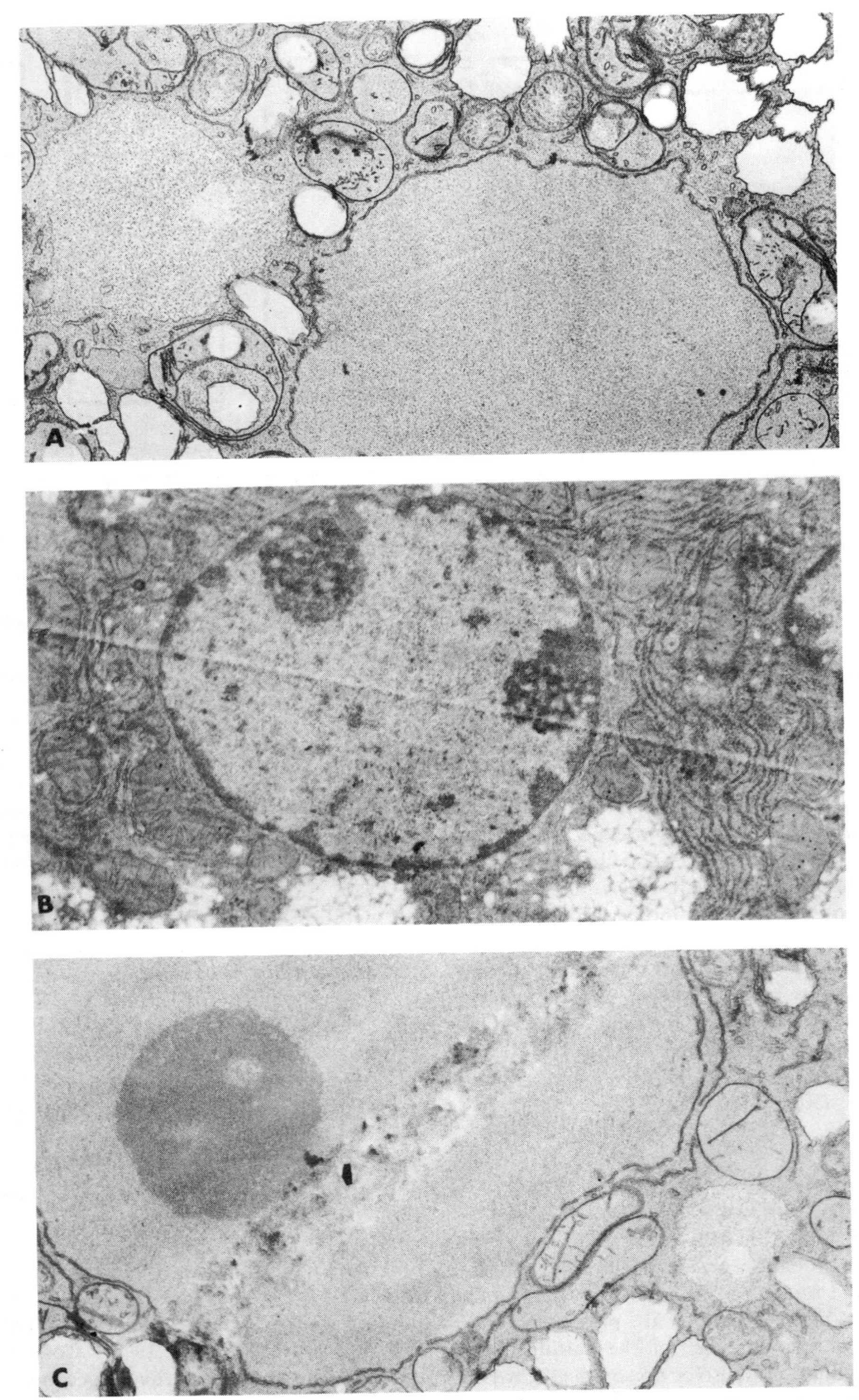

Figure 3.10 Knife marks running perpendicular to the cutting edge. A: Plant embryo cell showing a knife mark restricted to the nucleus. B: Mouse liver cell showing a knife mark throughout the section. C: Plant embryo showing a knife mark gouging the cell

filament which is bent from a 0.5 mm tungsten wire (Stang and Johansen, 1988). A circular holder with slots to accommodate the desired number of knives should be used. The holder has an inner radius of 39.5 mm, and the slots are milled in such a way that the knives are tilted 11° towards the filament. With the filament tip 75 mm above the centre of the holder, this geometry makes the line-of-sight between the filament and the knife edges bisect the cutting angle of the knives. Correct positioning gives an edge-to-filament distance of 63 mm.

To prevent evaporation of tungsten oxide, the filament should be glowed red hot for 1 min under high vacuum ($<10^{-2}$ Pa) before the glass knives are placed into the holder. The coating occurs with a current of ~30 A when the voltage selector is adjusted in the 10 V position. An almost linear film thickness is obtained as a function of the duration of evaporation. Freshly prepared knives are coated for 30 s to obtain a film of ~2 nm thickness. Films thicker than 2 nm are undesirable. Coated knives can be stored for several days before use, without adverse effects (Stang and Johansen, 1988).

DIAMOND KNIVES

Diamond is the hardest (10 on the Mohs scale) known material; its carbon atoms are bound together in a symmetrical array with greater binding energy per unit volume (~1375 kcal/cm^3) than in any other substance. Hardness is a measure of resistance to scratching or abrasion. Diamond is the most wear-resistant and the best heat conductor of all known materials. However, diamond has only moderate toughness, toughness being defined as resistance to breakage. Diamond edges are very brittle and so can easily be damaged by careless handling. Sections thicker than 1 μm should not be cut with a diamond knife.

The diamond knife consists of a single-crystal diamond. Although a natural diamond is the hardest known substance, it can be cleaved along its defined, cleavage planes, which are ~0.2 nm apart. It is suitably mounted and then abraded and polished, following precise crystallographic orientation by special microgrinding techniques first developed by Fernández-Morán (1953). The result is a wedge-shaped facet, which provides a stable and uniform cutting edge of molecular dimensions, with edge radii of ~2–5 nm (Fernández-Morán, 1986). Since diamond knives are made by grinding a single crystal with precision equipment, the angles of these knives can be controlled.

The diamond cutting edge usually is obtained by cleaving the natural diamond into slabs ~25 μm thick along the (III) plane. Taking account of the small size of the slab, it is soldered into a slotted shaft. Grinding begins by tapering one side of the shaft until the diamond just becomes visible; then the other side is tapered (Sawers, 1972). Grinding is continued until no nicks are visible at 50× magnification. This is followed by polishing until no nicks are seen at 600× magnification. However, fine nicks are always present to some degree. The shaft is fixed in a knife holder incorporating a trough. When ordering a new knife, the make of the ultramicrotome should be specified, because certain ultramicrotomes require special knife holders.

Diamond knives (Fig. 3.11) are, without a doubt, far

Figure 3.11 A: A diamond knife. B: Close-up of the cutting edge. (Courtesy E. I. du Pont de Nemours and Co.)

superior to glass knives. The crystalline diamond edge is sharper than the rounded glass edge, even if the latter shows no visible defects. However, the edge of a diamond knife is not exceedingly sharp, but looks more like a very sharp wedge. The radius of curvature of such a wedge has been estimated to be a few nanometres or a few dozens of atoms (Lickfeld, 1985).

The advantages of a diamond knife are many. Cellular structures in sections cut with a diamond knife are clearer and sharper than those seen in sections obtained with a glass knife. A considerable amount of time, invested in the preparation and 'setting up' of a glass knife on the microtome, can be saved by using diamond knives. Diamond knives can be used repeatedly without any adverse effect on the quality of sections, for they are not subject to flow or molecular rearrangement with time at room temperature. A theoretical cutting edge of 2–5 nm gives the diamond knife superior cutting characteristics.

Thin sections of materials (biological as well as non-biological) of almost all degrees of hardness or softness can be obtained with a diamond knife. It is indispensable for sectioning extraordinarily hard materials such as metal, wood, cartilage and bone. Because of minimal compression during cutting, sections show especially clear images of relatively delicate structures. For obtaining extremely thin sections (1–2 nm) for high-resolution electron microscopy, a diamond knife is preferred. A diamond knife is also invaluable in studies in which a large number of serial sections must be obtained. Because the knife is not replaced, the orientation of the specimen relative to the cutting edge remains unchanged. In the author's experience, epoxy resins and other relatively hard blocks are best sectioned with a diamond knife. Diamond knives with cutting edges ranging from 1 to 8 mm in length are available which can cut a section of almost any reasonable size.

Unfortunately, the quality of diamond knives commercially available is not consistent. This is a reflection of the problems involved in attaining a cutting edge of molecular dimensions. Two of the most common complaints are dullness and very fine nicks in the knife edge. These defects are presumably caused, in part, by the improper or uneven sharpening and polishing of the cutting facet of the knife. There is some evidence which indicates that a relationship exists between the orientation of the crystal planes of the diamond and the direction of polishing relative to the cutting facet. It is claimed that knives of good quality show the direction of polishing parallel to the cutting facet.

Steady progress has been made towards improving the quality of diamond knives. Currently, knives of fairly good quality with included angles of 40–60° are available from many sources. It is advisable to get a knife which suits the material to be sectioned. Knives with a relatively acute included angle (~45°) are entirely satisfactory for sectioning most biological materials. Metals, bones and other hard materials are sectioned best with knives having a less acute included angle (~56°). The length of the effective cutting edge is stated by the manufacturer. This is the only length which should be used for sectioning, although the total length of the cutting edge may be somewhat longer. The angle of the knife may vary by 1–2° along its length. When a diamond knife is of poor quality, compression can be avoided by reducing the size of the block face to less than $0.5 \times 0.2\,mm^2$, the specimen advance to less than 0.5 mm/s and the section thickness to ~50 nm.

A new diamond knife (Diatome-semi) has been introduced by DIATOME S.A. Company (Biel, Switzerland) for cutting semithin sections (0.2–2.0 μm) (Reymond, 1986). This knife costs about the same as a conventional diamond knife, and the former has no advantages over the latter for thin sectioning.

Examination of Diamond Knives

Even today the quality of diamond knives is uneven, for certain knives yield unsatisfactory sections from the outset. This is an almost inevitable result of inherent variations in the purity of different diamonds with no apparent bearing on the manufacturing source. However, the tests at present used by manufacturers to evaluate a diamond for ultramicrotomy need to be improved. More sensitive methods to detect the presence of impurities (e.g. silicon) should be used. Manufacturers do allow a reasonable amount of time (1–3 months) to the buyer for testing the quality of the knife; a knife of poor quality or with an unwanted knife angle can be returned to the manufacturer for replacement. This arrangement is also justified by the special demands of the user.

It is advisable to evaluate each knife when received, in order to ascertain whether it should be kept or returned for further sharpening or polishing. The most dependable and practical way to test a knife is by sectioning and examining the sections in the TEM. The chatter, compression, score marks and similar artefacts can be detected by such a test. This test can be combined with optical evaluation. A direct examination of the knife can be made under a dissecting microscope. With reflected light it is easy to see two polished cutting facets on each bevel; the one adjacent to the extreme edge is usually narrow.

If time and facilities are available, knives can be examined by dark-field microscopy. Knife marks and weak parts along the cutting edge can be detected by this method. The sharpness of the cutting edge and the quality of knife facets can be evaluated by examining the stress at the block face when the section is being parted from it. The cutting edge can be surveyed with a variable incident angle of light at a magnification factor ranging from 10 to 80, while detailed examination can be carried out with a fixed incident angle of light at a magnification of 600–850× (Persson and Persson, 1976).

The purchaser should take advantage of the manufacturer's offer to test the knife on a specimen block supplied to him. This testing ensures that the knife will cut the material supplied. Certain manufacturers (e.g. Du Pont) furnish knives with grids and electron micrographs of sections cut from each millimetre of the cutting edge. However, since electron microscope facilities may not be available to some of the knife manufacturers, microdefects and chatter may go undetected. Examination of the sections under the electron microscope by the buyer is, therefore, imperative.

All diamond knives eventually show nicks and chips in their cutting edge. If the damage is not drastic, they can be resharpened by the supplier. Debris adhering to the cutting edge of a used knife is extremely difficult to remove. In this case, also, resharpening is the solution. Another solution is to microtome soft copper at depths of 0.5–1.0 μm; such cuts have sufficient force to carry off the debris.

Care of Diamond Knives

It is advisable for a beginner to prepare and use glass knives before opting for diamond knives. Without some previous experience in sectioning with a glass knife, one is likely to damage the expensive diamond knife. Diamond knives cost between $2000 and $5000 (£1123 and £2809) per knife. Since the cutting edge of a diamond knife is extremely fragile, great care should be taken in mounting and removing the knife from the microtome. The edge can be chipped by accidental touching against a metal or other hard surface. One should be on the lookout for possible damage, especially by objects such as specimen block, chuck, syringe needle or micropipette, pair of forceps and grid. Particles of various kinds and microchips of razor blade that may become embedded in the block during trimming can damage the knife. To prevent this, the razor blade should be cleaned with acetone or another solvent before use and the block should be examined thoroughly with a binocular microscope before sectioning. It is advisable not to cut a specimen block with a diamond knife if the same block has already been exposed to a glass knife. Microchips of glass may adhere to the face of the block and damage the diamond edge. The diamond knife should not be used to cut sections thicker than 1 μm, especially of hard specimens. Also, diamond knives should not be used for trimming specimen blocks, although a discarded diamond knife can be used for this purpose.

Too much or too little clearance angle may cause chipping of the diamond edge. It is suggested, therefore, that the starting microtome setting of the clearance angle recommended by the manufacturer be observed. One should remember that even a slight pressure at the right angle to the diamond edge can damage it.

In view of the price of a diamond knife, it is economical only if its expected duration of service is utilised by handling it with utmost care. Experience has shown that only one person should be entrusted with the use and care of a diamond knife. Although some knives wear out and need to be resharpened or polished after a short period of time, a good knife can last for a long time if it is handled with care.

In order to obtain good sections, the cutting edge must be scrupulously clean. Therefore, it is recommended that the knife be cleaned immediately after use while the trough and the cutting edge are still wet. Once the leftover sections become dry, it is extremely difficult to remove them. The sections or their debris can adhere firmly to both the inner and outer surfaces of the cutting edge (Fig. 3.12). The debris on the inner surface may prevent proper wetting of the cutting edge. The debris on the outer surface may cause wetting of the back of the knife and block face. The debris may also cause striations in the sections.

Acids or other strong reagents should not be used for cleaning, for they may affect the finish of certain troughs, resulting in contamination. Toothpicks or other rigid wooden devices are also not recommended because they can be a source of contamination (e.g. oil) and possible damage to the delicate cutting edge. The best approach is to use a segment of Tygon tubing (R-3603, Fisher Scientific Co.) for cleaning the cutting edge (Gorycki and Oberc, 1978). A 2 mm length of the tubing (of 7.9 mm inner diameter and 1.6 mm wall thickness) is cut into four equal segments, one of which is impaled along its length on a needle in a holder or a standard dissecting needle (Fig. 3.13). The cutting edge is cleaned with a gentle unidirectional stroking motion of the wet squeegee parallel to the knife edge—never obliquely or at a right angle. This cleaning is done under observation through a binocular microscope. A small blast of compressed air (e.g. Freon) can also be employed to clean the cutting edge.

At the end of the work, it is advisable not to remove the knife from its adapter; instead, the adapter should be removed with its knife. The trough should im-

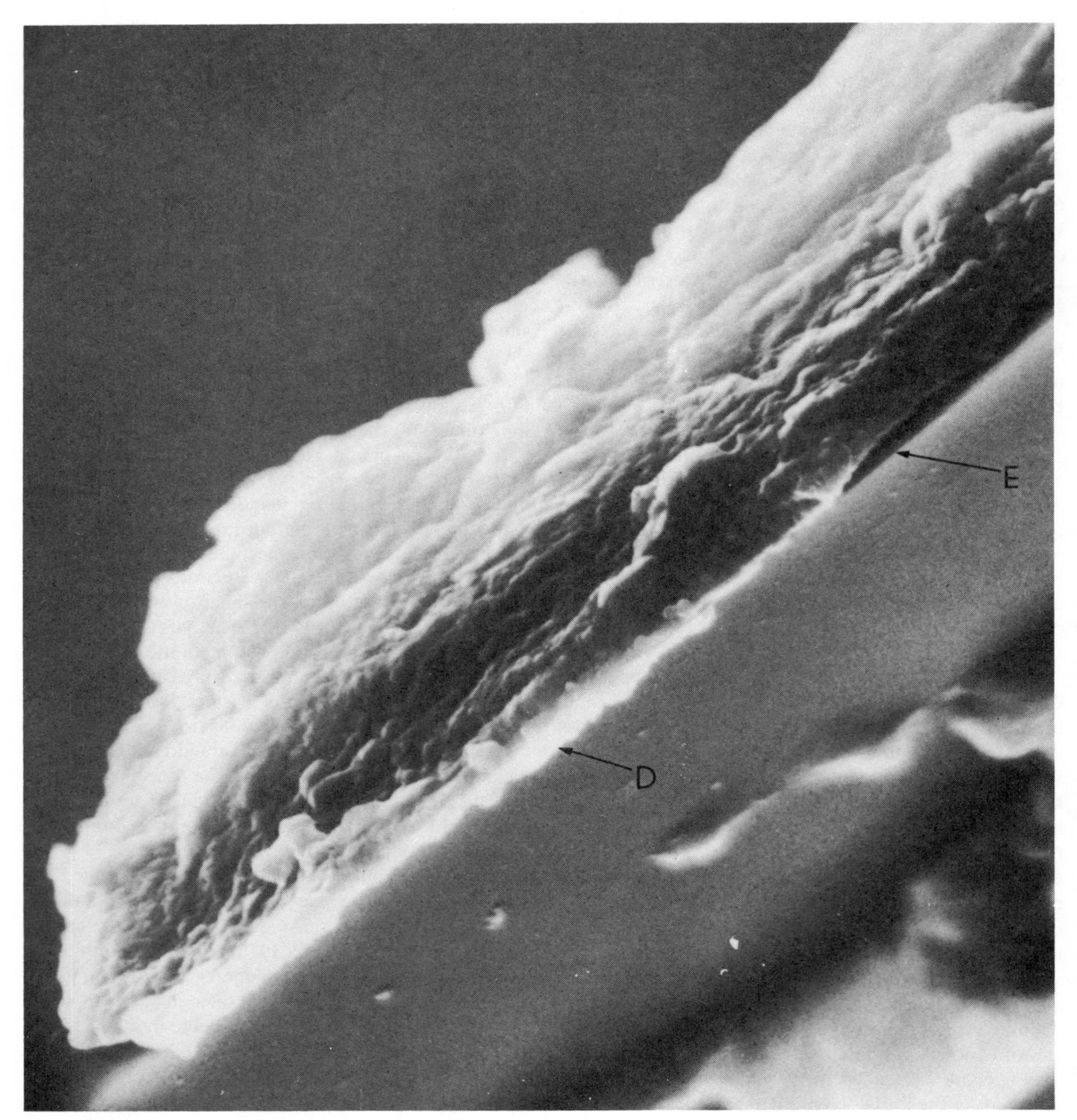

Figure 3.12 Edge of a used diamond knife as seen with the SEM (6804×). The debris (D) is strongly adhering to the edge and it is very difficult to remove. The debris develops during the cutting process and either envelops the edge or lies just below the edge. Nicks in the cutting edge invite the accumulation of debris. Edge of knife (E). (Courtesy J. T. Black)

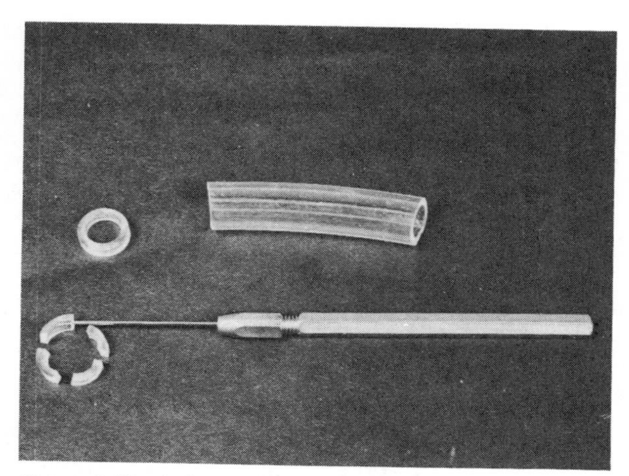

Figure 3.13 A 2-mm-long piece of Tygon tubing cut into equal segments (lower left) one of which is impaled on a needle to form a squeegee used to clean the surfaces and edges of diamond knives. The cutting edge is cleaned with a gentle unidirectional stroking motion of the squeegee parallel to the knife edge. (From Gorycki and Oberc, 1978)

mediately be rinsed with distilled water to remove the dust and clinging debris. To ensure cleaning, the trough should be immersed in a dilute solution of a suitable detergent for 10–15 min. Alternatively, Eastman Kodak lens cleaning fluid can be used. The final rinse should always be with distilled water. It is advisable to dry the knife under a warm-air dryer.

Ultrasonic cleaning accompanied by treatment with a detergent (e.g. Alconox) is quite effective in safely removing even tenaciously adhering debris (Wallstrom and Iseri, 1972). The knife is immersed in a 7% aqueous solution of a detergent contained in a 50 ml Tri-pour polyethylene beaker (Scientific Products). Because of the internal geometry of this type of beaker, the knife edge does not come into contact with the beaker wall. The beaker is placed in water in an ultrasonic cleaner (40 W, 60 Hz) (Heatsystems Ultrasonics Inc., 38 East Mall, Plainview, N.Y.) for 10–30 min. The knife is removed and rinsed thoroughly in distilled water. Evaporative drying of the cutting edge must be prevented. The knife is immersed in distilled water in a second 50 ml beaker and subjected to ultrasonic radiation for 10 min. The knife is rinsed again in distilled water and dried by displacement—i.e. the trough is filled to a convex meniscus with distilled water, and a sudden, forceful blast of compressed Freon from a Freon duster is applied to the cutting edge. Such drying will result in a clean, dry edge. Repeated cleanings by ultrasonic treatment will partially remove the black coating from the trough.

If all the above-mentioned treatments fail, a particularly dirty knife (e.g. one that produces many knife marks but is not chipped) can be cleaned with concentrated NaOH, which seems to release most stubborn debris. However, the corrosive effect of this reagent on the knife should be kept in mind. If even this treatment fails, the only resort left is resharpening.

SAPPHIRE KNIVES

A synthetic sapphire (Al_2O_3; synthetic corundum) has a hardness of 9 on the Mohs scale and is resistant to breakage. Like a diamond, synthetic sapphire is a single crystal. The knife is produced by cleaving or fracturing of the crystalline sapphire (Hodgson and Cunningham, 1987). The advantage of sapphire knives is their low price, about one-tenth of that of a diamond knife. The superior toughness and low cost of the former may make them a useful alternative to glass or diamond knives. As with a diamond, it is difficult to wet the surface of a sapphire with water. Further testing is needed to determine the usefulness of sapphire knives.

Sapphire knives with a knife angle of 30° and a clearance angle of 3–5° have been reported to produce satisfactory thin sections (Yamamoto and Maruyama, 1985). The block face should not exceed 0.5 mm on a side and sections thicker than 1 μm should not be cut with these knives. Soft blocks can be sectioned at a speed of 0.19–1.9 mm/s; hard blocks require slower cutting speeds. Sapphire knives with a knife angle of 30° are sharper but more brittle compared with those with a knife angle of 60°. Sapphire knives (Sapphotome) are commercially available (Sakura Ltd, Tokyo; LKB Instruments, Gaithersburg, MD).

MECHANISM OF THIN SECTIONING

Although thin sectioning of biological specimens is carried out routinely, an understanding of the process of sectioning is a basic requirement. At the outset it is pointed out that the process of thin sectioning of biological specimens embedded in a resin is completely different from metal machining, although, according to Lickfeld (1985), the mechanical process occurring around a cutting edge is related to that occurring when metal is lathed. The molecular structure of biological specimens and the elastic properties of the resin are different from those of metals. In this respect, resin properties also differ from those of paraffin. Additionally, the molecular structure of knives used for thin sectioning of metals is different from that of glass or diamond knives.

According to Acetarin *et al.* (1987), thin sectioning is essentially a cleavage process, whereas Sitte (1984) believes it to be a real sectioning process instead of a splitting or cleavage process. It is definitely not a scraping process. It seems that a relatively large facet angle produces sectioning by cleavage, while a small facet angle results in a real sectioning process. Scraping occurs when the facet angle approaches 90° (Sitte, 1984). Relatively hard resins appear to favour a cleavage process, while soft resins allow a real sectioning process. An excessively soft resin block may bypass the cutting edge without initiating the sectioning.

The mechanism of thin sectioning is exceedingly complex, and a number of factors control this process. To understand the sectioning process, knowledge of the physicochemical properties of the resin used is necessary. For example, since monomeric epoxy resins copolymerise with the specimen, the embedded block shows reduced heterogeneity, resulting in smooth sectioning (Acetarin *et al.*, 1987). This is not true in the case of Lowicryls, which cure by radical polymerisation, without copolymerisation with the specimen. Thus, epoxy resins produce sections with smoother surface relief, compared with those obtained from Lowicryls.

On the other hand, polar resins (melamines) require less kinetic energy for cutting a section, compared with non-polar resins (Epon 812) (Frösch *et al.*, 1985). The reason is that the inherent bonding energy is released more easily from the polar resin, resulting in less plastics flow before the rupture point is reached during sectioning. For this and other physicochemical reasons, the cutting qualities of polar resin seem to be better in that it can produce sections that are thinner (8 nm) and have smoother surfaces than those obtained from Epon. However, Epon sections as thin as 15 nm can be easily obtained with a diamond knife. Heat generated during sectioning has little effect on the quality of sections (Acetarin *et al.*, 1987). Such heat may soften the resin, but does not melt it at the cutting edge.

SECTION THICKNESS

At the outset it is appropriate to clarify the terminology used in this volume to indicate the thicknesses of resin sections. Sections as thick as 50 μm (Seliger, 1968) of epoxy and as thin as 8 nm (Frösch *et al.*, 1985) of melamine resins have been used. Sections as thick as 0.25 μm provide satisfactory resolution with the 100 kV transmission electron microscope, while sections as thick as 10 μm can be viewed in the high-voltage electron microscope. Sections as thin as 0.1 μm can be imaged in the light microscope (Rieder and Bowser, 1983). Thus, there is no boundary between section thicknesses that can be used for either light or electron microscopy. In other words, sections of the same thickness can be examined in the light or electron microscope.

The most accurate way to describe a particular section is by its actual thickness (determined by interference colour) or nominal thickness (determined from the ultramicrotome setting). For example, a section of 200 nm thickness should be referred to as a 200 nm section. A class or group of sections should be described as follows:

Thin sections	8–100 nm (0.1 μm)
Semithin sections	0.1–2.5 μm
Thick sections	2.5–10 μm

The use of words such as 'ultrathin' and 'semithick' should be avoided. It is the author's hope that, for the sake of consistency and international understanding, the above terminology will be accepted in the fields of light and electron microscopy. For somewhat different viewpoints, the reader is referred to Bowser and Rieder (1986) and Glauert (1987).

As stated above, most materials are too thick to be examined in the TEM. Therefore, the material must be cut very thin so that an electron beam can penetrate. When an electron beam passes through a section, it undergoes two kinds of collisions—elastic and inelastic. Electrons involved in elastic collisions do not give up their energy but change their direction, whereas inelastic collisions cause energy loss from the beam. This loss in energy involves a change in the associated de Broglie wavelength of the electron. Thus, the thicker the section, the more inelastic scattering and the wider the distribution of wavelengths. In other words, a loss of energy causes an extended spectrum of electron velocities in the beam, which, in turn, leads to chromatic errors in the image. This change in wavelength, therefore, determines the limit of resolution obtainable for a particular thickness of specimen. Because energy loss is considerably greater in the presence of relatively thick sections, chromatic errors in the image are a problem when dealing with these sections. In principle, the higher the resolution desired, the thinner the section should be, since this reduces scattering or loss of electrons as they pass through the section. The smallest detail resolved is limited to $\sim\frac{1}{10}$ of the section thickness.

On the other hand, a decrease in section thickness is accompanied by a decrease in contrast (Fig. 3.14). Furthermore, the resistance of some resins to differential sublimation under electron bombardment does not help to increase the contrast between the tissue and the background. To obtain the maximum contrast and the highest resolution, then, how thin should a section be cut? A section should be cut thin enough to eliminate

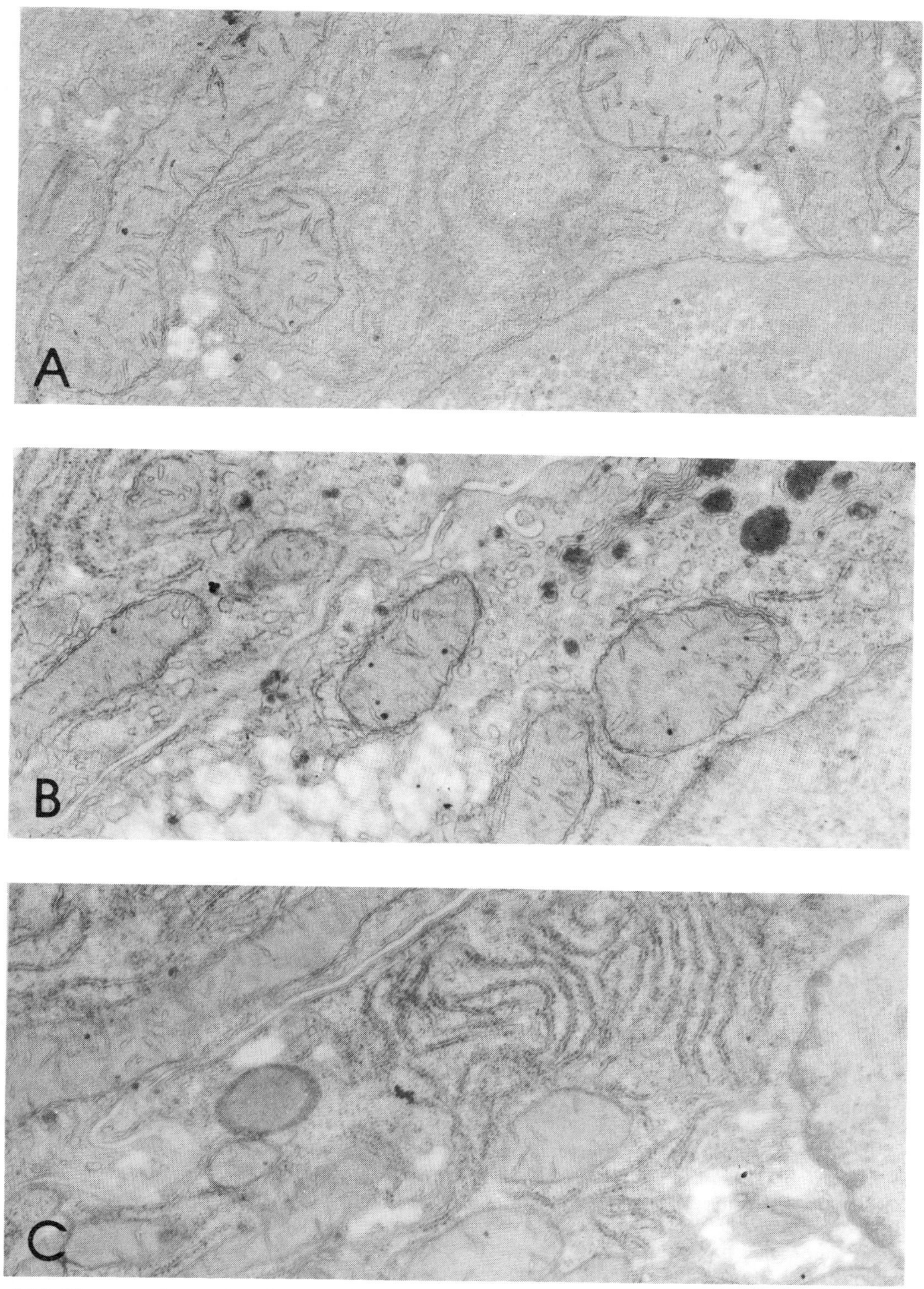

Figure 3.14 Electron micrographs of mouse liver sections of various thicknesses. A: ~40 nm thick. B: ~60 nm thick. C: ~80 nm thick. The uppermost micrograph shows the highest resolution but lacks adequate contrast

structural overlap, to be viewed in transmitted electrons and to be resolved by the optical system of an electron microscope. In general, although overlapping of cell components occurs in all thicknesses, it becomes a problem in sections thicker than 150 nm. The contrast can also be increased by using a small objective aperture or by post-staining the sections with heavy metals.

After the above three demands are satisfied, the most desirable thickness is determined by the objective of the study. It is erroneous to believe that the most desirable sections are always the thinnest sections. In some studies relatively thick sections are desirable from the standpoint of contrast and accurate interpretation of the distribution of structures within a tissue. The importance of relatively thick sections (with gold interference colour) is exemplified by studies of resorption of collagen associated with tooth eruption or periodontal disease. Collagen fibrils in the periodontal ligament are demonstrable in sections of gold interference colour, but not in grey sections (Melcher and Chan, 1978). A possible explanation is that the minimum volume of collagen necessary to react and stain with heavy metal salts is not present in very thin sections. The importance of section thickness in the localisation of enzyme activity is also apparent. Moreover, a thicker section is easier to obtain and handle than a thinner one.

Since cutting involves bond rupture and plastic flow, the surfaces of a section are damaged. Thus, the thinner the section, the larger the fraction of its volume that has been permanently distorted. Over-thin sectioning can, therefore, introduce mechanical distortion such as fragmentation of structural elements, especially of membranes.

Blurring of images of cellular structures, even in thin sections, is not uncommon. This is to be expected, since the electron microscope has a larger depth of field, which images structures above and below focus. Thus, the image of the structure at the true focal plane is blurred. This blurring limits the resolution obtained. When it is possible to section below the diameter of a structure, it is possible to gather information about the inside of the structure rather than be limited mainly to its surface. Methods for cutting sections with bevelled edges of infinite thinness are available (Colquhoun, 1980). These section edges allow improved resolution, which provides, for example, images of discrete protein units in the membrane and multilobed structure of glycogen granules. Loss in contrast of and damaging effects of the knife and electron beam on these edges are concomitant problems.

All sections of different embedding media undergo varying degrees of sublimation under the electron beam. Since embedding medium is partially removed as a result of sublimation, the effective thickness of the section is reduced. Thus, resolution is also aided by sublimation. The undesirable consequence of sublimation is the introduction of artefacts, which appear to be particularly severe in high-resolution work. The rate of sublimation is dependent upon the type of embedding resin and microscope used, and the operating conditions under which the sections are examined. Since an intense beam tends to increase the rate of sublimation, it is advisable to view sections under a beam of low intensity, especially in the early stages of viewing. The effect of sublimation is usually less pronounced in thin sections than in thick sections.

The loss of thickness occurs instantaneously when the sections are exposed to the electron beam. This reduction in thickness is accompanied by an increase in the refractive index of the sections. For example, Araldite sections show an increase in the combined refractive index from 1.54 to 1.9 when exposed for 2 min at a current density of 0.02 A/cm^2. The loss of mass and the increase in the refractive index are due to the fact that electrons knock out most of the lighter atoms from the section (hydrogen, oxygen and nitrogen), with the exception of carbon atoms.

RELATION BETWEEN SECTION THICKNESS AND INTERFERENCE COLOUR

The thickness of a section is commonly estimated by observing the interference colour in light reflected from the section while it is floating on the surface of a liquid in the trough (Peachey, 1958). The interference colours are similar to those exhibited by oil films on a water surface. These colours are produced by interference between the wave (ray of light) reflected by the top surface of the section and the wave that is reflected by the section–water interface. The appearance of the colour depends on the relative phase and amplitude of these two waves. The constructive and destructive interferences of waves follow certain fundamental laws, provided that they have the same light source. These colours are seen clearly only when the light is viewed reflected in the surface. It should be noted that the interference colour will vary slightly, depending upon the angle at which the section is viewed; this angle should not be greater than 45°.

The interference colour of the section serves as the index of its approximate thickness (Table 3.1); this scale is applicable to all embedding resins (epoxy polyester and methacrylates) with a refractive index close to 1.5. The colours produced are related to the section thickness and the refractive index of the sec

Table 3.1 Continuous interference colour index and thickness scale for thin sections

Interference colour	*Approximate thickness* (nm)
Grey	60
Silver	90
Gold	150
Purple	190
Blue	240

From Peachey (1958).

tioned resin and the specimen. Since the interference colours were observed by incandescent light (Peachey, 1958), fluorescent light will cause some shift of the colours towards the blue end of the spectrum because of the higher short-wavelength content of the latter light. However, in view of the subjective nature of the colour estimation, this small error can be ignored for routine studies. On the other hand, small differences in the refractive index of various resins are significant. The refractive index varies even from batch to batch of the same resin. Methacrylate sections of silver interference colour are one size category thicker than Vestopal sections of the same interference colour (Walter, 1961). This difference seems to be due to a difference in the refractive index of the two resins.

Since the interference colours form a continuous spectrum instead of clearly separate colours, the colour scale provides only a guide to the actual thickness. The estimation of section thickness based on these colours is reliable to within a range of 10–20 nm for sections thicker than 60 nm. However, the thinnest sections, described as grey, are produced by a wide range of thickness below 60 nm. Since each interference colour represents a range of ~30 nm, thickness given by ellipsometry is not accurate enough to be used for quantitative electron microscopy. Considerable variations in thickness occur between sections of the same colour.

Since it is difficult to measure accurately very thin sections by the ellipsometer used by Peachey (1958), Sakai (1980) reinvestigated the relationship between section thickness and interference colours. By studying resectioned thin sections, he indicated the following thickness measurements which differ from those estimated by Peachey (1958):

Silver interference colour	54–64 nm
Grey interference colour	26–29 nm
Dark-grey interference colour	20–21 nm

Yang and Shea (1975) also associate silver interference colour with 49–60-nm-thick Epon–Araldite sections.

INTRASECTION VARIATION IN THICKNESS

The surface of a section is not smooth—i.e. it is rough and is not uniformly thick. Since all embedded tissues are non-homogeneous, intrasection variation in thickness or surface roughness is common. The major reason for such variation is compression. A section undergoes an uneven thickening in the form of compression during cutting with a glass or diamond knife. Although when flattened with a solvent, the section reverts towards the shape it possessed before cutting, the recovery varies in different components of the section. The surface roughness is more pronounced over the specimen than over pure resin (Mollenhauer, 1987). Such roughness is greatest over mitochondria, lipid droplets and red blood cells.

Intrasection variation in thickness is a potential source of error in the interpretation of electron micrographs, especially in quantitative electron microscopy. Such variations affect section staining, contributing to variations in the image contrast (Mollenhauer, 1987). Thus, the informational content of the stained section is influenced by surface roughness. The effects of such variations are more pronounced at low magnifications than at high magnifications, because a relatively large portion of a section is observed at low magnifications. The extent of surface variations in thickness depends on manifold factors, including knife sharpness, type of resin used, resin formulation, tissue permeation by the resin, nature of specimen and sectioning parameters.

It is not possible to detect intrasection thickness variation in the interference colour by optical means, unless it is of a high magnitude. At present, no reliable method is available to measure such variation on the section itself before it is floated onto the water surface in the trough. A brief review of the available methods is given below.

Williams and Meek (1966) have reported that by interferometry and microdensitometry the thickness variation within a single section was found to be commonly of the order of ~10%. Carlemalm and Kellenberger (1982) estimate the depth variation at the surface of Lowicryl K4M-embedded tissue sections to be 3–5 nm. A thickness variation of as much as 7 nm or ~10% of the total Spurr section thickness has been reported by Mollenhauer (1987). Intrasection thickness variation for *E. coli* embedded in Epon is 1–2 nm, using the shadow of the polystyrene spheres as references (Acetarin *et al.*, 1987). On the other hand, intrasection thickness variation for rat kidney tissue embedded in epoxy resins is ~7%, except for grey or dark-grey sections (Yang and Shea, 1975). Cryostat sections (10–20 μm) of the adrenal gland showed intrasection thickness variability in the range 4–7% (Anthony *et al.*, 1984).

Different types of specimens embedded in the same resin display different degrees of intrasection thickness variations. Differences in the thickness variations for sections of similar resins are due to the differences in the methods used to obtain such data. In general, Epon yields smoother sections than those produced by other resins. Diamond knives produce smoother sections than those obtained with glass knives. Sapphire knives produce sections almost as smooth-surfaced as those obtained with diamond knives (Mollenhauer, 1987).

In addition to intrasection variations, there is often a deviation in the level of the section from that of a flat sheet. Such deviations are similar to the minor variations in the thickness observed when a slice of cheese is cut. After a few cuts, these variations in thickness build up to variations of sizeable level. As a result, for instance, the tenth slice of cheese is not flat. Variations in level ranging between 0.2 μm and 0.5 μm for Vestopal and Epon blocks and as much as several micrometres for unprepolymerised methacrylate blocks have been reported (Helander, 1969). The higher variation in level for the methacrylate blocks is related to the uneven polymerisation of this embedding medium.

One of the causes of variation in level is the circular course of the arm of some types of microtome. Since the arm of certain microtomes (e.g. Porter–Blum and LKB) moves along an arc of a circle, the centre of the cut face of the block usually rises above the periphery. Consequently, sections cut with these microtomes resemble segments of a cylinder wall prior to their flattening (Helander, 1969). It is emphasised that the curved path of the specimen during sectioning can explain only a minor portion of the intrasection variations in level. The main reasons for these variations must be attributed to the elastic and plastic properties of the embedding media.

A more serious source of error in making quantitative measurements on tilted sections is a reduction in section thickness on exposure to the electron beam. The reduction in thickness due to irradiation tends to be uniform throughout the thickness of the section, and it does not seem to be a surface effect. The change in thickness varies between 20% and 50%. There is little reduction parallel to the grid, presumably because the section is prevented from shrinking in this direction by its attachment to grid bars or the supporting film. Most of the reduction seems to occur transverse to the grid, since there is no retaining force in that direction (Bennett, 1974).

MEASUREMENT OF SECTION THICKNESS

For correct interpretation of an electron micrograph, it is necessary to have an accurate knowledge of section thickness. Knowledge of section thickness in high-resolution electron microscopy is as important as is knowledge of magnification. Since volume requirements are so crucial in the reconstruction of the three-dimensional structure of organelles and whole cells, an accurate knowledge of section thickness is invaluable. Information on section thickness is also important in estimating the efficiency of antibody labelling and the accessibility of antigenic sites. In electron autoradiography an accurate knowledge of section thickness is important for determining radioactive concentration.

Although the interference method (Peachey, 1958) is the most direct and the most popular approach for measuring section thickness, it is also the least accurate, because of its subjectivity. For example, since gold interference colour corresponds to a section thickness range of 90–150 nm, section thickness is estimated with an error of ~25%. An error of this magnitude is unacceptable in the interpretation of morphometric data and three-dimensional reconstruction of cell components.

The actual thickness of the section does not correspond to the thickness set on the ultramicrotome. Ideally, section thickness should be measured before and after irradiation of the section, since thickness decreases on exposure to the electron beam. The extent of thinning due to exposure to the electron beam is not known. A number of methods have been introduced for estimating section thickness with a fair degree of precision. Porter–Blum (1953) calculated section thickness by measuring the length of a heavy-metal shadow at known angle. This method has the drawback that only an extreme edge of the section is measured, and so the thickness of the remaining part of the section remains unknown.

Huxley (1957) compared section thickness with the regular spacing of muscle filaments within the section while viewing in the TEM. This approach cannot be used routinely. Electron scattering can also be used to determine mass thickness and, hence, specimen thickness (Zeitler and Bahr, 1962). Williams and Meek (1966) measured the thickness by determining the radioactivity of sections containing radioactive markers. Reedy (1968) measured section thickness relating to the kind of wrinkles produced when sections were mounted on the grids, while Adachi *et al.* (1968) made the measurements by the folding method.

Silverman *et al.* (1969) introduced a method based on quantitative electron microscopy (Bahr and Zeitler,

1965) for measuring section thickness. They used a loosely cross-linked anion exchange plastic, Dowex, stained with PTA as a standard for measuring absolute section thickness. This method has an advantage over interference microscopy in that it allows thickness measurement very close to the structure whose mass is to be determined.

Gillis and Wibo (1971) employed an interferometric method for measuring section thickness, whereby thickness measurement and examination can be performed on the same section, and standard objects need not be embedded together with the specimen. The accuracy of measurement is claimed to be within a margin of error of less than 1 nm. This method appears to be quite reliable and accurate for measuring the thickness of individual sections prior to exposure to the electron beam.

Edie and Karlsson (1972) presented yet another method for calculating thickness while the section is under observation in the electron microscope. The method involves essentially an electron beam attenuation measurement resulting from electron scattering in the plastics embedding medium. Section thickness is determined from an equation using the previously determined absorption coefficient for the embedding medium. This coefficient must be determined for a given accelerating voltage, objective aperture and magnification employed during the measurement. The accuracy of measurements of Formvar films has been claimed to be better than $\pm 5\%$, and similar accuracies are thought to be feasible for embedding media. Section thickness has also been estimated by quantitative electron microscopy (Casley-Smith and Crocker, 1975).

Since considerable variation in section thickness along a ribbon occurs, Gunning and Hardham (1977) estimated the average section thickness in a ribbon by measuring the block face. Berthold *et al.* (1982) obtained mean section thickness from the height of a trimmed specimen mesa of a known height and the number of sections cut from it. Using this mean value and the mean section electron-scattering value, the relative thickness of a single section ($\sim$100 nm) can be calculated with an uncertainty of $<10\%$. Although the method is tedious, it is useful for estimating the thickness of individual sections in a ribbon.

One of the most commonly used methods involves re-embedding and resectioning of sections (Yang and Shea, 1975; Ohno, 1980; Helander, 1983). The thickness is measured by viewing the section profiles in the TEM. A secondary section mounted on a Formvar-coated slot grid provides sufficiently normally cut segments for measurements yielding a precise estimate of mean thickness. The method presented by Bedi (1987) involves dividing a section into two portions, one of which is used for stereological counting and the other of which is re-embedded and resectioned through its thickness in order for its thickness to be measured. Because of intrasection variation in thickness and other reasons, the difference in thickness between the two portions of a section is $\sim$4%. This method differs from other resectioning methods in that it allows stereological counting and thickness measurement from the same section. The details of this method are given below.

Semithin sections are cut from a trimmed block to produce a smooth face. With a razor-blade, a shallow cut (score) is made into the block face such that it is at a right angle to the knife's cutting edge. The cut is made slightly off-centre of the block face so that each section produced by the knife yields two unequal portions; the larger portion is used for counting and the smaller for re-embedding.

A different, direct method consists of embedding in a resin, followed by electron microscopy of transverse sections (the grid sectioning technique) (Jésior, 1982). This method is suitable when the section is cut through a homogeneous material. For an inhomogeneous material, Jésior (1985) has proposed a method using non-central sections obtained from latex spheres. Berriman *et al.* (1984) described a trigonometric method using an image tilt series to obtain the thickness after marking the specimen surfaces with small gold particles. The electron scattering method (Weybull, 1970) is the best approach, in terms of ease and reliability, to estimate the thickness of thin sections. This method is carried out in the electron microscope and no additional equipment is required.

Robertson *et al.* (1984) have developed an interface method for measuring the thickness of semithin and thick sections in the range 0.3–45 μm. An incident illumination objective incorporating a beam splitter and adjustable reference mirror is employed to generate interference fringes by reflection from the upper surfaces of sections on glass slides. Cryostat sections are much thinner than the cryostat microtome settings (Pearse and Marks, 1976). The effective thickness of cryostat sections (10–20 μm) after being mounted onto glass slides is reduced by flash drying by 90% relative to microtome thickness setting (Anthony *et al.*, 1984).

SECTIONING ANGLE

The sectioning angle consists of the clearance angle and the knife angle. The quality of sectioning is affected, in part, by each.

CLEARANCE ANGLE

The clearance angle is that formed between the cutting edge and the face of the specimen block (Fig. 3.15). In other words, the cutting edge is inclined at such an angle to the face of the specimen block that it is not quite parallel to the face. This angle is necessary in order to avoid the block face scraping the back of the knife after cutting a section. The net result is that as soon as a section is cut, the block moves clearly away from the back of the knife. The clearance angle should be kept within the range 2–5°. In principle, the clearance angle should be as small as possible (provided that the block does not come in contact with the back of the knife after the section is cut), because a knife with a large clearance angle will tend to scrape the sections instead of cutting them, which will result in chatter. Too large a clearance angle can also cause chipping of the knife. The chips thus produced may not be visible under the binocular microscope, but their effect, in the form of lines perpendicular to the cutting edge, shows up in the micrographs.

Although knife holders have a marker indicating the clearance angle, this is only a rough approximation of the angle at which the holder is set. A difference of as little as 0.5° in the clearance angle may affect the quality of sectioning. Bucek and Arnott (1972) have introduced a device for accurately setting the clearance angle.

KNIFE ANGLE

The knife angle is critical for keeping the compression to a minimum. The knife angle is determined by the exact position of the score on the glass when preparing the knife. If the square of glass is bisected precisely, the knife angle will be the same as the scoring angle (gross angle). It is possible theoretically to obtain this angle of 45° (angle β in Fig. 3.15 and A in Fig. 3.8); however, in practice it is difficult, if not impossible, to achieve this angle, because of the final deflection of the diagonal break. This deflection is responsible for the formation of a facet at the cutting edge of the knife. The actual knife angle is, therefore, formed by this facet. The farther the deflection from the corner of the square of

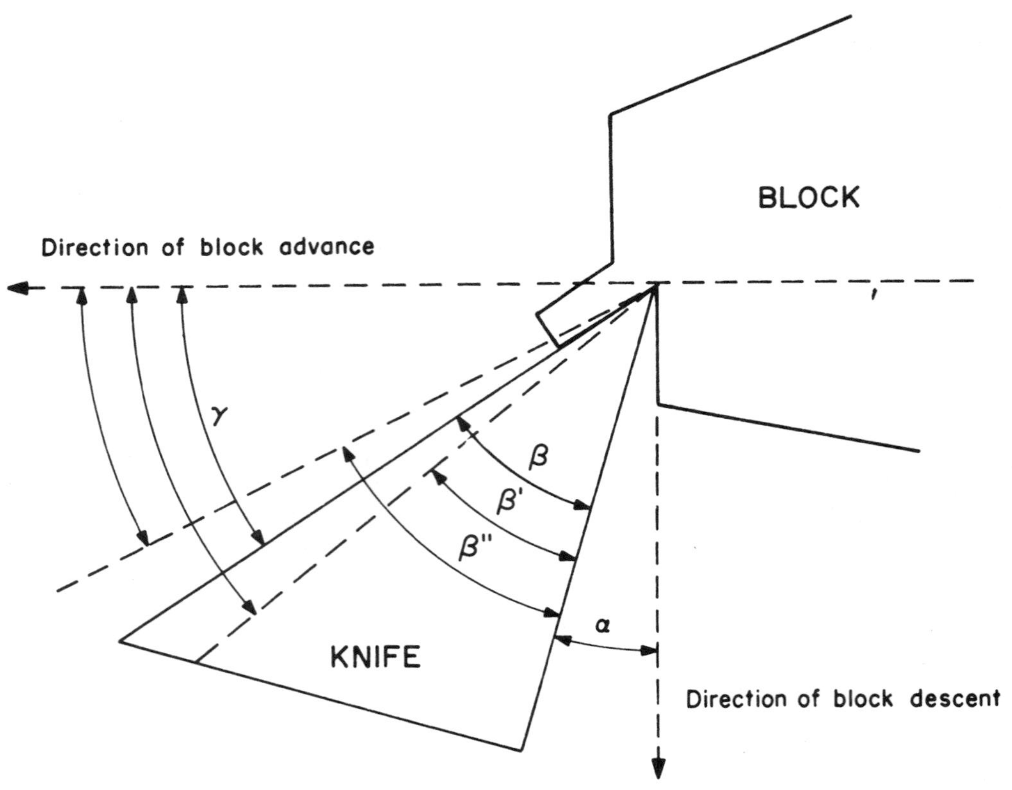

Figure 3.15 Diagram indicating various angles and directions involved in sectioning. α: Clearance angle; β: knife angle; β′: knife angle of less than 45°; B″: knife angle of greater than 45°; γ: shear (rake) angle. Dotted arrows indicate directions of block movement before and after cutting a section

glass, the less acute will be the facet, and also the less acute will be the knife angle. Thus, the facet determines the knife angle and the knife angle, in turn, determines the clearance angle. The more acute the facet, the smaller the knife angle it will make in order to obtain the optimal clearance angle. A smaller knife angle will result in decreased stress on the cutting edge of the knife, provided that the optimal clearance angle is maintained. In the preparation of triangular knives, therefore, the squares of glass should be bisected as precisely as possible.

As stated above, with pliers it is almost impossible to break the square of glass so precisely as to obtain knives with an angle of 45°. In practice, the most acute angle usually obtained is ~52°, on account of the curvature which always occurs at the end of the score. However, with a Knifemaker using scoring angles between 35° and 55°, it is possible to consistently obtain knives of good quality with knife angles ranging from 35° to 55°. This range is adequate to cut most biological materials. In the case of diamond knives, because the edge is sharpened and polished at an angle to the sides of the knife, the bevel formed by sharpening is known as the cutting facet. The clearance angle, therefore, is set by taking into account the angle formed between the cutting facet and the midpoint of the whole knife. To avoid the block face rubbing the back of the knife, the facet is tilted a few degrees away from the block face. Diamond knives are specified by the included angle at the cutting edge, taking into account the angle of the cutting facet. Most embedded biological materials can be sectioned with knives having included angles of 35–48°, whereas metals and other hard materials are best sectioned with knives having angles of 55–80°. It has been suggested that an included angle of 55° is better than one of 45°, because the former yields a longer usable cutting edge (Ward, 1977). The angle that is actually present at the cutting edge of a glass knife can be ascertained with the help of a special optical system devised by Fullagar (1966). Section deformation caused by pressure by the cutting edge can be reduced by lowering the knife rake angle (Fig. 3.15); such knives are expected to cut thinner sections (Lickfeld, 1985).

SPECIMEN BLOCK

TRIMMING AND PREPARATION OF THE BLOCK FACE

Small size and an appropriate shape of block face are prerequisites to satisfactory sectioning. In order to fulfil these two requirements, therefore, trimming of the specimen block is a necessity. In addition, it is desirable in most cases to centre the area of interest in the block face. Although sections as large as the specimen grids can be obtained using modern ultramicrotomes, the most desirable size of block face does not exceed 0.5 mm on a side. As a general rule, the smaller the block face, the less the amount of compression produced, and thus the quality of sections will be better. Smaller block face is especially important when very thin sections or serial sections are needed. In practice, the block face is trimmed to the size and shape consistent with the objective of the study. Experience indicates that when the block face becomes larger, the quality of sections is adversely affected. Since removal of each section results in the enlargement of the block face, it is desirable to retrim the face after an appropriate number of sections have been cut.

Retrimming of the block face interrupts the sectioning. An interruption of any length of time for retrimming, change of knife and resetting the advance system during sectioning changes the stability of the ultramicrotome. As a result, the first several sections of a new cutting sequence usually show several different interference colours ranging from grey to purple.

All the excess embedding medium around the tissue at the tip of the block should, whenever possible, be removed by trimming (Fig. 3.16). This will result in a block face of uniform density, and no part of the section would be without tissue. However, this may not be possible in the case of particulate specimens. It is emphasised that variations in the density of the block

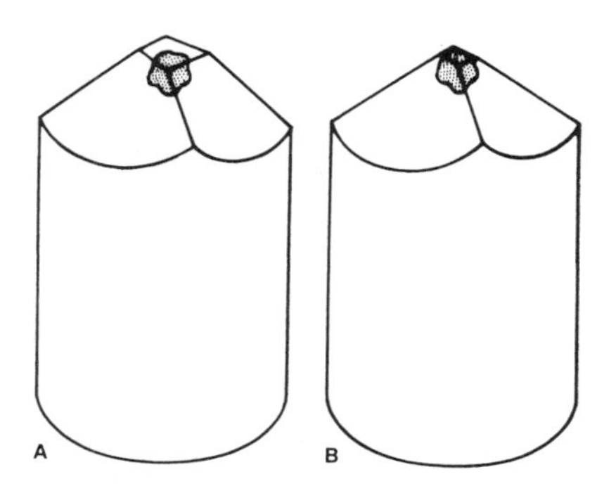

Figure 3.16 Diagrammatic representation of untrimmed (A) and trimmed (B) specimen blocks. (A) shows the resin around the specimen and at the tip of the block. The block face is large in size. In (B) excess resin has been trimmed and the block face is of a smaller size and contains only the specimen

Figure 3.17 An almost straight ribbon floating in the trough of a diamond knife; note that the specimen block is not extended more than a few millimetres from the front of the collet jaw. (Courtesy E. I. du Pont de Nemours and Co.)

face are a major cause of difficulty in sectioning. Embedded tissue should also be trimmed if it is too large to obtain a block face of a suitable size (0.1 mm or less). However, thick sections can be cut from a face of 5 mm^2 or even larger.

The ideal shape of the block face is trapezoidal, having parallel upper and lower sides and two sloped sides. During sectioning, the trapezoid-shaped face is supported much better than a face of any other shape, and the ribbon is formed more easily if the lower side is slightly longer than the upper side. While trimming, the upper and lower sides of the block face must be made parallel to each other. This will facilitate obtaining a straight, unbroken ribbon (Fig. 3.17), since such a face comes in contact and leaves the cutting edge evenly. A straight, unbroken ribbon is important in keeping serial sections in sequence.

Hand Trimming

The easiest and quickest way to trim a specimen block is with a sharp one-edged razor-blade which has been rinsed in acetone. If the block is very hard, a jeweller's saw, a fine file or a motor-driven abrasive wheel can be used. The selection of a particular tool is determined by the hardness of the block and also by availability and personal preference. However, most embedding media can be trimmed with a razor-blade. A block should always be trimmed after it has been mounted in the chuck (block holder). The chuck with the block is mounted on a special trimming block. Alternatively, the block in the chuck can be mounted on the ultramicrotome and then trimmed; the chuck can be clamped on some ultramicrotomes so that its axis is in a vertical plane and thus the specimen can be viewed from above.

The simplest way is to fit the chuck on the trimming block, which is placed on the stage of a stereomicroscope. The trimming block is sufficiently heavy and need not be held by hand. Fineran (1971) has described a similar device for holders from the LKB Ultrotome. An inexpensive device to hold a chuck for hand-trimming can be constructed by mounting a plastics tube cap vice from an ultracentrifuge on a wooden block (Mack, 1964). Holding the chuck in the hand during trimming is undesirable. The entire trimming operation should be carried out under a binocular microscope.

While viewing the specimen block from above under a stereomicroscope, a thin, horizontal cut is made to expose the specimen at the tip. Deep, sharp cuts should be avoided. The razor-blade is held at an angle of ~60° to the vertical and cuts are made down four sides of the specimen block (BEEM-type). The cuts should be short so that the block takes the form of a short pyramid, the semi-angle of which should be ~50°. The specimen face is supported much better if the sides leading to it are kept short. If the block tip is too thin and long, it will vibrate at a high frequency as a result of the impact of the cutting edge. Such vibrations are one of the major causes of periodic variations of thickness in a section.

It is emphasised that only thin slices should be cut with the razor-blade. This precaution will result in a trimmed block with smooth sides, which facilitate adequate adhesion between adjacent sections. The cuts should be made at such an angle that the angle of the sloping sides is sufficiently large to impart adequate rigidity to the specimen block during sectioning. Only a sharp razor-blade should be used for trimming, and better control of the trimming may be achieved if the blade is held with both hands. Great care should be taken while trimming, for the razor-blade may slip and cut the operator's fingers.

Some workers prefer to follow hand-trimming with fine trimming with the aid of an ultramicrotome. The advantage of such a practice is that the sides of the block tip are made smooth and clean. The smooth sides leading to the block face facilitate easy sectioning. This is especially true of the lower and upper side as the block is mounted for sectioning. If the lower side is not clean, the formation of a section may be hindered, since this is the side of the block face which first comes in contact with the cutting edge. If the upper side is not clean, the section may fail to be detached from the face of the block and may be dragged over the cutting edge. The use of ultramicrotomes and other mechanical trimmers will be described later.

During manual trimming, epi- (or oblique) illumination does not sufficiently display small specimens (e.g. single cells) for accurate trimming. This problem is largely alleviated by transmitting light through the resin block. According to one procedure, a hole, 3–4 mm in diameter, is drilled through the rear of any LKB chuck or through the base-plate and spherical base of the Porter–Blum Chuck (Scott and Thurston, 1975). The specimen block is placed in this modified chuck and viewed on a stereomicroscope capable of supplying transmitted illumination (Fig. 3.18). According to another procedure, the resin block is trimmed under a dissecting microscope using transmitted, reflected and dark-field illumination, either separately or in combination (Friedman, 1985). An acrylic block holder (wood may also be used) can be easily manufactured and fitted on the base of most microscopes; this holder can accommodate specimen blocks of most sizes and shapes. The T-holder shown in Fig. 3.19 (A, B) is sized to fit a Wild type of transmitted light stand (bright-field/dark-field). Slide clips, supplied with the

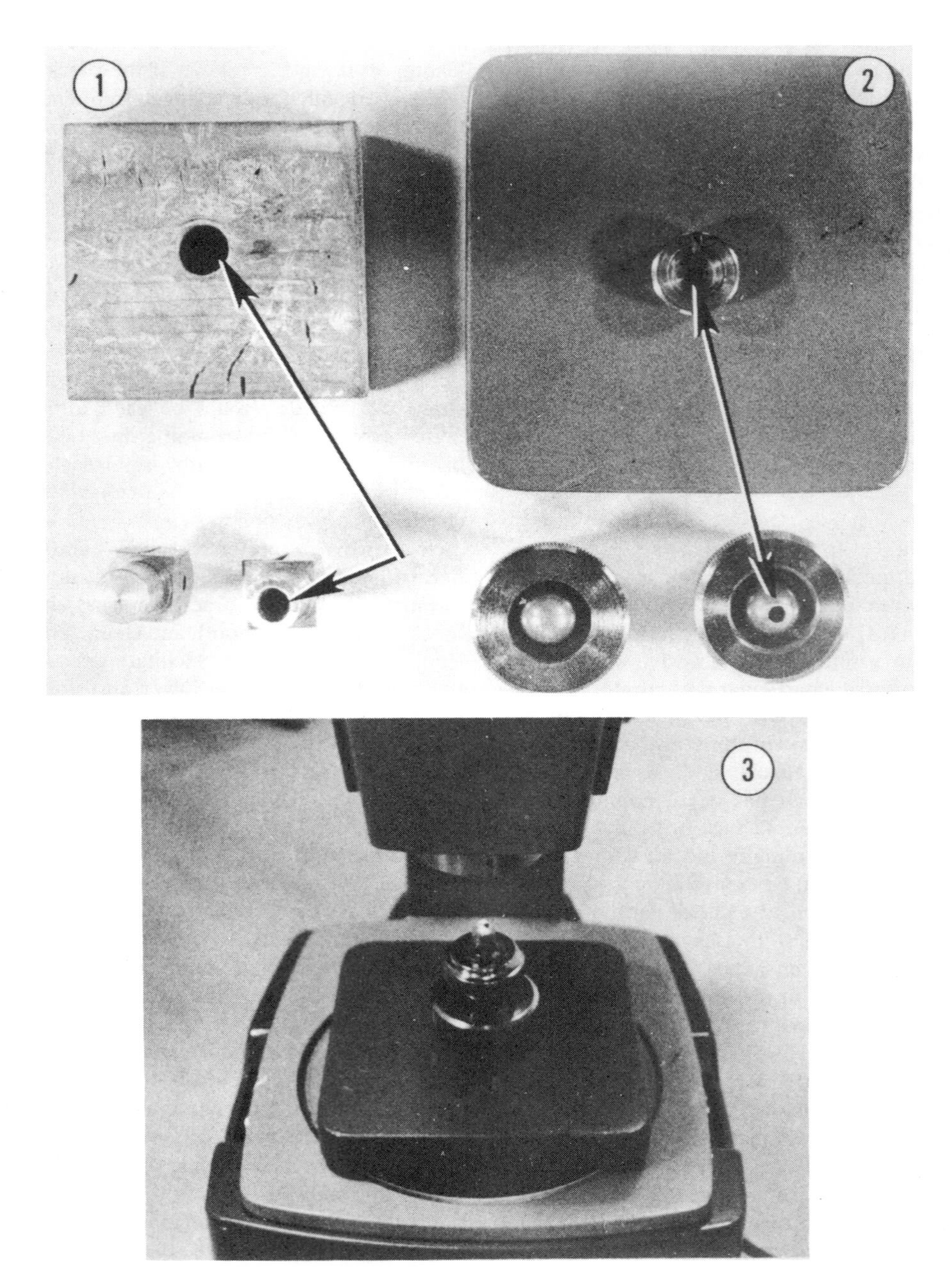

Figure 3.18 Specimen chuck modification for improved viewing during manual block trimming. Transmitted illumination through the plastics block or Lucite mounting rod accurately displays the specimen. A 3 mm hole is drilled through the rear of the LKB chuck (1) or the Porter–Blum chuck (2). Arrows indicate the modified chucks and their bases. (3), Specimen block is mounted in a modified Porter–Blum chuck and is being viewed by transmitted illumination using a dissecting microscope. (From Scott and Thurston, 1975)

stand, are removed. The holes are tapped, with an 8–32 threaded tap, to accommodate two long screws with knurled tops. The T-shaped holder permits one's hands to rest immediately adjacent to, and on either side of,

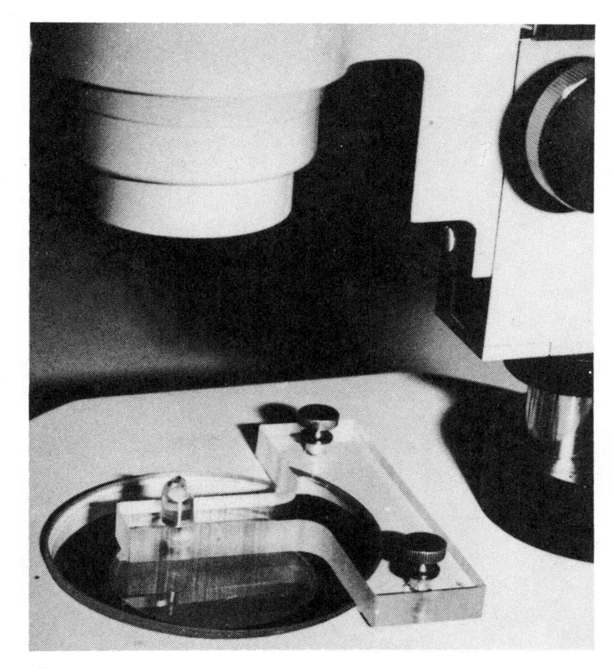

A

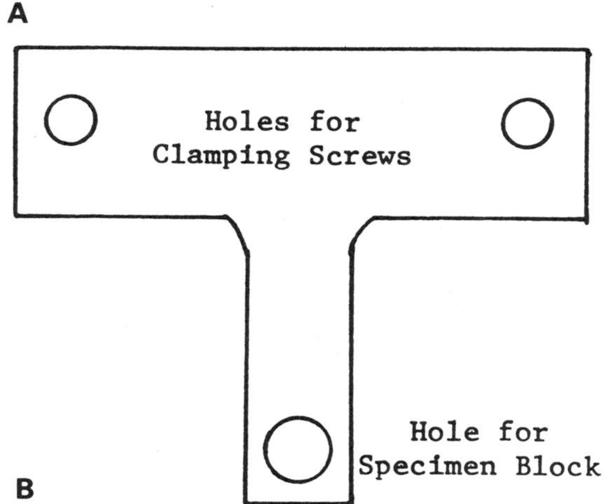

B

Figure 3.19 An acrylic block for trimming the specimen block. A: A view of the T-holder fastened onto the microscope base, with specimen block installed into the hole. B: A drawing of an overhead view of the T-holder shown above. (From Friedman, 1985)

the specimen block in the proper position for trimming with a single-edged razor-blade. The specimen block must fit tightly into its hole in the T-holder.

A simple approach is to paint reflective surfaces on the ends of the specimen block with white correction fluid of the type used by secretaries (Mollenhauer, 1976). For hand-trimming, the best approach is to paint the end of the block opposite where the specimen is embedded. For ultramicrotomy, the lower 45° facet of the block just below the specimen may also be painted. The painted surface reflects overhead light into the block and back around the specimen, which results in some improvement in visibility.

Hand-trimming can be rendered less susceptible to error by the operator if a razor-blade guide is used. Such a guide has been introduced by Butler (1974). The Butler Block Trimmer (Ernest F. Fullam Inc., Schenectady, N.Y.) (Fig. 3.20) is a free-hand microtome adapted for trimming the sides of a pyramidal block rather than the block face. The advantages of this device are: (1) the angle, location and thickness of the slices can be controlled; (2) the trimming process is quite rapid; (3) it can be used with any stereomicroscope that is provided with adequate epi-illumination; and (4) damage to the specimen during uncontrolled free-hand trimming can be avoided.

The specimen block (dowel-shaped) is inserted into the chuck (Fig. 3.20). The micrometer screw is rotated several full turns to back the knife guide away from the chuck. A flat object is laid across the tip of the specimen block, which is pressed against the ejector spring into the chuck until the block tip is level with the angled surface of the knife guide. The retaining screw is tightened with a screwdriver. The knife guide is oriented with respect to the specimen, and the micrometer screw is adjusted to position the guide so that a thin slice will be removed from the block when the blade, resting on the surface of the guide, is stroked downward.

The screw is readjusted and additional slices are removed from the block until one face of the pyramid is formed. The entire trimmer is turned 90° and the knife guide is rotated about the axis of the chuck, to establish the plane of the second pyramid face. Slices are removed to form the second face, after which the operations described above are repeated, to produce the two remaining pyramid surfaces.

Most trimming methods presuppose that the block face is perpendicular to the axis of the specimen holder. However, the Reichart TM-60 can produce parallel-face edges at any angle to the axis of the specimen holder only if the block face is formed on the trimmer as part of the trimming operation. Gorycki (1978) has introduced a device for use with the MT-2, which can produce a block face with parallel top and bottom edges, even if the block face is not perpendicular to the axis of the specimen holder. Guglielmotti (1976) designed a device for use with the LKB ultramicrotome, which produces a truncated pyramid from both flat and dowel embedments. Schneider and Sasaki (1976) introduced a razor-blade guide for trimming either flat-embedded or dowel-mounted specimens.

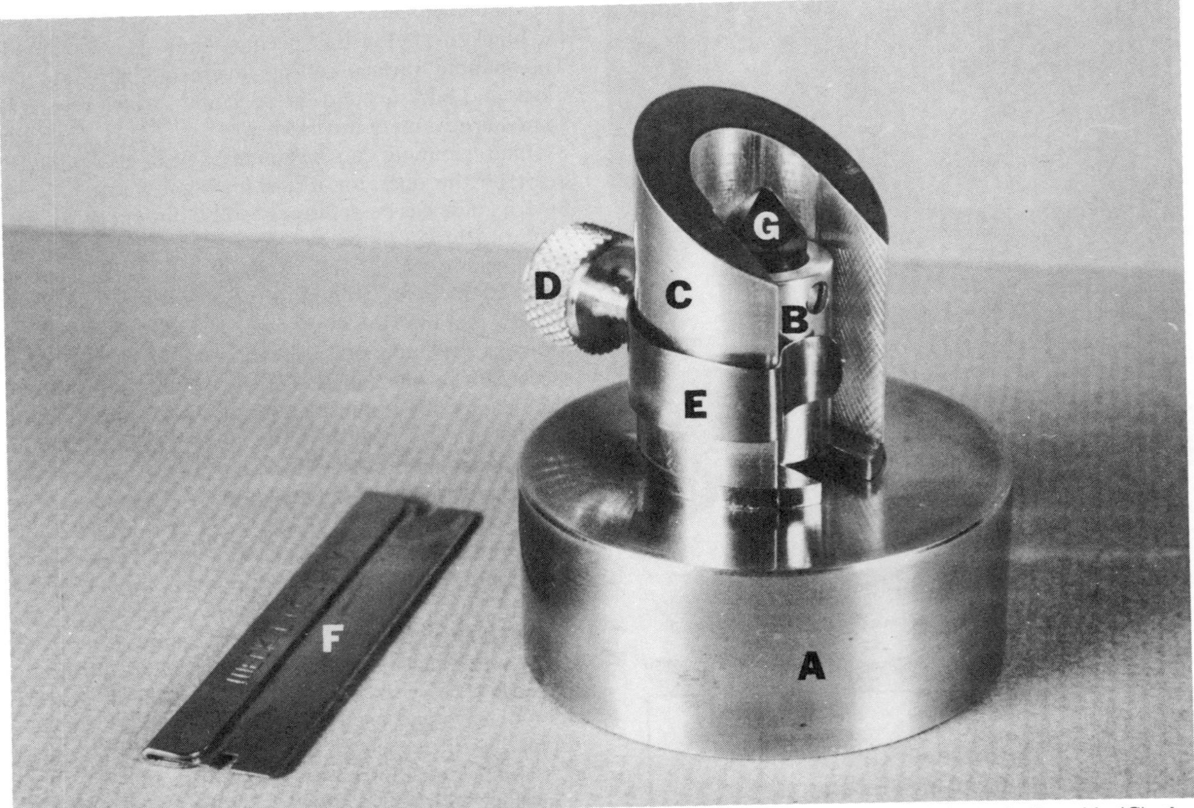

Figure 3.20 Precision hand trimmer showing the main parts: the base (A), the specimen chuck (B), the knife guide (C), the micrometer screw (D), the restoring spring (E), the hair shaper blade (F) and the specimen block (G). This trimming permits delicate control of the pyramid shaping of a specimen block face, and can be used with any stereomicroscope provided with epi-illumination. (From Butler, 1974)

Mechanical Trimming

Even the best trimming by hand is considered rough trimming because a hand-held razor-blade cannot cut slices sufficiently thin to produce smooth sides and precise enough to achieve upper and lower edges of the block face exactly parallel to each other. However, the quality of trimming is considerably improved by using razor-blade guides (e.g. the Butler Block Trimmer, described earlier). These devices are simple and may replace ultramicrotomes and other completely mechanised instruments used for trimming.

Smooth trimming can be obtained by utilising the advance and cycling mechanisms of certain ultramicrotomes (e.g. LKB). Glass or discarded diamond knives are used to trim the specimen block by cutting thick sections from each of the four sides in turn. The specimen block and the knife are mounted on the ultramicrotome, it being ensured that the tip of the block is perpendicular to the cutting edge. The specimen is exposed at the tip by cutting sections several micrometres thick. The knife stage is rotated through ~30° to the right and thick sections are cut; then 30° to the left, the block again being trimmed. Thick sections are cut at both times until the desired area of the specimen is exposed. This operation will result in the two edges of the face being parallel to each other.

The specimen block is rotated through ~90° so that the other two sides come in position to be trimmed. These two sides are trimmed in the same manner as the first two sides. This method can produce a block face as small as a sharp point. Some workers may find this method too time-consuming to be used routinely.

Three precision specimen block trimming machines are available commercially. The Cambridge Block Trimmer utilises a steel blade as the trimming edge. The blade is advanced by a feed-wheel between each stroke. The specimen block is illuminated from below. The Reichert TM 60 specimen trimmer uses a diamond milling cutter which revolves at a high speed. Specimen holders from most ultramicrotomes can be accommodated by this machine. An overhead fixed-focus light is used for illumination. The LKB Pyramitome uses a glass knife for trimming. The specimen is advanced by a hand wheel. A focused light can be rotated about the cutting point. Adaptors are available to accommodate specimen holders from most ultramicrotomes. This machine can be fitted with the Target Marker, which enables the image of a stained survey section on a glass slide to be superimposed on the block face. Thus, the desired area can be accurately localised on the block face for retrimming. The details of the operation of these three machines can be found in the manufacturers' manuals.

The Mesa Technique

When there is a need for survey sections throughout a specimen block, it can be trimmed to a mesa. If the specimen block is trimmed in the usual way, a lot of tissue in the block is cut away. This loss is unacceptable when only a very thin (shallow) piece of tissue is available; an example is a renal biopsy specimen, all of which may need to be examined. Sections at multiple depths from several areas on the same block face can be obtained by using this technique. The height of the mesa is usually kept under 50 μm and thus valuable tissue is saved.

The first step in the process is to cut a thick survey section, and then orient the corner of a glass knife to one side of the desired area. A few sections, each a few micrometres thick, are cut until a depth of ~30 μm is achieved (Fig. 3.21). The block is rotated through 90° and the next side is trimmed. The remaining two sides are trimmed in a similar manner. In other words, the tissue block is cut sequentially in the numbered regions as shown below. In the diagram on the right, the resin has been cut from faces 1–3. The resultant mesa obtained is then used to obtain thin sections. After the desired number of thin sections have been obtained, a new mesa can be made either on the same spot or in a different location. This process is repeated as often as is necessary to make certain that all the desired structures in one specimen are sectioned.

Mounting the Specimen Block

The precise alignment of the specimen block face in relation to the cutting edge is a prerequisite to satisfactory sectioning. The most important consideration in mounting the block is to minimise the forces of stress and strain produced during the process of cutting. Thus, the cutting edge is protected from excessive strain and the sections undergo minimum compression. The block should be oriented both in a vertical and in a horizontal plane relative to the cutting edge, so that the entire face of the block is cut. While aligning the block, the long axis of its face should be parallel to the cutting edge of the knife. As a result, the width, rather than the length, of the section is subjected to vertical compression during sectioning. Therefore, the stress resulting from the impact of the block is distributed over a longer portion of the cutting edge. Furthermore, the area along which any two sections come in contact with each other is longer, and thus mutual attachment of the sections is facilitated. If the block face is of trapezoidal shape, the shorter of the two parallel edges is mounted up (Fig. 3.17) to ensure that each section, as it is cut, will move the previous section away from the cutting edge. This procedure facilitates the formation of a straight ribbon (Fig. 3.17).

Ultramicrotomy of tissue culture monolayers presents special problems, for the monolayers possess an unusually restricted range of thickness for sectioning. Thus, in order to minimise cell loss during the first few strokes, special attention should be given to block alignment. Exceptional accuracy and consistency in the alignment of the block face are required in order to achieve a section of the maximum size after the first

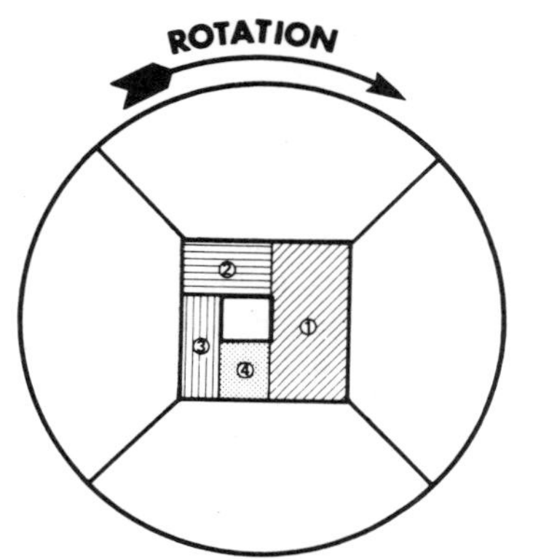

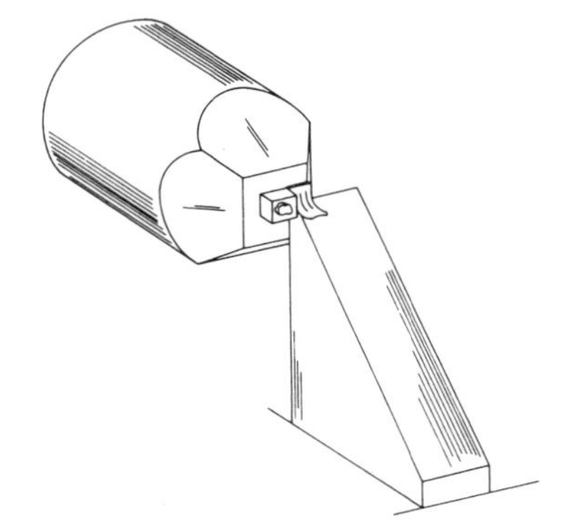

Figure 3.21 The mesa technique of specimen trimming. The block is trimmed sequentially in regions 1–4 (left). See text for further explanation

cutting stroke. The theory and mechanics of block alignment for sectioning narrow specimen layers (>15 μm) have been discussed by Grimley (1967).

The correct block orientation can be checked by making a lighting arrangement which casts a shadow of the cutting edge on the block face (Isler, 1974). A spotlight is directed towards the base of the knife. When the block face has the optimal orientation, the distance between the cutting edge and the shadow edge is the same at the right and the left, as well as at the upper and lower limit of the block face when the block is moved up and down. At the midpoint of the movement of the specimen arm, the height of the block face should be the same as that of the cutting edge. The front of the block face should be visible under the binocular microscope. This can be accomplished by rotating the specimen holder and suitably positioning the microscope and the illumination.

After proper alignment of the block, it is firmly clamped in the holder, to prevent any movement during sectioning. Very often a loose block is the cause of defective sections. The consequences of cutting a loose block include failure of sectioning, alternate thick and thin sections, section of uneven thickness and section wrinkling. Even a slightly loose block will make it difficult, if not impossible, to obtain a series of good-quality sections. Moreover, a loose block may damage the cutting edge.

To obtain a firm clamping, the block should not extend more than a few millimetres from the front edges of the collet jaw (Fig. 3.17). During sectioning, a longer block is not as well supported as a shorter one. Furthermore, when the major part of the block is firmly clamped, the block does not gradually squeeze out of its holder during sectioning. Periodic checking after preparing one or two grids is advisable, to ensure that the block continues to be held firmly. This checking will obviously necessitate realignment of the specimen block with the cutting edge. It is not uncommon for a firmly held block to become loosened after a number of sections have been cut. It is emphasised that a loose block is one of the most common errors of beginners and occurs very often when freshly polymerised or soft specimen blocks are being used.

Chucks are available for holding and orienting flat-embedded specimen blocks. Special chucks have been devised by Godkin and Keith (1975) for flat rectangular castings obtained from flexible silicone rubber moulds. Isler (1974) has modified the specimen holder of the MT-1, which allows specimen block orientation in both horizontal and vertical planes. An additional ring is mounted on the existing locking ring and is fitted with four radial screws which push the shaft of the chuck.

PREPARATION OF TROUGHS

Thin sections are extremely fragile, and adhere to the dry surface of a glass or diamond knife with such tenacity that it is almost impossible to preserve them. This difficulty is overcome by use of a trough and suitable flotation fluid. After a satisfactory glass knife has been prepared, a trough is constructed to collect the sections. Various methods are in use to prepare troughs, and they are discussed below. Alternatively, ready-made troughs can be bought.

A piece (~3 × 0.7 cm) of black electrical adhesive tape (Scotch No. 3) is wrapped from one side to the other of the cutting edge of the glass knife, as illustrated in Fig. 3.22. While the tape is being wrapped, its lower edge should preferably be parallel to the lower edge of the knife; also, the upper edge of the tape must reach exactly the tip of the cutting edge. This orientation results in a nearly straight trough tilted neither towards nor away from the cutting edge. A straight trough facilitates obtaining an optimal meniscus level, which is necessary in order for the sections to glide away smoothly from the cutting edge and to correctly show interference colours useful in judging section thickness. The black colour of the tape prevents the formation of reflections, which may otherwise interfere with viewing the sections on the water surface.

Excess tape extending from both sides of the knife face is removed by a single slashing cut with a sharp and clean razor-blade. The blade must not touch the back surface of the cutting edge. The trough thus obtained is ~0.8 cm long. The use of troughs of too large a size should be avoided, because of the difficulty in handling the floating sections properly in such a large space. To make the trough waterproof, the region of the heel of the trough is sealed with paraffin wax or beeswax, which is kept melted in a dish on a hot-plate or in any paraffin oven. In order to ensure a better seal, the paraffin wax or beeswax used for sealing the trough must be heated until it fumes. These fumes are considered to be carcinogenic; therefore, necessary precautions should be taken to avoid them. Care should be taken that the fingers do not touch the adhesive side of that part of the tape which comes in contact with the flotation fluid, as otherwise grease will be transferred from the fingers to the tape and eventually to the flotation fluid. Sealing wax is also a potential source of contamination, and therefore should be kept clean. A considerable degree of neatness is required to avoid oil and dirt contamination during trough preparation. Knives should always be kept covered when not in use.

The procedure described above has been satisfactory in that, with proper care, no leakage or contamination develops. However, a cleaner and faster seal can be

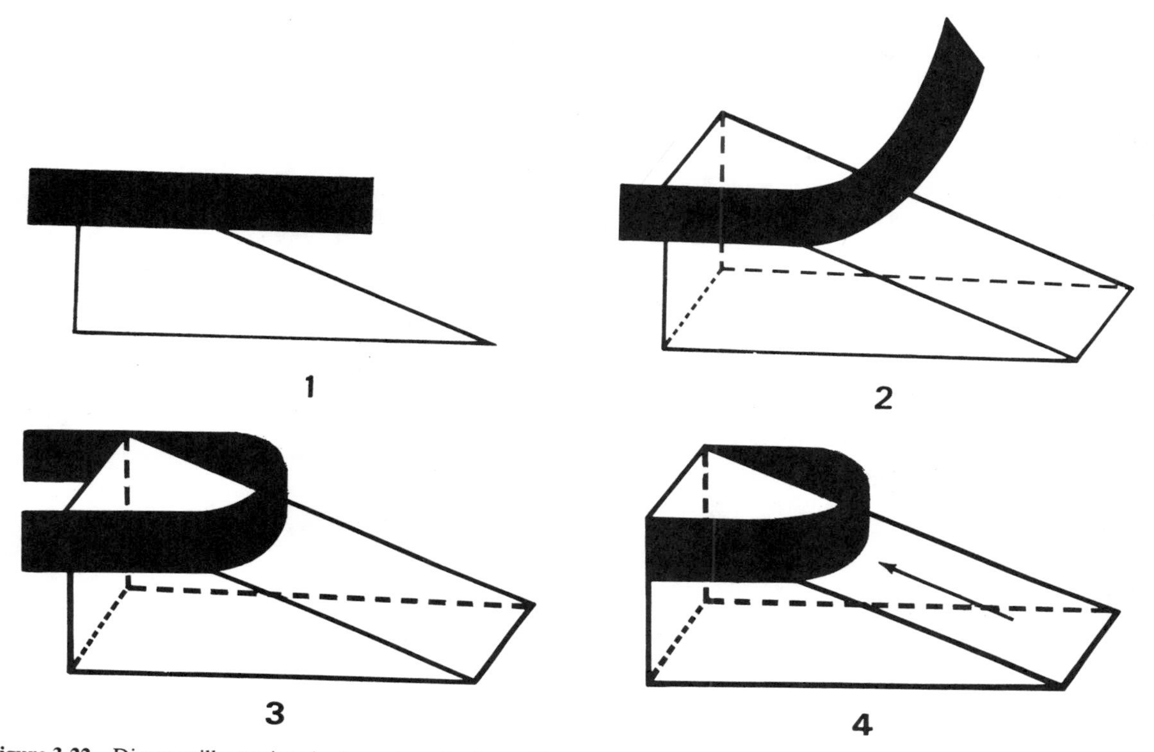

Figure 3.22 Diagram illustrating the four stages in the making of a tape trough; arrow indicates the heel of the trough, which is sealed with beeswax, nail polish or paraffin

obtained by using fingernail polish. The polish adheres easily to glass, metal or tape surfaces. Even a single brushing ensures a waterproof trough. The only disadvantage of this method is that the lacquer polish does not possess bridging capacity; therefore, the tape or metal must make a good fit with the glass edges.

Another type of trough is made of bronze, stainless steel or any other suitable material, and can be bought from the microtome manufacturing companies. This type of trough is sold ready to be sealed to the glass knife with fingernail polish, molten paraffin wax, beeswax or dental baseplate wax. A small piece of sealing material is placed inside the trough against the metal surface where it meets the glass and is melted with a match or microburner. McKinny (1969) suggested the use of a commercially available alcohol torch for sealing the metal trough. This torch provides an accurately controlled flame which helps to achieve good wax flow and seal near the knife edge. Wyatt (1972) has introduced an improved metal trough which avoids the use of adhesive tape or paraffin wax for sealing purposes, and thus possible contamination by these materials is eliminated.

Metal troughs, if already not black in colour, should be blackened in order to avoid reflections. A diamond knife is supplied sealed to a stainless steel or aluminium alloy trough. The inside of this trough is smooth and black, and is shaped in such a way relative to the cutting edge that the desired meniscus level is easily achieved.

MOUNTING THE KNIFE

Since the cutting edge is very delicate, the knife should be handled with the utmost care during mounting on the ultramicrotome. The first step in the use of a knife involves placing it in the proper holder without touching the cutting edge. If the glass knife leaks, it should be stopped by sealing with fingernail polish. It is preferred to mount the trimmed specimen block in the specimen holder before mounting the holder with its knife. Before mounting the knife, the knife stage is withdrawn a sufficient distance from the specimen block to ensure that the cutting edge will not hit the specimen accidentally. This distance is also needed to allow necessary adjustments between the specimen block and the cutting edge. The holder with its securely fastened knife is mounted on the ultramicrotome, and

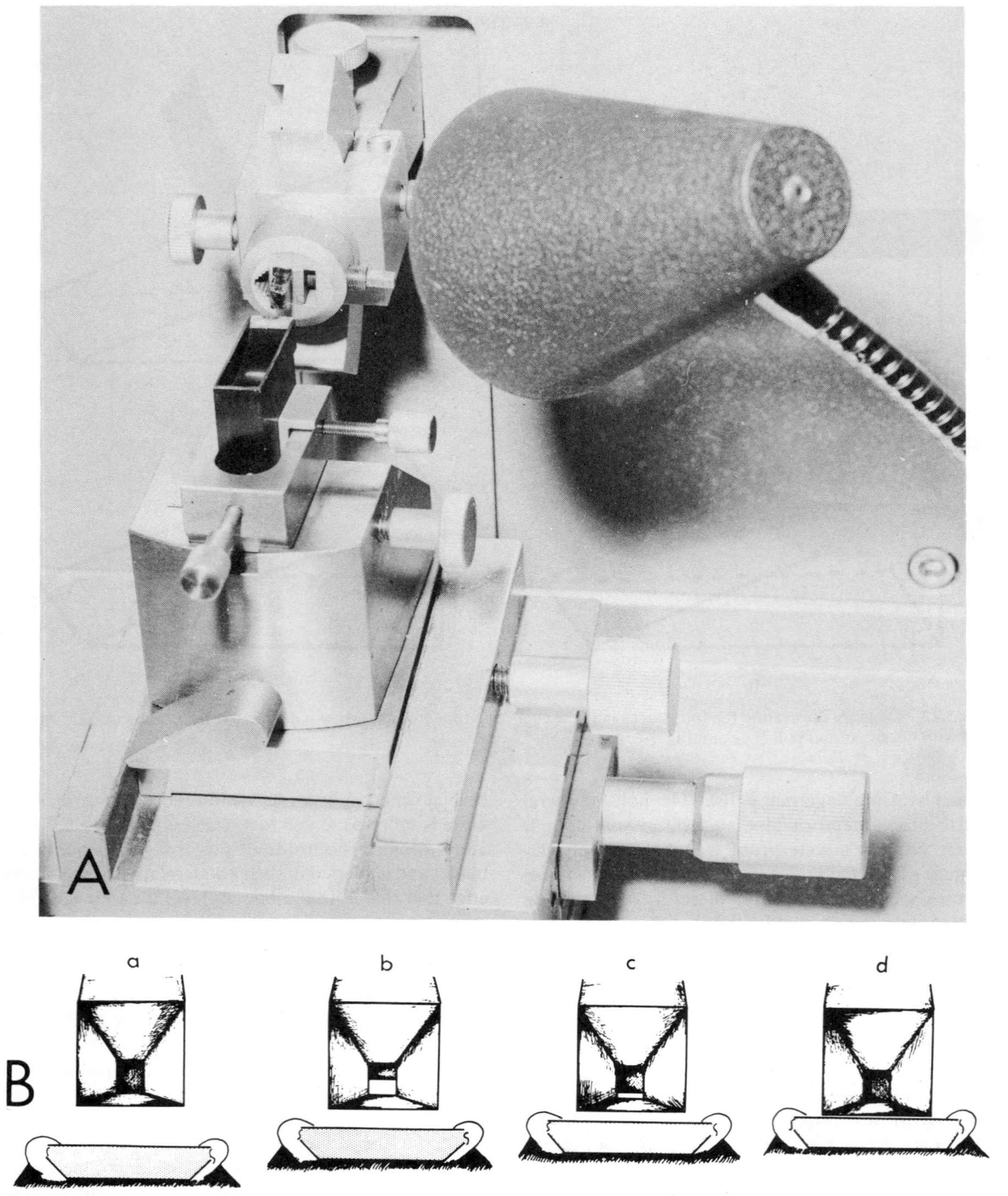

Figure 3.23 (A), Photograph of high-intensity lamp positioned so as to cast a reflection of the diamond knife on the face of a specimen block mounted in the chuck. This positioning facilitates the safe approach of the knife edge to the block, as explained here and in the text. (B), Schematic representation of steps during the approach of the knife to the block face when using the lamp: a, no reflection of the knife is visible on the block face when the knife and block face are separated by a distance greater than 200 μm; b, bright reflection appears on the lower portion of the block face as the knife approaches to within 100 μm of the block; c, reflection on the block face diminishes in height as the knife moves closer to the block; d, no reflection is visible on the block face when the knife is advanced to within 10 μm of the block. After this last step, the knife should be advanced extremely carefully. (From Nelson and Flaxman, 1973)

the knife angle is set to the desired sectioning angle, which is commonly in the range 2–5°. Knife holders have an engraved scale for determining this angle. The knife angle is determined primarily by the properties of the specimen block and the shape of the cutting edge.

The cutting edge should be parallel to the block face (Fig. 3.23B). In other words, the cutting edge should be perpendicular to the specimen arm. If the cutting edge is positioned obliquely to the specimen arm, lateral distortions would occur in the sections. The height of the knife should bring it level with the midpoint of the specimen arm movement. If the knife height is not optimal, the tangential relationship between the cutting edge and the radius of the arm movement will change. Ultramicrotomes are equipped with a reference mark for setting the knife height. The knife stage is moved laterally so that the useful part of the cutting edge comes in line with the face of the specimen block. The knife, the knife holder and the knife angle should be clamped firmly. Many problems encountered during sectioning are caused by loose connections. Even the smallest vibrations due to a loose connection between the knife and its holder, and between this holder and the ultramicrotome, will adversely affect the quality of sections and may lead to permanent damage to the knife.

At the start of sectioning, bringing the cutting edge sufficiently close to the block for it almost to touch its face is risky and time-consuming. This operation is carried out with the aid of a stereomicroscope at relatively high magnifications (30–40×) under the illuminating system of the ultramicrotome. The illumination that this system provides is unsuitable for observing the gap between the cutting edge and the block face. Controlling the approach of a diamond knife to the block face is more difficult than controlling the approach of a glass knife. The reason is that a glass knife produces a single bright image of its edge on the block, which facilitates a rapid and safe alignment of the edge with the block. A diamond knife, on the other hand, produces several reflections, and all of them are faint. It is not clear which of these reflections represents the actual edge. At magnifications higher than ~5×, they may become invisible.

The above-mentioned alignment of a knife with the block face is difficult. This difficulty can be minimised by concentrated illumination from a small, high-intensity lamp positioned laterally adjacent to the microtome (Fig. 3.23A) (Nelson and Flaxman, 1973). When the light is directed on the knife, a reflection of the knife edge is seen on the block face as a bright band, the upper edge of which may be used for the purpose of alignment. This procedure is useful for either glass or diamond knives, and provides good visibility at magnifications up to 40×. The reflection disappears when the knife edge has advanced to within 10 μm of the block face. At this stage, the lamp is withdrawn and the block is allowed to cool for at least 1 min before resuming the advance of the knife. This pause counteracts the thermal expansion of the block caused by the heat radiated from the lamp. This procedure is simpler than those using reflectors, which are described below.

It is well known that the visibility of the gap between the cutting edge and the block face can be improved by reflecting light upward towards the cutting edge from a white surface positioned below and in front of the knife. Several methods to improve the visibility of the gap are discussed below.

The simplest method consists of illuminating the back surface of the diamond knife with aluminium foil so that its reflection is visible on the block face Chaplin, 1972). A small piece of the foil is placed in the trough (Fig. 3.24) to produce a single image of the cutting edge as a band as bright as or brighter than that obtained with a glass knife. A piece of foil (25 mm^2) is folded in half, slightly crumpled to produce a multifaceted surface, and made concave by fingertip pressure. The foil is pressed gently into the trough near the cutting edge but not so close as to risk contact with the edge. It rests on the rim of the trough near the cutting edge and rises at the opposite end. After the foil is pressed down on the rim for support, it can be moved back and forth until the desired reflection of the cutting edge is achieved. The reflection usually consists of only one bright band, the lower edge of which is the line used to advance the cutting edge to within 1 μm of the block face. The knife is advanced until the space between the cutting edge and the bright reflection disappears. The foil is lifted out and can be re-used, and the trough is filled with the flotation fluid.

The block face can also be brought safely in contact with the diamond cutting edge before filling its trough with water. This procedure, 'dry diamond knife approach', has been explained by Purdy-Ramos (1987). Although this procedure is tedious and requires initial practice, it is worth trying when the above-mentioned methods are unsuccessful.

TROUGH FLUIDS

One of the major reasons for floating the sections on a liquid surface is to permit surface tension forces to restore them almost to their original dimensions after being compressed during cutting. The most desirable characteristics of a trough fluid are: (1) it should not damage cell ultrastructure; (2) it should easily wet the

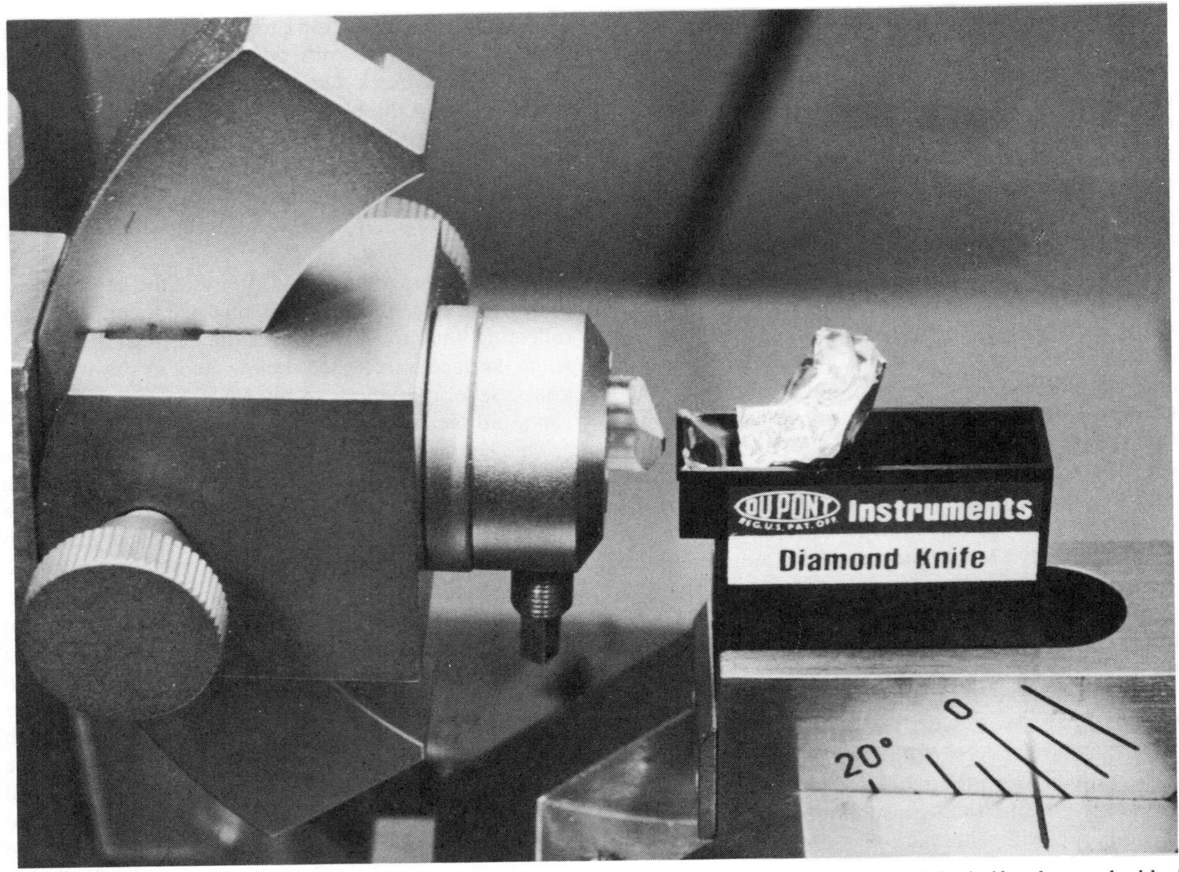

Figure 3.24 Diamond knife with aluminium foil in place, which produces a bright single band of the knife edge on the block face. This device facilitates rapid and precise alignment of the block with the knife edge. (From Chaplin, 1972)

cutting edge of the knife to the very end; (3) it should serve to release the section from the knife facet immediately before the next one is beginning to be cut; (4) it should relieve some of the gross compression caused by the forces necessary to move the section down the knife facet; (5) it should absorb the heat generated as the lower surface of the section rubs on the knife, (6) it should eliminate the possible electrostatic charge, for if the specimen block face picks up the static electricity charge, it is very difficult to obtain a section that is intact; (7) it should have a moderate surface tension; (8) it should not wet the sections; and (9) it should not corrode or in any other way damage the cementing material of the trough.

Distilled water is entirely satisfactory for sections of specimens embedded in various resins, provided that the sections are exposed very briefly to the vapours of a strong solvent, such as chloroform, or to the heat pen, to relieve compression. The water in the trough lubricates the sliding section by capillary movement and thus reduces the compression and shearing generated in the section. There are many advantages of using water: (1) sections can be kept floating for any length of time without subjecting them to the strong dissolving actions of solvents such as acetone; (2) the relatively high surface tension of water allows the sections to float; (3) the low viscosity of water facilitates free movement of the sections; and (4) the relatively low rate of evaporation of water enables the optimal meniscus to be maintained without frequent additions.

It should be noted that any concentration of a solvent (acetone) in the trough sufficient to remove the compression may also dissolve and destroy the cellular materials. Another reason for not recommending the use of acetone or ethanol solutions is that they may wet the block face more readily because of their lower tension. Furthermore, plain water is desirable for highly polar plastics such as glycol methacrylate. To prevent bacterial growth in the water container, only fresh water is used.

Some extraction and displacement of cell components occur during cutting and transferring of sections from the trough fluid onto the grid. Even materials of low solubility may be removed from the section by the aqueous solution, since the volume of water is large relative to the small amount of materials in a section of 80 nm or less in thickness. A case in point is sections of undecalcified bone tissue, which may be demineralised if allowed to float on water for longer than 3 min. Massive losses of extracellular calcitic mineral from earthworm calciferous glands during section flotation on water have been reported (Morgan, 1980). To minimise demineralisation, the pH of the water is maintained above 7.0. Water saturated with calcium and phosphate has been used to prevent demineralisation of sections of embryonic bone (Boothroyd, 1964). One solution is to keep sections dry during cutting and transferring to the grid. The dry transfer of sections can be accomplished in the presence of a strong electrostatic field (Koller, 1965).

Water is not used when sectioning at low temperatures. Collection of dry sections is preferred in the case of frozen specimens (see p. 404). Dimethyl sulphoxide solutions (40–50%) may be used for obtaining flat sections of frozen specimens. Solutions of anhydrous solvents such as ethylene glycol, hexylene glycol, glycerol and dibutyl phthalate have been tried as trough fluids. Freons, isopentane and liquid paraffin have also been used, without much success.

CUTTING SPEED

The rate at which the specimen block passes the knife during the cutting phase is called the cutting speed, which is expressed in mm/s. Automatic ultramicrotomes maintain a constant cutting speed over a specified distance. Success in obtaining thin sections of good quality depends partly upon the speed of cutting. As a general rule, sectioning should be performed at a relatively slow speed. Each time the cutting edge of a knife moves through the specimen block, the knife, the specimen block and the ultramicrotomes are placed under stress. The contact between the knife and the specimen causes the former to vibrate and the latter to compress. Therefore, sufficient time must elapse between two consecutive sections to enable the distortions to relax.

If the cutting speed is too high, the vibrations will overlap, resulting in variations in section thickness. Too high a cutting speed may also cause excessive compression, wrinkles and fine chatter parallel to the cutting edge in the sections, since, when cut at high speeds, resins tend to fracture as they are deflected by the cutting edge. It should be noted that cutting causes severe shear strain, which generates an elevated temperature as the lower surface of the section rubs on the cutting edge. Too high a cutting speed may cause an excessive rise in temperature, which, in turn, may cause permanent deformation of the fine structure. A very slow cutting speed is also undesirable, because changes in temperature (thermal fluctuations) and draught during a cycle will cause thermal drift. Too slow a cutting speed may also drag the trough fluid over the back of the knife. If the whole cycle from one cut to another is too long, the steady cutting motion will be disrupted, which will produce variations in the thickness of sections.

It is difficult to recommend an optimal cutting speed, because it varies according to the type and hardness of the resin, the specimen type, the knife and the ultramicrotome. Softer resins can be sectioned at a relatively high speed (~5 mm/s), whereas epoxy and polyester resins normally will section best at a speed of 1–2 mm/s. Very hard specimens such as bone and cellulose microfibrils require high speeds (>10 mm/s). Ultramicrotomes capable of operating at speeds ranging from 0.1 to 39.9 mm/s are available. Enough time as well as a constant time between the two consecutive sections are necessary in order to obtain sections of uniform thickness. It is, therefore, imperative to maintain a steady cycling rhythm during sectioning. This rhythm and the desired speed are mechanically maintained when using automatic ultramicrotomes. Considerable experience is, however, required to maintain a constant speed when using a hand-operated ultramicrotome. When operating these ultramicrotomes, the slow and constant speed is used only during the actual cutting phase and the rest of the cycle can be performed rapidly. In this way, the duration of a cycle is reduced, resulting in decreased thermal and other fluctuations. A relatively fast return speed (10–100 mm/s) also saves time. Automatic ultramicrotomes have a built-in mechanism which produces this variation in speed.

SECTION FLOTATION

Before sectioning is commenced, the specimen arm is taken at the lowest position and, with a syringe or a micropipette, distilled water is added at the rear of the trough. Usually enough water is added to obtain an almost flat (horizontal) surface. When using a diamond knife, especially a new one, more water than necessary is added; after the cutting edge has been wetted, the excess water is slowly withdrawn. The cutting edge of a diamond knife may have to be wetted by drawing the water up with a hair moistened with saliva or an appropriate wetting agent. The trough fluid can also be drawn up to the cutting edge with a small, soft wooden toothpick carrying some saliva.

Some workers prefer not to keep the diamond knife scrupulously clean, for the accumulating films on the face of the diamond knife facilitate proper wetting of the cutting edge. Another method of achieving wetting of the cutting edge involves immersion of the diamond knife in a 0.5% aqueous solution of Aerosol OT (dioctyl sodium sulphosuccinate) for 2–4 h, followed by thorough rinsing with distilled water. The treated knife will wet uniformly (Fig. 3.25) in daily use for several weeks before requiring renewal of the treatment (Garner and Steever, 1970). In general, because of the relatively hydrophobic character of the diamond surface, the meniscus level is kept higher than for glass knives. It is more difficult to wet the cutting edge of a diamond knife than that of a glass knife.

Before sectioning, the safest starting meniscus level is obtained by first developing a clearly convex meniscus; a concave meniscus is then obtained by withdrawing gradually the fluid with a syringe or micropipette until there is incipient drying of the cutting edge. The initial convex meniscus will ensure the wetting of the cutting edge. It is safe and desirable to start sectioning with a low meniscus rather than with a high one, and then to bring it up to the optimal level by adding more fluid. This procedure will almost eliminate the possibility of wetting the face of the block or the back of the knife face. The adjustment of the meniscus level is carried out while observing through a binocular microscope at a high magnification. If the knife facet is clean and the meniscus level is optimum, the sections should glide away smoothly from the cutting edge. It should be remembered that to maintain the optimal level of the meniscus, the fluid usually needs to be adjusted frequently during sectioning. The adjustment is also necessary for observing the interference colour of the sections.

After a suitable meniscus level has been achieved at the start of sectioning, extremely small quantities of flotation fluid are needed to be added or withdrawn subsequently during sectioning and collection of sections. A hypodermic syringe with a fine needle is most effective in adjusting the meniscus level. A glass micropipette can also be used. The fluid level can also be controlled by manipulation of a hypodermic syringe connected by a plastics tube to an opening in the base of the trough (Gay and Anderson, 1954). This and other devices (Westfall and Healy, 1962) introduced in conjunction with serial sectioning facilitate a more gradual and smoother adjustment of the fluid. However, these set-ups are rather elaborate, and may not be practical for routine sectioning. Finer control of the fluid level can be obtained by using the Reichert Reflexomat or a liquid dispenser.

It is emphasised that sectioning must not begin until the knife facet is wet to the very edge. This condition is necessary not only to obtain intact sections, but also to facilitate viewing of the sections and the release of a section from the cutting edge after the next one is just beginning to be cut. A relatively low surface tension impels the trough fluid to cover the space between the section and the knife facet and remain there. This is important, because the presence of a fluid between the surface of a section and the surface of a knife is necessary in order to obtain intact sections. On the other hand, under high surface tension, sections push the trough fluid away from the extreme end of the knife facet. This results in a dry cutting edge, which invites sections to adhere to it.

Wetting of the knife facet is related primarily to the shape of the meniscus of the trough fluid (Fig. 3.26). A convex meniscus generally tends to over-wet the cutting edge, which results in the wetting of the face of the specimen block. When this happens, the back of the knife face also picks up some trough fluid. Under these conditions, it is impossible to obtain sections. On the other hand, a concave meniscus may not wet the knife facet to the very end, which results in the sections sticking to the knife and crumpling. Therefore, the optimal wetting of the knife is achieved when the meniscus is almost flat, provided that the knife angle is neither acute nor obtuse to an appreciable degree. To obtain the optimum wetting of the facet of a knife having an acute angle, a somewhat concave meniscus is essential, whereas in the case of a knife having an obtuse angle the usual required meniscus shape is convex.

SECTION VIEWING

Since the appearance and interpretation of micrographs are dependent upon section thickness, keen observation and accurate evaluation of the quality and thickness of sections are essential steps prior to their being viewed in an electron microscope. Almost continuous viewing is required to ensure the maintenance of a desirable meniscus level in the trough during the cutting and transfer of sections. Thus, it is necessary that the sections be viewed not only during their transfer on to grids, but also during the cutting process. A low-power binocular microscope which can be set to a magnification range of 10–80×, without changing the working distance, is adequate for viewing sections. The microscope can conveniently be mounted on the front of the ultramicrotome, so that it can swing ~50° around

Figure 3.25 Mounted diamond knives. A: Untreated knife showing uneven wetting of the edge (single arrow) and accompanying distortion of water surface (double arrows). B: Uniform wetting of knife edge and undistorted water surface after treatment with Aerosol OT. (From Garner and Steever, 1970)

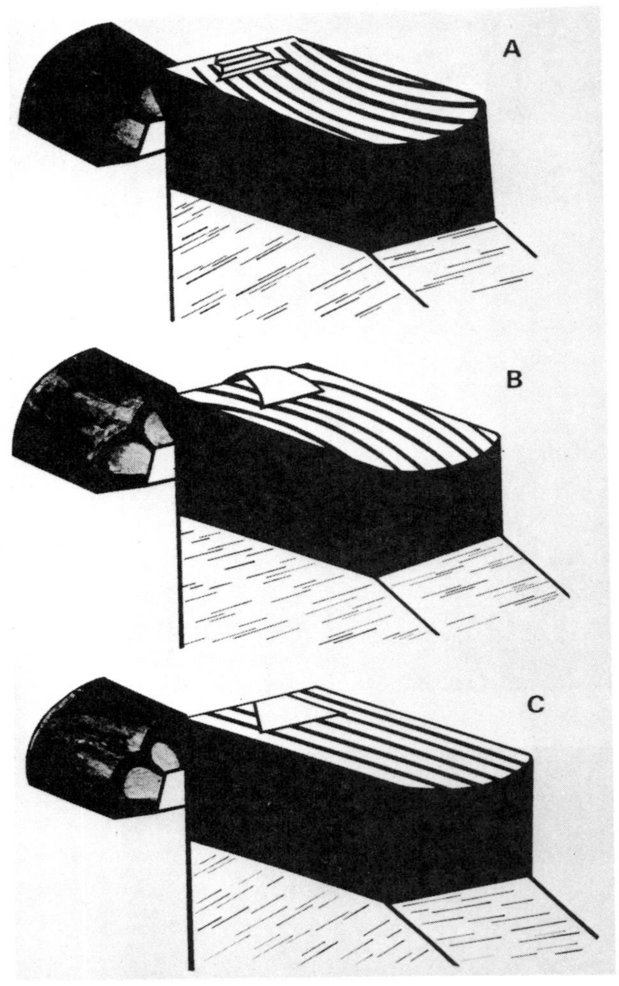

Figure 3.26 Diagrammatic representation of the appearance of various meniscus levels and their effects on the section. A: A concave meniscus may cause crumpling of the section. B: A convex meniscus may cause wetting of the face of the specimen block. C: An almost flat meniscus causes the optimum wetting of the cutting edge. Note the effect of varied meniscus levels on the section

the cutting edge without changing the microscope objectives so that the desired illumination angle can be obtained, and the bright reflection of light from the water surface can be seen through the microscope.

With a little experience, it is fairly easy to detect under a binocular microscope the presence of defects of even small size (e.g. fine lines, wrinkles and uneven thickness) in the sections. By manipulating the illumination angle, most of the possible flaws and the interference colours of sections can be observed. Proper illumination plus optimal meniscus level are necessary for the detection of the presence of contaminants such as debris and oil in the trough. As a general rule, as soon as a detectable flaw appears in a section, sectioning should be stopped and proper corrective steps should be taken (see p. 186). One should remember that the flaw that appears minor under a binocular microscope is enormously magnified in an electron micrograph. Experience indicates that even when sections appear perfect under the binocular microscope, they may prove to be imperfect in the electron microscope. Therefore, maximum effort should be made to eliminate, if possible, all the defects that appear during sectioning.

SELECTION AND HANDLING OF GRIDS

Copper grids are quite satisfactory for most studies. If oxidation of copper grids during prolonged exposure to aqueous solutions is a problem, it can be prevented by treating the grids with 0.05% collodion in amyl acetate, followed by quick drying. If copper is affected by a chemical treatment, then grids of other metals (e.g. gold, nickel, platinum and silver) can be used. Grids are either 2.3 or 3.05 mm in outer diameter, and the bar thickness is 6–50 μm. Many different types of grids, with openings of various shapes and sizes, are available (Fig. 3.27). The size of an opening is indicated by the 'mesh size', which refers to the number of openings to the linear inch. Mesh size indicates the amount of open area in each grid. For instance, a 100 mesh grid should have 100 bars per inch and ~65% open area, although the actual amount of open area in a given mesh size varies with the manufacturer. The open area decreases with increasing number of mesh per inch. Grids with thin bars provide a relatively large open area. For example, a conventional 200 mesh (square in shape) grid provides 46% open area, whereas a similar mesh grid with thin bars has 84% open area (Table 3.2).

Grids are available in mesh sizes ranging from 50 to 1000. Selection of the type of grid depends primarily upon the size of the section and the objective of the study. Single sections must be supported on all sides by the grid bars. In routine studies, a mesh size of 300–400 is satisfactory for single sections. Serial sections in the form of a ribbon are mounted on single-hole grids (Fig. 3.27) for an uninterrupted view of the sections. Very thin sections are supported better by hexagonal-shaped openings. Certain grid types have identification marks which help to locate areas of interest in the electron microscope. For high-resolution electron microscopy, specimen apertures are preferred to specimen grids, for the latter may render exact focusing difficult, because

Table 3.2 Grid reference chart (courtesy Electron Microscopy Sciences)

	STYLE	EMS CAT.#	BAR WIDTH (u)	HOLE WIDTH (u)	MESH* PITCH (u)	OPEN AREA (%)	ASPECT** RATIO
SQUARE MESH	50	0050	50	450	500	81	9.0:1
	75	0075	50	283	333	72	5.7:1
	100	0100	50	200	250	64	4.0:1
	150	0150	50	117	167	49	2.3:1
	200	0200	40	85	125	46	2.1:1
	250	0250	40	60	100	36	1.5:1
	300	0300	38	45	83	29	1.2:1
	400	0400	33	30	63	23	0.9:1
HEX MESH	75	H075	50	283	333	72	5.7:1
	100	H100	50	200	250	64	4.0:1
	150	H150	50	117	167	49	2.3:1
	200	H200	40	85	125	46	2.1:1
	300	H300	38	45	83	29	1.2:1
	400	H400	33	30	63	23	0.9:1
PARALLEL BAR	100	R100	50	200	250	80	4.0:1
	150	R150	50	117	167	70	2.3:1
	200	R200	40	85	125	68	2.1:1
	250	R250	40	60	100	60	1.5:1
	300	R300	38	45	83	54	1.2:1
PARAL. BAR W/DIVDR	100	R100D	50	200	250	78	4.0:1
	150	R150D	50	117	167	68	2.3:1
	200	R200D	40	85	125	66	2.1:1
	250	R250D	40	60	100	58	1.5:1
	300	R300D	38	45	83	52	1.2:1
SLOTTED	50/75	575	50	450×283	500×333	76	9:1/5.7:1
	75/300	753	40	293×43	333×83	46	7.3:1/1.1:1
	100/400	1040	38	212×25	250×63	34	5.6:1/0.7:1
THIN BAR SQUARE	200	T200	10	115	125	84	11.5:1
	300	T300	10	73	83	77	7.3:1
	400	T400	7	55	62	70	7.9:1
	1000	T1000	6	19	25	55	3.2:1
	Thin Bar Variable Mesh	TVM	10	155	165	88	15.5:1
			10	115	125	84	11.5:1
			10	73	83	77	7.3:1
			10	52	62	70	5.2:1
THIN BAR HEX	200	T200H	10	115	125	84	11.5:1
	300	T300H	10	73	83	77	7.3:1
	400	T400H	7	50	57	77	7.1:1
	600	T600H	7	30	37	65	4.3:1
SINGLE HOLE	ROUND	0600		600			
	ROUND	1000		1000			
	OVAL	0215		200×1500			
	OVAL	1020		2000×1000			
	SLOT	1002		200×1000			
	SLOT	0205		200×500			
SPECIAL PATTERN	VAR.	R12CA 1	130	120	250	23	0.9:1
			50	75	125	36	1.5:1
	VAR.	100SC	56/32	100	156/132	41/57	2.8:1/4.1:1

* BAR CENTER TO BAR CENTER ** RATIO OF HOLE TO BAR

Upon request, we will be happy to quote on your requirements, for special grids.

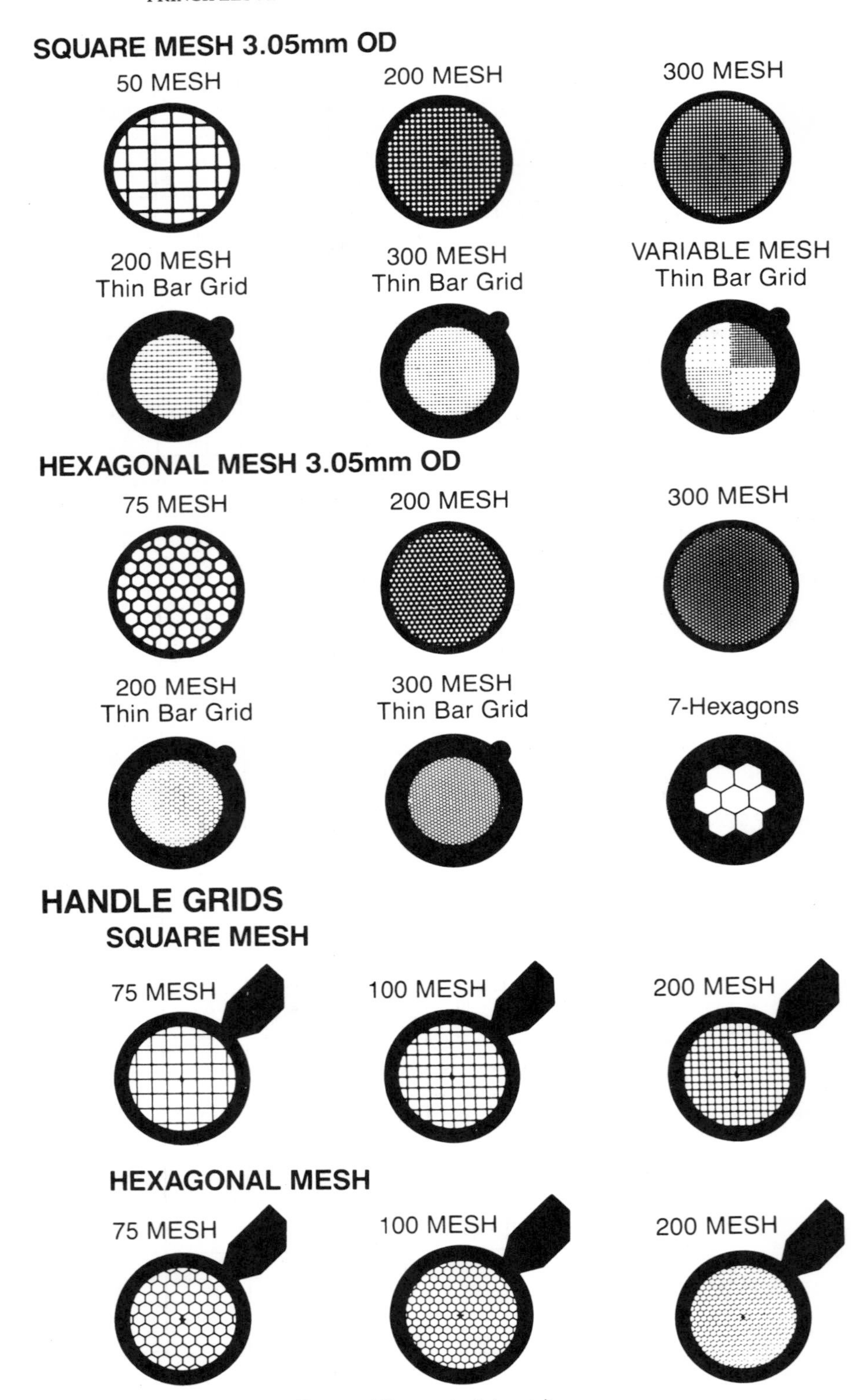

Figure 3.27 Various grid designs. (Courtesy Electron Microscopy Sciences)

PARALLEL BAR 3.05mm OD

200 MESH

WITH DIVIDER

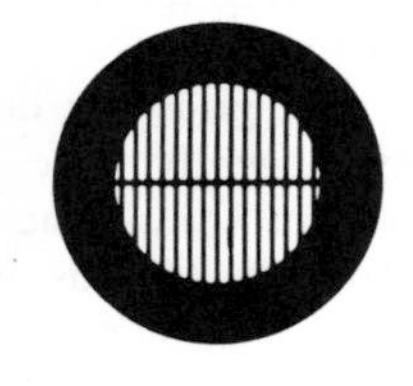

200 MESH

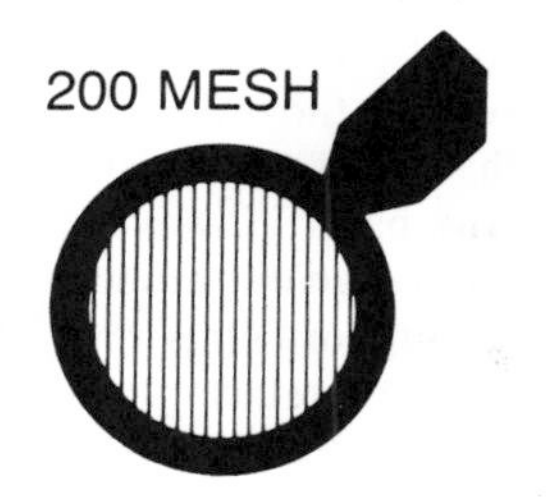

WITH DIVIDER

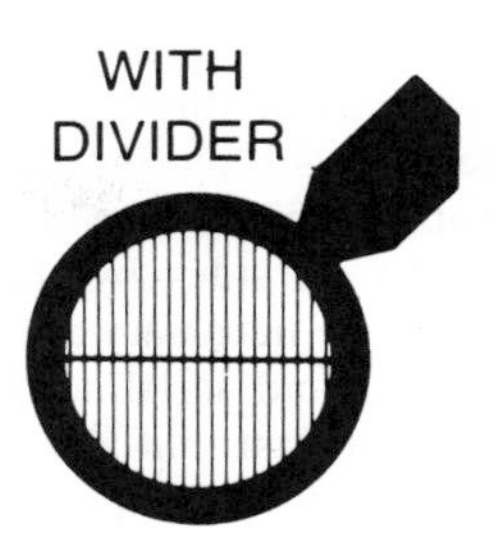

SLOTTED MESH 3.05mm OD

50/75 MESH

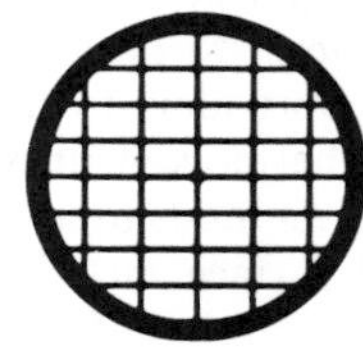

75/300 MESH

1.0mm Single Hole Grid

0.2 × 1.5mm Oval Hole Grid

1 × 2mm Oval Hole Grid

0.2 × 1mm Rectangular Hole Grid

PINPOINTER 100 MESH

COORDINATES 100 MESH

PINPOINTER 200 MESH

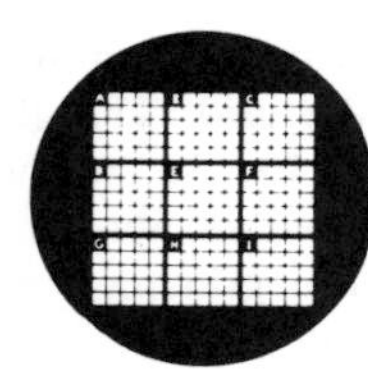

SPECIAL PATTERNS

SPECIAL PATTERNS

Oyster 100/200

of the limited depth of focus in such studies.

Since sections of three-dimensional polymer resins are sufficiently strong by themselves, they do not require support films, provided that the sections are supported on all sides by the grid bars. However, support films are necessary for examination of particulate specimens and serial sections. Support films not only reduce contrast, but also introduce artefacts. It is rather difficult, for example, to completely eliminate air bubbles, dirt and dust particles from Formvar films. Another type of artefact resembling plasmodesmata has been noticed when sections of plant tissues are mounted on Formvar-coated grids. This artefact seems to be wrinkles in either the section or the Formvar film. The wrinkling is probably caused by shrinkage of the cell wall relative to the cytoplasm (causing folds in the film) or by shrinkage of the cytoplasm relative to the wall (causing folds in the section) (Cox, 1968) or by the embedding medium (Epon or EMbed-812) being of high viscosity. No such artefact is observed in sections of plant tissues mounted on uncoated grids. This artefact can also be avoided by using embedding media (ERL-4206, LX-112) of low viscosity, even in the presence of Formvar film (Pihakaski and Suoranta, 1985).

The use of unsupported grids is, therefore, highly desirable unless the presence of film is absolutely necessary. If the sections do not stick firmly to the surface of the unsupported grids, rinsing the grids, immediately prior to use, in absolute acetone may help. The adhesiveness of the grid surface can be increased by burnishing. This is accomplished by placing the grid on the forefinger with the smooth surface up (if one side is polished); the handle surface of a pair of fine forceps is then rubbed firmly over the grid surface (Brown, 1969). This burnishing removes most of the oxidised surface from the grid. The grid is then immersed in 1 N HCl for ~10–15 s, washed thoroughly with distilled water, rinsed in absolute ethanol and dried on filter paper. The grid must be used within 10 min after this treatment.

If necessary, the adhesiveness of the surface of the unsupported grids can also be increased by treatment with chloroprene rubber (Neoprene W) (Fukami and Adachi, 1964). According to the original method, the grids are placed directly on a glass surface and covered with Neoprene solution. Grids coated in this way stick to the glass surface and are difficult to remove from the glass without damaging them. However, this difficulty can be overcome by attaching a thoroughly cleaned sheet of polythene (0.006 in thick) to the glass slide prior to placing the grids on it. Approximately 1 in wide by 5 in long sheet is pulled reasonably tight on the glass surface by taping it under the slide (Wyatt, 1970). The grids are placed on the surface of the polythene and 0.3–0.5% solution of Neoprene in toluene is dropped over them with a pipette. The excess solution is removed by tilting the slide, which is then placed in a desiccator for drying.

The coated grids are detached from polythene by stretching it gently. The stretching also breaks the Neoprene film around the grids, which can be picked up with forceps. If grids are damaged while handling with forceps, one can use a simple suction needle with a mouthpiece and plastics tube (Darley, 1972). Several grids can be picked up rapidly by employing stronger suction.

In addition to imparting adhesiveness, the coating described above protects the grids from the possible dissolving effect of strongly acidic or basic reagents. The coating can also be used to provide an adhesive surface for attaching support films to the grids, with consequent reduction in the movement of films under the electron beam. Other methods used for increasing adhesiveness of the grid surface include treating the grids with polybutene or a solution of tape adhesive in chloroform.

Grids should be handled with great care before, during and after mounting the sections, and should be protected from damage, such as bending or buckling, especially during post-staining. Any damage to the grid before or after mounting the sections may affect the quality of sections. It is apparent that sections are usually flat like the grid they are attached to. Buckling of even a segment of the grid may make the seating of the grid in its holder uneven. This is the reason why the deliberate bending of one edge of the grid to facilitate the transfer of sections is questioned by some workers. If the grid is not perfectly flat, it may become difficult to achieve an even focusing. Furthermore, bending or buckling of the grid after the sections have been mounted may result in section distortion.

Various methods have been suggested for handling the grids safely. Handling systems utilising magnetism are obviously not amenable to copper grids, while the use of fine stainless steel wire-mesh platforms has proved rather tedious. A simple device was introduced by Rowden (1968). In this method a glass plate with a surface having a series of ridges of ~2 mm in height permits the 3 mm type of grid to be picked up without any damage to the edges.

Grids most widely used today are of the Athene type, which usually have a rim and are characterised by both sides being flat, with one side polished and the other dull (matte) (Fig. 3.28). Sections should be collected on the dull side, because they get a better contact and are easily seen against a dull background by the naked eyes or under a light microscope. Support films also form a better contact with the dull surface. The reason is that

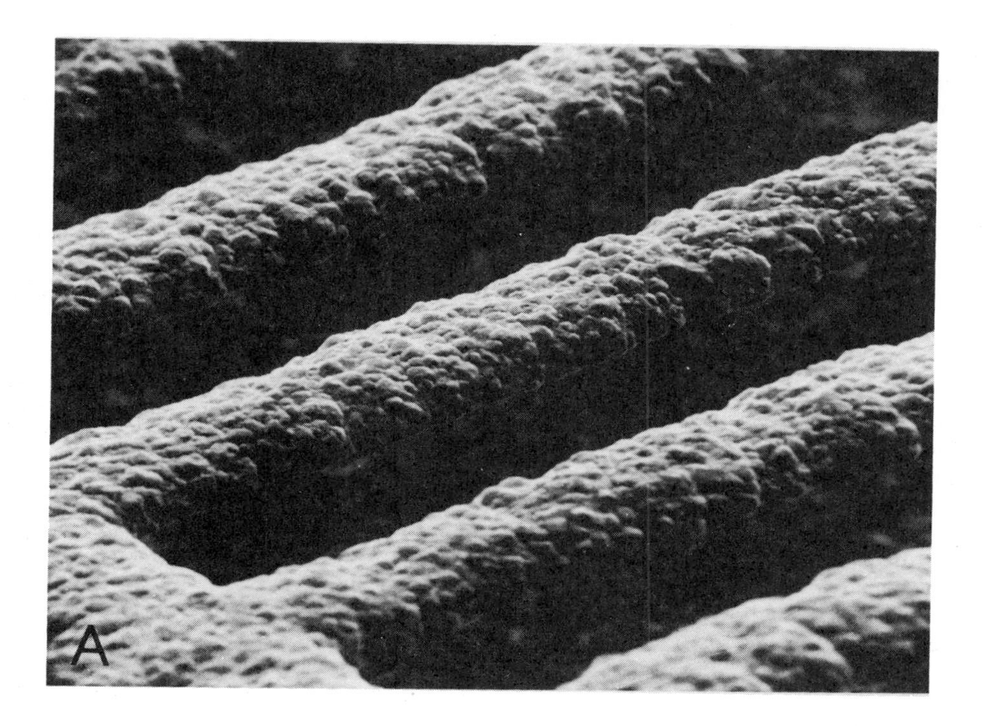

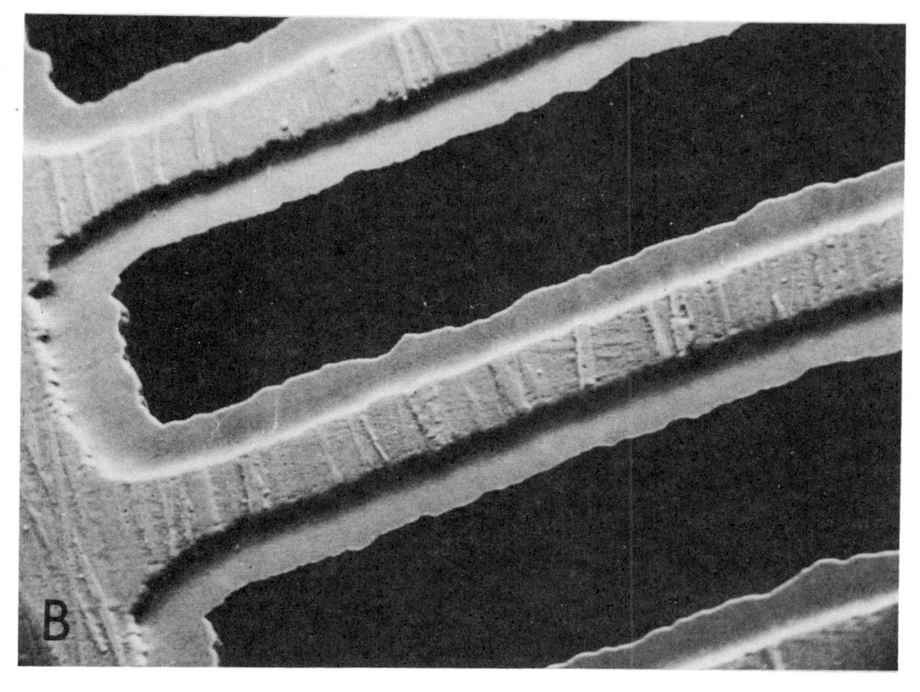

Figure 3.28 The surface structure of the Athene-type grid bars. The polished and dull surfaces are strikingly different in appearance. A: Dull surface has an extremely rough appearance. B: Polished surface is smooth, with only a few striations. The dull surface is preferred to the polished surface to facilitate section adherence. (From Summers and Rusanowski, 1973)

the rough surface of the dull side provides a larger amount of surface for contact. The tight adherence of both the sections and support films (if used) to the grids is essential in order for them to remain attached throughout the rigours of staining, washing and exposure to the electron beam.

SECTION COLLECTION

Section collection from the trough requires a considerable amount of care and skill. The two most important requirements of a successful collection are that the sections should be in the centre of the grid and should not overlap one another. Sections that stick to the rim of a grid are outside the field of view in the electron microscope; thus, these and overlapping sections are completely wasted. If sections form a ribbon, the possibility of overlapping during transfer is almost eliminated, provided that only one ribbon is picked. This is one of the main reasons why ribbon formation is so important in sectioning. If sections are scattered, they should be grouped together, by gently paddling the fluid with a stick carrying a single hair, towards the centre or at least away from the sides of the trough, especially away from the cutting edge. Sometimes sections have to be detached from the cutting edge and undesirable sections need to be separated from the desired ones. Although the paddling technique is rather risky in terms of damaging the sections, it lessens the possibility of overlapping and facilitates the transfer and centring of the sections on the grid. Additionally, it minimises any damage to the cutting edge and the grid. The manipulative ability to transfer and centre the sections on the grid is helped if the hand holding the stick or the forceps gets good support and the view through the binocular microscope is clear. While paddling the sections, they should be touched on their upper surfaces. If sections are moved by touching their edges, they may adhere to the hair.

If sections show a tendency to adhere to the human hair while being aligned, it can be replaced with a sliver of Teflon (Du Pont's fluorocarbon resin) (Reichle, 1972; Nyhlén, 1975). Teflon has a low coefficient of friction and is hydrophobic. Slivers are cut from Teflon foils 0.05–0.2 mm thick. Using a razor-blade, two parallel cuts, which converge at one end to form a fine point, are made in the foil. In other words, the cut is made from the base of the sliver to the tip. The sliver is glued to a suitable stick. Slivers can be made so thin and soft that they will penetrate the water surface. Thus, sections can be moved onto the wet grid without breaking the support film. Cutting the slivers at a less sharp angle makes them more rigid. Before gluing the sliver, it should be checked under the light microscope; if the surface is excessively rough, the sections may adhere to it.

If human hair or Teflon sliver is not working well, a hair shed from a Dalmatian dog may be helpful (Forbes, 1986). This white hair is long, is thick along most of its length and abruptly tapers to form a very fine tip. The dog hair is also recommended for handling cryosections (Barnard, 1982; Parsons *et al.*, 1984). A glass fibre drawn out from a micropipette into a fine thread has also been used for section manipulation (Davison and Rieder, 1985). The surface of the glass fibre is clean, smooth and hydrophilic.

The simplest and the easiest way to collect sections of relatively small size is by lowering the grid flat on to the upper surface of sections (Fig. 3.29). This approach is especially useful when more than a few sections need to be picked up. An alternative method of collecting sections is to dip the grid and then pick up the sections from below. The disadvantages of this method include: (1) the sections may be scattered by the upward movement of the grid in the trough and (2) the sections are difficult to centre on the grid. However, this method is recommended for large sections or ribbons of sections; it is discussed in more detail later.

It is easy to lay a grid flat on a ribbon if one edge of

Figure 3.29 Collecting a ribbon floating in the trough; note that one edge of the grid is bent to facilitate flat positioning of the grid over the ribbon

the grid is bent, making an angle of ~120° (Fig. 3.29). This can be accomplished by simply holding the edge of a grid with a pair of fine curved-tipped forceps and pressing it gently against a piece of filter paper. Grids with a handle or tab are also available. Sections can be easily picked up by bending the handle and lowering the grid flat over the ribbon; the bent handle forms an angle of 45° to the grid bars, which makes it easy to retrieve sections at a natural angle. The handle is also useful for transferring the grids during post-staining and rinsing. The handle is cut off before the grid is placed in the specimen holder in the TEM. Since the handle is held by the forceps and not the grid, there is less chance of damaging the grid or sections.

While picking up sections, the grid need not be pushed down into the trough fluid, because, as soon as the grid reaches across a ribbon, the sections along with a tiny drop of fluid stick firmly to the grid. In other words, the grid is pressed gently over the floating sections so that the fluid surface is depressed but not broken. This attraction of the ribbon is due to the presence of an electrostatic charge on the grid surface. It is advisable to rinse the grids with acetone just prior to use, since too much electrostatic charge will cause the ribbon to 'fly' towards the grid. Such violent attraction not only breaks the ribbon, but also is an obstacle in controlling the orientation and centring of the ribbon on the grid. In order to avoid contamination by grease, the tips of the forceps must be prevented from coming in contact with the skin. If necessary, the forceps can be cleaned with a solvent, or with lens paper or any other soft paper.

The grid with the sections is dried by placing it on a piece of filter paper, which absorbs the adhering fluid. Obviously, the side of the grid without sections is brought in contact with the filter paper. While placing the grid on a filter paper, if the forceps carry excess fluid, the grid is drawn up between the tines of the forceps. This can be avoided by inserting a piece of filter paper between the tines, to absorb any excess fluid. Although this procedure is effective, there is the possibility that the fluid may still remain trapped between the tines.

A more reliable method of overcoming these difficulties is hydrophobic coating of the forceps. The inner side of each tine is sprayed for ~1 s with polytetrafluoroethylene aerosol spray (e.g. Hiflon), a dry film lubricant and antistick agent (Wyatt, 1971). While spraying, the points of the forceps should be directed upward in a well-ventilated room; the spraying of the last 3 cm of the forceps is adequate. The hydrophobic plastics film formed on the surface of the forceps prevents retention of any fluid between the tines.

As stated above, large sections (~0.4 mm^2) are difficult to collect from above as they float on the water. Folding or wrinkling of the section may result because the grid may not touch all points of the section simultaneously. This method is also undesirable when a ribbon of sections needs to be precisely oriented on a large hole of the grid. Serial sections must be picked up from below.

Large sections can be collected by first transferring them to a small, support-film-free disc having a single hole with a diameter of 0.8–1.0 mm (Arbuthnott, 1974). The disc is placed over the floating section and is lifted clear of the water surface. The hole of the disc carries the section along with a drop of water, which is gently removed from below with a piece of filter paper. When sufficient water has been removed, the section is supported only by a thin film of water and is held taut by the surface tension of the water. The section is transferred to a Formvar-coated grid by superimposing the disc over the grid.

The method introduced by Rowley and Moran (1975) involves the use of a Formvar-covered aluminium supporting rack on to which slot grids that contain sections suspended in water are placed. Other methods for collecting sections have been discussed by Knobler *et al.* (1978).

MECHANICAL DEVICES TO COLLECT SECTIONS

Manual collection of sections may result in damage to the sections as well as to the grid. Transferring sections to the centre of the grid requires a great deal of experience. Therefore, efforts have been made to collect sections by using mechanical aids.

Westfall (1961) and Westfall and Healy (1962) introduced the Westfall-Healy Section Mounter (Western Instrument Scientific, Vacaville, Cal.), which consists of a diamond knife clamp, a grid container, a grid stage boat, a water control unit, and a cradle to adjust and position the grid stage boat. It is claimed that once this device is set up in the microtome, it need not be removed and can be used continuously for sectioning. The floating ribbon is guided over the grid, and the ribbon is then lowered onto the grid by lowering the trough fluid until it is level with the water control unit. The manufacturer emphasises that with the help of this device multiple ribbons can be mounted on a single grid. In the 'third hand' method of Behnke and Rostgaard (1964), the grid is positioned mechanically while the technician's hands are free for manipulating the sections. The LKB Ultrotome NOVA uses this method and is supplied equipped with a syringe system for raising or lowering the water level in the trough. Ward (1972) introduced a section lifter that can be attached to an ultramicrotome. For further information, the reader is referred to Knobler *et al.* (1978).

STATIC ELECTRICITY

The generation of static electricity at the specimen block–knife interface during sectioning is not uncommon, especially under certain environmental conditions; for example, excessively dry air in the sectioning room encourages the build-up of charge. Electrical charges are built up by the interaction of a block with a knife. Such static electricity can make sectioning difficult. The build-up of static electricity manifests itself as an attraction between the previously cut section still attached to the cutting edge of the knife and the descending block on its subsequent stroke. This attraction results in the lifting of the previously cut section towards the block face and its getting trapped between the block face and the cutting edge. The net result is that the previously cut section is damaged (e.g. excessively compressed) and the new section may not be cut at all. In addition, ribbon formation is hampered, which becomes a serious problem in serial sectioning. These problems increase as sectioning progresses.

Both the extent of charge generated during sectioning and the speed of charge dissipation affect the quality of sectioning. Charge neutralisation at the block–knife interface helps to ensure good sections, including ribbon formation. A number of approaches are in use to neutralise the static electricity at the time of its production. The water in the trough neutralises charge build-up on the inner (wet) face of the knife, while the charge on the block face can be dissipated by touching it with moistened tissue paper.

When this approach is ineffective, the use of a piezoelectric antistatic device is recommended (Fraser, 1976); this device (Zerostat) is commercially available (St. Ives, Huntingdon, U.K.; Ted Pella, Tustin, CA). The high voltage in this device causes the air around its needle to become ionised. Positive ions are released when the trigger is slowly squeezed, and negative ions are produced when the trigger is released. The block face is neutralised irrespective of whether it is charged negatively or positively, since the device produces both negative and positive ions sequentially. The device is hand-held and looks like a pistol. Figure 3.30(A) shows the damaged ribbon due to charged block face, and Fig. 3.30(B) shows a straight ribbon after using an antistatic device.

Another approach to dissipating the charge involves application of a strip of silver paint to the specimen block; the paint runs from just behind the block face to the chuck (Trett, 1981). Haskins and Nesbitt (1975) have used an ionising unit with a polonium element for the elimination of static electricity. Mollenhauer (1986c) has presented a modification of the polonium strip method in which a very small polonium strip is glued to the back (dry) face of the knife. It irradiates the block face before the beginning of each sectioning cycle. However, polonium is radiotoxic and its use is risky.

Figure 3.30 A: Sections cut from a charged specimen block. Note the badly damaged sections and the tendency of the ribbon to separate. B: The ribbon obtained while using the antistatic pistol. (From Fraser, 1976)

SECTIONING PROCEDURE

The ultramicrotome should be protected from external vibrations, draught, changes in temperature and sunlight, and should be housed in a room without distractions. It is essential to use a chair with adjustable seat height, because a comfortable seating position contributes to obtaining satisfactory sections. Before sectioning, the work table and other surfaces should be wiped free from dust with a wet sponge, and the forceps, pipette and other instruments should be absolutely clean. The points of the forceps must be in perfect condition; they should be periodically cleaned with a solvent and should be protected with a tip shield when not in use. Only freshly prepared solutions and fresh water should be used.

The materials needed are as follows: trimmed specimen block; several glass knives with a trough attached (or a diamond knife and a cleaning stick); fresh distilled water in a small bottle; grids recently rinsed in acetone in a Petri dish, which has a cover and the bottom of which is mantled by two layers of filter paper; a small camel's-hair brush reduced to a single hair (or dog's hair glued to one end of a small stick); a small-bore syringe or pipette; chloroform in a small bottle with a stopper or a heat pen; cotton swabs; a pair of fine forceps with a locking device; lens tissue; filter paper; a Petri dish, with two layers of filter paper at the bottom, for transferring grids after picking up the sections; and a wire transfer loop for transferring ribbons of sections directly to the stains.

It is assumed that the worker is familiar with the operation of the microtome. The following step-by-step procedure should be helpful in obtaining thin sections. Although this procedure has worked well in the author's laboratory, other procedures may, of course, work equally well or even better.

(1) Reset the advance mechanism of the microtome.

(2) Retract both the coarse and fine stage advances. This is an essential precaution, especially when using a diamond knife, against bumping the knife and the specimen block into each other.

(3) Mount the trimmed specimen block in its holder and screw tightly; make certain that the long axis of the block face is parallel to the cutting edge.

(4) Gently wipe off any resin debris present on or near the block face with a strip of lightly moistened lens paper.

(5) Mount a glass knife or diamond knife in its holder and screw tightly. The height of the knife should be adjusted.

(6) Set the knife angle between 2° and 5°, depending on the type of the specimen. Knife holders are provided with a scale to determine this angle.

(7) Cycle the microtome manually and stop when the face of the block is slightly above the height of the cutting edge.

(8) Position the binocular microscope and the illumination system so that the block face can be viewed through the microscope.

(9) Bring the mounted knife forward slowly and carefully until the cutting edge and the block face can be viewed through the microscope.

(10) Using a pipette, add distilled water in the trough until the cutting edge is wet. With a proper angle of illumination, a uniformly bright surface of the water normally indicates a slightly negative meniscus level.

(11) Adjust the angle of illumination so that some light is reflected on to the block face.

(12) Move the knife by laterally adjusting the knife holder so that sections will be cut by the best portion of the cutting edge.

(13) While viewing through the microscope, very slowly advance the knife towards the block face, by using first the coarse adjustment and then the fine adjustment. Advance the knife in increments of 1 μm between each stroke, until the first contact between the cutting edge and the block face is made. The importance of this first contact cannot be overemphasised, since a very thick section can damage both the knife and the block face.

(14) Adjust the desired cutting speed (2–3 mm/s can be used as a starting point) and section thickness (80 nm, pale gold in colour). If the need arises, the thickness should be changed only during the return part of the cycle, not during the cutting part.

(15) Adjust the meniscus level so that the interference colour of the sections can be seen clearly.

(16) Make sure that the ribbon is straight and that no knife marks are seen on the sections.

(17) After the desired number of sections have been cut, gently paddle them away from the cutting edge as well as from the sides of the knife and group them together in the middle of the trough by using a brush with a single hair.

(18) Expand the sections by holding a cotton swab dipped in chloroform close to but not in contact with the sections for 6 s.

(19) Lower the grid onto the sections and collect them on the dull side of the grid. Alternatively, bring the grid up under the floating sections. The author prefers the former.

(20) Place the grid on a filter paper (section side up) to dry, and cover the Petri dish.

Throughout the cutting process the quality of sections should be closely watched, and the correct meniscus level should be maintained by adding or removing small amounts of water with a syringe. Alternatively, one can use a commercially available device for precisely controlling the water level in a knife trough (Fullam

Inc., Schenectady, NY). The hand control allows micro-amounts of water to be pumped in or out of the trough to maintain the proper meniscus. Advanced ultramicrotomes have a built-in device for controlling the meniscus level. Only enough sections should be cut at any one time to be mounted on one grid. The same portion of the glass cutting edge should not be used for cutting more than one or two ribbons. Periodically checks ought to be made to ensure that the knife, specimen block and knife stage assembly are tightened firmly. After such checks, the block face will require realignment in relation to the cutting edge. Once the first section has been cut, every stroke must produce a section; otherwise, the sections will be either too thick or too thin, or the block face will rub against the clearance facet of the knife. For a detailed discussion on sectioning difficult specimens, see Hayat (1986b).

The problem of static electricity with grids kept in a plastics Petri dish can be solved by spraying the lids (inside up) with an antistatic spray such as Hansa (Agar Aids). The lids are sprayed lightly from a distance of 1 ft; this treatment lasts for the life of the dish.

Sectioning speeds of 1.5–2 mm/s for diamond knives and 1.5–3.5 mm/s for glass knives are recommended. Diamond knives should not be used at cutting speeds exceeding 2 mm/s. Sections of a large block face are easier to cut at slower speeds. To obtain a fairly constant section thickness for serial sectioning, sections should be cut at a low speed (~0.5 mm/s). When difficulties arise, lowering the sectioning speed helps. A pause in sectioning for several minutes is also helpful.

The durability of glass knives can be improved by coating the cutting edge with a film of evaporated tungsten metal (Roberts, 1975). Coated knives are thought to section hard specimens better than standard glass knives. Freshly prepared knives are attached to the base-plate of a vacuum evaporator with adhesive tape. They are hinged to allow tilting towards the evaporation source. The film is deposited on both faces of the knife simultaneously. The evaporation source is a 25-mm-long and 0.5-mm-thick tungsten wire (V-shape filament), which is mounted at a distance of 8–10 cm from the cutting edge. A pressure of 5×10^{-5} Torr for 3–5 min is required to provide a sufficiently thick film. A glass knife with an angle of 50° seems to be more durable than one with an angle of 45°; the former also has a longer cutting edge, but it may produce excessive section compression.

Glass knives are thought to section hard tissues embedded in epoxy resins with less difficulty when 1% Dow Corning 200 fluid silicone plastics additive (Dow Corning, Midland, MI) is added to the embedding medium (Langenberg, 1982). It has been claimed that glass knives last 5–15 times longer when cutting modified rather than unmodified resin. The silicone additive is added to the resin mixture while stirred and just before addition of the accelerator.

The thickness of a section is commonly estimated by observing the interference colour of light reflected from the section while it is floating on the surface of a liquid in the trough. Since the interference colours form a continuous spectrum instead of clearly separated colours, the colour scales in Table 3.1 are only a guide to the actual thickness.

Thin sections of epoxy resins are sufficiently strong to withstand the electron beam bombardment without support film. However, adhesion of sections to unsupported grids may be a problem in lengthy immunolabelling procedures or with serial sections. The following procedure will increase the adhesion of sections to these grids (Zelechowska and Potworowski, 1985). A 1% stock solution of Formvar is diluted in ethylene dichloride to a final concentration of 0.1–0.25%. The grids (precleaned in acetone) are dipped in this solution for 1 s and are immediately dropped onto a Whatman filter paper. After several seconds, they can be used immediately or stored horizontally in a covered container. A thin film of solidified Formvar is formed only on the bars, and the mesh spaces remain uncovered.

IMPROVEMENTS IN SECTIONING

The quality of sections can be improved by the means listed below.

(1) A small knife angle results in decreased stress on the cutting edge of the knife. However, a knife with an angle of 50° seems to be more durable than one with an angle of 45°; the former also has a longer cutting edge (see also p. 158).

(2) Thin sections of good quality can be obtained by coating the cutting edge of a glass knife with a film of evaporated tungsten metal (Roberts, 1975). Also, harder resin blocks can be sectioned easily with coated knives. Such treatment reduces the friction produced during cutting and bending of the sections, and the durability of the glass knife is improved.

(3) Fluorocarbon coating of the Ralph glass knife improves its cutting qualities and reduces section adherence of specimens embedded in paraffin (Richards, 1979).

(4) Relatively hard resin blocks have less intrinsic compression. Very thin sections are difficult to obtain from soft blocks.

(5) The quality of sections can be improved by reducing the heterogeneity of the specimen and the resin in the block. This is accomplished by using a resin

(e.g. an epoxy resin) which copolymerises with the specimen. However, this requires thermal polymerisation. Alternatively, heterogeneity can be minimised by hardening the specimen before infiltration and embedding (Jésior, 1985). This is accomplished by impregnating the specimen with a series of heavy metal salts such as OsO_4, PTA and uranyl acetate. The binding of heavy metals can be facilitated by using mordants (e.g. tannic acid). Precipitation of inorganic phosphate in the tissue can be accomplished by a rapid ethanol dehydration (Colquhoun and Rieder, 1980). The objective is to equalise the hardness of the specimen and that of the resin. Because compression diminishes as hardness increases, hardening of the specimen will limit the compression.

(6) Sectioning of the specimen block becomes easier when the resin contains a surfactant such as lecithin (see p. 98).

(7) Glass knives seem to section hard tissues with less difficulty when 1% Dow Corning 200 fluid silicon plastics additive is present in epoxy resins (see p. 184).

SECTION DEFORMITIES

The sources of and remedies for defects appearing during sectioning are listed in Table 3.3.

NORMAL SURFACE DAMAGE

It is well established that sections are damaged at both the lower and upper surface, regardless of the sharpness of the cutting edge. In other words, considerable damage to the section surface occurs during cutting with glass or diamond knives. The upper surface of the floating section is rougher than the lower surface. Glass knives produce a greater difference between the texture of the two surfaces of the section than do diamond knives. This damage clearly differs from knife marks (Fig. 3.10), which are usually present only on the bottom surface of a section. The occurrence of the surface damage becomes apparent when one realises that a section is not separated from the face of the specimen block as a result of fracturing, but as a result of shearing. The resin block is placed in compression prior to being sheared as it passes the knife edge.

The stress applied during cutting is not easily relieved, because polymerised resin is amorphous and possesses no crystalline plane which could provide a preferred path on which the stress would move. The stress is thus supported elastically in the interior and plastically in the surface of a section. The elastic strain is recovered, while the plastic strain, which is irreversible, remains in the section after cutting has been completed. The plastic strain is therefore primarily responsible for the section surface deformity, and the plastic strain, in turn, arises as a result of the energy expended to separate the section from the specimen block. Surface irregularities as great as 7 nm can be found in the epoxy sections (Mollenhauer, 1986b). Frösch *et al.* (1985) indicate that the depth of holes in the surface of Epon and Nanoplast FB 101 may be as great as 12 nm. The damage to section surface structure caused by diamond and glass knives was demonstrated by Black and Boldosser (1971). The images produced by cellular structures that extend to the section surfaces can be affected by the texture of the section surface. Relatively low plastic flow during cutting results in less damage to the two surfaces of a section.

This deformity is difficult to detect, because of the large depth of field characteristic of the TEM, which brings into focus dominantly the internal region of a section rather than the surfaces. The factors that increase this deformity include increased depth of cut, deteriorating cutting edge and increased knife angle, which decreases the rake angle. The information available concerning fundamental engineering of the cutting process and the nature of the material being machined is rather meagre. Black (1971b) has pointed out the major areas of investigation: density and hardness numbers of tissue–resin composite; sliding friction coefficient between diamond or glass and the composite materials; and plastic deformation curves.

SECTION COMPRESSION

To understand the sectioning process, it is important to know the nature of interaction of the specimen embedded in a resin with the cutting edge of a glass or diamond knife. In other words, an understanding of the origin of section compression and high-frequency vibration occurring during cutting is necessary. Compression seems to be an anisotropic deformation of an originally random pattern of macromolecules (Sitte, 1984). The deformation is accompanied by a specifically oriented polarisation phenomenon caused by pressure. This preferential orientation of the molecular structure may be due to the formation of bonds, which can be broken by an organic solvent or heat.

All embedding resins undergo compression of varying degrees during sectioning. Compression may or may not be easily detected during sectioning, but it will be observed in the electron microscope. Increased compression is produced by softer specimen blocks and thinner sections, especially when cut at a high speed. The result of such compression can be seen in that,

Table 3.3 Defects appearing during sectioning and their remedies

Defect	*Possible causes*	*Remedies*
Sections are not cut	(1) Face or block rubbing on back of knife	(1) Increase clearance angle
	(2) Knife, knife holder or block loose	(2) Tighten them up
	(3) Vibrations and/or change in temperature	(3) Check possible causes of vibration or change in temperature, such as draught, lights and air-conditioner
	(4) Advance mechanism reached to the end	(4) Reset advance mechanism
	(5) Blunt knife edge	(5) Replace knife
	(6) Face of block away from knife edge	(6) Bring face of block close to knife by cycling microtome while still advancing by fine adjustment
Only left or right portion of block face is cut	Cutting edge not parallel to block face in horizontal plane	Reorient knife or reface block
Only top or bottom of block is cut	Cutting edge not parallel to block face in vertical plane	Reorient block face or reface block
Sections vary in thickness	(1) Knife, knife holder or block loose	(1) Clamp them tightly
	(2) Face of block too large	(2) Reduce size by trimming
	(3) Knife tilted too much forward or backward	(3) Adjust knife angle (2–5°)
	(4) Cutting speed too fast	(4) Reduce speed
	(5) Blunt knife edge	(5) Try another portion of knife edge; get knife resharpened; replace knife
	(6) Advance setting reduced too much to give very thin sections	(6) Increase advance setting and after some time try again to reduce it
	(7) Interruption in microtome cycling	(7) Cycle microtome rhythmically
	(8) Mechanical vibration or thermal drift	(8) Check possible causes, including draught and heaters
	(9) Block too soft	(9) Heat block in oven at 60–80 °C for 24 h or change block
	(10) Defective mechanism or installation of microtome	(10) Check and reinstall or return microtome to maker
Ultramicrotome cuts on alternate strokes (skipping)	(1) Dull knife	(1) Use another portion of knife edge; replace glass knife; get knife sharpened
	(2) Knife angle too great for this block and section thickness	(2) Try lower knife angle
	(3) Clearance angle too high	(3) Reduce clearance angle

Problem	Cause	Remedy
Variations in thickness within section	(1) Dull knife	(1) Change knife
	(2) Uneven consistency of specimen block	(2) Retrim block and remove plain resin
	(3) Vibrations	(3) Remove cause of vibration
	(4) Hydrated hardener	(4) Use fresh supply
Block lifts section	(1) Meniscus too high	(1) Remove fluid from trough just enough to achieve bright reflection under binocular microscope near knife edge
	(2) Clearance angle too small	(2) Increase clearance angle by tilting knife backward
	(3) Upper edge of block face or knife dirty	(3) Replace glass knife; clean diamond knife with a stick; clean tissue block face with lens paper
	(4) Fluid drop on block face	(4) Clean block face with lens paper
	(5) Fluid drop on back of knife	(5) Replace glass knife; dry diamond knife with lens paper
	(6) Block too soft	(6) Heat block in oven at 60–80 °C for 24 h or change block
	(7) Face of block electrified	(7) Increase room humidity; ionise air with high-frequency charge
	(8) Knife, knife holder or block loose	(8) Tighten them up
Block face gets wet	(1) Drop of fluid on vertical back of cutting edge	(1) Remove drop with piece of lens paper
	(2) Incomplete infiltration	(2) Change specimen block
	(3) Too much fluid in trough	(3) Lower fluid level
	(4) Cutting speed too slow	(4) Increase cutting speed
	(5) Debris on back of cutting edge	(5) Clean by wiping with a hair or lens paper
Ribbon is not formed	(1) Meniscus too high, causing section to leave knife edge as soon as cut	(1) Lower meniscus level
	(2) Upper and lower edges of block face not parallel	(2) Retrim block face; coat sides of block with thin layer of Tackiwax
	(3) Upper and lower edges of block face not straight	(3) Retrim block face
	(4) Cutting speed too slow	(4) Increase cutting speed
	(5) Block face or knife area electrified	(5) Touch block face with moist lens paper to remove charge; increase humidity so that charge may leak out into moist air
	(6) Unsteady, or break in, cycling rhythm of microtome	(6) Cycle microtome steadily and without interruption
	(7) Upper edge of block face not parallel to cutting edge	(7) Reorient or retrim block face
	(8) Leading end of ribbon obstructed by side of trough or debris in trough fluid	(8) Centre ribbon in trough and remove debris
	(9) Debris adhering to cutting edge	(9) Clean cutting edge

Table 3.3 (cont'd)

Defect	*Possible causes*	*Remedies*
Ribbon is curved	(1) Upper and lower edges of block face not parallel	(1) Retrim block face
	(2) Knife not uniformly sharp, causing differential compression across section	(2) Try another portion of knife; get knife resharpened; replace knife
	(3) Upper and lower edges of block face not parallel to knife edge	(3) Reorient block face
	(4) Edges of block parallel to knife edge not straight	(4) Retrim block face
Sections difficult to see	(1) Too much fluid in trough	(1) Lower fluid level
	(2) Incorrect angle of illumination	(2) Adjust angle of illumination
Sections show scratches	(1) Fine nick in knife edge	(1) Try another portion of knife edge; get knife resharpened; replace knife
	(2) Dirty knife edge	(2) Replace glass knife; clean edge of diamond knife
	(3) Hard material in specimen	(3) Replace specimen or use diamond knife
Sections crumple or stick to knife edge	(1) Low meniscus level causes dry cutting edge	(1) Raise meniscus level
	(2) Dirty knife edge	(2) Replace glass knife; clean edge of diamond knife
	(3) Blunt knife edge	(3) Replace glass knife; get diamond knife resharpened
	(4) Knife angle too small	(4) Increase knife angle
Sections show wrinkles	(1) Poor embedding	(1) Heat block in oven at 60–80 °C for 24 h; get another block
	(2) Dull or uneven sharpness of knife edge	(2) Use different portion of knife edge; get knife resharpened; replace knife
	(3) Block face too large	(3) Reduce size of block face
	(4) Knife, knife holder or block loose	(4) Tighten them up
	(5) Normal compression	(5) Use vapours of organic solvent or heat
Irregular folds in section	(1) Knife angle too small	(1) Increase knife angle
	(2) Not enough fluid in trough	(2) Add fluid
Sections show tiny holes	(1) Tiny air bubbles in block face	(1) Eliminate bubble by retrimming block face
	(2) Debris on knife edge	(2) Try another portion of knife edge; replace knife; remove debris with moist lens paper
	(3) Piece of hard material in specimen	(3) Bring up another specimen, using a different preparatory procedure; use a harder block corresponding to hardest part in specimen

Specimen crumples and drops out of section	Imperfect embedding	Extend duration of infiltration
Sections show chatter	(1) Dull knife (2) Very hard resin (3) Cutting too fast or unsteady (4) Knife tilt too large (5) Chatter localised on section	(1) Change knife (2) Use softer resin mixture (3) Cycle microtome slowly and rhythmically (4) Reduce angle (5) Change area of knife or portion of block face
Blocks too brittle	(1) Excess hardener (2) Wrong ingredient mix	(1) Reduce proportion of hardener; include dibutyl phthalate as plasticiser (2) Prepare fresh embedding mixture
Blocks discoloured	Old catalyst	Use fresh supply
Blocks too soft	(1) Incomplete polymerisation (2) Insufficient catalyst (3) Wrong ingredient mix	(1) Repolymerise; increase polymerisation temperature (2) Increase proportion of catalyst (3) Prepare fresh embedding mixture
Section contaminated	(1) Dirty water in trough (2) Trough greasy or dirty (3) Dirty knife (4) Dirty forceps, grids or pipette (or syringe) (5) Block face greasy or dirty	(1) Use fresh distilled water (2) Replace glass knife with new knife and clean trough (3) Replace glass knife; clean diamond knife (4) Clean these accessories (5) Clean block face with distilled water or reface block

when a section is cut, it is shorter than the corresponding area of the block face. In other words, the section is thicker than the slice separated from the specimen block. The shortening is obvious only in the axis of the section perpendicular to the cutting edge, without significant widening.

The result of section shortening is ultrastructural distortion. Cell components showing shortening in the axis perpendicular to the cutting edge may be accompanied by an elongation in the axis parallel to the edge. These compressions may even result in an increased spacing among cell components in a direction perpendicular to the cutting edge. The net result is distortion in the spatial distribution of cell components, which is a serious problem in morphometric studies. This distortion of cell components may or may not be accompanied by a visible change in the shape or size of the section compared with the corresponding shape and size of the block face. Wrinkles and folding of the section may accompany the compression.

Specimen blocks undergo compression before sectioning at the onset of cutting forces. Sections undergo compression due to the friction between the resin and the cutting surface of the knife. Tension is introduced perpendicular to the plane of sectioning as the knife entry progresses during sectioning. Shearing stress is introduced when the resin is insufficiently flexible. The importance of resin flexibility becomes apparent when one considers that a thin section has to bend from the vertical position to the horizontal position (Fig. 3.31) at an angle of 90° before floating on the water surface. No irreversible events occurring during bending of the section are known. However, bending may cause plastic flow and section deformation. Such bending is smooth when the resin possesses flexibility.

Factors Affecting the Compression

Factors that influence section compression include the type of embedding resin and the quality of its polymerisation, the size and shape of the block face, knife angle, facet angle, clearance angle, quality of knife, type of specimen and cutting speed. Although sections of all types of resins used for embedding manifest a certain amount of compression, some are more prone to the stress of sectioning than others. Methacrylates are considerably more susceptible to this than are epoxy resins. Maraglas, Epon and Araldite sections normally show a compression of ~11%, 16% and 24%, respectively. Thin sections (100 nm thick, cut at a speed of 2 mm/s) of Vestopal W-embedded cat spinal roots showed a compression of ~5% in the direction of sectioning (Berthold *et al.*, 1982). Helander (1984) indicated that semithin glycol methacrylate sections showed compression ranging from 4% to 19%.

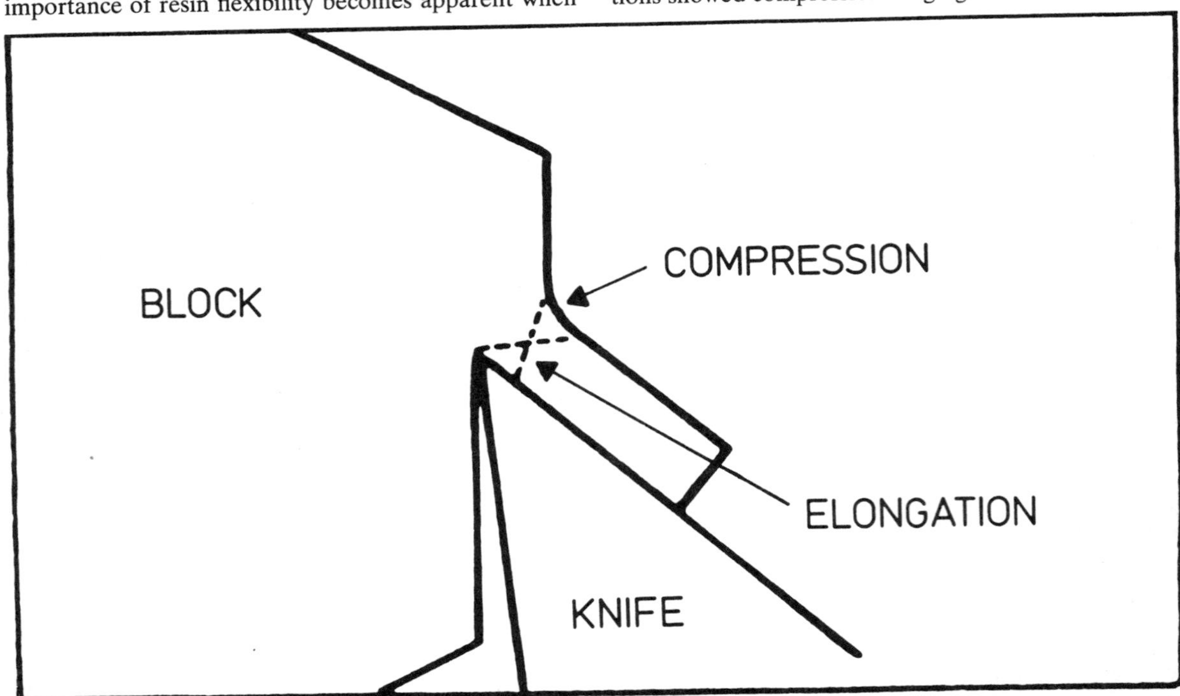

Figure 3.31 Diagrammatic representation of the section bending from the vertical position to an almost horizontal position (accompanied by compression) during the cutting process

Thin sections of different types of specimens embedded in the same resin may show compressions of different magnitude. The degree of compression varies in different components of a cell or a tissue. Various cell components seem to possess different plasticities and, as a consequence, compression varies. An example of this was provided by Nilsson (1964), who found no compression in the discs of Vestopal-embedded retina, whereas interspaces showed a compression of ~25%.

Since section compression is influenced by the angle of the cutting edge, knives with small angles should be used. However, knives with very small angles are brittle and easily wear out. Sections are difficult to produce when the knife edge angle is smaller than 35° (Helander, 1984). As stated above, cutting speed is a factor in determining the extent of compression. However, according to Jésior (1986), compression is independent of sectioning speed in a reasonable range (0.1–5 mm/s). The reason given is that, during sectioning, the fracture ridge propagates faster than the knife into the specimen block; the fracture propagates at a rate faster than 10 mm/s. Since compression occurs primarily in the section itself and not in the sectioned specimen block, there is little or no residual compression left in the block by the previous sectioning process. Compression of sections has also been discussed elsewhere in this volume.

Removal of Section Compression

Compression in thin sections is partially reversible, for most of it can be relaxed by vapours of an organic solvent (chloroform) or heat. As much as 90% of the compression can be relaxed. However, it is much better to prevent or drastically reduce compression during sectioning than to minimise it after the section has been cut. Therefore, prevention of compression will be discussed first. This can be accomplished by using a low-angle diamond knife (<30°), the lowest possible clearance angle (~3.8°) and cutting sections in the thickness range 50–90 nm. A difference of even 1° in the clearance angle will affect the amount of compression. The diamond knife is equipped with a built-in clearance angle—i.e. it is the angle at which the diamond has been mounted on its boat. This angle should be added to the clearance angle set on the knife holder to obtain the final, effective clearance angle (4–8°).

Very thin sections (<40 nm) show relatively more compression. Compression can also be minimised by increasing the hardness of the resin. Jésior (1985) has shown that the intrinsic resin compression decreases as resin hardness increases. Hardness of the resin primarily determines its cutting quality. However, the mechanical properties of the embedded specimen are more important than those of its surrounding resin in determining the compression. Compression is also decreased when better lubrication is provided by the water in the knife trough.

Usually sections of all types of embedding media require flattening, to eliminate foldings or crumples caused by compression forces. This is accomplished by exposing the sections to the vapours of organic solvents such as chloroform or trichloroethylene. Some of the compression is relieved when the sections are floated in the trough, but most of it is removed by the vapours. Alternatively, heat can be applied to remove compression.

It should not be assumed that because trough fluid may contain some solvent, subsequent application of a solvent in the form of vapours is not necessary. Although sections do imbibe some of the solvent present in the trough fluid, they remain compressed to some extent. In fact, solvent vapours alone are quite effective in removing the compression. Consistently good results are obtained with distilled water as the flotation fluid and chloroform vapours as the section flattener, using either glass or diamond knives. It is recommended that sections be exposed to vapours of chloroform, irrespective of the type of trough fluid and embedding medium used.

Vapours of various solvents have been used to flatten sections of different embedding resins. From a practical point of view, an ideal solvent for this purpose should possess a relatively high vapour density and a relatively low vapour pressure. The advantage of having a sufficiently high vapour density is that the solvent vapours will easily descend upon the sections, and the advantage of possessing a low vapour pressure is that the solvent will not evaporate before it is brought near the sections. However, these two characteristics do not always determine the usefulness of a solvent in vapour form: for example, chloroform is most effective in flattening sections of polyester and epoxy resins, although it has a higher vapour pressure than that of xylene. Chloroform should be used in sufficient quantity to compensate for its high vapour pressure. In this connection, it should be remembered that large amounts of vapours may cause breaking of the ribbon and scattering of sections. An overlong exposure to chloroform is also detrimental, for the solvent can easily dissolve most embedding resins. Trichloroethylene is another solvent that is equally effective for flattening sections.

Exposure of sections to vapours is accomplished by carefully bringing a wisp of cotton dipped in chloroform on a small stick close to the sections floating in the trough. Care should be taken that the applicator stick does not carry too much solvent, as otherwise it may drip into the trough. Care should also be taken not to let the stick touch either the sections or the flotation fluid. The solvent should be held a few centimetres above the sections. If it is brought too close, the sections will expand in both length and width to dimensions that are larger than those of the face of the block. This overexpansion will result in permanent distortion of cell ultrastructure. Furthermore, chances of a contact between the solvent and the sections or the trough fluid increase when the solvent is too close to the sections.

Sections can be exposed to vapours at any time, since exposure immediately after cutting is as effective as exposure after some time has elapsed. After being softened by the solvent vapours, the sections spread under the influence of surface tension forces. In fact, the flattening of sections occurs at a moderate speed, so that the process can actually be observed under a binocular microscope. The flattening of the section is accompanied by a decrease in thickness and a resulting change in the interference colour.

The vapours do not dissolve the embedding medium, since sections are not in direct contact with the solvent in a liquid phase. Nevertheless, sections should not be exposed for longer than is necessary to produce the initial rapid expansion, as vapours of the solvent can permanently damage the fine structure of the tissue. In addition, stretching of sections to remove compression is accompanied by variations in thickness within the section. Such variations may be greater than those that normally occur. Moreover, longer exposure is of no advantage. One method of determining the optimum duration of exposure is to compare the size of the flattened section with the corresponding dimension of the block face. Normally most of the compression is eliminated within 30 s of exposure. Variations in the amount of compression removed are presumably due to variations in the amount and form of the wrinkling produced during the process of cutting.

As stated above, application of heat is another method for removing section compression. Heat stretching eliminates the need for solvents used in solution or as vapours. Thus, the possible damaging effect of the solvents on the fine structure is avoided. Furthermore, slight shrinkage of sections immediately after the removal of the solvent vapours is not observed after heat stretching. This advantage is particularly important when cutting sections of plant materials containing cell wall or cuticle, which possess different cutting properties compared with the rest of the cell. Also, it has been suggested that heat-stretched sections are more stable under the electron beam and show less residual compression than those expanded by solvents (Roberts, 1970). Peachey (1958) demonstrated that a photo-flood lamp decreased section compression from 35% to 5–10%. Roberts (1970) introduced a V-shaped, heated wire filament (heat pen) (Polysciences or Polaron) for expanding sections, and indicated that the method had no apparent adverse effect on staining properties, section stability or resolution. It should be noted that heat-stretched sections of Maraglas, Araldite and Epon show a residual compression of ~2%, 4% and 10%, respectively. Solvent-stretched sections of these three embedding media tend to show a higher residual compression.

SECTION WRINKLING

Besides section compression, discussed earlier, another major deformity in sections is wrinkling or folding of some areas. The latter problem becomes serious in studies involving serial sectioning, low magnifications or stereology. Wrinkles may appear as electron-opaque, straight and narrow streaks or S-shaped. Scanning electron microscopy shows them to be tubular in shape (Abad, 1988). Water seems to play a part in the formation of such wrinkles. They may appear in the blank resin and/or in the specimen, but tend to be larger in the former.

Wrinkles do not show any particular orientation associated with the direction of the cut, nor are they present in every section. They occur erratically as regards both frequency and degree of severity from section to section. Their occurrence is not dependent on block hardness. Although Pihakashi and Suoranta (1985) indicate that wrinkles can be avoided by using a low-viscosity resin such as ERL-4206, their occurrence does not appear to be exclusively a function of the resin viscosity.

One type of wrinkle is formed during settling of the sections onto the support film. Such wrinkles are difficult to remove by chloroform vapour or heat treatment. Certain other wrinkles formed on uncoated grids are also difficult to remove by the above treatments. Since sections are not of uniform density, denser areas touch and bond to the film first, preventing the adjacent areas of the section from obtaining the surface required to settle flat (Abad, 1988). In other words, not every part of the section is provided with adequate film surface to expand and settle uniformly. Sections of a specimen with less variation of density tend to show fewer wrinkles. The structure of the specimen is primarily responsible for this type of wrinkle.

Abad (1988) has presented a method for preventing the formation of certain types of wrinkles in sections mounted on Formvar-coated single-hole grids. The film floating on water is picked up on an aluminium platform (2.5 × 5.5 × 1.5 cm in height) with holes 3.5 mm in diameter. The platform is placed on a slide warmer (62 °C) and covered by the bottom of a glass Petri dish, to keep the ambient air at the desired temperature. Sections are picked up from the trough on an uncoated, single-hole grid, which is deposited (dull side down) over the hole of the platform and allowed to dry. The grid is carefully pushed through the orifice onto a filter paper with the blunt head (3 mm in diameter) of a nail. The film is not affected by the heat and humidity under the Petri dish, and the sections are expanded by the heat. The adverse effect of heat on the fine structure is not known.

SECTION SHRINKAGE

The other major defect is section shrinkage (contraction) to varying degrees, caused by irradiation. The loss of mass by evaporation of an irradiated section is very rapid. Sections exhibit a rapid initial shrinkage even at a low electron dose, followed by a much slower phase of thinning. According to Cosslett (1961), however, terminal section thickness is reached instantaneously.

Various resins vaporise to different degrees under the electron beam. As a result, heat generated by the electron beam affects to different degrees the image contrast of sections of various resins. Thin sections of three resins were exposed to the electron beam (Mollenhauer, 1986b). They lost the mass in ascending order as follows: Ladd resin, Spurr mixture and Quetol. Losses as high as 20% of total resin mass have been reported (Luft, 1973). According to Cosslett (1960), section thinning may vary between 20% and 50%. Various resins treated differently will show different degrees of thinning. The thinning has a 'clearing' effect, resulting in rapid enhancement in contrast between the stained specimen and the background resin. The loss of mass described above is more critical in high-resolution studies than in routine work.

Lateral section shrinkage caused by irradiation may manifest itself as wrinkles, which predominate near the grid bars. Both the support film and the section may undergo shrinkage. It is not clear whether the resin and/or the specimens undergo the same degree of shrinkage as does the support film. The degree of shrinkage in sections containing the specimen differs from that occurring in pure resin. Section shrinkage is not a surface phenomenon, and is uniform throughout the thickness of the section (Bennett, 1974).

Section shrinkage is essentially a function of the electron dose. Colloidal gold particles have been used as markers of the section surface for monitoring the shrinkage (collapse) of sections as a function of electron dose (Luther *et al.*, 1988). The two surfaces of the section are separately coated with colloidal gold. The section is tilted to an angle of 45° in the electron microscope and a series of micrographs recorded, corresponding to increasing electron dose. When the tilted section is imaged, the shrinkage normal to the section manifests itself in the image as a relative movement of the two sets of gold particles. From the extent of this movement the degree of shrinkage can be estimated.

Displacement of tissue components occurs within the section under the electron beam. The initial section shrinkage is accompanied by movement of the specimen. Sections may move hundreds of angströms in spots immediately on irradiation. The primary cause of displacement is resin vaporisation; to a lesser extent electrical charges are also a cause. Displacement of tissue components also takes place within the resin matrix, both in the tissue block and in sections stored at room temperature.

CHATTER

One of the most common and bothersome section deformities encountered is chatter. Commonly, the term 'chatter' is used to describe all types of periodic variations in the contrast of a section (Fig. 3.32). Chatter manifests itself as a cyclic variation in optical density of an electron micrograph due to specimen thickness variations. When chatter appears only in certain regions of the section, the reason is the heterogeneous nature of the specimen. Chatter marks usually arise parallel to the cutting edge while the section is being cut—that is, they are oriented at right angles to the path of the moving section past the knife. This deformity is easily distinguishable from the marks present in the section as the result of a nick or a weak spot in the cutting edge. Characteristically, chatter marks tend to blend into each other, whereas knife marks are usually deep and abrupt (Fig. 3.10). In addition, knife marks may be accompanied by tearing of small regions of the section and are always parallel to the movement of the section past the cutting edge. Ordinarily, knife marks appear fewer in number than chatter marks, unless the cutting edge is completely damaged.

The chatter deformity is troublesome because it is very difficult to detect at the time of sectioning. When the variation in section thickness is of a wavelength of 1 μm or less, it is especially difficult to see. It is disappointingly discovered while viewing the section in

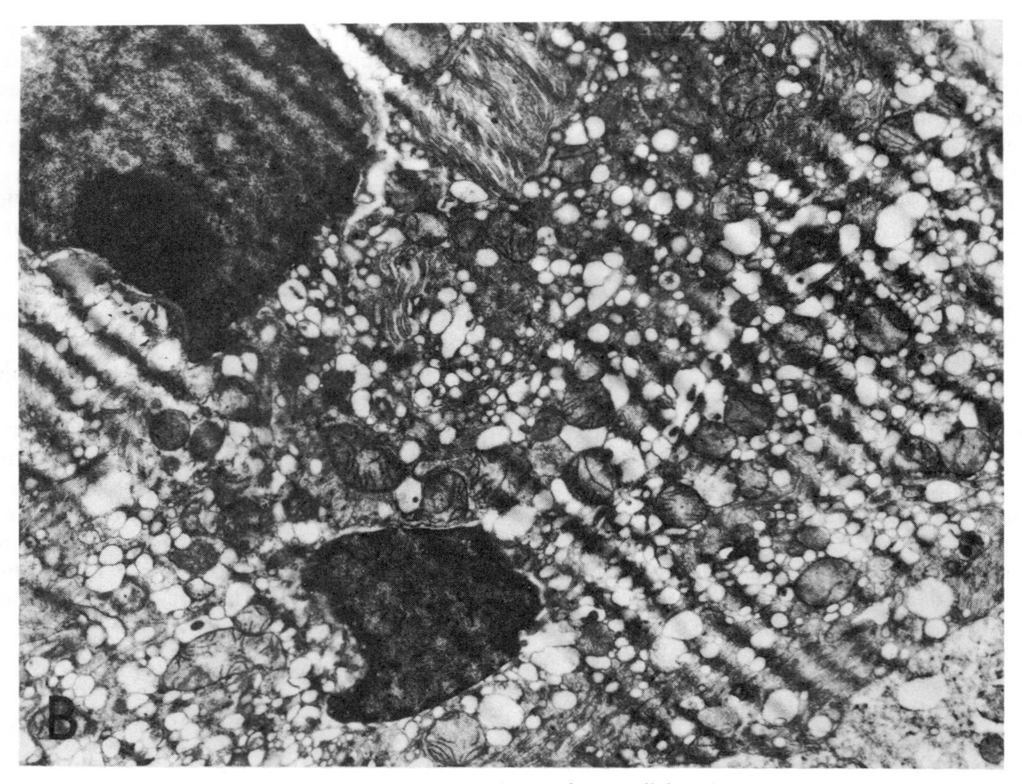

Figure 3.32 Human kidney cells showing strong chatter marks running parallel to the cutting edge

the TEM, and then it is too late to rectify the error. Very fine chatter is visible only at high magnifications in the TEM. Occasionally, in severe cases, it is observable with an optical microscope.

The vibrations that cause chatter can be picked up by a 'chatterbox' (McCutchen and Tice, 1973). Its output is amplified and played to the ear by a headphone. Thus, chattered sections can be detected as they are cut. A simple and sensitive vibration monitor has also been designed by Webster and del Cerro (1981). Some embedding media are more prone to this type of deformity than are others—e.g. methacrylates are more susceptible than are epoxy resins.

As stated above, the primary cause of chatter is considered to be the vibration of ultramicrotome components, including the cantilever (arm), the specimen block and the cutting edge. Vibrations <1000 c/s can result from vibrations of ultramicrotome components. For example, loose clamping of the specimen block, knife or knife holder can cause such vibrations. The alternating thick and thin bands resulting from mechanical vibration are of relatively long periodicity and cross the entire field without interruption. Bands of this type also manifest a considerable variation in thickness.

Simple mechanical vibrations are not the only cause of chatter. Another cause of this deformity is the impact of the knife on the embedding resin and the specimen. The vibrations responsible for this type of chatter are an immediate result of the release of stresses that had developed as soon as the specimen in block received the first impact from the cutting edge. This release of stresses takes place only after the specimen block passes its maximum elasticity or yield. When the resistance of the structures to cleavage equals the restoring force of their elasticity, the knife passes through the zone of distortion and produces a thick band, and the process is then repeated.

The first contact between the knife and the specimen block does not result simultaneously in sectioning, since a certain minimum force is needed to overcome the forces of cohesion in the resin and thus separate the section from the block. The first event is elastic bending of the resin by the force of the motion of the ultramicrotome arm. This results in vibrations that may exceed 1000 c/s (Sitte, 1984). Cutting forces can cause vibrations of as high a frequency as 10 000 c/s. Actual sectioning begins after the minimum force has been achieved. Subsequently, friction on both faces of the knife edge reduces the sectioning velocity, resulting in additional vibrations.

In the second type of chatter, it is most likely that the thickness and periodicity of the bands are related to the condition of the cutting edge of the knife as well as the toughness and elasticity of the embedding medium and the tissue. In this connection, it should be noted that during sectioning the surface layer of the block is compressed by the cutting impact of the knife, and from this compressed layer sections are cut. The extent of compression is dependent upon the plastic and elastic properties of the block. If the specimen block is relatively soft, it will be compressed to such an extent that no section can be cut. On the other hand, a specimen block that is too hard will induce greater impact and cutting forces. Rigid and highly electron-dense structures are relatively more susceptible to this type of chatter. As a general rule, the amplitude of vibration is proportional to the force of impact on the cutting facet. The bands or striations thus produced are of shorter periodicity than those resulting from a simple mechanical vibration. This type of band shows an average spacing of 0.2 μm.

It has been postulated that periodic distortion is partly dependent upon the deflection of the harder cell components in their softer matrix. This type of chatter is localised to some extent. According to Phillips (1962), the cutting process is accompanied by a severe shear deformation of each elemental width of the section. The degree of recovery is different in different components of the section. This difference in recovery is illustrated by lipid droplets, especially in plant cells, which revert towards the shape they possessed before cutting much less than do the rest of the cell components. This results in the localised presence of short-spaced contrast variations on denser structures such as lipid droplets. The degree of recovery in different components of the cell is apparently partly controlled by the mechanical properties of these components. The short-spaced bands are therefore caused by the deformation during sectioning. Chatter of very small amplitude is visible only in especially sensitive cell components such as nucleoli and cell walls and in certain cells such as erythrocytes (Sitte, 1984).

Not all the sources of chatter have as yet been determined. The known possible causes of this deformity are: (1) clearance angle too large, (2) knife angle too large, (3) cutting speed too fast, (4) block face too large, (5) room temperature too high, (6) specimen temperature too high (7) specimen block too long from the point of attachment, (8) undesirable height of mounted knife, (9) dull knife, (10) undesirable position, in relation to the cutting edge, of the mounted specimen block, (11) undesirable shape of block face, (12) specimen block loosely held in the vice, (13) specimen block too hard, (14) specimen block too soft and elastic, (15) knife not firmly held, (16) sudden change in temperature in the immediate vicinity of the instrument, (17) occurrence of static friction in any moving part of the instrument, (18) incomplete polymerisation of the embedding resin, (19) non-uniform density of tissue–resin composite, (20) debris adhering to the diamond cutting edge, (21) unstable microtome table and (22) building vibration caused by slamming a door, an air conditioner turning on, people walking down the hall or even outside traffic.

The above-mentioned causes of chatter should be helpful in finding ways to eliminate this problem. Chatter can be significantly minimised by reducing the cutting speed, the size of the block face and the knife angle. Reduced cutting speed also minimises the distance between the waves occurring in a section. Cutting speeds of <1 mm/s may eliminate some types of chatter. If chatter is encountered when sectioning a hard block, knife and clearance angles should be reduced, whereas, in the case of soft blocks, the reverse is normally recommended. Most importantly, the ultramicrotome should be mounted on a vibration-isolated table.

SECTION CONTAMINATION AND DAMAGE

In order to obtain electron micrographs of good quality, it is imperative that sections be kept free from every possible type of contamination. A section can pick up contamination from the moment it is cut until the time it is micrographed. The importance of maintaining clean sections becomes at once apparent when a dirt particle is found in the region of a section in which one is most interested. Many of the artefacts once attributed to fixation were evidently the results of post-section contamination. One should remember that a contaminated section cannot be decontaminated.

Sections can become contaminated not only during sectioning, but also during the subsequent handling of specimen grids. The sources of contamination are varied but the following are considered to be the major ones.

Dirt Extreme caution is necessary to prevent sections from picking up dust or dirt particles. The possible sources of this kind of contamination include: (1) the knife edge, (2) the flotation fluid, (3) the grids, (4) the grid container, (5) the forceps' tips, (6) the supporting film, (7) the trough, (8) the sealing wax and (9) the atmosphere. It is obvious that, to avoid or at least lessen this type of contamination, certain precautionary measures are necessary. As mentioned earlier, the glass strip should always be clean and dry before a knife is prepared, and it should be covered immediately after it is prepared.

The containers holding the flotation fluid and grids should be cleaned periodically, even if they appear to be clean. It should not be presumed that because grids appear clean they are dirt free. All grids should be thoroughly rinsed in absolute acetone and preferably again immediately prior to use. Grids should be kept covered at all times. Since supporting films, especially Formvar, are one of the major potential sources of contamination, their routine use should be avoided. However, in certain studies, such as those of viruses or those requiring serial sections, unsupported grids cannot be used. Sectioning should be carried out in a relatively dust-free room, since flotation fluids can easily pick up dirt from the surrounding air.

Oil films The sources of grease or oil film contamination, like those of dirt, are varied. Some of the more common are: the flotation fluid, the knife face, the forceps, the syringe or micropipette, the grid and its container, the trough, the fingers, the sealing wax and the wax applicator. It is advisable to use only acetone of the highest grade and deionised or double-distilled water. If the cutting edge of a glass knife appears even slightly contaminated with oily substances, the knife should be discarded. Some workers prefer to dip the tips of forceps in acetone each time they are used, since it is difficult to know when the pair of forceps has picked up greasy substances. If the hypodermic syringe or micropipette is not disposable, it should be rinsed in acetone before each cutting session. As stated earlier, the portion of electrical tape that comes in contact with either the knife edge or flotation fluid should not contact anything else. Metal troughs should be rinsed in acetone or xylene before each cutting session. It is good practice to clean the sealing wax container and replace the wax applicator periodically. As a general rule, the fingers must not touch any object or substance that comes in direct contact with the sections.

Staining artefacts Sections can be contaminated due to various kinds of precipitates that may form during the process of staining tissues either before or after sectioning. Dirt particles and oily films can also be introduced during the procedure of staining sections. These problems have been discussed in detail in the chapter on staining.

Electron beam contamination All sections, irrespective of their thickness, are contaminated immediately after they have been bombarded by electrons in a vacuum. The contamination layer is built up on the section at a rate of 0.1 nm/s or more. The contamination continues as long as the section remains exposed to the beam. Variations in the rate of contamination are dependent upon the types of microscope, the current density of the illuminating beam, the temperature of the specimen in relation to its surroundings, the section thickness, and the amount of oil and grease in the vacuum system. The rate of contamination is increased when the area exposed to the beam is decreased. In general, relatively thick sections are more susceptible to contamination. The contamination is caused when vapours of organic molecules form an adsorbed film by condensation on surfaces that are exposed to the beam (i.e. specimen, apertures and internal walls of the microscope). The contaminating vapours arise from grease around the vacuum seals and diffusion pump oil, remains of organic solvents used for cleaning parts of the microscope, the oil and grease used for lubrication and the rubber gaskets. The contamination films interact with electrons, and are transformed into graphite-like materials. The material of these layers is a hydrocarbon polymer, and is amorphous in nature.

The contamination is usually heavier near the grid bars than in the centre of a grid hole. The reason for this is that the areas near the bars are cooler than the centre, because grid bars are conductors of heat. Higher temperatures at the centre of the grid hole will cause evaporation of contamination layers before they get fixed permanently by the beam. It should be noted that the rate of contamination decreases significantly with increased temperature of the section; however, a high beam current may cause other types of damage to the specimen. Hart *et al.* (1970) have discussed the parameters affecting contamination rate.

The contamination layers reduce resolution and contrast, obscure the fine structural details and alter the dimensional relationships of cellular structures. Similarly, the heat produced in the specimen by its interaction with electrons causes changes in the thickness of the whole specimen or of cell organelles and thus reduces the resolution available. It should be remembered that not only the specimen but also the support film and the embedding resin are damaged by the electron beam.

It is emphasised that sections should always be irradiated first with a beam of low intensity, which causes carbon to be deposited on the sections. This deposition stabilises the sections before they are subjected to a more intense beam; thus, distortion of the fine structure due to heat is minimised. However, the best approach is to scan the grid at a low magnification of ~100–200×, using a beam intensity as low as is consistent with comfortable viewing, as rapidly as possible, and with sufficient photographs taken at the first examination. Another approach is to focus on one region of the section but photograph another. The reduction in contamination is helped by sweeping away organic compounds while cleaning the microscope column, by excluding organic vapours and by using an anticontamination device in the electron microscope, which reduces contamination by condensing the residual gases by cooling the space around the specimen.

Ultracentrifugation contamination In certain cases, when the pellet of centrifuged cell cultures and other particulate specimens is embedded and sectioned, score marks show up in the sections. Score marks may obscure details of the sections and may cause instability of the specimen under the electron beam. This contamination is thought to be due to the presence of hard, foreign particles (e.g. asbestos crystals) in the pellet causing nicks in the knife edge (Smith and Luther, 1976). These hard particles are ingested from the growth medium. The contaminants that cause this problem may come from certain filters (Seitz), glass-wool and glass containers. For a procedure to remove contaminants from the growth medium by filtration, the reader is referred to Smith and Luther (1976).

SERIAL SECTIONING

For most studies, the two-dimensional information yielded by study of individual sections is adequate. A third dimension may be added to individual sections by stereomicroscopy (Gray and Willis, 1968) or be inferred from stereological methods (Weibel and Bolender, 1973; Weibel, 1979). A review of the theory and practice of the technique of specimen tilting has been presented by Lange (1976). However, in order to analyse the form and structural relationships of cells or their components in greater depth, their structural details can be reconstructed in three dimensions from examination of serial sections. For example, 12 (~50 nm thick) consecutive serial sections might be required to observe the arrangement of cristae in a mitochondrion of 0.6 μm in diameter. Serial sections can also aid in locating specific constituents within cells and their relative numbers. Quantitative analyses of normal structures is important for comparison with pathological changes, and can be used to check the reliability of quantitative data obtained from analysis of single sections. Knowledge of the distribution of cellular organelles, such as lysosomes or mitochondria, provides information of value in interpreting physiological and biochemical parameters of cell function. Serial sectioning becomes a necessity when a cell or its component cannot be selectively stained or prepared as a whole mount, and when it is too large to be totally included within a section of practical thickness. Three-dimensional reconstruction is also required for correlative light and electron microscopy (Hayat, 1987); to determine the exact number of cell components, and their correct structure and spatial relationships; and to elucidate the structure of cell–cell relationships and interactions (Rieder, 1981).

Serial sectioning has been used to elucidate aspects of complex cellular arrangements in a variety of plant and animal specimens. Rieder (1981) has presented a detailed review of thick and thin serial sectioning for three-dimensional reconstruction. Stevens *et al.* (1980) have reviewed the reconstruction of microcircuitry in neurons by electron microscopy of serial sections.

There are at least three major steps in obtaining three-dimensional information: (1) standard fixation and embedding, (2) sectioning and transfer and orientation of ribbons of serial sections onto the grid, and (3) reconstruction of the three-dimensional image from the sections. The first step has been described in detail in earlier chapters and steps (2) and (3) are discussed below.

The procedure for conventional sectioning discussed earlier also applies to serial sectioning if the following additional comments are noted. A pyramid-shaped block face is not recommended for serial sections, because such a face keeps enlarging during sectioning and requires frequent trimming. Removal of the block from the ultramicrotome invites damage to the region of interest, because of inadvertent excessive trimming and loss of orientation. Therefore, the block face should be trimmed in such a way that the leading face and sides are vertical and only the trailing face is sloped to avoid bending of the block during sectioning (Fahrenbach, 1984). To improve cohesion of sections with each other to form a ribbon, the leading and trailing surfaces of the block should be coated with contact cement (Weldwood). A 1:1 mixture of contact cement and commercially available contact cement thinner is recommended. A heat pen is preferred for stretching a ribbon of sections still attached to the cutting edge. Vapour stretching, because of convection, may break a long ribbon. The ribbon can be detached from the cutting edge by increasing the water level in the trough and then gently stroking the cutting edge with a hair glued to an applicator stick.

TRANSFER OF SERIAL SECTIONS

The major difficulty in the preparation of serial sections for observation does not reside in the microtomy itself, since all the microtomes currently used can produce ribbons of section. It is the transfer of sections without wrinkling, and their alignment on the grid so that the sections are adequately supported and yet not obstructed by the grid bars, that is difficult. It is desirable, if possible, to accommodate the series of sections to be examined on one grid instead of more than one. The maximum length of a ribbon that can be accommodated by a standard grid and scanned in the TEM is ~1.5 mm. Most conventional grids have only ~50% of their total area open; the remainder is hidden behind the bars. Therefore, special grids having larger open areas rather than regular squares are used to mount the ribbons. These special grids, with openings of different shapes and sizes (Fig. 3.27), are commercially available. Desired size and shape of openings can also be obtained by modifying the conventional grids. A hole of the desired shape and size can be cut in the conventional grid by a sharp razor-blade; then the edges of the hole are flattened by pressing the grid between two clean glass surfaces. A single spherical hole (~1 mm in diameter) in the centre of the grid is better than a slit, since it does not require exact orientation of the ribbon.

A number of methods have been described for collecting and aligning ribbons of serial sections on grids. The simplest, one-step, method for controlled mounting of serial, thin sections obtained with glass knives was introduced by Couve (1986). A coated slot grid is placed on a square piece of Parafilm somewhat larger than the grid size, which is then attached onto the front, oblique, face of the knife, about 2–3 mm below the cutting edge (Fig. 3.33). The long axis of the slot is oriented perpendicular to the cutting edge. A collecting trough is constructed around the knife.

After cutting a ribbon, it is detached from the knife edge and allowed to float on the water surface in the trough. With the aid of a syringe or a water control unit, the water level is lowered and simultaneously the ribbon is guided, with a single hair, onto the top of the grid slot. Once the ribbon has been oriented onto the slot, the water is carefully withdrawn completely. After the water evaporates from the surface of the grid, it is removed with a pair of forceps. Since the grid remains stationary during withdrawal of water, the precise alignment of the ribbon onto the grid slot is facilitated.

In the method devised by Gay and Anderson (1954), the floating ribbon is picked up from below on a Formvar-coated wire loop (~4 mm in diameter) held in a self-locking forceps. The loop is then attached to a disc resting on the microscope stage. An unfilmed grid is placed on the microscope condenser, which can be elevated. By manipulating the disc, the ribbon is optically superimposed over the unfilmed grid. When the condenser is elevated, the grid passes through the loop, picking up the ribbon and the Formvar film. This multi-stage mounting technique is rather tedious and a fairly heavy Formvar film is required to both pick up and transfer the ribbon. Grids with a single circular hole with a diameter of 0.7–1.5 mm, made by drilling holes in 0.2-mm-thick copper plates, have also been used. Such Formvar-coated grids are reinforced with a carbon film.

Barnes and Chambers (1961) demonstrated a rather simple micromanipulator consisting of several glass microscope slides, bent tubing and a platinum loop. Two slides are oriented perpendicularly and glued in place as a base. A third piece of glass can be attached and slid onto the vertical portion of the base by lubrication with a viscous grease. A bent piece of tubing is attached to the third piece of glass with glue. A platinum loop attached to a glass rod is held in the bent tubing by friction. The ribbon is picked up on the Formvar-coated platinum loop before it is placed in the tubing. The sliding action of the glass provides the necessary control to orient the serial sections in the holder for transfer onto the grids.

Dowell (1959) introduced a one-step method in which a special grid with a single large hole (0.6–1.0 mm in diameter) is coated with a composite carbon–collodion film. This special grid can be made by punching 0.15 mm hardened copper sheet. The coated grid is dipped at a near-vertical angle in the water and then brought underneath the floating ribbon. This method has the advantage that the ribbon is picked up directly on the grid.

Westfall (1961) demonstrated a rather simple one-step method. A narrow wedge of black Bakelite is glued to the knife so that it is in a plane parallel with the binocular microscope. The wedge is attached 1 mm below the cutting edge in order for the sections to glide away easily. After a ribbon is cut, a Formvar-coated grid is slid below the ribbon. The ribbon is then detached from the cutting edge and oriented over the grid by an eyelash mounted on a stick. The black colour of the wedge surface helps in determining the interference colours of sections and in aligning the ribbon on the slotted grid. Later Westfall and Healy (1962) introduced a sophisticated device which can be easily clamped to the diamond knife holder. This device can also be attached to a glass knife by means of rubber cement. The coated grid is placed near the cutting edge and under the floating ribbon. When the ribbon is centred over the grid, the water is completely withdrawn from under the ribbon with the aid of a hydraulic device. This results in the dropping of the ribbon onto the grid surface. The complete device is available commercially.

Galey and Nilsson (1966) introduced a method which employs two slot grids, one coated and the other uncoated, to collect each ribbon. The uncoated grid is placed on the water surface so that the slot encloses the floating ribbon. When the grid is lifted, it carries a drop of water along with the ribbon. The ribbon can be

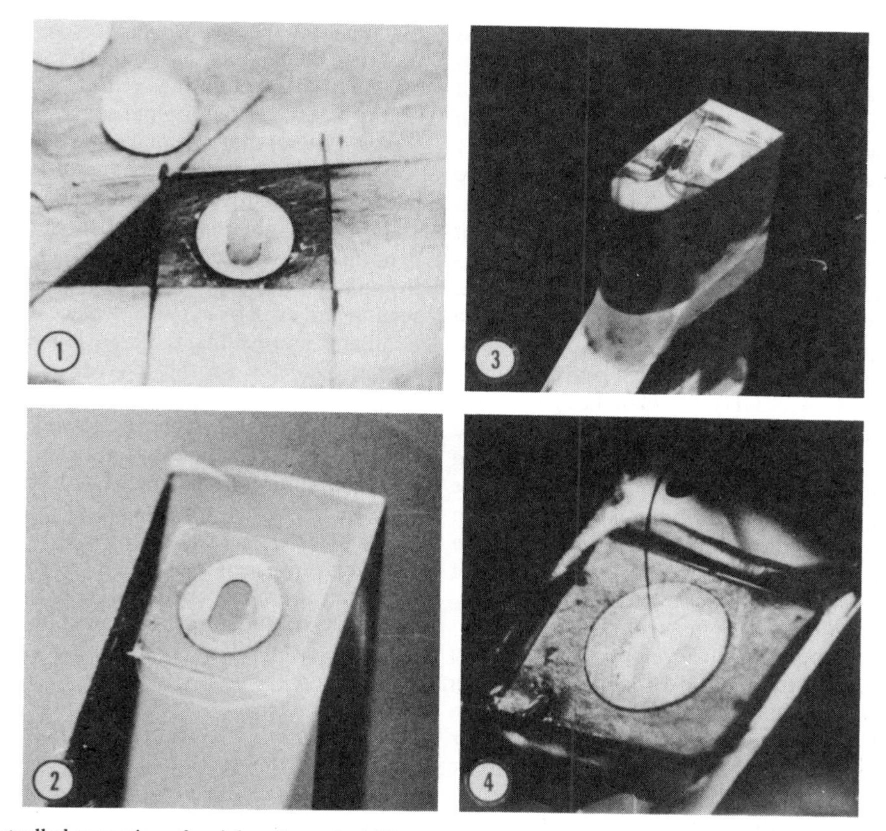

Figure 3.33 Controlled mounting of serial sections. 1: A Formvar-coated grid attached to a square piece of Parafilm is cut out. 2: The Parafilm is fixed near the cutting edge of the knife. Note that the long axis of the grid slot, which is face up, is perpendicular to the cutting edge. 3: A glass knife is ready for sectioning. Note the coated grid fixed on the bottom of the trough. 4: A single hair is touching a ribbon mounted onto a coated grid which is still attached by the Parafilm to the knife surface. (From Couve, 1986)

stained by transferring the grid from one liquid surface to another. Finally, the grid is placed on top of the coated grid and the slits are aligned. The water can be drained off with a piece of filter paper.

The above method was modified by Wegner (1971), because the last step is rather tedious. He designed a grid support which facilitates the transfer of ribbons from uncoated to coated grid. The grid support can be easily prepared in an ordinary laboratory. The method is also suitable for transferring large single sections.

In another technique gelatin is employed as a substratum to achieve the desired grid–section orientation. Ribbons of standard-size sections or large sections (1–2 mm) can be mounted on coated or uncoated grids. The method is applicable to both mesh and slot grids. With the aid of a loop, sections are transferred from the trough onto the surface of 3% viscous, but not set, gelatin in a Petri dish (Anderson and Brenner, 1971). The sections are expanded with chloroform vapours and then arranged with a hair to facilitate grid placement.

After the gelatin has been solidified in a refrigerator, the Petri dish is placed under a stereoscope at an angle of about 45°, to allow the sections to be viewed clearly. The section or the ribbon is centred in the field of the microscope and orientated by rotating the dish. With the aid of fine-tipped forceps the grid is placed on the gel and the desired grid–section orientation is achieved by manipulating the dish or the grid.

The gelatin is liquefied in an oven at 60 °C, and the grid is picked up and drained, and then floated, section side down, on a 2% solution of acetic acid for 30 min at 60 °C. The sections are air-dried and can be post-stained according to standard procedures. If necessary, sections can be stored for several weeks on the gelatin surface, provided that evaporation is prevented by placing the covered Petri dish in a refrigerator.

According to the method of Wells (1974), a glass

microscope slide is polished with fine alumina powder, dipped into a solution of 0.3% Formvar in dichloroethane, and drained vertically. The film is cast onto the water surface (Fig. 3.34A) and plastics rings (6 mm in diameter, 1 mm thick and 2 mm high) are carefully placed onto the film (Fig. 3.34B). The filmed rings are removed with a piece of coarse paper towel (Fig. 3.34C) and coated in an evaporator with a thin layer of carbon (Fig. 3.34D). A ribbon of sections is manoeuvred within the single slot of the grid (Fig. 3.34E, F), which is then placed on a large drop of water on a suitable hydrophobic surface (Fig. 3.34G). After staining and washing, the grid is placed on a ring containing the Formvar film (Fig. 3.34H); the filmed ring has been attached to a filter paper with double-sided adhesive tape for preventing surface tension from lifting the ring towards the gird. After air-drying in a Petri dish, the grid along with the supported ribbon is removed from the ring by inverting the assembly over a peg of the same diameter as the grid (Fig. 3.34I,J). The surface of the peg and the support film do not come into contact. This step is carried out with a dissecting microscope.

Holec and Ciampor (1978) developed an accessory for the transfer and orientation of serial sections on grids. This device is mounted on the knife holder of an LKB Ultrotome, and consists of a specimen loop for the transfer of sections and a specimen-grid holder for the orientation of sections on the grid. Heywood *et al.* (1977) introduced a method for applying photographic emulsion uniformly to serial sections which allows the ribbon to remain intact. The grid is attached to a copper tack prior to coating with emulsion, and is removed only after the autoradiograph has been developed.

Although a great deal can be learned about the details of cellular structure from the study of serial sections, in practice it is almost impossible to make a complete and detailed three-dimensional reconstruction of a cell or even of relatively large organelles. This difficulty becomes apparent when, on examination of serial sections, one finds that a given membrane, for instance, cannot be distinctly followed from one section to another. Several explanations have been offered for this anomaly. It was considered that the observed discontinuity of the membranes of adjacent sections was due to the fact that the two surfaces of a section are different with respect to how they are cut. The upper surface of a given section is in continuity with the lower surface of the previous section (as the sections float in a trough). As a consequence, the upper surfaces of two adjacent sections of a ribbon are not continuous. However, this explanation was discounted by Williams and Kallman (1955), for they did not find any systematic improvement in the continuity of membranes of the two adjacent sections which were reoriented in such a way that their upper surfaces represented the surfaces originally adjacent during sectioning. Another explanation is based on the original suggestion by Hillier and Gettner (1950) that sectioning is essentially a ripping or gouging action, and some material is usually removed between consecutive sections. Thus, even two adjacent sections will not match in their surface detail. This hypothesis was strengthened by the observation that the failure to match varies in degree from section to section and even from place to place in the same section (Williams and Kallman, 1955). However, no information is available on the average amount of material removed and its eventual location. Yet another explanation emphasises the role played by sublimation of the sections under the electron beam, since the process of sublimation is rather rapid and of sufficient magnitude to affect the appearance of cell structures. It is likely that the removal of embedded material between the adjacent sections combined with the effect of sublimation are responsible, at least in part, for the lack of visible continuity of structures of the adjacent sections.

SECTION THICKNESS FOR SERIAL SECTIONING

Ideal section thickness for three-dimensional reconstruction is determined by the structural detail desired. Whether thin or relatively thick sections should be used depends on the type of cell or its component under study. Both types of sections have advantages and limitations. Since the thickness of the consecutive sections determines the resolution of the reconstructions, the thinner the section, the more detailed the reconstructions can be made. Thinner sections have fewer tracing ambiguities. It is well known that ambiguities that arise when neuronal processes are reconstructed from thick sections are often eliminated by using thin sections.

In practical terms, section thickness is determined by the complexity and size of the structure under study. In general, sections should be thinner than the smallest detail of interest. The importance of this generalisation is indicated by the following example (Bundgaard, 1984). Reconstructions based on 50 nm sections indicate that free plasmalemmal vesicles often occur in the cytoplasm. However, reconstructions based on 15 nm sections show that the vesicular profiles do not represent true vesicles. The vesicles in reality are parts of the plasmalemmal invaginations. In the 50 nm sections, the continuity between vesicular profiles is often lost.

On the other hand, very thin sections have the

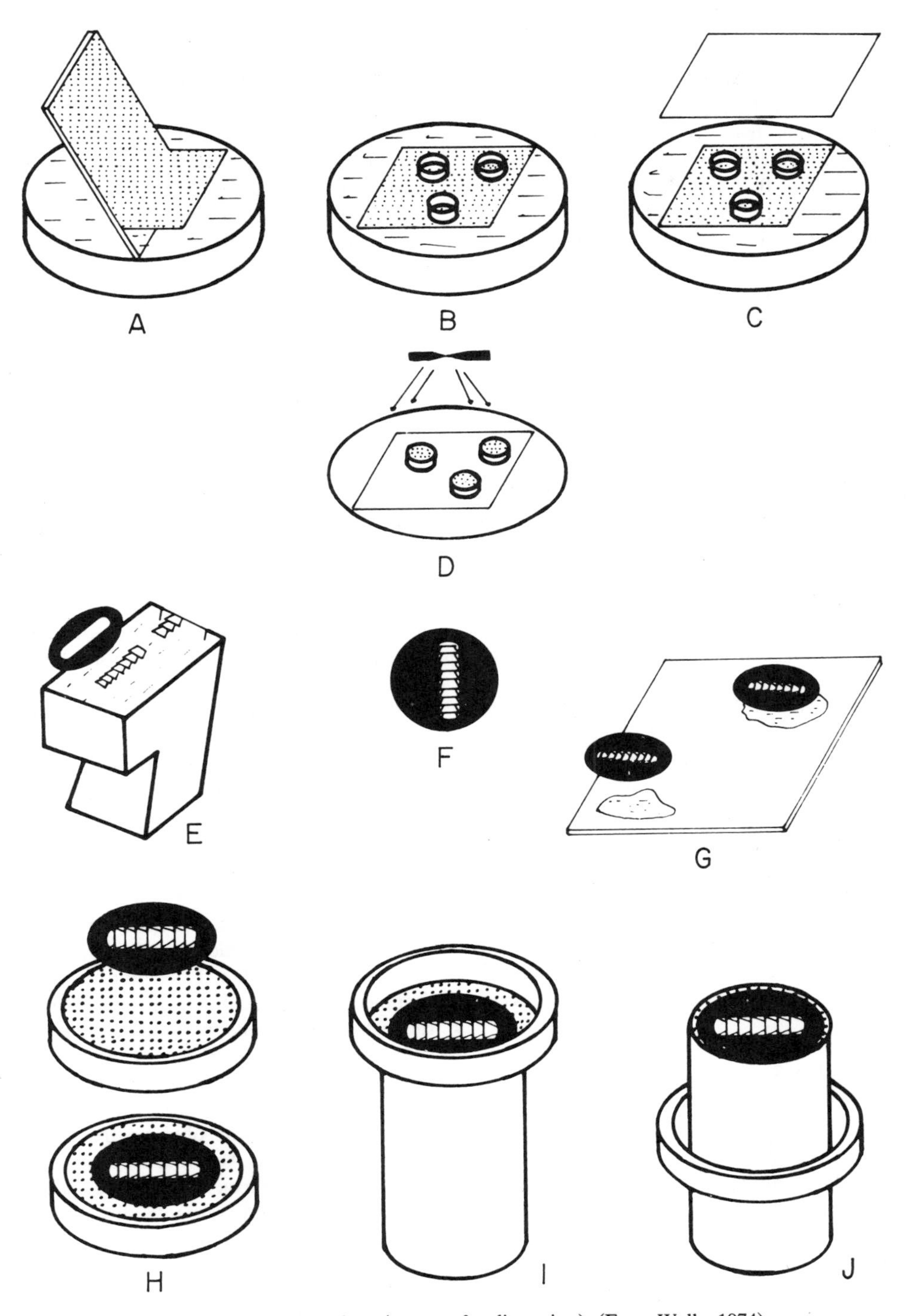

Figure 3.34 Method for collecting thin serial sections (see text for discussion). (From Wells, 1974)

following limitations: reductions in (1) image contrast, (2) mechanical strength and (3) spatial information. Moreover, a large number of very thin sections are required to obtain three-dimensional reconstruction, even of relatively small structures. Konig and Loos (1980) have accomplished three-dimensional reconstruction of neurons from relatively thick sections (0.25 μm) with a conventional TEM (80 V). In this study, the number of 0.25-μm-thick sections required was only a third of that of 80 nm sections.

In relatively thick sections (0.25 μm), more cells or cell components can be constructed in the same amount of time that it takes to complete a single reconstruction from thin sections. Certain structures which may be visible in relatively thick sections may go undetected in thin sections. This is exemplified by two components associated with the microtubule, which are seen in 0.25–0.60-μm-thick sections but are not detected in thin sections (Rieder, 1981). The reason is that thick sections capture enough of the component to be adequately stained. Similarly, stereo-viewing of thick sections often reveals relationships between cell components which cannot be readily elucidated from thin sections.

Relatively thick sections are easier to handle and show better resistance to the electron beam, which results in reduced section loss. The sturdiness of these sections allows the use of grids with a large opening, which may or may not have a support film. Konig and Loos (1980) have used grids (without support film) with 1000 × 500 μm openings for reconstructing a neuron as well as its processes and environment. Relatively thick sections, after staining, can be examined in the light microscope, the conventional TEM and the HVEM. These sections yield images of acceptable quality.

The limitations of relatively thick sections include low resolution, overlapping of cell components and blurring of obliquely cut membranes. The problem of overlapping of cell components can, in some cases, be eliminated by selective staining *in situ*. In any case, overlapping places constraints on section thickness.

THREE-DIMENSIONAL RECONSTRUCTION

Several methods are in use for visualising and reconstructing three-dimensional structures from a series of electron micrographs. One method involves tracings or line drawings. The drawings are essential to produce a three-dimensional reconstruction. Each micrograph can be traced onto transparent sheets of plastics (Barajas, 1970) or become part of a linear reconstruction (Dunn, 1972; Thompson and Gottlieb, 1972). As an alternative to making tracings, the negatives can be used directly (Fuscaldo and Jones, 1959). Another method uses various forms of model-making, which can be accomplished by cutting out a series of tracings on graph paper (Hoffmann and Avers, 1973), cardboard (Sotelo *et al.*, 1973) or polystyrene (Pedler and Tilly, 1966).

Since the above methods are rather laborious, photographic techniques have been introduced to overcome some of the difficulties. Preparation of serial sections is still a prerequisite. In one such technique, images from the negatives are enlarged onto sheet film, which is developed as a positive print (Brown and Arnott, 1971). The positive image on the film is in effect a transparency. These are stacked between glass or Plexiglas plates that approach a thickness of ~90 nm on the magnified scale. Usually 6–10 transparencies can be analysed simultaneously. In order for this technique to be used, it is essential that the original negatives be of high contrast.

A related photographic technique involves the use of stereophotographs of stacked transparencies made from serial sections (Ashton and Schultz, 1971). Using this method, introduced by Jordan and Saunders (1976), the information from up to 60 sections can be visualised. Electron micrographs are traced onto clear acetate sheets (250 mm^2) using coloured waterproof felt-tip pens. The sheets are supported by four stainless steel rods on a Perspex base. Machined Perspex spacers of thickness equal to the section thickness are used between the sheets. The rods are stabilised by a removable Perspex top, keeping the sheets taut. The whole set-up is immersed in water, making the sheets invisible; thus, the outlined structures are viewed clearly. The best stereophotographs are produced when the illumination is directed from below.

The above-mentioned methods are painstaking and time-consuming. The development of three-dimensional reconstruction programs for microcomputers has greatly reduced the time involved. Computer graphics allows visual evaluation of three-dimensional reconstructed results without building models. The computer-generated images can be rotated three-dimensionally on the video screen, and stereoimages produced. This rotation allows the visualisation of cell patterns that are not obvious from the analysis of individual sections. Accurate reconstruction is possible irrespective of the axis in which the specimen has been sectioned. In many respects the mutually exclusive advantages of trans- and cross-sections are obtained in a single reconstruction (Schierenberg *et al.*, 1986). The object can also be shaded to provide an illusion of a three-dimensional structure (Baba *et al.*, 1986). It is possible to observe an inner structure as seen through an outer one.

Computer-aided three-dimensional reconstruction of cell components in its simplest form consists of focusing a camera on the electron micrograph of thin or thick sections. The camera transfers the image to a computer graphics terminal, where diameters and co-ordinators of the cell components are entered into a computer and stored on magnetic discs. Using various display routines, the data can be displayed as three-dimensional reconstructions on a colour monitor. Inherent limitations include variations in section thickness, compression of sections during cutting and inaccuracy in alignment of the electron micrographs during input into the computer (Schierenberg *et al.*, 1986).

Future improvements consist of viewing thin sections with a TEM equipped with a video display. This would enable direct digitisation of selected cellular structures, using, for example, the Carl Zeiss-Kontron TV-overlay board system (Tuohy *et al.*, 1987). This approach would eliminate the costly process of producing first a negative and then a print for analysis. Ultimately, it should be possible to animate ultrastructural developmental processes in biomedical research. A summary of the published software available up to mid-1984 is presented by Howard and Eins (1984), and reviewed by Briarty (1986) for plant cells. Johnson and Capowski (1984) have reviewed computerised serial section reconstruction systems for animal cells.

SEMITHIN SECTIONING

The use of semithin resin sections is firmly established in histology and clinical diagnostic studies. Because embedding in resins preserves cellular components better than embedding in paraffin, the importance of semithin (thick) resin sections is obvious. During hardening, resins shrink less than does paraffin. Semithin sections of resin-embedded specimens provide greater definition of cellular details in light microscopy than that obtained in sections of paraffin-embedded tissues. Resin sections (1 μm thick) suffer from only ~20% of the image blurring that occurs in paraffin sections (5 μm thick). Semithin sections provide a bridge between the images obtained with a light microscope and those obtained with an electron microscope. These sections (1 μm thick) are about five times thinner than an average paraffin section, and many times thicker than a thin section. The image of a semithin section resembles a low-magnification electron micrograph. It therefore provides the identification of corresponding cellular regions in photomicrographs and electron micrographs.

Semithin sections can be examined under a phase microscope or, after staining, under a standard light microscope. The specificity of staining achieved in semithin sections by using chromatic stains has not been obtained in thin sections. Complex intracellular as well as extracellular structures are more easily and accurately studied by using semithin sections than by using serial thin sections. Semithin sections are more easily cut and handled and are physically more stable. Finally, spatial relationships between cellular components are better preserved, without significant overlapping of structures.

By determining precisely the area of interest in a semithin section prior to thin sectioning, one can enhance the accuracy of the study and save a considerable amount of time. One procedure is to locate the area of interest in a semithin survey section (stained or unstained) under the light microscope and then trim the block face so that it contains only the region of interest or mainly this region. This block is used for thin sectioning. A more accurate approach involves the use of the Target Marker, attached to the LKB Pyramitome. This allows the image of a stained semithin section to be superimposed over the face of the specimen block. Since cellular details are clearly seen in the semithin section, the block can be trimmed down precisely to the desired area. Another advantage is that the same semithin section can be used for both light and electron microscopy, allowing correlative studies.

A disadvantage of using semithin resin sections rather than paraffin sections is the small size of the specimen that can be viewed. Paraffin sections measuring ~60 mm^2 can be cut, whereas the size of semithin sections cut by an ultramicrotome is usually less than 1 mm^2. However, this difficulty can be circumvented by using a long-edged Ralph knife (discussed later) instead of a Latta and Hartmann triangular glass knife. The latter's cutting edge extends only up to 10 mm, whereas the former has a much longer cutting edge. Moreover, microtomes designed specifically for large-area sectioning of resins have been introduced: Sorvall JB-4A (Du Pont Instruments), HistoRange (LKB) and Autocut (Reichert Jung). With these microtomes, specimens as large as 12 × 16 mm embedded in resin blocks can be cut. These microtomes can cut block faces about 200 times larger than those used for electron microscopy, and a section thickness ranging from 0.25 μm to 10 μm can be obtained. They can also cut paraffin blocks up to 40 mm long and accept glass or steel knives.

How thick should a resin section be cut for staining for light microscopy? The desired thickness is determined primarily by the objective of the study, including the stain to be used, and by the type of embedding resin. In general, sections less than 0.2 μm thick are difficult to transfer from the trough to the glass slide, and contain a very small quantity of material that is available for reaction with the dyes. Thus, sections of less than 0.2 μm thickness are too faintly stained to be photomicrographed. A thickness range of 0.5–2.0 μm is suitable for most purposes. Since ultramicrotomes can cut thin as well as thick sections, it is rather easy to cut a semithin section before, after or during thin sectioning.

Semithin sections of almost any desired size can be cut with either a diamond knife or a glass knife (the size of the section is limited only by the length of the cutting edge), although a diamond knife is preferred. Glass knives develop nicks easily and consequently only a few sections can be cut before the cutting edge has to be changed. Substandard diamond knives are adequate for cutting semithin sections. Diamond knives with a cutting edge ranging from 1.5 mm to 4.0 mm in length are available; a diamond histo-knife with a cutting edge of 6.0 mm in length is also available for semithin sectioning (Diatome, Fort Washington, PA).

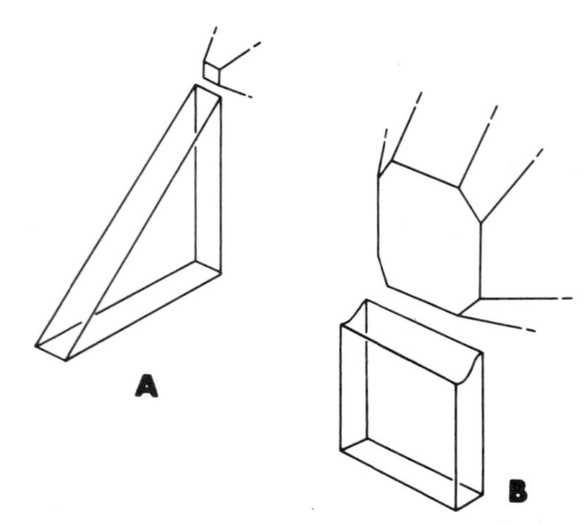

Figure 3.35 (A), Latta and Hartmann glass knife aligned with the specimen block face. (B), Ralph glass knife aligned with the specimen block face; the larger block face and longer knife edge are apparent

THE RALPH KNIFE

Bennett *et al.* (1976) have published a method for making long-edged glass knives, which they called Ralph knives in honour of the late Dr Paul Ralph, who invented the method. The width (not the thickness) of glass strips determines the length of the cutting edge. The glass strips commonly used have a width of ~25 mm and a thickness of ~6.5 mm (the same width is used for the Latta and Hartmann type of knife), although, theoretically, glass knives of unlimited edge length can be produced; an edge length of 4 cm has been recorded. The choice of the length of the cutting edge is based on the principle that the longest length of a block face to be sectioned must be shorter than the length of the knife edge. Approximate lengths of the cutting edges of the Latta and Hartmann and Ralph knives are compared in Fig. 3.35.

Both thin and semithin sections undergo compression during the cutting process. The compression of semithin resin sections ranges between 4% and 19% (Helander, 1984); nuclei show less compression than the cytoplasm. The degree of compression depends on many factors, including the angle of the cutting edge, the section thickness and the ambient temperature. In general, the larger the edge angle, the higher the compression. Thus, the use of knives with small edge angles is recommended. However, resin sections are difficult to cut when the edge angle is smaller than 35°. Figure 3.36 shows the best angle (shape) of the cutting edge of the Ralph knife. The edge angle of these knives varies between 12° and 58° (Helander, 1984). The right portion of the cutting edge is superior to the left portion; the latter has a large number of ridges that can be visualised with the SEM.

Ralph knives can be made by hand (Bennett *et al.*, 1976) or with one of the commercially available instruments: Histo KnifeMaker (LKB) and Longknife Maker (Polaron Instruments). The Histo KnifeMaker provides glass knives 25 or 36 mm long. Glass knives are superior to steel knives, even for cutting large sections. Satisfactory sections thinner than 2 μm are difficult to cut with steel knives. The Ralph knife can be used immediately after it has been made by attaching it to a glass knife holder (Szczesny, 1978); this assembly is then placed in a Sorvall JB-4 microtome steel knife holder for semithin sectioning. A modified holder for the Ralph knife for use with the Sorvall Porter–Blum MT-2 ultramicrotome was described by Gorycki and Sohm (1979), while details of the construction of the Ralph knife holder for use with the MT-1 ultramicrotome are given by Smith and Wren (1983). These methods have been discussed elsewhere (Hayat, 1986b).

Instead of using razor-blades with the Vibratome, as is generally done, one can use Ralph knives by attaching them to this instrument with a special holder (Electron Microscopy Sciences, Fort Washington, PA) (Johansson, 1983). The sectioning is carried out at a vibration rate of 7 scale units, a feeding of 2 scale and a thickness of 5–50 μm.

Glass knives are better than razor-blades for freehand trimming of large glycol methacrylate blocks. The former are especially useful for obtaining smooth block faces. Butler (1980) has described the procedure for specimen block trimming. A disadvantage of the

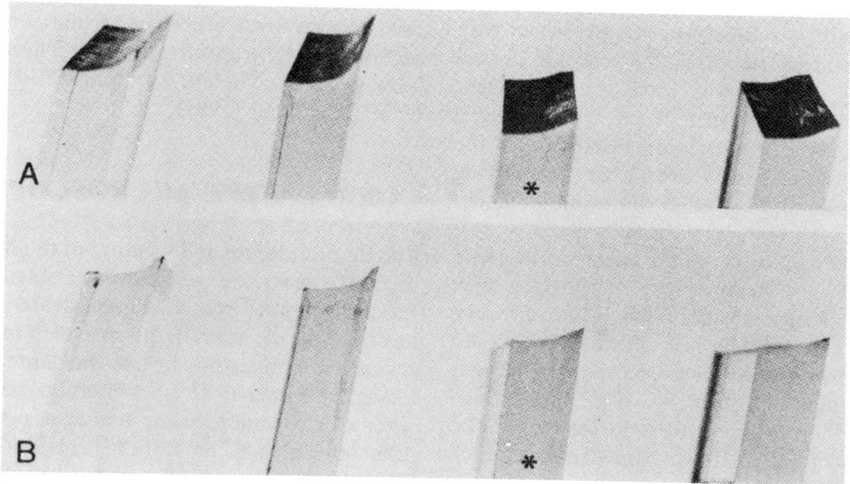

Figure 3.36 Photomicrographs of the Ralph knives. Different types of knives are shown with different angles. The knife marked with an asterisk in (A) and (B) has an angle which is optimal for the majority of specimens. The knives shown to the left are too extreme in their angles and those to the right are excessively flat. (From Johansson, 1983)

Ralph knife is that it tends to dull rather rapidly. The durability of this knife can be increased by reducing the clearance angle. Hard specimens such as undecalcified bone are sectioned better with tungsten carbide knives.

Sectioning
Usually semithin sections (0.5–2.0 μm) are cut from trimmed or incompletely trimmed specimen blocks, although they can be obtained from untrimmed blocks. When blocks are trimmed before semithin sectioning, enough resin should be left at the tip so that on the first few strokes the knife cuts through ~1 mm of resin before encountering the tissue. Ordinarily, semithin sections are handled individually except when serial sections are needed. As in the case of thin sections, semithin sections are floated on the trough fluid.

Superficial folds in the section can be removed by warming the section with one's breath. Deeper folds can be eliminated by floating the section on a drop of water at room temperature or briefly at 55 °C. If this fails, folds can be pulled apart with brushes or forceps while the section is on a glass slide wet with water. Alternatively, the section can be immersed in 100% ethanol for a few minutes and then floated on water. It is then mounted on a glass slide freshly cleaned with ethanol. Mounting is accomplished by inserting the slide at an angle into the water and bringing it up to the surface under the section. The slide with the section is removed from the water, and the section is positioned with a brush and allowed to dry on a hot-plate at 60 °C. Excessive heating should be avoided.

SECTION TRANSFER

As in the case of thin sections, semithin sections can be transferred individually or as a ribbon from the knife trough. A number of methods are available for transferring semithin sections. One method involves placing carefully a fine brush under the floating section and then raising it to pick up the section along with some water. The section is placed on a drop of water on a glass slide; if desired, more than one section can be transferred to the slide. Initially, the section wraps itself around the brush, but will unfold on the slide. Instead of a brush, a glass rod, whose end has been melted into a ball about the size of the section, can be used to pick up the floating sections. Alternatively, a floating section can be picked up with a fine wire loop or a thin foil annulus (Marinozzi, 1964) or a tungsten wire loop (Deer, 1983).

If desired, robust semithin sections can be cut with a dry knife using a slow sectioning speed. As the section begins to be cut, the edge of the section is lifted from the knife with a brush to minimise folding or wrinkling. Alternatively, the section is picked up from above with a moistened fine brush. If the dry section is of large size, it may be picked up from one edge with a fine forceps.

Merzel (1971) has introduced a simple method for transferring individual serial sections to glass slides. A simple method for transferring a large number of semithin sections to a glass slide has been described by Leknes (1985). This method is particularly useful for transferring 0.5–1.0 μm sections and when serial mounting is not required.

A versatile device, a tungsten or stainless steel wire loop (Fig. 3.37), to transfer individual semithin or thin sections or their ribbons was introduced by Deer (1983). This device is also useful for transferring a group of sections or removing floating debris from the trough. The device is equally useful for transferring sections from one liquid to another during staining and immunolabelling.

The following procedure is used to construct this wire loop (Deer, 1983). A 30 mm piece of tungsten wire (0.015 mm thick) is bent around a small nail, forming the 2-mm-diameter loop. The two ends of the wire are bent radially to form parallel support arms. Each arm is inserted into a 5 mm section of a hypodermic syringe needle which is then crimped to hold the wire. The distance from the loop to the proximal end of each needle is ~3 mm. These needles are spot-welded or glued, using epoxy resin, to the tips of a dissection-type stainless steel forceps which is ~12 cm long (Fig. 3.37). A hole is drilled with a carbide drill through the finger grip area, and a small bolt is inserted. A machine nut is mounted between the forceps arms, and a second nut is mounted on the outside. The inner nut is tightened until the bolt no longer turns; the outer nut is used for adjustments. A short piece of angled aluminium is glued with epoxy resin to the arm of the forceps nearest the operator, which serves as a finger grip. The loop is opened and closed by manipulation of the forceps to which it is attached. When the forceps is closed, the loop closes and its support arms touch each other. When the forceps is open, the support arms are separated by ~0.75 mm. A loop with a diameter of 4 mm is used for removal of debris, while semithin or large sections can be transferred with a 3 mm loop. A 2 mm loop is recommended for thin sections. For further details, see Deer (1983).

CORRELATIVE MICROSCOPY

Precise orientation and location of the desired area in a specimen block are well-known problems in sectioning for electron microscopy. A great deal of time is spent in locating specific areas in the specimen block. Especially severe sampling problems are encountered with pathological specimens. These limitations become obvious when electron microscopic analysis is needed of uncommon objects such as pathological lesions in a biopsy sample. These problems can be partially alleviated by examining in the light microscope a semithin section of relatively large size. After the desired area is located in this section with the light microscope, the specimen block is trimmed so that the final block face of very small size contains the desired area, preferably in its centre. This correlation is achieved when the orientation of the block face is maintained between the semithin and thinthin sections. An attachment, the Target Marker, to the LKB Pyramitome allows superimposition of the image of a stained semithin section over the block face, enabling a precise trimming to the desired area for thin sectioning. Alternatively, one and the same thin section (75–100 nm) mounted on a grid can be stained and examined in sequence with the light

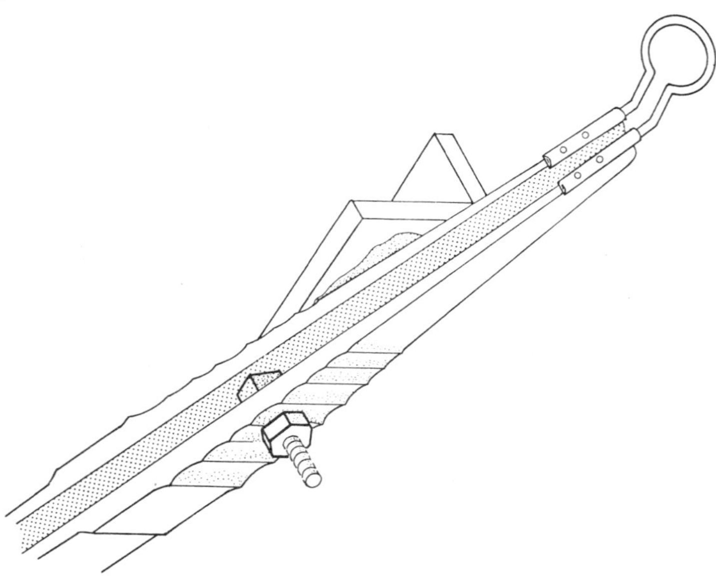

Figure 3.37 Forceps with a wire loop for transferring the floating thin or semithin sections. (From Deer, 1983)

and electron microscope.

The study of semithin sections can also provide a better understanding of the specimen as a whole. By examining such sections, unsatisfactory (e.g. poorly infiltrated) or uninteresting specimens can be rejected without cutting thin sections. The fact that the wide variety of differential staining used in light microscopy has not been adapted for electron microscopy clearly shows the advantage and desirability of using the evidence obtained by the former method in the interpretation of cell fine structure. Using specific staining, cellular materials can be recognised easily with the light microscope and then examined in the TEM; conversely, the nature of histochemical detail observed in the TEM can be evaluated and confirmed with the aid of light microscopy.

Semithin sections can be examined after staining or under a phase-contrast microscope. Procedures for staining these sections are presented in the chapter on staining in this volume and elsewhere by Hayat (1975, 1986b).

All types of microscopic systems can be used to obtain correlative data. Preparatory methods are available which allow examination of the same specimen with the light microscope and the scanning and transmission electron microscope. The specimens that have been studied with the SEM can be processed for light and transmission electron microscopy. The same paraffin section that has been examined in the light microscope can be studied with the TEM or the SEM combined with X-ray microanalysis. Correlative studies can be carried out in the areas of immunocytochemistry (Wilander and Lundqvist, 1987) and radioautography (Vuillet, 1987). Correlative methods also include scanning transmission electron microscopy and phase, polarising and ion microscopy. Instruments combining light microscopy with scanning electron microscopy (Wouters and Płoem, 1987) and with transmission electron microscopy (LEM-2000) have been developed. Confocal scanning laser microscopy provides unusual images that can be correlated with those obtained by light and electron microscopy (van der Voort *et al.*, 1987). These and other correlative methods have been discussed in detail elsewhere (Hayat, 1987).

4 Positive Staining

INTRODUCTION

Biological materials consist largely of molecules containing carbon, hydrogen, oxygen and nitrogen, with a few atoms of high atomic weight, and are therefore largely electron-transparent. It is difficult to differentiate among the electron opacities of various components in a cell. Such differentiation is rendered even more difficult in thin sections, because cell components are chemically similar to the embedding resins surrounding them. Hence, in order to determine the chemical composition, location, concentration, shape and dimensions of cell components, the application of selective staining is necessary.

Low-Z atoms in the specimen cause inelastic scattering of electrons accompanied by chromatic aberration. These effects result in unsharp images, with decreased contrast and low resolution. These problems can be minimised by employing stains of heavy metals such as osmium. These stains in thin sections (<80 nm) scatter the electron beam primarily elastically and the chromatic aberration is negligible. On the basis of theoretical and experimental studies, an understanding of the chemical methods employed for enhancing electron opacity selectivity has been emerging in recent years. The need for such understanding has become imperative, since electron microscopes capable of resolving structures at 0.2–0.3 nm spacings are at present available. These instruments can provide structural information that enables modes of function, including interaction properties of the molecule, to be inferred. With high-resolution electron microscopy, selective staining has already provided information, for instance, on the sequence of nucleotides of nucleic acids, and on amino and carboxyl groups and guanidine bonds in proteins. In addition, successful studies have been carried out for visualising single heavy atoms.

In optical microscopy various components of a tissue can be identified by a wide variety of selective staining techniques, which can give the different components characteristic colours. With respect to electron opacity, however, specific chromatin stains have no value. In terms of colour, the effects of two different stains are essentially indistinguishable by electron microscopy. Thus, in electron microscopy cell structures can only be identified by their morphology and electron opacity, and to some extent by their location. In principle, however, it is possible to distinguish colours with energy analysis, as in scanning transmission electron microscopy.

Different parts of an electron micrograph are recognisable because they differ in electron opacity, and contrast, in turn, is a result of differential electron opacity. In other words, contrast in the photographic image is produced by the differential scattering of electrons by a thin section of the embedded specimen. The loss of electrons from the beam is detected by a lighter image on the silver halide emulsion. It is obvious, therefore, that almost any substance that, when added to the specimen, increases the mass or density of certain sites selectively has a value as a 'stain' in electron microscopy. Although the scattering cross-section per atom for electrons varies at the operating voltages generally used for electron microscopy, it increases approximately linearly with atomic number. Image contrast, however, is a function of electron scattering cross-section per gram, which is practically independent of atomic number and atomic weight. Differential scattering for electron microscopy arises almost entirely from differences in density and thickness of the specimen. However, since a structure under study has a fixed number of sites at which the stain will bind, the stain having the greatest mass or density is most efficient. This is the reason why the most effective stains usually have the highest atomic number. Moreover, the staining is more pronounced in larger structures because, with a given density of sites, a thicker object binds more staining molecules in a given area. Other factors influencing the contrast include accelerating voltage and aperture size.

An appreciable increase in mass or density resulting from incorporation of heavy metal ions permits the use of thin sections, and consequently the attainment of high effective resolutions. By comparing the two electron micrographs in Fig. 4.1, it becomes obvious that the deposition of heavy metal ions of uranium and lead, in sufficient quantities, helps to obtain better contrast and resolution. However, deposition of heavy metal salts to visualise the specimen limits the ultimate resolution relative to the size of the metal ions.

Staining not only facilitates focusing, but also stabilises certain components so that they become more

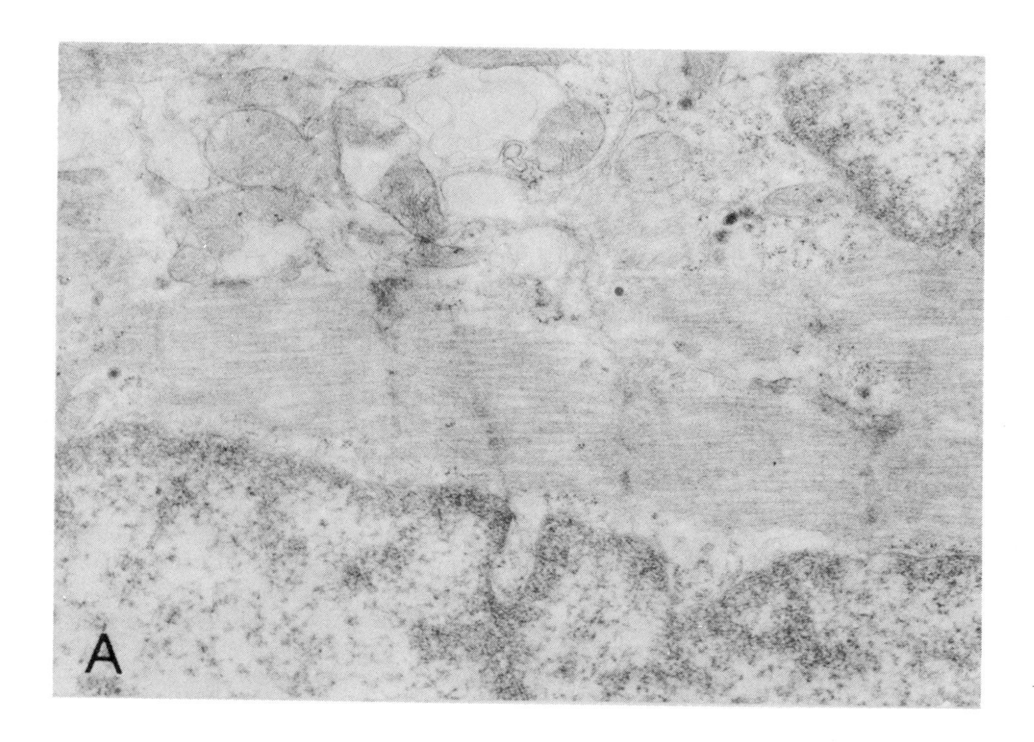

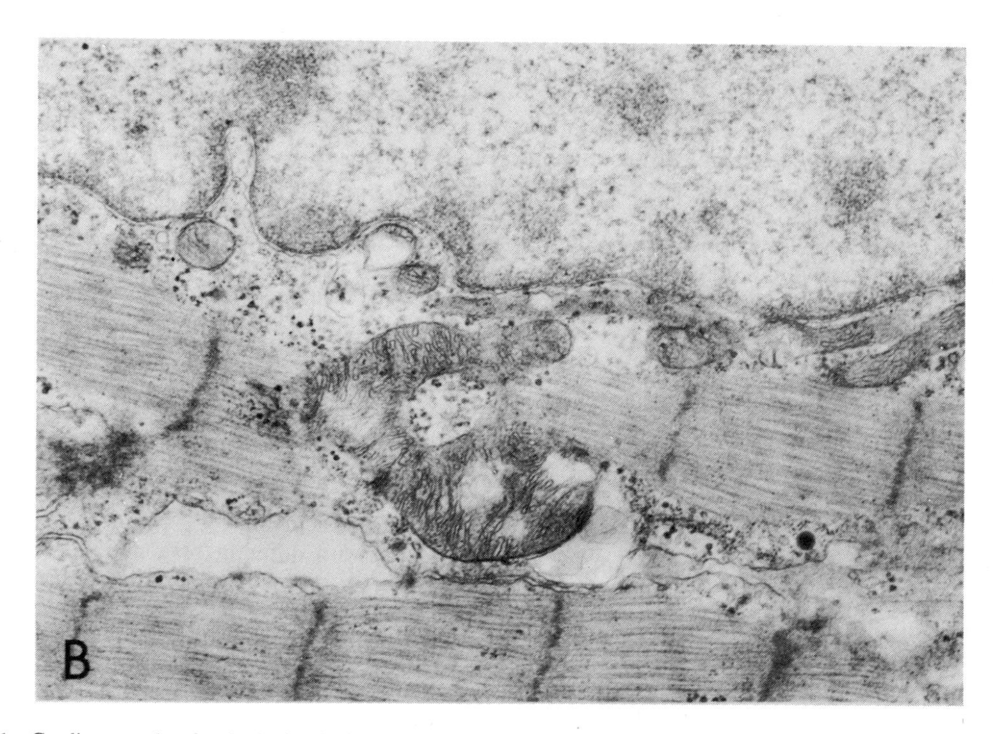

Figure 4.1 Cardiac muscle of tadpole fixed with glutaraldehyde followed by OsO_4 and exposed to 2% uranyl acetate prior to dehydration. A: No post-staining. B: Post-stained with uranyl acetate followed by lead citrate. Clarity of structural details is apparent in the lower micrograph

resistant to the damage caused by the electron beam. In other words, those cellular components that have been stained are less prone to sublimation under the electron beam than those that remain unstained. This selective stabilisation, in turn, contributes to an increased image contrast. The contrast is also enhanced as a result of differential extraction of tissue components as a result of staining, especially of long duration. Many metal chelates are soluble in organic solvents, and this could lead to such loss of tissue components.

The necessary requirements for a successful positive electron stain are: (1) it should increase the local electron-scattering power of the specimens sufficiently for an appreciable increase in image contrast to result; (2) it should be localised selectively, so that its localisation will result in positive identification of a given chemical group; and (3) its density should be higher than that of the embedding medium. For a significant overall increase in contrast, for instance, poliovirus must combine with almost its own weight of the stain.

The increase in the mass or density of the specimen is determined by the amount of stain deposited at a specific site, the total number of sites on each molecule and the frequency of occurrence or incidence of stainable molecules in the specimen. Whereas the number of sites per molecule is inherent in the specimen, the amount of stainable material is dependent on the specimen thickness. In practice, the electron opacity at a site is determined by the selective deposition of heavy metal atoms, or their complexes. Under suitable conditions it is possible to obtain selective high-contrast staining. Since the number of sites in a given region of the specimen is unknown, the increase in opacity caused by a stain is difficult to determine.

Staining can be performed during fixation and/or dehydration, or after embedding. In general, the staining carried out after fixation and embedding is decidedly superior to that accomplished before embedding. The reasons for this superiority are as follows:

(1) The extraction and displacement of the stain are minimal because the tissue has already been embedded.
(2) There is less chance distortion of fine structure.
(3) The increase in contrast is faster and greater.
(4) The staining is more uniform.
(5) Staining at this time is usually simple and can be controlled.
(6) The effect of different stains can be easily compared under almost identical conditions by treating serial sections of the same specimen block, although serial sections are not absolutely necessary to make useful comparison.

In addition, staining before embedding may make the tissue very hard, thereby rendering it more difficult to cut satisfactory sections.

The most effective and widely used method of introducing contrast into the tissue is by floating the grid-mounted sections, section side downward, on the surface of a drop of the staining solution, or by immersing the grid in the staining solution. Although staining by floating the grid is most commonly used, the immersion method is preferred for those stains that act only on the surface (e.g. phosphotungstic acid). In certain cases it is desirable to achieve staining by floating the sections prior to their being mounted on the grid. This method will prevent any adverse effect the stain may have on the metallic grid or on the support film.

In general, the specimen is stained more than once in order to achieve adequate differential electron opacity. For example, sections of a specimen fixed and stained with OsO_4 are usually post-stained with some other heavy metal such as lead. Post-staining increases the electron opacity of the structures that have already been stained with OsO_4, and permits identification of some fine structural details not easily detected in specimens stained with OsO_4 alone in connection with fixation. A case in point is nucleic acid-containing structures, which are not adequately stained with osmium and thus require uranyl acetate treatment (Fig. 4.2).

In any attempt to analyse the factors that may influence staining results, one must consider all the treatments the tissue has been given prior to or following staining. The use of stains is limited by the type of fixation and dehydration employed prior to staining. It is known, for instance, that lipids will not be visible when the tissue is fixed with aldehyde alone, since they are extracted during dehydration. Double fixation with an aldehyde and OsO_4 before dehydration, on the other hand, facilitates the staining of at least some of the lipids. Osmium tetroxide, of course, acts not only as a stain for unsaturated lipids, but also as a general fixative. Besides OsO_4, other stains known to act as fixatives are potassium permanganate, uranyl acetate and lanthanum nitrate. Potassium permanganate preserves and stains phospholipid–protein complexes, and uranyl acetate and lanthanum nitrate stain and stabilise mostly nucleic acids.

Fixation with different reagents alters the staining characteristics of the tissue. Lowering of the isoelectric point of proteins as a consequence of fixation is bound to alter the physical nature of proteins, with a resulting increase or decrease in the number of sites available for stain uptake. Furthermore, since bound or free fixative in the tissue may react with the stain, the staining results may vary, depending on the type and concentration of the fixative present, and on whether the staining

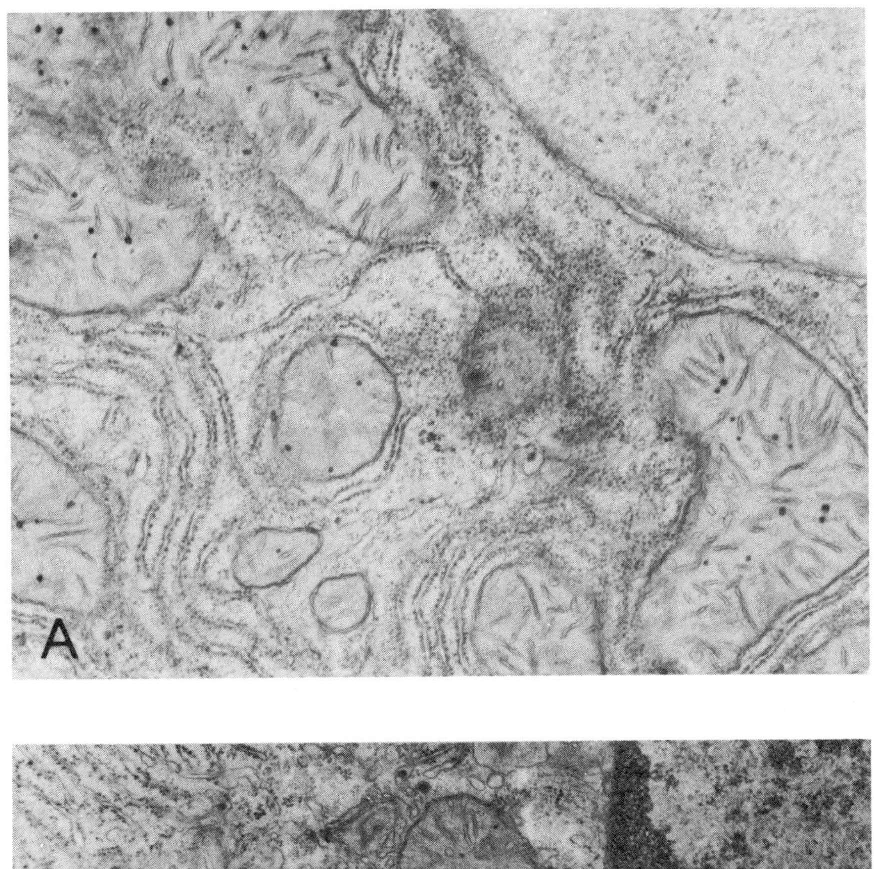

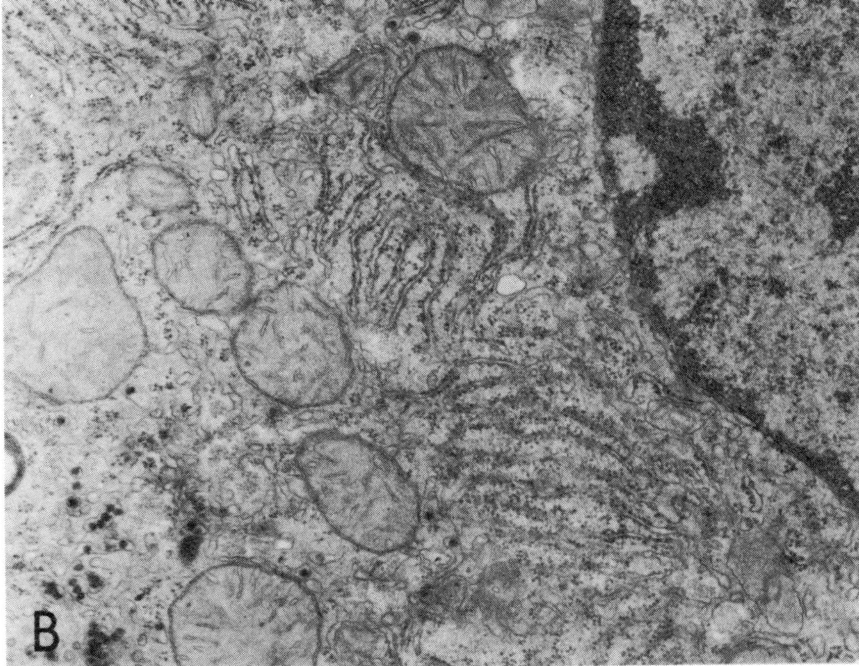

Figure 4.2 Mouse liver fixed with glutaraldehyde followed by OsO_4. A: Post-stained with lead citrate. B: Post-stained with uranyl acetate followed by lead citrate. It is apparent that nucleic acid-containing structures (chromatin and ribosomes) are stained more intensely when uranyl acetate is employed

precedes or follows the embedding. The reaction between the fixative and the stain may affect the quality of fixation. For this reason every effort should be made to wash out the unbound fixative before staining. Also, a fixative may modify the physicochemical state of the tissue, and thereby change the reactivity of the tissue components with the stain. It is conceivable, for example, that binding sites may become blocked through reaction with active groups in the fixative and the embedding medium. Thus, preparatory procedures could lead to increased or decreased staining. For a detailed discussion on the effects of fixation on staining, the reader is referred to Hayat (1989a).

Available evidence indicates that, at least in certain tissues, staining of fixed and dehydrated specimens does not result in any apparent morphological change. The studies by Brody (1971) on human epidermis have indicated that the morphology of tissue fixed and stained conventionally was similar to that observed in stained and unstained sections of unfixed and dehydrated tissue. Nevertheless, dimensional changes introduced by the stains into molecules and macromolecules, especially at the level of high-resolution electron microscopy, cannot be overlooked. In fact, certain cell structures appear larger after staining than before. One explanation of this phenomenon is that staining improves the visibility of the perimeter of these structures, but no actual increase in size takes place. It is conceivable, however, that certain stains cause topographical changes in cell components, including an increase in their size. Staining with excess stain, and with stains that tend to form polymers, may increase size unless the staining reaction is controlled. It must be admitted that a great deal of experimental and theoretical investigation of this aspect of the staining mechanism remains to be done.

IMAGE CONTRAST

Image contrast in the transmission electron microscope (TEM) is due to selective electron scattering by the atoms composing the biological specimens. When an electron beam passes through a specimen, it can interact with the nuclei of the specimen atoms and/or with electrons of the atomic shell, but the interaction with nuclei is more important. Thus, electrons of the beam are scattered as they approach positively charged atomic nuclei. The scattered electrons follow a hyperbolic trajectory around the atomic nucleus as a focal point. Electrons of a given velocity which pass at a given distance from the atomic nucleus will be more strongly deflected by a nucleus whose positive charge is high (high atomic number Z). Therefore, image contrast at any given beam potential and aperture size is proportional to the atomic number. The size of the atomic nucleus is directly related to its atomic number and therefore, approximately, to its density. Consequently, the greater the specimen density, the greater the electron scattering. If less than ~5% of the electron beam is scattered, the specimen cannot be seen in the TEM.

Another factor which influences the degree of electron scattering is specimen thickness. The degree of scattering is proportional to the specimen thickness because a large number of atoms encountered by the electron beam during its passage through the specimen results in increased scattering. Therefore, the total electron scattering is dependent upon the product of specimen density and thickness. This effect is termed the mass thickness of the specimen.

Using a conventional TEM containing a standard objective aperture, a specimen is reasonably well-defined if the product of its density and thickness is greater than ~400 times that of its surroundings. For a specimen to be visible at all, the value of this difference must be greater than ~100. Generally, since biological specimens have a density of ~1, unstained specimens must have a minimum thickness of ~10 nm to be visible.

FACTORS AFFECTING CONTRAST

Since contrast in an electron micrograph depends primarily upon differences in the ordinary density and thickness of different parts of the section, the efficiency of an electron stain is determined by the mass of the stain attached to biological substrates. Thus, if reagents of high atomic number are used, generally fewer molecules are required to impart a given increase in density. Furthermore, since a fixed number of binding sites are available in the tissue, the reagents of greatest mass are most efficient as electron stains. These are the reasons why the majority of the electron stains used are heavy metals.

Employing heavy metals renders certain structures in a cell more opaque through combination with elements of higher atomic number than their surroundings. This becomes clear when one considers that the atoms of major cellular organic substances (carbon, hydrogen, oxygen and nitrogen) are of lower atomic number and ordinary density than those of inorganic heavy metals (e.g. osmium and lead) commonly employed as electron stains.

However, in some cases atoms of metals with moderate atomic number are as effective in imparting opacity as are atoms of metals with high atomic number or density. This is exemplified by manganese (atomic number 25), which usually enhanced tissue opacity as much as does osmium (atomic number 76). This indicates that factors other than atomic number or density are also important in increasing the tissue electron opacity. One such factor is the amount of stain that can be bound by the substrate. The large amount of manganese attached to the substrate imparts as much increase in mass density as do stains of high atomic number. Manganese is thus equally effective as an electron stain. In general, large quantities of the majority of the electron stains are deposited in the tissue to achieve adequate electron density or opacity.

The speed with which local increases in mass per unit area can be obtained is an important factor in producing increased contrast. As a rule, the higher the incidence of given staining sites, the more staining molecules are bound. Consequently, a thicker section binds more stain molecules in a given time. Variations in contrast between stained sections can easily arise from a difference in thickness and not from any additional deposition sites. However, section thickness is apparently of less importance, in the case of a few stains which penetrate only the surface of a section. Intrinsic differences in mass of the tissue components contribute little to these variations in contrast. In this connection, differences in the ease of penetration of a stain into various embedments should also be considered.

The pH of the staining solution is another important factor responsible for the degree of increase in tissue contrast. It is known that, in general, heavy metal salts are hydrolysed at increasing pH levels and form polynuclear ions. These polynuclear complexes are formed as a result of the aggregation of ions when they lose protons during hydrolysis. These ions eventually precipitate as either hydroxides or hydrous oxides, which leads to increased contrast. The significance of this factor in staining with various heavy metals will be discussed later.

The advantages of post-staining of sections compared with *en bloc* staining have been mentioned earlier. In general, heavy metal stain aggregates on the surface of thin sections impart higher contrast than those embedded within the sections (Haydon, 1969a). Stereographic examination at lower magnifications suggests that tangentially sectioned membranes stain more densely on the surface than through the section (Gray and Willis, 1968). Maximum contrast is apparently achieved when a stain is applied both *en bloc* and as a post-stain for thin sections.

As mentioned earlier, much of the contrast in the TEM image may arise from variations in mass of the different parts of a specimen. Such variations not only result from the incorporation of heavy metal ions, but are also due to differential losses of parts of the tissue specimen and the embedding medium on exposure to the electron beam. Both types of losses will increase the contrast.

In tissue specimens some components are much more resistant to the effects of radiation than the others—nucleic acids and some proteins are relatively resistant. The greater the overall resistance to radiation damage, the less the differential contrast. The variations in contrast which are due to loss of embedding media have been discussed in some detail by Bahr and Zeitler (1965).

Contrast in the TEM is affected not only by the number of electrons that are eliminated from the beam as a result of the scattering ability of the specimen, but also by the angles through which the electrons are scattered and the size of the objective aperture. Small apertures and low voltages give greater contrast. When the electrons are scattered through an angle that is too great, they fall outside the objective aperture opening. These electrons are lost from the beam and become unavailable for image formation, which results in an increase in the local image density. Although the fraction of electrons scattered by any specimen can be almost doubled by simply halving the operating voltage, a number of difficulties arise with low voltages; one of these is that the photographic emulsions are not penetrated by the slower electrons and thus are very slow, despite the high contrast in the electron image. A long exposure time would be expected to increase the contrast with low voltage, but because of the image movement the resolution would be seriously lowered.

DURATION OF STAINING

The effect of the duration of staining is a complex function of the rate of adsorption of metal ions, the alterations in the staining solution and the possible extraction of cellular materials. Since very little is known regarding the actual mechanism of staining, it is difficult to predict the optimal duration of staining for various tissue types. However, available data indicate that the optimal staining duration is determined by the type of tissue and the embedding medium, the coated or uncoated surface of the grid, the pH, the type and concentration of the staining solution, the type of fixation used prior to staining, the section thickness and the objective of the study. In general, plant tissues require less staining time than do animal tissues.

No evidence is available to indicate that any digestion and extraction of cellular materials or embedding resin occurs when the sections are stained with heavy metals for short periods of time. However, it is advisable to keep the staining time as short as possible, for prolonged exposure will cause extraction of cellular materials or even destaining. Most lead salts stain most effectively in a few minutes. On the average, a range of 2–15 min is adequate. The thinner the section, the longer the duration of staining required. Comparatively, a shorter duration is required to obtain adequate staining of methacrylate sections, presumably because stains penetrate methacrylates more rapidly than they penetrate epoxy resins. However, epon-embedded specimens stain more rapidly than do Araldite-embedded specimens. It is cautioned that in electron microscopy, as in light microscopy, overstaining is a pitfall, which can result in an overall increase in contrast with little differentiation.

SIZE OF STAIN AGGREGATES

The formation of large stain aggregates in the specimen is undesirable, primarily because they obscure cellular fine details. The size of the stain aggregates is particularly important in high-resolution electron microscopy. In this case, the usefulness of a stain is limited by the size of its aggregates in the cell, macromolecule or molecule. For example, for base localisation in nucleic acids, stain aggregates larger than 0.6 nm in diameter are undesirable (Highton and Beer, 1963). In such a small space it is difficult to accommodate many heavy atoms.

Very little information is available on the size of the aggregates formed by various electron stains. A resolution of ~1 nm has been achieved after staining thin sections with uranyl acetate (Sjöstrand, 1963). According to Haydon (1969b), Reynold's lead citrate forms aggregates measuring ~1 nm in diameter, whereas Sjöstrand (1969) indicates a diameter of ~5 nm for lead. However, Sjöstrand did not specify the staining conditions responsible for such large aggregates. Silver is known to form relatively large aggregates (4–6 nm in diameter). Alcian blue has been estimated to form aggregates ranging in diameter from 2 nm to 8 nm.

The size of the stain aggregate is dependent upon many factors, including duration of staining, concentration of staining solution, type of stain, type of tissue, temperature at which staining is carried out, pH of staining solution, type of fixative, the presence of other ions in the staining and/or fixative solution, mode of staining and section thickness. Duration of staining and concentration of the staining solution are probably the most important factors which determine the aggregate size. How these factors interact among one another is not known. Admittedly, aggregate size for particular stains, under specific conditions, have not yet been determined.

STAIN PENETRATION

The great depth of field of the electron microscope allows simultaneous focusing of cellular structures present at all levels of a thin section. The depth of field of an electron microscope with a resolution of 0.6 nm is ~400 nm. This depth of field allows focusing of the entire depth of a thin section (80 nm). Hence, the accuracy of the image of a cellular structure depends on the penetration and distribution of the stain related to that structure.

The penetration of a stain into a thin section is controlled by manifold factors, including the embedding medium, the type of the stain and the vehicle in which it is dissolved, the concentration and pH of the staining solution, temperature and duration of staining, and the section thickness. Depth of penetration for any given stain is also dependent on the nature and location of the cellular component.

The penetration of heavy metal stains is often incomplete and non-uniform. Lack of penetration by the stain throughout the depth of the section becomes a serious problem for thick sections (0.5–2 μm) for high-voltage electron microscopy. According to Peters *et al.* (1971), however, cellular structures in thin sections are stained throughout the section thickness. This conclusion is based on the evidence that thicker sections show increased concentration of homogeneously distributed cell components (ribosomes and synaptic vesicles) whose diameter is less than the section thickness. If the staining were a surface phenomenon, the concentration of these components would be independent of section thickness. This observation is somewhat at variance with that of Shalla *et al.* (1964). Even when a section is uniformly stained throughout its depth, only a small fraction of the available binding sites may be occupied by the stain (Richardson and Davies, 1980).

The embedding medium is a major obstacle in stain penetration, since most embedding resins are non-polar, whereas most stains are ionic or polar. Tissue structures poorly infiltrated by a non-polar resin are readily penetrated by polar stains (PTA and uranyl acetate) (Horobin and Tomlinson, 1976). However, according to Richardson and Davies (1980), the cytoplasm is easily penetrated by aqueous PTA compared with the penetration of chromatin in intact nuclei, even though the resin content of cytoplasm is greater than that of chromatin bodies. It must be indicated that PTA does not readily penetrate chromatin bodies even in the absence of an embedding resin. Staining atoms can build up in the surface regions, which may deny its access to the interior of the section. Since aqueous uranyl acetate seems to diffuse through the hydrophilic pathways in the epoxy resins, its penetration is not affected by the resin content of a cellular structure.

Lead citrate penetrates into the embedding resin faster than does PTA or uranyl acetate. Penetration by aqueous PTA is limited to ~100 nm from the surface of the section. This fact may be related to the relatively large size of the PTA molecule, associated with its high molecular weight. Penetration by a stain varies from one tissue component to another. Differential stain penetration plays a part in determining somewhat selective staining. Lead citrate penetrates more rapidly collagen fibrils compared with chromatin (Richardson and Davies, 1980). Uranyl acetate penetrates more readily glycogen-rich regions than myofibrils (Horobin and Tomlinson, 1976). In general, penetration by the stain is aided by using elevated temperatures and alcoholic vehicles for the stain. The effect of temperature and ethanol is presumably to increase the movement and separation of the resin polymer chains, facilitating stain diffusion (Richardson and Davies, 1980).

STAIN SPECIFICITY

Specific staining is probably the most important technique for the determination of the chemical composition of cell structures. The majority of the popular electron stains are general-purpose stains, so they suffer from the disadvantage of not being very specific in action. In fact, there is probably no electron reaction which is absolutely specific for a single type of biological polymer. However, some staining reactions involving heavy metals are specific to some degree under controlled conditions of pH, temperature, concentration of staining solution, duration of staining, and methods of fixation and embedding. A few examples will be given. Uranyl acetate, under very stringent conditions of staining, has been reported to differentiate between RNA and DNA (Monneron and Bernhard, 1969). Ribonucleic acid and DNA may also be differentiated by using uranyl acetate and lead citrate under rigidly defined conditions (Daems and Persijn, 1963). Some organic chemicals also impart specific staining, although these reactions usually do not result in a very high density. For example, organic iodine derivatives stain faintly and specifically the cell membrane of sperm heads (Silvester and Burge, 1959), but then high contrast may not be absolutely necessary where high chemical specificity is the primary objective.

Specificity of the staining can be improved by employing solutions of relatively low concentrations. The usual practice in staining is to employ solutions of rather high concentrations, resulting in the saturation of most reaction sites, and thus no discrimination between them would be possible. Since it is very likely that the staining of two substances bearing different ligands with different affinities occurs over different concentration ranges, the specificity can be improved by selecting the optimal concentration of the stain for a given substance. One method of choosing the stain concentration for optimum specificity is to stain the specimen with a series of solutions of different concentrations.

Another method of improving specific staining is strict control of the duration of staining. A stain may react with more than one binding site, but the rate of reaction with various sites is likely to be different. Thus, by controlling the duration of the staining reaction, it might be possible to terminate the reaction after the desired ligand has been adequately reacted.

Specific staining of certain structures can be obtained by controlling the pH of the staining solution. A well-known example is variable staining with PTA when employed at different pH levels. When employed at pH levels below 2.0, PTA shows a high affinity for polysaccharides, while at pH 3.0–3.5 it shows affinity for proteins and nucleoproteins. In this case, the vehicle in which the stain is dissolved is also important.

Specific staining can also be improved by employing the proper method of fixation. In general, fixation with glutaraldehyde, rather than OsO_4, improves the specificity of stains, because the former is not a stain and thus does not obscure the specificity of subsequently used stains. Uranyl acetate, for instance, stains intensely and specifically nucleic acid-containing and membranous structures in the aldehyde-fixed tissue (Fig. 4.3). Lead citrate, on the other hand, stains selectively RNA-containing structures and hydroxyl groups of carbohydrates in similarly fixed tissues. If the tissue must be fixed with OsO_4, the specificity can still be enhanced by employing dilute solutions of strong oxidising agents, such as hydrogen peroxide or periodic acid, which are capable of removing lower oxides of osmium from the fixed tissue. The reason for using OsO_4 is to take advantage of its excellent fixation properties.

Specificity can also be enhanced by employing specific blocking agents to prevent binding of the stain at secondary sites. Watson and Aldridge (1961) have shown that specific staining of nucleic acids could be obtained with indium trichloride after secondary reactive groups in proteins and carbohydrates were blocked by acetylation or reduction. Specific digestive enzymes can also be used to remove the cellular materials which

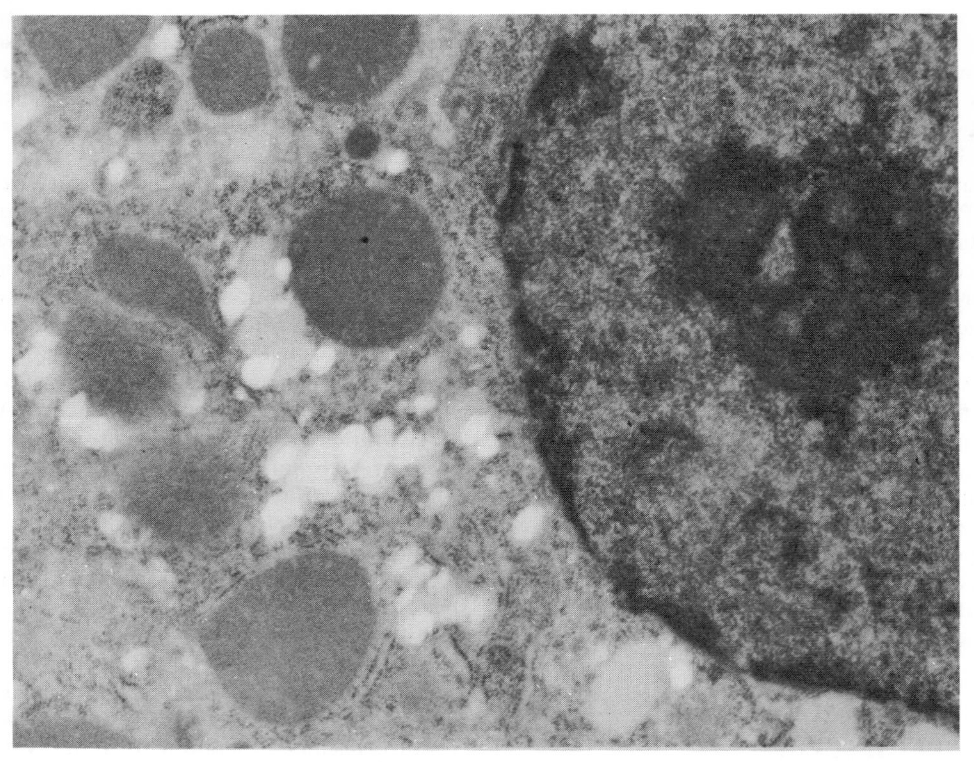

Figure 4.3 Mouse liver fixed with glutaraldehyde and the thin section post-stained with uranyl acetate. Chromatin and nucleolus are stained intensely and ribosomes, mitochondrial matrix are stained adequately, and lipids, glycogen and lipoproteins (membranes including cristae) are not stained

may hinder specific staining. These techniques have been effectively employed at the histochemical and cytochemical levels, and they are beginning to be employed in electron cytochemistry. For a detailed discussion the reader is referred to Hayat (1975, 1989a).

STAINS

As stated previously, inorganic reagents are most commonly employed as electron stains. These reagents generally lack the specificity which can be achieved in light microscopy by employing organic dyes. The use of organic dyes for electron microscopy is limited because they are composed of atoms of low atomic weight and ordinary density, and thus do not impart sufficient electron opacity. Also, the diffusion of dyes into the embedding media causes difficulty in interpretation. However, non-metallic dyes can yield some visible contrast, especially at low operating voltages. Another approach is to use a combination of organic dyes and metal salts as staining compounds.

Regardless of the staining method used, the usual criteria for a high degree of contrast, the integrity of fine structure and the preservation of enzyme activity (specificity, sensitivity and precision of localisation) must be satisfied. The need for controls such as omitting the stain, pre-digestion of the substrate, etc., in specific staining cannot be overemphasised. A detailed discussion on staining techniques for electron enzyme cytology is outside the scope of this volume; the reader is referred to a six-volume treatise on electron cytochemical methods (Hayat, 1973–77).

In summary, the staining techniques most commonly employed to obtain structural and cytochemical information from which might be inferred the modes of functions are: (1) direct attachment of heavy metal salts, (2) attachment of reaction products of heavy metals, (3) attachment of organic reagents, (4) attachment of organic reagents substituted by heavy metals and (5) precipitation of antibodies by their homologous antigens. Only heavy metal stains will be given detailed consideration here. For immunocytochemical staining methods, the reader is referred to Sternberger (1979) and Hayat (1989b).

Knowledge of the specific reactions involved in staining is necessary in order to determine the chemical properties of specific regions through their reaction with various stains. Although some progress has been made in recent years in understanding the nature of electron staining, the mechanism of attachment of stains to the reacting sites still remains largely unknown.

ALCIAN BLUE

Alcian blues are a family of polyvalent basic dyes, of which alcian blue 8G (1300 mol. wt.) (formerly designated 8GX) is the most reliable and commonly used member. Since its introduction by Steedman in 1950 as a selective stain for mucins, alcian blue has become one of the most important dyes in histochemistry and cytochemistry. It has been widely used to demonstrate glycosaminoglycans at the light microscopy level, using paraffin sections (Quintarelli and Dellovo, 1966) or semithin resin sections (Mallinger *et al.*, 1986). This copper phthalocyanin dye shows a high degree of specificity for polyanionic substances such as hyaluronic acid, sialic acid and chondroitin sulphates at the subcellular level. The critical electrolyte concentration requirement for alcian blue staining is indicative of its affinity for polyanionic substrates (Scott and Dorling, 1965; Scott, 1972a,b). That acidic polyanions can sequester large quantities of metallic cations has been well established. By changing the pH or ambient salt concentration, characteristic patterns of staining can be obtained; such patterns are easily interpretable in biomolecular terms.

The width of alcian blue molecules is 2.5–3.0 nm (Scott, 1972b). It has a relatively high solubility in salt solutions. It stains more slowly than other basic dyes. Solutions of alcian blue at pH 5.6 are unstable at room temperature or raised temperatures (56 °C). However, at pH 5.6 and 56 °C the stability is increased in the presence of $MgCl_2$.

PURIFICATION OF ALCIAN BLUE

Commercially available lots of alcian blue contain about 49% dye; the remaining constituents are boric acid, sulphates, dextrin and other unknown organic substances. It can be purified by the following method, presented by McAuliffe (1983). One gram of alcian blue is placed in a flask to which is added 100 ml of a 9:1 mixture of acetone and distilled water. This mixture is stirred on a magnetic stirrer for 1 h and then filtered through No. 2 filter paper. Small quantities of additional acetone are used to wash the dye from the flask and down the sides of the filter paper. The filtrate has a faint yellow-green colour. The residue is recovered from the filter paper and dried. Small quantities of the dye should be purified as they are needed.

The above method yields alcian blue containing about 78.7–83.5% dye, depending on the commercial source. If needed, this purified alcian blue can be further purified by the solvent extraction method of Horobin and Goldstein (1972). Scott (1972a) has also introduced a method for removing the contaminants from alcian blue. In this method one part of a 2–5% aqueous solution of alcian blue is added to 5–10 parts of acetone. The purified precipitate contains about 80% dye. The modification introduced by McAuliffe (1983) is simpler and more reliable, and yet yields dye of a purity comparable to that obtained by the Scott method.

CHEMICAL COMPOSITION

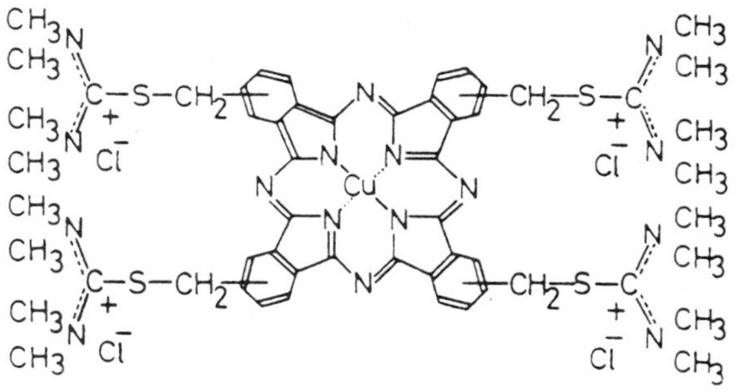

Chemical structure of alcian blue

Alcian blue is composed of geometrical isomers of copper phthalocyanin which contains at least two positively charged isothiouronium groups linked to the phthalocyanin core by methylene bridges. Elemental analyses and electrophoresis studies suggest a phthalocyanin with three *S*-methylene tetramethylisothiouronium side-chains per molecule (Scott, 1972c). Since copper phthalocyanin ring is an exceedingly insoluble and chemically inert substance, isothiouronium groups are introduced into the molecule for the preparation of aqueous solutions. These onium groups impart cationic character to the dye and serve to carry the phthalocyanin nucleus to the site of deposition in aqueous solutions.

MECHANISM OF STAINING

Alcian blue has electron density sufficiently high for it

to be capable of being used as an electron stain. The deposition of large quantities of this dye at the reaction sites is also responsible for obtaining adequate electron density. Owing to its positive charge, alcian blue is intensely attracted by a negative charge carried by the substrate molecule. Thus, the staining reaction is due to electrostatic linkages (salt linkage) between the polyvalent basic dye molecules and the anionic sites on the substrate. The anionic sites are generated in the tissue usually by sulphated, carboxylated and phosphated polyanions. The dye has especially strong affinity for both sulphate and carboxyl groups. Acid mucopolysaccharides are polyanions containing sulphate esters and carboxylic acid residues.

When used in the primary fixative at physiological pH, alcian blue binds the extracellular glycoproteins and glycoaminoglycans. Owing to its cationic nature, alcian blue is strongly held in the negative electrostatic field associated with glycoconjugate polyanions such as chondroitin sulphate, hyaluronic acid, heparin and acid glycosaminoglycans. The precipitates formed by polyanions with alcian blue are soluble only in strong salt solutions. The dye–carbohydrate complex forms an insoluble precipitate with OsO_4, yielding a staining pattern identical in many ways with that seen when ruthenium red is employed in the fixative. However, alcian blue–OsO_4 precipitate is less electron-opaque compared with ruthenium red–OsO_4 precipitate. It is thought that alcian blue in the presence of $MgCl_2$ preserves the proteoglycan *in situ* and prevents its clumping that occurs with glutaraldehyde or OsO_4 alone.

Biochemical studies have indicated that approximately one molecule of alcian blue combines with each dissacharide repeating unit of chondroitin 4-sulphate (Whiteman, 1973a,b). As stated earlier, the alcian blue molecule has up to four cationic isothiouronium groups and as an average two of the groups are linked with chondroitin 4-sulphate when complexes are formed in the presence of 50 mM $MgCl_2$. Hyaluronate appears to bind more stain per anionic charge than does chondroitin 4-sulphate. Heparin, with its high charge density, shows more binding per repeating unit than does chondroitin 4-sulphate.

Treatment of the tissue with hyaluronidase, chondroitinase or keratinase prior to staining results in decreased alcianophilia. Vigorous methylation of tissue sections removes sulphate esters from acid mucopolysaccharides and esterifies carboxyl groups. Thus, methylation removes the anionic charge from reactive groups. The binding of acidic groups of acid glycoproteins including sialomucus to alcian blue can be further tested by identifying an acid group by its removal from the tissue. For instance, the tissue is subjected to sialidase digestion to identify sensitive sialomucins. Sialic acid can be identified by acid hydrolysis which removes all sialic acid residues, which, in turn, enables the identification of sulphated substances.

The specificity and degree of interaction of organic cations or polycations with polyanions are influenced by the pH and electrolyte concentration of the medium. The degree of binding of anionic groups (e.g. —COO—, —OSO_3— and $C_6H_4SO_3$—) in mucosubstances is also influenced by the concentration of the dye and electrolytes, pH and duration of staining. In the absence of added electrolytes, staining is non-specific. Low concentrations of inorganic cations, especially Mg^{2+}, displace the dye from non-specific binding sites, resulting in not only increased specificity but also enhanced staining.

REACTION WITH NUCLEIC ACIDS

Alcian blue shows a very low affinity for nucleic acids under standard conditions of staining. The four bulky *S*-methylene tetramethyl isothiouronium side-chains are located almost exclusively in the peripheral positions of the dye molecule (Scott, 1972b). This position of the side-chains prevents its intercalation with the sugars and phosphates constituting the polynucleotide backbone. Consequently, the dye fails to intercalate into the stacked base-pairs of native nucleic acids. In this case the electrostatic binding of alcian blue is restricted to the outside of the helix.

However, at elevated temperatures alcian blue shows intense staining reaction with nuclei, depending on the DNA content of chromatin (Stockert and Juarranz, 1980). This temperature-dependent staining reaction of chromatin is thought to be based on hydrophobic and van der Waals interactions of the dye with unpaired bases of denatured DNA. Thus, alcian blue stains the DNA component of chromatin under conditions that are known to cause DNA denaturation in fixed cells. Nucleic acid-containing sites can also be stained with alcian blue (pH 4.0) after prolonged staining (a few days) at 25 °C (Goldstein and Horobin, 1974a).

CRITICAL ELECTROLYTE CONCENTRATION

The phrase 'critical electrolyte concentration' means the characteristic concentration of salt at which a given polyanion ceases to stain with a given dye. The higher the critical electrolyte concentration, the fewer substrates are stained—i.e. the specificity of the reagent is increased (Scott, 1973a; Meek *et al.*, 1985). The structures that are strongly stained with this dye may represent binding to sulphated polyanions, since they have a high critical electrolyte concentration. The critical electrolyte concentration method of alcian blue staining of tissue polyanions requires staining to be carried to a true (reversible) equilibrium (Scott and Dorling, 1965). The word 'critical' is used in this phrase to indicate that when the concentration of salt is steadily increased in a solution of polyions, the change in staining occurs abruptly. Overwhelming evidence indicates that the two conditions, reversibility and attainment of equilibrium, can be met using alcian blue.

The specificity of alcian blue for polyanions is primarily due to underlying specific ion effects. Thus, staining with this dye is controlled by the ionic character of the predominant tissue polyanions, such as chondroitin sulphate in cartilage matrix and phosphate in DNA. At a critical concentration of an electrolyte, the dye is displaced from its complex with the substrate. The reason for this displacement (dissolution) is that the cations of the electrolyte compete with the cationic (isothiouronium) groups on alcian blue for binding to the negative sites on the polyanions. The concentration of $MgCl_2$ required to displace the alcian blue–carbohydrate complex depends on the relative affinity of the competing cation for the polyanion. The concentration of $MgCl_2$ required to accomplish this displacement is also a measure of the affinity of alcian blue for the substrate. Such affinity, in turn, is determined by the type of anionic groups on the biomolecules. In general, a cationic dye bound to anionic sites of a substrate can be displaced by other cations at concentrations which will vary according to the dye–substrate affinity.

The critical electrolyte concentration values derived from model systems by Scott (1972b) correlate well with the staining of tissue sections. The critical electrolyte concentration allows a semiquantitative assessment of affinities of dyes for substrates. Dyes can be compared with each other for their affinity for a substrate. Also, substrates can be compared with each other for their affinity for a dye. It is interesting to note that hyaluronic acid has a critical electrolyte concentration of about 0.1–0.2 M $MgCl_2$. The critical electrolyte concentrations for a number of polysaccharides in model systems and histological preparations have been determined (Scott and Dorling, 1965; Scott and Willett, 1966).

A different point of view has been expressed by Goldstein and Horobin (1974a). According to these workers, alcian blue stains a few tissue components to equilibrium and the staining is reversible after only the shortest staining times. They indicate that the reversibility of staining with this dye is significantly affected by the duration of staining and that the dye is largely or completely irremovable from most substrates after any except the shortest staining times. Nevertheless, even these workers show that a large proportion of the dye is removable from cartilage matrix and mast cell granules with 1.0 M $MgCl_2$, even after prolonged staining. Since alcian blue stains relatively slowly and is chemically unstable, it has been suggested that staining equilibrium is unlikely in many tissue sites (Goldstein and Horobin, 1974a; Tas, 1977b). However, a number of other studies with the electron microscope strongly support the method of critical electrolyte concentration.

Alcian blue is displaced from mucopolysaccharides and staining ceases at a characteristic and narrow range of Mg^{2+} concentration (the critical electrolyte concentration). Alcian blue in 0.1 M acetate buffer (pH 5.8) stains hyaluronic acid, chondroitin sulphate and keratin sulphate in the presence of up to 0.2 M, 0.8 M and 1.0 M $MgCl_2$, respectively (Mallinger *et al.*, 1986). Bernfield and Banerjee (1972) demonstrated uniform staining of epithelial–mesenchymal interface of mouse embryo salivary gland in the presence of Mg^{2+} concentrations of 0.2–0.4 M. At Mg^{2+} concentrations of 0.6 M or 0.7 M the staining was markedly reduced. On the other hand, Mg^{2+} concentrations as low as 0.1 M reduced staining of intra-axonal structures with alcian blue (Hinkley, 1973); this staining was completely abolished at Mg^{2+} concentrations higher than 0.2 M. Increasing the $MgCl_2$ concentration from 0.4 M to 0.9 M resulted in the loss of chondroitin sulphate activity (Scott and Stockwell, 1967). An $MgCl_2$ concentration of 0.4–0.5 M produced intense staining of chondroitin sulphate in the epiphyseal cartilage matrix granules at the subcellular level, whereas such staining was drastically reduced when the $MgCl_2$ concentration was increased to 0.9–1.0 M (Schofield *et al.*, 1975). In general, addition of 0.1 M $MgCl_2$ to the alcian blue solution reduces background staining due to the masking of weakly charged anionic sites.

In summary, electrolyte concentration not only affects alcian blue binding, but also influences the morphology of both intracellular and extracellular alcian blue-positive materials.

ROLE OF pH

At a very strongly acidic pH of 1.0 or below, only sulphate moieties appear to retain their negative charges and bind to alcian blue. When alcian blue is used at pH 2.6 (in 3% acetic acid), some polysaccharides are masked by proteins, because at this pH polysaccharides and proteins form salt links. This masking will prevent an interaction between glycosaminoglycans and the dye. This problem can be minimised by staining at a pH of about 5.8. However, at this pH some proteins and glycoproteins may become weakly ionised polyanions capable of forming complexes with the dye. Moreover, at this pH carboxyl groups associated with proteins could contribute to the dye-binding. The staining is completely general at this pH and thus useless for ordinary purposes. These undesirable effects can be largely abolished and maximum reaction between glycosaminoglycans and alcian blue can be achieved at pH 5.8 in the presence of low concentrations of a salt (e.g. 0.05 M $MgCl_2$) (Scott and Dorling, 1965). In the presence of low concentrations of $MgCl_2$ the background staining is considerably minimised, as would be expected if weakly charged polymers containing $—COO^-$ groups were involved.

Various tissues behave differently in regard to optimal pH necessary for staining with alcian blue. Presumably, differences in steric arrangement are reflected in the pH at which the acidic groups ionise and stain with the dye. For example, in epithelial cells glycoprotein is strongly stained at pH 2.6, whereas the chondroitin sulphate of connective tissue does not stain at pH 2.0 or below, because of its binding with proteins (Szirmai, 1963; Scott *et al.*, 1968). This difference in response to pH may be due to the difference in the arrangement of acidic groups in the two types of tissues. A repeating arrangement of densely packed acidic groups is characteristic of the latter, whereas in the former the acidic groups are more irregularly distributed, with looser arrangement. It has been shown that the former type possesses more affinity, in terms of staining and retaining the dye, than does the latter (Goldstein and Horobin, 1970).

In human tissues, sulphomucins tend to stain over a wider range of pH than does the sialomucin content. Tissues containing sialomucin alone fail to stain with alcian blue at any pH level below 1.5, whereas sulphomucins stain even at pH 1.0 (Jones and Reid, 1973a). The ionisation of sulphomucins at a strongly acidic pH level suggests a sulphate group with a very low pK_a value. The bound protein of this sulphomucin does not prevent its staining with alcian blue at less than acidic pH levels, since the addition of a salt (0.1 M $MgCl_2$) to the staining solution fails to affect its staining response. Thus, alcian blue used at pH 1.0 can distinguish epithelial sulphomucins from sialomucins (Jones and Reid, 1973b).

RATE OF STAINING

The rate of staining with alcian blue is dependent primarily on the speed of diffusion of the dye into the substrate. The diffusion, in turn, is affected by the permeability of the substrate, including the plasma membrane, and the effective size of the aggregates in the staining solution. Alcian blue is present in aqueous solutions almost entirely as aggregates containing two or more molecules. These aggregates can be decreased at higher temperatures or by adding urea, resulting in increased diffusion of the dye, especially into relatively slow-staining substrates (Goldstein and Horobin, 1974a). Another effect of higher temperatures may be to increase the permeability of the substrate. These factors may exert an almost opposite, final effect on the rate of staining of readily penetrable substrates such as mucin. Other factors that affect the rate of staining include the pH, the presence of salt, the stain concentration, the duration of staining, the nature of the solvent and the age of the solution. Some of these factors influence the size of the stain aggregates. For example, the addition of salts tends to increase the size of dye aggregates. The role of pH in staining with alcian blue has been discussed earlier. It suffices to mention that the rate of staining can be increased by lowering the pH of the dye solution. The affinity of the substrate for the dye *per se* is of relatively less importance in determining the rate of staining.

Studies using model systems indicate that the time required to reach the optimal level of staining depends upon the concentration of the substrate as well as the stain (Tas, 1977a). Although the dye aggregates increase in a more concentrated alcian blue solution, optimal staining is obtained more rapidly with 1% than with 0.5% staining solution. It is not clear whether only alcian blue monomers penetrate the model system or also dimers or still higher aggregates.

With short durations of staining, relatively selective staining can be achieved with alcian blue. The inability of large aggregates of the dye to penetrate into dense tissue components plays a part in the selective staining. This selectivity is influenced by the rate of staining of various basophilic components. The rate of staining in descending order is: epithelial mucin, mast cell granules, pericapsular cartilage matrix, interstitial cartilage matrix, nuclei and cytoplasmic chromidial substance (probably RNA) (Goldstein and Horobin, 1974a). By choosing an appropriate duration of staining and other conditions, some of these components can be selectively stained.

The intensity of alcianophilia at the light microscope level can be considerably increased by treating the collagen-bound glycosaminoglycans with urea prior to fixation (Banerjee and Yamada, 1984). This treatment unmasks the carbohydrates bound to proteins. The PAS reaction is likewise increased in intensity following treatment with urea. This indicates unmasking of 1,2-glycol groups of protein-bound neutral glycoproteins.

FIXATION AND STAINING PROCEDURES

The maximum staining of cell coat and intercellular materials is obtained when alcian blue is used in combination with glutaraldehyde during fixation. The alcian blue–mucosubstance–glutaraldehyde complexes are osmiophilic, while individually the components are not. It is pointed out that this behaviour of alcian blue is almost identical with that shown by ruthenium red. Unlike ruthenium red, alcian blue is not very toxic to living organisms.

According to one procedure, tissue specimens are fixed with 4% glutaraldehyde containing 1% alcian blue for 1–18 h. After a brief rinse with a buffer, the specimens are post-fixed with 2% OsO_4 for 1–4 h. Glutaraldehyde and OsO_4 solutions are buffered at pH 6.5.

Acid mucopolysaccharides can be demonstrated by perfusing tissue specimens intravitally with alcian blue. The electron-opaque aggregates range from 3.3 to 50.0 nm in diameter in the intercellular matrix. Perfusion fixation was employed to fix and stain gap junctions in mouse liver by 3% glutaraldehyde containing 0.1% alcian blue (Goodenough and Revel, 1971). In certain cases in which the electron density introduced by alcian blue is inadequate, the density of the stained tissue components can be increased by silver sulphide impregnation. The method was used by Geyer *et al.* (1971) to demonstrate acid mucosubstances in various types of cells, including goblet and Paneth cells of rat. Sections of glutaraldehyde-fixed cells are stained with 0.1% alcian blue in 3% acetic acid for 30 min at room temperature. The excess dye is removed in distilled water. The sections are then developed for 10–15 min at room temperature in the dark with the following medium.

40% gum arabic	5.0 ml
1.2% boric acid	1.0 ml
2% hydroquinone	1.2 ml
12% silver nitrate	0.3 ml

The sections are washed with distilled water and picked up on coated grids. Post-staining with uranyl acetate and lead citrate can be obtained, if desired. The sections are handled most easily with the aid of Marinozzi's rings. Mucosaccharides stained with alcian blue become impregnated by tiny granules of metallic silver. Since different lots of alcian blue vary considerably, one of the criteria used to ensure the quality is that it should be easily soluble in water to at least 5% w/v.

BISMUTH

Bismuth (atomic number 83) is one of the heaviest metals used as an electron stain. It tends to bind with basic proteins (histones, protamines, polylysine and polyarginine) and to other molecules containing free amino groups (tryptamine and dopamine). It also binds to structures rich in phosphates (ATP). Structures rich in lipids or polysaccharides usually show a weak reaction with bismuth. Cell components strongly stained with bismuth include nucleoli, ribosomes and nuclear granules. It has been shown that bismuth staining of nucleoproteins is sensitive to proteinase K digestion (Brown and Locke, 1978). The lipid moiety of membranes and glycocalyx show low contrast. According to Locke and Huie (1977), specialised membranes such as the postacrosomal dense lamina in mouse sperm and the inner alveolar membrane in *Paramecium* are stained with bismuth. Alkaline bismuth shows affinity for sugars, which will be discussed later. The stain shows great promise in the study of chromatin-associated proteins.

Bismuth staining is sensitive to the type of aldehyde fixative used to fix the specimen. Formaldehyde fixation allows bismuth to react with amino groups and some phosphorylated molecules. Fixation with glutaraldehyde inhibits reaction of bismuth with amino groups, although guanidyl and phosphate groups remain insensitive to this dialdehyde (Brodie, 1982). In the formaldehyde-fixed tissues, bismuth stains interchromatin and perichromatin granules and the entire nucleolus. In this case, bismuth strongly reacts with nuclear basic proteins such as histones and protamines. In the glutaraldehyde-fixed specimens, bismuth also stains interchromatin and perichromatin granules, but the nucleolar staining is limited to its specific regions (Takeuchi, 1987). The reaction with the granules is due to their high phosphate content, while the staining of nucleolar components is due to the presence of a highly phosphorylated protein, 100 kD (Gas *et al.*, 1984). This protein is thought to be similar to the acidic nucleolar phosphoprotein (C23) that is stained with the Ag–NOR method (Fakan and Hernandez-Verdun, 1986). The differential effect of the two aldehydes may be explained on the basis of irreversible cross-linking of amino groups in proteins by glutaraldehyde, and at least the initial reversible reaction of formaldehyde with proteins. In other words, bismuth staining is blocked by cross-linking the amino groups and is increased by exposing these groups. It seems that bismuth staining of specimens which is due to amino groups is glutaraldehyde-sensitive, while the staining that is due to guanidine groups is less sensitive to the dialdehyde, and the staining due to organic phosphates is insensitive to glutaraldehyde.

Bismuth iodide (BiI_4^-) impregnation of glutaraldehyde-fixed tissue has been used to obtain detailed information on the structural organisation of synapses. When the stain is used in combination with post-staining with uranyl acetate and a lead salt, termed BIUL procedure, the fine structure of internal and external fuzz coats of nerve cell membranes and a double-layered subunit within the synaptic cleft becomes quite distinct (Pfenninger, 1971). In some respects the BIUL reaction is similar to that of ethanol-PTA (E-PTA), which is discussed later in this chapter.

It should be noted that the staining of synapses is accomplished only when impregnation by bismuth iodide is followed by post-staining with both uranyl and lead. It is likely that the uranyl–lead complex reacts not only with acidic groups in the tissue, but also with the bismuth iodide complex present in the tissue. According to Barrett *et al.* (1975), however, maximum general staining is obtained when bismuth staining follows treatment with salts of uranium and lead. A thorough rinsing of sections after each staining step is necessary. Acetylation and methylation experiments indicate that the BIUL technique demonstrates both the acidic groups and the basic amino acids except the guanidino groups of arginine. Guanidino residues are known to be condensed with aldehyde fixatives. Partial selective staining of basic groups can be achieved by blocking carboxyl residues by methylation.

When thin sections are exposed to bismuth, the staining is non-specific. In contrast, *en bloc* staining is somewhat specific. Periodate-reactive mucosubstances and polysaccharides containing 1,2-glycols in thin sections of conventionally fixed tissues can be stained with alkaline bismuth (Ainsworth *et al.*, 1972). The sugar residues are oxidised by periodic acid and the resulting aldehydes presumably reduce chelated bismuth subnitrate to metallic bismuth, which, in turn, appears as a fine-electron-opaque precipitate at the sites of the reducing sugars. This staining reaction can be prevented by omitting periodate oxidation or alkaline bismuth treatment, or by aldehyde blockage with a blocking agent such as *m*-aminophenol (1 M) or sodium borohydride (1%); the latter is preferred.

Alkaline bismuth is effective in staining cell walls and Golgi vesicles in plant cells (Park *et al.*, 1987). This method is also useful for studying glycoproteins in the extracellular matrix of elastin-rich tissue (Fanning and Cleary, 1985).

In the absence of periodate oxidation, glycogen, liver lysosomal dense bodies and occasionally ribosomes show marked staining. This staining is presumably through chelation of bismuth by hydroxyl groups, since it is blocked by acetylation (45 min at 60 °C) with 40% acetic anhydride in anhydrous pyridine. Methylation, on the other hand, does not affect this non-specific staining.

The above method is considered to be simpler and more specific than the majority of the other methods for the ultrastructural detection of periodate-reactive substances. The periodic acid–silver methenamine technique, for instance, produces large grains of reduced silver, which usually restrict the usefulness of the technique to relatively low magnifications. Furthermore, silver produces non-specific reactions and its solutions are unstable.

Alkaline bismuth intensifies the electron density and increases the size of ferritin by staining thin sections (Ainsworth and Karnovsky, 1972). Ferritin is a protein containing more than 20% iron. Bismuth increases the contrast of ferritin used alone as a tracer or to label antibodies. This increase in contrast probably arises through their phosphate. Uranyl and lead salts are unable to enhance ferritin contrast. By using bismuth, ferritin can be easily detected at low magnifications and concentrations. This metal simultaneously imparts some contrast to the tissue fine structure. Although counterstaining is unnecessary, if additional contrast is desired, uranyl and lead can be used; these do not interfere with the staining of ferritin by bismuth. The increase in the size of ferritin may be due to the addition of the metallic bismuth molecule to the normally undetectable spherical protein shell, apoferritin. The ferritin size may increase about twofold.

MECHANISM OF STAINING

The reactive form of bismuth is the free metal, Bi^{3+}. The trivalent charge seems to be important, since Bi^{3+} and the tetravalent polynuclear lead complex, but not the divalent lead ion, stain polyarginine (Brodie, 1982). Polynuclear complexes of bismuth with a higher valency than that of the bismuth ion do not bind to polyarginine. This implies that the distance between charges on the metal salt may influence the stability of the bond between the metal ion and its ligand.

Bismuth can localise amine, amidine and phosphate groups. Certain amino groups such as guanidyl are particularly reactive with this metal. Bismuth binds to nucleoproteins by reacting with primary phosphates and exposed amino groups. Nucleic acids do not react with the metal (Locke and Huie, 1977; Brown and Locke, 1978). Bismuth strongly binds to creatine and

ATP as well as the inorganic phosphate of mitochondrial granules (Brodie *et al.*, 1982a). X-Ray microanalysis indicates that granules in palm seedling tissue stained with bismuth are rich in phosphorus (DeMason and Stillman, 1986). Bismuth intensifies the staining already obtained with salts of uranium and lead. General contrast is increased with bismuth through its interaction with reduced osmium.

As stated earlier, some selective staining can be achieved by selecting an appropriate fixation procedure. After fixation with formaldehyde, bismuth binds to some amine, amidine and primary phosphate groups (Locke and Huie, 1976a,b; 1977). In contrast, after fixation with glutaraldehyde, bismuth stains only phosphorylated proteins. Glutaraldehyde seems to block the reaction of this metal salt with basic reactive groups by cross-linking the amines. Amino groups, but not phosphate groups, are sensitive to glutaraldehyde fixation.

Bismuth stains only certain types of phosphorus-containing molecules. The metal shows affinity for very reactive non-esterified phosphate groups of protein, but not with phosphate groups linked to nucleic acids. Thus, phosphate groups linked to proteins can be distinguished from phosphate groups in nucleic acids. Bismuth stains interchromatin granules, immature ribosomal precursor granules and mitochondrial granules, all of which contain phosphate reactive with bismuth (Brodie *et al.*, 1982a). These structures are also stained with uranyl acetate. On the other hand, other structures such as ribosomes, heterochromatin and mature ribosomal precursor granules, which also contain phosphorus, do not stain with bismuth; these structures are stained with uranyl acetate. It is likely that bismuth reacts with cellular groups in addition to those mentioned here. For example, Golgi complex beads produce a weak phosphorus signal in electron spectroscopic images, although they stain intensely with this metal (Brodie *et al.*, 1982b). This and other evidence indicates that the intensity of staining of a structure obtained with bismuth cannot be a measure of its phosphate content.

When sections are stained with bismuth, the staining is relatively non-specific. On the other hand, *en bloc* staining with this metal is highly specific. Using sections, general staining of glycogen, lysosomes, ribosomes and aldehyde groups (periodate-reactive sites) of polysaccharides (Ainsworth *et al.*, 1972) and dextrans (Thorball, 1982) have been demonstrated. According to Shinji *et al.* (1974), 1,2-glycol groups of polysaccharides (e.g. glycogen) can be demonstrated with bismuth without oxidising agents. The implication is that carboxyl groups rather than aldehyde groups may be involved in such staining.

Bismuth staining of formaldehyde-fixed tissues reveals nucleoli, nuclear nucleoprotein granules and ribosomes. Digestion with DNase causes the chromatin to stain; presumably, histone amino groups are released which react with bismuth (Locke and Huie, 1977). On the other hand, bismuth staining of glutaraldehyde-fixed tissues reveals only nuclear nucleoprotein granules and a part of the nucleolus (Locke and Huie, 1980). These results indicate that after fixation with glutaraldehyde bismuth binds to phosphate groups of phosphorylated compounds other than nucleic acids. Wassef (1979) has also shown that after such fixation bismuth staining of interchromatin granules occurs through its phosphate groups.

FIXATION AND STAINING PROCEDURES

General Staining (Riva, 1974)

Sodium tartrate	400 mg
Sodium hydroxide (2 N)	10 ml
Bismuth subnitrate	200 mg

Sodium tartrate is dissolved in NaOH. Drops of this solution are stirred into the bismuth subnitrate. After the addition of 6–8 ml of this solution, the mixture clears, and after the addition of all of this solution, all the bismuth is chelated. Conventionally fixed tissues are stained *en bloc* with a saturated solution of uranyl acetate in water for 20 min. Thin sections on grids are stained under cover for 3 min by immersion in the above bismuth solution and then rinsed with distilled water.

Staining of Mucosubstances, Glycoproteins and Polysaccharides (Ainsworth *et al.*, 1972)

Solution A

Periodic acid	0.8 g
Ethanol (100%)	70 ml
Sodium acetate (0.2 M)	10 ml
Distilled water	20 ml

The above solution is prepared by dissolving periodic acid in ethanol and then adding sodium acetate and water.

Solution B

Sodium tartrate	400 mg
Sodium hydroxide (2 N)	10 ml
Bismuth subnitrate	200 mg

The solution is prepared by adding drops of sodium

tartrate and NaOH mixture to bismuth subnitrate. After 6–8 ml has been added, the solution will clear. Thin sections on nickel grids are floated on solution A for 10–30 min. After being thoroughly washed 20–30 times for a total period of ~10 min with distilled water to remove periodate, the grids are floated on solution B for 30–60 min. They are again thoroughly washed with distilled water.

Staining of Polysaccharides without Periodic Acid (Shinji *et al.*, 1975)

Polysaccharides, including glycogen, can be stained with alkaline bismuth without oxidising agents. Thin sections of specimens fixed with glutaraldehyde followed by OsO_4 are stained with the following solution.

Stock Solution

Sodium hydroxide (10%)	50 ml
Potassium sodium tartrate	2 g
Bismuth subnitrate	1 g

The final staining solution is prepared by diluting 1 ml of the stock solution with 40 ml of distilled water. Thin sections are stained for 40 min at 40 °C.

Selective Staining of Nucleoproteins (Locke and Huie, 1977)

Sodium hydroxide (1 N)	10 ml
Sodium tartrate	400 mg
Bismuth oxynitrate	200 mg

Sodium tartrate and NaOH are added dropwise to bismuth oxynitrate to make the stock solution. The final staining solution is prepared by mixing the stock solution (3 parts) with 0.2 M triethanolamine buffer (pH 7) (1 part). The pH is adjusted to 7.0 with HCl. Formaldehyde-fixed specimens are rinsed in a 0.1 M triethanolamine buffer and then stained *en bloc* for 1 h. They are rinsed and then post-fixed with 1% OsO_4 in 0.05 M cacodylate buffer containing 4% sucrose.

Staining of Synapses (Pfenninger *et al.*, 1969)

Bismuth carbonate	0.5 g
Potassium iodide	2.5 g
Formic acid	50 ml

The above mixture is heated to 50 °C and filtered. It should not be exposed to bright light.

Thin sections of aldehyde-fixed tissue, mounted on unsupported grids, are stained in the above solution by flotation for 2 h. After careful rinsing in 2 N formic acid and in distilled water, the sections are post-stained with uranyl acetate and lead hydroxide (Karnovsky, 1961).

COLLOIDAL GOLD

Gold (atomic number 79) is sufficiently heavy to impart electron-density to biological macromolecules. It is widely used in a variety of receptor and tracer applications. Immunoelectron microscopy using colloidal gold promises to advance morphological diagnosis, including virological, by a quantum leap. The major objective in using colloidal gold is the *in situ* localisation of cellular macromolecules. This information is used to elucidate biochemical properties and functions of cellular compartments and components. The colloidal gold method is applicable to most of the microscopical systems. Methods for multiple labelling of the same specimen with colloidal gold particles of different sizes are available. The silver-enhanced colloidal gold method can be used for both light and electron microscopy. Although post-embedding labelling is most common, it can be used for pre-embedding immunolabelling. Thin conventional sections as well as thin cryosections can be labelled with colloidal gold.

Adsorption of macromolecules to gold particles is not based on chemical covalent cross-linking but rather on complex electrochemical interactions. Therefore, most bound macromolecules essentially retain their biological activity. Colloidal gold has been adsorbed to a wide variety of molecules, including proteins A and G, immunoglobulins, lectins, toxins, glycoproteins, lipoproteins, dextran, enzymes, streptavidin and hormones (e.g. insulin). Colloidal gold labelling can be quantified and an approximation of the relative density of antigenic determinants at different sites can be made. Principles, methods and applications of colloidal gold have been presented in a two-volume series edited by the author (Hayat, 1989b).

ADVANTAGES OF COLLOIDAL GOLD

(1) Colloidal gold particles are electron-dense and are easily detected in the TEM, even at relatively low magnifications. In contrast, the iron core of ferritin has a low electron-density.

(2) Compared with ferritin, gold particles are more easily visualised in relatively thick sections or on structures with a high inherent density.

(3) Non-specific adsorption by gold particles is low, which is difficult to achieve with thin sections floating on ferritin markers.

(4) The gold method alleviates the time-consuming chemical coupling of probes to ferritin.

(5) Because of the particulate and very dense nature of gold particles, quantification of labelling can be

carried out by direct counting of the number of particles in a given area of the section. This cannot be accomplished with the amorphous reaction product of the immunoperoxidase method.

(6) Higher concentrations of antisera than those acceptable for the immunoperoxidase method can be used for the gold method.

(7) Affinity cytochemistry via the avidin–biotin complex is feasible with the gold method.

(8) Specimens marked with gold can be quickly visualised without microscopic observation, because the marked specimens show red to black colour, depending on the size and density of the gold particle.

(9) The fluorescent gold method allows optimisation of labelling conditions by observing the fluorescence before electron microscopy.

(10) Because of good emission of secondary and back-scattered electrons by the gold particles, they are easily visualised in these two modes of the SEM.

(11) Gold particles can be located on cell surfaces with the SEM without metal coating (Horisberger, 1989).

(12) Monodisperse colloid gold particles of various dimensions (2–150 nm in diameter) can easily be produced for multiple labelling, which allows the visualisation of more than one macromolecule simultaneously on the same section.

(13) Easy conjugation of proteins and polysaccharides to colloidal gold particles can be accomplished.

(14) Macromolecules adsorbed to the gold particles generally maintain their bioactivities.

(15) The Protein A–gold complex is extremely immunosensitive on frozen sections.

(16) Gold particles display little non-specific adsorption to embedding media, enabling their use in post-embedding labelling.

(17) Both surface and intracellular antigens can be localised with gold particles.

(18) Gold particles can be used for microinjection and as a molecular probe for living cells as well as for resin and paraffin sections.

(19) Gold particles of small size are less susceptible to steric hindrance.

(20) Colloidal-gold-labelled sections can be counterstained, allowing the study of the ultrastructure.

(21) Colloidal gold can be used as an electron-opaque cytochemical probe for direct or indirect labelling techniques.

(22) The post-embedding immunogold method does not require permeabilisation of membranes, for the reactants have direct access to cellular structures.

(23) Since immunogold particles can identify a specific protein, the information on its presence, distribution and quantity can be obtained in the presence of other types of proteins. Very small quantities of a protein can be detected with the immunogold method.

(24) Gold particles absorb or reflect light, and can be used for subsequent silver intensification for both light and electron microscopy.

(25) The colloidal gold method allows correlative studies using various microscopic systems. The same gold-labelled reagents can be used for light and electron microscopy. Semithin and thin sections can be cut from the same specimen block for staining with colloidal gold.

(26) Since the binding constants of gold markers to cell surface are very high, specimens can be processed for scanning and transmission electron microscopy without significant loss of bound gold particles.

(27) Colloidal gold is easy to prepare or can be obtained from commercial sources. Stabilised colloidal gold can be stored, without adverse effects, for months at 4 °C.

(28) Colloidal gold can also be used for non-microscopical procedures, which are discussed in the two volumes edited by Hayat (1989b).

STABILISATION OF COLLOIDAL GOLD

The energy of interaction between two approaching particles determines the stability of a colloid. The primary force of attraction seems to be the van der Waals force, and the major force of repulsion appears to be the electrostatic double-layer force (Verwey and Overbeek, 1948). A colloid flocculates once the energy barrier has been overcome. The extent of the barrier depends on ionic strength and electrostatic charges (Horisberger and Clerc, 1985). The barrier decreases in the presence of electrolytes, which compress the ionic double layer surrounding the colloidal gold particles. As a result, the radius of repulsive forces is reduced, which causes the flocculation of the colloid. This flocculation is accompanied by a change in colour from red to violet. However, such flocculation in the presence of electrolytes does not always occur; this is discussed later.

Colloidal gold sol can be stabilised by increasing the surface potential and/or charge density by adsorbing the surface-active long-chain ions of a molecular weight <100. The stabilisation can also be achieved by reducing the van der Waals forces by adsorbing relatively rigid hydrophil macromolecules of molecular weight >1000 (Horisberger and Clerc, 1985). These two approaches can be carried out by using proteins. Besides proteins, adsorbing substances such as hormones conjugated to protein, glycoproteins, enzymes, antisera, lectins, fluorescent probes, polysaccharides and polyethylene glycol prevent electrolyte-induced flocculation of colloidal gold (Faraday, 1857; Horisberger and Vauthey, 1984).

In addition to the stabilising effects depending on Coulomb and van der Waals forces discussed above, a third type of stabilisation, termed steric stabilisation, has been suggested (Heller and Pugh, 1960). It indicates that the adsorption of flexible polymers of sufficiently high molecular weight (polyethylene glycol) causes polymer chains to protrude from the particle surface. The slight interpenetration of these chains keeps colloid particles at a distance too large to give a van der Waals interaction sufficient for coherence (Horisberger, 1981).

The phenomenon responsible for preventing the flocculation is very complex and not completely understood. A coat of protein is not always sufficient to prevent flocculation in the presence of electrolytes. The stabilisation of colloidal gold by protein depends on manifold factors, such as pH, temperature, electric charge, molecular weight and shape and configuration of the macromolecule.

Contrary to the opinion expressed in the past, it does not seem necessary to avoid commonly used buffers containing electrolytes during the preparation of stable gold complexes with protein A. No gross release of protein occurs after complexing to gold in the presence of electrolytes. Besides protein A, a wide range of other proteins (lectins and immunoglobulins) can be complexed to gold particles (6–22 nm) in the presence of buffers such as 0.1 M cacodylate and PBS containing up to 0.15 M NaCl (Lucocq and Baschong, 1986). The inhibition of electrolyte-induced gold particle flocculation in the presence of buffers seems to be due to binding of protein A to gold and not some other mechanism. The adsorption of protein to gold is more rapid than induction of flocculation by electrolytes (Lucocq and Baschong, 1986).

The advantages of complexing protein A to gold in the presence of buffers are: (1) the ionic environment of protein A is maintained during complexing; (2) pH can be controlled, obviating the need for pH adjustment of the gold colloid prior to complexing; (3) proteins such as certain antibodies supplied in buffers (PBS or saline) can be complexed directly to gold particles without prior dialysis against low-molarity buffers; and (4) protein A–gold complexes remain cytochemically active for months when stored at 4 °C (Lucocq and Baschong, 1986). The importance of maintaining physiological ionic environment becomes apparent when one considers that biological activity and/or solubility of a protein would be affected at a low ionic strength.

Various approaches have been used to determine the optimal protein concentration required to stabilise gold particles. These approaches include spectrophotometric determination of adsorption isotherms (Geoghegan and Ackerman, 1977), isoelectric focusing (Geoghegan *et al.*, 1978) and radioassay (Goodman *et al.*, 1979). The most practical approaches are microtitration assay, a visual estimation of gold flocculation by colour change, and visualisation with the TEM (Fig. 4.4). Representative examples of protein requirements are given here. For the stabilisation of 10 ml of colloidal gold (14 nm particle size), ~30 μg of protein A is needed. For the stabilisation of 10 ml of colloidal gold of 3 nm particle size, ~60 μg of protein A is required. For the stabilisation of 10 ml of heterodisperse gold particles, ~50 μg of protein A is needed.

SIZE OF COLLOIDAL GOLD PARTICLES

Spherical gold particles with diameters ranging from 2.6 nm to 150 nm can be prepared by varying the reductant used (Faulk and Taylor, 1971; Frens, 1973; Horisberger, 1981; Mühlphordt, 1982; Slot and Geuze, 1985; Baschong *et al.*, 1985; van Bergen en Henegouwen and Leunissen, 1986). In general, large-size particles (20–45 nm) are ideal for scanning electron microscopy. Such particles can easily be visualised at a low magnification, so that a clear overall picture of the labelling pattern is obtained. In addition, particles of this size appear more distinct over electron-dense cell structures than do those with small diameters. However, relative labelling efficiency and selectivity by colloidal gold particles decrease with an increase in their size. This is attributed to steric problems.

Another disadvantage of large size is that the number of gold particles usually decreases by a factor of approximately 8 for each doubling in size. In other words, an increase in particle size results in decreased label density. However, this trend is not universal. The decrease in labelling efficiency might also be due to poor penetration and/or repulsion forces (Slot and Geuze, 1984). Chemical permeabilisation of resin sections with a saturated solution of NaOH in ethanol for 1 min or 10% hydrogen peroxide for a few minutes facilitates the penetration by the primary antibody and the gold-labelled immunoreagents to the antigenic sites (Lackie *et al.*, 1985). However, this treatment damages the ultrastructure.

On the other hand, gold particles of small size allow not only larger number of particles per unit volume, but also a higher protein content (De Mey, 1985). Small gold particles require more protein to be stabilised than

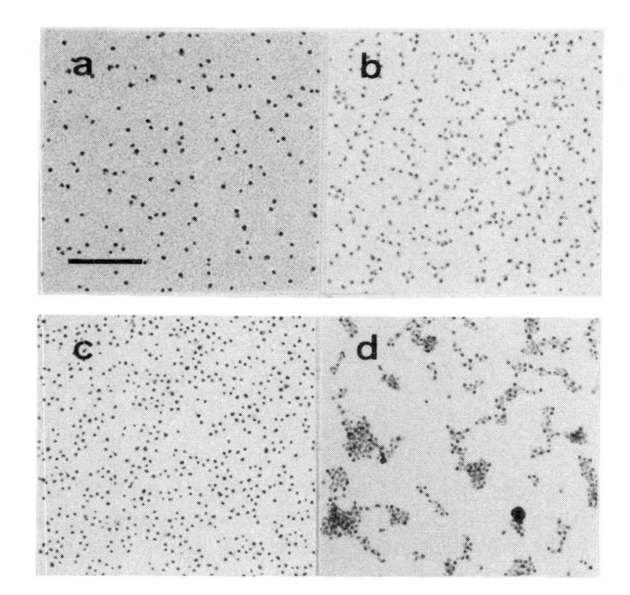

Figure 4.4 Serial dilutions of protein A complexed to colloidal gold particles. Various amounts of protein A dissolved in 100 μl of distilled water were mixed with 0.5 ml of Au_{SCN} solution (pH 6.1) for 2 min. Later 0.5 ml of aqueous 10% NaCl was added. After 30 min at room temperature, these complexes were adsorbed to Parlodion–carbon-coated copper grids for 5 min. Grids were rinsed with distilled water, air-dried and examined for aggregates in the TEM. Protein A: a = 25 μg, b = 12.5 μg, c = 6.25 μg and d = 3.125 μg. Turbidity was assessed visually in diffuse light and by measuring light scattering at $OD_{550\,nm}$. Turbidity was observed only in d and lower concentrations of protein A. Relatively small aggregates were seen in b and c. Bar represents 50 nm. From Baschong *et al.* (1985)

is required to stabilise large particles (van den Pol, 1986). Some proteins appear to adsorb better to small gold particles than to large ones. Some substances of low molecular weight have been found to stabilise only gold particles smaller than 40 nm (Roth and Binder, 1978; Goodman *et al.*, 1979). For example, cholera toxin, useful for tracing neuronal pathways, adsorbs poorly to 40 nm gold particles, but stabilises 20 nm gold particles with little problem (van den Pol, 1986). Gold particles ~5 nm in diameter (prepared by white phosphorus) bind more efficiently to Ig–Fc than gold particles of ~12 nm and 16 nm diameters prepared by sodium ascorbate and sodium citrate, respectively (van Bergen en Henegouwen and Leunissen, 1984). In addition, gold particles of small size offer a moderate increase in resolution and multiple labelling (Baschong *et al.*, 1985). Small gold particles (5 nm in diameter) also give better labelling following silver enhancement (Lackie *et al.*, 1985).

It is worth noting that the 3 nm gold particle is one of the smallest particles available for cytochemistry and allows a lateral resolution of ~16 nm (Roth, 1982). Apparently, small gold particles are ideal for high-resolution transmission electron microscopy. Indeed, even a resolution of the order of 1.5 nm is possible with very small (0.82 nm) undecagold probes (Safer *et al.*, 1982). Surface tagging with these probes has been used for examining avidin–biotin interactions. Thus, gold probes are available to elucidate biological activity at the molecular level by visualising functional sites. However, very high resolving power is not necessary for immunocytochemical studies, since interacting molecules (i.e. antigens and immunoglobulins) are of relatively large size. Table 4.1 lists gold particles ranging from 5.3 to 63.3 nm in diameter, depending on the type and concentration of the reductant used.

Determination of Gold Particle Size

The exact size distribution of gold particles can be determined by attaching colloidal gold sol to Formvar-coated grids. Artefactual clustering of gold particles can be prevented by using grids coated with Formvar–poly-L-lysine (De Mey, 1984). Negatively stained catalase crystals can be used to calibrate the final magnification of the prints. Ideally, about 95% of the gold particle population should be of the desired size.

PROCEDURE (De Mey, 1984)

About 10^{-2}% solution of poly-L-lysine (mol. wt. 70 000) is prepared in distilled water, and the pH is adjusted to 7.4. It is centrifuged at 10 000 *g* just before use, and can be stored at −20 °C. After placing a drop of this solution on a Formvar-coated grid for 10 min, the grid is rinsed in distilled water and the excess water is removed with a piece of filter paper. The grid is allowed to air-dry, and then one drop of colloidal gold is placed on it for 10 min. It is rinsed in distilled water, excess water is removed with filter paper, and then the grid is allowed to air-dry completely. The final magnification (electron microscope and photographic magnifications) is calculated by using a negatively stained catalase crystal. Gold particles can be measured on the electron micrographs with a magnifying glass or an electromagnetic instrument.

PREPARATION OF COLLOIDAL GOLD

The preparation and use of colloidal gold is rather easy. Relatively rigid, spherical gold particles can be prepared in colloidal sols by various reducing agents: white phosphorus, sodium citrate, tannic acid, sodium ascorbate and sodium thiocyanate. Gold sols are prepared

Table 4.1 Colloidal gold sols obtained by various methods

Reducing agent	*pH*	*Particle diameter* (nm)	*Particles*/ml	*Colour*	*Absorbance* (nm)
Phosphorus	7.4	5.3	3466	Red	518
Sodium ascorbate	7.9	13.3	157	Orange-red	523
Sodium citrate (1%), 2 ml/100 ml	4.8	23.7	33.1	Red	526
Sodium citrate (1%), 1.6 ml/100 ml	4.4	30.3	14.1	Red	527
Sodium citrate (1%), 0.8 ml/100 ml	3.3	46.6	3.76	Dark red	539
Sodium citrate (1%), 0.6 ml/100 ml	3.3	63.3	1.47	Red-violet	542

Modified from Horisberger (1985).

by condensing metallic gold from a 0.01% solution which is obtained by reducing Au^{3+}, usually in the form of gold chloride. Gold chloride is completely reduced to element gold during colloidal formation (Horisberger and Rosset, 1977). The conditions under which reduction and subsequent sol formation occur determine the size and dimensional variability of the resulting gold particles (Slot and Geuze, 1985). The size is expressed as the average particle diameter, and the dimensional variability is indicated by the coefficient of variation. When the coefficient exceeds 15%, sols are considered to be heterodisperse or polydisperse, and when it is less than 15%, sols are designated as homodisperse. The homogeneity of the colloidal particle diameter in the final suspension is controlled by the rate of icosahedral nuclei formation compared with shell condensation (Horisberger, 1981). Monodisperse sols are useful for multiple labelling, in which two or more probes of different sizes are employed to mark different target molecules in a single specimen (Horisberger and Rosset, 1977; Geuze *et al.*, 1981; Bendayan, 1982).

Colloidal gold sols containing particles ranging from 2 nm to 150 nm in diameter can be prepared by using the above-mentioned reducing agents (see p. 228). Gold particles as small as 0.82 nm, composed of 11 gold atoms, have been prepared by reducing triaryiphine gold complexes with sodium borohydride (Bartlett *et al.*, 1978; Wall *et al.*, 1982).

Determination of Optimal pH for Preparing Gold Sol

The colloidal gold preparation is passed through a membrane filter (pore size 0.22 μm) (Hodges *et al.*, 1984). About 7 μl of 100 mM buffers with a pH range of 4.0–11.0 is placed into each well of duplicate rows of a 96-well conical-bottom microlitre tray. The recommended buffers are citrate–phosphate (pH 4.0–6.5), phosphate (pH 7.0–7.6), Tris–HCl (pH 8.0–9.0) and carbonate (pH 9.5–11.0). To these wells is added 93 μl of the colloidal gold, resulting in the final buffer concentration of 7 mM. The two solutions are mixed by tapping the tray, and allowed to remain at room temperature for 5 min. Stability (absence of flocculation) of the colloidal gold at a given pH is demonstrated by its red colour. This colour will change to violet and then to pale blue with destabilisation of the colloidal gold (increasing flocculation). The pH range that allows the colloidal gold to remain stable is determined from the range of wells showing red colour. This pH range will include the isoelectric point of the protein used.

Before the pH of colloidal gold can be checked, it needs to be stabilised. This is done by adding two drops of 1% polyethylene glycol (mol. wt. 20 000) to 10 ml of colloidal gold sol before inserting the pH electrode. Failure to stabilise the gold before measuring pH results in the porous plug or wick of the electrode becoming plugged with colloidal gold, presumably flocculated by the KCl in the electrode (Geoghegan and Ackerman, 1977). Estimates of pH made with pH paper may vary by as much as 0.5 units.

White Phosphorus Methods

METHOD 1

White phosphorus can be used for the chemical reduction of gold chloride ($HAuCl_4$) to produce 5–12 nm gold particles (Faulk and Taylor, 1971). Three millilitres of 1% aqueous solution of $HAuCl_4$ is added to 240 ml of distilled water in a thoroughly cleaned flask, and the solution is neutralised with 5.4 ml of 0.2 M potassium carbonate. Then 2 ml of a saturated ether solution of white phosphorus is added, and the mixture is shaken for 15 min. The mixture is turned brownish red, and after heating over an open flame for 5 min, it becomes wine-red in colour. The pH of the colloidal gold sol is adjusted according to the pI of the protein that is going to be tagged with the gold. The ether solution consists of four parts of diethyl ether and one part of ether-saturated white phosphorus. Some workers prefer always to use the same glass flask for the preparation of the colloids, as gold granules may be deposited on wall impurities (Horisberger and Rosset, 1977).

White phosphorus is volatile and sublimes *in vacuo* at room temperature when exposed to light. It is inflammable and ignites at about 30 °C in moist air; the ignition temperature is higher when the air is dry. When exposed to air in the dark, it emits a greenish light and gives off white fumes. The fumes and the element itself are poisonous. It should be handled with plastics forceps and must be stored under water. Phosphorus and phosphorus-containing ether solution should be discarded after adding an excess of copper sulphate.

METHOD 2

Roth (1982) has introduced a modification of the method proposed by Zsigmondy (1905) which gives initially a monodisperse gold sol with a particle size of 2–3 nm and can be used directly for the preparation of protein A–gold complex. To obtain this sol, 1 ml of 1% aqueous $HAuCl_4$ solution is mixed with 100 ml of distilled water, adjusted to pH 7.2 with 0.2 M potassium carbonate and heated just to boiling. The reduction is initiated by rapid addition of 0.5 ml of diethyl ether saturated with white phosphorus, resulting in a change of colour to dark red. After boiling for 5 min, the colour starts changing to reddish orange. After additional boiling for 2 min, the reduction is complete. The pH of this sol is about 8.0–8.2, and must be adjusted to the pI of the protein to be tagged with 0.1 M HCl. Since non-stabilised colloidal gold plugs the pores of the electrode, five drops of polyethylene glycol (mol. wt. 20 000) are added to 3 ml of the sol, which is then adjusted to the pI of the protein.

METHOD 3

The white phosphorus method was modified by van Bergen en Henegouwen and Leunissen (1986) for preparing gold particle populations with controlled average diameters up to 12 nm. The initial gold particle population, with an average diameter of 5.6 nm (±0.9 nm), is prepared by reduction of $HAuCl_4$ with white phosphorus. A controlled increase in the diameter of these particles is obtained by additional reduction of $HAuCl_4$ with white phosphorus in the presence of colloidal gold particles. After one additional reduction cycle of $HAuCl_4$, potassium carbonate and white phosphorus the average particle diameter is increased from 5.6 nm to 6.7 nm, while after six additional cycles the diameter is increased to about 12.1 nm During additional cycles no new particles with the diameter of 5.6 nm are formed. Thus, addition of white phosphorus results in the growth of particles and not in the formation of new particles. Similarly, the particle concentration remains constant during the growth process.

Sodium Citrate Method

Trisodium citrate ($C_6H_5Na_3O_7 \cdot 2H_2O$) can be used for the reduction of $HAuCl_4$ to produce monodisperse colloids with particle diameters ranging from 10 nm to 150 nm (Frens, 1973). The desired size of the particles can be obtained by varying the proportion of trisodium citrate to a constant volume of the $HAuCl_4$ solution (Table 4.1). As the proportion of trisodium citrate increases, the particle size decreases to a minimum of 10 nm.

To obtain particles of 15 nm in diameter, 100 ml of 0.01% aqueous $HAuCl_4$ is heated in an Erlenmeyer flask. When the solution begins to boil, 4 ml of 1% aqueous trisodium citrate is added as quickly as possible (Roth, 1982). Immediately, the boiling solution becomes light blue. After boiling for 5 min, the reduction is complete, as shown by a colour change to reddish-orange. After the colloidal gold sol has cooled to room temperature, the pH is adjusted to the pI of the protein to be tagged with 0.2 M potassium carbonate.

Sodium Citrate–Tannic Acid Method

A mixture of sodium citrate and tannic acid is used as reducing agent for Au^{3+} (Mühlphordt, 1982). Under optimal conditions, homodisperse sols with gold particle size ranging from 3 nm to 15 nm can be prepared (Slot and Geuze, 1985). This method is probably the best for making gold particles in this size range. Tannic acid causes faster reduction compared with that caused by sodium citrate. When excess of tannic acid is used to obtain gold particles of small size, the reduction is accomplished almost exclusively by tannic acid. Even when very small amounts of tannic acid are used, it is the tannic acid concentration that determines particle size and, hence, the number of nuclei formed. Sols are prepared by adding Au^{3+} solution and a mixture of tannic acid and sodium citrate (the reducing mixture) quickly together at 60 °C. At higher temperatures the sols become heterodisperse, while at lower temperatures the size of the particles is increased. The maintenance of optimal pH (7.5–8.0) in the reducing mixture is also necessary; Au^{3+} solution is very acidic. Higher pHs cause heterodispersity, while lower pHs induce large particle size. Since addition of tannic acid to the reducing mixture lowers the pH (4 ml of 1% tannic acid lowers the pH by 1.5 units), enough potassium carbonate is added to compensate for the acidifying effect of tannic acid. When 0.5 ml or less 1% tannic acid is added, the pH shift is insignificant and the addition of potassium carbonate is omitted. If potassium carbonate is added to the gold solution rather than to the reducing mixture, the sol formation is slower and gold particles are larger and much more heterogeneous.

The sodium citrate–tannic acid method has many advantages compared with other methods. The particle size can be controlled accurately, so these particles are useful in multiple labelling procedures. When bound to various proteins (e.g. protein A and IgG), the complexes are stable for long periods. The coefficient of variation is 10–15% for sols containing gold particles >5 nm in diameter and below 10% for sols with gold particles <5 nm in diameter. This means that these sols are more homogeneous than those reported for other methods. The sodium citrate–tannic acid method avoids the danger involved in the white phosphorus method.

Gold solution is prepared by adding 1 ml of 1% $HAuCl_4$ to 79 ml of distilled water. The reducing mixture consists of 4 ml of 1% trisodium citrate ($C_6H_5Na_3O_7 \cdot 2H_2O$), 10 µl–5 ml of 1% tannic acid (low molecular weight, Mallinckrodt, St. Louis, Mo.), 25 mM potassium carbonate (same volume as tannic acid) and distilled water to bring the total volume to 20 ml. The two solutions are separately brought to 60 °C on a hot-plate, and then the reducing mixture is quickly added to the gold solution with stirring. The colloidal gold sol is formed within a second when large amounts of 1% tannic acid are added. The reaction time increases gradually when small amounts of 1% tannic acid are used. In the absence of tannic acid, the reaction is completed in ~1 h at 60 °C. After the sol formation is completed, as evidenced by the red colour, it is heated to boiling. If in this procedure the tannic acid concentration increases, the average gold particle diameter will decrease (Fig. 4.5). For example, if 0.015 ml, 0.1 ml, 0.5 ml or 3 ml of 1% tannic acid is added to the reducing mixture, average gold particle diameters of 14 nm, 9 nm, 6 nm or 3.5 nm, respectively, can be expected. Such sols are sufficiently homogeneous so that they can be used together as markers in multiple-staining studies. Catalase crystals can be used in the TEM for calibrating average particle diameter and the coefficient of variation. For further details, see Slot and Geuze (1985).

Sodium Ascorbate Method

Sodium ascorbate ($C_6H_7NaO_6$) can be used for reducing $HAuCl_4$ to obtain gold particles with a diameter ranging from 5 nm to 20 nm (± 4 nm) (Stathis and Fabrikanos, 1958). The most stable colloids with this method are formed near pH 7.0. The aqueous solutions of sodium ascorbate are unstable and subject to rapid oxidation by air at pH >6.0.

Horisberger (1984) modified this method to produce gold particles of 12–13 nm in diameter. To prepare the colloid, a mixture of 10 ml of 0.1% $HAuCl_4$ and 1.5 ml of 0.1 M potassium carbonate is added as rapidly as possible at room temperature to 10 ml of 0.07% sodium ascorbate, which is vigorously agitated with a clean magnetic bar. The colloid forms almost immediately. Distilled water (78.5 ml) is then added to adjust the volume to 100 ml. The colloid has an orange-red colour.

According to a modification by Slot and Geuze (1984), 10 ml of aqueous 0.07% sodium ascorbate solution is added as fast as possible to a rapidly stirred solution containing 1 ml of aqueous 1% $HAuCl_4$, 1.5 ml of 0.2 N potassium carbonate and 25 ml of distilled water at 0–4 °C. Higher temperatures tend to increase the particle size. The colloid is formed immediately, as evidenced by purple-red colour. The solution is heated until boiling so that the colour becomes red. The considerable variation in particle size of the ascorbate gold colloid particles restricts their suitability for cytochemical studies, especially for multiple labelling. However, Slot and Geuze (1981) have introduced a procedure for the purification of uniform subfractions from ascorbate gold preparations.

Sodium Thiocyanate Method

Sodium thiocyanate (NaSCN) can be used for reducing Au^{3+} to prepare monodisperse and homodisperse sols with gold particles having a mean diameter of 2.6 nm (de Brouckère and Casimir, 1948; Baschong *et al.*, 1985). These particles are used directly for cytochemical labelling without further separation. They have a mean coefficient of variation of about 15%, which is comparable with that of the sodium citrate–tannic acid method. Thiocyanate gold sol is suitable for multiple labelling without further fractionation by size. The small size of the particles is ideal for high-resolution transmission electron microscopy.

To prepare the sol, 0.3 ml of 1 M NaSCN is added with stirring to 50 ml of double-distilled water containing 0.5 ml of 1% $HAuCl_4$ and 0.75 ml of 0.2 M potassium carbonate. The reaction is accomplished in a siliconised conical flask. The yellowish solution is kept overnight (12–15 h) at room temperature, preferably in

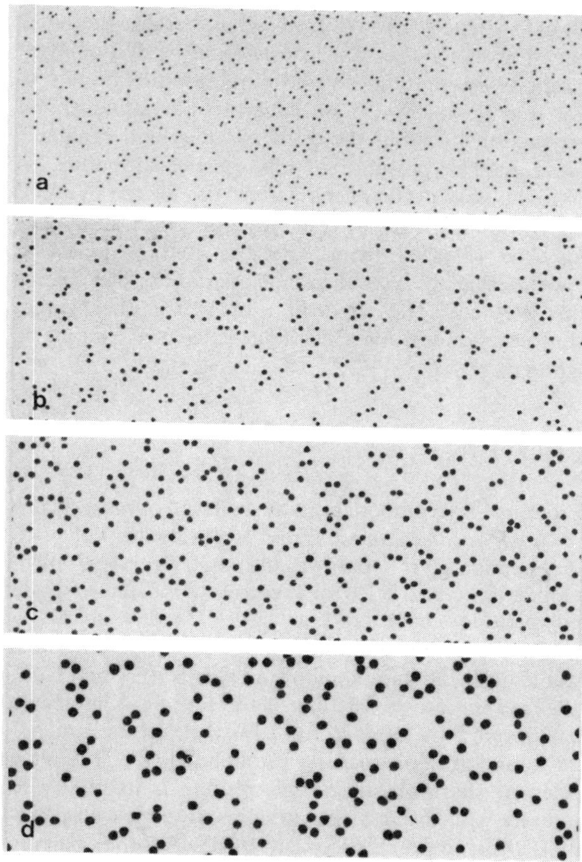

Figure 4.5 Colloidal gold particles prepared by the sodium citrate–tannic acid method. The diameter of the gold particles is increased with decreased proportion of tannic acid. Gold particles of 4 nm (±11.7%) (a), 6 nm (±7.3%) (b), 8.2 nm (±6.9%) (c) and 11.5 nm (±6.3%) (d) in diameter were obtained by using 1% tannic acid in the amounts of 2 ml, 0.5 ml, 0.125 ml and 0.03 ml, respectively, in the reducing mixture. 100 000×. From Slot and Geuze (1985)

a dark place. Storage for several days results in the formation of larger gold particles, showing reddish-brown colour. These aggregates can be removed by centrifugation at 10 000 *g* for 10 min.

Sodium Borohydride Method

Sodium borohydride ($NaBH_4$) can be used to prepare gold particles with a diameter of about 2 nm by reduction of $HAuCl_4$ (Bonnard *et al.*, 1984). To prepare the gold particles, 150 μl of a 4% $HAuCl_4$ solution and 200 μl of 0.2 M potassium carbonate are added to 40 ml of double-distilled water that has been cooled to 4 °C. With rapid stirring, 400 μl aliquots (3–5) of freshly prepared $NaBH_4$ (0.5 mg/ml) are rapidly added until no further colour change from bluish purple to reddish orange is observed. The gold solution is additionally stirred for 5 min.

INFLUENCE OF EMBEDDING MEDIA ON THE COLLOIDAL GOLD METHOD

Effects of dehydration and embedding on staining with colloidal gold in immunocytochemistry, especially in the post-embedding method, cannot be overlooked. In this connection, two factors are most important: (1) the effect on the preservation of antigenicity and (2) the rate and extent of penetration by staining solutions, immunoglobulins and other media into the section. Dehydration and embedding reagents significantly influence the preservation of antigenicity. Although epoxy resins provide adequate preservation of many antigenicities, these resins give less reliable results than those obtained with hydrophilic resins such as Lowicryl K4M and LR Gold. The reason is that dehydration and infiltration in epoxy resins are usually carried out at room temperature, and embedding is accomplished at ~60 °C. These temperatures adversely affect the preservation of antigenicity due to denaturation. Loss of tertiary structure in proteins is significantly reduced in Lowicryl compared with that in epoxy resins. Low-temperature dehydration and embedding in some acrylic resins result in better preservation of antigenicities. Stain penetration is discussed on p. 214.

Processing of specimens at low temperatures enhances the staining by the immunocytochemical protein A–gold method. Embedding at low temperatures also lowers the background staining, resulting in increased specificity. In other words, at these temperatures non-specific labelling of the resin (e.g. Lowicryl K4M) by protein A–gold is suppressed, increasing the sensitivity of this labelling method. Most of the lipids are extracted in organic solvents and resin monomers used for dehydration and embedding, respectively, unless the temperatures are maintained below ~220 K (Weibull *et al.*, 1983, 1984). Ion retention in the tissue appears to follow the same rule (Wroblewski and Wroblewski, 1984). Some solvent-induced precipitates are smaller in size at low temperatures.

Recently introduced acrylic resins that can be used at low temperatures are: the Lowicryl series (Chemische Werke Lowi GmbH, FRG) and the LR series (London Resin Co., P.O. Box 29, Woking, Surrey, UK). Lowicryl K4M and K11M are hydrophilic. By virtue of their low viscosities, the Lowicryl resins can infiltrate the tissue at the following low temperatures:

Lowicryl K4M (polar and hydrophilic)	230 K

Lowicryl HM20 (non-polar and hydrophobic)	220 K
Lowicryl K11M (polar and hydrophilic)	210 K
Lowicryl HM23 (non-polar and hydrophobic)	190 K

These resins can also be photopolymerised at the same low temperatures, provided that no long-wave (360 nm) UV-absorbing pigments or osmium deposits are present (Armbruster *et al.*, 1982; Carlemalm *et al.*, 1985). One disadvantage of Lowicryl resins is that OsO_4 cannot be used with them. In spite of claims to the contrary, the quality of ultrastructure preservation is not as good as that provided by epoxy resins. For example, cellular membranes show negative contrast in specimens embedded in Lowicryl K4M in the absence of treatment with OsO_4. This limitation can be significantly minimised by including picric acid in the glutaraldehyde fixative and embedding in less complex acrylic resins such as LR White.

Some of these acrylic resins are ideal for immunocytochemistry with the colloidal gold labelling method. Dehydration and embedding in Lowicryl K4M can be carried out at −35 °C. LR Gold cures at −25 °C in 20 h. LR White cures either at 50 °C or at room temperature using an aromatic tertiary accelerator. Since oxygen terminates the curing reaction of acrylic resins, it must be excluded during their curing. Gelatin capsules should be used as moulds for specimen embedding. Another excellent hydrophilic embedding medium is melamine (Nanoplast) (Rolf Bachhuber, Drosselweg 10, 7902 Blaubeuren, FRG). Specimens can be infiltrated with melamine without dehydration with organic solvents (Bacchuber and Frösch, 1983). This resin is better suited for the localisation of certain diffusible ions. For further information on the use of various embedding resins with colloidal gold, see Hobot (1989), Newman (1989) and Newman and Hobot (1989).

INFLUENCE OF OTHER FACTORS ON THE COLLOIDAL GOLD METHOD

The degree of colloidal gold labelling depends on the concentration of gold markers, temperature, medium viscosity, size of gold markers and duration of staining. The most important factor is the concentration of gold particles (Park *et al.*, 1987). The degree of labelling is inversely related to the square root of the size of gold particles. The orientation of the antibody on colloidal gold particles may also influence the labelling. Although all gold particles are covered by antibodies, the latter may not show the same orientation. The dominant antibody orientation on gold particles with respect to Fab and Fc portions of an antibody may influence the interaction between the gold particles and antigenic proteins. The presence of non-specific antibodies on colloidal gold particles may affect the degree of staining. In the presence of competing proteins, antibodies may be released from colloidal gold particles. It appears that the use of monoclonal antibodies avoids this problem. The effect of a change in the protein conformation on the labelling efficiency cannot be ignored.

PROTEIN A

Protein A is a component of the cell wall of most strains of *Staphylococcus aureus*, and is covalently linked to the peptide-glycan part of the wall (Verwey, 1940; Goudswaard *et al.*, 1978). It consists of a stable, single polypeptide chain with a very extended shape, a molecular weight of 42 000 and pI 5.1. This protein is stable to heat and denaturing agents (Björk *et al.*, 1972). The native structure of the protein A molecule has about 50% α-helix, 10–20% β-structure and 30–35% randomised structures (Sjöholm, 1975). The most relevant characteristic of this protein is its ability to interact with the immunoglobulins of almost all mammals (Forsgren and Sjöquist, 1966). Another equally important property is that this interaction does not involve the antigen–antibody region of the immunoglobulin molecule and, therefore, the antigen–antibody reaction remains unaffected. In other words, the interaction is a pseudoimmune reaction. These two unique properties of protein A qualify it to be the most extensively used probe in immunocytochemical techniques.

Protein A interacts specifically with the Fc fragment of immunoglobulin G and in some species with IgA and IgM (Goudswaard *et al.*, 1978). One molecule of protein A contains four Fc-region binding sites, and is able to bind two immunoglobulin molecules (Sjödahl, 1977). However, when protein A is released from the bacterial wall, only two binding sites are expressed. Therefore, protein A is functionally bivalent (Langone, 1982). Tyrosine and histidine amino acids in protein A seem to be important for binding to Fc (Sjöhalm *et al.*, 1973). Thus, procedures that modify these two residues may decrease the reactivity of protein A. The reaction characteristics of protein A allow double labelling with two sizes of colloidal particles. For additional details, the reader is referred to the two-volume series edited by the author (Hayat, 1989b).

Preparation of Protein A–Gold Complex

Protein A in combination with gold particles was first used by Romano and Romano (1977). Colloidal gold particles are prepared by a simple two-step procedure given on p. 227. This is followed by adsorption of protein A onto the surface of these particles due to electrostatic interaction between the negatively charged surface of gold particles and positively charged groups of protein A. This protein is rapidly and irreversibly bound to the gold particle, involving electrostatic van der Waals forces, resulting in the formation of a stable complex, protein A–gold. This binding is non-covalent and does not interfere with the bioactivity of the protein except that of catalase (Horisberger, 1978).

Studies by Horisberger and Clerc (1985) indicate that gold sols can bind much more protein A than is indicated by the gold number. However, this 'excess' protein is bound less tightly and may dissociate during the first weeks after the preparation. The dissociated protein A may compete for reactive sites with the protein A–gold and thus lower the immunolabelling intensity. Therefore, protein A in excess of its stabilisation concentration should not be added to the gold sol. This stabilisation concentration is determined by the coagulation test.

However, a consensus on the above-mentioned point is lacking. According to Goodman *et al.* (1981), with increasing protein A concentrations, increasing amounts of protein are adsorbed to gold particles as multilamellar shells, whereas, according to Horisberger and Clerc (1985), this protein is adsorbed as a monolayer on gold particles. Tokuyasu (1983a) also indicates protein A–gold (particle size 5 nm) complexes give significantly higher density of labelling when they are prepared using more than 10 times the concentration of protein A than that shown by the coagulation test. Tinglu *et al.* (1984) also demonstrate that IgG–gold complexes obtained by using the IgG–colloidal gold ratio indicated by the coagulation test leads to poor labelling. Nevertheless, the coagulation test is widely used to determine the concentration of protein A necessary to prepare stable gold complexes.

pH

The binding of protein to gold is pH-dependent. Specifically, the pH of the gold sol controls to a large extent the optimal concentration of protein A needed to stabilise the colloidal gold (Geoghegan and Ackerman, 1977). The adsorption of protein to colloidal gold at pH values acidic to the isoelectric point causes the colloid to flocculate. The flocculation is abolished when the pH is raised to the isoelectric point of the protein. Raising the pH still higher may or may not cause flocculation by the addition of electrolytes. In general, stable complexes can be achieved at a pH equal to or slightly higher than the isoelectric point of the protein involved.

A gold sol binds more protein A near its isoelectric point (pH 5.1) (Horisberger and Clerc, 1985). This is due to the surface charge of the gold particles becoming less negative, as a result of which mutual repulsion due to Coulomb forces decreases. Another desirable effect of this pH is the increased binding of colloidal gold sol to IgG. On the other hand, at pH 7.2 fewer than 4.5 protein A molecules are adsorbed per gold particle with a diameter of ~11.2 nm (Horisberger and Clerc, 1985). In general, the number of protein A molecules adsorbed per gold particle depends on the size of the particle. The smaller the size of the particle, the fewer the number of protein molecules adsorbed. A seemingly desirable side-effect of pH 7.2 is that the protein A binding to gold is more stable. But the more tight binding at this pH may have a denaturing effect on the protein, making the probe less reactive. Therefore, the binding of protein A to gold should be carried out at pH 6.0. Besides pH, many other factors, including temperature, ionic strength, and size and configuration of the macromolecules, affect the adsorption of proteins to gold.

DETERMINATION OF OPTIMAL PH FOR PREPARING PROTEIN A–GOLD COMPLEX

Protein stock solution is prepared at a concentration of 5–10 mg/ml distilled water. This stock solution is diluted to obtain a protein concentration of 50–1000 μg/ml of distilled water, depending on the protein; the protein concentration should be in excess of the stabilising level required for the gold marker system (Hodges *et al.*, 1984). About 7 μl of the working protein dilution is added to the pH-buffered colloidal gold in the range of those wells showing a red colour (showing colloidal gold stability). The two solutions are mixed by tapping the tray and are allowed to remain at room temperature for 10–15 min. Stability of the protein gold mixture with pH change is determined visually: a red colour indicates that the mixture is stable, unflocculated, whereas a blue colour indicates that it is flocculated.

Determination of Optimal Stabilising Amount of Protein A

Each protein has a specific optimal concentration at which it stabilises colloidal gold. Before preparing the protein A–gold complex, it is necessary to determine the optimal amount of protein or other macromolecules that fully stabilises a given volume of colloidal gold against aggregation. This step must be carried out under optimal conditions of pH and ionic concentrations. The optimal amount of protein necessary to stabilise colloidal gold can be assessed from the resistance of gold to salt-induced aggregation as a function of protein concentration. The aggregation can be monitored spectrophotometrically (Geoghegan and Ackerman, 1977) or by microtitration assays (Goodman *et al.*, 1981). The latter method (coagulation test) is outlined below (Hodges *et al.*, 1984).

Exactly 20 μl of double-distilled water is placed in each of the wells of one series of duplicate rows of a 96-well conical-bottom microtitre tray. To the first wells of each row is added 20 μl of a 500–1000 μg/ml protein-working dilution (prepared in double-distilled water from a 5–10 mg/ml stock protein solution). Twofold serial protein dilutions are prepared, leaving a total volume of 40 μl in the final wells. To the serially diluted protein in each of the wells (except the last wells) is added 100 μl of the buffered colloidal gold sol. Mixing is accomplished by tapping the tray. The mixture is allowed to remain at room temperature for 15 min.

To assess resistance of the buffered colloidal gold–protein A mixtures to salt-induced aggregation, 20 μl of 10% NaCl (in double-distilled water) is added to all except the last wells of the duplicate rows. The mixtures are allowed to remain at room temperature for 5 min. Stabilisation of the gold is determined visually by the absence of aggregation. The last well in the sequence of serial protein dilutions that maintains red colour represents the dilution endpoint at which the protein stabilises 100 μl of buffered colloidal gold. For example, an endpoint in well No. 6 indicates that 20 μl of protein diluted to 1/64 stabilises 100 μl of buffered gold sol. For 1 ml of buffered gold sol, the protein-stabilising concentration is 3.21 μl of the working protein dilution.

Alternatively or additionally, stabilisation can be confirmed with the TEM by the presence of aggregates containing five or fewer particles (Fig. 4.4d). In the case of gold sols prepared by the sodium citrate–tannic acid method, the colour change is slow and is masked by the brownish colour contributed to the sol by tannic acid. This is a problem, especially with the finest sols, when a relatively high concentration of tannic acid has been used to prepare the sol. This problem can be obviated by adding low concentrations of hydrogen peroxide (~0.2%) to the test sol (J. W. Slot, personal communication).

Procedure for Preparing Protein A–Gold Complex

Before preparing the protein–gold complex, three parameters are established: (1) the pH range at which colloidal gold is stable, (2) the optimal pH for a stable protein–gold complex and (3) the optimal protein concentration that stabilises the colloidal gold. The procedures to determine these conditions are given elsewhere in this volume. Protein is dissolved in distilled water at 1 mg/ml. The pH of the colloidal gold sol is adjusted to 6.0 with 0.1 N NaOH. The pH of the unstabilised sol should not be checked with the pH meter, for it will clog the electrode. The best approach for adjusting the pH is to take less than 5 ml of the sol (pH 5.0–5.5), stabilise it with excess protein A (5 μg/ml of sol) and determine the concentration of NaOH required to bring the pH to 6.0. Then this concentration of NaOH is added to the unstabilised sol and the stabilisation concentration of protein A in small samples is determined as described above. The protein A–gold complex is now prepared by adding this concentration of protein A to the sol at pH 6.0.

The protein A–gold complex is centrifuged for 45 min at 125 000 *g* and 50 000 *g* for 5 nm and 10 nm gold particles, respectively. The resulting pellet consists of a large loose part and a small tightly packed part. The supernatant is completely removed without disturbing the pellet, and the loose part of the pellet is resuspended in PBS (pH 7.3); the volume of PBS to be added is ~1/25 of the original amount of the sol used. This is layered over a 10–30% continuous glycerol gradient in PBS containing 0.1% Carbowax 20-M. The gradient is centrifuged for 45 min at 41 000 rev/min and for 30 min at 20 000 rev/min for 5 nm and 10 nm gold particles, respectively, in a SW 41 rotor (Beckman Instruments). The dark-red band is collected, which contains protein A–gold essentially free of clumps (Slot and Geuze, 1981).

Protein A–gold complex may lose its activity within weeks, owing to the dissociation of the protein from the gold. However, the complex can be stored in 20% glycerol frozen in small samples at −70 °C or in 45% glycerol at −20 °C; in the latter case the complex remains fluid. Alternatively, the complex can be stored with 0.02% sodium azide in a refrigerator. Although protein A–gold stored under these conditions remains viable for several months for immunocytochemistry, the signal-to-noise ratio seems to be lower compared with that obtained with freshly prepared complexes. Commercial colloidal gold adsorbed to proteins varies considerably in concentration, and appropriate working dilutions can be prepared after spectrophotometric analysis (De Mey, 1983). Before use, the content of colloidal gold can be checked by measuring the optical density at 520 nm, in order to ensure that a comparable amount of gold is present in the incubation medium.

Labelling
Nickel grids with mounted thin sections are floated on a drop of 1% chick ovalbumin in PBS (pH 7.4) for 5 min at room temperature to block non-specific attachment of antibodies to residual glutaraldehyde. The grids are transferred (without rinsing) onto drops of the antibody solution, at a concentration previously determined to be optimal, for 1–2 h at room temperature or overnight at 4 °C in a moist chamber. The grids are thoroughly rinsed in four changes of PBS for a total period of 5–10 min, and dried by holding them edgewise on filter paper. They are transferred to drops of protein A–gold solution for 30–60 min at room temperature in a moist chamber. The grids are extensively rinsed in several changes of PBS, and then in distilled water. Post-staining is carried out with 5% aqueous uranyl acetate followed by lead citrate. The background staining is usually the result of high concentrations of antisera or gold. Inadequate rinsing or accidental drying during one of the incubation steps may cause contamination (Childs *et al.*, 1986). For additional information the reader is referred to the three volumes edited by Hayat (1989–90).

Considerations in the Use of Protein A–Colloidal Gold
The use of an aldehyde as the sole fixative is often necessary to adequately preserve the immunoreactivity. In the absence of post-fixation with OsO_4, the preservation of ultrastructure is less than satisfactory. This drawback can be minimised by adding ~50 mM lysine to the aldehyde fixative. When OsO_4 is used, thin sections usually are oxidised with sodium metaperiodate. However, this treatment should be carried out with care, since it may damage the ultrastructure. For post-embedding immunocytochemical techniques, Lowicryl resins give better results than epoxy resins.

Before applying protein A–gold to localise bound immunoglobulin, a test should be made to determine whether or not given cells or tissues contain components that bind gold (Behnke *et al.*, 1986). This can be carried out by using gold stabilised with PEG or Tween 20 at the same pH and ionic strength as those planned for the experiment. If such binding does not occur, the protein A–gold complex can be used. If such binding does occur to the putative 'naked' gold surfaces, cell components responsible for this binding are identified; an 'inert' competing protein (with a higher affinity for gold than for the components) should be determined and included in the incubation mixture. For example, gelatin has a higher affinity for gold than that shown by fibronectincollagen fibres in human fibroblast cultures (Behnke *et al.*, 1986).

In the preparation of protein A–gold complex, free protein is completely removed, to avoid competition between the complex and the free protein for the binding site during labelling. The presence of free protein results in a low efficiency. The specificity of antibodies as well as the purity of proteins (protein A, immunoglobulin and enzymes) to be tagged with gold particles significantly affect the quality of results, especially the level of non-specific labelling (Bendayan, 1986). During labelling, drying of thin sections and evaporation of the reagents should be avoided, to prevent artefactual adsorption and clustering of gold particles.

Background staining can be minimised by: (1) preincubating the tissue sections with a solution of albumin or Tween 20; (2) using Lowicryl instead of epoxy resin; and (3) performing all incubations by floating the grids on droplets of various reagents rather than immersing them (Bendayan, 1986). Antisera of high concentrations should be avoided, since they usually cause background staining, formation of electron-opaque precipitates on the section and clustering of gold particles. Since gold particles may aggregate during storage, the protein A–gold complex can be centrifuged at a low speed immediately before use.

Unsatisfactory fixation, embedding and sectioning can cause non-specific adsorption of gold particles. Cellular debris and sectioning imperfections may lead to unwanted adsorption of gold particles. Rinsing of thin sections during various steps of fixation and labelling should be sufficiently thorough to remove unbound antibody molecules as well as gold particles non-specifically adsorbed to thin sections (Bendayan, 1986). Too harsh a rinsing may dislodge specifically adsorbed gold particles.

Although gold particles of very small size (e.g. 5 nm) allow high resolutions and intense labelling, the choice of the size of the gold particles is determined by the objective of the study. Very small gold particles require visualisation at high magnifications, limiting the field of view. Gold particles ~15 nm in diameter facilitate easy identification of labelled structures at medium magnifications, allowing relatively large field of view (Bendayan, 1986). The resolution achieved also depends on the distance existing between the gold particle and the section surface.

Possible Limitations of the Protein A–Gold Method
The application of protein A–gold is based on the assumption that the protein is irreversibly adsorbed and that the surfaces of the gold particles are constantly and completely coated with the protein. If these conditions are not fulfilled, then uncoated gold surface may be

available for binding to proteins in cells and tissues (Behnke *et al.*, 1986). It has been indicated that stabilisation of gold sol with protein A against salt flocculation is not tantamount to saturation of the gold surface with the protein (Goodman *et al.*, 1980; Horisberger and Vauthey, 1984). It is possible that the stabilisation of gold sol occurs with only partial coating of the sol's surface. If uncoated gold surfaces on protein-stabilised gold particles are present, these surfaces will be negatively charged and might electrostatically bind to the cell and tissue proteins that carry a net positive charge at the pH and ionic strength used (Behnke *et al.*, 1986). Only some proteins may show such non-specific binding.

It is also thought that steric hindrance is the main mechanism responsible for the stabilisation of colloidal solutions (Heller and Pugh, 1960). This means that the extended molecule of protein A may be attached to gold particles at one end, with the other end protruding free into the solvent, preventing collision between the particles. That the same amount of protein stabilises colloidal solutions of widely different particle diameters corroborates the involvement of steric hindrance (Bontoux *et al.*, 1969).

The assumption that proteins are irreversibly adsorbed to the surface of gold particles has been questioned, on the basis that adsorbed proteins can be desorbed (Schwab and Thoenen, 1978; Goodman *et al.*, 1981; Warchol *et al.*, 1982). Since adsorption of proteins is pH-dependent, desorption of proteins might be similarly affected (Behnke *et al.*, 1986). The latter is possible because the adsorption is accomplished at ~pH 5.5–6.0, but the protein A–gold complex is stored and used at pH 7.0–7.2. However, protein A adsorption to gold particles can be carried out at pH 7.2 (Lucocq and Baschong, 1986).

Non-immunological binding of stabilised gold particles to cell and tissue components cannot be ignored (Behnke *et al.*, 1986). This non-specific binding seems to be independent of the particular stabiliser used. In other words, this binding is not due to protein–protein interactions, but rather is the result of an interaction between the gold particles and cell and tissue components, in which the stabiliser is not involved. This non-specific binding can be significantly reduced by including competing proteins that have a higher affinity for the putative 'naked' gold surfaces than that possessed by the cell and tissue components (Behnke *et al.*, 1986). These competing proteins do not appear to interfere with the specific protein–protein interactions. Gelatin is one such competing protein. Other limitations of colloidal gold are discussed elsewhere (Hayat, 1989b).

PROTEIN G

Protein G is a protein isolated from the cell walls of human group G streptococcal strain (G 148) (Björck and Kronvall, 1984). Like protein A, protein G binds specifically to the Fc fragment of IgG molecules (Åkerström and Björck, 1986). The binding properties and avidity of protein G for several polyclonal and monoclonal antibodies are superior to those shown by protein A (Reis *et al.*, 1984). Bendayan (1987, 1989a) introduced the application of protein G–colloidal gold to the ultrastructural localisation of antigens on tissue sections. Protein G molecules surrounding the gold particles bind to the Fc region of the IgG, and the gold particles enable the localisation of the antigenic sites on thin sections.

The specificity of protein G–gold is indicated by the observation that it is not adsorbed spontaneously to the surface of the tissue section and is unable to bind directly to the immunoglobulins included in the tissue. A specific primary antibody molecule is needed for obtaining an immunolabelling.

Procedure

Protein G is isolated from group G staphylococcal strain by enzymatic digestion with papain and purified by chromatography, and is commercially available (Amersham Corp., Arlington Heights, Ill.). The complex is prepared by dissolving 0.25 mg of protein G in 100 μl of double-distilled water. Exactly 10 ml of the gold (14 nm) suspension (pH 5.0) is added to this protein solution with continuous stirring. After 2 min, 100 μl of 1% polyethylene glycol (mol. wt. 20 000) is added and the mixture is centrifuged at 25 000 rev/min for 30 min in a Beckman ultracentrifuge, using the Ti-50 rotor (Bendayan, 1987). The clear supernatant containing the free protein G is discarded, and the dark-red sediment is resuspended in 1 ml of 0.01 M PBS (pH 7.2) containing 0.2 mg/ml PEG. This protein G–gold complex can be stored at 4 °C.

Thin sections of the aldehyde-fixed tissue are mounted on nickel grids, which are floated for 5 min on drops of PBS containing 1% ovalbumin. They are incubated on drops of the specific antibody for 2 h, rinsed in PBS to remove unbound antibody molecules and then transferred to the PBS–ovalbumin solution for 5 min. The grids are incubated on drops of the protein G–gold complex for 30 min. The complex has an optical density of 0.5 at 525 nm. After incubation, the grids are thoroughly washed with PBS, rinsed in distilled water and air-dried. The grids are post-stained with salts of uranium and lead. For further details, see Bendayan (1989a).

MULTIPLE IMMUNOGOLD STAINING METHODS

Multiple immunostaining at the subcellular level is a revealing method, especially in studying antigen colocalisation. More than one type of cellular antigen can be identified simultaneously on the same tissue section (Fig. 4.6). This is possible because of the availability of colloidal gold particles of varied dimensions. Double labelling with protein A–gold complexes containing gold particles of different sizes can be performed on the same surface of the section. However, since protein A reacts only with Fc fragments of IgG molecules of many species and Fab, interactions of both labellings can occur, leading to codistribution of both small and large gold particles. This confusion is avoided by performing each labelling separately on each of the two surfaces. Since the reaction of protein A with IgG (Fc fragments) of some species (e.g. sheep and goat) is considerably weaker than that with IgG of other species, in such instances it seems advantageous to use protein G–gold probes (Bendayan, 1989a). Each of the following four methods has advantages and limitations, and the reader can choose the one that suits the objective of the study.

(1) When the primary antisera are raised in different species, the following conventional method can be used (Varndell and Polak, 1984). Thin sections mounted on grids may be etched with 10% solution of H_2O_2 for 10 min. This treatment is supposed to permeabilise the epoxy resin to facilitate antibody penetration. However, since H_2O_2 tends to damage the ultrastructure, this treatment should be avoided by using incompletely cured resin blocks (12–18 h cure at 60 °C). The grids are thoroughly rinsed in filtered (0.45 μm pore size) distilled water, and transferred onto droplets of normal goat serum (1:30 dilution in PBS containing 0.1% BSA and 0.01% sodium azide, pH 7.2) for 30 min. After removing the normal goat serum with a piece of filter paper, the grids are incubated in droplets of primary antisera for ~1 h at room temperature. A mixture of two antisera (one raised in a rabbit and the other in a guinea-pig) is used; the antisera should not cross-react. The final concentrations of the two antisera are equal to their predetermined optimal titre.

The grids are thoroughly rinsed with 50 mM Tris buffer (pH 7.2), followed by three changes for 15 min each with 50 mM Tris buffer containing 0.2% BSA. They are incubated in a mixture of gold-labelled anti-primary species immunoglobulin at optimal titres for 1 h. Typically, anti-guinea-pig immunoglobulin labelled with gold (20 nm) and goat anti-rabbit immunoglobulin labelled with gold (10 nm or 40 nm) can be used. The grids are thoroughly rinsed with Tris buffer containing 0.2% BSA, followed by Tris buffer and finally distilled water. The grids can be post-stained with salts of heavy metals.

(2) When the primary antisera are raised in the same species, the simplest method is the two-side immunolabelling introduced by Bendayan (1982). Each surface of a tissue section mounted on a grid (without a support film) is stained separately, using colloidal gold particles of two different sizes following incubation with a specific antibody. The section is sufficiently thin to allow simultaneous visualisation of the gold particles present on both surfaces. Caution must be observed to prevent sinking of the grid and wetting both sides of the section. Such a wetting is possible when long incubations are required. The staining of the first face of the sections presents no problems at room temperature, but low temperatures can cause water condensation on the upper surface of the grid. The staining of the second face of the section may present difficulties because of its contamination during the first half of the immunostaining method. A schematic description of the principles of the two-side method is presented in Fig. 4.7.

Procedure 1 (Bendayan, 1982) Face A (Fig. 4.7) of the section is exposed to 0.5–1% ovalbumin in PBS for 5 min. The grid (face A down) is transferred to a drop of the first antibody solution and incubated for 1 h at room temperature. The grid is rapidly rinsed by floating it successively on several drops of PBS, and then incubated (face A down) on a drop of protein A–gold (~19 nm particle size) solution for 30 min at room temperature. The grid is thoroughly washed in PBS, rinsed in distilled water and air-dried. The grid is turned over (Fig. 4.7) and face B is incubated with 0.5% or 1% ovalbumin in PBS for 5 min. The grid (face B down) is transferred to a drop of a second antibody and incubated for 1 h at room temperature. The grid is rinsed on several drops of PBS and incubated (face B down) on a drop of the protein A–gold (~12 nm particle size) solution for 30 min. The grid is thoroughly washed in PBS, rinsed in distilled water and dried. Care should be taken not to wet the opposite surface during the incubations. Counterstaining is carried out with uranyl acetate followed by lead citrate on face A of the sections.

The two-side method described above can be modified to demonstrate three antigenic sites (Doerr-Schott and Lichte, 1986). It is necessary that antisera be used at maximal dilutions. First, face A is incubated with a mixture of two antisera and then with a mixture of two corresponding species-specific immunoglobulins adsorbed, respectively, to 5 nm and 20 nm gold particles. Second, face B is incubated with a third antiserum

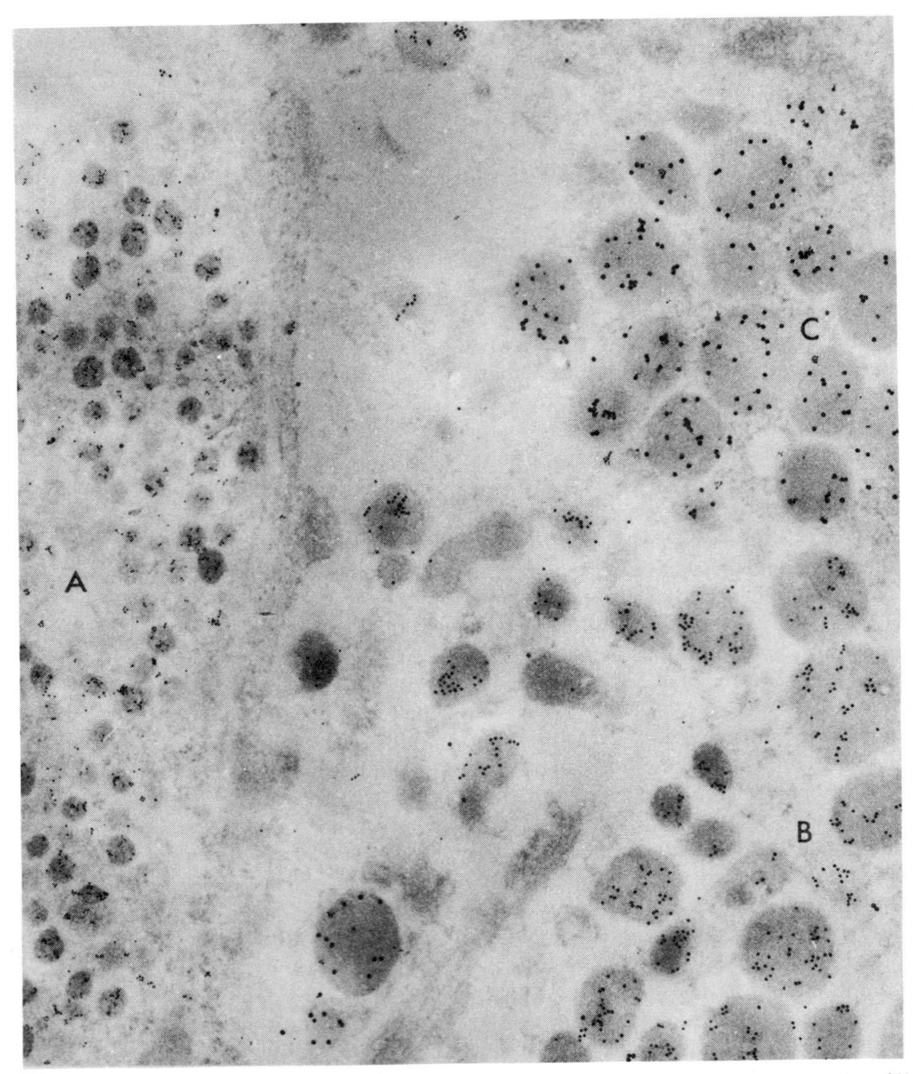

Figure 4.6 Simultaneous demonstration of three antigens on the same thin section of the amphibian (*Xenopus laevis*) hypophysis. Three cell types that synthesise three different hypophyseal hormones are seen. Labelling is specific and cross-reaction between the primary antisera and the non-corresponding secondary labelled antibodies is absent. A: ACTH cells containing small secretory granules (100–200 nm in diameter) labelled with 5 nm gold particles. B: LH cells containing irregularly shaped secretory granules (200–600 nm) revealed with 10 nm gold particles. C: LTH cells containing round secretory granules (200–800 nm) covered with 20 nm gold particles. 31 200×. Courtesy J. Doerr-Schott

and then species-specific secondary antibodies adsorbed to 10 nm gold particles. The results of this triple immunogold staining are shown in Fig. 4.6. For additional details, see Doerr-Schott (1989).

(3) As stated earlier, in the two-side method contamination or cross-reactivity of the second surface of the sections may be a problem. Such a cross-reaction, even when the grid becomes submerged, can be minimised by coating the section surface, after the first antigen has been stained, with a thin film of carbon (Beesley *et al.*, 1984; Powell, 1987). In other words, the side of the section on which the first immunolocalisation has been accomplished is blocked by a carbon film. However, this method still risks some contamination of the second surface with the first antibody and may result in false-positive localisations. This problem is avoided in method 4, given later.

Procedure 2 (Powell, 1987) All incubations are carried out on pieces of sealing film on top of damp filter paper inside a Petri dish. Thin sections are mounted on a nickel grid (without a support film). The grid is

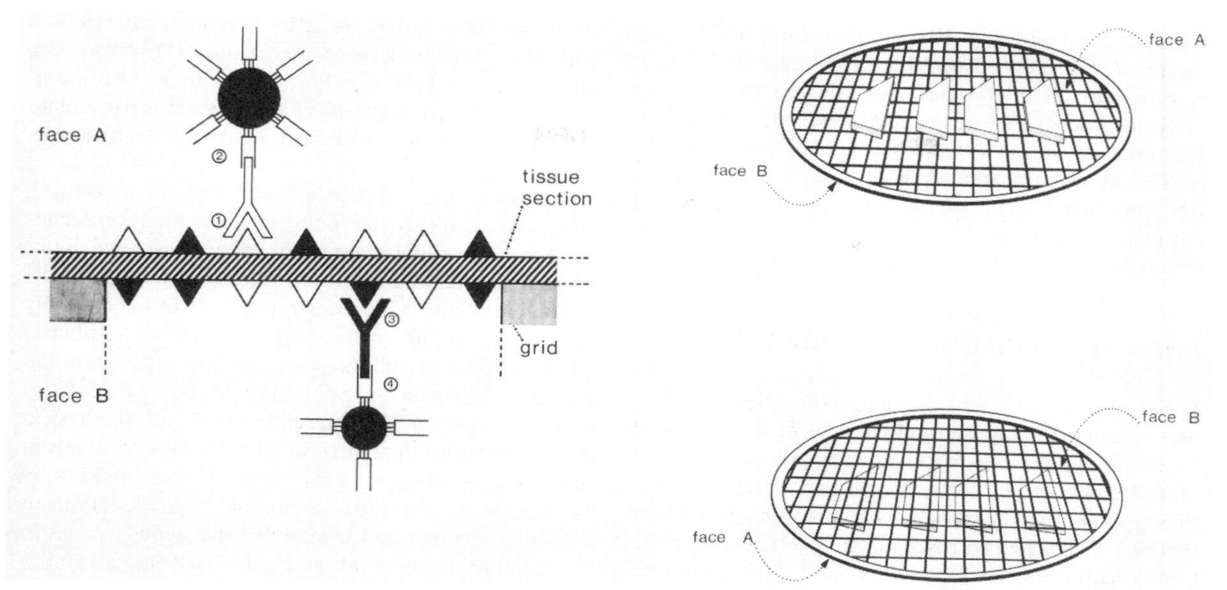

Figure 4.7 Schematic illustration of the principles of the double-labelling, two-side protein A–gold method. In the first step (1), face A of the tissue section is exposed to a specific antiserum. The immunoglobulin (IgG) interacts specifically with the antigen present at the surface of the section. In the second step (2), protein A molecules surrounding the large, spherical gold particle interact with the Fc portion of the same IgG molecule. In the third step (3), face B of the same tissue section is exposed to a specific antiserum and the IgG molecule interacts with its specific antigen. In the fourth step (4), protein A molecules surrounding the smaller gold particle interact with the Fc portion of the second IgG. With this indirect method the gold particles of each protein A–gold complex enable the localisation of their respective antigenic sites. From Bendayan (1982)

incubated by floating (section side down) on a 20 μl drop of 5% normal serum diluted in 0.05 M Tris buffered saline containing 0.1% BSA (TBS–BSA; pH 8.2) for 30 min. The serum is drained off by touching the edge of the grid with a piece of filter paper, and the grid is transferred to a drop of primary antibody diluted with 1% normal serum in TBS–BSA and incubated overnight. The grid is rinsed on drops of TBS–BSA.

The grid is incubated on a drop of the colloidal gold conjugated secondary antibody diluted 1:20 with TBS–BSA for 2–4 h. It is rinsed in a series of drops of distilled water. After drying, the grid is attached (immunostained side up) to a glass slide by sticking the edges to a piece of double-sided sticky tape. The grid is coated with a thin film of carbon (5 nm), which shows a light buff colour on a piece of broken, white tile. A carbon fibre evaporation device gives a reproducible thin film. After separating the grid from the slide, the second side of the section is treated the same way as the first side, using secondary antibody labelled with gold particles of a different size. Incubations are carried out at room temperature, because damp filter paper is necessary to provide a humid atmosphere which prevents reagents drying out. Copper grids should not be used. Although the grid can be post-stained with heavy metals, the side of the section coated with carbon will not take up the staining solution.

(4) The contamination of the upper surface of a section with the inappropriate antibody can be prevented by coating one side of the grid with a Formvar film before incubating the other side (Shore and Moss, 1988). This film is then removed with a solvent, and a second Formvar film is coated on the reverse side to protect the first immunolocalisation before immunostaining the second antigen. The embedding resin used should be resistant to chloroform (e.g. Lowicryls). The possible adverse effect of chloroform on the physical properties of the resin or on the antigenic sites is not clear. Although this technique should give good results using protein A–gold with the second antibody, only gold-labelled specific secondary antibodies against the two primary antibodies applied on either side of the section have been used.

Procedure 3 (Shore and Moss, 1988) Thin sections are mounted on a Formvar-coated nickel grid. Incubation in the first antibody solution and treatment with protein A–gold are accomplished by immersing the grid (section side up). The grid is rinsed with distilled water and dried on a filter paper. The Formvar film is removed by briefly immersing twice the grid in 100% ethanol in a vial for a total period of 10 s, in chloroform for 30 s and then twice in 100% ethanol for a total period of 20 s. It is rinsed in distilled water and allowed to dry. The opposite side of the grid is coated with Formvar film. Immunolocalisation for the second antigen is accomplished by using protein A–gold of a different particle size from that used above for the localisation of the first antigen. If necessary, the second Formvar film can be removed and the sections post-stained.

IMMUNOGOLD–SILVER METHOD

Silver enhancement of colloidal gold probes is a sensitive immunostaining method capable of detecting low levels of immunoreactivity on conventional thin sections (Lackie *et al.*, 1985; Scopsi and Larsson, 1985), thin cryosections (Bastholm *et al.*, 1986), semithin resin sections (Danscher and Nörgaard, 1983), paraffin sections (Scopsi and Larsson, 1986) and thick tissue sections (30–50 μm) before they are embedded in resin for electron microscopy (van den Pol, 1985). Use of small colloidal gold particles coupled with silver enhancement is a valuable tool for pre-embedding labelling studies (van den Pol, 1986). Since small gold particles cannot be detected directly in the light microscope, a silver physical enhancement is necessary for visualising immunogold-stained cell compartments (Holgate *et al.*, 1983). Silver-enhanced colloidal gold complexes have also been used as markers for both scanning electron microscopy (Scopsi *et al.*, 1986) and photoelectron microscopy (Birrell *et al.*, 1986). The immunogold–silver staining allows correlative localisation studies with light and electron microscopy.

As stated above, the advantage of labelling with colloidal gold particles of small size (5–10 nm in diameter) is the achievement of relatively high labelling efficiency and density. However, a disadvantage of using such particles is that they require comparatively high magnifications, to be visualised. Furthermore, these particles are difficult to visualise when thin sections are post-stained with salts of uranium and lead. These disadvantages can be circumvented by silver enhancement. Increasing the size of small gold particles with silver enhancement enables them to be detected at low magnifications. With controlled silver enhancement, the size of the colloidal gold particles can be increased to the desired dimension. A beneficial side-effect of the silver treatment is increased positive staining of certain proteins.

It appears that the silver precipitated around the colloidal gold particle covers not only the gold, but also the adsorbed protein A or immunoglobulin, thus reducing or eliminating further cross-reaction when the silver–gold method is used for the first antigen and non-intensified horseradish peroxidase (HRP) for the second antigen (van den Pol, 1986). Silver also intensifies HRP reaction product, and gives it a particulate quality, recognisable with both light and electron microscopy.

The gold–silver method provides superior resolution compared with that yielded by the immunofluorescence method. The silver staining of sections is permanent, whereas immunofluorescence fades with time.

The possibility of demonstrating gold in the tissue by physical development has long been known (Roberts, 1935). This silver enhancement method is based on the reaction employed for photographic physical development (James, 1977). Danscher (1981) used physical development for detecting minute amounts of metals in biological specimens. This method was applied by Holgate *et al.* (1983) to immunogold-stained sections to achieve a strong intensification of the signal. It uses in its original form a photographic developer and a silver donor—silver ions dissolved from the silver halide in the photographic emulsion. During development, the latent image in the photographic film catalyses the precipitation of silver atoms on its surface, and it thereby becomes the final, developed silver grain (Bienz *et al.*, 1986). When this procedure is applied to the immunogold method, the colloidal gold serves as catalyst, and so the size of the gold particle (label) enlarges during the development. After the gold particles are deposited at the site of immunoreactivity, they catalyse the reduction of silver nitrate to metallic silver by depositing a shell of metallic silver around them. This results in the enlargement of the size of the gold particles (Fig. 4.8). The metallic silver shell formed around the gold particles also catalyses this reduction; thus, the reaction is autocatalytic.

The silver deposition is determined by the number of gold particle nuclei per unit area and not by their mass. Thus, the size of the gold particles has little effect on the progress of silver deposition.

The developer contains silver ions (silver nitrate or silver lactate) and a reducing substance (hydroquinone), has a pH of 3.5–3.8, and usually contains a protective colloid such as gum arabic, polyethylene glycol, polyvinylpyrrolidone or dextran (Scopsi and Larsson, 1986). The nitrate salt of silver is preferred because of its high dissociation coefficient, and, hence, rapid rate of reaction. The colloid inhibits the interaction between silver ions and hydroquinone, allowing fine control of the reaction product. Such control seems to be necessary because the reaction occurs quickly (3–6 min) and the interval between optimal specific staining and undesirable background is very short. However, some workers avoid using protective colloids

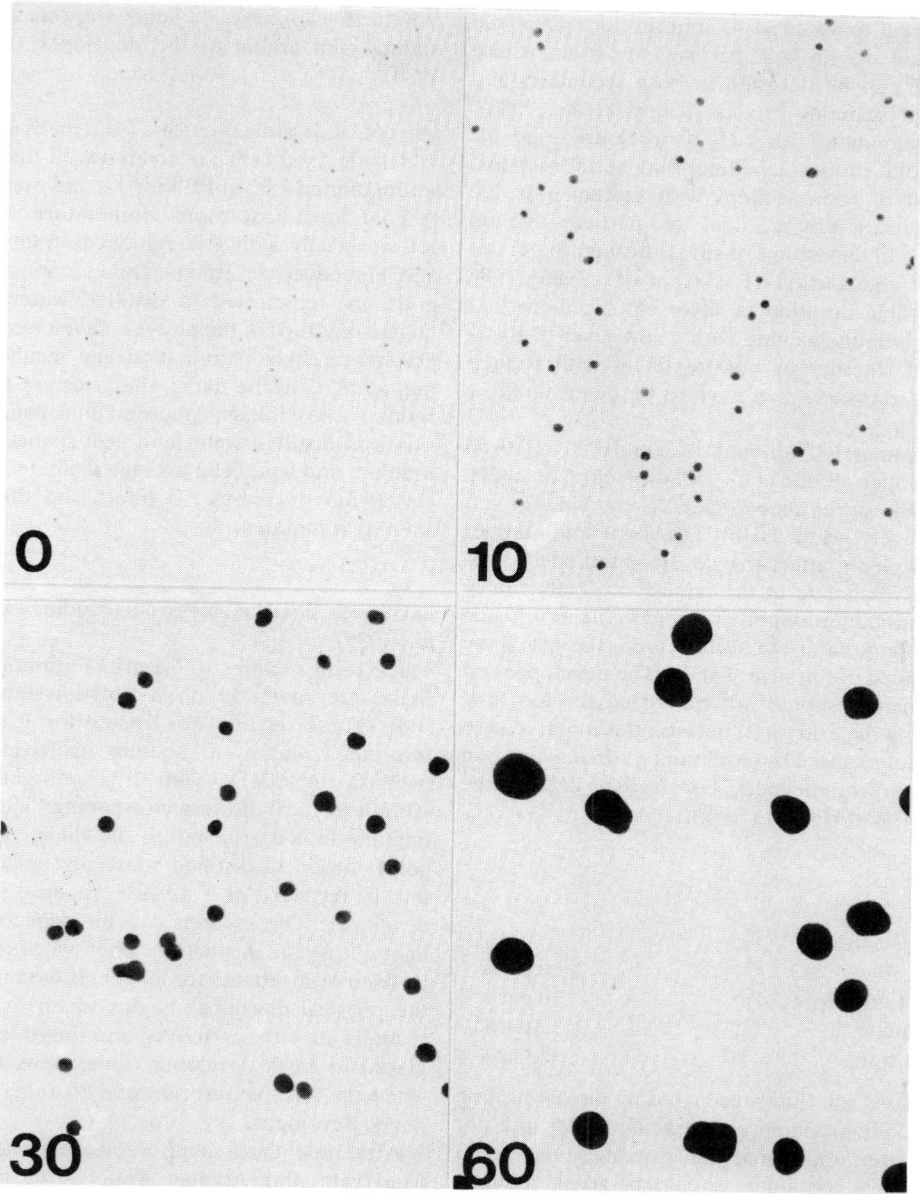

Figure 4.8 IgG-coated colloidal gold particles of 5.7 nm in diameter spread on a Formvar and poly-L-lysine-coated nickel grid. Silver enlargement by physical development in the silver lactate–gum arabic developer for 10, 30 and 60 min. Courtesy L. Scopsi

(Hacker *et al.*, 1985). Non-specific precipitations from the developer can be minimised by using double-glass-distilled water and coating the slides (for light microscopy) just before development with a thin layer of gelatin (Danscher and Nörgaard, 1983); gelatin can be removed by a prolonged rinse in warm, running tap-water.

The enhancement of staining by silver is due to the increased size of individual gold particles. This increase is progressive and linear with the duration of development. In other words, the longer the silver development, the larger the size of silver deposits (Fig. 4.8). To obtain a reproducible and predictable increase in particle size, a relatively slow development speed should be used, which minimises the effect of inevitable small technical fluctuations (Bienz *et al.*, 1986). A 2–3-fold increase in the size of the gold particle (5–10 nm) seems desirable for most studies.

Development times of 30–45 min produce 40–50 nm particles when ~5 nm gold particles are used, whose strong signal can be detected in both secondary and backscattered scanning modes (Scopsi *et al.*, 1986). Shorter development times (5–10 min) are used for enlarging gold probes labelling thin resin sections. Pretreatment of resin sections with sodium ethoxide aids the penetration by colloidal gold particles of 5 nm in diameter and deposition of silver through the entire thickness of the section (Lackie *et al.*, 1985). The shortest possible duration of silver enhancement that yields clear immunostaining with a low level of background, and whether or not treatment with sodium ethoxide is necessary, may have to be determined by trial and error.

The recommended amount of emulsion is 20–50 mg/ml developer (Bienz *et al.*, 1986). Below or above this range, the development speed is too slow or too fast, respectively, to be useful. The shape and number of emulsion pieces affect the development speed, because the silver halide in the emulsion is only slowly soluble in the sodium sulphite present in the developer. Therefore, the greater the surface area, the faster the dissolving speed of the silver halide. The developer and emulsion mixture should not be stirred for too long before placing the grid on it, because this might lead to irregular results caused by oxidation; a stirring duration of 5 min is recommended. For further details, see Scopsi (1989) and Hacker (1989).

Procedures

Physical developer

Gum arabic (50%)	60 parts
Citrate buffer (pH 7.4)	10 parts
Hydroquinone	15 parts
Silver nitrate	15 parts

The gum arabic solution is prepared by dissolving 1 kg of commercially prepared gum arabic powder in 2 l of deionised water under magnetic stirring. Only clean glass or plastics containers should be used; metallic containers and tools should be avoided. Several hours later the solution is filtered through multiple layers of gauze and stored in plastics tubes at −20 °C. Citrate buffer is prepared by dissolving 25.5 g of citric acid and 23.5 g of trisodium citrate (dehydrated) in deionised water to make up 100 ml. Hydroquinone solution is prepared by dissolving 0.85 g of hydroquinone in 15 ml of deionised water. Silver nitrate solution is prepared by dissolving 0.11 g of silver nitrate in a final volume of 15 ml deionised water. The silver nitrate is highly sensitive to light and is added immediately before use in the dark. The physical developer is filtered through Whatman paper No. 4. Some workers advise against adding gum arabic to the developer (Birrell *et al.*, 1986).

PROTOCOL 1: THIN SECTIONS (Marchetti *et al.*, 1987)
Aldehyde-fixed cells are treated with the specific antiserum diluted 1:30 in PBS for 1 h and then with protein A–gold for 1 h at room temperature. The cells are refixed briefly with glutaraldehyde followed by OsO_4 and embedded in Epon. Thin sections mounted on grids are rehydrated in distilled water for 15 min, floated on drops of the physical developer under photographic safelight illumination and incubated for 5–15 min at 28 °C in the dark. The grids are fixed with 5% sodium thiosulphate (photographic fixer) for 1 min, rinsed in distilled water and post-stained with salts of uranium and lead. The average diameter of gold particles (5 nm) increases ~3–6-fold and the background staining is minimal.

PROTOCOL 2: THICK SECTIONS (Modified from Lackie *et al.*, 1985)
Thick resin sections (0.5 μm) of glutaraldehyde-fixed tissue are mounted on a poly-L-lysine-coated glass slide. These sections are treated for 1–10 min with a saturated solution of sodium hydroxide in ethanol (sodium ethoxide) or with 10% hydrogen peroxide for 10 min to facilitate immunostaining. The sections are immunostained (including labelling with colloidal gold), rinsed in distilled water and incubated for ~7 min in the dark in a freshly prepared silver physical developer. The sections can be inspected under dim light during the incubation, after which they can either be fixed or incubated for longer. If the brown colour of the physical developer begins to turn grey or black, sections are either removed and rinsed in buffer or are placed in fresh developer. Overintensification occurs when the sections are allowed to remain in the darkened developer.

After adequate silver intensification, sections are fixed with photographic Amfix for 5 min, rinsed in buffer and post-osmicated. After rinsing in distilled water, they are mounted in PBS–glycerine (1:1) for easy re-embedding for thin sectioning. The sections on the slide are dehydrated and embedded in a resin by placing a drop of it on the slide. The sections are separated from the slide and embedded or re-embedded by the pop-off method for thin sectioning.

After incubation for 7 min, the volume of silver deposited around 5 nm gold particles is ~5000 times that of the original gold particle. This increase is significantly more than that observed by Marchetti *et al.* (1987).

IMMUNOGLOBULIN–COLLOIDAL GOLD METHOD

The first application of colloidal gold as a label in immunocytochemistry was carried out in combination with immunoglobulins (Faulk and Taylor, 1971). The preparation of immunoglobulin–gold complex is more complex than that of protein A–gold complex, for polyvalent immunoglobulins consist of a mixture of molecules with a wide range of isoelectric points (Slot and Geuze, 1984). The variation in the isoelectric point becomes a hindrance in determining the optimal pH for the preparation of the complex. The following procedure is based on recommendations by De Mey *et al.* (1981) and Slot and Geuze (1984).

Exactly 4 ml of antiserum is adsorbed to 2 ml of CNBr-activated Sepharose gel, to which rabbit immunoglobulin has been bound, for 2 h at room temperature. The gel is washed with PBS and anti-rabbit immunoglobulin is eluted with 4 ml of 3 M KCNS. The protein concentration is adjusted to 1 mg/ml in 3 M KCNS, and the solution is dialysed against 2 mM borax (pH 9.0) at room temperature. Immunogold preparations readily form aggregates, some of which precipitate after centrifugation at 50 000 *g* for 30 min, but the smaller ones remain in suspension. The degree of aggregate formation depends on the immunoglobulin used. The aggregation can be minimised by diluting the immunoglobulin before dialysis or by slightly raising the pH of the borax solution.

Coupling is carried out by adding a sufficient amount of immunoglobulin to 30 ml of gold sol (12 nm) at pH >9.0 for the concentration of the former to exceed the stabilisation point by 10%. After 1 min, a sufficient volume of 10% aqueous bovine serum albumin is added to obtain a final concentration of 0.25% to stabilise gold particles. This suspension is centrifuged at 50 000 *g* for 45 min. The pellet consists of a large loose portion and a small compact portion. The supernatant is removed without disturbing the pellet, and the loose portion of the pellet is resuspended in the residual supernatant. This is layered over a 10–30% continuous sucrose or glycerol gradient (volume 10.5 ml, length 8 cm) in TBS (0.01 M Tris, 0.15 M NaCl, pH 8.2). The gradient is centrifuged in a SW 41 rotor (Beckman) at 20 000 rev/min for 30 min. The immunoglobulin–gold complex can be stored for months with 0.02% sodium azide at 4 °C. For longer storage, it can be frozen in small samples at −70 °C.

Immunolabelling of Thin Cryosections (Geuze *et al.*, 1981)

Thin sections mounted on carbon-coated copper grids are exposed to 2% gelatin in PBS for 10 min at room temperature; all other treatments are also carried out at room temperature. The grids are treated with 0.02 M glycine in PBS for 10 min, and then with a pure preparation of immunoglobulin (~25 μg/ml) in drops as small as 5 μl for 30 min. They are rinsed in four changes of PBS for a total period of 4 min, exposed to the immunoglobulin–gold complex (which has been diluted just before use with PBS and 1% bovine serum albumin to a protein concentration of 0.05%) for 30 min, and again rinsed in four changes of PBS for a total period of 20 min. Procedures for post-staining and embedding are given below (Tokuyasu, 1980a; Griffiths *et al.*, 1982).

Grids are rinsed in three changes of distilled water, and stained for 10 min with 2% uranyl acetate in 0.15 M oxalic acid brought to pH 7.0 with 5% NH_4OH. After rinsing three times in distilled water, they are again stained with 3% uranyl acetate for 10 min. The grids are rapidly passed over three drops of 1.5% methylcellulose in distilled water; this solution is prepared by dissolving 1.5 g of methylcellulose in 100 μl of distilled water for 24 h or longer at 4 °C and stored in the refrigerator. The grids are picked up from the drops within 30 s with a wire loop, and the excess solution is removed by touching them with a piece of filter paper. The air-dried grids should show gold or gold-blue colour. For further details, see van Bergen en Henegouwen (1989).

LECTIN–COLLOIDAL GOLD COMPLEX

Lectins are valuable tools for the localisation of sugar sequences in oligosaccharide units of glycoconjugates at the light and electron microscope levels. Some lectins can be used to determine differences in the carbohydrate composition of normal and malignant cells. Although lectins have been widely used for cell surface studies and internalisation experiments, their use for the localisation of intracellular binding sites is limited, owing to their poor penetration into cells and tissues. Lectin-binding sites in thin sections can be detected by using either lectin–gold complexes directly (Roth and Binder, 1978) or native lectins followed by glycoprotein–gold complexes. The use of the indirect label has the advantage of facilitating the unimpeded interaction of an unconjugated macromolecule with the cell surface (Geoghegan and Ackerman, 1977). Upon binding, the macromolecule can protrude like an arm from its binding site, providing greater stereochemical accessibility to the gold–protein, which can lead to greater sensitivity (Temmink *et al.*, 1975).

Since lectins are neither enzymes nor electron-opaque, they must be conjugated to an electron-opaque molecule or a metal particle (e.g. gold) for use in electron microscopy. Alternatively, lectins can be conjugated to enzymes (e.g. HRP) that produce electron-opaque reaction products. Colloidal gold, with a relatively large surface area, adsorbs irreversibly a wide variety of lectins through a non-covalent process (Horisberger, 1981). Lectins adsorbed onto gold particles retain their sugar-binding capacity, although their precise specificity may be modified (Debray *et al.*, 1983). Lectin–gold complexes have a general application to the localisation of macromolecules.

Gold particles labelled with lectins can be used in pre-embedding or post-embedding techniques. In the former the penetration of gold–lectin complex becomes a problem, because of steric hindrance when gold particles of relatively large size are used. In this case a meaningful relationship between the number of lectin-binding sites and the number of bound gold particles is lacking, because the accessibility of gold particles is strongly regulated by their size (Horisberger, 1984). The binding of lectin-labelled gold particles to lectin-binding sites also depends on ionic strength and pH. In general, the binding density tends to increase when the size of the gold particle is reduced (Horisberger and Vonlanthen, 1979). Both in the pre-embedding and post-embedding techniques, almost total inhibition of binding can be achieved by adding specific monosaccharides to the lectin-labelled gold marker. However, in the case of concanavalin A and wheat germ lectin, total inhibition is achieved only when specific polysaccharides are incorporated (Nurden *et al.*, 1980). Properties of commonly used lectins and their varied applications are described elsewhere (Hayat, 1989b).

To prepare the complex, 8 ml of gold sol (7 nm particle size) is added to 1 ml of concanavalin A (conA) solution (20 μg/ml of distilled water) and the mixture is stirred for 5 min at room temperature. To this mixture is added 1 ml of polyethylene glycol (0.1% w/v), and the mixture is stirred for 5 min and then centrifuged at 100 000 *g* for 30 min at 4 °C. The supernatant is discarded and the pellet is resuspended 1:1 in 0.9 M mannitol in 10% 'White's stock solution' containing 0.1 M Na–phosphate buffer (pH 7.4). This is the lectin–gold solution for use. Lectin–gold particles can be checked for effective conjugation by negative staining. For further details of the colloidal gold–lectin methodology, see Benhamou (1989).

LECTIN–HORSERADISH PEROXIDASE–COLLOIDAL GOLD METHOD

The electron-opaque reaction product formed with the lectin–HRP–diaminobenzidine method is flocculated rather than particulate, and while capable of showing relative surface binding characteristics, is incapable of showing subtle differences in the concentration of lectin-reactive sites on plasma membranes (Ackerman and Freeman, 1979). Although the ferritin-conjugated lectin method allows the direct demonstration of lectin-reactive sites seen as fine particles at high magnifications, permitting the quantification of these sites, covalent bonding between lectin and ferritin appears to alter the physicochemical binding characteristics of the lectin. As a result, the lectin–ferritin complex may not have access to some of the lectin receptor sites on the plasma membrane, owing to steric hindrance (Temmink *et al.*, 1975).

The above-mentioned limitations can be avoided by using colloidal-gold-labelled HRP as an indirect marker for lectin-binding sites on the plasma membranes (Geoghegan and Ackerman, 1977). Binding of colloidal gold to HRP is electrostatic and the reaction product is particulate and larger than ferritin, allowing quantification at relatively low magnifications. Also, the lectin molecule (con A) is unmodified when reacted with the plasma membrane. Horseradish peroxidase–colloidal gold complex is prepared as follows (Ackerman and Freeman, 1979).

About 8 mg of HRP is dissolved in 5 ml of triple-distilled water, which is then added dropwise with gentle stirring to 200 ml of colloidal gold. The resulting complex is mixed for 1 min and 2 ml of 1% aqueous polyethylene glycol (mol. wt. 20 000) is added to prevent aggregation. This complex is centrifuged at 1500 rev/min for 20 min to remove any large aggregates. The sediment is discarded and the supernatant is centrifuged at 11 500 rev/min for 1 h at 4 °C. The supernatant is discarded and the red HRP–gold pellet is suspended in 0.1 M phosphate buffer (pH 7.4) with 4% polyvinylpyrrolidone (mol. wt. 10 000) and 0.2 mg/ml polyethylene glycol and centrifuged at 11 500 rev/min to remove any unlabelled HRP from the HRP–gold complex. The supernatant is discarded and the HRP–gold complex is diluted to 10 ml in phosphate buffer containing polyvinylpyrrolidone and polyethylene glycol.

LECTIN–HORSERADISH PEROXIDASE–COLLOIDAL GOLD–RUTHENIUM RED METHOD

Two- and three-dimensional distribution of cell surface receptors can be demonstrated in thick sections by employing the lectin–gold–ruthenium red method in conjunction with high-voltage electron microscopy (Takata *et al.*, 1984). In the image obtained, colloidal gold particles represent lectin-binding sites and ruthenium red enhances the staining of the cell surface, including the plasma membrane. In the absence of staining with ruthenium red, the plasma membrane is not easily discerned. Lectin-binding sites are randomly distributed on the cell surface at 0 °C, whereas the redistribution and endocytosis of lectin–gold complexes occurs at 37 °C (Takata *et al.*, 1984).

Macrophages are suspended in Eagle's minimum essential medium and allowed to adhere to plastics tissue culture dishes (35 mm) (Takata *et al.*, 1984). Concanavalin A binding sites are visualised by the indirect colloidal gold labelling method (Ackerman and Freeman, 1979; Benhamou, 1989). Briefly, cells are washed with PBS and incubated with con A (50 μg/ml) in PBS for 1 h at 0 °C. After washing with PBS, the cells are incubated for 1 h at 0 °C in an HRP–colloidal gold medium. The cells are washed with PBS containing 4% polyvinylpyrrolidone and then with PBS alone. The cells are fixed with 1.2% glutaraldehyde in 0.067 M PIPES buffer (pH 7.3) containing ruthenium red (0.5 mg/ml) for 1 h at 4 °C. After washing with 0.15 M PIPES buffer, the cells are post-fixed for 30 min at room temperature with 0.67% OsO_4 in 0.067 M PIPES buffer containing ruthenium red (0.5 mg/ml). The cells are thoroughly washed with PIPES buffer and dehydrated, after which they are detached from the culture dish by adding propylene oxide and embedded. Thick sections (0.1–3.0 μm) are viewed without counterstaining.

ENZYME–COLLOIDAL GOLD METHOD

Enzyme–gold complexes can be used to localise various cellular or tissue substrates in thin sections of glutaraldehyde- and OsO_4-fixed and resin-embedded tissues (Bendayan, 1981). It is a direct, one-step post-embedding method. Since this method is based on the affinity existing between an enzyme and its substrate, the results are specific for a given substrate. It appears that during incubation the enzyme molecules at the surface of the gold particles interact with their specific substrate molecules exposed at the surface of thin sections. This is possible because the activity of the enzyme molecules adsorbed onto gold particles is retained. Although the exact mechanism responsible for the attachment of the enzyme–gold complex to the substrate is not known, the following comments may be of some help. Several types of interactions can occur between the enzyme-active sites and the substrate molecule. These interactions include electrostatic interactions, hydrogen bonding and cohesion between non-polar portions of the enzyme molecule and similar regions on the substrate surface (Hirs, 1964).

The activity of the enzyme molecule is affected by the pH, and so labelling must be carried out at pH values compatible with the activity of the enzyme used. The detection of some substrates is strictly dependent on the compatible conditions of fixation and embedding. RNase– and DNase–gold complexes have been used for localising nucleic acids (Bendayan, 1981). Other macromolecular systems that have been localised with this method include elastin, collagen, glycogen, and xylan and chitin proteins (Bendayan, 1984b).

Preparation of Enzyme–Gold Complex (Bendayan, 1984b, 1989b)

To prepare an enzyme–gold complex, 0.1–0.5 mg of enzyme is dissolved in 0.2 ml of double-distilled water in a centrifuge tube and mixed with 10 ml of the colloidal gold suspension, which has been brought to the optimal pH. The optimal pH is determined by the isoelectric point of the enzyme. The enzyme adsorption to the surface of the gold particles occurs upon mixing. Two drops of a 1% solution of polyethylene glycol (mol. wt. 20 000) may be added to further stabilise the colloidal gold. The complex is centrifuged at 25 000 rev/min with a Ti-50 rotor for 30 min at 4 °C, to remove the excess unbound enzyme. Three phases are obtained in the tube: a clear supernatant containing free enzyme; a dark-red sediment at the bottom, which is the enzyme–gold complex; and a black spot along the side near the bottom, which is metallic gold not stabilised by the enzyme.

The supernatant must be completely removed, for it contains free enzyme which, if left, will compete with the enzyme–gold complex during labelling. The enzyme–gold complex is recovered and resuspended in 1.5 ml of 0.01 M PBS (Na_2PO_4/Na_2HPO_4, NaCl: 0.14 M) containing 0.02% polyethylene glycol; the pH is adjusted to the value of the optimal enzyme activity. The metallic gold not stabilised by the enzyme remains at the bottom of the tube. The stock solution of enzyme–gold complex can be stored for up to 10 days at 4 °C; the enzyme activity decreases with time. It can be diluted immediately before use.

Labelling (Bendayan, 1984b)
Thin sections of glutaraldehyde- and OsO_4-fixed tissues are mounted on nickel grids. To unmask the substrate, the grids are floated on a drop of saturated solution of sodium metaperiodate for 1 h at room temperature and then rinsed in distilled water followed by 0.01 M PBS (pH is adjusted according to the enzyme under study). Oxidation with sodium metaperiodate is not done when the tissue has not been post-fixed with OsO_4. The grids are transferred onto a drop of the enzyme–gold complex for 1 h at room temperature. Duration and temperature of incubation may vary, depending on the enzyme–substrate system, and the fixation and embedding procedures used. After incubation, the grids are thoroughly rinsed with PBS followed by distilled water, and then post-stained with salts of uranium and lead. Although post-fixation with OsO_4 results in excellent preservation of the ultrastructure, it may irreversibly alter the molecular conformation of some substrates. During labelling, drying of the sections or evaporation of reagents should be prevented, otherwise artefactual adsorption and clustering of the gold may occur. If aggregation of gold particles is suspected, the enzyme–gold complex can be centrifuged at a low speed for a short period to remove these aggregates.

COLLOIDAL GOLD–DEXTRAN METHOD

Gold–dextran particle preparation combined with pre- and post-embedding labelling techniques can be used to investigate the surface and intracellular binding patterns of lectins indicating the distribution of carbohydrates (Hicks and Molday, 1985). Diaminoethane-derivatised dextran T-10 (Pharmacia, Upsala, Sweden) is prepared by causing a solution of 10 g of dextran in 50 ml of 0.1 M sodium acetate (pH 5.0) to react with 1.6 g of sodium periodate for 90 min (Hicks and Molday, 1985). This solution is dialysed against 6 l of water and then caused to react with 0.5 M diaminoethane (pH 9.5). After 1 h , 0.72 g of sodium borohydrate is added to stabilise the diaminoethane dextran. The solution is thoroughly dialysed and lyophilised. Gold–dextran particles are prepared by adding 100 mg of diaminoethane dextran to 100 ml of colloidal gold (~18 nm) in 20 mM phosphate buffer (pH 7.0). Excess diaminoethane dextran is removed by centrifugation at 17 000 rev/min for 45 min. The gold–dextran pellet is washed again by centrifugation in 30 ml of 20 mM sodium phosphate buffer, and then the pellet is resuspended in 3 ml of Tris-buffered saline (TBS) (0.15 M NaCl, 0.02 M Tris, pH 7.4) containing 1 mg/ml bovine serum albumin (BSA).

For preparation of glutaraldehyde-activated gold–dextran particles, 100 mg of diaminoethane dextran in 4.5 ml of 20 mM sodium phosphate buffer (pH 7.0) is caused to react with 0.5 ml of 25% aqueous glutaraldehyde (Hicks and Molday, 1985). After 10 min, the dextran solution is added with stirring to 100 ml of colloidal gold. The mixture is stirred for 1 h and washed twice by centrifugation as described above. The final pellet is resuspended in 3 ml of 20 mM sodium phosphate buffer (pH 7.0) and caused to react with a lectin at a final concentration of 1–2 mg/ml. The reaction is allowed to proceed overnight and then terminated by adding TBS containing 1 mg/ml BSA. The conjugate is washed twice with 10 ml of TBS and the pellet is resuspended in 3 ml of TBS containing 1 mg/ml BSA and 10 mM NaN_3. The stock gold–dextran conjugate diluted 1:100 with water has an absorbance in the range of 0.15–0.25 at 520 nm.

Labelling of Intact Tissue
Glutaraldehyde-fixed tissue blocks are incubated in con A (100 μg/ml in 0.1 M cacodylate buffer, pH 7.4) for 18 h. After a thorough wash in the same buffer, the conA is detected by incubating the specimens in gold–dextran particles (1:10 dilution in TBS containing 1 mg/ml BSA) for 18 h. The specimens are thoroughly rinsed in 0.1 M cacodylate buffer (pH 7.4), post-fixed with 1% OsO_4 for 1 h and embedded in a resin. Thin sections are post-stained with salts of uranium and lead. Control experiment is carried out by adding α-methyl mannoside (0.2 M) in the lectin incubation medium.

Labelling of Thin Sections
Thin sections of glutaraldehyde-fixed and Lowicryl K4M-embedded tissue are treated with TBS for 10 min, followed by incubation for 30 min in con A, as above. Grids are carefully rinsed in filtered buffer, followed by incubation in gold–dextran particles, as described above. Grids are rinsed carefully in buffer followed by distilled water, and post-stained as above.

LIGHT AND ELECTRON MICROSCOPIC IMMUNOCYTOCHEMISTRY ON THE SAME SECTION (Mar *et al.*, 1987)

The following method involves post-embedding immunostaining of intracellular antigens at the subcellular level. The method allows correlative light and electron microscopy using the same section. Epon is removed-from a semithin section (2 μm), and this is followed by immunostaining with colloidal gold and staining with haematoxylin for light microscopy. The same section is re-embedded in Epon using the pop-off technique for thin sectioning for electron microscopy. Since membranes tend to be disrupted during sectioning, intracellular penetration of immunoreagents, including antibodies, substrates and colloidal gold, is not a problem. The use of chemical reagents that augment antibody penetration is unnecessary. However, the application of strongly alkaline reagents to deplasticise the section may adversely affect the antigenicity. Embedding and re-embedding procedures might cause some diffusion and/or extraction of the antigenicity. Weak immunolocalisation at light as well as electron microscope levels can be optimised by the silver enhancement method described on p. 240.

Semithin sections (2 μm) of fixed tissues are transferred to a glass slide and dried at 45 °C on a hot-plate for 4–6 h. The slide is immersed in a 1:1 mixture of saturated solution of sodium hydroxide and 100% ethanol for ~20 min to remove the resin. The saturated solution of sodium hydroxide is prepared by dissolving 15 g of NaOH pellets in 100 ml of 100% ethanol on a stir-plate for 1 h, which is kept at room temperature (capped) for 5 days before decanting of supernatant for use. The slide is thoroughly rinsed four times (5 min each rinse) in 100% ethanol, hydrated in an ethanol series of descending concentrations, placed in 0.3% aqueous hydrogen peroxide and rinsed in running water for 10 min.

Immunogold staining is carried out by covering the sections with a 100 μl drop of each of the following preparations:

(1) 10% normal goat serum in 0.05 M Tris–HCl-buffered saline (TBS) plus 2% BSA for 20 min.
(2) TBS rinse.
(3) Primary antiserum for ~12 h at 4 °C.
(4) TBS rinse.
(5) Goat anti-rabbit IgG conjugated to colloidal gold diluted 1:5 with TBS for 1 h.
(6) TBS rinse.
(7) PBS rinse.
(8) Fixation with 3% glutaraldehyde in PBS for 10 min.
(9) PBS rinse.
(10) Staining with haematoxylin for 5 min.
(11) Rinsing in water.
(12) Mount with a drop of glycerol under a cover-slip for light microscopy.

The sections should not be allowed to dry throughout the entire procedure.

Re-embedding is carried out by the following procedure. The coverslip is removed by soaking the slide in water, and sections on the slide are post-fixed with 1% OsO_4 for 5 min. The sections are dehydrated in an ethanol–propylene oxide series, and then covered with a 1:1 mixture of propylene oxide and Epon for 1 h. A drop of Epon is applied to the section for 1 h, followed by another drop of Epon for 1 h. Excess resin is wiped as close as possible from around the sections. BEEM capsules filled with Epon are inverted over the sections. Slides with mounted capsules are polymerised at 60 °C for 24–48 h. The capsules are removed, along with sections, from the slides by warming on a hot-plate. Thin sections are post-stained with salts of uranium and lead. For additional details, see Mar and Wight (1989).

DIAMINOBENZIDINE–OSMIUM TETROXIDE

Diaminobenzidine is effective in demonstrating sulphated mucopolysaccharides. This reagent, when used at an acid pH, forms a specific linkage with sulphated groups, and can be subsequently oxidised with OsO_4. As a result, these mucopolysaccharides are localised as heavy, electron-dense precipitates. The precipitate is partly due to oxidised DAB, a phenazine polymer, and partly to an osmium black. The precipitate is of 'non-droplet' type and thus yields a finer ultrastructural definition than that obtainable by using colloidal iron or colloidal thorium. This method also has the advantage of specificity for sulphate groups. Another advantage is that it can be applied as a post-stain to sections.

Thin sections of the tissue fixed with aldehydes and embedded in glycol methacrylate are collected on gold grids and washed in 5% (saturated) boric acid in distilled water for 5 min. The grids are then floated on a freshly prepared 1% solution of DAB in 5% boric acid for 20–30 min. After two washes in 5% boric acid (1–2 min), the grids are floated on 2% OsO_4 in distilled water for 10–20 min. Finally, the grids are washed in distilled water. For the methods using DAB for localising enzymic activity, the reader is referred to Essner (1974).

IODIDE–OSMIUM TETROXIDE MIXTURES

The zinc iodide–OsO_4 technique (Maillet, 1963) and the sodium iodide–OsO_4 technique (Champy, 1913) have been employed for light microscopic investigation of the autonomic nervous system for many years. These techniques, especially the former, have also been used at the subcellular level.

ZINC IODIDE–OSMIUM TETROXIDE (ZIO)

Mechanism of Staining

Although the mechanism responsible for the deposition of this stain in specific structures remains obscure, it is suggested that the ZIO reaction results in the deposition of reduced osmium at the positive sites. It has been indicated that deposition of the sodium iodide–osmium stain within cell organelles depends upon some type of reduction reaction (Garrett, 1965). In the ZIO reaction, zinc does not seem to be specifically necessary. The iodine probably catalyses deposition of the osmium rather than contributing directly to the reaction product. This concept is strengthened by evidence obtained by using electron probe analysis (Ostendorf *et al.*, 1971) and energy-dispersive X-ray microanalysis (Osborne and Thornhill, 1974). These studies showed that the iodine was absent in the ZIO reaction product. However, a remote possibility, that iodine is an integral part of the reaction product but does not survive the fixation and embedding treatments, does exist. According to Ostendorf *et al.* (1971), the ratio between osmium and zinc in the precipitate ranged from 3:2 to 4:1, while Osborne and Thornhill (1974) indicated that the precipitate was composed primarily of osmium and only a trace of zinc.

It has been postulated that the ZIO mixture uncouples lipid moieties from lipoprotein complexes, and then newly exposed groups of the lipid become available for interaction with the metal. It was, therefore, suggested that the material preferentially stained was a lipid. This concept is strengthened by the fact that ZIO reactivity is strongly inhibited by exposure to lipid solvents (e.g. methanol). On the other hand, the non-involvement of lipids has been suggested on the basis that leukocyte granules and erythrocytes fail to yield a positive ZIO reaction, but do exhibit positive histochemical reactions for lipids and phospholipids (Ackerman, 1964). Joó *et al.* (1973) have also suggested that chemical groups responsible for the osmiophilia of membranes are not likely to be involved in the ZIO reaction.

The staining of the Golgi apparatus and the gastrodermal surface coat by ZIO indicates the involvement of acid mucopolysaccharides, since pretreatment with hyaluronidase or neuramidase prevents staining. The presence of these carbohydrates in the gastrodermal surface coat is further indicated by its staining with colloidal iron, alcian blue and toluidine blue. Since lipids, carbohydrates and glycolipids are present in the Golgi apparatus and cell surface coats, the sensitivity of ZIO staining to both the digestive enzymes and lipid solvents is expected.

The cellular reactivity with ZIO has also been attributed to certain reducing substances such as catecholamines and ascorbic acid (Stockinger and Graf, 1965). Compounds containing reactive sulphhydryl groups have been suggested as being responsible for mitochondrial activity with ZIO; *N*-ethylmaleimide is known to inhibit mitochondrial staining. However, structures such as leukocyte granules, erythrocytes and skin epithelial cells rich in sulphhydryl groups fail to yield a positive reaction with ZIO. Neurotransmitter compounds have been suggested as reacting with ZIO (Niklowitz, 1972).

The observation that when a small amount of phosphate buffer is added to the ZIO mixture, the staining intensity is augmented (Osborne and Thornhill, 1974) is quite interesting. This augmentation of staining intensity may be due to the chemical reaction between phosphate ions and ZIO. Some evidence is available to strengthen this assumption. For instance, it has been suggested (Joó *et al.*, 1973) that ZIO positivity of the Golgi complex and synaptic vesicles may be correlated with the presence of thiamine pyrophosphatase in these organelles (Griffith and Bondareff, 1972). Also, other ZIO-positive sites (e.g. mitochondria, lysosomes, endoplasmic reticulum and perinuclear cisternae) are known to contain phosphatases.

There is evidence to suggest that the ZIO reaction is independent of glycolytic and respiratory enzyme activity. Dinitrophenol, iodoacetate, malonate, fluoride and azide do not inhibit ZIO reactivity. However, other enzymes may be involved in ZIO staining. That heat or cyanide treatment inhibits blood and bone marrow staining by ZIO (Clark and Ackerman, 1971) suggests an enzymatic basis for the staining of these tissues. Since *N*-ethylmaleimide does not affect the staining of the cisternal systems but completely inhibits the staining of mitochondria, it is apparent that different mechanisms are involved in the positive reactivity of ZIO with various cellular structures.

Staining Specificity for Synaptic Vesicles

If properly applied, mixtures of OsO_4 and soluble iodides yield a relatively specific blackening of synaptic vesicles. Factors which affect the quality of staining with ZIO include pH, type and osmolarity of the buffer solution, presence of Ca and Mg ions, and pre-fixation solutions used. Any deviation from the optimal parameters will affect the uniformity and specificity of staining. Even a small shift in the pH results in variations in the staining quality. The optimal pH of ZIO solution is considered to be 7.2–7.6. However, it should be noted that the influence of pH on the staining properties depends on the type of pre-fixation and its quality. The presence of Ca and Mg ions seems to be an important prerequisite for proper staining. Therefore, Tris–HCl buffer yields best staining, for it contains Ca and Mg ions. However, the buffering capacity of this buffer is weak at low pHs, and thus subject to fluctuations. Addition of a small amount of phosphate buffer to the ZIO mixture augments staining intensity. In general, hyperosmolarity of the buffer solution is desirable (Vrensen and de Groot, 1974b).

The initial claim of an absolute specificity of ZIO for acetylcholine has been disproved. Besides cholinergic vesicles, aminergic vesicles show positive reaction with ZIO. Moreover, a number of other organelles (e.g. Golgi complex, cisternae of endoplasmic reticulum, mitochondria and myelin sheath) in a variety of mammalian and non-mammalian cell types also show staining with ZIO. Endomembranes, especially rough endoplasmic reticulum, nuclear membrane and Golgi, in thin and semithin sections of plant cells, can be selectively stained with this method (Stephenson and Hawes, 1986).

In the past, many studies have indicated that not all synaptic vesicles in a given section yield positive reaction with ZIO. In the nerve terminals of spinal grey matter of adult rats, for instance, only half of the vesicular population was stained (Kawana *et al.*, 1969). Dennison (1971) also demonstrated that in the goldfish spinal cord there were no synaptic terminals in which all the vesicles were ZIO-positive, nor were there any that were entirely ZIO-negative. A number of controversies concerning the specificity of ZIO staining can partly be attributed to differences in the staining conditions described above. At present there is no one ZIO staining method which can be universally applied to nervous tissue with consistent success.

Staining Solutions

Before staining solutions are presented, a brief comment on the role of glutaraldehyde fixation in ZIO staining is in order. Several earlier studies have suggested that pre-fixation with glutaraldehyde tends to reduce the affinity of the ZIO stain for synaptic vesicles, whereas it increases the affinity for other organelles (e.g. Owman and Rudeberg, 1970). According to Joó *et al.* (1973), Purkinje cells in the rat brain were ZIO-negative unless they were pretreated with an aldehyde. However, recent studies show that glutaraldehyde neither hinders nor significantly enhances the staining of synaptic vesicles with the ZIO mixture. After pre-fixation with glutaraldehyde the success or failure of staining is partly dependent upon the properties of the buffer in which the ZIO mixture is dissolved.

The ZIO technique has been employed to obtain preferential staining of the central periodic lamella of the Langerhans cell granule, the Golgi region and the nuclear envelope of epidermal Langerhans cells (Niebauer *et al.*, 1969). The technique has been successfully used for visualising Langerhans cells in the epidermis of patients with histiocytosis. The procedures for the preparation of solutions, fixation and staining are presented below.

Zinc iodide solution	
Metallic iodine	5 g
Metallic zinc	10–15 g
Distilled water	200 ml

The combined powders are slowly added to distilled water in a beaker and after ~5 min the solution is filtered. Caution should be observed while dissolving the powder, since this is an exothermic reaction.

Final solution	
Zinc iodide solution	8 ml
Unbuffered OsO_4 solution (2%)	2 ml

The solution should have a pH of 5.9 and an osmolality of 218 mosm. It should be kept in the dark prior to and during use. Tissue specimens are placed in the OsO_4–ZnI_2 solution in the dark for 24 h at room temperature. Rinsing in distilled water is followed by dehydration and embedding in a resin according to standard procedures. Caution must be used in exposing the tissue specimens to uranyl acetate, for this stain is capable of dissolving the osmium–zinc precipitate. Exposure of the precipitate to uranyl acetate for longer than 2 min results in gradual dissolution, which is complete after 10 min.

The ZIO technique has been employed for staining different regions of the Golgi apparatus distinctively in the root apex cells (Dauwalder and Whaley, 1973). The staining patterns obtained differ from those of a number of other cytochemical procedures. Tissue specimens are treated with ZIO mixture in small vials wrapped in aluminium foil for 24 h at room temperature; the vials should be protected from light. The ZIO mixture is prepared by combining 12.5 g of zinc dust (Mallinckrodt Analyzed Reagent) with 5.0 g of iodine crystals, and slowly adding this to 200 ml of deionised water. This stock solution can be stored in tightly capped bottles, which should be protected from light, in a refrigerator for several months. For use, the stock solution is Millipore-filtered (0.45 ± 0.02 μm pore size) and mixed with 1% aqueous, unbuffered OsO_4 in a 4:1 ratio. The final concentration of the OsO_4 is thus 0.2%. The osmolality of this mixture ranges from 210 mosm to 213 mosm, and the pH varies from 5.8 to 6.0. This solution is rapidly transferred to vials for fixation.

As commonly employed, the zinc iodide–OsO_4 technique results in poor preservation of ultrastructure. However, this problem can be alleviated by pre-fixing the tissue with an aldehyde. Pre-fixation of the tissue with aldehydes has been employed for obtaining ZIO-positive reaction of synaptic vesicles (Krstić, 1972). The animals are perfused with 4% paraformaldehyde in 0.05 M phosphate buffer (pH 7.4). Tissue blocks are immersed in 6.25% glutaraldehyde in 0.1 M phosphate buffer (pH 7.4) for 2 h at room temperature. The specimens are washed three times in a stock buffer solution for 5 min at room temperature, and this is followed by impregnation with ZIO mixture (pH 6.25) for 16 h at 4 °C. The solutions are prepared as given below.

Stock buffer solution (1200 mosm)	
Distilled water	100 ml
NaCl	3.3 g
$CaCl_2$	0.06 g
$MgCl_2 \cdot 6H_2O$	0.31 g
Tris–aminomethane	0.605 g

The pH is adusted to 7.4 with HCl.

ZIO mixture is prepared by combining the zinc dust with the iodine and slowly adding the mixture to the distilled water.

Distilled water	40 ml
Zinc (powder)	6 g
Iodine resublimate or bisublimated iodine	2 g

Filter and add 4 ml of the filtrate to a mixture of 4 ml of Tris–HCl buffer and 2 ml of a 2% OsO_4 solution immediately prior to use.

According to another modification of the standard ZIO procedure, zinc iodide alone is supplied during pre-fixation with glutaraldehyde; this is followed by post-fixation and staining with OsO_4. This procedure yields the same staining of cell components as that obtained with zinc iodide–OsO_4. In addition, the reaction products are more intense, both qualitatively and quantitatively. Moreover, when Zn^{2+} salts are added to glutaraldehyde, some degree of specificity can be obtained. By using this procedure, Niklowitz (1971) obtained staining of perikarya and postsynaptic regions, but mossy fibre boutons remained unstained. The recommended solution is 6.2% glutaraldehyde in 0.1 M cacodylate buffer (pH 6.5) containing 0.13 M sucrose. To this solution is added 3% zinc acetate or zinc sulphate. Tissue specimens are fixed for 20 h and then washed for 1 h in cacodylate buffer. The specimens are stained for 3 h with 2% OsO_4 in cacodylate buffer and then washed in the buffer for 12 h.

When ZIO solution is adjusted to pH 7.4, Golgi materials stain distinctly, but staining of the cell surface coat is not visible (Elias *et al.*, 1972). This approach may serve as a marker for the Golgi apparatus (Fig. 4.9). The staining of Golgi materials and surface coat can also be increased or decreased by changing the duration of treatment; shorter durations result in selective staining of Golgi apparatus and certain cell surfaces.

In addition to zinc iodide, other iodides are also effective when used in combination with OsO_4 to reveal different kinds of synaptic vesicles. In fact, the appearance of the two components of the vesicles, the matrix and the core, is dependent upon the iodide used and the pH of the mixture. It has been demonstrated, for instance, that in the pineal glands of rats, the matrix is more reactive with ZIO (pH 5.5), while the core is more reactive with cadmium iodide–OsO_4 (pH 6.0) (Pellegrino de Iraldi and Suburo, 1970). Potassium iodide–OsO_4 at pH 7.2 stains the core most prominently but the matrix is stained scantily, whereas the same mixture at pH 5.5 stains more densely both the matrix and the core, and also brings out tubular structures, either isolated or connected to a vesicle. The reasons for the differences in the electron density of the matrix or the core of the synaptic vesicles due to variations in the pH are not clear.

Temperature and duration of staining also influence the extent and intensity of the reaction. In general, an increasing number of synaptic vesicles and other cell components are stained when the temperature and duration are increased. This is clearly seen in the nerves. When the staining is carried out at 4 °C for 2 h, the matrix of the granulated vesicles of the pineal nerves appears electron-dense, whereas the matrix of the granulated vesicles of the vas deferens nerves remains electron-lucent (Pellegrino de Iraldi, 1974). On the other hand, when the staining is carried out at 20 °C for 15 h, the matrix of the granulated vesicles of the vas deferens nerves is also deeply stained.

SODIUM IODIDE–OSMIUM TETROXIDE

The application of sodium iodide–osmium tetroxide mixture in electron microscopy has indicated that this method is not specific for adrenergic fibres in light microscopy. Electron microscopic studies have shown that a wide variety of cells and cell components show positive reaction with this mixture (Garrett, 1965; Breathnach and Goodwin, 1965). Little is known of the

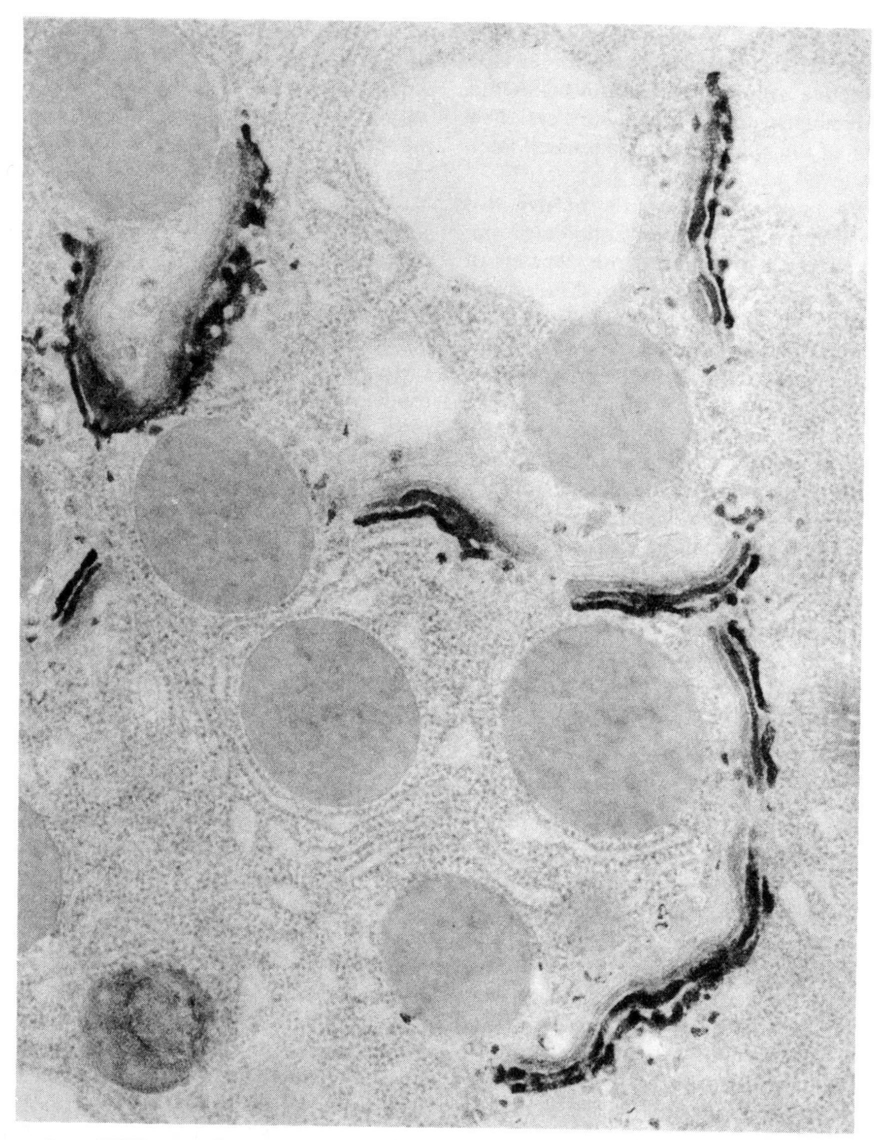

Figure 4.9 Thin section of ZIO-reacted mouse pancreatic exocrine cell. Many Golgi stacks are selectively stained; the *cis* region is stained more intensely than the *trans* side. 38 000×. Courtesy T. Noda.

histochemical basis of the reaction of the staining mixture.

It has been claimed that the mixture stains selectively type II alveolar epithelial cells and fibroblasts (McNary and El-Bermani, 1970). However, not all of the type II cells react with the same intensity. This difference in the staining intensity is probably related to the metabolic state of the cell. The fixing and staining mixture is given below.

Osmium tetroxide (1%)	1 part
Sodium iodide (3%)	3 parts

The tissue specimens are treated for 24 h at room temperature. No additional staining of the sections is required.

IRON

Iron (atomic number 26) is moderately heavy. Positively charged colloidal solutions of this metal are employed for specific staining of acid mucopolysaccharides and sialic acid at the cell periphery. The negatively charged form of the colloidal iron, on the other hand, possesses great potential as a specific stain for surface proteins; this selective affinity can be utilised in the elucidation of chemical structure of cell surfaces. The staining behaviour of the negative sol is discussed later in this section.

Iron is effective in locating the sites of negative charges on the membrane surface. Colloidal ferric oxide has been employed to obtain dense deposits of positive iron particles on the surface of red blood cell membranes (Marikovsky and Danon, 1969). Such a deposition is a result of strong attraction of neuraminic acid (carrier of the negative charge on the outer surface of the red blood cell membrane) for positively charged colloidal particles. Iron has also been used to stain pectin in plant cell walls (Albersheim and Killias, 1963a,b). Aqueous ferric ammonium sulphate solution has been employed to preferentially stain myofilaments and interfilamental structural material of the I-bands in glutaraldehyde-fixed insect flight muscle (Allman, 1971). That Fe^{3+} forms complexes with amino acids (e.g. glycine), has been demonstrated (DeVore *et al.*, 1971). However, iron is employed most effectively in the form of a positively charged colloid for specific staining of acid mucopolysaccharides, protein polysaccharides or glycosaminoglycans.

MECHANISM OF STAINING

Iron binds to polysaccharides, mucopolysaccharides and glycoproteins. Heparin, chondroitin sulphate, amylopectin sulphate, hyaluronic acid, *N*-acetylneuroaminic acid and sialic acid react with iron. It seems that a variety of macromolecules which possess negatively charged hydroxyl, carboxyl, sulphate and phosphate groups have an ability to chelate with iron. However, iron binding may not be totally a function of the charge present in the molecule. This is consistent with the observation that although amylopectin sulphate binds iron, dextran alone shows significant binding ability (Bella and Kim, 1973). Furthermore, the binding is not necessarily proportional to the charge, because a negatively charged polygalacturonate binds less iron than does dextran. Obviously, factors other than charge (such as accessibility and conformation) can be very important in the binding.

The positive staining of chromatin, ribosomes and phospholipids of membranes is attributed to the binding of free ferric ions to the phosphate groups. The staining of basement membranes is thought to be due to the binding of iron to their numerous sulphate groups. The staining of myofibre cell coat is due to the binding of iron to its carboxyl groups. Similarly, the staining of the surface of human erythrocytes is due to the binding of iron to the carboxyl groups of *N*-acetylneuraminic acid in the sialoglycoprotein (glycophorin). Glycoproteins contain accessible carboxyl groups such as those of sialic acid residues in sialomucins. These groups are Schiff-negative without prior oxidation. It has been suggested that the staining of glycoproteins with iron may also be due to hydroxyl groups.

Treatment with acidified methanol (methylation) prior to staining can block iron staining of acid mucopolysaccharides in two ways: (1) by converting the free carboxyl groups to their methyl esters, methylation renders these acid groups inaccessible to cations of the iron; and (2) sulphate groups (which bind the iron) in the acid mucopolysaccharide molecule are detached by methylation. Methylation for 6–18 h at 60 °C blocks most of the iron staining. Acidified methanol is prepared by adding 0.8 ml of concentrated HCl to 100 ml of methanol. Nitrous acid deamination also prevents iron uptake by lysyl groups.

The staining of acid mucopolysaccharides can also be blocked by treating the tissue with barium hydroxide or neuraminidase; both treatments are known to extract sialic acid. It is apparent, therefore, that when sialic acid is removed enzymatically or is blocked chemically, the positive particles of the iron do not attach to the cell. The fact that incubation with neuraminidase results in the removal of iron-staining components, but not the material that is stained with PAS, indicates that sialic acid residues in sialomucins are responsible for iron staining, and that galactose and n-acetylgalactosamine are stained by PAS.

In the localisation of pectic substance (polyuronide) within cell walls with iron, the reaction proceeds in two steps: (1) reaction between pectin and hydroxylamine results in the production of pectic hydroxamic acids; and (2) these acids form insoluble, electron-dense complexes when treated with ferric ions. Albersheim and Killias (1963a) have indicated that hydroxylamine produces pectic hydroxamic acids via nucleophilic substitution at the carbonyl carbon. The amount of iron deposited is dependent upon the concentration of reactive pectin substances, since pectic hydroxamic acids are formed by substituting hydroxylamine for the methoxyl groups of pectin.

It is known that the negatively charged form of iron colloid (according to R. D. L. Lillie, iron binding does not occur in the colloidal state, and at least temporary passage through an ionic state is needed) differs from the positively charged colloid in its affinity for the ionogenic groups of the cell surface. The former type stains positively ionised groups such as basic amino groups present in the protein side-chains. It has been stated earlier that ferric and ferrous salts of iron react selectively with protein amino and guanidyl groups. This reaction will occur when ferric salts are protected from hydrolysis to basic salts and ferrous salts are protected from oxidation in the absence of excess acid. The specific groups that may possibly react are side-chain amino groups of arginine, lysine and hydroxylysine; imino and imidazolic groups of histidine residues; and phospholipid and glycolipid amines. Most of the binding takes place at the side-chains of lysine, hydroxylysine, arginine and histidine.

The stainability of surface proteins by negative colloids is dependent upon the removal of blocking sialic acid. Prior to staining, therefore, either sialic acid is removed chemically by neuraminidase or carboxylic groups of this acid are blocked by methylation. The interference by sialic acid with the staining may be due to a steric hindrance effect, or to electrostatic repulsion of the negative colloidal stain by the negative charges of sialic acid.

The positively charged colloidal form of iron stains acid and carbohydrate hydroxyl sites. This will occur when iron is hydrolysed to basic ferric salts or, in the case of ferrous salts, oxidised to the ferric state in the absence of the acid required to satisfy the higher valency. It appears that the complex ion $FeOH^{2+}$ is involved in this type of reaction. It has already been stated that the binding of this form of iron is prevented by methylation, impaired variably by acetylation and unaffected by deamination. The binding of Fe^{2+} and Fe^{3+} to lysyl residues, on the other hand, is prevented by both methylation and deamination. Arginyl reaction is prevented by methylation but only moderately impaired by deamination.

Staining by iron may be electrostatic in nature, or it may be dependent upon chelation or complexing mechanisms, or both. However, the strong dependence of iron binding upon pH and ionic strength indicates that electrostatic forces play an important role in the staining mechanism. Furthermore, theoretical considerations favour the view that the staining is governed primarily by electrostatic forces. Polysaccharides frequently carry fixed negative charges and are polyelectrolytic chains, which enable them to interact and bind to cations of heavy metals under favourable cytochemical conditions.

Almost all methods for demonstrating acid mucopolysaccharides are based upon the ionic interaction of cationic reagents with the acid groups (carboxyl, sulphate) of the complex carbohydrates. The specificity of the reaction seems to be due to the ionic charges of the interacting molecules rather than to the chemical type of the end-groups involved.

Colloidal iron is known to react with nucleic acids at weakly acidic pH values. The available evidence indicates that iron has a higher affinity for DNA than for RNA. Microphotometric analysis of nuclear and cytoplasmic binding of iron in fibroblasts showed that 3–4 times more colloidal iron was bound per unit amount of DNA than per unit amount of RNA (Auer, 1972). It is emphasised that the binding of iron polycations to chromatin is dependent not only upon the amount of 'free' phosphate groups, but also upon their spatial arrangement—i.e. the charge density.

An increased iron binding to deoxyribonucleoprotein (DNP) complex can be obtained by pretreating the tissue block with solutions of increased alkalinity or increased ionic strength, and by blocking the amino groups in histones with acetic acid anhydride or treatment with trypsin. This increased binding is, therefore, interpreted to be due to an increased electronegativity of the DNP complex as a result of decreased interaction between the DNA and histones. This relationship becomes apparent when one considers that the majority of the negatively charged phosphate groups are bound to positively charged residues in histones. The release of negatively charged groups of DNA results in an increased iron binding.

pH

At a pH of ~1.8, positively charged colloidal ferric hydroxide binds to acidic groups—i.e. groups capable of becoming negatively charged. Three types of groups can theoretically bind the positive colloid iron to the cell surface: carboxylic groups of sialic acids and amino acid residues; phosphate groups of phospholipids; and sulphate groups of glycolipids and glycosaminoglycans. At pH 1.8 primarily sialosyl and sulphate groups ionise and bind to the iron.

At pH values between ~0.7 and 1.0 negatively charged colloidal iron reacts with basic amino groups in proteins. At this pH range the majority of the acidic groups are not ionised (except some sulphate groups in glycosaminoglycans), and thus do not repulse negative particles of the iron. The staining of proteins is not affected by blocking or removal of acidic groups at this pH range. At pH values lower than 0.7 the colloidal iron becomes very unstable.

The pH of the iron solution has a profound effect, not only on the appearance of fine structure (especially when *en bloc* staining is employed), but also, as stated above, on the specificity of the staining. The damaging effect of iron on the fine structure during *en bloc* staining at acid pHs is well documented. It is also known that the pH of the staining solution is a critical factor in obtaining specific staining of acid mucosubstances. A pH range of 1.4–2.0 yields a greater specific staining of mucosubstances, because at this pH range the majority of the anionic groups remain unchanged except acidic groups such as *N*-acetylneuraminic acid and its derivatives, sulphate, and ribonucleic acid or phosphate. In other words, ionisation of weakly acidic radicals will be suppressed so that only carboxyl or sulphate groups will react with positively charged iron particles. For example, the reaction of iron at pH 1.7 appears to be restricted to carboxyl groups of sialic acid and sulphate groups of monoesters of sulphuric acid. At pH values higher than 2.0 iron also reacts with phosphate groups and carboxyl groups of sialic acid. Apparently, then, in order to achieve a specific staining of mucosubstances and satisfactory preservation of the fine structure, the best approach is to post-stain the section at low pH values.

The optimal staining of other cell components is also influenced by the pH of the staining solution. Ferric chloride solution is most effective as a post-stain for chromatin at pH 3.0–6.5. In *en bloc* staining, ferric ammonium sulphate shows strong affinity for chromatin, mitochondria and basement membranes at pH 4.5, whereas at higher pH values chromatin staining decreases, although nucleolar staining remains constant (Sprumont and Musy, 1971). Collagen fibres are best stained at pH 4.5 and below. The binding of iron at acid pH values is primarily due to the availability of free ferric anions to the phosphate, sulphate and certain groups with low p*K* values.

At higher pHs there are more free electrostatic radicals in the tissue to react with iron anions and eventually with cations. At pH 8.0 and above iron also acts as an anionic stain. At these pHs amine and ammonium groups in the tissue probably react with iron. The staining of cytoplasmic ground material at pH 7.2 and above has been attributed to the participation of protein groups in iron complexing. According to Geyer *et al.* (1973), iron anions show a higher binding rate with positively charged sites on the surface of human erythrocytes at pH 4–6.

How the variations in the pH of iron solution (employed *en bloc*) influence the staining intensity of various cell components is shown in Table 4.2. It is important to note that a change in the pH results in an immediate alteration of the electrostatic properties of the cell component, which, in turn, induces conformational changes. Therefore, increased or decreased binding cannot be completely explained by a simple change in the electrostatic properties of the cell component: other factors, such as accessibility and conformation, are also important.

Nicolson and Painter (1973) reported the effect of pH on the appearance of the iron reaction product.

Table 4.2 Staining intensity of cell components at various pH levels

Component	*pH of the iron solution*					
	1.9	*3.0*	*4.5*	*6.0*	*7.2*	*8.0*
Chromatin	+++	+++	+++	+	+	+
Mitochondria	+++	+++	+++	++	+	+
Basement membranes	++	++	++	+	+	+
Nucleolus	++	++	++	++	++	++
Ribosomes	++	++	++	++	++	++
Cytoplasmic matrix (proteins?)	+	+	++	++	+++	+++

They showed that the topographic distribution of colloidal iron hydroxide on ghost membranes of human erythrocytes was sensitive to pH in low ionic strength buffers. At pH 6.5–7.5 the colloidal iron hydroxide appeared in a dispersed state, while at pH 5.5 the stain appeared in an aggregated state. It follows from the above considerations that colloidal iron can be a useful cytochemical instrument when its two electrical forms are used knowledgeably, with caution, at carefully controlled pH levels.

RATE OF PENETRATION

The penetration of the iron stains into the tissues is rather slow. Poorly fixed tissues are more difficult to penetrate. The type of fixation also influences the rate of penetration. According to Wetzel *et al.* (1966), iron stains penetrate less efficiently into glutaraldehyde-fixed tissues than into formaldehyhde- or OsO_4-fixed tissues. The rate of penetration is apparently also dependent upon the type of tissue. Frozen tissues show relatively rapid penetration. The staining is carried out at room temperature, for higher temperatures seem to retard penetration of the stain into sections.

MODE OF STAINING

Both *en bloc* staining and post-staining of sections have proved successful. Some workers prefer *en bloc* staining, since direct treatment of sections is technically rather tedious. Furthermore, *en bloc* staining allows the examination of the tissue in the light microscope. On the other hand, stained tissue blocks are more difficult to cut than the unstained blocks.

Since the stain does not penetrate readily, extremely small segments (~0.5 mm^3) of the tissue are used for *en bloc* staining, and cells near the surface are stained best. In general, a duration of ~3 h for *en bloc* staining is adequate. The duration of post-staining of sections varies considerably, depending primarily upon the staining solution and the objective of the study; a duration of 1 h is recommended. Shorter durations and lower concentrations may be necessary to achieve specific staining.

STAINING SOLUTIONS

The three commonly used iron formulations are: (1) Hale's colloidal iron (1946), as modified by Mowry (1958); (2) dialysed solution of anhydrous ferric chloride, glycerin and ammonium hydroxide, as described by Rinehart and Abul-Haj (1951); and (3) ferric chloride solution (0.1–0.4%). Treatment with Mowry's colloidal iron stain results in the deposition of iron in the form of fine, regular, dense particles ~3 nm in diameter. Heavy staining may cause these particles to aggregate, forming spherical masses of up to 30 nm in diameter. Staining with ferric chloride or dialysed iron, on the other hand, may yield particles of various sizes. In general, non-dialysed colloidal iron solutions give finer deposits than the dialysed ones.

The coarse deposits given by dialysed iron are due to the decreased stability of its micelles, and this decreased stability is due to the removal of stabilising positive ions Fe^{3+} and H^+ during dialysis. Positive ions Fe^{3+} and H^+ are formed when the sol is prepared by hydrolysing $FeCl_3$. The net result of the removal of these ions is the aggregation of smaller micelles into coarser particles of varying sizes. The most common iron deposition, however, is in the form of very fine grains. Although a low concentration of iron particles is present throughout the cytoplasm and nuclei, this rather homogeneous distribution is clearly different from that seen in positively reacting sites. Details of the preparation of various staining solutions are given below.

Colloidal Ammonium Ferric Glycerate

Ferric chloride (granules)	275 g
Distilled water	1000 ml
Glycerine	400 ml
Ammonium hydroxide (28%)	220 ml

Glycerine is added after the ferric chloride is completely dissolved in water. Ammonium hydroxide is added gradually in amounts of 50 ml or less at a time. The mixture is stirred vigorously each time to dissolve the precipitate before adding more ammonium hydroxide. The final colloidal solution should be clear, and deep reddish brown in colour. This solution is then dialysed by pouring into Cellophane bags which have been previously softened by immersion in water (dialysis is necessary in order to remove acid formed by the hydrolysis and non-colloidal iron salts). The bag should not be filled to more than 40% of its capacity, since the volume of the solution inside the bag is increased during the process of dialysis. The bag is tied at both ends and placed in a large container filled with distilled water. The completion of dialysis requires 8–10 changes of distilled water over a period of 72 h. The final dialysed ammonium ferric glycerate solution is stable at room temperature. Glacial acetic acid is added to this stock solution in the ratio of 1:3 just before use to achieve the desired pH.

Positive Ferric Oxide Solution

Dissolve 6.75 g of $FeCl_3 \cdot 6H_2O$ in 50 ml of distilled water, and add 50 ml of this solution (0.5 M) in a slow stream to 600 ml of boiling distilled water. Dialyse this solution against ten volumes of distilled water for 5–6 days, with two changes of water each day. In order to adjust the concentration of the solution, iron is determined by reduction to iron(II) with stannous chloride (after acidification) and titration with potassium dichromate, using diphenylamine sulphonate as an indicator. Add distilled water to the iron solution to obtain a final concentration of 1.2 g of iron per litre. Mix 100 ml of this solution (pH 5–6) with 100 ml of 0.0012 N HCl. This mixture (pH 3.5) is the stock solution. The final positive ferric oxide solution is prepared as follows.

Stock solution	10 ml
Glacial acetic acid	10 ml
Distilled water	20 ml

This solution has a pH of ~1.8 and contains 0.15 g of iron per litre.

Negative Ferric Oxide Solution

Negative ferric oxide solution is prepared by recharging the positive ferric oxide sol with ferrocyanide ions (Hazel and Ayers, 1931). This is accomplished by slowly pouring 100 ml of the stock solution of the positive ferric oxide (see above) into a beaker containing 100 ml of 8 mM potassium ferrocyanide which was previously dissolved in boiling distilled water with constant stirring. The 1:1 ratio of the two chemicals results in the maximum stable negative charge on the Fe_2O_3 colloid. This stable stock solution contains 0.3 g of iron per litre, and can be stored for 1–2 weeks without coagulation. The final staining solution is prepared by combining 20 ml of this stock solution with an equal volume of boiled distilled water. The pH of this mixture is ~6.0. A pH of 3.4–4.2 can be obtained by mixing 20 ml of the negative solution with 20 ml of 0.01 M acetate buffer (pH 3.4). It should be noted that at a pH lower than 6.0 these solutions become unstable; the degree of instability is related to the increase in hydrogen ions. These unstable solutions show green or greenish-blue colour, which indicates partial conversion of the negative colloid into Prussian blue or other iron blues.

Negative Colloidal Ferric Hydroxide

Negatively charged colloidal iron solution is prepared by dissolving 6.53 g of ferric chloride in 100 ml of distilled water. This solution is added fairly rapidly (almost in a stream) to 1 litre of boiling distilled water. This final solution has a pH of ~1.8 and contains 1.7 g of iron per litre. Prior to use, the pH is lowered to 0.8 by adding HCl. Chloride ions at pHs lower than 1.0 bring about the recharging of the conventional positively charged colloid to a negative form. The physicochemical factors involved in this recharging phenomenon have been discussed by Blanquet and Loiez (1974).

This negative colloid has been used for localising with relative specificity positively ionised groups such as basic amino groups of protein side-chains in the inner and outer hydrophilic leaflets of the cell surface membrane (Blanquet and Loiez, 1974). In other words, the negative charge of the colloidal iron (at pH 0.8) permits the identification of positive ionogenic groups. This staining solution possesses considerable potential for aiding the study of the chemical organisation of intracellular components. It should be noted that the strong acidity of this solution is bound to cause some damage to cell structures.

Positive ferric oxide solution (pH 2.0) was employed by Matukas *et al.* (1967) to demonstrate sulphated mucopolysaccharides in the granules of cartilage; fibres did not show uptake of colloidal iron. Colloidal iron was visualised as electron-opaque particles 3–5 nm in diameter. Sections of glutaraldehyde- and OsO_4-fixed tissues are mounted onto a supported or unsupported grid and stained on a drop of the staining solution for 1 h at room temperature. The grid is rinsed by dipping in three changes of 12% acetic acid and two changes of distilled water. Sections can be examined with or without post-staining.

Colloidal iron hydroxide has been employed to localise sialic acid in plasma membranes isolated from rat liver and hepatoma (Benedetti and Emmelot, 1967). Channels in microspore walls have been rendered visible by immersing the anthers, whose stamen filaments were cut, for ½–3 h in colloidal iron solution prior to fixation (Rowley, 1971). Colloidal iron hydroxide was used to label the topographic distribution of acidic anionic residues on human erythrocyte membranes (Nicolson, 1973). This electrokinetic negative surface charge is probably due to acidic oligosaccharide anionic residues such as *N*-acetylneuraminic acid. Glutaraldehyde-fixed erythrocyte ghosts are spread flat at an air–water interface and mounted on a collodion–carbon-coated grid. The mounted membranes are exposed to 5% solution of bovine serum albumin in distilled water for 3 min.

The excess albumin is removed by touching the grid once to a drop of distilled water. While wet, the grid is floated on a drop of the freshly prepared staining solution for 2 min. Then it is floated on several drops of 12% acetic acid (pH 2.0) and washed in distilled water. The staining solution is prepared by rapidly adding 5 ml of 0.5 M $FeCl_3$ to 60 ml of boiling distilled water. After cooling, 10 ml of glacial acetic acid is added and the pH is adjusted to 1.8. Ultracentrifugation of this solution prior to the staining procedure prevents the formation of coarser deposition particles.

As stated earlier, iron can also be employed to study the distribution of pectic substances (polyuronide) within cell walls. The specific staining of these substances can be obtained by employing alkaline hydroxylamine and ferric chloride ($FeCl_3$). For staining, the tissue is fixed with glutaraldehyde and post-fixed with OsO_4. The fixed tissue is washed and taken gradually into 60% ethanol, and then treated with alkaline hydroxylamine solution for 1 h. The tissue is washed in dilute HCl for 15 min, and then immersed in 2% ferric chloride solution for 1 h. Now the tissue is ready for dehydration and embedding according to standard procedures.

Ferric chloride (Schmörl reaction) has been employed to stain lipofuscin pigment (Hendy, 1971). The stain is prepared by mixing equal volumes of 3% ferric chloride and 1% potassium ferricyanide; only freshly prepared solutions should be used.

Erythrocyte membrane ghosts take up a moderate number of negatively charged iron particles from a modified colloidal iron solution (Gasic *et al.*, 1968). The average amount of binding is ~3100 iron particles/μm^2 (Linss *et al.*, 1971, 1972). Ghosts of human erythrocytes are prepared by osmotic haemolysis (Geyer *et al.*, 1972b). They are fixed with 3.5% glutaraldehyde for 15 min at 4 °C, and then stained with the modified colloidal iron reaction at pH 6.0 for 60 min. After washing in distilled water, the specimens are post-fixed with 1% OsO_4 for 15 min and washed again. The specimens can be transferred to coated grids as whole mount preparations. Alternatively, the specimens are dehydrated, embedded and sectioned. Post-staining is unnecessary.

Negatively charged colloidal iron has also been employed for staining basic amino groups of protein side-chains in the cell surface membrane (Blanquet and Loiez, 1974). Electron-dense particles are distributed in the outer and inner leaflets of the triple-layered structure of the membrane. Isolated membranes are washed thoroughly with Sorensen's buffer, fixed with buffered formaldehyde for 2 h at 4 °C and washed again with the buffer. The membranes are rinsed in 1% HCl, homogenised very gently in the staining solution and left in it with constant swirling for 1 h at 4 °C.

Iron haematoxylin, a standard stain in histology, has been employed for staining granules in enteroendocrinic cells at the subcellular level (Nichols *et al.*, 1974). Iron was deposited on spherical and non-spherical granules measuring 200–500 nm in diameter. This staining is probably due to the presence of basic protein residues in the granules, which are capable of binding ferric ions. The staining procedure involves two solutions—iron alum (ferric ammonium sulphate) and haematoxylin. Five per cent iron alum is prepared in distilled water. This solution should be ripened for at least 24 h and filtered before use. The sections mounted on grids are floated (section face down) on the iron alum solution in a dish, which, in turn, is floated on a water-bath at 85–90 °C for two 30 min intervals. The same procedure is repeated with the iron haematoxylin solution, followed by rinsing in distilled water.

IRON DIAMINE

The high iron diamine (HID) and the low iron diamine (LID) methods are useful for ultrastructural localisation of carboxylated and sulphated glycoconjugates in extracellular and intracellular sites. These methods were originally introduced for the demonstration of acid glycoconjugates at the light microscope level (Spicer, 1965).

The HID method demonstrates specifically sulphated complex carbohydrates at the light microscope level as well as ultrastructurally. It does not show reactivity with sialmucins, hyaluronic acid and neutral mucosubstances. Thus, the HID reagent can distinguish between sulphate and carboxyl groups in complex carbohydrates.

The diamine oxidation products complex with iron in the HID reagent and thus impart electron density to reactive sites. This method has the limitation of imparting a low degree of electron density to some sites. Such low electron density, however, can be enhanced by using it in conjunction with the thiocarbohydrazide–silver proteinate (TCH–SP) procedure (Sannes *et al.*, 1979). Thiocarbohydrazide was originally introduced as a bridge between periodate-engendered tissue aldehydes and either OsO_4 (Seligman *et al.*, 1965) or silver proteinate (Thiéry, 1967) for detection of vicinal glycols at the ultrastructure level. Since TCH also complexes with heavy metals (Seligman *et al.*, 1966), iron in the HID reagent is thought to complex with TCH. The bound TCH presumably then reduces the silver proteinate to yield an electron-dense reaction product (Sannes *et al.*, 1979).

In contrast to the HID method, the LID method seems to localise both sulphate and carboxyl groups of acid glycoconjugates. Alcian blue or dialysed iron also shows affinity for these two groups. Low iron diamine stains hyaluronic acid, sulphated glycosaminoglycans, sulphated glycoprotein and sialylated glycoprotein (Takagi *et al.*, 1982). Neither LID nor HID reagents stain phosphate groups of nucleic acids. The LID staining requires post-osmication to obtain adequate staining. The electron density imparted by the HID method is also enhanced by post-osmication. The staining of LID-reactive sites is also enhanced by treatment with TCH–SP. The LID method is a useful adjunct to the HID method in assessing the nature of acidic glycoconjugates.

High iron diamine	
N,*N*-Dimethyl-*m*-phenylenediamine $(HCl)_2$	120 mg
N,*N*-Dimethyl-*p*-phenylenediamine HCl	20 mg
Distilled water	50 ml

Add 1.4 ml of 40% (w/v) ferric chloride to the above solution. The pH of the solution is 1.3–1.6. Because of the low pH, specimens exposed to the staining solution show poor cytological preservation.

Low iron diamine	
N,*N*-Dimethyl-*m*-phenylenediamine	30 mg
N,*N*-Dimethyl-*p*-phenylenediamine	5 mg
Distilled water	50 ml

Add 0.5 ml of 40% ferric chloride to the above solution, freshly prepared. If TCH–SP staining is used, the SP solution should be filtered twice through Whatman filter No. 2 to remove background staining. Post-osmication is accomplished in 1% OsO_4. Control specimens are incubated for about 18 h at 22 °C in $MgCl_2$ solution, which is prepared by adding 0.5 ml of 40% $MgCl_2$ to 50 ml of distilled water and adjusting the pH to 1.8 with HCl.

LANTHANUM

Lanthanum has an atomic number of 57, and trivalent

lanthanum cation (La^{3+}) possesses electron scattering power sufficiently high to produce contrast in electron microscope images. The diameter of the lanthanum colloidal particle is less than 2 nm. When tissue specimens are treated with lanthanum prior to or during fixation, or during washing in a buffer, the extracellular space becomes filled with an electron-opaque precipitate. This staining indicates the presence of anionic molecules, possibly mucopolysaccharides or glycoproteins. Also, it is assumed that the fine tissue spaces filled with electron-opaque aggregates of lanthanum, under physiological conditions, are occupied by interstitial fluid or water. Thus, physiological pathways of tissue fluid can be traced by permeating the tissue with lanthanum. Lanthanum has the disadvantage of being toxic to living tissues and must be applied to fixed specimens.

In addition to its role as a marker for extracellular materials (both at the cell surface and in the intercellular space), lanthanum is taken up intracellularly by the axoplasm of certain neurons (Lane and Treherne, 1970) which contain numerous neurotubules. In cross-section, the entire core of the stained neurotubule appears electron-dense. The axons near the cut surface of the tissue specimen take up the stain because they get severed during dissection. According to Lane (1972), lanthanum penetrates the neural lamella and lacunae, clefts and gap junctions between adjacent perineural cells, but no further.

Evidence is available indicating that this tracer is capable of penetrating across the undamaged sarcolemma of skeletal muscle (Forbes and Sperelakis, 1979). The tracer is also known to be able to penetrate septate junctions. It may penetrate by intercellular diffusion into the regions of intercellular channels containing septate desmosomes. The dye seems to be accessible to the intercellular space (~2 nm) delimited by adjacent perineural membranes in the gap junctions. Extracellular spaces of small peripheral nerves are also accessible to externally applied lanthanum.

On the other hand, sufficient data are available to indicate that lanthanum does not penetrate tight junctions (e.g. Neaves, 1973). This difference in the penetration seems to be due to the difference in the manner in which the tracer is administered. In general, the tight junctions are more permeable to both ionic (La^{3+}) and colloidal ($La(OH)^{2+}$) or $La(OH)_2^+$) lanthanum when the tracer is administered *in vivo* prior to fixation. The difference in penetration may also be due to the presence of leaky junctions in some tissues, while tight junctions are present in other tissues. It is concluded that in using lanthanum for monitoring the permeability barriers, one must consider the manner in which the tracer is administered to the tissue.

Lanthanum is able to penetrate various organelles in a cell to which it has gained access. In mouse and fish olfactory mucosa, lanthanum has been reported to penetrate endoplasmic reticulum, mitochondria, microvilli and microtubules, outlining their substructure in negative contrast.

However, it is known that, in general, lanthanum does not enter cells unless the plasmalemma is damaged. Henrikson and Stacy (1971) have shown, for instance, that lanthanum was unable to penetrate more than a few millimetres into the ruminal epithelium by way of the cut surface at the perimeter of the tissue block. Lanthanum, in fact, has a stabilising effect on membranes, since it prevents the membrane leakiness induced by metabolic depletion (Casteels *et al.*, 1972). However, cell walls seem to be freely permeable to lanthanum.

MECHANISM OF STAINING

Although the mechanism of lanthanum staining is not known, it has been suggested that, because of its trivalency, La^{3+} displaces bound Ca^{2+} and binds to the same sites more strongly. Lanthanum ions are considered to form complexes with nucleic acids, nucleoproteins and phospholipids. Under certain conditions, lanthanum seems to act primarily as a selective stain for certain cells and extracellular structures. This suggestion is strengthened by the results of studies on the mucosa of mouse caecum (Henrikson, 1974). The tissue was treated briefly with lanthanum and then exposed to large volumes of lanthanum-free solutions for long periods of time. Lanthanum staining persisted. If the staining had been simply an inert probe of the cellular environment, one would expect to find minimal amounts of the metal in the tissue specimen.

The other possibility is that lanthanum follows simple diffusion pathways. Probably colloidal particles of the stain are trapped non-specifically in small interstices between cells, which accounts for much of the staining effect. Lack of staining specificity in some studies indicates that lanthanum follows simple diffusion pathways. However, the available evidence is insufficient to decide whether lanthanum follows diffusion channels with little true chemical binding or whether it actually reacts with polysaccharides and other cell materials.

It is pointed out that lanthanum may be hydrolysed or transformed into a non-dissociated or slightly dissociated form in tissue. Possibly lanthanum is complexed or chelated by a biological carrier, probably a macromolecule. On the other hand, lanthanum may be available in tissue in the ionic form (La^{3+}). These different chemical forms may possess different diffusion rates through the plasma membrane barrier, and consequently a higher cell membrane permeability could be reflected in a higher concentration of the tracer in the cell fluid. Either carrier-mediated transport or a simple diffusion process, or both, may be in operation.

In certain tissues lanthanum staining is difficult to achieve, perhaps because of leaching of the stain or tissue components, or both. Inequalities of stain distribution probably reflect various degrees of mechanical damage caused by the preparatory procedures. Differential fragility of the plasma membranes due to metabolic differences may also play a part in the sporadic staining.

FIXATION AND STAINING PROCEDURES

It should be noted that fixation with OsO_4 is necessary to retain lanthanum in tissues, because lanthanum deposits are not apparent when specimens are fixed with aldehydes alone. In specimens fixed with aldehydes, lanthanum is washed out during dehydration. Lanthanum should be used at room temperature, because it is precipitated in the cold. Phosphate buffer is undesirable for use with lanthanum, for mixing them results in precipitation. For general purposes, the following procedures of fixation and staining of intercellular materials are recommended.

(1) Tissue specimens are fixed for 2 h in 2.5% glutaraldehyde in cacodylate buffer (pH 7.2) to which 1% lanthanum nitrate ($La(NO_3) \cdot 6H_2O$) has been added. They are then rinsed for 30 min in the buffer containing 1% lanthanum nitrate, and post-fixed for 30 min in 1% OsO_4 in the buffer containing lanthanum nitrate in the final concentration of 1%. The optimal duration of fixation and the pH are determined by the objective of the study.

(2) Lanthanum nitrate is employed in combination with $KMnO_4$:

Lanthanum	1 g
$KMnO_4$	1 g
Veronal acetate buffer	20 ml
Zetterqvist salt solution	6 ml
0.1 N HCl to attain pH 7.8	
Distilled water to make	100 ml

Tissue specimens are fixed in the above solution for 1 h at 4 °C, followed by treatment with several changes of Tyrode's solution,

(3) Lanthanum can also be employed in combination with OsO_4 in a different procedure. According to this method, tissue specimens are pre-fixed with glutaraldehyde containing alcian blue or cetylpyridinium chloride (a cationic substance). Tissues can be pre-fixed by either immersion or vascular perfusion. For immersion fixation, tissue specimens are fixed with 3% glutaraldehyde, buffered with 0.1 M cacodylate, containing 0.5% alcian blue or cetylpyridinium chloride, for 2 h at room temperature.

After rinsing in the buffer, the specimens are post-fixed with a 1:1 mixture of OsO_4 (1%) and lanthanum nitrate (1%) buffered with 0.1 M collidine (pH 8.0) for 2 h at room temperature. It should be noted that the pattern of staining of the surface coat obtained with this method is similar to that achieved by using colloidal iron, dialysed iron, colloidal thorium, ruthenium red, silver methenamine and glutaraldehyde–alcian blue fixation.

(4) Tissue specimens are fixed in glutaraldehyde and post-fixed in 1% OsO_4 (pH 7.2) containing 0.5–1.0% lanthanum hydroxide for 1–2 h at room temperature. Lanthanum hydroxide is prepared by bringing a 3% solution of lanthanum nitrate to pH 7.6 by gradually adding 0.01 N NaOH. During this titration the solution should be stirred vigorously. At pH 7.8 lanthanum hydroxide becomes insoluble, which results in its becoming flocculated.

(5) *Permeation technique* Tissue specimens are fixed in cacodylate-buffered (pH 7.2) glutaraldehyde (2.5%) containing 1% lanthanum nitrate for 12–72 h at 4 °C. The specimens are washed in the same buffer containing 1% lanthanum nitrate for ~6 h, and then post-fixed in buffered OsO_4 (1%) containing 1% lanthanum nitrate for 1 h at 4 °C. This is followed by rinsing in the same buffer containing 1% lanthanum nitrate for 2 h. After dehydration with 50% ethanol containing 1% lanthanum nitrate, the specimens are stained with 1% uranyl acetate and 1% lanthanum nitrate in 50% ethanol for 30 min. Further dehydration is carried out with 60% and 90% ethanol containing 1% lanthanum nitrate and 100% ethanol and propylene oxide without lanthanum nitrate.

(6) *Vascular perfusion technique* Fixatives containing lanthanum, when applied by immersion, do not show a consistent penetration of the tracer into certain tissues. In order to alleviate this problem, lanthanum is perfused in combination with aldehydes. In general, vascular perfusion by a mixture of paraformaldehyde, glutaraldehyde and lanthanum (final concentration 1.12%) for ~30 min is recommended. The specimens are then rinsed and post-fixed with OsO_4. The sections can be examined without additional post-staining.

LEAD

Lead (atomic number 82) is slightly heavier than osmium, but the latter is twice as dense. At present lead salts are the most widely used stains for electron microscopy. These stains have a high electron opacity and show affinity for a wide range of cellular structures, including membranes, nuclear and cytoplasmic proteins, nucleic acids and glycogen. Lead salts increase contrast much more intensely in the presence of reduced osmium than do uranyl salts, although uranium has a higher atomic number than that of lead. In the absence of reduced osmium, the best overall staining is obtained with uranyl salts. Although lead is employed most commonly as a post-stain, it can be applied *en bloc* during fixation for certain studies (see later).

Lead stains penetrate through the entire section. The effective application of lead stains is somewhat limited by the fact that upon exposure to air all solutions of lead salts become clouded by the formation of lead carbonate. Lead ions readily form insoluble crystals with most anions, and these crystals are extremely hydrophobic. The insoluble contaminant is deposited on the surface of sections as electron-opaque fine needle-like crystals, small granules or large amorphous polygonal precipitates (Fig. 4.10). However, by observing utmost care and using improved formulations, this contamination can be avoided. When even a slight precipitate appears in the flask, or when excessive contamination is found on a number of stained grids, the staining solution should be discarded without any hesitation. The toxicity of lead solutions should be kept in mind and necessary precautions taken at all times to avoid contact with the skin.

MECHANISM OF STAINING

Various cell components differ in the degree of their affinity for lead stains, depending upon the differences in the mechanism of attachment of the stain to the reacting groups of these components and the pH of the staining solution. Furthermore, the appearance of some cell components after lead staining is independent of the type of fixative used, whereas others show either an increased or a decreased affinity for the stain, depending upon the type of fixative; for instance, after lead staining the appearance of glycogen in the tissue fixed with formaldehyde or OsO_4 is similar, while after lead staining ribosomes and nucleoli are denser when the tissue is fixed with formaldehyde than when OsO_4 is employed as a fixative (Daems and Persijn, 1963). On the other hand, the constancy of the results obtained with different alkaline lead stains indicates that the basic chemical mechanism of staining is similar in each instance. In general, with prolonged treatment, the staining becomes more general and less specific. Staining with lead citrate for longer than 30 min results in destaining and loss of cellular materials.

According to available evidence, staining by lead salts at a higher pH is much more rapid and intense than at a lower pH. Also, alkaline lead solutions are relatively stable. At increasingly higher pH values, phosphate, sulphhydryl, tyrosyl and carbonyl groups in tissue become more ionised, which results in increased binding of Pb^{2+} ions. For this reason lead hydroxide (pH 8.15) stains more intensely than monobasic lead acetate (pH 7.0). Ionisation of phosphate, sulphhydryl, tyrosyl and carboxyl groups of proteins occurs following fixation, and these anions become available for combining with lead cations. This mechanism is also operative in staining solutions of very high pH (>11.5).

Basic salts are formed when divalent lead salts are combined with lead hydroxide or other alkalis in aqueous solutions at higher pH values. These compounds, of the general type $Pb(OH)_2PbX_2$, ionise as shown in the following equation (Reynolds, 1963):

$$Pb(OH)_2PbX_2 \rightleftharpoons [Pb(OH)_2Pb]^{2+} + 2X^-$$

Divalent cations containing two atoms of lead impart more density at pH 12.0, since tissue would bind twice as much lead as would be bound by the divalent lead ions (Pb^{2+}) at pH 7.0.

It is generally thought that cationic rather than anionic salts of lead are responsible for staining. This assumption is supported by the demonstration that staining is prevented in the presence of ethylene diaminetetraacetate (EDTA) (Reynolds, 1963). Since EDTA is a powerful chelating agent, it tends to sequester only cationic salts of lead while anionic salts are left free in solution (however, EDTA action on cations could lead to new equilibria in which anionic lead may also be reduced in concentration). Furthermore, the lessening of the staining intensity with lead citrate at pH 14.0 is related to the conversion of cationic salts of lead present at pH 12.0 to anionic hydroxyplumbites. Thus, it is unlikely that these anionic salts of lead are responsible for staining at pH 12.0. It also seems unlikely that positively charged groups are available in the tissue for ionic binding of the negatively charged plumbite ion at high pH levels.

On the other hand, some workers (e.g. Karnovsky, 1961) believe that at high pH levels anionic forms of lead (plumbite ion) are responsible for staining. Several workers have suggested the formation of plumbite in the alkalinised solutions of lead salts (Lever, 1960; Millonig, 1961). Both lead hydroxide and lead monox-

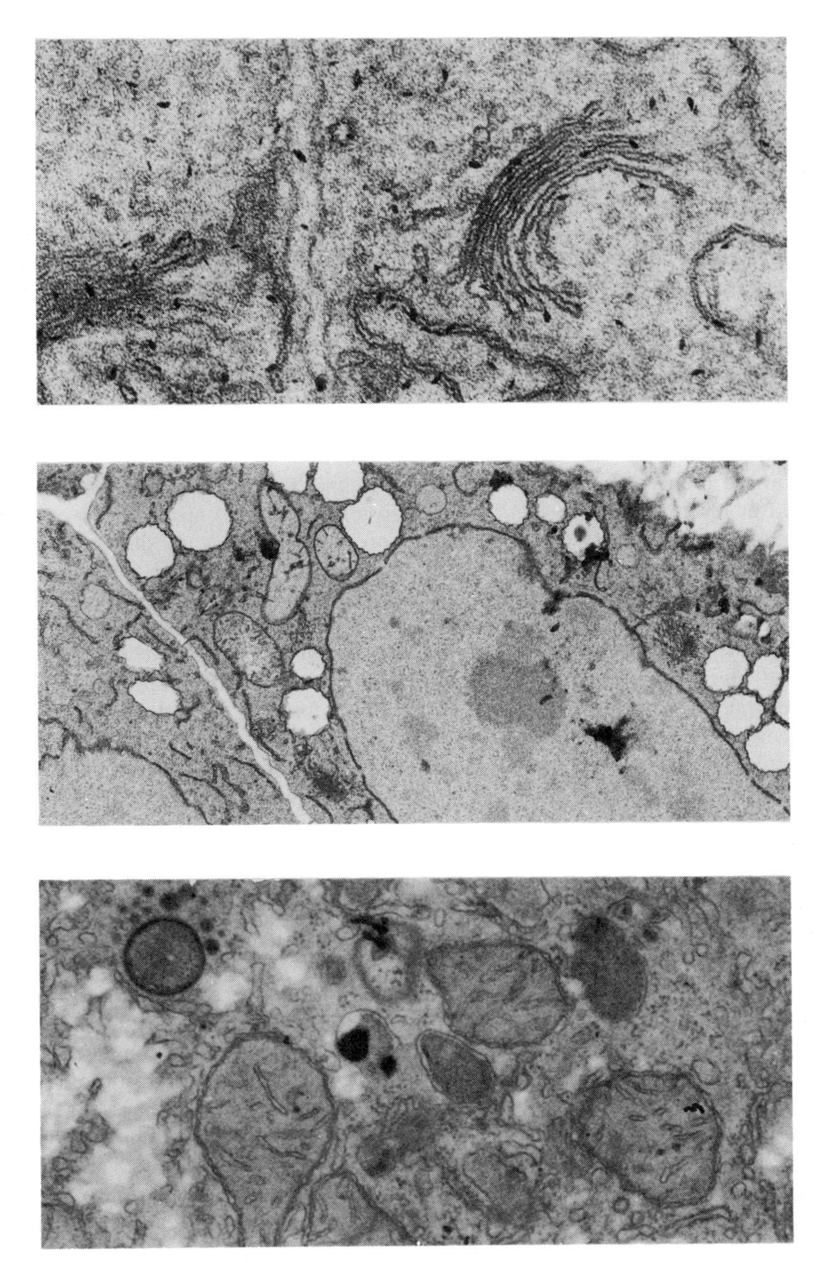

Figure 4.10 Various shapes and sizes of electron-opaque precipitates deposited on the surface of sections post-stained with lead citrate

ide are amphoteric and act as acid anhydrides, forming plumbite ion on treatment with alkali. Accordingly, it is expected that at high pH levels (>11.5) lead is present as the plumbite ion, $Pb(OH)_3^-$ or $Pb(OH)_4^{2-}$.

It was suggested by Karnovsky (1961) that staining is accomplished through hydrogen bonding of the plumbite ion to groups in the tissues. This hypothesis was supported by the observation that stained sections lost their stain when they were treated for a short time with 8 M urea (however, urea could chelate). When staining was preceded by urea treatment, no decrease in staining intensity was observed. Nevertheless, overwhelming evidence supports the position that in practice both anionic and cationic forms would be well hydrated and hydroxylated, and capable of extensive hydrogen bonding.

The increase in contrast with lead citrate after staining with uranyl acetate depends primarily on the amount of the latter bound to the specimen. It appears that lead citrate binds more strongly to the stain or fixative already bound to cellular materials than directly to these materials (Cattini and Davies, 1983). Variations in lead staining of various cell components and of cell types are due to differences in both their affinity for the stain and the rate of their extraction during staining. The mechanisms involved in the staining of individual cell components are discussed below.

Reaction with Membranes

With aldehyde fixation the membranes remain almost invisible before and after lead staining. In fact, membranes are not revealed even with OsO_4 and lead treatments when they follow dehydration by alcohols. Thus, membranes are stained by OsO_4 treatment only when it precedes the dehydration. This indicates that at least unsaturated lipids formerly present in the lipoprotein complexes of the membranes are extracted or irreversibly changed during dehydration.

Lead staining of membranes does occur after fixation with OsO_4, presumably because reduced osmium is indispensable for the attachment of lead to the polar groups of phosphatides. It is thought that OsO_4 reacts primarily with unsaturated lipids followed by cleavage of the diester and migration of osmium oxide to polar groups of the phosphatides. Since the osmium oxide thus bound is acidic, it exhibits a strong affinity for positive dye ions. This probably is the reason why the attachment of lead ions to membranes can only proceed at completely negatively charged polar groups. According to Hoogeveen (1970), lead binds firmly to phosphotidylcholine membranes *in vitro*, but organic Pb compounds in this system have not been identified as yet.

Reaction with Glycogen

The typical pattern of lead staining of glycogen is characterised by coarse, precipitate-like granules. This type of pattern can be understood by assuming that, through chelation, lead ions attach themselves to hydroxyl groups of carbohydrates. This initial, fine impregnation is followed by the accumulation of lead hydroxide around the primarily attached lead. Such an aggregation of lead results in the characteristic coarse, granule-like appearance of glycogen. Glycogen aggregates which stain intensely in thick sections appear as finely granular deposits in thin sections (~40 nm thick). Glycogen, in sections showing grey interference colour, usually stains less intensely than do ribosomes. This transition from densely staining aggregates to less dense, finely granular deposits suggests that the former are formed as a result of superimposition of lead-sequestering glycogen subunits of the latter.

The persistence of the positive reaction of glycogen with lead salts after OsO_4 and aldehyde fixation is probably related to the inertness of carbohydrates towards the former and their incomplete reaction with the latter. Since the glucose polymer of glycogen is not known to form coordination complexes with metal ions, it is difficult to visualise glucose as an ionic lead-binding site. It is conceivable that the staining of the glucose polymer depends upon the formation of a stable lead–glucose complex through hydrogen bonding.

It is well known that the density and configuration of glycogen particles are strongly influenced by the fixation and staining methods. Depending upon the method of preparation, glycogen may appear black, grey or light in colour. The metabolic state of the cell at the time of fixation may also influence the appearance of glycogen. Liver fixed with OsO_4 and stained with lead shows rosette-shaped glycogen, while glutaraldehyde-fixed and lead-stained specimens show glycogen in the form of irregular masses. The chemical factors responsible for this difference in morphology are not known. However, it is hypothesised that in the glutaraldehyde-fixed tissue, residual alcohol molecules may react with lead, thus producing an insoluble lead precipitate which blurs the image of glycogen. On the other hand, OsO_4 fixation prevents the subsequent interaction between residual —OH groups and lead. This hypothesis is supported by the fact that if the dehydration precedes fixation with OsO_4, glycogen stained with lead appears in irregular masses.

Another possible explanation of the rosette-shaped glycogen is that it is the result of extraction of a certain material associated with glycogen. It is known that fixation with concentrated solutions of paraformaldehyde causes rosette-shaped glycogen. Vye (1971) has studied the configuration of glycogen in β-particle form by employing periodic acid oxidation in conjunction with thiosemicarbazide, OsO_4, silver protein, sodium chloride, uranyl acetate and lead citrate. The range of variation in the appearance of liver glycogen due to different methods of fixation and staining has been discussed by Minio *et al.* (1966).

The pH of the OsO_4 solution also influences the appearance of glycogen. It has been shown that after direct immersion in OsO_4 (pH below 7.0) and post-staining with lead, the β-granules (small particles) have indistinct margins and show clumping, while α-granules (large particles) are pale with dark rims (Bhagwat and

Wong, 1972). On the other hand, at pH 7.0 or above, the β-granules show more sharp margins and less clumping, while α-granules are uniformly dark. A pH below 7.0 apparently is detrimental to the ultrastructure of glycogen.

The type of buffer employed may also influence the appearance of glycogen regions. It has been suggested that in the specimens exposed to cacodylate buffer glycogen regions seem to appear light following lead staining, whereas in the phosphate-buffer-treated specimens lead stains glycogen intensely (Leskes *et al.*, 1971).

Reaction with Other Cell Components

Biochemical studies indicate that lead forms mercaptides with the —SH group of cysteine and less stable complexes with other amino acid side-chains. Proteins with large numbers of free —SH groups (e.g. thionein) bind Pb^{2+} firmly both *in vitro* and *in vivo* (Ulmer and Vallee, 1969). Asparagine, cysteine, aspartate and citrate form insoluble complexes with Pb, which remain soluble in the presence of phosphate. It has been shown that the staining of zymogen granules (which are rich in free amino groups) with lead is abolished when the tissue is subject to oxidative deamination by dinitrofluorobenzene (Ponzo *et al.*, 1973). Lead tends to enhance the activity of some enzymes, while it inhibits the activity of other enzymes. A list of both types of enzymes has been presented by Vallee and Ulmer (1972).

Lead forms complexes with the phosphate groups of nucleotides and nucleic acids, and catalyses a non-enzymatic hydrolysis of nucleoside triphosphates (especially ATP) (Rosenthal *et al.*, 1966). Comparative studies indicate that RNA-containing nucleoproteins show a strong affinity for uranyl acetate. From the available evidence it can be assumed that the action of OsO_4 depresses the staining of RNA-containing nucleoproteins with lead salts, and strongly depresses that of DNA-containing nucleoproteins. The possibility of the attachment of lead to negative groups of the nucleoproteins has been suggested by Daems and Persijn (1963).

Lead shows a strong tendency to stain ground substance, and this is much more prevalent with tissues that have been fixed either primarily or secondarily with OsO_4. In connection with their studies on the fine structure of α-keratin, Filshie and Rogers (1961) demonstrated that the staining effect of lead hydroxide is not elicited unless the keratin has been previously stained by reduction with OsO_4. The affinity of this stain for cryptococcal polysaccharides was indicated by Hirano *et al.* (1964).

Although lead has no known chemical affinity for cellulose, cellulose microfibrils of cell walls in thin sections can be stained with lead citrate if preceded by uranyl acetate (Cox and Juniper, 1973). Sections are stained with 2% uranyl acetate followed by lead citrate for 30 min each. This staining is considered to be physical in nature, since the stain is removable by washing with water. Lead stains have been used for selective staining of certain components of the Golgi system (McClintock and Locke, 1982). Lead stains different cisternal components from those stained by uranyl acetate.

LEAD ACETATE

In the method described by Dalton and Zeigel (1960) lead acetate rather than lead hydroxide is used. It is believed that the formation of lead hydroxide occurs directly in the sections, which reduces the formation of lead carbonate to a minimum.

A saturated solution of lead acetate is prepared in boiled distilled water and kept in a glass-stoppered bottle containing some undissolved crystals of the stain. To prevent the formation of lead carbonate, the bottle is kept completely filled; whenever a few drops of the stain are removed, the bottle is refilled to the top with boiled distilled water. The sections on the grids are stained for 3–20 min at pH 5.9, followed by an immediate thorough rinse in boiled distilled water. The grids are blotted dry and then exposed for 5 s to the vapours of 1–5% ammonium hydroxide. This treatment imparts a general increase in contrast. It is believed that at higher pH levels greater numbers of anionic sites are made available in most proteins for binding the excess lead present in the sections. In general, lead acetate imparts less density than does lead hydroxide. A saturated solution of monobasic lead acetate ($Pb(C_2H_3O_2)_2 \cdot Pb(OH)_2 \cdot H_2O$) selectively stains ribonucleoprotein granules, since more carboxyl groups become available for combining with lead in solutions from nucleoproteins than from other proteins at pH 7.0.

Another modification of the lead acetate method was introduced by Björkman and Hellström (1965). With this method the precipitation of lead carbonate, on exposure to atmospheric CO_2, is eliminated and the staining is obtained at a low pH level. Thus, this technique differs from most other methods in that the staining is accomplished at a nearly neutral pH, rather than at a very high pH level. Since lead salts are soluble in ammonium acetate, and form soluble, slightly dissociated lead acetate compounds, a stable staining solution can be prepared by combining ammonium acetate (NH_4Ac) and lead acetate ($PbOAc_2$). When ammonium acetate is combined with lead acetate in the molar proportion of 2:1, a possible resulting ionising compound is $2(NH_4)^+(PbOAc_4)^{2-}$.

The staining solution is prepared as follows:

Ammonium acetate	18.5 g
Saturated lead acetate solution	100 ml

Saturated lead acetate solution is prepared by dissolving 39 g of lead acetate in 100 ml distilled water. The ammoniated concentrated lead solution remains clear for several weeks, even after it is insufflated with expired air for 30 s. On the other hand, the staining solution loses its stability on the addition of as little as 2% water. It is important, therefore, that the solution always be saturated with respect to lead acetate. In general, tissue specimens are fully stained after immersion for 45 min in the solution. After staining, the specimen is rinsed with distilled water.

The staining obtained with ammoniated lead is not as intense as that obtained with Reynolds' lead citrate, although the formation of contaminants is almost eliminated in the former method.

Since lead acetate is soluble in ethyl alcohol and this solution is stable on exposure to air, sections can be stained suitably and efficiently with this solution (Kushida, 1966a). To prepare the stain, lead acetate is added in excess to 100% ethyl alcohol. The mixture is gently stirred for ~10 min and then filtered. This saturated filtrate is the staining solution, which can be stored in a closed container. There is some evidence that exposure of tissue to acetone enhances the staining activity of the lead. It is therefore advisable to expose the tissue specimen to acetone after dehydration with alcohols. The grids are stained by immersion, washed thoroughly in 100% alcohol and blotted dry.

Kushida (1966b) demonstrated that tissue blocks can be suitably stained with lead acetate before embedding, and that the stained tissue blocks section as well as unstained ones, because the hardness of the former is not changed by staining. In addition, the tissue blocks are not overstained and section contamination is thereby avoided. The staining solution is prepared by saturating 100% ethyl alcohol with lead acetate, followed by the addition of an equal volume of 100% acetone; the mixture is stirred for ~10 min and filtered. This filtrate, which is the staining solution, is most effective when used immediately after filtration. Tissue blocks are stained adequately in 1–2 h. After staining, the blocks are washed in two changes of 1:1 mixture of 100% alcohol and 100% acetone for 10–20 min. The staining quality is not affected if the stained tissue is exposed to reactive solvents such as propylene oxide.

LEAD ASPARTATE

Lead aspartate was introduced for *en bloc* staining by Walton (1979). Contamination of sections with lead carbonate is avoided by this method. Sections can be viewed immediately after they are cut. Contrast enhancement by lead aspartate is slower than that obtained with lead citrate. Thus, staining with the latter is easily controllable. Lead aspartate seems to stain nucleic acids more intensely than does lead citrate in the absence of staining with uranyl acetate (J. M. Yoshiyama, personal communication). If *en bloc* staining with uranyl acetate is required, it should precede the treatment with lead aspartate. A thorough rinse with distilled water is necessary between the two staining treatments.

Since lead aspartate is used at pH 5.3, ionisable hydroxyl groups in glycogen are unavailable for Pb^{2+} binding. Consequently, lead aspartate does not stain glycogen. Lead citrate, on the other hand, is used at pH 12, providing binding sites for Pb^{2+}. The overall quality of ultrastructural preservation is less than satisfactory with lead aspartate. *En bloc* staining should be attempted as a last resort.

LEAD CITRATE

At present, Reynolds' (1963) lead citrate preparation is the most widely used stain. This preparation differs from other alkaline stains in that the chelating agent employed is citrate. Since citrate is added in sufficient excess to sequester all of the lead present, lead is prevented from combining with carbonate, which results in an appreciable reduction in the formation of contaminants. Citrate does not appear to interfere with staining, because anionic tissue binding sites apparently have a greater affinity for the lead cation than for citrate. Most probably, citrate forms stable complexes with cationic alkaline lead salts. Thus, it is likely that lead citrate transfers its lead component to tissue binding sites such as cysteine, orthophosphate and phosphate groups, and to other lead-sequestering sites formed during fixation. The staining solution is prepared as follows:

Lead nitrate ($Pb(NO_3)_2$)	1.33 g
Sodium citrate ($Na_3(C_6H_5O_7)\cdot 2H_2O$)	1.76 g
CO_2-free distilled water	30 ml

In order to complete the conversion of lead nitrate to lead citrate, the mixture is shaken vigorously at intervals for ~30 min in a 50 ml volumetric flask. The completion of the conversion is superficially indicated by the appearance of a uniform milky suspension. To this suspension is added 8 ml of 1 N NaOH. It is then diluted to 50 ml, and mixed by inversion until lead citrate dissolves and the suspension clears up completely. For staining, this solution is used either as it is or diluted 5–1000 times with 0.01 M NaOH, depending upon the contrast desired. The author prefers to use the solution in its concentrated form. After this solution has been prepared, a small amount is decanted and the pH is tested. It should be between 11.9 and 12.1. If the pH is too low, 1 N NaOH is added dropwise until the desired pH is reached. If the pH is too high, the solution should be discarded.

It is important that the NaOH used to prepare the solution should be fresh and carbonate-free. Only freshly purchased reagents should be used. Reagents such as sodium citrate and NaOH pellets that have been stored for about 6 months should not be used. Sodium hydroxide pellets used from frequently opened containers are a major source of carbonate contamination. Carbonate-free, boiled distilled water should be used for preparing lead solutions as well as for rinsing the grids. Water is boiled for 10 min to remove CO_2. It is then covered and allowed to cool, which may take ~1 h. Alternatively, water can be degassed in a vacuum desiccator. When the desiccator is evacuated, water boils. This water can be used immediately after air is admitted into the desiccator, because it has not been heated. Stock and staining solutions will keep for several months if tightly stoppered; if precipitates appear, they should be discarded. Lead solutions should be stored in plastics containers rather than in glass, for alkaline lead citrate solution leaches Si-containing compounds from glass in sufficient quantities to form precipitates during the staining procedure. The staining procedure is described on p. 313.

A simplified method for the preparation of lead citrate staining solution was recommended by Venable and Coggeshall (1965). To 10 ml distilled water is added 0.01–0.04 g of commercially available lead citrate and 0.1 ml of 10 N NaOH in a screw-topped vial. The vial is closed tightly and the mixture shaken vigorously until all the lead citrate is dissolved. To minimise the formation of lead carbonate, one should use carbonate-free NaOH and avoid exposing the solution to atmospheric CO_2. A staining time of ~1 min is adequate for most tissues. Longer staining times should be avoided, to lessen the chance of the formation of contaminants.

Another simplified method using commercially available lead citrate was used by Fahmy (1967). Freshly prepared staining solution is used each time and then discarded. Previously boiled, cooled, double-distilled and Millipore-filtered water, which can be stored in the refrigerator, is used. One pellet of NaOH (0.1–0.2 g) is dissolved in 50 ml of the above water. Approximately 0.25 g of lead citrate is added, and the solution is ready for use. The pH is ~12. Sections are stained for 3–15 min.

Another modification of lead staining was introduced by Sato (1967). This modification was refined by Hanaichi *et al.* (1986). The procedure of preparing this staining solution is simple, and it can be stored for months at room temperature without the production of lead carbonate precipitates. The formulation is given below.

Calcined lead citrate ($Pb(C_6H_5O_7)_2$)	0.2 g
Lead nitrate ($Pb(NO_3)_2$)	0.15 g
Lead acetate ($Pb(CH_3COO)_2 \cdot 3H_2O$)	0.15 g
Sodium citrate ($Na_3(C_6H_5O_7) \cdot 2H_2O$)	1.0 g
Distilled water	41.0 ml

The calcined lead citrate is obtained by heating lead citrate crystals for several hours at 200–300 °C until the colour changes to light-brownish yellow. Overheated lead citrate with a dark-brownish or black colour should not be used. The above reagents are placed in a 50 ml flask and mixed well, to produce a yellowish milky solution. About 9 ml of 1 N NaOH solution is added while stirring, to obtain a light yellow colour. A staining time of 2 min is recommended. The author finds this staining solution superior to all other lead solutions.

LEAD HYDROXIDE

Lead hydroxide does not stain when dissolved in neutral or acidic solutions. In alkaline solutions, however, it is a very effective stain. Since it is effective as a stain only in alkaline solutions, it can be assumed that a chemical reaction occurs when sodium or potassium hydroxide is added to lead hydroxide, which results in the formation of one or more new salts in the mixture. In fact, in such a mixture the presence of compounds such as lead oxides and hydroxides, sodium plumbite and hydroplumbite has been indicated (Meller, 1927), although their isolation has not as yet been accomplished. Saturated solutions of lead hydroxide stain more intensely than either lead acetate or monobasic lead acetate. Presumably the reason for this is that more anions become available for combining with lead at increasingly higher pH levels.

Although Watson's original formulation is highly effective and has probably not been surpassed as a general stain, its preparation is rather tedious and has the disadvantage of forming lead carbonate immediately on exposure to CO_2 of the air and breath, resulting in contamination of the section. This reaction takes place so quickly that, even though staining has been carried out in a filtered solution and in a CO_2-free atmosphere, a precipitate may form while the grid is being transferred from the staining solution to water for washing. In other words, if the grid comes in contact with lead hydroxide and the air at the same time, even for a few seconds, the precipitate may form. The formation of lead carbonate can be avoided by preventing the stain from coming in contact with air. To accomplish this, several elaborate procedures have been introduced (see, e.g. Normann, 1964). There is no doubt that some of these methods are satisfactory, at least when carried out with skill.

Another approach to lessen the formation of insoluble lead carbonate is the modification of Watson's lead hydroxide stain. Several modifications of the original formulation have been introduced (Lever, 1960; Karnovsky, 1961; Reynolds, 1963; Venable and Coggeshall, 1965; Björkman and Hellström, 1965; Sato, 1967; Hanaichi *et al.*, 1986), and these improved preparations have largely replaced Watson's original method. The tartrate and citrate preparations of Millonig (1961) and Reynolds (1963), respectively, contain the lead in a chelated form, and thus result in far less contamination. The risk of contamination is also reduced with Karnovsky's hydroxide. All of the variants listed above are safer to handle and easier to prepare, but the degree of contrast yielded by Watson's lead hydroxide is not always obtained.

Lever (1960) prepared the staining solution by bringing to the boil 1 g of lead hydroxide in 100 ml of distilled water, after which the solution is cooled and filtered to remove any possible contaminants. Then 2 N potassium hydroxide is added drop by drop to the filtrate until upon agitation it clears completely. This is the staining solution, which probably contains some lead in the form of plumbite. After staining the grids by immersion in drops of the staining solution in a Petri dish for 5 min, they are rinsed in 1% solution of potassium hydroxide, after which they are thoroughly rinsed in distilled water. If the grids show some contamination, it can be removed by further immersion in potassium hydroxide. It is emphasised that the duration of exposure to potassium hydroxide is critical, since the alkali tends to destain or cause uneven staining.

Karnovsky (1961) believed that highly alkaline solutions of lead hydroxide (pH >11.5) yield stable solutions which stain rapidly and intensely, without necessitating excessive precautions against lead carbonate contamination. He also thought that in these solutions the plumbite ion is responsible for the staining. He introduced two methods for preparing the staining solutions.

In the first method 270 mg of lead monoxide (PbO) is added to 20 ml of 1 N sodium hydroxide in a flask, and the lead monoxide dissolves when the mixture is gently boiled for 2 min with continuous stirring. The flask is stoppered, and the mixture is cooled rapidly to room temperature and then filtered through Whatman No. 1 filter paper. The resultant filtrate is the concentrated stock solution, which can be stored for months. The final staining solution is prepared by diluting the stock solution 1:20 or 1:50 with distilled water. The diluted solution is also stable. Immediately before use, the solution is centrifuged in a stoppered tube for 5–10 min.

In the second method lead monoxide is added in excess to 15 ml of 10% sodium cacodylate solution in water in a flask. The flask is stoppered and the mixture is stirred thoroughly for 10–15 min (a magnetic stirrer is preferred) and filtered, since some lead monoxide remains undissolved after the stirring. The filtrate is diluted 2:8 with 10% sodium cacodylate with continuous stirring, and then 1 N sodium hydroxide is added drop by drop from a pipette. A faint cloudiness may gradually form, but the addition of more alkali drops will clear the solution. Only enough drops are added to clear the solution once more. If the cloudiness does not appear, a maximum of six drops is required. The clear solution is the final staining solution, which can be stored for months.

The stain prepared by the first method stains more rapidly and intensely than that prepared by the second method, although the latter is more stable on exposure to air than is the former.

LEAD TARTRATE

Millonig (1961) achieved stabilisation of an alkaline lead hydroxide solution by the addition of sodium potassium tartrate. The addition of tartrate does not greatly reduce the staining capacity, and the solution stays clear even when not protected from the carbon dioxide of the air and breath. Among various lead salts, tartrate has the advantage of greater stability. Tartrate is, however, a relatively weak complexing agent. The stock solution is prepared as follows:

NaOH	12.5 g
K–Na tartrate	5.0 g
Distilled water to make	50.0 ml

To prepare the staining solution, 0.5 ml of this stock solution is diluted to 100 ml with distilled water and heated, and 1 g of lead hydroxide is added. After cooling, the solution is filtered and should stay clear. The pH should be approximately 12.3. A staining time of 5–15 min is adequate.

Millonig (1961) suggested adding tartrate to stabilise

alkaline lead acetate solution in the same manner that he proposed for lead hydroxide solution. The stock solution is prepared as follows:

NaOH	20 g
K–Na tartrate	1 g
Distilled water to make	50 ml

To prepare the staining solution, 1 ml of this stock solution is added to 5 ml of a 20% lead acetate ($Pb(CH_3COO)_2{\cdot}3H_2O$) solution. This solution is stirred, diluted 5–10 times with distilled water and filtered. The resulting solution is clear and ready for staining. A staining time of 5–15 min is adequate.

EN BLOC STAINING WITH LEAD

Lead citrate can be used for *en bloc* staining. The formulation is given below.

Solution A	
Formaldehyde (6%)	25 ml
Glutaraldehyde (20%)	12 ml
Phosphate buffer (0.1 M, pH 7.4)	13 ml
Final solution	
Solution A	12 ml
Osmium tetroxide (2%)	8 ml
Lead citrate solution (saturated)	4 ml

All solutions should be prepared in phosphate or arsenate buffer and kept at 4 °C; at higher temperatures the solution turns brown, although no precipitate is formed. Solution A should be freshly prepared, while others can be stored in the cold for several weeks.

A mixture of 0.14% lead citrate and 0.006 M $MnCl_2$ has been used for *en bloc* staining of the vesiculotubular system in the gill epithelium (Pisam *et al.*, 1987). Tissue blocks are further stained with ferrocyanide-reduced osmium for 1 h to increase the contrast of cytoplasmic membranes without staining the background.

Lead acetate also has been employed for *en bloc* staining (Kushida, 1966b).

Lead aspartate has been used for *en bloc* staining. The preparation of the staining solution is as follows:

Aspartic acid stock solution	
Aspartic acid	0.998 g
Glass-distilled water	250 ml

The aspartic acid dissolves more rapidly when the pH is raised to 3.8. The stock solution is stable for 1–2 months.

Staining solution	
Stock solution	10 ml
Lead nitrate	0.066 g

The pH is adjusted to 5.3 with 1 M KOH. The solution contains 0.02 M lead nitrate and 0.03 M aspartic acid. It should be heated to 70 °C before use, to ensure its stability. Tissue blocks fixed with glutaraldehyde and OsO_4 are placed in a stoppered vial and covered completely with the staining solution for 30–60 min at 70 °C (Yoshiyama *et al.*, 1980).

DOUBLE LEAD STAINING METHOD

The double lead staining method improves the contrast of tissue components that show inadequate contrast with conventional staining (Daddow, 1986). This inadequate contrast may be due to prolonged fixation with glutaraldehyde or the embedding medium (e.g. Spurr mixture) used. Thin sections of aldehyde- and OsO_4-fixed specimens are stained with Reynolds' lead citrate for 30 s, rinsed in distilled water, stained with saturated uranyl acetate for 1 min, rinsed in distilled water and stained again with lead citrate for 30 s. The lengthy duration of staining needed by sections of Spurr medium can be reduced by pretreatment of sections with 10% H_2O_2 for 1 min (Pfeiffer, 1982). However, this treatment may render certain osmiophilic tissue components vulnerable to extraction during subsequent staining with salts of uranium and lead (Parker, 1984). Hydrogen peroxide can be replaced with 0.02 M NaOH (Daddow, 1986).

GLYCOGEN STAINING

It has been stated earlier that lead at an alkaline pH stains many cytoplasmic components, including glycogen. Glycogen and ribosomes can be stained selectively with lead, provided that bound osmium is removed from tissue sections prior to staining by oxidation with either periodic acid (PA) or H_2O_2. However, the available evidence indicates that this method is only moderately specific for glycogen. Overoxidation of thin sections should be avoided, for it causes extraction of glycogen. Staining with uranyl acetate may obscure the fine details of glycogen particles.

Thin sections are exposed to 0.8–3.2% PA for ~30 min or to 1.5% H_2O_2 for 10–15 min. The sections are rinsed in distilled water and then stained with lead citrate for 10–15 min.

TRICOMPLEX FIXATION AND STAINING

The acellular layer lining the alveolar walls is a complex

of lipids, proteins, carbohydrates and inorganic substances. The highly surface-active components of the layer are called surfactants. They are considered to be primarily dipalmitoyl lecithin containing large amounts of saturated lipids. Neither glutaraldehyde nor OsO_4 preserves the surfactants.

The method of tricomplex fixation has been claimed to be effective in preserving the surfactants. It employs cations and anions which form a link between the phospholipid amphoions, and the resulting electrostatic interactions provide cohesion between the phospholipid molecules. The essential components of the fixative are lead nitrate ($Pb(NO_3)_2$) and potassium ferricyanide ($K_3Fe(CN)_6$).

Contrary to the claim found in the literature, amorphous precipitates formed on the alveolar surface of the lungs, followed by treatment with tricomplex fixatives, are non-specific and cannot be regarded as being related to surfactant phospholipids. It is possible that these precipitates may be artefacts. The possibility that these precipitates are non-specific has been strengthened by the evidence that identical precipitates are found even after lipid extraction with chloroform–methanol. Furthermore, identical precipitates have been found within the cytoplasm of red blood cells (Gil, 1972).

Elbers *et al.* (1965) first employed the method of tricomplex fixation for phospholipid in the brain, and Dermer (1970) used this method to preserve pulmonary surfactant. These studies were carried out with tissues fixed by immersion. Adamson and Bowden (1970) applied this method to demonstrate purportedly saturated phospholipid in the lung. They used various methods, including intratracheal fixation and vascular perfusion with glutaraldehyde, to preserve the tissue. The procedure described below is most effective and was employed by Finlay-Jones and Papadimitriou (1972).

The trachea of the anaesthetised animal (e.g. rabbit) is exposed and clamped, and the thorax is opened. Vascular perfusion into the right ventricle is carried out with 3% glutaraldehyde buffered to pH 7.4 with 0.1 M sodium cacodylate and HNO_3 at 4 °C. After an initial influx of the perfusate, the left atrium is opened and 1 ml of glutaraldehyde per gram body weight is injected slowly. The heart, lungs and trachea are dissected from the animal and immersed in fresh glutaraldehyde for 1 h at 4 °C.

The lungs are then filled to full expansion by injecting a mixture of $Pb(NO_3)_2$ and $K_3Fe(CN)_6$. The mixture is prepared by dissolving 0.826 g $Pb(NO_3)_2$ and 0.549 g $K_3Fe(CN)_6$ in 100 ml distilled water. The trachea is clamped and the whole lung incubated in the mixture for 30 min at 37 °C. Small pieces of tissue are excised, washed in the buffer and post-fixed in buffered 1% OsO_4 for 1 h. The success of this method depends upon satisfactory fixation of the tissue. Therefore, tissue blocks should be chosen from those areas of the lung which have been fully permeated by the salt mixture.

The advantage of this procedure is that by vascular perfusion the fixative comes in contact with the cell surface from the aqueous side and thus leaves the surfactant largely intact. In contrast, with immersion fixation the aerated lung tissue is forced to change from air to aqueous phase. This treatment destroys the bimolecular leaflet of phospholipid orientated at the cell surface, which results in the formation of micelles, which may be extracted.

OSMIUM TETROXIDE

Osmium (atomic number 76) is slightly heavier than tungsten. The primary role of OsO_4 in electron microscopy is as a reliable fixative rather than as a stain. Nevertheless, it does stain membranous structures, ribosomes, Golgi complex and multivesicular bodies. The staining is due to the deposition of lower oxides of osmium, although a small degree of density may be contributed by organically bound but unreduced osmium. The mechanism of osmium reaction with unsaturated lipids and proteins and its attachment to the lipoprotein membranes has been discussed in Chapter 1. Little is known concerning the chemical nature of reaction between osmium and ribosomes (see Hayat, 1981). This reagent has little direct staining action on viruses. In fact, it tends to damage many isometric plant viruses, such as turnip yellow mosaic and wild cucumber mosaic (Hills and Plaskitt, 1968).

Osmium deposition and the resultant staining can be increased by prolonging the duration of treatment. Osmium staining is usually further intensified by post-staining with lead, PTA or uranyl acetate. Staining by osmium can also be increased by treating the tissue with *p*-phenylenediamine (PPD); this procedure will be discussed later. It can also be dramatically increased by treatment with thiosemicarbazide or thiocarbohydrazide and additional osmication (Hanker *et al.*, 1966).

The addition of ions to fixation or staining solutions of OsO_4 influences the staining intensity of specimens. For instance, the addition of divalent cations to the OsO_4 solution enhances the staining intensity of heterochromatin of photoreceptor nuclei (Ladman, 1973). One possible explanation for this enhanced staining is that several divalent and monovalent cations bind to nucleotides to produce heterochromatin staining. Increasing the sucrose concentration to 5% or higher also results in enhanced heterochromatin staining. Cytoplasmic membranes of Gram-negative bacteria have also been reported to show enhanced staining when the specimens are fixed with OsO_4 in the presence of Ca^{2+} (Silva and Sousa, 1973). It is emphasised that very little information is available on the role of ions in the staining mechanism.

The intensity of overall staining produced by OsO_4 can be enhanced by PPD with bound but unreduced osmium. Fixed tissue blocks are post-treated with 1% PPD in 70% ethanol for 15–25 min; sections do not require additional staining (Ledingham and Simpson, 1972). Semithin sections of the tissue thus treated can be examined with a light microscope, without additional staining. This method eliminates the need for post-staining the sections for both light and electron microscopy. Caution is warranted in handling PPD, for it can cause skin irritation, contact dermatitis and bronchial asthma.

Available data indicate that certain cellular structures fixed in glutaraldehyde acquire the ability to bind stains only after OsO_4 post-fixation, whereas some other structures fixed in glutaraldehyde can bind the same stains without OsO_4 post-fixation. An example of the former type is the surface coat of intestinal microvilli; and of the latter, myelin sheaths (Napolitano *et al.*, 1968). In this connection, the solvent medium for OsO_4 is also important. For instance, it has been demonstrated that the acid mucopolysaccharide-rich surface coat of the intestinal epithelium can bind heavy metal stain when the tissue is post-fixed with phosphate-buffered OsO_4, whereas the surface is incapable of binding heavy metal stains after the tissue is post-fixed with OsO_4 in carbon tetrachloride (Pratt and Napolitano, 1969).

Osmium tetroxide has been used successively to stain the Golgi complex and the vesicles found within multivesicular bodies in the rat epididymal tissue. Smooth- and rough-surfaced endoplasmic reticulum are usually stained much less intensely by OsO_4 than are the elements of the Golgi complex. Thus, it is possible to distinguish, on the basis of the intensity of staining, the cisternae of these two organelles after controlled OsO_4 treatment. It is pointed out, however, that cell structures other than Golgi complex and endoplasmic reticulum show large concentrations of reduced osmium compounds after prolonged exposure to OsO_4 solution. When parietal cells of the stomach were exposed to OsO_4 for 60 h, in addition to Golgi complex and endoplasmic reticulum, perinuclear cisternae, mitochondrial cristae, the surface coat of the microvilli of intracellular canaliculus and multivesicular bodies were stained (Winborn and Seelig, 1974). In this connection, it is pertinent to remember that the sites of the deposition of reduced osmium compounds may not necessarily be the sites of reduction. Apparently, correlative biochemical studies are needed to explain the chemical composition of the cellular materials which produce the stain deposition. A somewhat preferential staining of the Golgi complex can be obtained by the following method (Friend and Murray, 1965).

Small fragments of fresh tissue are fixed with unbuffered 2% OsO_4 (pH 6.2–6.8) in foil-wrapped vials in an oven at 40 °C. After 24 h, the solution is replaced by fresh 2% OsO_4, and the tissue is returned to the oven for another 16–24 h. The tissue is then treated with 0.5% uranyl acetate for $1\frac{1}{2}$ h at room temperature, prior to dehydration and embedding.

In another procedure OsO_4 can be deposited selectively within the Golgi apparatus and rough- and smooth-surfaced endoplasmic reticulum of zona fascicula cells of the rat adrenal cortex active in steroid synthesis and secretion. In this technique osmium is not deposited in lipid droplets, in mitochondrial matrix inclusions or in the outer mitochondrial spaces. It has been suggested that steroids, rather than cholesterol or neutral lipids, are at least partially responsible for this staining reaction. It is likely that a prolonged immersion in OsO_4 is responsible for retention and staining of steroids. The staining procedure used by Friend and Brassil (1970) is given below.

Immediately after mincing, the tissue is immersed in 2% aqueous OsO_4 solution (unbuffered) in vials wrapped with aluminium foil to exclude light for 40–48 h at 40 °C. The solution is decanted after 24 h and replaced by fresh solution. The pH of the solution is lowered to 3.0 during this procedure.

The OsO_4–ethanol technique has been employed to reveal osmiophilic periodic lamellae of synthetic dipalmitoyl lecithin, and the surface film of acellular alveolar lining layer of the mammalian lung (Kaibara and Kikkawa, 1971). This technique has also been used to examine the distribution of osmiophilic lamaellae within the alveolar and bronchiolar walls of mammalian lungs (Kikkawa and Kaibara, 1972). According to this technique, tissue specimens are fixed with glutaraldehyde followed by OsO_4. After rinsing in a suitable buffer, the specimens are post-osmicated in a mixture containing 2% OsO_4 and 100% ethanol in equal parts. Some of the osmiophilic content observed after the second osmication is not seen after initial osmication.

Osmium textroxide in vapour form has been employed for staining thin frozen sections (Werner *et al.*, 1974). After drying in an incubator at 50–60 °C, grids with sections are placed in a small vial containing OsO_4 crystals for $\frac{1}{2}$–1 h. This treatment results in the general staining of cell organelles, including nuceli, mitochondria, granular endoplasmic reticulum and microtubules in flagella.

Osmium tetroxide can also demonstrate calcium-binding sites in neurons. The central nervous system fixed with s-collidine-buffered OsO_4 (1.5%) containing 5 mM $CaCl_2$ shows electron-dense granules associated with synapses; these granules probably represent calcium binding sites in neurons (Sampson *et al.*, 1970). The OsO_4–ruthenium red method has been used to study cell wall structure of both cultured plant cells (Leppard *et al.*, 1971) and native plant tissues (Leppard and Colvin, 1972). The details of this method are given in the section on ruthenium red.

OSMIUM TETROXIDE–IMIDAZOLE COMPLEXES

The kinetics of the oxidation of imidazoles by OsO_4 in both the presence and the absence of ligands (e.g. pyridine) have been studied by Kobs and Behrman (1987). Imidazoles can function as ligands as well as olefins (Deetz, 1980). Three reactions of OsO_4 with imidazoles are given below (Kobs and Behrman, 1987):

Reaction 1

OsO_4 + HN N ⇌ HN N-OsO_4

1

Reaction 2

1 + 2 Im → [cyclic osmate ester with two imidazole ligands]

2

Reaction 3

2 —H^+ or OH^-→ ring cleavage products + Os(VI) species

The first reaction indicates OsO_4–ligand complexation, which is a relatively fast reaction. The second reaction shows addition of OsO_4 to the 4,5 double bond of the imidazole to form the cyclic osmium(VI) ester, which is a slow reaction. The third reaction shows degradation of the osmate ester via ring cleavage of 4,5-dihydroxydihydroimidazole, which is very slow. The final reaction product seems to be an osmate(VI) ester of a 4,5-dihydroxydihydroimidazole with two imidazole molecules occupying the remaining coordination sites.

Like tannic acid, imidazole compounds act as mordants for heavy metals and improve the retention and staining of certain cellular components. Addition of imidazole to OsO_4 results in improved staining of lipids compared with that yielded by conventional post-fixation of glutaraldehyde-fixed specimens with buffered or unbuffered OsO_4 (Angermüller and Fahimi, 1982). Lipids containing unsaturated fatty acids as well as lipoproteins are strongly stained with this mixture. Lipid droplets appear well circumscribed, with no clear evidence of diffusion.

Treatment of cells with a mixture of OsO_4 and 3-amino-1,2,4-triazole produces improved staining of membranes that stain less clearly with the conventional double fixation method (van Emburg and de Bruijn, 1984). The contrast of endomembranes of these cells can be further improved by post-staining thin sections with salts of uranium and lead. The staining of apolar regions of membranes is increased with a mixture of OsO_4 and triazole, while the contrast of polar regions of membranes is enhanced with a mixture of potassium osmate and aminotriazole (de Bruijn *et al.*, 1984). The latter mixture also increases the staining of glycogen.

The increased staining of polar regions of membranes seems to be due to the presence of an amino group in the heterocyclic nitrogen compounds, such as triazole. It is thought that these nitrogen compounds stabilise certain osmium valency species (Os^{VI}, Os^{IV} and Os^{III}), which are responsible for increased contrast (de Bruijn *et al.*, 1984). As a result, imidazole compounds stabilise hexavalent osmium in the resulting glycogen osmate and cyclic oxo compounds of membrane lipids. These methods have the potential of differentiating various types of lipids, and could be used in conjunction with selective lipid extraction techniques.

The staining of the apolar membrane regions is increased by a reaction with mixtures of OsO_4 and 1,2,3-triazole or tetrazole. The increased contrast-staining of the apolar regions is due, it is suggested, to the stabilisation of the cyclic osmic esters (resulting from the OsO_4 reaction with membrane lipid alkenes) with 1,2,3-triazoles or tetrazoles. This reaction product might be similar to the product in reaction 2 (see left), proposed by Kobs and Behrman (1987). In reactions by mixtures with osmate ($K_2OsO_2(OH)_4$) plus triazoles this cyclic oxo compound is absent and, hence, the membrane contrast (de Bruijn *et al.*, 1984).

The latter mixture also increased the electron-scattering capacity of the glycogen. This was explained by the assumption that osmate had reacted with hydroxylic groups as shown before for the complex formation with cyanides. ESCA measurements revealed that the amino-triazole was unable to reduce the osmate to lower oxides within 24 h (de Bruijn *et al.*, 1984). The reaction of osmate and 1,2,3-triazole was confirmed spectroscopically, *in vitro*, by a shift in absorption maximum in osmate–triazole mixtures (Riemersma *et al.*, 1984).

OSMIUM TETROXIDE—IMIDAZOLE (Angermüller and Fahimi, 1982)

Slices of glutaraldehyde-fixed tissues about 50 μm thick are placed in 0.2 M imidazole and then 4% aqueous OsO_4 is added; the final concentration of OsO_4 is 2% and the pH is 7.5. This post-fixation is carried out for 30 min at room temperature. Osmium tetroxide–imidazole penetrated tissue slices 50 μm thick and 1 mm^3 tissue blocks in 5 and 30 min, respectively.

OSMIUM TETROXIDE–3-AMINO-1,2,4-TRIAZOLE (van Emburg and de Bruijn, 1984)

Glutaraldehyde-fixed cells are thoroughly rinsed in 0.1 M cacodylate buffer containing 0.03 M $CaCl_2$ and 0.25 M sucrose for 1 h at 4 °C. They are again rinsed twice in the above buffer containing 0.02 M 3-amino-1,2,4-triazole (Sigma). Secondary fixation is carried out with 1% OsO_4 in 0.1 M cacodylate buffer (pH 7.2) containing 0.03 M $CaCl_2$ and 0.1 M 3-amino-1,2,4-triazole for 90 min at 4 °C.

POTASSIUM OSMATE–3-AMINO-1,2,4-TRIAZOLE (de Bruijn *et al.*, 1984)

Glutaraldehyde-fixed tissue specimens are rinsed in 0.1 M cacodylate buffer (pH 7.3) for 1 h at 4 °C, and then post-fixed with freshly prepared 1.27% potassium osmate ($K_2OsO_2(OH)_4$) in the buffer containing 0.05 M 3-amino-1,2,4-triazole for 24 h at 4 °C. The final pH of this mixture is ~11.4. Potassium osmate is prepared by dissolving 1 g of KOH in 10 ml of distilled water, to which is added 1 g of OsO_4. After dissolving the OsO_4, the solution is cooled to 0 °C and poured into 10 ml pure ethanol (0 °C) while stirring. The precipitate is collected, washed with pure ethanol (0 °C) and dried over solid KOH in a desiccator.

OSMIUM TETROXIDE–POTASSIUM FERRICYANIDE OR FERROCYANIDE

Osmium tetroxide in combination with ferricyanide ($Fe(CN)_6^{3-}$) or ferrocyanide ($Fe(CN)_6^{4-}$) has been used for enhanced contrast-staining of a wide variety of cellular components. De Bruijn (1968) first demonstrated the improvement of membrane and glycogen staining by addition to OsO_4 of complex iron cyanides, ruthenium and osmium cyanides. Karnovsky (1971) confirmed the usefulness of ferrocyanide-reduced OsO_4 to enhance membrane contrast.

Osmium tetroxide–potassium ferricyanide or ferrocyanide mixtures (OsFeCN) are used as a post-fixative for aldehyde-fixed specimens. The OsFeCN mixture yields enhanced electron-scattering capacity of many cellular components compared with that obtained by OsO_4 alone, and reduces the electron-scattering capacity of ribosomes and nuclear chromatin. Both osmium and complex cyanides must be present simultaneously (de Bruijn, 1973) to produce the enhanced contrast in the aldehyde-fixed specimens. This suggests that intermediate compounds are formed by the reactants, which are responsible for the enhanced staining.

The mixture contains $Os^{VIII}O_4$, $Fe^{II}(CN)_6^{4-}$, $Fe^{III}(CN)_6^{3-}$ and $Os^{VI}O_2(OH)_4^{2-}$, and labels cyano-bridged Os–Fe species containing osmium in nominal oxidation states of VIII, VII and VI valency (White *et al.*, 1979). These osmium compounds are thought to be chelated by donor atoms in the tissue macromolecules, which results in the reduction of osmium to lower (water-insoluble) oxidation states. The greater reactivity and concentration of Os^{VII} and Os^{VI} intermediates was considered to lead to additional deposition of osmium in both membranes and glycogen (White *et al.*, 1979). The chelation seems to be primarily responsible for the stabilisation of osmium and the enhanced staining. However, X-ray microanalysis established (de Bruijn and van Buitenen, 1980, 1981) that in aldehyde-fixed liver glycogen, osmium–iron or osmium–ruthenium complexes (when $K_2Ru(CN)_6$ was used) were present with a rather constant weight ratio of Os:Me = 1:3 when $K_2OsO_2(OH)_4$ was the osmium compound in the reaction mixture (rather than OsO_4). In the remaining cellular components such complexes were not established (Os:Me = 1:0). Different weight ratios were obtained with similar combinations with OsO_4 in the mixture (in the glycogen, Os:Fe = 1:7; and 1:5 for the cellular components and 1:2 and 2:1 for Os:Ru).

Quantitatively, using unfixed isolated glycogen as target material, ESCA measurements revealed that in the aqueous phase of the treatments with osmium mixtures, predominantly (80–100%) Os^{VI} species were present in the glycogen, and mainly Os^{IV} and some Os^{III} species after ethanol dehydration (de Bruijn *et al.*, 1984).

Initially the enhanced staining of glycogen with the OsFeCN method was explained on the basis of the involvement of C_2–C_3 hydroxyl groups (de Bruijn and den Breejen, 1974). These ligands were presumed to immobilise the osmate ($Os_2^{VI}(OH)_4^{2-}$) by chelation. The resulting chelated lower-valency osmium component would thus be responsible for the subsequent accumulation of the osmate–cyanide complexes, leading to enhanced glycogen staining. Two models of the first chelate and the subsequently added complex have been proposed (de Bruijn and van Buitenen, 1981).

By means of spectrophotometry and conductometric titration it has recently been shown that glycogen and other glucose derivatives in aqueous medium react with added potassium osmate in forming osmate esters. Absorbance and conductivity data show characteristic stoichiometry. Four glucose monomers are combined with one osmium atom in glycogen and in the cyclodextrins used as model compounds of known conformation. In contrast to the polysaccharides examined, glucose and derivatives such as 1-methylglucose bind to osmium in a 2:1 ratio. The data indicate that glycogen and certain other polysaccharides react with osmate ions by forming a primary chelate in which osmium is bound in such a way that adjacent glucose molecules in different parts of the chain (or different chains) are involved; a molecular model of the reaction product has been proposed (Riemersma *et al.*, 1984). Under fixation and staining conditions, of course, other substances play a role and the reaction mechanism is more complex.

As stated earlier, the OsFeCN method effectively enhances the staining of membrane in general. To stain membranes, the formation of Os^{VI} seems to be necessary, which can be provided when OsO_4–potassium ferricyanide ($K_3Fe(CN)_6$) or OsO_4–potassium ferrocyanide ($K_3Fe(CN)_6$) is used. In this regard, the reaction of OsO_4 (present in the mixture) with unsaturated lipids in the membrane cannot be disregarded.

The OsFeCN mixture sharply delineates membranes of both the endoplasmic reticulum and the nucleus in plant and animal cells. It also strongly stains plasma membranes, for example, in neural tissues (Rivlin and Raymond, 1987). The OsFeCN mixture may fill the lumen of membranes of the endoplasmic reticulum and nucleus with an electron-opaque deposit. The presence of Ca^{2+} seems to play some role in determining which membrane systems are selectively stained. Omission of Ca^{2+} from the aldehyde fixative or addition of PO_4^{2-} (which precipitates Ca^{2+}) abolishes selective staining of the membranes of the endoplasmic reticulum and the nucleus (Hepler, 1981). It is possible that divalent or trivalent cations bind selectively to certain cell stuctures during fixation, which enables them to react with OsFeCN. It is thought that staining with OsFeCN is based on the presence of Ca^{2+}-binding proteins. However, since not all stained elements of the smooth endoplasmic reticulum actively accumulate Ca^{2+}, it has been suggested that any connection between staining with OsFeCN and Ca^{2+} is coincidental (Walz, 1982).

The presence of Ca^{2+} might be explained from the models proposed for the complexes by the residual overall negative charge of the complex, in which the presence of coincidental calcium could be established by X-ray microanalysis (de Bruijn and van Buitenen, 1980).

The OsFeCN mixture can be used as an extracellular tracer. Freshly prepared mixture fills extracellular spaces in the muscle when phosphate buffer is used and Ca^{2+} is omitted from the aldehyde fixative (Forbes *et al.*, 1977). On the other hand, in the presence of Ca^{2+} and with omission of PO_4^{2-}, selective staining of intracellular tubular membrane systems in the muscle can be accomplished. Under favourable processing conditions, a dense precipitate is formed in the extracellular spaces in the neural tissue, serving to outline individual cells (Rivlin and Raymond, 1987).

The OsFeCN mixture also stains myelinated nerve fibres (Langfold and Coggeshall, 1980), microfilaments in animal cells (McDonald, 1984), mitotic apparatus (Hepler, 1980), smooth endoplasmic reticulum (Walz, 1982), Golgi complex and secretory vesicles (Schnepf *et al.*, 1982), cell surface coat (Neiss, 1984) and cochlear tissue (de Groot *et al.*, 1987).

Ruthenium cyanide additions can be advocated for contrast enhancement, especially when quantitative X-ray microanalytical ferritin iron estimation is aimed for (de Bruijn and Cleton, 1985; Cleton *et al.*, 1986).

A disadvantage of the OsFeCN method is its capriciousness. Certain regions of a tissue block are well penetrated by the mixture, whereas others are not. This results in uneven staining of membranes and variable filling of extracellular spaces. Staining irregularities may also be due to functional differences among various regions of a cells or its components. This method does not appear to be specific for Ca^{2+}-sequestering membrane systems. Detailed studies by Schnepf *et al.* (1982) demonstrate the variability and sometimes contradictory staining of different cell types with OsFeCN.

This might be explained by the fact that the initially advocated duration of the post-fixation reaction (18 h) is ignored. Moreover, the inhibition of the complex formation below pH 5.5 has been reported earlier (Saxena, 1967). Such low pH values sometimes occur—e.g. after cytochemical reactions. Rebuffering at pH 7.4 mostly cures the problem. Moreover, from the chemical point of view the optimal pH conditions for osmate are 10.5 (Riemersma *et al.* 1984). Nevertheless, this method is useful for enhancing the contrast of various membrane systems and even occasionally for differentiating between different membrane types.

FIXATION AND STAINING PROCEDURES

According to a general method, cells or thin tissue slices are fixed with 2% glutaraldehyde in 0.05 M cacodylate buffer (pH 7.2) containing 5 mM $CaCl_2$ for 90 min at 4 °C. They are rinsed in buffer containing $CaCl_2$, and post-fixed with a mixture of 1% OsO_4 and 0.8% potassium ferricyanide or ferrocyanide for 1–2 h at room temperature. Although post-staining is unnecessary, thin sections can be stained with salts of uranium and lead. The $CaCl_2$ is added to these solutions to prevent dissolution of cellular calcium crystals to be investigated. Solutions of OsO_4–ferrocyanide may have a brownish precipitate, whereas OsO_4–ferricyanide shows a clear amber colour.

Osmium–potassium ferrocyanide can be used as a selective stain for cardiac sarcoplasmic reticulum (Forbes *et al.*, 1985). However, occasionally mitochondria and often nuclear membranes are also stained. The hearts are fixed by retrograde perfusion of the coronary arteries with 1.6% glutaraldehyde in 300 mosm cacodylate buffer (pH 7.6) containing sodium chloride to a total milliosmolality of 300. After a 5 min wash, the perfusion is switched to Tris containing 100 mg% DAB and 0.01% hydrogen peroxide. After 20 min the perfusion is terminated and the heart tissue is left in fresh fixative overnight at 0–4 °C. The tissue is post-fixed in 2% OsO_4 buffered with cacodylate and containing 0.8% potassium ferrocyanide for 2 h. Thin sections can be counterstained with uranyl acetate and lead citrate. Selective staining of sarcoplasmic reticulum is shown in Fig. 4.11.

The above-mentioned staining is accomplished only in the tissue exposed to both DAB and potassium ferrocyanide. No staining is seen in the tissue perfused with Tris alone or with Tris and hydrogen peroxide. Although staining occurs when DAB without hydrogen peroxide is perfused, it is less intense. This staining method has considerable potential application in high-voltage microscopy.

OXALATE–GLUTARALDEHYDE

As stated earlier, oxalate can be used as a precipitation agent, mostly for calcium, since other plentiful physiological divalent ions do not form complexes as insoluble as calcium oxalate. According to this procedure, tissue is fixed with an oxalate-containing glutaraldehyde solution or variations of this formulation.

Constantin *et al.* (1965) reported a procedure for perfusing muscle fibres with calcium followed by oxalate which produced dense deposits in terminal sacs of sarcoplasmic reticulum. Podolsky *et al.* (1970) later identified these deposits as calcium oxalate, using electron probe microanalysis. These investigators began the procedure by 'skinning' the fibres, which involves removing the muscle membrane. This is necessary to permit oxalate to enter the cell, since the unfixed muscle membrane is normally impermeable to this ion. After perfusion with 10 mM $CaCl_2$ plus 80 mM sodium citrate, 10 mM sodium oxalate plus 140 mM KCl is added. Five minutes after the perfusion the preparation is fixed with 6.5% glutaraldehyde in 0.2 M cacodylate buffer (pH 7.2), washed in the buffer and post-fixed in 2% OsO_4. All of these solutions contained 10 mM oxalate. Tissue specimens are rinsed in 10 mM oxalate and dehydrated according to standard procedures. Both stained and unstained sections should be examined, since lead staining tends to remove the oxalate deposits.

According to another procedure, tissue specimens are fixed with buffered 3% glutaraldehyde containing 0.0125 M ammonium oxalate for 2 h. The specimens are then treated with buffered 0.025 M ammonium oxalate solution for $2\frac{1}{2}$ h, followed by post-fixation with 2% OsO_4 for 2 h.

PHOSPHOTUNGSTIC ACID

Phosphotungstic acid (PTA) is an anionic stain. Its use dates from the early days of electron microscopy; for instance, Hall *et al.* (1945) pointed out that PTA is particularly suitable for staining specific regions in the fibrils of mollusc muscles, although the staining specificity was not utilised effectively until much later. A possible reason may be that in the past aqueous solutions of PTA were employed which tended to cause distortion of the fine structure, especially at a low pH, where the uptake of the tungsten is greatest. However, large amounts of the metal are also taken up by the tissue from solutions of PTA in 100% ethanol without any apparent deleterious effect on the fine structure. In fact, treatment with PTA seems to exert a protective effect against the damage caused by polymerisation and distortion during sectioning. Phosphotungstic acid is an excellent negative stain. For a comprehensive discussion on the theory, mechanism and methods of negative staining with PTA and other stains, the reader is referred to Hayat and Miller (1989).

Metallic tungsten has an atomic number of 74, is brilliant white and melts at 3400 °C. It is among the least fusible and least volatile of metals, and is stable in air at room temperature, but on heating it is oxidised by air to WO_3. Tungsten is very stable to acids, because it easily becomes passive. Heteropolyanions such as

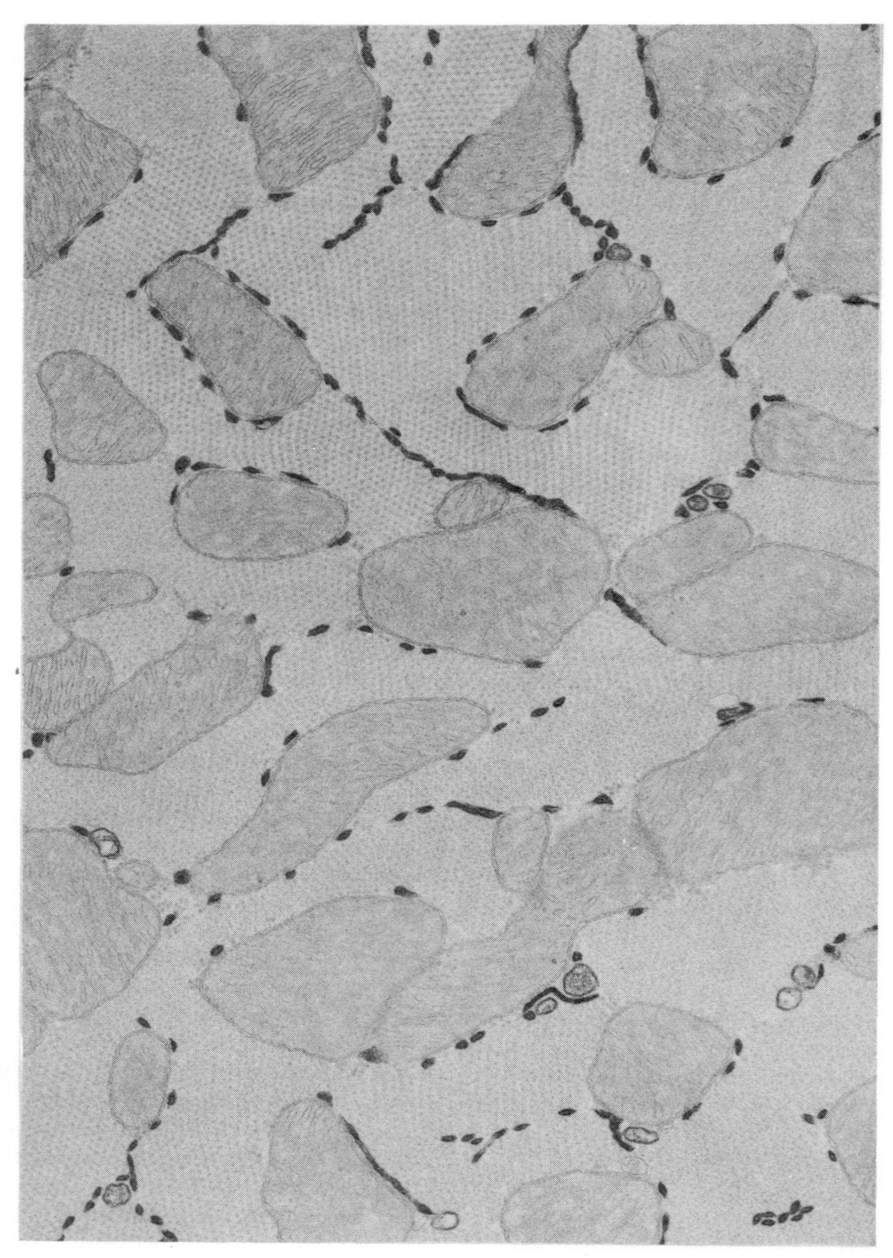

Figure 4.11 Thin transverse section of right ventricular wall of mouse heart post-fixed with ferrocyanide-reduced OsO_4 so as to selectively opacify the sarcoplasmic reticulum (SR). Deposition of osmium allows ready identification of SR profiles. 30 000×. Courtesy M. S. Forbes

phosphotungstate and tungstate can exist in more than one form, and their composition and charge may change with a change in pH. In the phosphotungstate ion ($PW_{12}O_{40}^{3-}$), at a low pH, 12 tungsten—oxygen octahedra surround a single PO_4 tetrahedron and the anion is roughly spherical, with a diameter of 1.1 nm (Bradley and Illingworth, 1936). At pH 5.5–7.0, the polyanion is present as $PW_{11}O_{39}^{7-}$. The presence of 12 or sometimes 24 atoms in each molecule gives PTA a high molecular weight (3313.5), and every attached molecule imparts high density to the tissue. Large quantities of this dense stain can be deposited in the tissue

(Fig. 4.12 on p. 276). The amount of tungsten taken up may be equal to the dry weight of the tissue, and the tissue density may rise from ~1.6 to 3.0 (Huxley, 1959).

MECHANISM OF STAINING

At present, an agreement on the type of interaction and affinity of PTA for organic substrates is lacking. From the available data two interpretations can be advanced: (1) acidic PTA in an aqueous medium forms a complex with highly polymerised carbohydrates; and (2) PTA precipitates organic cations, including proteins, and the interaction is ionic in nature. The evidence supporting the first interpretation will be presented first.

It has been indicated that PTA in aqueous solutions has a special affinity for polysaccharides at strongly acidic pH values, and that only the carbohydrate moiety of glycoproteins is responsible for PTA uptake, while the basic groups of protein are not involved. The fact that digestion with hyaluronidase removes considerable PTA-positive material supports this view. Furthermore, acetylation or sulphation to block hydroxyl groups tends to prevent PTA binding, while deacetylation results in the recovery of PTA staining. In addition, methylation to block carboxyl groups and deamination carried out on sections does not alter PTA staining.

Pease and Bouteille (1971) demonstrated that the carbohydrate-containing matrix, but not the filaments, of native collagenous fibrils was vigorously stained with acidic PTA. Dermer (1973) showed that the surface coats of cells in normal and neoplastic duct epithelium can be stained with PTA; the tissue was fixed with glutaraldehyde and embedded in glycol methacrylate, and staining was carried out on thin sections. When sections were treated with 0.1 N NaOH at 100 °C, staining of the cell surface coat material was reduced. This treatment is known to release ~89% of the carbohydrate groups from glycoproteins. Furthermore, a reduction in the staining of the cell surface coat occurred when the sections were pretreated with 0.05 M H_2SO_4 for 1 h at 80 °C. The results of these treatments, and experiments with pronase, trypsin, α-amylase, hyaluronidase and neuraminidase, indicated that the staining of surface material was due to the presence of sialic acid residues within surface coat glycoprotein.

Ponzo *et al.* (1973) showed that infusion of 2,4-dinitrofluorobenzene into the tail vein of rats *in vivo* did not reduce the staining of zymogen granules of the pancreas with PTA. It is known that zymogen granules contain a large amount of proteins which are rich in free amino groups, and 2,4-dinitrofluorobenzene produces oxidative deamination of amino groups. It is emphasised that only thin-section staining with aqueous solutions of PTA at an acid pH shows a high degree of specificity for carbohydrates; unconjugated proteins are not stained with an aqueous solution at pH 3.0 or below.

Binding of aldehydic groups to PTA has been indicated. According to Palladini *et al.* (1970), when aldehyde-fixed specimens are treated with aqueous solutions of PTA, an oxidative action on the PAS-positive groups (double bonds, *vic*-glycols) takes place, while anhydrous solutions of PTA tend to oxidise amino groups. This oxidation results in the production of aldehydic groups which bind to PTA molecules. Phosphotungstic acid can be directly demonstrated, using either microradiography or stannous chloride reduction to the heteropolyblue. On the other hand, studies of glycol splitting power of PTA by Scott (1973b) indicate that, although PTA shows Schiff reactivity, it is not a glycol-splitting reagent. In this connection, it should be noted that PTA is unstable at pH above 1–1.5 and that hydrazines are reducing agents. These factors may play a part in the destruction of PTA. In the case of native collagen, potential binding sites for acidic PTA are blocked after fixation with glutaraldehyde (Pease and Bouteille, 1971). The circumstances under which aldehydes may block PTA staining are unclear.

Experiments by Marinozzi (1968) and Rambourg (1967) indicate that all blocking reactions that affect hydroxyl groups in the tissue also affect the staining with PTA. Also, the same blocking reactions affect adversely the staining by PAS as well as by PTA. In other words, PAS-reactive materials are also PTA-positive. It has been shown that PTA-containing areas are Schiff-positive and that treatment with a phenylhydrazine after PTA staining prevents the Schiff reactivity (Palladini *et al.*, 1970). These data indicate that hydroxyl groups could be involved in staining by acidic PTA. Studies of aldehyde blockage, methylation and acetylation suggest that both the hydroxyl and carboxyl groups are involved in the PTA reaction at acidic pH values (Tsuchiya and Ogawa, 1973). Pease (1970) has reviewed in detail the reasons justifying this interpretation. Rambourg (1974) has reviewed the reasons justifying the involvement of hydroxyl and aldehyde groups in the PTA reaction whose specificity can be compared with that of the PTA–Schiff technique.

On the other hand, several studies fail to support the view that PTA has a special affinity for polysaccharides. On the basis of chemical and histochemical experiments, Silverman and Glick (1969) indicated that PTA selectively stains tissue proteins, while no apparent reactivity with acid polysaccharides was observed. By using various stains (thorium, iron, ruthenium red,

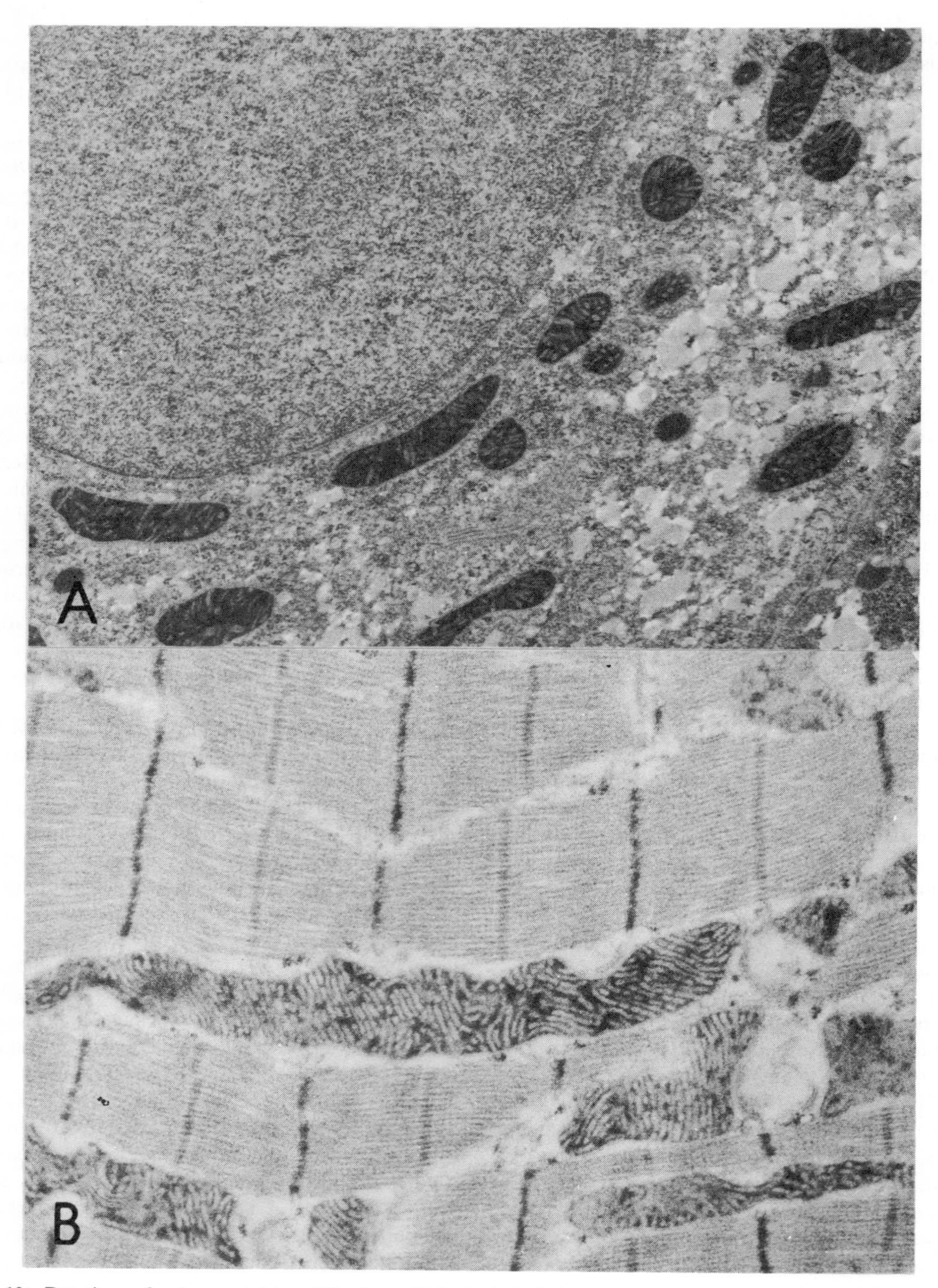

Figure 4.12 Rat tissue fixed overnight in 1% glutaraldehyde in 0.067 M potassium phosphate buffer (pH 7.4) with added sucrose to give a final osmolality of 300±10 mosm, and stained for 3 h in 5% PTA in 6.25% Na_2SO_4 prior to dehydration in acid alcohol (pH 2.0). Cell membranes either are not discernible or are demonstrable only as negative images against their positively stained surroundings, as in the case of mitochondria. Neither glycogen nor lipid droplets are discernible. A: Liver parenchymal cell showing a uniform general staining with marked staining of mitochondrial matrix and desmosomes. B: Heart muscle showing Z-bands as densely stained as mitochondrial matrix, and myofibril stained with moderate intensity. Courtesy L. Silverman and D. Glick

alcian blue and PTA), Zacks *et al.* (1973) concluded that in myofibre external lamina, proteins, rather than polysaccharides, were stained by PTA in strongly acid solutions. Studies by Quintarelli *et al.* (1971a) showed that PTA did not interact with chondroitin sulphate, sialic acid and hydroxyl groups of the sugar moieties of these macromolecules. Studies of the interaction between aqueous solutions of PTA and solutions of various organic compounds indicated that the interaction took place only between the metal and the desulphated hexosamine amino groups, and that when the N-desulphated product was acetylated, no PTA binding occurred (Quintarelli *et al.*, 1971b).

Acetylation of sections in a mixture of 10% acetic anhydride and pyridine results in the blocking of chemical sites that react with PTA; pyridine when used alone does not affect PTA staining. However, not all sites that react with PTA are blocked by acetylation. For example, although nucleoli of microsporocytes in the lily react avidly with ethanolic PTA, this reaction is not affected by acetylation (Sheridan and Barrnett, 1967). Acetylation treatment is thought to be sufficient to block most free amino groups.

It has been indicated that under test-tube conditions peptide bonds and amide groups in protein react with PTA (Scott and Glick, 1971). *In vitro* studies of bovine plasma albumin indicate that deamination decreases the binding of PTA with protein by as much as 85% (Silverman and Glick, 1969). The treatment of isolated plasma membranes with protease can remove the stainable material with PTA (Beneditti and Bertolini, 1963). A study of the specialised interneural contact zones in the nervous system of several animals has indicated that digestion with trypsin and pepsin removes the material having affinity for ethanolic PTA, and the effects of trypsin digestion are completely suppressed by soybean inhibitor (Bloom and Aghajanian, 1968).

Phosphotungstic acid binding is selective for positively charged groups. The presence of negative charges inhibits PTA uptake in their immediate vicinity. Several recent studies have shown that PTA is an effective anionic stain for positively charged groups of collagen (Tzaphlidou *et al.*, 1982a). Steric factors play a significant role in the binding of PTA to collagen (Nemetschek *et al.*, 1979). The density imparted by PTA appears to be related primarily to the concentration of positively charged groups of proteins.

Phosphotungstic acid has been used as a precipitating agent in the isolation of amino acids. In fact, the use of heteropolyacids as precipitating agents is one of the oldest methods for the isolation of amino acids from natural sources. It is most likely that PTA reacts with protein rich in basic amino acids such as lysine, histidine and arginine, although other amino acids (e.g. cystine) also form insoluble complexes with PTA. *In vitro* studies indicate that PTA also reacts with serotonin, histamine and epinephrine (Silverman and Glick, 1969). It has been shown that some bands of collagen containing relatively large amounts of arginine are selectively stained with PTA (Latta, 1962). Purified collagen binds ~850 mg of PTA, of which 350 mg is linked to lysine, while ~380 mg is taken up by arginine (Kühn *et al.*, 1957). Positively charged groups which may interact with PTA anions include side-chain ε-amino, guanidino and imidazole groups of lysine, arginine and histidine, respectively; these groups are the principal electron donors in these interactions.

The stainability of proteins with PTA is somewhat restricted to certain specialised reaction sites. In the lactating mammary gland, for instance, the stainability is limited to the lumen of the glandular alveoli, whereas the other regions of the plasma membrane remain unstained (Beneditti and Bertolini, 1963). The desmosomes and terminal bars in the rat liver plasma membranes are stained by PTA, while the plasma membrane of the endothelial cells lining the blood capillaries is not stained. Certain areas of the plasmalemma and pinocytotic vesicles in the isolated tomato fruit protoplast can be stained selectively with PTA (Mayo and Cocking, 1969). Intracellular membranous structures of the acinar cells in the rat submaxillary gland can be stained with aqueous solutions of PTA (Vidić, 1973). The secretory vesicles containing milk precursor droplets can also be stained with PTA. The stain is particularly useful in outlining the shape of unsectioned virus particles.

Basic proteins associated with nuclear DNA and nucleoli also show affinity for PTA. The affinity of nuclear material for PTA is probably due to the association of basic proteins with nuclear DNA. This preferential affinity for basic proteins was skilfully employed by Bloom and Aghajanian (1968) to obtain specific staining of the synaptic material, without increasing the density of the synaptic membranes. This material is thought to be proteinaceous, containing a high proportion of basic amino acids such as lysine, histidine and arginine.

Since PTA shows a strong affinity for positively charged groups in proteins, its specificity for polysaccharides has been questioned. In fact, it has been suggested that PTA could stain the protein moiety of a mucoprotein molecule, the polysaccharide moiety of which could give a periodic acid–Schiff reaction (Glick and Scott, 1970).

It has been proposed that any polymer including polysaccharides can, by protonation of hydroxyl groups at low pH, form polycations which are capable of precipitating PTA in a salt-like complex (Scott and

Glick, 1971). Even sialic acids (common components of glycoproteins), under acidic conditions, may partially degrade and release part of their nitrogen in the form of ammonia, which reacts strongly with PTA. The implication is that any polymer which dissolves in an aqueous acid to produce a polycation at pH values above 1.0 will interact electrostatically with PTA, and may be precipitated. Thus, increased specificity by PTA would be accomplished at higher pH values rather than at a lower pH, as suggested by Pease (1966b). The biochemical evidence and theoretical reasons supporting the ionic interaction between positively charged biological material and negatively charged PTA have been discussed by Scott (1971) and Scott and Glick (1971).

The controversy over staining specificity of PTA has arisen primarily as a result of contradictory evidence obtained from biochemical experiments and histological studies. A metal ion–macromolecular interaction in solution is significantly different from that occurring after the tissue has been fixed and embedded. Fixation may alter the steric relations to such an extent that staining may be either increased or decreased. The embedding medium may influence staining, since its active groups may block the available sites in the biological substrates and thus prevent metal binding. The situation is further complicated because of differential leaching of PTA during the final stages of dehydration and infiltration of the embedding medium, so that very few discrete patterns emerge which are simply interpretable. Another complication is that PTA has been used in aqueous medium as well as in organic solvents. Aqueous solutions of PTA probably stain differently compared with those prepared in organic solvents.

Furthermore, conformational changes of the biological molecule as a result of physical or mechanical stress can hinder metal binding. In other words, the spatial distribution of the active groups in the biological molecule may be so drastically altered during processing that they become unavailable for metal binding. This phenomenon may explain in part, for instance, the conflicting data on the staining of elastic fibres with PTA. In the unextended elastin, the spatial distribution of aspartic and glutamic acid side-chains could be so drastically reduced that their ionised groups may not be available for metal binding (Quintarelli *et al.*, 1973a). All the evidence presented above is probably valid, and the controversy can be resolved by finding a satisfactory explanation that bridges the gap between biochemical and histological findings.

It is pertinent to point out that digestion treatments cannot be considered valid in all cases. It is a well-known fact, although not often admitted, that PTA-positive material present at various sites shows different sensitivities to digestion treatments. The exact implication of these different sensitivities remains to be determined. In addition, the role played by impurities present in commercially obtained digestive enzymes and other reagents cannot be overlooked.

FIXATION AND STAINING PROCEDURES

Fixation procedures involving OsO_4 are not recommended, for PTA staining is masked by the overall staining; polysaccharides and lipoproteins show non-specific interaction with bound osmium. However, the desired preferential staining in the sections of the OsO_4-fixed tissue can be obtained by dissolving the lower osmium oxides with 20% hydrogen peroxide, or with 1% periodic acid, which acts as a strong oxidant (Silva, 1967). Nevertheless, this technique is not always successful and may cause distortion of organelles. Consequently, the tissues are fixed with aldehydes, or prepared by freeze-substitution or 'inert dehydration' with glycol.

A general increase in contrast of the sections of OsO_4-fixed tissue can be obtained by treatment with PTA solution for 20–30 min. A prolonged treatment, especially at high concentrations, with PTA extracts the reduced osmium from the sections. This apparently results in the progressive loss of the staining of those cellular components which had stained as a result of the presence of reduced osmium. It is evident that the reaction product of reduced osmium and PTA is soluble in the staining solution. A prolonged treatment, on the other hand, also results in the gradual increase in the staining intensity of certain cellular components.

Glutaraldehyde seems to be the ideal fixative for tissues to be stained with PTA. After glutaraldehyde fixation, PTA in aqueous acidic medium intensely stains the mitochondrial matrix, cisternae of endoplasmic reticulum, desmosomes, myofibrils and the Z-band of muscles, whereas glycogen, lipid droplets and the majority of the membranous structures are not stained (Fig. 4.12). Mitochondrial cristae and the membranes of endoplasmic reticulum stand out in negative image. Glycogen and phospholipids are probably extracted during the preparatory procedures. It is important to note that the glutaraldehyde–PTA method imparts distinctly greater density to the mitochondrial matrix and the cisternae of endoplasmic reticulum than that imparted by the glutaraldehyde–OsO_4–uranyl acetate–lead citrate method. This suggests that the former method probably retains more proteins than does the latter method.

Since PTA reacts with monomer epoxy resins, a considerable extraction of the stain occurs when the stained tissue is embedded in these resins. For this reason, the tissue is usually embedded in glycol methacrylate or in polyester resins. However, staining of the tissue block can be retained by removing the excess protons from the bound PTA with an alkaline solution (0.01% NaOH) prior to dehydration with propylene oxide and embedding in epoxy resins. This problem is encountered only when the tissue is stained prior to embedding, for the polymerised resins apparently do not react with the stain. It is emphasised that when PTA is allowed to mix with propylene oxide for longer than 5 min, the result is an explosive exothermic reaction. To avoid this, the excess PTA solution is removed from the tissue by a quick rinse in cold absolute ethanol, prior to the longer final dehydration with propylene oxide. Acetone presents no such problems.

The effect of the stain is considered to be of an additive nature. Since higher contrast is achieved by exposing only one side by floating the grid on the staining solution, it is likely that the staining is largely a surface phenomenon unless prolonged treatment is employed. During *en bloc* treatment, PTA penetrates relatively slowly. Applied to 1 mm^3 blocks of aldehyde-fixed tissue, PTA penetrates up to 100–200 μm from the outer surface of the block (Vrensen and de Groot, 1974a).

Staining intensity can be increased by exposing the specimen to absolute ethanol prior to staining, or by repeatedly refreshing the staining solution during staining. A slight increase in the water content of the ethanol–PTA (E–PTA) staining solution by the addition of 95% ethanol enhances selective staining. The optimal E–PTA concentration, pH, duration of staining and water content for a tissue has to be determined by trial and error. Commonly observed variability in the selectivity of E–PTA staining is due primarily to the water content and pH of the staining solution.

Ethanol–PTA is especially useful for obtaining selective staining of synaptic contact zones. Ethanol tends to extract phospholipids from membranes in the aldehyde-fixed tissue. As a result, it is assumed that the protein moiety of membranes is liberated, which results in minimum background staining. The net result is that synaptic contact zones are stained more selectively with E–PTA than with aqueous PTA. This phenomenon may be one of the reasons for the difference between staining by E–PTA and that by aqueous PTA. Enhanced selectivity and uniformity of staining of synaptic contacts is achieved when 50–100 μm tissue slices are stained with E–PTA at 60 °C.

The glutaraldehyde E–PTA method is quite effective in discerning ultrastructural differences between synaptic junctions, differences which remain undetected when conventional procedures are used (Jones, 1973). Gordon and Bensch (1968) have employed 2% E–PTA to differentiate the cortex from the interior of the coarse fibres and to further differentiate the cortex into densely stained and pale-stained segments in glutaraldehyde- and OsO_4-fixed guinea-pig sperm. In general, for en bloc staining, during the final dehydration stage of glutaraldehyde-fixed tissue, 1% E–PTA for 1 h is adequate. For epoxy sections, 1% E–PTA for 5 min is recommended. For specific staining of elastin in sections of Durcupan-embedded tissue, treatment by 10% PTA in absolute methanol for 30 min is recommended. Elastin and collagen fibres in the aorta can be stained with 0.5% PTA in 0.1 N HCl.

Unsectioned virus particles are stained by immersing the grid in a drop of 2% aqueous PTA for 1 min, while the virus to be observed in sections can be stained either during dehydration or on a specimen grid. Selective staining of the complex carbohydrate matrix of native collagenous fibrils can be obtained with 2–5% PTA solution made up in 0.5–1.0 N HCl at pH 8.0 (Pease and Bouteille, 1971). Staining of surface coats can be achieved by treating thin sections of glutaraldehyde-fixed tissue with 3% PTA (pH 1.5) for 15 min. Staining of intracellular membranous structures of the acinar cells in rat submaxillary gland can be obtained with aqueous 10% PTA solution (pH 1.0–1.5) applied for 1 h (Vidić, 1973). Intracellular membranes can also be revealed by post-staining the sections with 10% aqueous PTA (pH 1.0) for 1 h; the desired pH can be obtained with HCl. Tissue specimens are fixed with an aldehyde mixture containing trinitro compounds (Ito and Karnovsky, 1968), followed by post-fixation with OsO_4.

Periodic acid–PTA procedures can be employed for obtaining selective staining of glycoproteins (Tsuchiya and Ogawa, 1973). Thin sections of glutaraldehyde and OsO_4-fixed tissue are treated with 0.1–1.0% periodic acid for 30 min at room temperature. After rinsing, sections are stained with 1% PTA in 10% chromic acid (pH 0.3) for 5–30 min at room temperature.

A periodic acid–Schiff–PTA procedure can be used for staining glycogen. Periodic acid oxidises 1,2-glycol groups of glucosidic molecules, exposing two aldehydes. The reason for using periodic acid is that it does not further oxidise dialdehydes, whereas other oxidants, such as chromic acid, $KMnO_4$ and hydrogen peroxide, continue oxidising. The exposed dialdehydes react with fuchsin–sulphurous acid Schiff reagent to form a coloured product. Since this product is not electron-opaque, PTA is employed as a source of heavy atoms. Other heavy metal salts (e.g. silver nitrate and lead citrate) can also be used for staining the dialdehyde–Schiff complex.

It is pointed out that this procedure is not specific for glycogen, since other tissue components (e.g. collagen) also stain intensely. Furthermore, the fine details of glycogen particles are not discerned. For this procedure, thin sections are exposed to Schiff reagent for 1–2 h at room temperature; light should be avoided. This step is followed by rinsing in five changes (2 min each) of freshly prepared sulphurous acid solution; one of the two solutions given below can be employed.

(1) Sodium metabisulphite (10%)	10 ml
HCl (1 N)	10 ml
Distilled water	200 ml
(2) Sodium or potassium metabisulphite	1 part
HCl (0.1 N)	1 part

The sections are then rinsed in three changes (2 min each) of distilled water, followed by staining with 2% aqueous PTA for 30–60 min. The pH of the PTA solution may be adjusted to 4.5 or 7.0. The sections are rinsed three times in distilled water.

It should be noted that staining with PTA is influenced not only by the pH of the staining solution, but also by the concentration and source of PTA. These parameters are known, for example, to affect the size and distribution of presynaptic dense projections and the arrangement of cleft material. The impurities present in PTA include Cl^-, NH_4^+, NO_3^- and Na^+. High levels of impurities apparently cause unsatisfactory staining.

POTASSIUM PERMANGANATE

Potassium permanganate is employed in electron microscopy not only as a fixative, but also as a stain. Its role as a fixative has been discussed elsewhere (Hayat, 1981), and as a stain it also has a rather limited use. Potassium permanganate forms insoluble electron-dense reduction products by reaction with tissue constituents, thus giving rise to image contrast by electron scattering. It was introduced as a stain by Lawn (1960), and subsequently other workers used it as a stain, with varied degrees of success. Permanganates should not be used for post-staining sections of tissues embedded in Epon containing the hardener nadic methyl anhydride (NMA), because this hardener retains its capacity to react with the stain, even in the polymerised Epon. However, sections of cured Epon mixture without NMA can be stained with permanganates. Also, post-staining of sections of other epoxy resins, polyester resins and methacrylates presents no problem.

The deposition of reduction products of permanganate in granular form is considered to be responsible for the gross darkening of tissue observed after permanganate treatment. High contrast of the cytoplasmic membrane system is probably due to the 'unmasking' of the protein and lipid components of the phospholipoprotein, for this protein reduces the permanganate and thus increases the electron density.

Concentrations of $KMnO_4$ higher than 1% do not affect staining intensity, whereas concentrations lower than 0.8% decrease the final staining density. Solutions having a pH lower than 5.5 do not influence staining intensity; however, an increase in pH above 6.8 results in the deposition of precipitates. Overstaining is avoided by employing dilute solution for brief durations.

Osmium tetroxide-fixed tissues stained with $KMnO_4$ show intense staining of tonofibrils, myelin sheath, basement membranes, terminal bars, desmosomes and cytoplasmic membranes in general. Monolayer tissue cultures of blood leucocytes (Sutton, 1968), feather keratins and α-keratins (Rogers and Filshie, 1962) have also been successfully stained with permanganate. This stain can be very useful for showing the true surface viruses forming in infected cells and the lignin components of cell walls. In comparison with some other stains, permanganate imparts uniform staining. However, this stain has the limitation that the zymogen granules, cytoplasmic RNP particles and some ground proteins are not stained. Glycogen is poorly stained, a characteristic which may be due to the oxidation of polysaccharide hydroxyl groups. However, the staining of glycogen can be enhanced by longer exposures (20–30 min) to $KMnO_4$.

Sections mounted on a grid are stained by immersion in 1% aqueous $KMnO_4$ solution. The staining container is completely filled with the stain and sealed to exclude air, for the exposed surface invites a rapid formation of a film of reduced compounds (a weak solution of citric acid will remove the surface deposits). Adequate staining is achieved within 30 min at room temperature; however, duration and concentration may be varied, depending upon the objective of the study. The stained grids are thoroughly rinsed in distilled water. If absolutely necessary, the precipitated stain can be removed by exposing the grid to 0.025% solution of citric acid for $\frac{1}{2}$–1 min. With prolonged treatment this solution can completely destain the sections. Cytoplasmic membranes may be damaged if the sections are exposed to high concentrations of the acid, or to lower concentrations for excessive durations.

Barium permanganate is a useful stain for increasing

the contrast of cellulose. Sections mounted on a grid are post-stained with lead acetate followed by aqueous 0.5% solution of barium permanganate for 2 min. This technique was employed by Hohl *et al.* (1968) to achieve strikingly increased contrast of cellulose fibrils in the stalk of slime and mould cells.

Staining intensity of sections treated with $KMnO_4$ can be enhanced by post-staining with uranyl acetate followed by lead citrate. This increase in staining may be due to the $KMnO_4$-induced alterations of protein, which, in turn, may lead to the formation of an increased number of available binding sites for uranyl acetate and lead citrate. When $KMnO_4$ is substituted by PA, the staining is hampered (Soloff, 1973). This evidence indicates that the facilitatory action of $KMnO_4$ is not due solely to its oxidising capacity.

According to the triple staining procedure, grids with sections are floated on a drop of $KMnO_4$ (0.9%) in phosphate buffer (pH 6.5) for a few minutes, and then thoroughly rinsed in distilled water. The grids are washed in 50% ethanol for ~15 s prior to being floated on a drop of saturated solution of uranyl acetate in ethanol. The grids are then floated on a drop of lead citrate (0.2%) for 1–2 min and rinsed in distilled water.

The contamination that accompanies staining with $KMnO_4$ results partly from interaction of copper grids with the staining solution. Such contamination can be minimised by using stainless steel or gold grids, or by acid washing of copper grids and buffer rinse immediately before exposure to $KMnO_4$; dilute formic acid can be used for rinsing the grids prior to use. Contamination by the precipitated stain is minimised by employing freshly prepared staining solution and by keeping the grid immersed throughout the staining procedure. The staining solution should never be filtered to remove crystals, since precipitation of stain is augmented by filtration. Adequate agitation ensures complete dissolution of permanganate crystals within 15–30 min.

POTASSIUM PYROANTIMONATE

It is known that inorganic cations (e.g. Ca^{2+}, Mg^{2+}, Na^+ and K^+) and anions (e.g. Cl^- and orthophosphate) are important constituents of all living animal and plant cells. Essential roles played by ions in cellula metabolic mechanisms are well established. For example, the informational role of Ca^{2+} in eukaryotes has been recognised. This cation is an important apoplastic nutrient and serves as a structural component of the cell wall. Most higher organisms maintain an intracellular concentration of ionic calcium of about 10^{-6} M, while the extracellular concentration is about 10^{-3} M. The importance of methods for studying subcellular distribution of Ca^{2+} and other diffusible ions is apparent. To study these ions, methods have been developed utilising cytochemical, autoradiographic and X-ray microanalytical approaches; only cytochemical approaches employed in conjunction with electron microscopy will be described here.

The potassium pyroantimonate (antimonate) ($K_2Sb_2O_7 \cdot 4H_2O$)–OsO_4 method was originally developed by Komnick (1962) for the localisation of Na^+ at the subcellular level, and it was employed for this purpose in many laboratories for about 10 years. The method has also been used for the localisation of other cations such as Ca^{2+}, Mg^{2+}, K^+, Mn^{2+}, Zn^{2+}, Fe^{2+} and Ba^{2+}. However, in recent years it has been almost exclusively used for the detection of Ca^{2+} in plant and animal tissues. Lists of the subcellular components with which Ca^{2+} is associated, as determined by antimonate precipitation, are given by Simson and Spicer (1975) and Wick and Hepler (1982). In comparison with the oxalate method, the antimonate method exhibits somewhat greater sensitivity but lesser specificity for Ca^{2+}. Improvements in the classical antimonate–OsO_4 method include the use of glutaraldehyde, high concentrations of phosphate or the addition of oxalic acid; these improved methods are described later.

MECHANISM OF STAINING

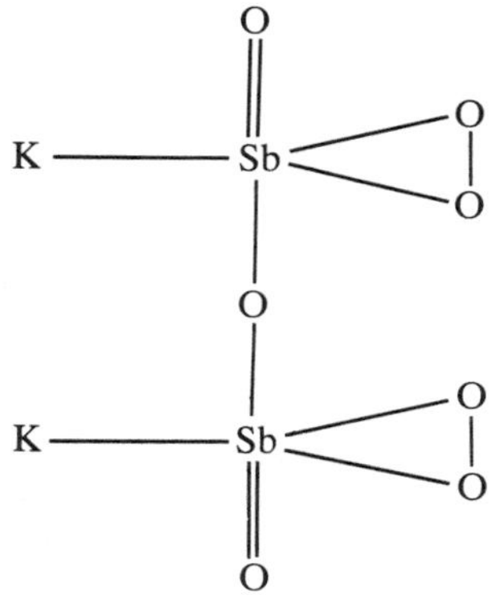

When one considers the chemical formula of the potassium pyroantimonate molecule, it becomes apparent that the reagent is a double salt possessing many unstable bonds, at which monovalent, divalent and trivalent ions can be added. Both *in vitro* and *in situ* studies indicate that antimonate forms insoluble, electron-opaque precipitates with inorganic (Ca^{2+}, Mg^{2+}, Zn^{2+}, Mn^{2+}, Ba^{2+}, K^+ and Na^+) and organic (—NH_3) cations. Pyroantimonate anion ($Sb_2O_7^{2-}$) readily reacts with endogenous cations, especially Ca^{2+}. Cations precipitate as electron-opaque Sb salts. For example, Ca^{2+} combines with antimonate to form $CaH_2Sb_2O_7$.

The reactive sites on or in cell components (e.g. plasmalemma and mitochondria) represent loci of cation-binding sites. Published reports of the reaction and staining of various cell components with antimonate are most likely due to the presence of cations in these components rather than because of any interaction between this salt and the cell components *per se*. For example, metallic cations bound to the phosphate groups of nucleic acids and histones are responsible for nuclear staining with antimonate. Most of the free cations are believed to be extracted during immersion in the aqueous pyroantimonate solution as it is in conventional preparatory procedures such as fixation, rinsing and dehydration. Cations bound to cell macromolecules are likely to be retained and react with antimonate. Tightly bound cations as such may not be available for reaction with this salt, unless their concentration or pH is increased.

Pyroantimonate precipitate formation is thought to involve the formation of an initial nucleus containing a minimal number of pyroantimonate–cation complexes (Mentré and Halpern, 1988). The precipitate grows by coalescence of more of these complexes. The size of the precipitate indicates the concentrations of cations. Large precipitates may be formed in high-concentration and low-mobility regions of cations as well as in low-concentration and high-mobility regions.

EFFECTS OF FIXATION

Osmium tetroxide–antimonate has been extensively used as a primary fixative in the antimonate method. This fixative has the advantage of penetrating somewhat faster than glutaraldehyde–antimonate mixture into the tissue, and facilitates rapid penetration of the antimonate anions into the specimen. Osmium–antimonate retains most of Ca^{2+} *in situ* in single cells, but most of K^+ and Na^+ are lost (Chandler and Battersby, 1976). In tissues, about 80% of Ca^{2+} is retained after a $2\frac{1}{2}$ h osmium–antimonate fixation (Van Iren *et al.*, 1979). Osmium–antimonate fixation may result in the formation of deposits that are not seen after treatment with antimonate followed by glutaraldehyde or OsO_4 (Yarom *et al.*, 1974a). Calcium is the dominant cationic component of such deposits.

The quality of ultrastructural preservation is far better with antimonate–glutaraldehyde fixation than that with antimonate–OsO_4 fixation. As in the case of OsO_4, glutaraldehyde does not affect significantly the specificity of antimonate reaction with cations. However, both fixatives may influence relative distribution of certain cations, because of differential loss or retention of the cations (Wick and Hepler, 1982). Calcium is the major cation found in precipitates formed during both antimonate–osmium fixation and antimonate–glutaraldehyde fixation. Most of the Ca^{2+} bound to macromolecules is retained during antimonate–glutaraldehyde fixation. This Ca^{2+} is precipitated when antimonate is present in the glutaraldehyde solution. Post-fixation with OsO_4 is not necessary to accomplish Ca^{2+} precipitation in antimonate-glutaraldehyde fixative (Weakley, 1979). Calcium precipitation in the presence of glutaraldehyde is more sensitive to buffer composition than in the presence of OsO_4.

The following variations of fixation and antimonate treatment can be used.

(1) Primary fixation with a mixture of glutaraldehyde and antimonate, followed by post-fixation with antimonate–OsO_4.

(2) Primary fixation with a mixture of glutaraldehyde and antimonate, followed by post-fixation with OsO_4.

(3) Primary fixation with glutaraldehyde, followed by post-fixation with antimonate–OsO_4.

(4) Primary fixation with a mixture of glutaraldehyde and potassium oxalate, followed by post-fixation with antimonate–OsO_4.

(5) Primary fixation with a mixture of antimonate–OsO_4 when the quality of ultrastructural preservation is not a high priority.

SPECIFICITY OF REACTION

Although antimonate anion precipitates a variety of cations, careful choice of reaction parameters can make this method specific for Ca^{2+} localisation. The affinity of antimonate for Ca^{2+} is about 10 000 times higher than that for Na^+. Precipitation of Ca^{2+} by antimonate is initiated at a lower concentration compared with that of Na^+ or Mg^{2+}. The preferential precipitation of Ca^{2+} is facilitated by buffering glutaraldehyde with a high concentration of phosphate (100 mM) at a slightly alkaline pH (7.2).

The specificity of the reaction can be checked by determining the concentration of Ca^{2+} in the antimonate precipitate by X-ray microanalysis. Advantages and limitations of this approach have been reviewed by Wick and Hepler (1982). Laser microprobe mass analysis has also been used, both to detect Ca^{2+} in freeze-dried specimens (Kaufmann, 1980) and to verify the composition of Ca^{2+}–antimonate precipitate obtained with the oxalate–antimonate technique (Van Reempts *et al.*, 1984). This approach involves vaporisation of very small regions of the tissue specimen by a high-energy laser pulse (De Noyer *et al.*, 1982). The elemental and molecular ions that are produced in this way are mass analysed. The precise location of the evaporated region after analysis can be verified in the TEM. Unlike X-ray microanalysis, this approach does not allow interference of the Ca^{2+} signal with the antimony signal. There are other advantages, including its high sensitivity (about 10^{-20} g).

Another technique for analysing the composition of antimonate precipitate is microincineration of thin sections followed by selected-area electron diffraction (Mizuhira *et al.*, 1971; Mizuhira, 1973; Hohman, 1974). Although individual precipitates can be analysed with this technique, not all deposits give clear diffraction patterns.

The specificity of the method for Ca^{2+} localisation can also be determined by employing ethylenediaminetetraacetic acid (EGTA) or ethyleneglycoltetraacetic acid (EGTA). These chelators prevent Ca^{2+} precipitation with antimonate as well as dissolving Ca^{2+}–antimonate precipitate. Electron-opaque deposits of Ca^{2+}–antimonate can be removed from both tissue blocks and thin sections with EGTA or EDTA. After treatment with EGTA, a complete extinction of peaks for Ca^{2+} results. This chelator may not penetrate thin sections of Araldite-embedded tissues. These chelators, on the other hand, have no effect on Na^{+}–antimonate deposits *in vitro* or in the embedded material. Also, other monovalent cation–antimonate salts are not removed by the chelators. Mg^{2+}–antimonate in thin sections remains essentially unaffected by EDTA, while it is slightly diminished by EGTA.

Thus, the above-mentioned differential effects of the chelator can be used to confirm the presence of Ca^{2+} in antimonate precipitates in the tissue. Apparently, the best approach to confirm the presence of Ca^{2+} is to use both chelator and wavelength-dispersive mode of X-ray microanalysis. When expensive equipment for X-ray analysis is not available, EGTA is equally reliable in confirming the presence of Ca^{2+}–antimonate. Usually 0.2 M EGTA (pH 6.8) is used for destaining of thin sections. Although improved procedures to achieve selective localisation of Ca^{2+} and vastly diminished coprecipitation of other cations as well as to obtain enhanced reproducibility are available, the results achieved with antimonate must be interpreted with caution.

REPRODUCIBILITY OF RESULTS

As stated earlier, the results of the antimonate method cannot always be reproduced. Inconsistencies in the localisation of cations are attributable to the details of the method, to the tissue employed and to the physiological state of the cells. The results are affected by all aspects of the methodology, including type of buffer and its molarity, duration of rinsing, pH, antimonate concentration, duration of incubation with antimonate, mode of antimonate application, type of fixative and its concentration and mode of application, and staining method. In the past insufficient attention has been paid to the optimal concentration of antimonate and duration of incubation. Careful selection and control of the reaction conditions can render the antimonate method highly reproducible.

The following discussion should help to optimise some reaction conditions. Since antimonate solutions tend to precipitate spontaneously above a certain concentration, optimal concentrations have not been used by most workers. According to the *Handbook of Chemistry and Physics*, the solubility of antimonate is 2.82 g/100 ml water at 20 °C. This is true at pH 7.0. However, 7% solutions of antimonate have been stored at 4 °C without spontaneous precipitation (Kashiwa and Thiersch, 1984). Eight grams of antimonate can be dissolved in 100 ml of boiling water without addition of KOH. Above this concentration, KOH must be added to dissolve the antimonate crystals completely (Kashiwa and Thiersch, 1984). Antimonate crystals from different commercial sources show different solubilities.

The conjugate acid of antimonate is insoluble, whereas the base moiety is soluble; the latter determines the pH of the solution. The pH but not the temperature reflects the maximum amount of the base dissolved or saturated in solution. The term 'saturation' is defined as the maximum concentration of antimonate in solution at any given pH. When the pH is lowered below the saturation pH, the insoluble conjugate acid is precipitated. Sucrose is most effective in stabilising the antimonate solution at an alkaline pH, and is abolished at neutral pH (Kashiwa and Thiersch, 1984). The pH of a 7% antimonate solution, for instance, cannot be lowered below 9.0 without sucrose, but with equimolar sucrose the pH can be lowered to 7.8 without spontaneous precipitation.

To optimise antimonate concentration and incubation time, information on the sensitivity of precipitation is desirable. The minimum concentration at which cations precipitate with unbuffered 2% antimonate (pH 7.8) at 4 °C is 10^{-6} M Ca^{2+}, 10^{-5} M Mg^{2+} and 10^{-2} M Na^{+} (Klein *et al.*, 1972). 10^{-1} M K^{+} does not precipitate with antimonate under these conditions. Simson and Spicer (1975) report that 1.25% and 2.5% antimonate precipitate 10^{-3} M Ca^{2+} in 1 h and 2.5% antimonate precipitates 10^{-4} M Ca^{2+} in 18 h. Lower concentrations of antimonate and Ca^{2+} require longer incubation times. Reaction of antimonate with divalent cations is linear, and even at lowest reacting concentrations efficiency of precipitation is near maximal (70% for Mg^{2+} and about 100% for Ca^{2+}) (Wick and Hepler, 1982).

Some information on whether cations alter each other's precipitation threshold is available. A 10:1 ratio of K^+:Na^+ inhibits Na^+ precipitation (Torack and LaValle, 1970). However, according to Klein *et al.* (1972), same ratios do not cause such inhibition. Precipitation of 1 mM Ca^{2+} with antimonate is somewhat increased in the presence of 25 mM Na^+ (Saetersdal *et al.*, 1974). Precipitable levels of Na^+ may coprecipitate K^+ at the same threshold level of K^+ present in antimonate (Torack and LaValle, 1970). Antimonate itself may contribute K^+ to other cation–antimonate deposits (Daimon *et al.*, 1978; Weakley, 1979). This could occur during fixation and dehydration (Tisher *et al.*, 1972; Wick and Hepler, 1982).

The role of buffers in antimonate reaction with endogenous cations is controversial. It appears that pH and buffer components affect the reaction sensitivity of antimonate for cellular cations. Phosphate buffers tend to severely inhibit the reaction of antimonate with monovalent cations such as Na^+ and K^+. Thus, the use of phosphate buffer may increase the specificity for divalent cation (Ca^{2+}) precipitation. Phosphate buffer may also stabilise intracellular Ca^{2+} against its extraction during fixation.

LIMITATIONS OF METHOD

An inherent drawback of the antimonate method is lack of reproducibility of results, including inconsistency in the specificity of Ca^{2+} precipitation. Slow access of the antimonate anion to the intracellular sites of cation precipitation seems to be the major cause. Glutaraldehyde fixation appears to limit access of antimonate to intracellular and intraorganellar sites in certain cases (Simson and Spicer, 1975). Treatments that increase the membrane permeability will facilitate the subcellular localisation of cations. Fixation with OsO_4 is one such treatment.

Another drawback lies in the possibility of relocation of elements during the preparative procedures and even prior to fixation following death or excision. Since antimonate precipitation takes place in an aqueous environment, movement of some ions will occur. Although most free, soluble cations, including cellular K^+, Na^+, Mg^{2+} and Ca^{2+}, are extracted during fixation with glutaraldehyde and OsO_4, elements bound firmly to molecular components are expected to be retained. Most of the intracellular Ca^{2+} is thought to be retained.

Antimonate precipitates formed in the tissue may undergo dissolution and redistribution during subsequent treatments such as *en bloc* staining, dehydration, embedding and post-staining (Yarom and Meiri, 1973). Uranyl acetate is one of the heavy metal salts that may reduce or dissolve the antimonate precipitate. Uranyl or lead ions may replace cations originally complexed with antimonate. If Ca^{2+}–antimonate begins to dissolve, it can be replenished by adding antimonate to rinsing, fixation and post-fixation solutions. The antimonate will preferentially complex with Ca^{2+}. Rapid dehydration in small volumes of solvent minimises dissolution of cation–antimonate precipitates.

Conditions that immobilise the cation prior to its precipitation with antimonate would be expected to facilitate its *in situ* localisation. To immobilise diffusible elements and establish their *in vivo* distribution, rapidly frozen tissues have been studied in conjunction with antimonate treatment and cryoultramicrotomy (Timms and Chandler, 1984). *In vivo* distribution of endogeneous ions such as Ca^{2+} and Zn^{2+} have been studied with this technique, although the quality of organelle preservation is poor. In any case, the relationship between the observed precipitates and the *in vivo* distribution of the ions is not clear. The presence of antimonate precipitates might present a problem in sectioning. A variety of improved modifications of the original Komnick method have been introduced to either overcome or minimise the above-mentioned drawbacks.

FIXATION AND STAINING PROCEDURES

A 5% solution of potassium pyroantimonate ($K_2Sb_2O_7 \cdot 4H_2O$) is prepared by boiling 5 g of the reagent in 100 ml of distilled deionised water. The solution is allowed to cool at room temperature, diluted to 100 ml with distilled deionised water, and filtered through Whatman No. 5 filter paper to remove pyroantimonate precipitate.

When 25 ml of 2% aqueous OsO_4 solution is added to 25 ml of 5% K-pyroantimonate solution, the final concentrations are 1% and 2.5%, respectively. The pH of this mixture is 9–9.5, which is lowered to 7.6–7.8 by adding up to 5 ml of 0.1 N acetic acid. The fixative has about 0.04 N acetic acid. Before use, the fixative is passed under pressure through a Millipore (0.22 μm) to remove any fine particles or precipitated pyroantimonate. At all stages the fixative is maintained at 4 °C. Tissue specimens are fixed for 1 h at 4 °C and then rinsed twice (for a total of 5–10 min) with 0.05 M potassium acetate buffer (pH 7.8 at 4 °C). Thin sections are post-stained with uranyl acetate and lead citrate. Several hours after mixing OsO_4 and antimonate solution, the mixture becomes metallic bluish-black in colour, which indicates the reduction of osmium.

Alternatively, a mixture of glutaraldehyde and K-pyroantimonate can be used to improve ultrastructural preservation. The fixative is prepared by breaking a 2 ml vial of purified 70% glutaraldehyde into 3 ml of distilled water, yielding a 28% glutaraldehyde solution (Weakley, 1979). Sufficient 2% aqueous K-pyroantimonate is added to 2.9 ml of the solution to yield a total volume of 20 ml. The pH is adjusted to 8.5–10.0 with 0.2 ml of 1% acetic acid. The final concentrations of glutaraldehyde, K-pyroantimonate and acetic acid are 4%, 1.7% and 0.01 N, respectively. The fixation time is 4 h at 4 °C.

RUTHENIUM RED

Ruthenium red (ammoniated ruthenium oxychloride) is an inorganic, synthetically prepared, intensely coloured, crystalline compound. Ruthenium (atomic number 44) is moderately heavy. It has long been used as a standard stain for pectins in plant tissues for light microscopy. For electron microscopy, it was probably introduced by Reimann (1961). The reagent was successfully used as an electron stain by Luft (1964) in animal histology to demonstrate a variety of extracellular materials (glycocalyx), the majority of which seem to be acid mucopolysaccharides. It has been demonstrated that staining with this metallic dye is prevented by prior treatment with either hyaluronidase or quaternary ammonium compounds, which bind to mucopolysaccharides.

Ruthenium red is a hexavalent cation (diameter ~1.13 nm) and reacts with a large number of polyanions. It gives a strong reaction with polyanions having a high charge density. However, a high charge density alone is not sufficient for a reaction to occur. Luft (1971a) has demonstrated that 1,3,5-benzene trisulphonic acid yields no precipitate with ruthenium red. Highly polymeric substances such as insulin or polyvinyl alcohol, even though extensively substituted with alcoholic hydroxyl groups, also produce no precipitate with ruthenium red.

Moderate reactivity is shown by substances of a low charge density (e.g. polyglutamic acid or hyaluronic acid), even though they are highly polymerised. In tissue, hyaluronic acid and chondroitin sulphuric acid are thought to react with ruthenium red. As stated above, although hyaluronidase generally removes ruthenium red-reactive materials, this enzyme has little effect on certain types of tissues. It has been shown that hyaluronidase does not extract ruthenium red–OsO_4 staining materials, for instance, from the myofibre and Schwann cell external lamina (Zacks *et al.*, 1973). Consequently, ruthenium red is an effective stain for studying the surface layer of cells infected with Rous sarcoma viruses, since the layer has been shown to contain acidic mucopolysaccharides, belonging to the hyaluronic acid and chondroitin sulphuric acid groups. It does react with carboxyl groups of neuraminic acid present in tissue and with glycosaminoglycans. It has been shown that glycosaminoglycans stained with ruthenium red during fixation are resistant to elastase digestion, and persist as a distinct meshwork in the regions of digested aggregates (Kádár *et al.*, 1973).

Ruthenium red does not bind specifically to sulphated mucopolysaccharides, although, according to Eisenstein *et al.* (1971), it seems to display a higher affinity for chondroitin sulphate, heparin sulphate and keratin sulphate than for hyaluronic acid. Diaminobenzidine, on the other hand, does bind specifically to sulphate groups of mucopolysaccharides, although it may also bind to phosphate groups of nucleic acids.

Ruthenium red reacts with certain lipids. Luft (1971b) has shown that the stain did not precipitate phosphatidyl ethanolamine, phosphatidylserine or phosphatidylinositol, while phosphatidylcholine (i.e. lecithin) did not react. The stain has affinity for certain intracellular lipids. For instance, it has been demonstrated that ruthenium red reacts with lipid inclusions and tonofilaments of the acinar cell in the rat maxillary gland (Vidić, 1973).

Although the observed density in tissue following staining with ruthenium red is due to the presence of acid mucopolysaccharides, these substances are invariably associated with proteins. The binding of ruthenium red to proteins is likely. Also, the possibility that some of the staining may be due to the leakage of cytoplasmic materials from inside the cell cannot be overlooked.

Ruthenium red, being a strong oxidant, probably acts not only as a stain, but also as a fixative. When this dye is applied during fixation, membranes and myofilaments are well preserved and sharply defined without post-staining. Alternatively, ruthenium red and OsO_4 together may form RO_4, which can react with some of the more polar lipids, proteins, glycogen and common oligosaccharides. For a discussion of the applications and effects of ruthenium red on cell surfaces, its role in ion transport and the impurities present in commercial lots, the reader is referred to Hayat (1975, 1989a).

PENETRATION

Some information on the penetration of ruthenium red has been obtained in connection with the study of electrical uncoupling and 'healing over' phenomena (Baldwin, 1973). Sheep heart false tendons containing strands of Purkinje cells were isolated and maintained in oxygenated Tyrode's solution until processed for electron microscopy. The dye penetrated ~100 μm from the cut surface when applied 10 min after cutting. The penetration increased to ~150 μm from the cut surface when applied 10 s after cutting. When the tissue was cut 15 min after applying the initial fixation, the dye diffused to 300–400 μm from the cut surface.

Because of its large molecular size, ruthenium red seems not to freely cross the undamaged plasma membrane. When the outer leaflets of the plasma membrane are fused together, ruthenium red cannot bind, nor can it pass such a fusion. The poor penetration of the dye may also be due to its precipitation and non-specific adsorption onto cell surfaces; the precipitate formed results from the interaction between the dye and acid mucopolysaccharides.

It is conceivable that a general lack of staining of the cytoplasm may be due to the very low concentration of the dye going across the interior of the cell and not to a lack of affinity of cytoplasmic components for the dye. It is known that ruthenium red does not penetrate and stain mast cell granules, lipids within certain cells, nuclei in chrondrocytes, certain cheek epithelial cells and the sarcoplasmic reticulum in the T-system. Gap junctions are ruthenium red-positive. Shelf-type septa are also permeable to the dye when applied in the fixative solution. However, certain septate junctions act as barriers to this tracer (Fig. 4.13) (Hori, 1986). Ruthenium red also penetrates agranular reticulum of the axon, first at the node of Ranvier and later in the internodal region in peripheral nerve fibres (Singer *et al.*, 1972). The dye also sequesters in the Schwann cell body and in adaxonal and paranodal Schwann cell cytoplasm. Better penetration is obtained when the tissue is immersed in ruthenium red–Ringer solution prior to fixation.

MECHANISM OF STAINING

The intensity of the ruthenium red reaction is dependent primarily upon the number of ionisable carboxylic acid groups available, provided that the molecular weight of the carrier is sufficiently high. Ruthenium red binds to its substrates electrostatically (salt linkage). If non-coulombic forces are also involved in its selective binding, then they will require steric conformation of the dye and binding site.

The studies of mast cell granules have provided some information on the nature of the staining mechanism. It is known that ruthenium red reacts selectively with mast cell granules by binding with acid glycosaminoglycan (heparin). This staining reaction seems to be stoichiometric and the stain binds to acid glycosaminoglycan in a molar ratio which results in one ruthenium red–cationic complex per sulphate group of the acid. The treatment of cells with neuraminidase or hyaluronidase significantly reduces the ruthenium red-stained material. This is in favour of the proposed specificity of ruthenium red for hyaluronic and/or chondroitin-like sites on the cell surface, although trypsin also somewhat affects the staining.

As a pectin stain, the active group in ruthenium red is considered to be the ruthenium ion and the associated complex of four ammonia molecules. The staining is considered to be accomplished when the host molecule has two negative charges 0.42 nm apart to accommodate the staining group (Sterling, 1970). Pectic materials possess such staining sites and are thus readily stained with ruthenium red. It is thought that in the case of pectins the staining site is intramolecular. De-esterification intensifies staining of cell walls, for it re-establishes the carboxyl group with a strong negative charge on the oxygen atoms.

Although pectins are easily stained red with ruthenium red in light microscopy, very little increase in density is observed in the electron microscope when staining solutions of usual concentrations are employed. However, when the stain is accompanied or followed by treatment with OsO_4, a selective increase in the electron density becomes obvious without additional heavy metal staining.

Before discussing the interactions among OsO_4, ruthenium red and cell components, it is pertinent to point out that when a sufficient amount of ruthenium red is deposited on a cell, it can be visualised by its own mass. This can be accomplished by employing solutions of higher concentrations or by staining the sections. Single cells can also be stained with ruthenium red alone. Ruthenium red alone can produce adequate contrast. Ruthenium red-positive substances may occur as amorphous or granular bodies. The primary advantages of using OsO_4 with ruthenium red are higher contrast and *en bloc* staining.

It is interesting to note that the reaction products of acid mucopolysaccharides and ruthenium red correspond in location and appearance to those observed after positively charged colloidal iron or thorium dioxide staining. Furthermore, the staining behaviour of ruthenium red is almost identical with that shown by alcian blue.

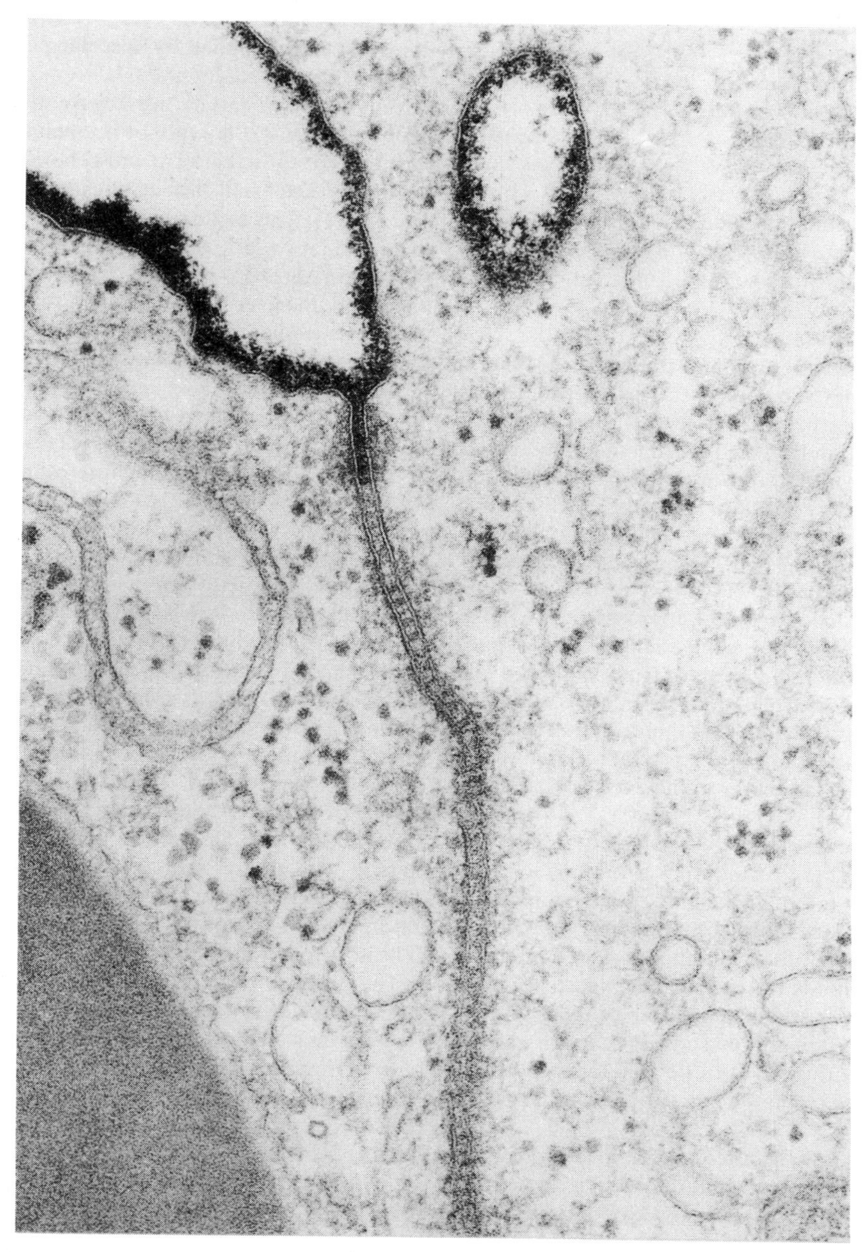

Figure 4.13 Apical portion of the gastrodermal cell stained with ruthenium red. Penetration of the dye is blocked at the boundary portion between the intermediate junction and the septate junction. From Hori (1986)

FIXATION AND STAINING PROCEDURES

Maximum contrast is obtained when ruthenium red and OsO_4 are applied together. The stain can be applied by adding it to the fixation solution, water or buffer solutions. All standard buffers can be used. However, phosphate buffers are not recommended for ruthenium red–OsO_4 procedures. Chloride-free cacodylate buffer is considered the most suitable. Inconsistent staining may result if the tissue is exposed to ruthenium red after primary fixation. For general purposes the following method of fixation and staining for acid mucopolysaccharides is recommended.

Solution A

Aqueous glutaraldehyde (4%)	5 ml
0.2 M cacodylate buffer (pH 7.3)	5 ml
Ruthenium red stock solution (1500 ppm in water)	5 ml

Solution B

Aqueous osmium tetroxide (5%)	5 ml
0.2 M cacodylate buffer (pH 7.3)	5 ml
Ruthenium red stock solution (1500 ppm in water)	5 ml

Tissue specimens are fixed and stained with solution A for 1 h at room temperature. The specimens are rinsed briefly with the buffer and then post-fixed and stained with solution B, which is prepared immediately before use, for 3 h at room temperature. This is followed by a brief rinse with the buffer, and dehydration and embedding according to standard procedures. Post-staining of sections with uranyl acetate and lead citrate is not necessary.

The staining reaction requires appropriate pH values for its optimum effects on each tissue type. A pH range of 7.2–7.4 is considered desirable. Ruthenium red and OsO_4 should be mixed immediately prior to use, for the mixture is not very stable. The application of this dye during fixation may result in the formation of artefacts such as myelin-like structures in the extracellular space and/or within cells.

There are disadvantages of *en bloc* staining with ruthenium red. These include (1) a prolonged treatment is necessary to obtain staining of cell components other than acid mucopolysaccharides; (2) there is a reduction in contrast during embedding in Epon; and (3) a useful reaction in the extracellular space is limited to ~100 μm in the tissue block. Some of these difficulties and the irregularity of the staining reaction, due, in part, to the poor diffusion of ruthenium red into certain tissues, can be overcome by employing vascular perfusion.

In some tissues (e.g. hamster adrenal medulla) ruthenium red yields positive results only when applied by vascular perfusion. The animals are perfused with ice-cold saline (NaCl, 0.9 w/v) for 2 min to remove most of the blood, and then with ice-cold 2.5% glutaraldehyde containing ruthenium red (3000 ppm). The fixative is introduced via a needle into the left ventricle of the heart, and is allowed to escape through a cut in the right auricle. After perfusion, the tissue is removed and immersed for 1 h in the same fixative. This is followed by rinsing for 3–10 min in cold cacodylate buffer (pH 7.4) containing 7.5% sucrose. The tissue is then post-fixed in 2% OsO_4 containing ruthenium red (3000 ppm) for 3 h in the dark at room temperature (Benedeczky and Smith, 1972). Alternatively, ruthenium red can be applied by vascular perfusion prior to fixation (Kimble *et al.*, 1973).

Cells grown in culture are apparently easier to stain than are tissues. A mixture of ruthenium red and OsO_4 was used to study the form and general composition of the external cell wall surface of cultured cells (Leppard *et al.*, 1971). This technique demonstrated the presence of a fribrillar coating on the external surface of suspension-cultured plant cells. The composition of these fibrils has been described as lignin-like and they are therefore termed lignofibrils. The staining of lignofibrils is assumed to be due to the binding of ruthenium red cations to the negatively charged acid groups in lignin. These lignofibrils have not as yet been demonstrated in normal plant tissues. The presence of a lignofibril coat poses the interesting question of whether or not this coat is related to the glycocalyx in animal tissues. The procedure employed by Leppard *et al.* (1971) is given below.

Cells taken directly from culture are fixed for 1 h at room temperature in 6% glutaraldehyde at pH 7.0 in 0.01 M phosphate buffer containing 0.4 M sucrose. They are washed in buffered sucrose solution followed by buffer solution. This is followed by post-fixation in a filtered, saturated solution of ruthenium red in 0.05 M phosphate buffer at pH 6.8 containing 1% OsO_4. The specimens are washed in cold buffer, dehydrated by methanol and propylene oxide, and embedded in Epon according to standard procedures. Sections are post-stained with uranyl acetate followed by lead citrate. This method was also employed to delineate the non-amorphous nature of the middle lamella between plant cells in suspension culture.

The problem of poor penetration and absorption of ruthenium red can also be alleviated by post-staining thin sections of specimens embedded in water-miscible or water-immiscible resins; the staining solution easily penetrates glycol methacrylate is well as Epon sections. Specific staining of mast cell acid glycosaminoglycans has been obtained by using ruthenium red as a post-stain for glycol methacrylate sections (Gustafson and Pihl, 1967). In fact, the maximum uptake of the stain by cartilage acid glycosaminoglycans takes place when the stain is employed in the fixative as well as a post-stain. In this method, specimens are fixed in 2.5% paraformaldehyde in cacodylate buffer (pH 7.4) containing 0.001 M ruthenium red for from 1 h to 14 days in the cold (only freshly prepared solutions should be used). Dehydration and embedding are carried out in glycol methacrylate. Thin sections are post-stained with 0.001 M ruthenium red in 0.1 M phosphate buffer (pH 7.4) containing 7.5% sucrose, followed by a brief rinse in distilled water. Filtered staining solution can be stored in the refrigerator, but should be discarded when it turns brownish.

SILVER

Silver (atomic number 47) is sufficiently heavy to be employed as an electron stain. It differs slightly from the other cationic stains in that it is also an oxidising stain and is a redox reagent. On application, the metallic silver is precipitated from the staining solution in the form of fine granules, primarily by the reductive activity of aldehyde radicals. The granules of the silver precipitate have a minimum diameter of 3–5 nm; after longer treatment these small granules fuse to form large aggregates of 15–50 nm in diameter. The average diameter of the granules is ~10 nm. In OsO_4-fixed tissues the density of these granules is usually so high that they completely mask the outline of the structures on which they are deposited. Furthermore, in these tissues the specificity of the silver stain is lost. Much better results are obtained when the tissue is fixed with glutaraldehyde.

In general, silver is deposited in regions containing mucopolysaccharides and diffuse proteins, including glycoproteins. Thus, the stain has been employed to study materials present in the Golgi apparatus, and at the surface of cells and basement membranes. Ammoniacal silver reaction has been employed for detecting arginine-rich proteins, especially histones (MacRae and Meetz, 1970), while silver nitrate selectively stains proteins associated with nucleic acids (Smith and Stuart, 1971). Silver hexamine is useful in the investigation of the distribution of disulphide-containing proteins in plant cell walls (Leppard *et al.*, 1971).

MECHANISM OF STAINING

The major staining mechanism involves a reaction in which metallic silver is precipitated from the staining solution by the reductive activity of aldehyde radicals. To free the aldehydic groups, the tissue is usually oxidised with periodic acid (PA) prior to staining. It is known that when aldehyde groups in the oxidised sections are blocked by chlorous acid ($HClO_2$), staining with silver is abolished.

In light microscopy carbohydrate macromolecules, which are oxidised by periodic acid, can be detected by the PA–Schiff techniques, while in electron microscopy the Schiff reagent may be replaced by alkaline silver solutions to detect the aldehydic groups released by PA oxidation. In fact, the staining of cell surfaces with the PA–Schiff reaction is more or less similar to that by the PA–silver or by PTA.

The PA–silver reaction is accomplished in two steps: (1) periodic acid oxidises 1,2-glycol and α-amino alcohol groups and transforms them into aldehydic groups; and (2) these aldehydic groups reduce silver tetramine contained in methanine silver solution, with the release of free silver. Since glycoproteins are rich in 1,2-glycol and α-amino alcohol groups, they show strong affinity for silver stains. This is the reason why PA–silver has been employed to stain the material at the cell surface, which is considered to be rich in carbohydrates. Other structures which show specific reaction with PA–silver include mucus, thyroid colloid, lysosomes, cartilage matrix, starch granules, cell walls, kidney basement membrane and glycogen (in hepatocytes and leucocytes).

The available evidence indicates that reducing groups such as aldehydes are present in the tissue for reaction with silver, even without periodic acid oxidation. Additional aldehyde groups may be added to the tissue as a result of fixation with aldehydes. It is quite possible that when tissues are fixed with aldehydes, certain structures may absorb the aldehyde and then reduce the silver. Thus, fixation with aldehydes may increase the number of structures unspecifically stained. An example of the role of aldehydes in silver staining is discussed below.

It will be mentioned later in more detail that noradrenalin as well as 5-hydroxytryptamine can be localised in glutaraldehyde-fixed tissues by using ammoniacal silver. To explain the specific staining of noradrenalin, it has been suggested that glutaraldehyde reacts with the primary amine group of noradrenalin, forming a yellow-coloured compound. The chemical structure of this compound seems to be such that one of the carbonyl groups of glutaraldehyde remains unreacted and free. This carbonyl group is capable of reducing an ammoniacal solution of silver hydroxide to form a metallic precipitate.

On the other hand, *in vitro* studies to elucidate the cytochemical basis of the localisation of 5-hydroxytryptamine indicate that free carbonyl groups are almost completely absent, while free OH and NH groups are present (Gardner and Silver, 1971). However, at least one of the two aldehyde groups is probably present in 'crypto' form, and can be readily released under appropriate reaction conditions. When silver stain is added, the aldehyde groups can re-form and reduce the silver ion. This reduction results in the precipitation of metallic silver at the sites of storage of 5-hydroxytryptamine in the tissue.

In glutaraldehyde-fixed tissue non-specific silver deposition can be obtained without introducing periodic acid oxidation. In these tissues the reacting sites are located on chromosomes, nucleoli, ribosomes, collagen fibres, plasma and red blood cells, certain pigments, the majority of cell granules, etc. (Rambourg and Leblond, 1967). On the other hand, the specificity of the silver stain for certain tissues is not affected in the absence of periodic acid oxidation. For instance, omission of pretreatment with periodic acid does not affect the staining specificity for the cytoplasmic granules of glutaraldehyde-fixed adrenal medulla (Chang and Bencosme, 1968).

Since PA oxidation does not affect specific or non-specific staining of certain structures of glutaraldehyde-fixed tissues to any significant degree, it is assumed that pre-existing aldehyde groups are responsible for reaction with the silver and/or that reducing groups other than aldehydes are also involved. It has been suggested that silver is taken up by histones after formaldehyde fixation (Black and Ansley, 1966). Thus, aldehyde groups from formaldehyde may account for the non-specific staining of chromatin, although the staining of chromatin may be due to the reaction of a protein moiety with silver.

It has been suggested that the staining of nucleic acid-associated structures is dependent upon the binding of silver to a protein (histone) moiety (Smith and Stuart, 1971). This reaction may involve the formation of chelate complexes between silver ions and the peptide backbone. However, it should be noted that treatment with 10% perchloric acid (which extracts most of the RNA) abolishes the staining of the nucleolus with silver, whereas the staining of cytoplasmic and mitochondrial ribosomes is not affected. A discussion of the binding of silver ions by DNA will be presented later. The selective reaction of silver with protein is probably dependent upon a particular protein conformation which satisfies the steric requirements of the reaction. It has been suggested that silver interacts with reactive centres in arginine.

According to Mishima (1964), the non-specific staining of pigments rich in melanin may be due to the reducing groups other than aldehydes. That free sulphhydryl groups (—SH and —S—S) in the cell wall of *Pityrosporum ovale* react with silver was demonstrated by Swift (1966). Studies of the effects of relatively low concentrations of Ag^+ on transport and permeability properties of frog skin suggest that the most likely sites of Ag^+ reaction are sulphhydryl groups; however, other possibilities, such as the imidazole moiety of histidine, cannot be excluded (Curran, 1972). By using a bovine γ-globulin (which lacks sulphhydryl groups but possesses disulphide groups which may be reduced to sulphhydryl) and Sephadex G25 (a cross-linked dextran rich in 1,2-glycol groups) complex, it has been shown that silver methenamine reacts with both sulphhydryl and 1,2-glycol groups (Burr, 1973).

Swift (1969) demonstrated that disulphide groups in the cystine residues of human keratin are oxidised by silver with concomitant deposition of metallic silver. According to Jessen (1973), staining of keratohyalin granules with silver is attributed to the presence of cysteinyl (—SH) groups and not cystine (—S—S) groups. Studies by Swift (1973), however, indicate that this staining is due to the latter groups, which results in the formation of thiosulphate-insoluble silver deposits. It has been proposed that when the silver methenamine treatment lasts for more than 1 h, disulphide bonds are oxidised and $AgNO_3$ is reduced as silver grains (Thompson and Colvin, 1970). It is certain that groups other than sulphhydryl, disulphide and aldehyde will reduce the silver salts (or ion). It has been demonstrated, for example, that reactive groups in xylem wall thickenings are not blocked by either iodoacetate or dimedone and bisulphite (Pickett-Heaps, 1967). Further work is needed to establish the identity of groups other than those mentioned above which react with silver.

Silver ion is one of the few cations which forms complexes with nucleic acids by interacting with purine and pyrimidine bases, rather than by interacting with phosphates. Yamane and Davidson (1962) have suggested that silver ions react with purine and pyrimidine bases. Biochemical studies indicate that there are at least three types of binding of silver ion by DNA. Denatured DNA binds more strongly than does native DNA (Jensen and Davidson, 1966). In native DNA, silver is probably bound between two nitrogens of complementary base-pairs. The complex of Ag^+ with DNA is characterised by the quantitative relationship: 1 ion to 4 bases (Ivanov *et al.*, 1967). The high strength of the binding suggests that the bound silver ions are somehow chelated. The Ag–DNA complexes have almost the same high intrinsic viscosity as the uncomplexed DNA, which indicates that the helical structure is not disrupted.

ROLE OF FIXATION

The type of fixation greatly influences the affinity of the tissue for silver stain. In the past, OsO_4 has been commonly employed to fix tissues prior to staining with silver. All cytoplasmic membranes show intense staining with silver after such fixation.

The staining of the membrane is considered to result from the reduction of the silver by the osmium bound to the unsaturated lipids of the phopholipids, for this staining can be prevented by eliminating the bound osmium by oxidising the sections with hydrogen peroxide or periodic acid. This oxidation removes the bound reduced osmium by making it water-soluble in the tetroxide form. Furthermore, the membranes do not regain their affinity for the silver stain if the sections are again treated with OsO_4 after oxidation.

As with cytoplasmic membranes, the staining of nucleoproteins is prevented when the reduced osmium is eliminated by oxidation with hydrogen peroxide. However, nucleoproteins are stained with silver if the sections are again treated with OsO_4 after oxidation. It is known that nucleoproteins retain their affinity for osmium even after the sections have been oxidised. It is important to note that double fixation with an aldehyde followed by OsO_4 enhances the intensity of chromatin staining with silver.

On the other hand, glycogen, mucin and elastic fibres in the osmium-fixed tissue do not lose their affinity for silver after oxidation of the sections. The persistence of this affinity is probably due to the reducing capability of the aldehyde groups which are released during fixation by the oxidising action of OsO_4.

In tissue fixed with acrolein alone silver is deposited heavily on the nuclei and ribosomes. Chromatin is stained more intensely than is the nucleolus. The strong affinity of chromatin for silver is probably conditioned by the reaction between the nuclear histones and acrolein. It has been claimed that fixation with a mixture of $AgNO_3$ and glutaraldehyde preserves and stains the chromosomal matrix, which is difficult to discern following standard fixation (Bobak and Herich, 1969). Although silver ions have been shown to bind to both DNA and RNA in solution (Ivanov *et al.*, 1967), the affinity of ribosomes for silver is probably related to their protein component rather than their RNA, because ribonuclease digestion does not affect the staining reaction. Collagen fibres and secretory granules of the pancreatic endocrine cells also stain strongly. Basement membranes and zymogen granules, however, stain very weakly.

As mentioned above, although a number of structures, including all cytoplasmic membranes, show intense staining with silver after OsO_4 fixation, the staining is non-specific. Thus, fixation with OsO_4 is not recommended if the objective is to achieve specific staining. On the other hand, a certain degree of selective staining with PA–silver can be obtained by employing glutaraldehyde as the sole fixative. Of all the fixatives, glutaraldehyde probably yields a relatively small number of non-specific reactions with this stain.

FIXATION AND STAINING PROCEDURES

Ammoniacal Silver

Thiéry and Bader (1966) demonstrated that in glutaraldehyde-fixed pancreatic islets, β-cells were selectively stained with silver methenamine, whereas δ-cells showed staining with ammoniacal silver. Tramezzani *et al.* (1966) showed that in aldehyde-fixed tissues, heterochromatin, ribosomes, certain secretion granules and the area around mitochondria were selectively stained with ammoniacal silver carbonate. According to this method, tissues are fixed in phosphate-buffered 6% glutaraldehyde for 6–24 h at 4 °C, washed in distilled water for ~10 min, and then stained *en bloc* with ammoniacal silver carbonate for 30 min at room temperature. After washing in distilled water for 15 min, the specimens are immersed in a 1% solution of paraformaldehyde for 30 min at room temperature.

Sections mounted on grids other than copper can be stained with ammoniacal silver, a procedure which results in general staining of cell components. The advantage of the method is that the stain can be seen to darken the section, and thus the intensity of staining can be easily controlled. Non-copper grids (e.g. stainless steel) with sections are floated on drops of the staining solution, heated until the desired contrast is achieved, and rinsed in distilled water. Staining for 2–3 min at 99 °C has been recommended. If necessary, silver salt crystals can be removed by rinsing with dilute ammonium hydroxide. The staining solution is prepared by adding 10% silver nitrate to a small amount of ammonium hydroxide (58%) until a faint turbidity remains; this is diluted with an equal amount of distilled water.

Ammoniacal silver has been employed for localising histones in root tip cells (Haapala and Nygrén, 1973). Silver grains with mean grain size of ~40 nm were located in the chromatin and nucleoli. The staining reaction was absent in the cytoplasm. Tissue specimens are fixed in 10% formaldehyde, neutralised with sodium acetate at pH 7.0, for 2 h. After washing in distilled water for 25 min, the specimens are immersed in ammoniacal silver solution prepared according to the method of Black and Ansley (1969). The specimens are washed in distilled water and then transferred to 3% formaldehyde solution for 3 min to develop the staining reaction.

Silver Methenamine

Chang and Bencosme (1968) differentiated three types of secretory granules in rat adrenal medulla with silver methenamine. It is important to note that these granules were not specifically stained when the tissue was fixed with formaldehyde or post-fixed with OsO_4 only. It is apparent that glutaraldehyde is able to preserve the argentaffinity of norepinephrine-storing granules.

The reaction between glutaraldehyde and noradrenalin is pH-dependent and the product is a polymer. The polymer can be selectively stained in tissues after months of storage at 4 °C. On the other hand, adrenalin does not combine with glutaraldehyde and is washed out before sections are exposed to the silver solution. The duration of staining is an important factor in obtaining this specificity. It should be noted that the glutaraldehyde–noradrenalin polymer can be rendered more electron-dense not only by silver, but also by OsO_4, potassium dichromate, sodium molybdate or potassium iodate.

The silver methenamine technique of Chang and Bencosme (1968) has been employed to distinguish between specific atrial granules, which were silver-negative, and lysosomal granules which showed intense impregnation with silver (Berger and Bencosme, 1971). This technique is presented below.

Sections of the glutaraldehyde-fixed tissue are transferred by wire loop to 1% periodic acid solution for 15 min at room temperature. Silver methenamine solution is prepared by adding 2 ml of 5% silver nitrate solution to 18 ml of 3% hexamethylenetetramine solution. The turbidity which appears will clear by continuous stirring. To the silver methenamine solution is added 2 ml of 2% borax solution, and the mixture is filtered twice through No. 42 Whatman paper.

Following oxidation with periodic acid, sections are rinsed in distilled water and then transferred to the staining solution. Staining dishes are immediately placed into an oven for 1–1½ h at 55–60 °C. At the end of the first 45 min, floating sections can be selected at 15 min intervals for examination on a glass slide, and the progress of staining reaction can be assessed with a light microscope. After sections are adequately stained, they are transferred by a platinum wire loop to 5% sodium thiosulphate solution for 1 min to stop the silver from reacting further. Sections are rinsed thoroughly in distilled water. Sodium tetraborate, periodic acid and thiosulphate solutions can be kept as stock solutions, but other solutions should be prepared immediately before use.

It has been suggested that silver methenamine (methenamine is also known as hexamethylenetetramine, hexamine or urotropin) interacts with cystine-containing proteins (Swift, 1968) under alkaline conditions. The preparation of the staining solution is given below.

Solution A	
Silver nitrate (5%)	5 ml
Methenamine (3%)	100 ml
Solution B	
Boric acid (1.44%)	10 ml
Borax (1.9%)	100 ml
Final solution	
Solution A	25 ml
Solution B	5 ml
Distilled water	25 ml

The final staining solution should have pH 9.2. It can be stored in the dark at 0 °C for up to 1 week.

Thin sections are stained by immersing the grid in the solution in a covered Petri dish placed in a light-tight oven at 45 °C. The duration of staining varies from 30 s to 2 h. Staining for longer than 75 min is considered to demonstrate the location of cystinyl residues in hair fibre. This staining can be eliminated by preliminary reduction and alkylation of sections with iodoacetate, which results in almost complete conversion of cystine to *S*-carboxymethyl cysteine in keratins (Maclaren and Sweetman, 1966).

Reduction can be accomplished by immersing the sections in 0.3 M benzyl thiol in 20% *n*-propanol solution for 90 min at room temperature, followed by a brief rinse in 20% *n*-propanol solution. Alkylation can be carried out by immersing the sections in iodoacetate solution for 4 h at room temperature, followed by washing with 50% *n*-propanol solution. The iodoacetate solution is prepared by dissolving 18.6 g iodoacetic acid and 12.35 g boric acid in 400 ml distilled water. The pH is adjusted to 8.0 with potassium hydroxide solution. This solution is diluted to 500 ml with distilled water, and finally 500 ml of *n*-propanol is added. Sodium thiosulphate solution (10%) can be used to remove covalently or loosely bound silver from the silver-impregnated sections.

The silver methenamine schedule of Swift (1968) was used to stain thin sections for visualising proteinaceous wound healing material (Burr and Evert, 1972). This schedule was modified in that sections are pretreated with sodium bisulphite or dimedone, and that alcoholic solutions of benzyl thiol and iodoacetate are replaced by solutions of dithiothreitol and iodoacetate in 0.2 M borate buffer (pH 8.0). In addition, the alkylation step is started on ice and subsequently allowed to warm to room temperature. The presence of proteins can be confirmed by extraction with various proteases. If the tissue has been post-fixed with OsO_4, its sections should be treated with hydrogen peroxide to remove the osmium, which otherwise may inhibit protease activity.

Silver methenamine has been used to demonstrate

sulphhydryl groups in keratohyalin granules and the peripheral envelope of cornfield cells in epithelia (Jessen, 1973). The details of the staining technique are given above. Sections mounted on gold grids are floated on drops of freshly prepared, cold (4 °C) staining solution, and incubated in the dark for 2–3 h at 45 °C. After a thorough rinse on distilled water, the grids are floated on 10% sodium thiosulphate for 15 min at room temperature. Blocking of sulphhydryl, disulphide and aldehyde groups prior to staining can be performed with freshly prepared reagents as given below.

Sulphhydryl groups Sections are immersed in iodoacetate reagent for 4 h at 45 °C.

Disulphide groups Grids are floated on 0.6 M mercaptoethanol in distilled water for 90 min at 45 °C. After a thorough rinse in distilled water, the reduced sections are alkylated with iodoacetate reagent for 4 h at 45 °C.

Aldehyde groups Sections are treated with a solution of dimedone (saturated at room temperature) in 1% acetic acid for 1–2 h at 60 °C.

The presence of cystine-containing proteins in hair follicle can be detected when thin sections of the unfixed fibres are treated with Gomori's silver methenamine reagent at pH 9.2 at 45 °C for 2–3 h (Swift, 1969). This treatment results in the deposition of globular metallic silver in the components containing cystine. It is a useful technique for investigating the detailed distribution of disulphide bonds in cell walls. This technique is successful only under carefully controlled conditions of pH, duration and temperature.

Silver methenamine has been used for localising the distribution of protein in plant cell walls (Colvin and Lepard, 1971). This technique, in a modified form, was used to demonstrate cystine-containing proteins in the marine alga *Bryopsis hypnoides* (Burr, 1972). Sections on gold grids are treated with 2% aqueous sodium bisulphite (sodium metabisulphite can also be used) for 30–40 min at room temperature in order to reduce possible free aldehyde groups. This is followed by a thorough rinsing with distilled water. The sections are incubated in a fresh solution of silver methenamine prepared according to the method of Thiéry (1967) for 1 h at 60 °C in a darkroom, using a safe light (Kodak Wratten Series OC filter).

After incubation the sections turn golden brown in colour. They are rinsed in distilled water and then floated on 5% sodium thiosulphate for 30 min in the dark. The sections are thoroughly washed with distilled water and dried by touching the grid to a filter paper.

The silver methenamine technique has also been employed for selective staining of exposed polysaccharides in the cell surface of bacteria (Dhir and Boatman, 1972). The specificity of this staining is dependent upon the production of aldehyde groups after periodate oxidation of the polysaccharide. The exposed aldehyde groups react with silver methenamine, causing a deposition of silver on specific sites. The deposition is proportional to the number of reactive groups present. Prior treatment with sodium deoxycholate (1%) results in a significant reduction in staining.

Sections on nickel grids are immersed in 0.005 M periodic acid for 20 min at 22 °C, rinsed twice in distilled water and stained with freshly prepared silver methenamine solution for 70 min at 50 °C. The grids are rinsed thoroughly in distilled water, treated with sodium thiosulphate (0.5%) for 2 min at 22 °C and washed in distilled water. The preparation of the staining solution is given below:

Silver nitrate (0.25%)	10 ml
Hexamethylenetetramine	0.3 g
Sodium borate (5%)	8.0 ml
Distilled water	12.0 ml

The non-specific background staining after periodic acid–silver methenamine treatment may be due to glutaraldehyde, acrolein contaminant in glutaraldehyde, sulphhydryl groups or other reducing compounds pre-existing in the tissue. This background staining may occur in glutaraldehyde–OsO_4-fixed tissues as well as in tissues fixed with glutaraldehyde only. In some cases the background staining can be eliminated by removing osmium with periodic acid oxidation. Only those staining reactions which appear after periodic acid oxidation should be considered specific in these staining procedures.

Silver nitrate methenamine has been employed for staining of argyrophilic cells (alpha cells) by using semithin sections of OsO_4 and potassium dichromate-fixed pancreatic tissue (Lee, 1967). The advantages of this method are a short incubation period and the opportunity to correlate light and electron microscopy. This method has also been used for staining the granules of certain mouse gastric endocrine cells (Weinshelbaum and Pittman, 1972). Although maximum staining was obtained in the tissue fixed with OsO_4–potassium dichromate, glutaraldehyde–OsO_4 fixation resulted in less interference from diffuse silver impregnation and better preservation of the fine structure. The procedure is given below.

Tissue specimens are fixed with OsO_4 containing potassium dichromate (Dalton's fixative) for 1 h at 4 °C. Semithin sections are placed on a glass slide and immersed in aqueous 4% sodium bisulphite for 2 min. The sections are washed with distilled water for 5 min and then incubated in freshly prepared silver nitrate solution at 60 °C in an oven for 1 h or longer. The intensity of staining should be checked at 15 min intervals until the desired intensity of staining is achieved. The staining may have to be continued for periods up to 4 h. The silver solution is prepared by adding 7.5 ml of aqueous silver nitrate to 85 ml of aqueous 3% methenamine. This solution is alkalinised by adding 15 ml of basic borax buffer (pH 9.0).

After incubation, the sections are rinsed in distilled water and exposed to 3% aqueous thiosulphate for 2 min to remove excess silver. This step is followed by washing in tap-water for 5 min, dehydration and mounting. Granules in argyrophilic and endocrine-like cells stain brown-black.

Highly selective staining of nucleic acid-associated proteins can be obtained with silver nitrate. Tissues are fixed in 2% glutaraldehyde in 0.1 M phosphate buffer (pH 7.3) containing 0.22 M sucrose. Sections are mounted on non-copper grids (e.g. titanium) and stained by immersion in freshly prepared 5% silver nitrate in distilled water for 30 min at room temperature. The grids are washed thoroughly in distilled water.

Silver Nitrate

Silver nitrate (Fontana reaction) has been employed for the staining of lipofuscin pigment (Hendy, 1971). The stain is prepared by adding ammonia dropwise to 25 ml aqueous 10% silver nitrate solution until the precipitate which first formed disappears. By adding distilled water, the solution is made up to 50 ml. The solution is left in the dark for 24 h, stored in a dark bottle and filtered before use.

A special formulation containing silver nitrate was applied to human pancreatic islet cells by Grimelius (1969) to obtain specific staining of secretory granules in α-cells; secretory granules in β-cells were not stained. This technique was successfully employed to obtain selective staining of endocrine granules (Vassallo *et al.*, 1971). The selective staining is claimed to be partly due to the presence of 5-hydroxytryptamine in the endocrine granules.

For staining, tissue specimens (less than 1 mm^3) are fixed in a mixture (1:1) of 2.5% glutaraldehyde and 2.5% paraformaldehyde in 0.1 M phosphate buffer (pH 7.4) for 3 h at 4 °C. The specimens are washed in the buffer for 15 min and then cut into slices 100–150 μm thick, preferably with an automatic tissue sectioner; larger specimens are inadequately impregnated with the stain. The tissue slices are collected in 0.9% NaCl, soaked in 0.1 M acetic acid–sodium acetate buffer (pH 5.6) for 15 min, and impregnated with 0.03 M $AgNO_3$ in 0.1 M acetate buffer (pH 5.6) in darkness for 24 h at 37–40 °C. The impregnated tissue slices are developed in a mixture containing 1% hydroquinone and 5% Na_2SO_3 for 1–2 min at 45 °C.

The specimens are rinsed in distilled water for 5 min and then transferred to 5% aqueous Na_2SO_3 for 5 min. This step is followed by a rinse in distilled water for 5–10 min and reimpregnation with $AgNO_3$ in darkness for 10 min at room temperature. The specimens are redeveloped for 2–3 min at 45 °C. After rinsing in distilled water for 5 min, they are post-osmicated in 1% OsO_4 for 30 min at room temperature. The sections are post-stained with uranyl acetate followed by lead citrate.

Impregnation Techniques

Silver nitrate has been employed in the Golgi impregnation method commonly applied for studying intercellular relationships in the neuronal tissue. This method was employed by Pinching and Brooke (1973) for studying the same cell in the rat olfactory bulb with both light and electron microscopy. The animal (e.g. rat) is perfused with a solution containing 5% glutaraldehyde and 2.4% potassium dichromate in 0.1 M phosphate buffer. After several hours olfactory bulbs are removed and cut into small pieces. These specimens are then further fixed in the same solution for 5 days. After washing in distilled water, the specimens are left in 0.75% silver nitrate solution for 5 days. The specimens are rinsed and placed in the same fixative for 2–3 days. After rinsing, the specimens are again transferred to the silver nitrate solution. The specimens are rinsed again and post-fixed with 2% OsO_4 in 0.1 M phosphate buffer (pH 7.3) for 1 h. All solutions should be freshly prepared, and during processing the specimens should be kept in the dark. All steps are carried out at room temperature.

A modification of the silver impregnation technique used by Fernandez-Gomez *et al.* (1969) for light microscopy has been employed for staining the fibrillar part of the nucleolus both in plant and animal tissues (Risueño *et al.*, 1973). This selective staining is thought to be due to the affinity of silver for long chains of polypeptides. Tissue specimens are fixed in a 1:1 mixture of 10% formaldehyde and 1% hydroquinone for ~15 h at 4 °C. After washing in distilled water for 30 min, the specimens are impregnated with 2% silver nitrate for 4 h at 70 °C in the dark. The specimens are washed in distilled water for 30 min and transferred to the formaldehyde and hydroquinone mixture for 1 h at room temperature to develop the staining reaction.

Modifications of silver and chromium impregnation techniques have been used to classify amine-producing endocrine cells. The argyrophil, argentaffin and chromaffin reactions can be performed directly on thin sections (Hakanson *et al.*, 1971). It is pointed out that the following techniques are primarily of cytodiagnostic value, because the morphological details are not very clear.

For the argyrophil reaction, tissue specimens are fixed with 10% neutral formaldehyde but not with glutaraldehyde; post-fixation with OsO_4 does not affect the staining. Sections are placed on a nickel grid which is allowed to float on a drop of 0.03% solution of silver nitrate in 0.02 M acetate buffer (pH 5.6) for 6–12 h at 60 °C in a humid chamber (a Petri dish covered on the inside with wet filter paper suffices). The grid is then developed for 1–1½ min in a freshly prepared aqueous solution of 1% hydroquinone and 5% sodium sulphite, followed by rinsing with distilled water.

For the argentaffin reaction, tissue specimens are fixed with glutaraldehyde; post-fixation with OsO_4 is undesirable. Sections on a grid are floated on a drop of ammoniacal silver solution for 1–4 h at 60 °C, and then rinsed in distilled water.

For the chromaffin reaction, tissue specimens are fixed as stated above, and sections on a grid are floated on a drop of 2.5% potassium dichromate solution containing 1% sodium sulphite in a 0.2 M acetate buffer (pH 4.0) for 2–8 h at 60 °C.

Silver Proteinate

The method for the demonstration of periodate-reactive carbohydrates by using thiosemicarbohydrazide in addition to the periodic acid–silver proteinate (PA–TCH–SP) (Thiéry, 1967) has been presented earlier. This silver proteinate method yields a very fine reaction product and is almost free of non-specific precipitates. It is more specific and yields better staining than that given by the periodic acid–silver methenamine method. For instance, the latter method cannot be employed to localise starch in plant amyloplasts, since starch grains are washed away when the sections are floated on the silver methenamine solution at 60 °C. The specificity of silver methenamine is limited, since many structures in non-oxidised sections show high reactivity with this stain.

It is pointed out that positively stained structures should not be assumed to be of carbohydrate nature unless they are found unstained in non-oxidised sections treated with thiocarbohydrazide–silver proteinate. It is, therefore, imperative that adequate controls be used in order to obtain a correct interpretation of the PS–TCH–SP test. The following control treatments are suggested.

(1) Non-oxidised sections should be treated first with thiocarbohydrazide and then with silver proteinate; free aldehyde and other groups reacting with thiocarbohydrazide and silver proteinate or with silver proteinate alone are stained.

(2) Periodic acid-oxidised sections should be treated with silver proteinate; chemical groups such as 1,2-glycol or α-aminoalcohol groups reacting with silver proteinate without modification by the periodic acid oxidation are stained.

(3) Non-oxidised sections should be stained with silver proteinate; chemical groups reacting with silver proteinate are stained. Similar results can be obtained by replacing silver proteinate with Protargol. By using this method, Geyer *et al.* (1972a) obtained satisfactory staining of mucous secretions of goblet cells and glycocalyx at the brush border, and Zagon (1970) obtained selective staining of kinetosomes, cilia, nuclear material and myonemes in ciliates. The staining method is given below.

Thin sections of aldehyde-fixed tissues are oxidised with 1% periodic acid for 30–90 min at room temperature. After rinsing in distilled water, the sections are treated with 1% thiosemicarbohydrazide for 1 h at 50 °C. The sections are washed in warm water (50 °C) and then incubated in 1% Protargol (previously warmed to 60 °C) for 5–15 min at 60 °C. This is followed by a rinse in distilled water, treatment with 5% sodium thiosulphate for 1 min, and a final wash in distilled water. Fixation procedures used and the pH of the staining solution are important factors in achieving selective staining by this method.

Various components of the cell wall which contain 1,2-glycol groups can be differentiated by employing silver proteinate (Thiéry, 1967) instead of silver hexamine at a specific pH level (Jewell and Saxton, 1970). According to this method, sections are oxidised for 40 min on a 1% solution of periodic acid in distilled water. The sections are then rinsed with distilled water and placed on the surface of a 1% solution of TCH in 10% acetic acid at room temperature; the duration of this coupling ranges from ½ to 24 h. After repeated rinsing, the sections are stained with 1% solution of silver proteinate (pH 8.4) for 30 min in the dark. This is followed by a thorough rinse and mounting on a grid.

When the pH of the silver proteinate solution is raised to 9.2, the staining of cell walls is enhanced. As usually employed, silver proteinate is an aqueous solution with a pH of 6.4. It should be remembered that the selectivity of silver staining is significantly conditioned by many parameters, including fixation procedures, constituents of fixing solution, pH of staining solution and duration of staining.

Periodic Acid–Silver Method

Silver staining is carried out usually before the sections are mounted on copper grids, since coarse precipitates form when copper grids come in contact with silver solutions. However, sections mounted on non-copper (e.g. nickel) grids can be stained. Silver solutions are very unstable, especially at higher temperatures, and thus may cause artefactual precipitates if the sections are allowed to be exposed for much longer than 60–80 min. It is therefore recommended to use only freshly prepared solutions. The staining solution is prepared as follows:

Hexamethylenetetramine (3%)	18 ml
Silver nitrate (10%)	2 ml
Sodium borate (2%)	2 ml

Prior to staining, sections are usually oxidised by floating on a 1% aqueous periodic acid solution for ~15 min at room temperature, and they are then washed thoroughly with distilled water (three changes). The washed sections are floated on a freshly prepared silver methenamine solution in an oven at 60–70 °C, preferably in the dark. After ~1 h the sections show a yellow tinge which is indicative of the completion of staining. Certain tissues may require longer than 1 h to achieve the desired intensity. The optimal duration of staining is dependent upon the section thickness and the temperature of the staining solution. After several rinses with distilled water, the sections are treated with 3% sodium thiosulphate for 5 min at room temperature, followed by a final thorough washing with distilled water. No post-staining is required.

Periodic Acid–Chromic Acid–Silver Method

In order to obtain better staining of polysaccharides for light microscopy, Mowry (1959) introduced a chromic acid step between PA oxidation and silver staining. This method was successfully used by Hernandez *et al.* (1968) and Rambourg *et al.* (1969) for electron microscopy. The PA–CrA–silver technique differs from PA–silver in that the non-specific staining of ribosomes and nuclei is minimised. In addition, this newer technique stains glycogen more intensely, a situation which is probably due to the production of additional aldehydic groups by the chromic acid oxidation of glycogen. This technique was successfully employed to study the Golgi apparatus in a number of cell types. To stain, the grids are treated as follows.

(1) Aqueous PA solution (1%) for 20 min
(2) Distilled water for 30 min
(3) Aqueous chromic acid solution (10%) for 5 min
(4) Aqueous sodium bisulphite solution (1%) for 1 min
(5) Distilled water for 30 min
(6) Silver methenamine solution for 30 min at 60 °C
(7) Distilled water for ~20 min

The desired intensity of staining can be obtained by adjusting the duration of staining with silver methenamine from 25 min to 40 min.

Staining of Nucleolar Organiser Region

A nucleolus organiser region (NOR) is a specific region of a chromosome which is responsible for the formation of a nucleolus in interphase. The NOR is the region of the chromosome which contains the main rRNA genes (18S and 28S rDNA) (Schwarzacher and Wachtler, 1983). The silver-positive material in the NOR is an acidic protein. Such a protein has a molecular weight of 195 000 and is a protein subunit of polymerase I (Williams *et al.*, 1982). According to Ochs *et al.* (1983), the NOR-associated protein is C23 phosphoprotein. This stainability is attributed to silver affinity for sulphhydryl and disulphide groups in the NOR proteins (Buys and Osinaga, 1980).

In interphase cell nuclei these proteins are numerous in the fibrillar centres and the dense fibrillar components of nucleoli. It is thought that the amount of silver grains deposited in the Ag–NOR staining represents the transcriptional activity of rRNA genes, and that the Ag–NOR proteins may have a regulatory role in the transcription of rDNA (Hubbell, 1985). However, this view has been questioned, since the inhibition, spontaneous or induced, of transcriptional activity does not abolish the staining of NOR proteins (Medina *et al.*, 1986). An alternative hypothesis is that the Ag–NOR method detects those proteins that are responsible for decondensation of nucleolar chromatin.

Hernández-Verdún *et al.* (1980) adapted the method of Goodpasture and Bloom (1975) for selectively staining the NOR at the ultrastructural level. A brief pre-fixation with glutaraldehyde minimises the extractive effect of subsequently used Carnoy's solution.

SILVER STAINING *in situ* (Hernández-Verdún *et al.*, 1980; Medina *et al.*, 1986)

Small root pieces (1 × 0.5 × 0.5 mm) are fixed with 3% glutaraldehyde in 0.025 M cacodylate buffer (pH 7.2) for 30 min at 4 °C. After thorough rinsing in the buffer, they are post-fixed with Carnoy's solution (3:1, ethanol and acetic acid) for 10 min at 4 °C. The specimens are rehydrated in a descending series of concentrations of ethanol (100% to 30%) and then rinsed in distilled water (pH 4.5) for 10 min. They are flooded with 50% silver nitrate solution (pH 4.5) and placed under a photoflood (2800 K bulb) for 6 min, so that the temperature of the solution reaches 50–55 °C. The specimens are rinsed in cold double-distilled water (pH 4.5), and developed for 2–15 min in a 1:1 mixture of 1.5% formaldehyde and ammoniacal silver solution (Vio-Cigna *et al.*, 1982) for 90–120 s until a pale-yellow colour is observed. They are rinsed in double-distilled water. Thin sections can be stained with salts of uranium and lead. Only freshly prepared solutions should be used. The results of this method are shown in Fig. 4.14.

SILVER STAINING ON SECTIONS (Moreno *et al.*, 1985)
TG cells (an established human cell line) are washed twice with PBS and fixed by immersion as above. Free aldehyde groups are blocked by incubation in 0.5 M ammonium chloride in buffer for 2–4 h at room temperature. They are embedded in Lowicryl K4M at a low temperature. Thin sections on gold grids are floated for 5 min at room temperature on a mixture of 2% gelatin in 1% formic acid and 50% silver nitrate solution. Grids are rinsed in distilled water, immersed in 5% thiosulphate solution for 10 min and thoroughly rinsed in distilled water.

SILVER LACTATE–OSMIUM TETROXIDE

This method was introduced to localise intracellular anions such as Cl^- (Komnick, 1962). After this treatment, chloride ions are precipitated as AgCl, as demonstrated by selected-area electron diffraction. The fixation and staining mixture is prepared by mixing 0.5–1.5% silver lactate ($AgC_3H_5O_3 \cdot H_2O$) with 1–2% OsO_4 in equal parts in a suitable buffer (pH 7.2). Tissue specimens are fixed for 2 h in this mixture in the cold, using a red safe light. During dehydration, the tissue blocks are treated with nitric acid (50% acetone

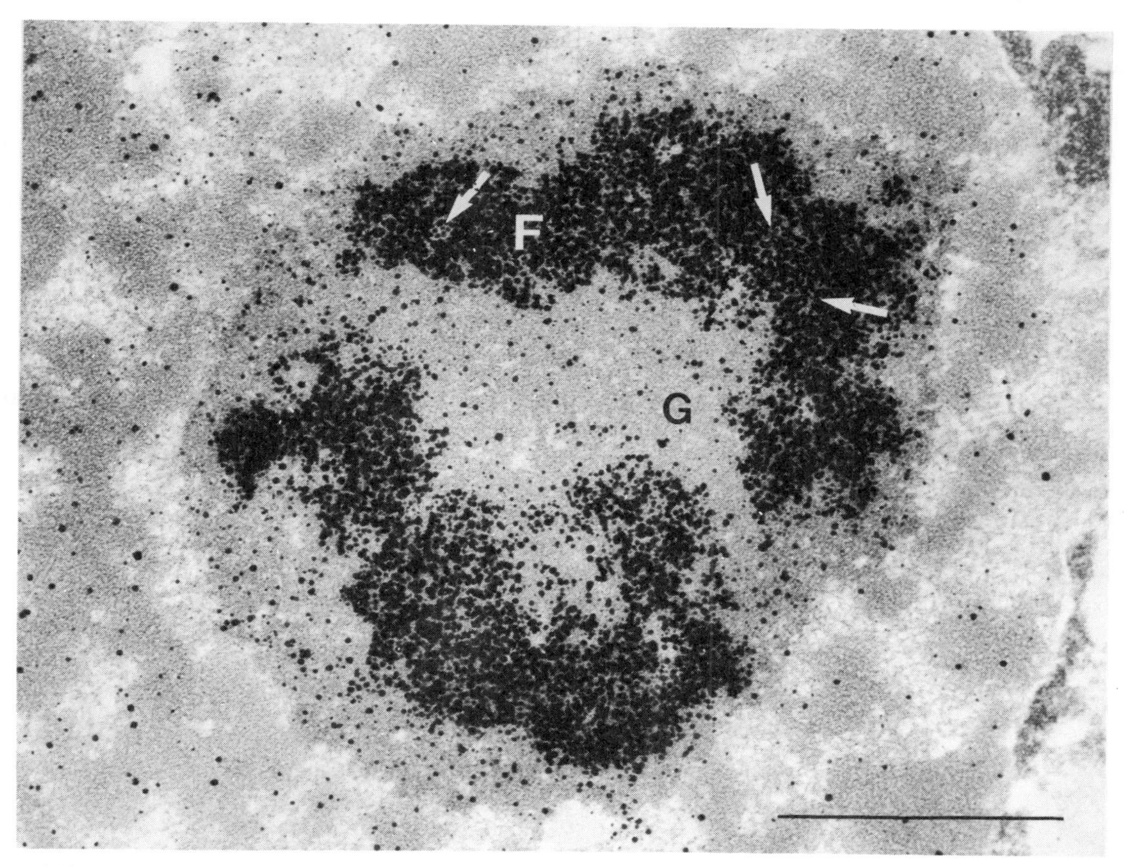

Figure 4.14 An onion root meristematic cell prepared with the Ag–NOR staining method. The section was post-stained with uranyl acetate and lead citrate. The nucleolus shows the interphasic localisation of the genetically 'unlocked' NOR chromatin, either engaged in active transcription, located in the dense fibrillar component (F) or transitorily inactive, lying in homogeneous fibrillar centres (arrows). These nucleolar structures appear densely covered by silver grains, whereas the granular component (G) is devoid of them. Bar = 1 μm. Courtesy F. J. Medina

containing 0.1 N HNO_3) in order to dissolve non-specific precipitates.

SODIUM TUNGSTATE

Sodium tungstate (Na_2WO_4) has a molecular weight of 293.83, and is available as a hydrate, colourless crystals or white crystalline powder. It is soluble in 1.1 parts water, but is insoluble in alcohol. The aqueous solution is slightly alkaline (pH 8–9). Although it was first used by Stevens and Swift (1966) and Swift and Adams (1966) as a stain for nucleic acid-containing structures, for a long time its use has been scarcely explored. According to Marinozzi (1968), it stains polysaccharides and glycoproteins at pH values of 1.0–4.5. It has been shown that in plant tissues sodium tungstate strongly stains cell walls and ribosomes at pH 1.5, while chromatin and nucleoli stain slightly (Stockert, 1977). At pH 5–6 it intensely stains chromatin. Chromatin staining is more intense after fixation with glutaraldehyde only and embedding in Epon than after post-osmication. At pH 5.0 this reagent differentially stains nucleoli and chromatin in animal cells (Takeuchi, 1981). At pH 9.0 the cell is not stained.

TANNIC ACID

Tannic acid ($C_{76}H_{52}O_{46}$) is a compound of glucose and gallic acid, and contains phenolic groups interspersed with carboxyl groups which are expected to be ionised at neutral pH. It is a natural substance found in a wide variety of plants, especially in their pathological growths known as galls. It is available from various commercial sources and consequently is of varying chemical structure. Tannic acid of low molecular weight is prepared from Aleppo nutgalls (Mallinckrodt, St. Louis, Mo.). Commercially available tannic acid (corilagin) consists of polyphenolic compounds with a galloxylated glucose structure:

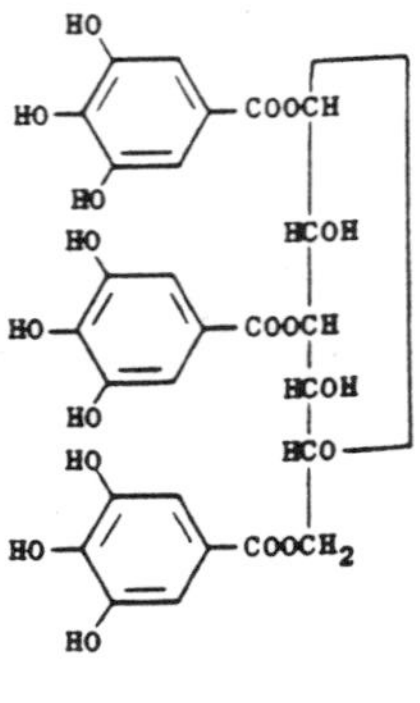

One gram of tannic acid can dissolve in 0.35 ml of water. It gives insoluble precipitates with albumin, starch, gelatin, and most alkaloidal and metallic salts. Tannic acid produces a bluish-black precipitate with ferric salts. It gradually darkens on exposure to air and light, and should be kept in a closed bottle protected from light. A historic perspective of tannic acid in histology has been provided by Chaplin (1985).

Tannic acid (2%) is compatible with phosphate and cacodylate buffers. Only freshly prepared mixtures of tannic acid and glutaraldehyde should be used; a delay of as little as 5–10 min may give poor results. For general ultrastructural studies, tannic acid of a low molecular weight should be used to facilitate its penetration and avoid the formation of precipitates of large size. If an excessive precipitation inside and outside of cells is a problem, it should be used at a low concentration (1%).

Tannic acid acts as a supplemental fixative and improves fixation with glutaraldehyde. Being hydrolysable, its molecule is potentially capable of multiple ligand-mediated interactions. Tannic acid reacts with bound osmium in membranes, and facilitates lead and uranium binding. Membranes of tannic acid-treated cells bind more uranyl acetate than do those of untreated cells. Tannic acid fixation accentuates the trilaminar aspect of membrane structure. Low-molecular-weight galloylglucose constituents of tannic acid act as mordants by reason of their polyphenolic structures. They readily form strong, highly cross-linked binary complexes with heavy metals (Akey and Edelstein, 1983).

Tannic acid-treated cells show relatively less shrinkage when critical-point dried for scanning electron microscopy (McCarthy *et al.*, 1985). Also, this compound renders plasma membranes of cells more conductive when treated subsequently with heavy metal salts. Tannic acid seems to protect certain cellular materials such as proteins from the adverse effect of dehydration. This acid appears to form a non-crystalline 'glass' upon drying on protein crystals (Akey and Edelstein, 1983). Specimens embedded in tannic acid are less prone to damage by the electron beam in the electron microscope.

Tannic acid, when used in conjunction with glutaraldehyde and OsO_4, interacts with many different types of proteins; carbohydrate polymers, including extracellular matrix; phosphatidylcholine; membranes; membrane junctions; acetylcholine receptors; sarcoplasmic reticulum; microtubules; microfilaments; elastic fibres; and internal cytoplasmic milieu. The remarkable ability of tannic acid to interact with diverse cellular components may be attributed in part to the many active groups present on its molecule.

Tannic acid is highly specific for acetylcholine receptors at known cholinergic synapses (Sealock, 1980). Tannic acid (1%) has been used in combination with methylamine tungstate (3%) for positive staining of plasma membranes (Skaer, 1981). Tannic acid in combination with ferric chloride and uranyl acetate can be used to visualise sulphated complex carbohydrates and anionic complex carbohydrates, respectively (Takagi *et al.*, 1983).

Conventional fixation and staining procedures are not suited to capturing the fleeting process of exocytosis. Tannic acid is effective in providing better visualisation of this process. Since tannic acid is usually unable to penetrate intact, living or aldehyde-fixed plasma membranes and remains mostly in the extracellular space, it serves as both the fixative and the mordant for staining extracellular proteins, including exocytosed secretory products. Methods using tannic acid to study exocytosis are described later. For additional details, see Buma (1989).

REACTION WITH PROTEINS

Tannic acid is thought to react with basic amino acids, although it is not clear whether ε-amino groups are involved. This reaction is regulated by both the pH and the concentration of the tannin. At pH 3.5 and a concentration of 2.5% tannic acid reacts predominantly with arginine (Meek and Weiss, 1979). At the same concentration, but at pH 7.5, tannic acid reacts with lysine. Under these conditions, probably non-electrostatic binding occurs between tannic acid and the protein. When the concentration at this pH is reduced to 0.01% (0.06 mM), only electrostatic interactions occur. High concentrations of tannic acid encourage its binding with several types of amino acids. At neutral pH tannic acid precipitates generally basic proteins.

The charged carboxyl groups in the tannic acid molecule may interact with favourably located ionic sites on the protein. Alternatively, phenolic groups in the gallic acid moiety of the tannin molecule may coordinate on non-ionic sites (e.g. peptide bond), resulting in a multipoint attachment of tannic acid to the protein (Mizuhira and Futaesaku, 1972).

Tannic acid possesses a large number of available hydroxyl groups for hydrogen bonding. These groups seem to play an important role in the formation of the protein–tannic acid–heavy metal complexes by functioning as ligands for salts of heavy metals (Akey and Edelstein, 1983). Tannic acid may also interact with proteins via non-bonded hydrophobic interactions. Thus, tannic acid has the unusual capability of interacting with both cellular materials and atoms of heavy metals, thereby facilitating the formation of stable ternary complexes.

Tannic acid enhances the preservation-staining of the proteinaceous fibrillar network underlying the human erythrocyte membrane (Fujikawa, 1983). This improved visualisation is due to interactions between tannic acid and exposed membrane proteins, resulting in an increase in their size. The increase in size is restricted to the membrane proteins that are located outside the membrane and thus are exposed directly to the fixative. The intramembranous particles seen on the fracture faces do not change in size. The diameter of microfilaments and microtubules preserved with tannic acid tends to increase in proportion to the concentration of this reagent used.

Tannic acid reacts with collagen (Meek and Weiss, 1979) and elastic fibres (Kageyama *et al.*, 1985). At an alkaline pH tannic acid aids in contrast enhancement of elastin, which is thought to be a reflection of the binding of anionic oxygen in alkaline tannic acid at a cationic group such as ammonium in elastin (Anwar and Oda, 1966). After reaction with tannic acid, proteins seem to lose most of their osmiophilic properties (Nehls and Schaffner, 1976). Proteins precipitated with tannic acid do not appear to be covalently cross-linked. It is apparent from the above discussion that the exact mechanism involved in the interactions between tannic acid and proteins at different pH values is unknown.

REACTION WITH CARBOHYDRATES

Mayer (1896) first indicated staining of tissue mucus sites with a mixture of tannic acid and ferric chloride. It has been proposed that tannic acid binds to mucins in complex and multiple ways (Pizzolato and Lillie, 1973). According to Sannes *et al.* (1978), tannic acid has selective affinity for tissue components rich in carbohydrates. Tannic acid seems to form complexes with some components of the extracellular matrix, which bind osmium and other heavy metals. Hyaluronic acid is one such component (Singley and Solursh, 1979). Tannic acid seems to bind to glycoproteins regardless of their electrical charge. The binding of tannic acid to complex carbohydrates is thought not to depend on electrostatic forces (Sannes *et al.*, 1978). At an acidic pH tannic acid facilitates staining of anionic glycoconjugates through hydrogen bonding of carboxyl, sulphate and phenol groups in the tannin. Prior methylation to block acid groups does not impair tannic acid affinity for carbohydrates. Considerable hydrogen bonding seems to be involved in the binding between tannic acid–metal salts and carbohydrates, and this variability is reflected in differing affinities for metal salts. Sannes *et al.* (1978)

have used tannic acid–metal salt combinations for staining tissue components (secretory granules and Golgi cisternae) rich in mucosubstances.

REACTION WITH LIPIDS

A direct reaction of tannic acid with the choline moiety in saturated phosphatidylcholine and sphingomyelin rather than a reaction with OsO_4 has been demonstrated (Kalina and Pease, 1977a,b). The resulting complex of reaction product is stabilised with OsO_4. Significant intercellular translocation of labelled choline phosphatides during dehydration and embedding is prevented when the labelled tissue has been treated with tannic acid prior to osmication (Saffitz *et al.*, 1981). Thus, tannic acid stabilises saturated fatty acids of phospholipids, rendering them resistant to extraction in lipid solvents such as acetone.

FIXATION AND MORDANTING EFFECTS OF TANNIC ACID

Tannic acid functions not only as a fixative, but also as a mordant. The fixation effect of tannic acid may result from its ability to bind soluble proteins either by hydrogen bonding or by chelation. Either the phenolic radicals or the electrostatic charges of the tannic acid may be responsible for the fixation effect (Futaesaku *et al.*, 1972). By using a mixture of tannic acid (2%) and glutaraldehyde (2%) as a pre-fixative, membrane antigens can be visualised by immunofluoresence and immunoelectron microscopy (Roholl *et al.*, 1981). Tannic acid-treated tissues seem to be more resistant to extraction during dehydration and embedding. It has been proposed that Ca^{2+}-binding proteins, possibly membrane-bound, are precipitated by tannic acid prior to or in conjunction with their immobilisation through cross-linking with glutaraldehyde during fixation (Wagner, 1976). Since tannic acid stabilises certain cellular structures, it can be termed a fixative.

It has been suggested, on the other hand, that tannic acid is not a true fixative but is a multivalent agent that acts as a mordant between osmium-treated structures and lead stains. A mordant facilitates the binding of a heavy metal stain to biological structures, enhancing their contrast. Only ~18% of the phospholipid was lost from the tissue that had been exposed to tannic acid and osmium, as compared with a 71% loss when tannic acid was used alone (Stratton *et al.*, 1982). It has been reported that tannic acid binds two reactive sites on choline and with osmium, resulting in increased membrane contrast (Kalina and Pease, 1977a,b). Such increased contrast is observed in tissues exposed to both osmium and tannic acid.

The galloylglucose moiety of tannic acid acts as a mordant between osmium-treated structures and lead stains. Tannic acid also acts as a mordant for other heavy metals, such as uranyl acetate. The reactive groups required for the mordanting effects are a carboxyl group and at least one hydroxyl group on the tannic acid molecule. The mordanting effect does not depend on the residual aldehyde groups in the aldehyde-fixed tissue, but does depend on osmium.

In addition to its being a mordant, tannic acid may impart electron density to reactive sites. This observation is consistent with studies indicating that tannic acid used *en bloc* enhances the electron density of elastin and collagen, even in the absence of OsO_4 fixation (Thyberg *et al.*, 1979). Subsequent binding of uranyl and lead salts to tannic acid further increases the electron density. However, treatment with glutaraldehyde–tannic acid results in negative image of cell membranes, which are similar to those observed in cells first fixed with glutaraldehyde without post-osmication.

PENETRATION

In general, intact tissue blocks are impermeable to tannic acid except near the cut surfaces. That penetration by tannic acid is confined to 0.3 mm from the surface of even a damaged skeletal muscle fibre has been demonstrated (Fig. 4.15) (Cottell and Hooper, 1985). The presumed fixation effect of tannic acid on the plasma membrane probably plays a part in its impermeability. The plasma membrane of cultured or isolated cells tends to allow penetration by tannic acid when it is used in combination with glutaraldehyde. The penetration can be facilitated by using tannins of low molecular weight ($C_{14}H_{10}O_9$) rather than those of high molecular weight ($C_{76}H_{52}O_{46}$). The penetration can also be helped by rendering the plasma membrane permeable by treatment with reagents such as digitonin, saponin or dimethyl sulphoxide. It has been suggested that for good penetration, tannins should be applied after osmication (Simionescu and Simionescu, 1976a,b).

NEGATIVE STAINING WITH TANNIC ACID

Fixation with tannic acid (pH 7.4) tends to alter the positive-staining pattern of certain tissues, reflecting a change in the charge profile. Tannic acid can impart either positive or negative staining to the cellular

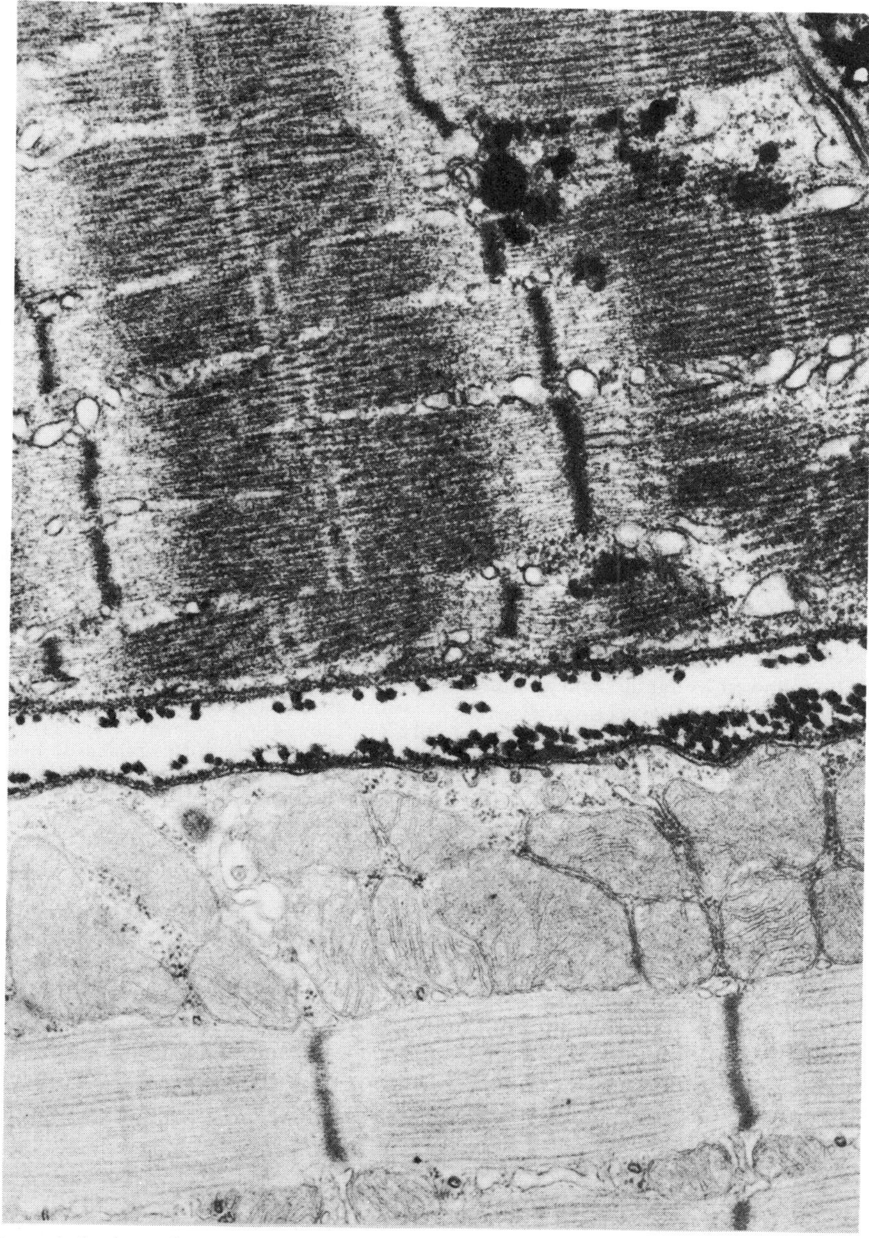

Figure 4.15 Mouse skeletal muscle treated with low-molecular-weight tannic acid (4%) in 2.5% glutaraldehyde for 3 h. Tannic acid penetration was confined to the damaged portion (above) and the site of experimental injury. It does not penetrate intact cell membranes (below). Thin section was post-stained with salts of uranium and lead. 13 000×. From Cottell and Hooper (1985)

components; the latter is especially pronounced on proteinaceous materials. That either positive or negative staining of catalase crystals with tannic acid under specific preparatory conditions can be achieved has been demonstrated (Akey *et al.*, 1984).

Non-electrostatic binding is involved in the negative-staining effect of tannic acid. Intense staining of the cell background substance with tannic acid results in negative images of structures such as microtubules. Relatively high concentrations of tannic acid at pH 7.5 and damaged plasma membranes encourage negative staining. This type of negative staining differs from conventional negative staining: the former is a specific phenomenon, whereas the latter is not. This view is, admittedly, a simplistic explanation of the mechanism of negative staining produced by tannic acid. Akey and Edelstein (1983) have explained the mechanism of negative staining in tannic acid-fixed sections of catalase crystals.

FIXATION AND STAINING PROCEDURES

General Procedures

The most common method consists of tissue fixation with a mixture of glutaraldehyde and tannic acid, followed by osmication. Tannic acid can also be applied with glutaraldehyde by vascular perfusion. *En bloc* staining with tannic acid after fixation with glutaraldehyde and OsO_4 is not recommended, since this approach may result in uneven penetration and artefactual precipitates. According to another general method, single cells can be fixed with glutaraldehyde and then with OsO_4, followed by tannic acid. Tannic acid (2%) can also be used to stain thin sections. Use of alcoholic uranyl acetate in conjunction with tannic acid staining should be avoided, since alcohol is known to destain the staining by the latter. Grid staining with aqueous uranyl acetate, followed by immersion in 2% tannic acid and subsequent staining with lead citrate is recommended (Dae *et al.*, 1982). Intracellular staining with tannic acid is facilitated when the plasma membrane has been disrupted or weakened. Methods of fixation and staining with tannic acid for visualising specific cellular components are presented below.

Visualisation of Mucosubstances (Sannes *et al.*, 1978)

Thin sections of the glutaraldehyde-fixed tissue are treated with filtered 5% aqueous tannic acid for 10 min, rinsed in distilled water, and stained with 2% uranyl acetate for 5 min.

Visualisation of Elastin (Kageyama *et al.*, 1985)

Thin sections of glutaraldehyde and OsO_4-fixed specimens are treated with 5% aqueous tannic acid solution (pH 7) for 10 min, rinsed in three changes of distilled water (pH 5.5) for 1 min and stained with 1% uranyl acetate (pH 4.1–4.3) for 5 min. Figure 4.16 shows intense staining of elastin, moderate staining of collagen and relatively weak staining of nuclei in the tunica media of the rat aorta.

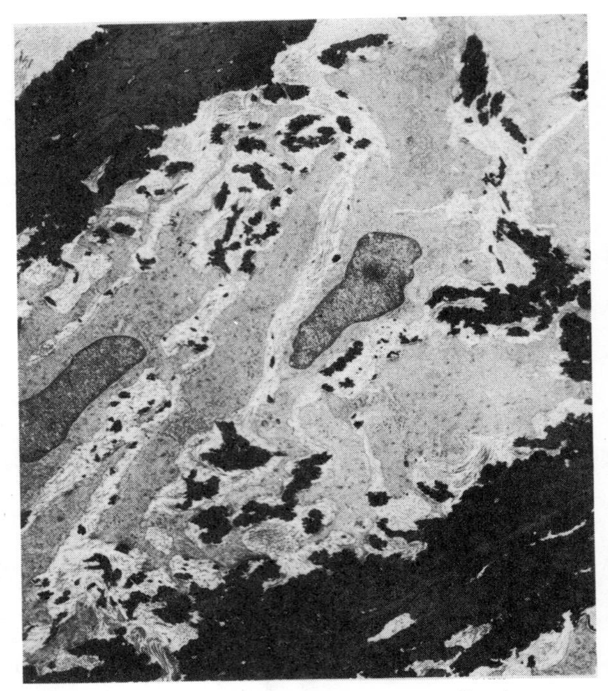

Figure 4.16 A thin section of the tunica media of rat aorta stained with freshly prepared and filtered 5% tannic acid (pH 7.0) followed by 1% uranyl acetate (pH 4.1) for 1 min. The tissue was fixed with glutaraldehyde followed by OsO_4. The section was not counterstained. The intense elastin staining is assumed to be due to complexing of tannic acid with uranyl acetate and/or OsO_4. 6000×. Frim Kageyama *et al.* (1985)

Visualisation of Cholinergic Synaptic Junctions (Bird, 1984)

Cells are fixed with 4% glutaraldehyde containing 2% tannic acid for from 90 min to 4 h. They are rinsed in distilled water and post-fixed with 2% OsO_4 for 15 min. The cells are stained with a saturated solution of uranyl acetate for 30 min.

Visualisation of Exocytosis

TANNIC ACID–GLUTARALDEHYDE–OsO_4 (TAGO) METHOD (Roubos and van der Wal-Divendal, 1980)

The tannic acid–glutaraldehyde–OsO_4 method involves fixation with a mixture of tannic acid and glutaraldehyde, followed by osmication. It is used for staining exocytotic products. Tissue specimens are fixed with a mixture of 1% tannic acid and 0.8% glutaraldehyde in 0.2 M cacodylate buffer (pH 7.3) for 2 h at 20 °C. The specimens are rinsed for 15 min at 20 °C in the same buffer, and then post-fixed with 1% OsO_4 in the same buffer for 1 h at 4 °C.

TANNIC ACID IN RINGER SOLUTION (TARI) METHOD (Buma *et al.*, 1984)
The tannic acid in the Ringer solution method seems to be somewhat more effective in visualising the exocytotic process compared with the TAGO method. Also, the latter requires more time. Tissue specimens are incubated for 2 h at 20 °C in Ringer solution containing the following ingredients:

Tannic acid	0.5%
NaCl	30 mM
$CaCl_2$	4 mM
$MgCl_2$	2 mM
KCl	1.5 mM
NaH_2PO_4	0.25 mM
$NaHCO_3$	18 mM

The pH is adjusted to 6.8 with concentrated NaOH. After incubation and rinsing, the specimens are fixed with glutaraldehyde and OsO_4 according to standard procedures. Thin sections are post-stained with salts of uranium and lead. This method can also be applied by vascular perfusion, the details of which are given below.

TARI METHOD BY VASCULAR PERFUSION (Buma and Nieuwenhuys, 1987)
The rat (200–250 g body weight) is anaesthetised with sodium pentobarbitone and perfused for 5 min with Ringer solution containing:

NaCl	154 mM
KCl	5.6 mM
$CaCl_2$	2.2 mM
$MgCl_2$	1.0 mM
$NaHCO_3$	6.0 mM
Tris	2.0 mM
Glucose	10 mM

The pH is adjusted to 7.0 with NaOH or HCl, oxygenated for 30 min and heated to 37 °C before use.

This is followed by perfusion with the same Ringer containing 1% tannic acid for 20 min, and then again with Ringer without tannic acid for 5 min. (To induce membrane depolarisation and calcium-dependent exocytosis, a second rat can be perfused in the same way with Ringer containing 50 mM KCl in place of NaCl.) Fixation is carried out by perfusion with a mixture of 1% formaldehyde and 1.25% glutaraldehyde in 0.1 M phosphate buffer (pH 7.4) for 20 min at 20 °C. The skull is opened and the whole brain is further fixed *in situ* overnight at 4 °C. Small pieces of the tissue are post-osmicated in buffered 1% OsO_4 for 1 h at 4 °C.

MODIFIED TARI METHOD (Brooks and Carmichael, 1987)

Locke's solution	
NaCl	154 mM
KCl	2.6 mM
K_2HPO_4	2.15 mM
KH_2PO_4	0.85 mM
$CaCl_2$	2.2 mM
$MgCl_2$	1.0 mM
Glucose	10 mM

The pH and osmolality are adjusted to 7.4 and 298 mosm, respectively.

Tannic acid medium 0.5% tannic acid solution is prepared in Locke's solution and pH is adjusted to 7.0 with 1 M NaOH with constant stirring. Since tannic acid slowly precipitates after ~40 min, this solution should be used immediately after its preparation at 37 °C. At lower temperatures or at pH values higher than 7.0, precipitation occurs more rapidly.

Procedure Cells are maintained in culture for 7–10 days prior to processing. Cells are grown as spots of 200 000 cells/spot on Lux Thermanox membranes (Miles Lab.). These membranes are prepunched, before the addition of cells, with a punch press (Detail Machine and Tool, Rochester, N.Y.) so that each spot of cells correspond to a well position in a 10-well Micro-Slide Chamber (Bellco Glass, Vineland, N.J.). The cells on the membrane are rinsed in three changes (5 min each) of Locke's solution at 37 °C in plastics reservoirs (Flow Lab., McLean, Va.). The individual spots of cells are transferred to small reservoirs of 0.2 ml, milled into a Lucite plate which contains the incubation solution. (Alternatively, the incubation solution is placed in troughs and the entire Thermanox membrane is transferred to the trough.) The incubation solution consists of 0.5% tannic acid in Locke's solution, and the incubation lasts for 10 min at 37 °C.

Spots of cells on tags on Thermanox membrane are transferred to a freshly prepared solution of 3% glutaraldehyde in 0.1 M cacodylate buffer (pH 7.2) at room temperature and then at 4 °C for 1 h. After rinsing three times in ice-cold 0.3 M sucrose in buffer, the cells are post-osmicated in 1% OsO_4. Thin sections are cut in a plane perpendicular to the membrane, and post-stained with salts of uranium and lead.

THIOSEMICARBAZIDE AND THIOCARBOHYDRAZIDE

Thiosemicarbazide (TSC) and thiocarbohydrazide (TCH) are sulphur-containing osmiophilic reagents which have been employed for detecting PAS-positive substances. In this method the vicinal hydroxyl or amino groups of certain PAS-positive substances are oxidised by periodic acid to release aldehyde groups. The aldehyde thiosemicarbazones reduce OsO_4 readily. The thiocarbamyl group of the reagent is responsible for this reduction. A black, electron-opaque deposit is formed at the reactive sites after TSC is substituted for the Schiff reagent and the specimens are exposed to OsO_4. As mentioned above, these reagents were introduced for the localisation of polysaccharides (e.g. glycogen), mucopolysaccharides and glycoproteins (e.g. mucin) (Hanker *et al.*, 1964; Seligman *et al.*, 1965), and for enhancing the contrast of lipid components (Seligman *et al.*, 1966).

Several modifications of TSC and TCH methods have been introduced for the localisation of specific carbohydrates. These procedures are given below.

PERIODIC ACID–THIOSEMICARBAZIDE OR THIOCARBOHYDRAZIDE–SILVER PROTEINATE (PA–TSC–SP or PA–TCH–SP)

As stated above, thiocarbamyl groups of TSC and TCH reduce OsO_4. These groups also react with silver salts and the reaction product is metallic silver, which replaces osmium blacks in the original PATO or PATCO procedures. In other words, this method depends upon the oxidation of 1,2-glycol linkages of certain carbohydrates to aldehyde. Thiosemicarbazide or TCH reacts with such aldehyde groups, but at the same time retains its own capacity to reduce silver proteinate. Treatment of the sections with silver proteinate solution releases silver at the sites of the original glycol linkages.

When silver nitrate or silver methenamine is used, some artefactual staining occurs, although these salts react strongly with thio groups. Silver proteinate is the best silver salt, for it causes very little background precipitation. This procedure is one of the most reliable methods for the staining of glycogen in thin sections. The major credit for its development goes to Thiéry (1967).

According to this method, thin sections mounted on inert gold or platinum grids, or carried in plastics rings, are exposed to 1% periodic acid solution for ~30 min at room temperature. The sections are rinsed in three changes of distilled water and then treated with either 1% TSC in 10% acetic acid or 0.2% TCH in 20% acetic acid solution. This is followed by rinsing thoroughly for 30 min (10 min in 10% acetic acid solution, 5 min in 5% and 1% acetic acid solutions, and 15 min in several changes of distilled water). The sections are treated with 1% aqueous silver proteinate solution for 30 min at room temperature under minimum conditions of lighting. The sections are rinsed in distilled water.

The intensity of the PA–TCH–SP method can be improved by using hot silver proteinate (50 °C) and enhancing the reaction product with silver (physical development). Minute concentrations of periodate-generated aldehyde groups that do not react in the conventional PA–TCH–SP procedure at room temperature can be revealed by the improved method. For details see Lo *et al.* (1987) and Neiss (1988).

PERIODIC ACID–THIOSEMICARBAZIDE OR THIOCARBOHYDRAZIDE–OSMIUM TETROXIDE (PATO or PATCO)

As stated earlier, aldehyde groups produced by periodic acid oxidation condense with hydrazine groups of thiosemicarbazide or thiocarbohydrazide. On reacting with OsO_4, the thiocarbamyl moiety of the aldehyde thiosemicarbazone yields an osmium black. Although the osmium black produced at the site of the reaction is intensely electron-dense, it does not obscure the fine details of glycogen particles.

Either the PATO or the PATCO procedure stains glycogen particles intensely, and very little non-specific staining is apparent. The α- and β-particles are easily recognisable, although fine details of these particles are difficult to discern.

According to this method, which was originally developed by Hanker *et al.* (1964), thin sections mounted on inert gold or platinum grids or carried in plastics rings are incubated in 1% periodic acid for 30 min, followed by a rinse in distilled water. The sections are then exposed to either 1% TSC in 10% acetic acid for 30–40 min or in 0.2% TCH in 20% acetic acid for the same duration. After a thorough rinse for 30 min (10 min in 10% acetic acid solution, 5 min in 5% and 1% acetic acid solutions, and 15 min in several changes of distilled water), the sections are exposed to OsO_4 vapours at 60 °C for 1–3 h. This is accomplished by mounting the grids on coverslips, which are placed in small jars containing OsO_4 crystals. The jars are sealed with grease and kept in a water-bath at 60 °C to volatilise the OsO_4.

Modified Method

The PA–TCH–SP method has the drawback of yielding a granular and non-homogeneous reaction product, which may hamper the visualisation of the structural organisation of carbohydrates. If this problem is encountered, a modification of the original PATCO or PATO method can be used (Derenzini *et al.*, 1986). This approach confers a very thin, non-particulate (homogeneous) reaction product on the reactive substances. In this approach, TCH or TSC is dissolved in distilled water (instead of acetic acid), the solution being bubbled with SO_2. The SO_2 treatment of TSC results in more intense staining than that obtained with the original PA–TCH–OsO_4 method. Whether this intense staining with SO_2 is due to extensive linking of TSC to aldehyde groups or increased binding of osmium to TSC is not clear. However, it is known that non-specific stains may become Schiff-like reagents after SO_2 treatment (Kasten, 1960). The effect of SO_2 does not seem to be related to the low pH of the TSC–SO_2 solution. The reaction is carried out on thin sections of formaldehyde-fixed tissues. The method selectively visualises the aldehyde groups liberated by oxidation of glycoconjugate complexes in thin sections.

Thin sections of formaldehyde-fixed and Epon-embedded specimens are mounted on gold grids (without a support film), which are floated on 1% periodic acid for 15 min. After rinsing, the grids are floated on a 1% TSC in distilled water (pH 2.0) (previously bubbled with SO_2 for 20 min) for 90 min at 60 °C. The grids are thoroughly rinsed in distilled water and dried. Osmication is carried out with 2% OsO_4 in a moist chamber for 1 h at 60 °C, followed by rinsing in distilled water.

SODIUM PERIODATE–THIOSEMICARBAZIDE–OSMIUM TETROXIDE

Stastňa and Travnik (1971) introduced this modification, which is sufficiently sensitive even in the presence of only a small amount of PAS-positive substances. The specificity of staining is achieved by substituting glutaraldehyde with paraformaldehyde, since some of the aldehyde groups of the former remain free and react with thiosemicarbazide, resulting in non-specific staining. The free aldehyde groups of glutaraldehyde can be blocked by exposing the tissue specimens to anilin hydrochloride.

The above method is effective for both thick and thin sections. Paraformaldehyde-fixed tissue specimens are oxidised with 0.05 M sodium periodate for 15 min, and then washed with 0.15 M phosphate buffer (pH 4.2) for 45 min. This step is followed by treatment with a saturated solution of thiosemicarbazide in 0.28 acetic acid for 2 h and post-fixation with 1% OsO_4 for 1 h.

URANYL PREPARATIONS

Uranium (atomic number 92) is the heaviest metal used as an electron stain. Uranyl ions or complex ions involving uranyl groups apparently contain atoms of sufficiently high atomic weight to be very effective in scattering electrons. In fact, macromolecules bound to even a few uranyl groups can be visualised in the electron microscope. It has been demonstrated that antibody molecules stained with about 30–100 uranyl ions are visible in electron micrographs (Sternberger *et al.*, 1965). Uranyl acetate is also an effective negative stain. For a comprehensive discussion of negative stains, the reader is referred to Hayat (1986b) and Hayat and Miller (1989).

Uranyl ions combine in large amounts with nucleic acids via phosphate groups and complex at cell surfaces with phosphate and carboxyl groups, but not with sulphydryl groups. Treatment with aqueous uranyl acetate following double fixation with glutaraldehyde and OsO_4 markedly improves the appearance of membranous and other structures. In order to obtain higher specificity in staining, uranyl acetate should be used in water solutions rather than in organic solvents, because of ionisation in the former solvent. Treatment with uranyl acetate, especially *en bloc* staining, adversely affects glycogen staining.

The word 'uranyl' is used loosely here, since it refers to one of a number of uranium-containing ions present in solution at pH 4.0. Uranyl acetate ($UO_2(CH_3COO)_2 \cdot 2H_2O$) is the usual source of uranium compounds; its properties are discussed later. Uranyl magnesium acetate has been used as a substitute for uranyl acetate in the double-staining technique and also as a single stain (Frasca and Parks, 1965). The main advantages of uranyl magnesium acetate are claimed to be reduction in the amount of the fine precipitate and less diffuse staining of the background material. A 7.5% solution in triple-distilled water (pH 5.2) is recommended; it should be filtered before use. The staining solution is stable at room temperature for several months. Uranyl formate (0.5%) has been reported to be effective in staining purified DNA (Brack, 1973). Uranyl magnesium acetate (0.5%) in 0.15 M NaCl has been used for *en bloc* staining of osmiophilic inclusion bodies in type II alveolar cells (Douglas *et al.*, 1973). Zinc uranyl acetate (50–100 mM) was employed as an electron-opaque sodium precipitant in conjunction with freeze-substitution (Harvey and Kent, 1981).

URANYL ACETATE

Uranyl acetate ($UO_2(C_2H_3O_2)_2$) has a molecular

weight of 388.15. It is a yellow, highly fluorescent, crystalline powder with slightly acidic odour, which decomposes at 275 °C. This reagent is slowly and incompletely soluble in water, owing to the presence of basic salt, but is freely soluble in water acidulated with acetic acid; the solubility also increases at higher temperatures and in alcoholic solutions, especially in methanol. A 7.7% solution of this salt in water is saturated at 15 °C. The saturated solution has a pH in the range 3.5–4.0. At this pH uranyl acetate is a weak electrolyte in water and is mostly un-ionised. The uranyl ion (UO_2^{2+}) is associated mainly with the acetate ion, and this association increases with increased concentration of the salt. The charged uranyl ions that are present in solution at this pH range exist not as a single species (UO_2^{2+}) but as a variety of complexes, cationic as well as anionic. These complexes have the general formula $UO_2(OUO_2)_{n+1}^{2+}$. These complexes are capable of binding with both negatively and positively charged groups in the specimen (Chapman and Hulmes, 1984). The relative uptake by the charged groups depends mainly on concentration and pH. That staining with uranyl acetate reflects only the negative charge distribution is not possible. Uranyl ion (UO_2^{2+}) will be the predominant ion only at pHs below 2.5.

Uranyl acetate, like lead citrate and strontium permanganate, completely penetrates through a thin section. When a grid is floated, section side down, on a drop of the uranyl acetate solution, the stain permeates the entire thickness of the section, and cell components are stained throughout the section depth. In other words, the staining is not confined to the surface of a thin section. For this reason, the effects of superimposition should be taken into account when analysing electron microscopic images (Peters *et al.*, 1971). For different views on stain penetration, see p. 214.

Uranyl acetate is an excellent general stain and fixative, because its various species react with both negatively and positively charged side-chains on proteins. However, at a high concentration (normally used for staining) positively charged groups show slightly stronger affinity than that exhibited by negative groups (Tzaphlidou *et al.*, 1982b). In fact, uranyl acetate, along with lead citrate, is the most widely used general counterstain; the former is also universally employed as a negative stain.

URANYL NITRATE

Uranyl nitrate ($UO_2(NO_3)_2$) has a molecular weight of 394. It is formed by dissolving an oxide of uranium in nitric acid. This salt is freely soluble in water, ethanol, acetone and ether, but not in benzene, toluene or chloroform. An aqueous solution can hold up to 56% anhydrous salt (w/v) at 25 °C. The aqueous solution is acid. Its solutions are more stable than are uranyl acetate solutions.

Uranyl nitrate shows different staining behaviour from that exhibited by uranyl acetate. Unlike uranyl acetate, uranyl nitrate is a strong electrolyte and completely dissociates at low pH into uranyl ions (UO_2^{2+}) and nitrate ions (Rothstein and Meier, 1951). As the pH is raised, hydrolysis occurs, resulting in the formation of polynuclear complexes in which the electrostatic charge is reduced by the addition of hydroxyl ions (Zobel and Beer, 1961). However, anionic and neutral complexes similar to those present in uranyl acetate solution seem to be absent (Tzaphlidou *et al.*, 1982b).

Uranyl nitrate, unlike uranyl acetate, reacts primarily with negatively charged groups in the absence of extraneous phosphate ions. It has been shown that uranyl nitrate binds primarily to the negatively charged side-chains of aspartic and glutamic acids in collagen under suitable conditions (Chapman and Hulmes, 1984). Uranyl acetate, on the other hand, as stated earlier, binds to both negative and positive charges. The reaction of uranyl nitrate with negatively charged groups is sensitive to the presence of phosphate and other ions. In the presence of phosphate, positively charged groups show increased uptake of uranyl nitrate. This increased uptake is thought to be via the negatively charged phosphate groups. Lysine residues, rather than arginine and histidine residues, seem to be responsible for the residual uptake of uranyl nitrate by positively charged groups (Tzaphlidou *et al.*, 1982b). It is concluded that, in general, staining with uranyl nitrate is also due to its affinity for both negatively and positively charged groups. Uranyl nitrate is, in general, a less efficient overall stain, because of its reduced uptake by positively charged groups.

MECHANISM OF STAINING

It appears that at low pH uranyl acetate solution contains only uncomplexed uranyl ions, while at higher pH levels, uranyl acetate complexes of the type UO_2Ac^+, UO_2Ac_2, $UO_2Ac_3^-$, $UO_2(OH_2)$ $UO_2Ac_3^-$, etc., are formed. The transition from uncomplexed to complexed uranyl ions can take place even at lower pH levels in the presence of higher concentrations of uranyl acetate or in the presence of added electrolytes. Thus, the effect on staining by uranyl acetate owing to the addition of electrolytes to the fixation vehicle cannot be overlooked. It is known that hydrolysis (proton release) occurs in solutions of uranyl acetate, especially at higher pH levels.

Uranyl ions form complexes of relatively high affinity with anionic groups, especially with phosphoryl and carboxyl. It has been indicated that UO_2^{2+} forms reversible but stable complexes with phosphoryl and carboxyl ligands on the outer surface of membranes (Rothstein, 1970). Uranyl ions bind ionically to the phosphate moieties of phospholipid head-groups of membranes. That UO_2^{2+} binding is reduced by as much as 75% in phosphate-depleted cells (Van Stevenick and Booij, 1964) suggests a high affinity of this salt with phosphoryl. Furthermore, it has been demonstrated that uranyl acetate reacts with the phosphate groups in lecithin monolayers (Shah, 1970). The negative charges of phosphate groups interact with uranyl ions at pH 3.9. At similar negative charge densities, uranyl ions show the strongest affinity for phosphate groups, less strong affinity for carboxyl and sulphate groups, and negligible affinity for sulphhydryl groups.

Studies of the infrared spectra of the DNA–uranyl complex suggest that the uranyl ion is bound at the phosphate loci (Zobel and Beer, 1961), and that this ion forms a simple salt with the phosphate groups in DNA. Such binding is expected on the basis of known electrostatic effects as well as coordination principles. The stoichiometric relations indicate that at pH 3.5 one uranyl ion is attached for every two phosphate groups in DNA. Such binding reduces the net charge of DNA molecules, which, therefore, tend to stick to each other. Also, the possibility of the formation of bridges between the phosphates of two molecules cannot be ruled out.

Uranyl acetate is soluble in esters and forms addition compounds with them. The attachment of uranyl salts is ionic in nature, and the possibility of chelation by the bases is remote. Uranyl acetate (2% aqueous solution) can combine with nucleic acids in amounts sufficient to increase the dry weight of purified DNA by a factor of almost 2 (Huxley and Zubay, 1961). Uranyl acetate staining correlates strongly with the phosphorus content of nucleic acids, proteins and inorganic deposits (Brodie *et al.*, 1982a). Inorganic phosphate of mitochondrial granules and nucleic acid phosphate of interchromatin granules and immature ribosomal precursor granules are stained by uranyl salts. Reactions of uranyl acetate with nucleic acids, proteins and lipoproteins are discussed in detail later.

REACTION WITH NUCLEIC ACIDS

The staining and stabilisation of bacterial DNA by uranyl acetate was proposed by Ryter and Kellenberger (1958). Valentine (1958) and Epstein (1959) used uranyl acetate to enhance the contrast of the central region of the unsectioned and sectioned adenovirus, respectively, which contains a large amount of DNA. Stoeckenius (1960) demonstrated that the extraction of DNA from its protein shell is prevented when DNA preparations are treated with uranyl acetate before dehydration, and that the density imparted by uranyl acetate is adequate for the visualisation of DNA strands in protein films. In their *in vitro* experiments Zobel and Beer (1961) convincingly showed the preferential staining of DNA over protein solutions. They recognised the advantage of using lower pH and concentration of the stain. Marinozzi (1963b) suggested that uranyl acetate acts as a 'stabiliser' of deoxyribonucleoproteins. The possible structures and reactions of the complexes formed between uranyl ions and adenine nucleotides have been discussed by Feldman *et al.* (1970).

These and other data indicate without a doubt that nucleic acid-containing structures are strongly and preferentially stained with uranyl salts, provided that conditions of pH and concentration are controlled. The binding of uranyl salts with nucleic acids is at the phosphate loci and is ionic in nature. It is known that uranyl cations coordinate preferentially with oxygen sites such as those in the phosphate groups of nucleic acids. The view that the staining reaction is based on ionic binding is supported by the evidence that staining is affected by a change in pH in aqueous solutions as well as in organic solvents with different dielectric constants (Derksen and Willart, 1976). There is no clear evidence for chelation by the bases.

Uranyl salts exhibit much stronger affinity for nucleic acids than for proteins. *In vitro* studies show that, on a weight basis, DNA binds seven times as much uranyl as an equivalent amount of the protein bovine serum albumin (Zobel and Beer, 1965). The mol ratio between phosphate and uranium in the DNA is 4:1 (Rothstein and Meier, 1951). The attachment of one uranyl ion per base-pair in the DNA molecule results in a detectable increase in contrast (Zobel and Beer, 1961). The greatest specificity for nucleic acids occurs at a pH of 3.5 and with a concentration of 1×10^{-5} M (Wolfe *et al.*, 1962). That DNA binds uranyl ions at a lower concentration of the stain than does protein is expected, since chemical studies indicate that an association constant of 8×10^6 is found for the DNA–uranyl complex, whereas that for the protein–uranyl complex is only about 1×10^4 (Zobel and Beer, 1961). As expected, solutions of higher concentrations (1×10^{-2} M) stain both the DNA and the protein. Relatively short durations of fixation also appear to result in increased specificity for nucleic acids.

Increased and rapid staining of nucleic acids can be obtained with uranyl acetate in an organic solvent instead of water. Increased staining of DNA can be achieved in a shorter time with uranyl acetate in acetone (Gordon and Kleinschmidt, 1968). That uranyl acetate in acetone acts both as a fixative and stain for cell materials, including nucleic acids and proteins, has been shown (Hayat, 1969). The staining intensity of the ribonucleoprotein complex is higher when uranyl acetate is dissolved in methanol than when it is dissolved in ethanol (Derksen and Willart, 1976). It is known that carbonic acids fail to enter into ionic interactions in media with a dielectric constant lower than 25. Stability constants for metal binding are usually higher in organic solvents. It is likely that the staining of nucleic acids with uranyl acetate in acetone or alcohol results in a comparatively high uranyl:phosphate ratio.

The effects of preparatory procedures on the structure, concentration and distribution of nucleic acids in a specimen are not known. However, Feulgen staining experiments indicate that a substantial part of the DNA that is originally present survives the preparatory procedures and is available in the specimen. It seems most likely that the attachment of uranyl to nucleic acid-containing structures is primarily responsible for the observed contrast. However, the role of nucleic acid-associated proteins in this contrast is not clear.

REACTION WITH PROTEINS

Uranyl salts bind primarily to the negatively charged carboxyl side-chains of aspartic acid and glutamic acid residues (Rothstein and Meier, 1951; Zobel and Beer, 1961). Positively charged amino groups also participate in binding with uranyl ions (Lombardi *et al.*, 1971). This is expected, since uranyl acetate is only weakly dissociated and a variety of ionic uranyl complexes (cationic, anionic and neutral) coexist in aqueous solution (Tzaphlidou *et al.*, 1982b). Thus, uranyl acetate stains not only negatively charged but also positively charged side-chains on the protein molecule. In fact, under the usual staining conditions, uranyl acetate does not react exclusively with negatively charged groups. The relative uptake of uranyl ions by negatively and positively charged groups is controlled by many factors, including the concentration and pH of the staining solution; the duration of staining; the ease of access to the reactive groups; and the presence of extraneous phosphate groups introduced, for example, by the buffer. The presence of other ionic species in solution or bound to protein can also affect staining with uranyl salts. Additionally, ionic groups introduced during fixation and embedding, and other previous treatments, can also influence this staining.

Although uranyl acetate shows a preference for nucleic acids over proteins, the binding of the stain to various proteins, including histones, ribonucleoproteins (RNP) and phosphoproteins, is likely. The amount of uranyl acetate bound to proteins seems to be considerably smaller than that attached to nucleic acids. These reacting proteins may stain intensely, or weakly, or not at all. Collagen fibres, however, stain intensely. The binding of the stain to proteins apparently varies with the charge on the protein. Studies of the interaction between uranyl ions and bovine serum indicate that the protein is precipitated by the stain. The protein–uranyl precipitate is soluble in strongly acidic or in strongly basic solutions. The precipitate seems to be stable only in the pH range 4–6.

Studies on the interaction between membranes and uranyl acetate indicate that the protein rather than the phospholipid part of the membrane is responsible for the staining. Studies of rat brain also indicate that the amount of protein extracted during dehydration with ethanol was four times less when the treatment with uranyl acetate was employed (Silva *et al.*, 1968). Free amino groups seem to play an important part in the binding of this stain to the tissue. Deamination with dinitrofluorobenzene, nitrous acid, or Chloramine T and $—NH_2$ blocking techniques applied to aldehyde-fixed tissues indicate that the stainability of zymogen granules by uranyl acetate is strongly reduced (Ponzo *et al.*, 1973). As stated below, other functional groups (—COOH) of proteins also bind uranyl acetate.

REACTION WITH LIPIDS AND MEMBRANES

Uranyl ions significantly influence the electrophoretic mobility of cells and phospholipid colloids. These ions can penetrate lipid bilayers as neutral complexes, such as uranyl acetate. It has been shown that UO_2^{2+} binds to model membranes (Ginsburg and Wolosin, 1979) as well as to biomembrances (D'Arrigo, 1975). The reactivity of UO_2^{2+} with biological and model membranes is due to its high affinity for the phosphate moieties of the polar head-groups of phospholipids. Uranyl ions show especially strong affinity for a membrane where phosphate groups are closely spaced. They react with the phosphate groups of both saturated and unsaturated lecithin. That uranyl ions readily interact with phosphate groups in lecithin monolayers, but not those in phosphatidic acid monolayers, has been demonstrated (Shah, 1969). At pH 4.8 the phosphate group in lecithin has one ionic oxygen, whereas in phosphatidic acid it has two. Steric factors probably prevent the interaction of uranyl ions with the divalent phosphate groups in phosphatidic acid monolayers. Uranyl ions require an optimal steric arrangement at the membrane surface in order to exert full effect. Consequently, uranyl ions bind to only a portion of the phospholipids present in the membrane.

In contrast to OsO_4 and $KMnO_4$, as stated above, UO_2^{2+} reacts with the phosphate groups in both saturated and unsaturated lecithins. The presence of double bonds is necessary for both OsO_4 and $KMnO_4$ to interact with fatty acyl chains of phospholipids. Since OsO_4 does not react with the polar group of egg lecithin, it may cause degradation of these monolayers. On the other hand, uranyl acetate reacts only with the phosphate group of lecithin and not with the fatty acyl chains. Thus, uranyl acetate does not cause disruption of phospholipid films. In fact, uranyl acetate is one of the few stains that do not cause degradation of phospholipids.

Since uranyl ions have a high affinity (10^5–10^6 mole^{-1}) for the phosphate group of phospholipids, UO_2^{2+} readily binds to the membrane surfaces. This binding is electrostatic in nature. Upon binding, UO_2^{2+} changes the surface charge density and, hence, the surface potential; the maximum change in surface potential is 88 mV, which corresponds to one uranyl ion per 31 nm^2 surface area (Ting-Beall, 1980). Uranyl ions, even at a low concentration, bind effectively to negative surface charges of a membrane, causing the surface charge to become either largely neutralised or even positive (D'Arrigo, 1975).

Uranyl ions absorb to the interface of lipid bilayer membranes made of phosphatidylcholine with a dissociation constant of about 3:10^6 M, and thereby charge the interface of the membrane and attain almost stoichiometric binding of one molecule of uranyl acetate per molecule of phosphatidylcholine at 1 M ionic strength and 20 μM of the uranyl salt (Ginsburg and Wolosin, 1979). Uranyl treatment results in increased surface charge density.

Uranyl ions are known to decrease membrane conductance. Alternatively, or in addition to the electrostatic effect of uranyl ions on membrane conductance, they may decrease the fluidity of the membrane core as a result of enhanced hydrophobic interactions between the alkyl chains of the phospholipids, since UO_2^{2+} have an effect of compression on the membrane (Hayashi *et al.*, 1972; Ginsburg and Wolosin, 1979). This may also result in increased membrane viscosity. The lateral diffusion of phospholipids in the membrane seems to be hampered as a result of UO_2^{2+} binding. Membrane stability is increased immediately after exposure to the UO_2^{2+}. The extraction of tissue lipids is minimised by the action of uranyl salts. Treatment of the tissue with uranyl acetate prior to osmication results in the stabilisation of phospholipids. In the absence of this treatment, phospholipids may leak out and form artefactual myelin figures. Uranyl acetate is a better fixative for lipoproteins than is formaldehyde. Also, the presence of uranyl in lipoproteins increases their dye-binding capacity (Lózsa, 1974).

Comparative studies indicate that uranyl acetate is superior to uranyl nitrate in terms of membrane stabilisation and staining. Some uranyl ions, when used in the form of uranyl acetate, diffuse across the membrane and bind to the phosphate groups on the opposite side (Ting-Beall, 1980). On the other hand, uranyl nitrate penetrates only rarely. This difference seems to be due to the existence of uranyl acetate as neutral complexes, whereas uranyl nitrate completely dissociates into uranyl and nitrate ions. It can be inferred that the movement of uranyl ions through the lipid bilayer involves the formation of uranyl acetate neutral complexes, which then diffuse across the lipid bilayer (Ting-Beall, 1980).

The extraction of tissue lipids is minimised by the action of uranyl acetate. Maleate-buffered 0.5% uranyl acetate has been employed during the perfusion of lung to preserve *in situ* lipids (Gil and Weibel, 1969/70). That uranyl treatment prior to osmication results in the stabilisation of phospholipids, which form myelin figures in hepatocytes, has been demonstrated by Mumaw and Munger (1971). They also showed the preservation of lipid components associated with mitochondrial membranes following uranyl treatment prior to osmication.

FACTORS AFFECTING URANYL STAINING

pH

The pH of the staining solution has a profound effect on the binding capacity of uranyl to various substrates of biological origin. As stated earlier, both phosphate and carboxyl groups bind uranyl ions. At relatively high acidic pH values (4.0–5.0) uranyl ions bind to both phosphate and carboxyl groups, while at lower acidic pH values (lower than 4.0) the binding is more specific for phosphate groups. The studies of DNA and protein solutions indicate that, at pH 3.5, attachment of uranyl ions to DNA is very strong and specific; the association constant for the substrate–metal ion complex is large (Zobel and Beer, 1961). At this pH uranyl ions show considerable preference for DNA over proteins, and even low concentrations of the stain are able to interact with DNA.

At pH values higher than 3.5 there is an increase in binding of uranyl ions to DNA. The likely reason for this increased binding is that the raised pH increases the effective negative charge on the DNA molecule, making possible the binding of additional uranyl ions. Another effect of the raised pH may be the formation of polynuclear complexes of uranyl and hydroxyl ions, which would also enhance the attachment of uranyl ions to the DNA. However, the binding at a higher pH is weaker than at pH 3.5. Such weak associations are accompanied by a decreased selectivity. An increase in uranyl concentration of the staining solution also results in an increased attachment of uranyl cations. It should be noted that both higher pH levels and higher uranyl concentrations lead to decreased selectivity.

On the other hand, lowering the pH from 3.5 to 2.2 results in a decrease in the amount of stain attached to the DNA molecule. This decrease is probably related to the structural changes that the molecules undergo on acidification of the solution, and to the changes in the extent of ionisation of the phosphate and base groups.

Buffer Types

Uranyl acetate reacts strongly with cacodylate and phosphate buffers, and readily causes the precipitation of the component salts of these buffers, whereas no precipitate is visible with veronal acetate. Thus, to avoid precipitation with the former buffers, specimens should be rinsed in dilute ethanol or acetone, after fixation to wash off the excess buffer, before employing this stain. Cope (1968) indicated variations in chromatin staining in mouse pancreas embedded in water-miscible methacrylates by uranyl acetate as a result of variations in the buffer media. In this connection, it should be noted that uranyl acetate forms complexes with various buffers, and thus a variety of ions of different complexity and different charge can be expected. Detailed data are lacking concerning the effect of various buffer systems on the staining and fixing capacity of uranyl acetate.

Fixation and Embedding Methods

Fixation and embedding methods also influence uranyl acetate staining, especially with regard to chromatin appearance. Although chromatin fixed with glutaraldehyde only shows maximum uptake of uranyl, prolonged rinsing in distilled water causes significant loss of uranyl stain (Cattini and Davies, 1984). On the other hand, loss of uranyl stain from the post-osmicated specimens after a lengthy rinse is insignificant. Uptake of uranyl ions is, in general, proportional to section thickness whether or not specimens have been post-osmicated. Post-staining with lead citrate results in increased retention of uranyl.

Chromatin staining by uranyl acetate is also affected by the embedding medium. For example, more intense staining with uranyl acetate is obtained when the tissue is embedded in glycol methacrylate–*n*-butyl methacrylate, as compared with the staining achieved after glycol methacrylate–styrene embedding. Uranyl ions rapidly penetrate, via hydrophilic channels, cell organelles (e.g. nucleus) exposed at the surface of the section, but reach only slowly to cell components small enough to be completely surrounded by the resin (Richardson and Davies, 1980). Information on stain penetration is given on p. 214.

OVERALL EFFECT ON TISSUES

Uranyl acetate treatment of tissues prior to dehydration, irrespective of the type of fixative used, results in an increase in contrast as well as stabilisation of the fine structure (Hayat, 1969). This improvement is definitely observed in the OsO_4-fixed tissues. Silva (1967) demonstrated that uranyl acetate treatment of OsO_4-fixed mouse liver cells prevents the removal of the density of cytoplasmic membranous structures by oxidation with hydrogen peroxide or periodic acid, while the same oxidation treatment causes a complete loss of the density in the absence of stabilisation by uranyl acetate. He also showed that even a stronger oxidation fails to remove the density of mesosomes when OsO_4-fixed Gram-positive bacteria were treated with uranyl acetate prior to dehydration.

It was further demonstrated by Silva *et al.* (1968, 1971) that the amount of phospholipid phosphorus extracted during the dehydration of *Bacillus subtilis* fixed with OsO_4 and treated with uranyl acetate was eleven times less than that extracted from bacteria of the same culture not treated with uranyl acetate. Similar results were obtained with rat brain. It is quite likely that uranyl acetate stabilises the fine structure by combining with the reduced osmium (lower oxides) that has been deposited in the cells. It should be noted, however, that this is only one of the factors responsible for better stabilisation, since treatment with uranyl acetate without previous fixation with OsO_4 also reduces the amount of extraction. Uranyl acetate increases contrast in almost all components of a plant cell fixed with $KMnO_4$. This is especially true in the case of cell walls.

De Petris (1965) demonstrated that when uranyl acetate is used before dehydration, a considerable amount of structure can be seen in the cell envelope, between the trilaminar cell wall and the cytoplasmic membrane in phage-infected *E. coli* fixed with glutaraldehyde and OsO_4. Uranyl acetate probably acts both as a stain and as a fixative. Uranyl acetate followed by lead citrate has proved effective in staining cellulose microfibrils of the cell wall in thin sections (Cox and Juniper, 1973).

Terzakis (1968) showed that dense components of malarial parasite membranes can be specifically preserved by employing uranyl acetate after double fixation with glutaraldehyde and OsO_4 and before dehydration. Aqueous uranyl acetate not only stained DNA and membranes in this tissue, but also markedly improved their preservation when the tissue was treated with the stain before dehydration. The improvement in the preservation of organisation of the membranous structures was indicated by their continuity and trilaminar appearance. Silva (1971) has demonstrated ultrastructural damage to bacterial membranes in the absence of post-fixation with uranyl salts.

Uranyl acetate is also quite effective in staining preferentially the protein shell and nucleic acid core of isometric viruses in the infected plant tissues (Hills and Plaskitt, 1968). The infected tissue is fixed with glutaraldehyde and then exposed to uranyl acetate during dehydration; post-fixation with OsO_4 is omitted. The advantage of this method over the conventional method is that ribosomes are not stained strongly, and thus the virus particles are easily recognisable in the infected plant tissue. After double fixation, however, the electron density of ribosomes is greatly augmented by this stain. Aqueous uranyl acetate also stabilises globular subunits in the bacterial cell wall and mesosomes.

Acid uranyl acetate (pH 2.0) stains selectively micropapillae present on the surface of human blood monocytes (Clawson and Good, 1971). The acidity of the staining solution is increased by adding HCl. The micropapillae became visible only when Ca^{2+} and Mg^{2+} were added to the fixative. It is assumed, therefore, that UO_2^{2+} preferentially replaces these divalent cations bound by the micropapillae, and thus enhances the electron density of the site.

It is clear from the above discussion that uranyl acetate is capable of preserving certain cellular materials by preventing their dissolution duruing dehydration. Available evidence is sufficient to recommend the introduction of aqueous uranyl acetate prior to dehydration in order to achieve better preservation of cellular components, particularly of DNA and membranes. It should be noted that prolonged treatment with unbuffered aqueous solution of uranyl acetate, prior to dehydration, is undesirable, for it may cause extraction and alteration in the appearance of glycogen in certain tissues. Vye and Fischman (1970) have reported extraction and aggregation of glycogen in the glycogen body of the chick embryo, and in skeletal and cardiac muscles stained *en bloc* with unbuffered aqueous solution of uranyl acetate (0.5%) for 1 h prior to dehydration. Studies by Mumaw and Munger (1971), on the other hand, indicated that uranyl acetate (0.5%) treatment for 1 h at 4 °C resulted in excellent preservation of glycogen in rat hepatocytes. They employed uranyl acetate in the pH range 2.0–8.0 prior to osmication of the aldehyde-fixed tissue.

STAINING SOLUTIONS

Uranyl acetate can be employed for staining thin sections as well as for *en bloc* staining prior to or following embedding; stain is not lost during dehydration. *En bloc* staining can be employed either prior to or after osmication. Various concentrations of the stain and various staining durations and temperatures have been employed to obtain the desired intensity of staining of different tissues. Although aqueous solutions of uranyl acetate (0.5–1.0%) employed briefly (1–10 min) stain chromatin, the nucleolus and ribosomes intensely at room temperature, application of heat or the use of alcoholic solutions further increases the effectiveness of the stain in certain cases. Thus, the duration of staining can be reduced by using alcoholic solutions and increased temperatures.

Sufficient contrast can be obtained by prolonged *en bloc* staining at higher temperatures (e.g. 60 °C), so that subsequent post-staining with uranyl acetate, lead, etc., becomes unnecessary, unless counterstaining is desired for achieving staining specificity. Prolonged *en bloc* staining can be utilised in high-voltage electron microscopy for viewing sections as thick as 1 μm. The advantage of this procedure becomes apparent when one considers that conventional staining procedures on thin plastics sections do not allow stains to penetrate deeper than ~100 nm.

Since uranyl acetate dissolves more readily in methanol than in water or ethanol, the use of methanolic uranyl acetate was suggested by Stempak and Ward (1964). Also, methanolic uranyl acetate penetrates the tissue faster and deeper than is possible with aqueous solutions, because of the imbibition of methanol by plastics sections. For epoxy sections, treatment with 0.1% solution for 30 min at room temperature is recommended. Aggregated virus particles centrifuged onto carbon-coated grids were successfully counted after staining with 0.4% uranyl acetate in methanol for 10 min (Ellis *et al.*, 1969). Methanolic uranyl acetate has a limitation in that collodion-coated grids cannot be used, because methanol readily dissolves collodion. Formvar-coated grids, however, can be used, although even Formvar is solubilised by too long a staining time, or by very strongly concentrated solutions.

Locke *et al.* (1971) obtained good general contrast in insect tissues by *en bloc* staining with 2% solution in 100% ethanol for 12 h at 60 °C. They did not use any subsequent staining and claimed to have obtained less granular staining. Although they claimed the absence of extraction due to the prolonged treatment, the effects of heat, absolute ethanol and prolonged staining on the decomposition and extraction of cellular materials should be taken into account when using these procedures.

A significant increase in contrast of actin filaments in thin sections can be obtained by treating isolated membranes with 1% aqueous uranyl acetate for 1 h after fixation with OsO_4 and before dehydration (Pollard and Korn, 1973). Increased contrast of mitochondrial inner membrane spheres can be achieved by treating the pellet with 2% uranyl acetate in methanol for 24 h at 4 °C, after fixation with OsO_4 and before dehydration (Telford and Racker, 1973).

Enhanced staining of cytoplasmic membranes of Gram-negative bacteria can be obtained by treating the specimens with 0.5% uranyl acetate solution (pH 3.9) for 30–120 min at room temperature, after fixation with Ryter–Kellenberger solution and before dehydration (Silva and Sousa, 1973). The staining is also enhanced by adding Ca^{2+} to the OsO_4 fixative. However, this treatment with uranyl acetate has an adverse effect on the R-layer and the material covering the surface of the cell wall. This adverse effect can be eliminated by section staining rather than staining prior to embedding.

En bloc staining can also be accomplished at neutral pH. The staining solution consists of 0.01 M uranyl acetate and 0.01 M oxalic acid, in 0.1 M collidine buffer adjusted to pH 7.2, with 6% ammonium hydroxide. This solution is claimed to give enhanced contrast, accompanied by little extraction of cellular materials, compared with *en bloc* staining with uranyl acetate at acidic pHs (Mumaw and Munger, 1969). Glycogen rosettes, for instance, are well preserved with uranyl acetate–oxalate.

The use of uranyl acetate, following double fixation with glutaraldehyde and OsO_4 and before dehydration, is strongly recommended for increasing the overall contrast and further stabilising membranous and nucleic acid-containing structures (Fig. 4.1). The author routinely uses 2% uranyl acetate in acetone or ethanol for ~15 min in the first step of dehydration for *en bloc* staining. Tissues can also be stained with equal success in 0.5% aqueous uranyl acetate at pH 3.9 for 10 min immediately prior to dehydration. The sections of the tissue thus treated are post-stained with 2% aqueous uranyl acetate for 2 min, followed by lead citrate for 5–10 min. Prolonged treatment with uranyl acetate should be avoided.

Tissue blocks can also be stained with uranyl acetate following dehydration and embedding. Adequately cured blocks are trimmed and faced, and then placed in tightly stoppered, oil-wrapped vials containing 1% uranyl acetate in 95% ethanol for 12 h at 60 °C. Sections can be cut from the exposed face after the block has been washed in ethanol; the first few sections should be discarded from the contaminated surface. If the block is softened by the hot ethanol treatment, it can be rehardened in an oven at 60 °C. By this method, satisfactory contrast is obtained to a depth of many micrometres from the surface of the block. This method is especially useful for obtaining a large number of sections from tissue blocks that were not stained *en bloc* prior to embedding.

As stated earlier, preferential staining of RNP can be obtained in the glutaraldehyde-fixed tissues by prestaining the sections with 5% uranyl acetate for 1 min, and then treating with 0.2 M EDTA for 20 min, followed by lead treatment for 1 min. Bernhard (1968, 1969) has given the optimal concentration and duration of treatment for various cell and tissue types.

Microfibrillar structures of cell walls can be seen in thin sections of plant tissues after post-staining with 2% uranyl acetate for 30 min, followed by lead citrate for the same duration (Cox and Juniper, 1973). Deoxyribonucleic acid samples prepared by the basic protein film technique (Kleinschmidt and Zahn, 1959) can be stained with uranyl acetate (Wetmur and Davidson, 1966). The grid with the DNA is dipped into a 95% ethanol solution of uranyl acetate for 30 s, followed by exposure to isopentane for 10 s.

Uranyl acetate has been used in combination with periodic acid and sodium hypochlorite to demonstrate glycogen. Exposed aldehyde groups are converted to carboxyl groups with hypochlorite. Uranyl ions are caused to react with the acidic carboxyl groups, producing an electron-dense stain for glycogen. The staining is probably due to the reaction of protein as well as polysaccharide components of the glycogen complex with uranyl ions. Several methods for staining glycogen at the subcellular level have been evaluated by Vye (1971), Anderson (1972), Rybicka (1981) and Cataldo and Broadwell (1986). Detailed discussion of glycogen staining is given on p. 304.

Thin sections of glutaraldehyde and OsO_4-fixed tissues are mounted on gold or platinum grids or carried in plastics rings and exposed to 1% periodic acid for 30 min and then rinsed in distilled water. This step is followed by incubation in a 2.5% solution of sodium hypochlorite dissolved in 10% acetic acid for 15 min. The sections are rinsed with distilled water and treated with a saturated solution of uranyl acetate for 10 min. The sodium hypochlorite solution should be freshly prepared and used in the dark.

Care should be taken in interpreting the structure of chromosomal fibres that have been stained with uranyl acetate, for the stain itself may form structures which closely resemble the 23 nm chromosomal fibre (Griffith and Bonner, 1973). These uranyl complexes exhibit a variety of forms and have an electron density similar to that of stained biological specimens. This artefact is formed in aqueous solutions that have been aged for 1 week or more. The formation of this artefact is enhanced by the presence of salts such as NaCl and CsCl. This artefact is not formed in non-aqueous solutions of uranyl acetate. Uranyl precipitates can be removed by exposing the sections to 0.5% oxalic acid for 30 s.

URANAFFIN REACTION

The uranaffin reaction as an ultrastructural cytochemical stain for the localisation of adenine nucleotides in organelles storing biogenic amines was introduced by Richards and Da Prada (1977). This reaction, when used under specific ionic conditions, stains neurosecretory granules (Payne *et al.*, 1984). It seems that the uranaffin reaction is specific for neuroendocrine cells and their neoplasms. It has been proposed that the hormonal content of neuroendocrine granules is bound to a protein–nucleotide core and that the uranyl ions bind to the phosphate groups of these nucleotides. At pH 3.9 the uranyl ions exist predominantly in the uncomplexed ionic form (UO_2^{2+}); they appear not to interact with the phospholipid present in the membranes of neuroendocrine cells but instead are deposited in areas where the negative charges of phosphate groups are highly concentrated (i.e. the polyphosphates of nucleotides) (Payne *et al.*, 1984). Uranyl ions bind exclusively to phosphate groups at concentrations below 10^{-5} M.

The uranaffin reaction seems to be specific for cell organelles containing high concentrations of nucleotides. These organelles include nuclei, ribosomes and neurosecretory granules. The staining of nuclei is due to the high concentration of DNA; that of the nucleoli and ribosomes is due to the high concentration of RNA; and that of the granules is due to the high concentration of ATP, ADP or AMP. Uranaffin-positive structures are easily differentiated from uranaffin-negative organelles in sections that are not counterstained (Hopkinson *et al.*, 1987). If the tissue is rinsed with cacodylate buffer instead of NaCl, non-specific staining will increase.

Procedure

Tissue specimens fixed in glutaraldehyde and OsO_4 are treated with 0.9% NaCl for 72 h at 4 °C (with several changes of fresh NaCl) to thoroughly wash out the buffer salts used during fixation. The specimens are exposed to 4% aqueous uranyl acetate (pH 3.9) for 48 h at 4 °C, and then rinsed in three changes of 0.9% NaCl (15 min each). The number of neurosecretory-like granules stained by this reaction can be statistically increased by increasing the time of exposure to uranyl acetate.

STAINING PROCEDURES

Thin sections mounted on grids can be stained individually or simultaneously.

Double Staining with Uranyl Acetate and Lead Citrate

The simplest method used for staining the grids individually with uranyl acetate and lead citrate is as follows. Several individual drops of a saturated solution of uranyl acetate are placed on a clean sheet of dental wax in a Petri dish. If dental wax is unavailable, a clean plastics Petri dish can be used; drops of suitable size and shape are not easily obtained on glass surfaces. The grid with mounted sections is floated (section side down) on a drop of the uranyl acetate for 5–15 min. The grid is rinsed in boiled distilled water and kept in a covered Petri dish for subsequent staining with lead.

A small quantity of fresh NaOH pellets is placed at one side of an absolutely clean Petri dish (Fig. 4.17) to produce a CO_2-free atmosphere. A piece of a clean sheet of dental wax is placed in the dish, which is covered as shown in Fig. 4.17. The preparation of this set-up should be done before staining the grids with uranyl acetate in a second dish, so that by the time the first staining is completed, the atmosphere in the dish will be free of CO_2.

With the aid of a clean pipette, a small quantity of lead citrate stain is drawn up from below the surface of the staining solution (freshly prepared solution is preferred), and 3 drops, each slightly larger than the grid, are placed on the wax in the dish. Each drop should be small enough to allow the grid to float on top of the dome of the drop instead of sliding down to the sides. Immediately the grid, already stained with uranyl acetate and wetted with boiled distilled water, is placed (section side down) on the drop of the staining solution (Fig. 4.17); alternatively, the grids can be stained by immersion. The dish is covered immediately after the grid has been placed in it. In fact, the dish should be covered during the interval when more than one grid

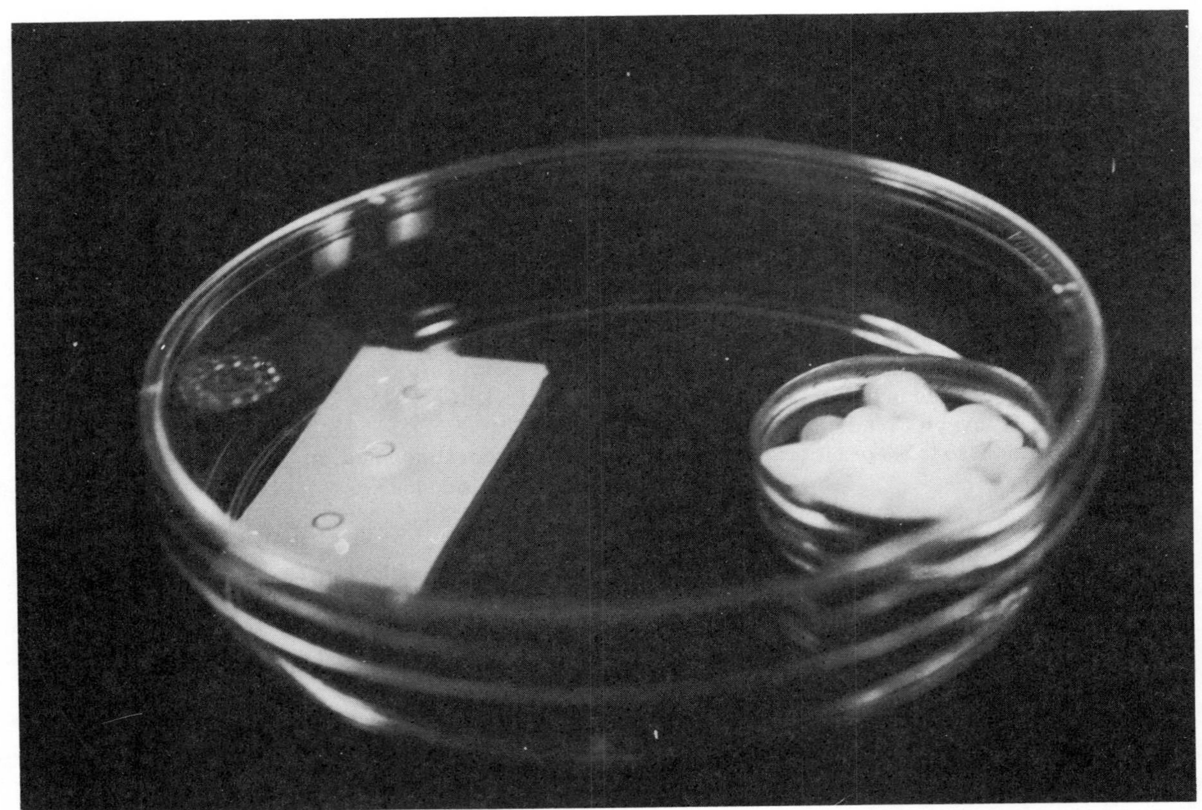

Figure 4.17 A Petri dish used as a chamber for staining grids with Reynolds' lead citrate. The dish contained pellets of NaOH and a small sheet of dental wax. Three drops of the staining solution are placed on the sheet, each showing a grid floating on it.

are placed in it. No time should be wasted between placing the drops of the stain in the dish and placing the grids on them. The duration of staining is 3–15 min. Coated grids, very thin sections and thick sections are stained for longer.

After the grid has been stained, it is quickly and thoroughly washed consecutively in 0.02 N NaOH solution and distilled water. While the grid is being washed, the lid should be replaced, since the other grids are still being stained. The washing can be accomplished by holding the grid at its edge with a forceps and dipping it rapidly in the fluid by agitation in a small beaker, or by washing under a jet from a plastics wash bottle. Irrespective of the method of washing used, care should be taken that the grid is quickly immersed in the washing fluid and all of the stain washed off the grid and the forceps. Any left-over stain between the tines of the forceps is a potential source of contamination. The washed grid is then blotted dry, making certain that the section side does not touch the filter paper. While placing the grid on a filter paper, if the forceps carry excess water, the grid is drawn up between the tines of the forceps. This can be avoided by placing a piece of filter paper between the tines of the forceps and advancing it towards the grid.

More than one grid can be easily stained simultaneously in one Petri dish. However, each grid should be floated on a separate drop of the staining solution. In fact, simultaneous staining by the method described above not only saves time, but also provides the opportunity to determine the optimal duration of staining. It takes about 2 min to wash a grid after it has been stained, so that when four grids are stained simultaneously and the first grid is stained for ~3 min, the second would be stained for ~5 min, the third for ~7 min and the fourth for ~9 min. The desired intensity of staining can be selected from this series. It is advisable not to stain more than five grids simultaneously. If lead precipitates are formed, they can be removed by exposing the sections to 10% acetic acid for 1 min.

Some stains penetrate more efficiently when the sections are not allowed to dry between the time of sectioning and staining. Another advantage of using wet sections is that they do not pick up contaminants from the surface of staining and washing solution. Contaminants usually adhere to a dry section. The first and last solutions to come in contact with the sections must be clean (e.g. boiled distilled water from a wash-bottle).

MULTIPLE GRID STAINING

A large number of methods have been introduced for holding and staining groups of grids (see, e.g., Hiraoka, 1972; Thaete, 1979; Giammara, 1981; Kobayasi, 1983; Brown, 1983b). Many of these devices provide closed staining systems with filters to remove the contaminants. The device now described is simple to construct and is disposable and inexpensive (Cardamone, 1982). It is constructed from Tygon tubing, rings and a vial (Fig. 4.18). The specifications of the tubing and rings are as follows:

Quantity	*Length*	*Width (outer diameter)*
1	50 mm	6 mm
2	8 mm	10 mm
2	8 mm	12.2 mm

First the 10 mm rings are placed on each end of the 50 mm tubing, and then the 12.2 mm rings are fitted over the 10 mm rings. These rings function as spacers that keep the main body of the tubing away from the wall of the vial. A razor-blade is used to cut a series of 10–12 slits in the tubing, just deep enough to reach the hollow centre of the tubing. These slits, when squeezed laterally, open to accommodate the edge of a grid. Conversely, when the lateral pressure is relaxed, the tubing firmly grips each grid (Fig. 4.18). To keep track of grid placement, a notch is cut at one end of tubing, thus denoting position 1. Once the grids are in place, the entire holder can be immersed in the staining solution in a vial (Fisher No. 03-340-2D) which is stoppered.

MULTIPLE STAINING

The desirability of post-staining after tissue has been fixed with an aldehyde followed by OsO_4 is now fully recognised. In fact, the best general increase in contrast is obtained when the tissue is exposed to several heavy metals. The advantage of multiple staining is obvious in that each of the stains used either enhances the contrast already imparted by another stain or increases the contrast of a very-low-density structure which does not react with another stain, or both. In other words, in general, multiple staining enhances the density of all cell components beyond the degree that can be obtained with any single stain alone. Specificity of staining is obviously not the primary objective of this technique.

Another advantage of multiple staining is that many of these heavy metals act as fixatives to various degrees, especially when employed prior to dehydration. For example, the stabilising effect of uranyl acetate on certain cell structures has been discussed earlier. Interaction between heavy metals appears to play an important role in better preservation and enhanced contrast of cellular structures. It is known, for instance, that lead is bound by reduced osmium. New methods using several stains are likely to be employed more and more in the future. Additional work will have to be done to elucidate the mechanism involved in the interaction of metal stains. The following method has proved the most effective for general morphological staining.

The tissue is fixed with glutaraldehyde followed by OsO_4, and then immersed in a 2% solution of uranyl acetate in 10% acetone for ~10 min. After dehydration and embedding, the sections are post-stained by floating the grids on a drop of 1% aqueous uranyl acetate for 5 min, followed by Reynolds' lead acetate for 6 min. The grids should not be allowed to dry completely between the different staining treatments.

SECTION CONTAMINATION AND ITS REMOVAL

Section contamination means the presence of any extraneous substance in or on a thin section. Typically, it is sporadic in appearance. In most cases, section contamination can be avoided by carefully evaluating and changing the fixation, rinsing, embedding and staining procedures (see later). However, it is important to understand the cause of contamination and how to avoid it, and, if present, how to eliminate it. Mollenhauer and Morré (1978) and Kuo *et al.* (1981) have described the following types of section contamination. The sources of these contaminations overlap one another.

Surface contamination An extraneous substance is introduced during post-staining, and is present on the surface of a section. The contaminant may be both large and dense, which makes the section look dirty. The sources of such a contamination are oil, dirt, lint, and dried residues of uranyl and lead salts. Careful post-staining minimises this contamination.

Fixation contamination Some defect in the fixation procedure produces fine, dark precipitate, which may be scattered throughout the section. Incomplete removal of glutaraldehyde from the specimen before post-osmication is one of the causes of this artefact. The presence of reduced osmium may also be respons-

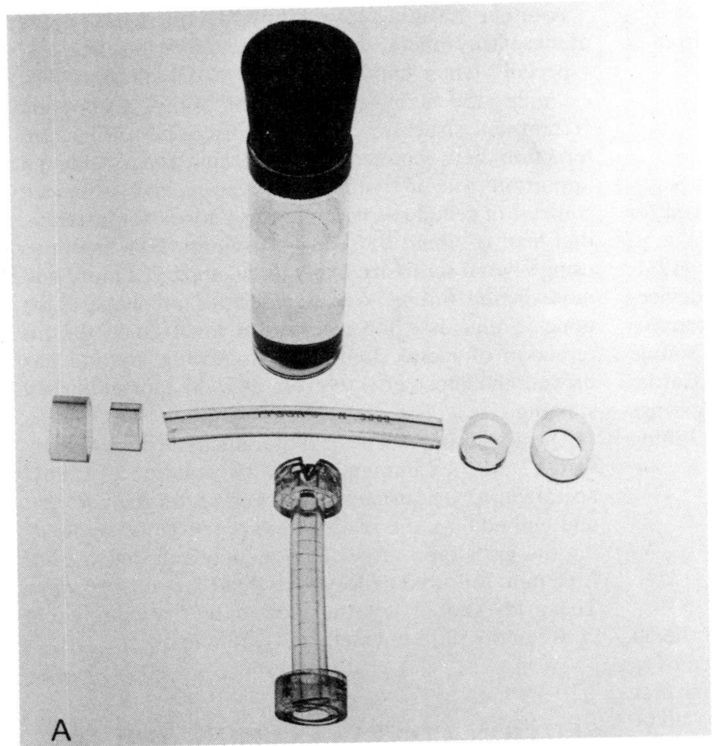

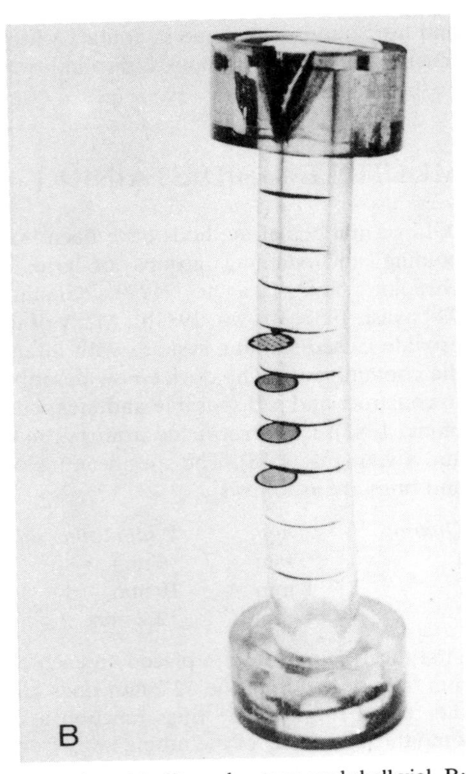

Figure 4.18 Multiple-grid staining device. A: Tygon tubing, rings and assembled device with slits and a stoppered shell vial. B: Close-up of the device, showing grids firmly held in slits. From Cardamone (1982)

ible for this type of contamination. Osmication at 4 °C and addition of 0.05 M sucrose to the fixative and buffer may reduce this contamination in certain specimens. This artefact can be removed by treating the section with freshly prepared 1% periodic acid or 3% H_2O_2 for 5–10 min (Ellis and Anthony, 1979).

Staining contamination Uranyl acetate alone rarely forms precipitates. If uranyl precipitates appear, they can be easily removed by rinsing the sections with warm distilled water or by filtering the staining solution before use. Prolonged staining (15–60 min) with lead citrate alone may form electron-opaque, spherical particles. These particles consist of lead carbonate formed by lead reacting with CO_2 from the air; this artefact is water-insoluble. It can be removed by treating the sections with aqueous uranyl acetate for ~7 min or with 10% aqueous acetic acid for 1 min; the former treatment is preferred, for acetic acid may damage the sections. Treatment of sections with acetic acid for a few minutes is also effective in the general removal of stains. Such sections can be restained.

If lead citrate is applied after staining with uranyl acetate without thorough rinsing, electron-dense, amorphous, netlike precipitates may form on or in thin sections. Fixation with potassium ferricyanides increases this artefact. It may be common in sections from tissue that has been stained *en bloc* with uranyl acetate. The density of this precipitate varies from one section to another. Occasionally it may cover the section completely. This precipitate appears to consist of lead acetate or uranyl citrate complexes and is water-insoluble. It can be removed by treating the sections with 0.5% oxalic acid for 15 s, 10% acetic acid for 1 min, or either aqueous or ethanol solutions of uranyl acetate for 1 min. The last treatment is preferred. Treating the sections with 1% EDTA before staining may help to avoid such artefacts.

Prolonged staining with alcoholic uranyl acetate solution followed by lead citrate may cause the production of crystalline needles. Unlike the precipitates mentioned above, this artefact is difficult to remove. The solution is to avoid using alcoholic uranyl acetate unless absolutely necessary, as well as prolonged staining. Staining artefacts generally are difficult to remove from the areas of a grid that have been subjected to prolonged exposure to the electron beam.

Embedding contamination Embedding contamination is usually associated with organelles such as mitochondria and microbodies and red blood cells. The source of this artefact appears to be improper preparation of the embedding resin. Incomplete infiltration of the specimen by the resin plays a role in the formation of the artefact. It can be minimised by preparing the embedding formulation carefully, ensuring complete infiltration of the tissue, and post-staining by immersing the grids instead of floating them.

Electron beam contamination The electron beam can cause a fine granular precipitate of uniform size; excessive granularity of membranes may also be seen. The contamination is a result of abrupt heating or charging effects of the electron beam on the section. Cell components of the section may be evaporated and redistributed over the section or aggregated into granular structures. Although this artefact cannot be eliminated, it can be diminished by exposing the section to low-dose irradiation.

PRECAUTIONS TO MINIMISE SECTION CONTAMINATION

A thorough rinsing between aldehyde and OsO_4 fixation is necessary. The entire staining procedure should be carried out extremely carefully. Artefactual staining precipitates can be minimised by following the guidelines below.

(1) Make sure all containers used during staining are absolutely clean and free from dust and lint.

(2) Use only clean, boiled distilled water for preparing lead and rinsing solutions.

(3) Avoid staining for longer than necessary.

(4) Wet the grids before transferring them to any staining solution.

(5) Do not allow the grids to dry between various staining steps.

(6) Wash grids thoroughly with *warm* distilled water after each staining step.

(7) If desirable, submerge sections into the staining solution instead of floating them.

(8) Add one drop of glacial acetic acid to 10 ml of uranyl acetate solution in water or 50% ethanol. Acetic acid prevents photodegradation of uranyl acetate solution and such solutions are less likely to produce artefactual fine granules on the sections.

(9) If possible, use freshly prepared staining solution; allow undissolved crystals of uranyl acetate to settle down before using the solution.

(10) If desirable, filter uranyl acetate solution before use.

(11) Use CO_2-free NaOH pellets from a freshly opened container to prepare 1 N or 0.02 N NaOH in a bottle with a rubber cap. If the container of NaOH pellets is opened frequently, it becomes contaminated with carbonate.

(12) Prepare and store lead solutions in plastics rather than glass containers.

(13) Maintain a CO_2-free environment during lead staining. Avoid breathing on sections during lead staining.

(14) If possible, eliminate the air–stain interface during lead staining.

(15) If a white precipitate ($PbCO_3$) is visible in the bottle containing the lead solution, discard the solution instead of filtering it. However, this precipitate can be sedimented by centrifugation.

The contamination of lead stains during storage can be significantly reduced by a simple procedure introduced by Richardson *et al.* (1983). The staining solution can be stored for at least 5 months without contamination, and stain from the container can be used repeatedly without contaminating the remaining stock solution. About 30 ml of lead citrate solution is filtered with a syringe-operated Millex GV-filter (Millipore Corp.) into a 60 ml serum bottle. The bottle is stoppered with a flange rubber, and a 25-gauge disposable Luer-lok needle is inserted into the rubber stopper (Fig. 4.19). A plastic tube attached to a can of compressed inert gas (Freon) is inserted into the hub of the needle. A few short bursts of gas are applied to purge air from the serum bottle through the same needle. The plastic tube inserted in the hub does not completely seal it. While the needle is being removed, gas is slowly released to replace most of the atmospheric air. This step reduces the amount of CO_2 that can form lead carbonate. Immediately before use, the staining solution is removed from the bottle with a disposable Luer-lok needle (20–25-gauge, ~4 cm) and a 3 ml syringe (Fig. 4.19).

REMOVAL OF SECTION CONTAMINATION

If lead precipitates are present, they can be removed by treating the grid with 2% aqueous uranyl acetate for 2–8 min at room temperature. Alternatively, lead precipitates can be eliminated by treatment with 10% aqueous acetic acid for 1–5 min. Precipitates resulting from post-staining with both uranyl and lead salts can be removed by treating the grid with 0.5% oxalic acid for 12 s, 10% acetic acid for 1 min or 2% aqueous (or ethanol) solution of uranyl acetate for 1 min. Aqueous uranyl is preferred to acetic acid. Uranyl acetate precipitation can be eliminated by rinsing the grid in *warm* distilled water. Acetic acid (10%) is also effective in

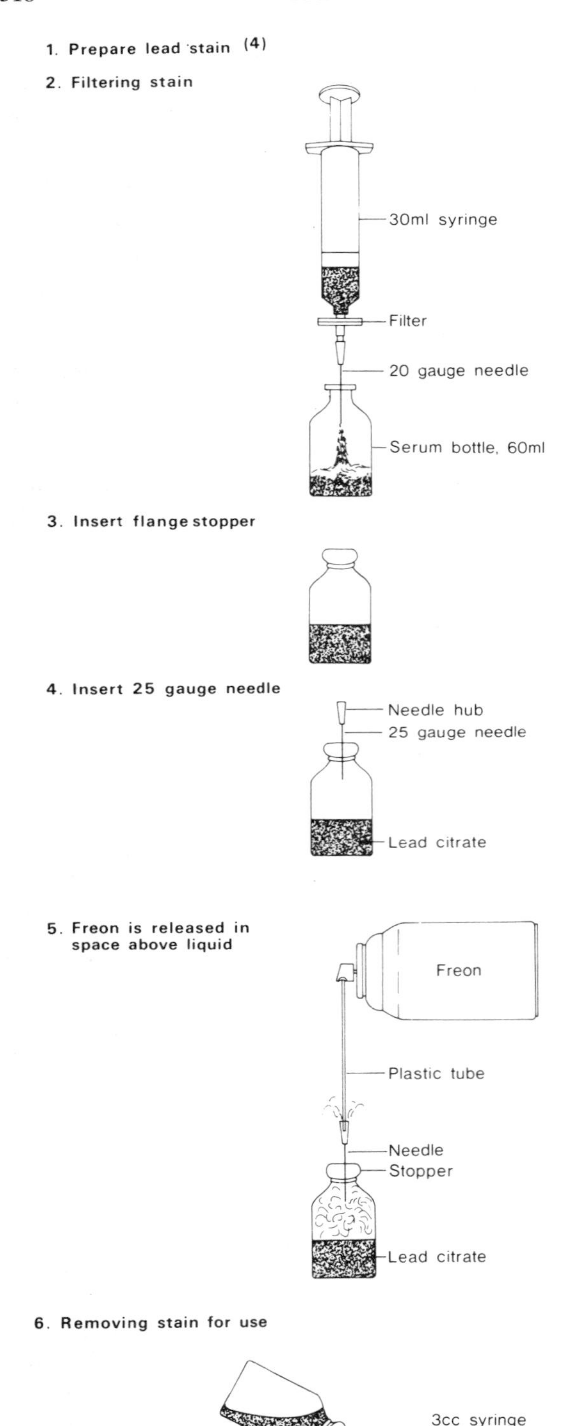

Figure 4.19 Schematic diagram of the storage and removal of lead staining solution without contamination. From Richardson *et al.* (1983)

general removal of stains. All the above-mentioned treatments may damage the fine structure.

SELECTIVE HEAVY METAL STAINING FOR HIGH-RESOLUTION ELECTRON MICROSCOPY

Since single atoms of heavy metals in a number of small model compounds have been directly visualised with the transmission electron microscope (TEM) and with the scanning transmission electron microscope (STEM), molecular structure can be determined by using heavy metal staining. The structure of a macromolecule can be elucidated provided that it can be assembled from its components. The basic approach is to label the component of interest with heavy atoms. After assembling the macromolecular complex, the disposition of that component can be inferred from the observed positions of the labels bound to it. It is now possible to determine the positions of chemical groups within macromolecules by attaching heavy atoms selectively to the sites of interest. This approach is also effective in determining the location of a particular macromolecule within a complex of several. For example, the position of histones in platinum-labelled nucleosomes has been determined by using chloroglycol-1-methioninatoplatinum(II) (Stoeckert *et al.*, 1984). However, the accuracy of these structural data is limited by possible damage to the specimen.

Two types of specimen damage are: (1) distortion of the specimen during deposition on the carbon film and subsequent dehydration, and (2) damage by irradiation during electron microscopy. Structural damage can be minimised by fixation. Drying from glycerol is an acceptable substitute for fixation. The migration of heavy atoms from their original site during viewing in the electron microscope is a problem. However, this problem can be minimised by low-dose electron microscopy. The electron dose can be precisely controlled in the STEM. This is possible because a given number of electrons scans a given area of the specimen, and thus the electron dose can be controlled by changing either the magnification or the beam current.

Using electron dosages of 10^3 e/A^2 to 10^4 e/A^2, it was found that 60% of the osmium atoms moved less than 0.5 nm between successive scans from their attachment sites in a synthetic polynucleotide as a result of radiation damage (Cole *et al.*, 1977). In this case a resolution of 0.5 nm might be achievable using osmium as a label. Similarly, little migration of the platinum atoms occurred from their original binding site in the Pt–GLM-labelled collagen (Beer *et al.*, 1979). Chemical analysis indicated that ~27 atoms of Pt were bound per collagen molecule. The problem of the signal/noise ratio can be improved by Fourier techniques.

STAINING FOR HIGH-VOLTAGE ELECTRON MICROSCOPY

Most of the staining methods used for high-voltage electron microscopy are modifications of the staining techniques employed for conventional electron microscopy. Uranyl and lead preparations penetrate thick sections without much difficulty. However, under ordinary conditions of staining, thick sections are difficult to stain with those metals that are primarily surface stains (e.g. PTA) except in the case of unicellular organisms. In general, longer durations and/or elevated temperatures are needed to stain thick sections. To facilitate thorough penetration into the section, certain staining solutions (e.g. uranyl acetate and PTA) are prepared in acetone or alcohols. In some cases higher concentrations of stains are needed. Both sides of the section should be exposed to the stain by immersing it in the staining solution.

En bloc staining is also used for high-voltage electron microscopy. Such staining with uranyl acetate is especially useful for the study of membranous structures. Only some of the stains can be employed before embedding and sectioning. *En bloc* staining can be used alone or followed by section staining. Usually two or more stains are employed to achieve adequate staining. However, this approach diminishes selective staining.

Aqueous staining solutions require rather long durations to penetrate throughout a thick section. Aqueous uranyl acetate solutions penetrate more slowly than does lead citrate. Thick sections are difficult to stain with aqueous uranyl acetate solutions, because only superficial parts of the section are stained, even after prolonged staining. Thick sections, on the other hand, are adequately stained with uranyl acetate in alcohol solutions. Uranyl or PTA solution in ethanol penetrates to depths of up to 15 μm in 24 h, especially when staining is carried out at 60 °C.

It has been estimated that, in general, after 2 h of staining the penetration for liver tissue is ~4 μm, and after 16 h the penetration is completed into a 19 μm section (Favard and Carasso, 1973). By adding a detergent (Tween 80, 0.3%) or DMSO to the staining solution, the penetration is made easier and more homogeneous. Apparently, the rate of penetration is dependent upon many factors, including temperature, type of tissue and embedding medium, concentration and vehicle of the staining solution, and mode of staining. If precipitates cling to the surface of the section as a result of prolonged staining, they can be dissolved by exposing the section for 1 min to 0.05% citric acid solution. That the stain has penetrated throughout the section can be verified by cutting the section after staining according to a technique devised for measuring section thickness (Favard and Carasso, 1973).

The highest overall contrast is achieved by *en bloc* staining with uranyl acetate before dehydration, followed by prolonged post-staining of the sections with uranyl acetate and lead citrate. However, the disadvantage of staining with both uranyl acetate and lead citrate is that many different structural elements are stained equally, and thus are not distinguished from each other. For relatively selective staining, either uranyl acetate or lead citrate should be used. Staining with uranyl acetate emphasises the membranous components and nucleic acid-containing structures, especially chromatin, while lead citrate alone results in the staining of ribosomes and cytoplasmic ground substance. Lead staining is essential for visualising certain structures such as bone crystallites (Glauert and Mayo, 1973).

Selective staining of certain structures can be achieved with Gomori's lead method without any counterstaining. By using this method, Golgi saccules (where acid phosphatase or thiamine pyrophosphatase activity is located) and endoplasmic reticulum (where glucose-6-phosphatase activity is located) have been stained (Favard and Carasso, 1973). The osmium impregnation method has been used for staining the forming faces of the Golgi complex and in some cases membranes and matrix of mitochondria (Favard and Carasso, 1973). Bismuth subnitrite has been employed for staining cell surfaces (Peachey *et al.*, 1974).

A silver impregnation technique has been employed to reveal the deposition of neurofibrils (Favard and Carasso, 1973). A zinc–iodine impregnation method (proposed by Maillet, 1968) has been used to reveal saccules of the forming face of the Golgi complex after short duration (4–24 h) and all saccules of the Golgi complex after long durations (48 h–7 days) (Favard and Carasso, 1973). The rapid Golgi method (Fregerslev *et al.*, 1971) has been employed for morphological analysis of the CNS (Palay and Chan-Palay, 1973).

Selective staining can also be achieved with molecular tracers. Lanthanum colloid is useful for selective staining of the T-system in skeletal muscle (Peachey and Franzini-Armstrong, 1977). Ruthenium red has been employed to examine the basal region of kidney tubule epithelial cells (Yamada and Ishikawa, 1972) Peroxidase has been used to examine the T-syster cardiac muscle (Yamada and Ishikawa, 1972). T dioxide has been employed to study cellula ments associated with endocytosis phen *et al.*, 1971). In general, it is desirable to achieve intense staining of certain structures and light staining of other cellular structures in order to facilitate a better interpretation of the micrographs. This can be accomplished by *en bloc* staining with uranyl acetate or staining of sections with lead citrate in addition to specific intense staining.

STAINING PROCEDURES

(1) *Uranyl acetate (1–2%) in 95% ethanol or acetone* Thick sections are stained for 20 min to 24 h at 60 °C, depending on the tissue type, the embedding medium used and the objective of the study. This treatment can be followed by staining with standard lead citrate for 20–90 min. The grids carrying the sections are immersed in lead citrate solution to stain both sides of the section. Alternatively, sections (0.5 μm) are stained in a 7:3 mixture of 5% uranyl acetate and 100% acetone for 10 min. They are rinsed and stained with a 7:3 mixture of Reynolds' lead citrate and acetone for 10 min.

(2) *Uranyl acetate–lead nitrate–copper sulphate (Thiéry and Rambourg, 1976)* Whole intracellular membranous system is stained in sections 0.5–2.0 μm thick. The staining solution is prepared by adding 2 ml of a 1 M lead nitrate solution (33%) to 0.5 ml of a 10% aqueous copper sulphate solution. A white precipitate is formed, which is mixed with 19 ml of a 4.6% aqueous sodium citrate solution. Then 4 ml of 4% NaOH is added to obtain a clear, bluish solution; it can be stored for several days.

Glutaraldehyde-fixed tissue blocks are immersed in a 5% aqueous uranyl acetate (pH 3.5) for 1 h at 37 °C. They are rinsed and then stained with the above solution for 1 h at 37 °C. The specimens are rinsed in distilled water and post-fixed overnight in 1% OsO_4 at 4 °C.

(3) *Uranyl acetate (0.2%) and critical point drying* Cells growing on coated grids attached to coverslips are fixed with glutaraldehyde and OsO_4 according to standard procedures. They are rinsed and then stained with 0.2% uranyl acetate in a mixture of 15% acetone and 5% DMSO for 15 min. After rapid dehydration, the cells are dried by the critical-point drying method and coated with carbon.

(4) *Uranyl acetate and phosphotungstate* *En bloc* staining is carried out by immersing the tissue blocks in 2% alcohol uranyl acetate for 12 h at 60 °C. Sections are stained with 2% PTA in 95% ethanol for 30 min at 60 °C.

(5) *Phosphotungstate and uranyl acetate* Hot alcoholic PTA solution yields excellent contrast to depths of ~15 μm, and thick sections thus stained can be observed without additional staining. Thick sections of glutaraldehyde-fixed specimens are stained with 2% PTA in 95% ethanol for 30 min at 60 °C. If needed, staining can be intensified by post-staining the sections with 2% alcoholic uranyl acetate for 12 h at 60 °C.

(6) *Osmium tetroxide* Selective staining of the Golgi complex and endoplasmic reticulum is obtained by impregnating the tissue with an unbuffered aqueous solution of OsO_4 (1%) for periods as long as 4 days. The staining solution should be renewed after 24 h. Thick sections are examined without additional staining.

(7) *Osmium tetroxide and potassium dichromate* The rapid Golgi method is useful for staining the central nervous system. Tissue blocks are fixed with glutaraldehyde and then immersed in a mixture of 0.19% OsO_4 and 2.33% potassium dichromate in distilled water for 4 days at room temperature in the dark. After a brief rinse in 0.75% silver nitrate, the blocks are immersed in this solution for 1–2 days in the dark at room temperature. The blocks are gradually transferred to pure glycerol at 4 °C.

Tissue blocks are embedded in 7% agar and cut at 200 μm. These sections are gradually transferred to distilled water. Gold toning is accomplished by placing the sections in 0.05% solution of yellow gold chloride ($HAuCl_4 \cdot H_2O$) for 10–15 min in an ice-bath. Effective agitation is carried out every 2 min. Sections are rinsed in three changes of cold distilled water. While the vials are still in the ice-bath, 0.05% oxalic acid is poured into them to reduce gold chloride to metallic gold; this reduction takes 2 min. Deimpregnation of silver chromate is done by treating the sections with 1% sodium thiosulphate for 1 h at 20 °C. Sections are rinsed in distilled water to remove sodium thiosulphate. They are post-stained with 2% OsO_4 for 30 min, rinsed in buffer and stained *en bloc* with 1% uranyl acetate in 70% ethanol for 2 h.

(8) *Horseradish peroxidase (Wilson, 1987)* Identified neurons can be stained with 4% HRP injected by a microelectrode. The peroxidase activity is reacted with DAB as a chromogen.

(9) *Silver nitrate* Specific staining of neurofibrils in nerve cells can be achieved by simultaneously fixing and impregnating tissue slices (2–4 mm) with silver by immersion in an aqueous solution of silver nitrate at 37 °C for periods as long as 14 days. After reduction by either pyrogallol or hydroquinone, the specimens are dehydrated and embedded. Thick sections are viewed without further staining.

(10) *Lathanum* The T-system in skeletal muscle can be stained by fixing the tissue with glutaraldehyde followed by OsO_4 according to standard procedures.

The specimens are infiltrated by lanthanum. Sodium hydroxide is added to form lanthanum hydroxide.

(11) *Cationised ferritin (Kawakami and Hirano, 1986)* Cationised ferritin has been used to label platelets for viewing in a 1000 kV TEM. Surface-bound ferritin is rapidly internalised at 37 °C by the platelets.

(12) *Ammonium paramolybdate (Smith, 1984)* Ammonium paramolybdate was used for viewing cytoskeleton in sections 1 μm thick.

STAINING OF THIN CRYOSECTIONS

Thin cryosections are usually stained with uranyl acetate. Tokuyasu (1980b) introduced the 'adsorption staining' method using uranyl acetate. Thin frozen sections mounted on a grid are washed four times on drops of distilled water for a total period of 4 min to remove most of the ions from the sections; uranyl acetate can precipitate in the presence of high concentrations of ions. The staining solution is prepared by mixing equal volumes of 0.3 M oxalic acid and 4% aqueous uranyl acetate, and the pH is adjusted to 7.0–7.4 with 5% NH_4OH, using pH paper instead of an electrode. The staining solution (2%) should be centrifuged for ~4 min at 14 000 *g* before use. The grid is stained on a drop of this solution for 10 min, and washed three times on drops of distilled water for a total period of 15 min. Since this treatment contributes only slightly to the staining intensity, the sections are stained again and embedded simultaneously as follows.

The grid is floated on a mixture of 1.8% polyethylene glycol (mol. wt. 1540), 0.2% methyl cellulose (400 cP) and 0.005–0.1% acidic uranyl acetate for 10 min. The grid is picked up on a loop of 2.5–3.0 mm in diameter, and after the excess mixture has been removed, it is allowed to air-dry. The excess mixture should be removed to such an extent that the dried film of the mixture shows a silver-gold to gold-blue colour (Tokuyasu, 1986). The reason for using polyethylene glycol as the embedding medium is that it penetrates rapidly into the frozen sections. The methyl cellulose stock solution is prepared by suspending its powder in hot water (55 °C, 333 K) and then cooling the suspension to 1 °C (277 K).

The above method was modified by Griffiths *et al.* (1982, 1984). This method results in a mostly negative and partially positive staining. The grid is treated with neutral uranyl acetate solution as described above, and then washed three times for a total period of 3 min on drops of distilled water. The grid is floated on a drop of a 1.8–2% methyl cellulose (25 cP) solution containing 0.1–0.4% acidic uranyl acetate on ice for 10 min. It is picked up on a loop and, after the excess solution has been removed, dried rapidly in a low humidity chamber.

This method is suitable for colloidal gold-labelled sections, but not for ferritin-labelled ones. The strong negative staining obtained facilitates identification of membrane profiles, even when obliquely sectioned. In the case of positive staining obtained, only membrane profiles that are transversely sectioned are visualised. The use of thin cryosections in conjunction with immunogold staining is described by van Bergen en Henegouwen (1989).

The above methods are suitable for staining membranes. To visualise cell filaments, thin cryosections are stained with 1% aqueous OsO_4 for 15 min, washed twice in distilled water, dehydrated in two changes of ethanol and washed twice in amyl acetate (Tokuyasu, 1983b). Each washing or dehydration is carried out for 2 min. The grid is left in a 1% ethyl cellulose solution in amyl acetate for 10 min, picked up with a forceps, touched on the surface of filter paper to remove the solution from the back of the grid and allowed to dry. The grid is stained with 2% acidic uranyl acetate for 10 min and washed with distilled water.

STAINING OF SEMITHIN SECTIONS

As stated earlier, it is desirable to study counterstained thick sections with the aid of a light microscope for the purpose of comparison, orientation and identification of structures. Alternate thick and thin sections can be employed to obtain the desired orientation of the tissue for electron microscopy. Also, unknown tissue components visualised in the electron microscope can be identified by comparing the thin sections with the thick sections stained by well-established histochemical methods and viewed in the light microscope.

Prior to staining, sections are transferred from distilled water onto a standard glass microscope slide. Sections are transferred onto the slide with a fine brush. It is advisable to transfer more than one section onto a slide. Large-size troughs (55 mm × 30 mm × 25 mm) made from stainless steel are useful for transferring ribbons directly to a glass slide. The end of the slide is immersed into the trough at an angle while the sections are manoeuvred with an eyelash.

Various methods have been used for attaching the sections to a glass slide. The easiest method is to float the sections on a drop of water in the centre of the slide and heat the preparation to 70 °C for 1 min on a hot-plate. Sections easily and quickly stick to the glass surface. Another method is to freshly coat the clean slide with a thin layer of Mayer's albumin adhesive or gelatin–chrome alum adhesive. The latter adhesive is prepared by dissolving 5 g of gelatin in 1000 ml of warm distilled water and then adding 0.5 g of chromium potassium sulphate. The wet-mounted section is air-dried either at room temperature or on a hot-plate at 50–60 °C. To obtain reliable adherence of sections to the slide, complete drying is imperative.

Sections can be stained either intact or after the embedding resin has been removed with a suitable solvent. It is recognised now that the sections from which the resin has been removed stain brilliantly and reveal cellular detail better. Furthermore, the structure of resin-free sections remains undamaged by the solvent action. It has also been suggested that polar groups remain available to cationic stains after resin extraction, although some plumbiphilic lipid is extracted from the tissue.

Because polymerised epoxy resins are not soluble in standard organic solvents, a special solvent is needed for use in the tissue–epoxy system. One such solvent consists of a mixture of sodium methoxide, benzene and methyl alcohol, and has proved an effective remover of the resin without causing detectable cytological distortion. Small pieces of 2.5 g of metallic sodium are slowly added to 25 ml of methyl alcohol at ~55 °C under a hood. Since the alcohol will evaporate at this temperature, the level of the solution during preparation should be maintained at 25 ml. After the sodium is completely dissolved, 25 ml of benzene is added. Alternatively, 2–5 g of commercially available sodium methoxide can be dissolved in 25 ml methyl alcohol and 25 ml benzene added. The clear mixture can be stored in a dark bottle for several months without any apparent deterioration.

The above mixture is used either at full strength or suitably diluted in a mixture of equal parts of methyl alcohol and benzene, depending on the thickness of sections. Floating sections which are to be stained are transferred to a drop of water in the centre of a glass slide with a fine brush, a wire loop or a sharpened applicator stick. Some workers prefer to use albumin-coated slides to ensure adhesion. The slide is heated on a hot-plate at 55–60 °C, and allowed to dry at this temperature. An exposure of 1–3 min is adequate to remove resin from the sections. The slides with sections are then rinsed in a mixture (1:1) of benzene and methyl alcohol, followed by two changes of acetone and water. At this stage the slides are ready to be stained. It is not a good practice to delay the staining after the embedding medium has been removed.

The above mixture is also useful for differentiating lipids from other tissue components in OsO_4-fixed tissues. Since the mixture is a powerful lipid-solvent, it can extract lipids from thick as well as from thin sections. The material removed by this mixture is assumed to be lipid.

A saturated solution of potassium hydroxide in absolute ethanol is another reliable solvent for Epon sections. Ethanol is poured over excess KOH. It is allowed to stand overnight, and the supernatant, dark-yellowish brown in colour, is used as a solvent. Glass slides with sections are immersed for 10–15 min in this supernatant, followed by rinsing in four changes of absolute ethanol in order to completely remove the potassium hydroxide.

Sodium methylate is also an effective remover of most types of resins. Sodium methylate (1.5 g) is dissolved in 25 ml of methyl alcohol. To this solution is added 25 ml of benzene. Slides are transferred to this solution for 10 min, to 70% methyl alcohol in 30% benzene for 5 min, to acetone (two changes, for 3 min each) and to water (three changes, for 1 min each). Slides can be exposed to 1% PA for 15 min to remove the osmium.

Iodine or bromine can also be employed for removing epoxy resin from the thick sections. In this method, sections are treated with a 10% solution of iodine in absolute acetone for 12–18 h to soften the resin, followed by washing with absolute acetone to remove the medium prior to staining. The embedding medium can also be softened and removed by exposing the dry sections to bromine vapour in a special, safe container for 30–60 s, followed by washing with 100% acetone.

The possible mechanism involved in rendering the epoxy resin soluble in acetone was described by Yensen (1968). Infrared spectroscopy studies of bromine-treated and untreated Araldite indicate that this halogen is progressively substituted onto the aromatic rings and/or a structural change occurs in these rings. These changes are considered sufficient to render the epoxy soluble in acetone. However, it is admitted that very little is known concerning the actual mechanism involved in rendering various polymerised resins soluble in organic solvents by the different methods discussed above.

Semithin sections may be stained either to obtain overall staining of the tissue components or to achieve differential staining of the components. The overall monochromatic staining is obtained by employing stains such as Sudan black B, Nile blue A, basic fuchsin, toluidine blue, paraphenylenediamine or cobalt sulphide. On the other hand, differential staining of tissue components can be obtained by using combinations of two or more stains. Haematoxylin and eosin, for instance, is a well-known combination that differentiates acidophilic and basophilic structures.

Since haematoxylin has an affinity for lipoproteins and phospholipids preserved by osmication, more distinct staining of certain animal tissues can be obtained by substituting celestin for haematoxylin. Snodgress *et al.* (1972) have presented differential staining methods based on aniline acid fuchsin–aniline blue, ammoniacal silver hydroxide, methenamine silver, periodic acid–Schiff, alcian blue, colloidal iron, etc.

It is apparent from the above discussion that polychromatic staining combinations can provide satisfactory results after polyester or epoxy embedment. In other words, differential staining of major cell components can be obtained in resin sections by using various staining combinations. However, the success in obtaining differential staining is dependent upon the type of fixative and buffer, pH, temperature, concentration of the staining solution, duration of staining, section thickness and type of embedding medium. The binding capacity, for instance, of basic stains for imparting differential staining appears to be pH-dependent. Thus, differential staining can be obtained by employing a combination of two or more basic stains which show maximum binding at different pH levels.

The maintenance of optimum temperature during staining is also important in order to achieve differential staining. Fluctuation from the optimum temperature can cause overstaining, understaining or destaining. Various staining methods apparently differ in their sensitivity to changes in temperature. Since cell structures do not all stain with equal rapidity, the intensity of staining should be critically evaluated and staining stopped when the appropriate intensity of staining has been achieved. For instance, collagen is stained by basic fuchsin much faster than are brush borders. The duration of staining is determined in part by the concentration of the staining solution, section thickness and degree of differentiation desired.

The kind of buffer used as a vehicle with a fixative is also important in staining. While the importance of pH in staining has been recognised in the past, little attention has been paid to the possible effect of a certain type of buffer on the staining reaction. It is evident that much more information is needed to elucidate and establish the exact role of a buffer in the staining reaction.

As stated earlier, the type of fixative employed is one of the factors which influence the intensity of staining of thick sections. It is apparent that tissues fixed with aldehydes will show negative reaction for lipids, since they are extracted during dehydration. On the other hand, differential staining of thick sections of OsO_4-fixed tissue may be difficult to achieve, because the fixative may be present, in excess, in the form of reduced osmium. Moreover, the presence of reduced osmium may slow down the penetration of stains into the sections. Osmium may also block certain reactive groups in the tissue. Furthermore, the presence of excess osmium may result in a cloudy, irregular staining, especially when haematoxylin is employed. Fixation with glutaraldehyde is therefore preferred, provided that lipid retention is not important. It is known that glutaraldehyde fixation, with or without OsO_4, enhances haematoxylin staining.

When the use of OsO_4 as a post-fixative is necessary, the reduced osmium is usually removed from the tissue sections to facilitate staining. A brief exposure to solutions of strong oxidants such as hydrogen peroxide, periodic acid or performic acid is sufficient to remove reduced osmium from tissue sections.

Although the exact role of oxidation in permitting staining of resin sections is not clear, several explanations can be offered: (1) the removal of reduced osmium bound to basophilic structures may make these structures available for subsequent staining; (2) peroxidation may attack the polymeric structure of the embedding resin and thus increase the porosity of the sections; (3) increased stainability is due to oxidative uncoupling of chemical bonds between the embedding resin and reactive groups in the tissue specimen.

Peroxidation by PA is carried out by exposing the sections to its aqueous solution (0.5%) for 10 min at 60 °C. Peroxidation by performic acid is accomplished by exposing the sections to its solution for 10 min prior to staining; only freshly prepared solutions should be used. The solution is prepared as follows:

Hydrogen peroxide (30%)	15 ml
Sulphuric acid (conc.)	1.5 ml
Formic acid (97%)	133.5 ml

Peroxidation with performic acid has been employed to facilitate differential staining by alcian blue–PAS–orange G or by aldehyde fuchsin–orange G.

Peroxidation by H_2O_2 is carried out by covering the section with 1–2 drops of the oxidant for 1 min at room temperature. The solution should have a pH of 3.2, and it should be checked for proper pH prior to use. The solution is prepared as follows:

Distilled water	30 ml
Hydrogen peroxide (30%)	15 ml
Sulphuric acid (0.01 N)	1.8 ml

The concentration and duration of treatment with the above-mentioned oxidants are critical, since other sensitive groups present in the tissue may be destroyed by the oxidation. Haematoxylin, for example, is thought to stain poorly following H_2O_2 treatment. A mild oxidant, such as paracetic acid, is therefore preferred. Another mild oxidant, Oxone (a monopersulphate of undefined composition used in laundry soaps), can also be employed to obtain specific staining. Sections are oxidised in 5% Oxone in distilled water for 1–4 h at room temperature. The $KMnO_4$–oxalic acid sequence is yet another method used to remove the excess of reduced osmium. To accomplish oxidation, sections are exposed to 0.1% solution of $KMnO_4$ for 1–3 min, followed by bleaching with 1% solution of oxalic acid.

Double fixation with an aldehyde and OsO_4 is most commonly employed to process tissues for electron microscopy. The following procedure (Snodgress *et al.*, 1972) is quite reliable in removing the polymerised resin as well as reduced osmium.

Solutions

A: 0.5% potassium hydroxide in 100% methyl alcohol

B: 10% picric acid in acetone

C:	Solution A	10 ml
	Acetone	20 ml
	Benzene	20 ml
D:	Solution A	20 ml
	Benzene	20 ml
E:	30% H_2O_2	1.5 ml
	80% ethanol	40 ml

Procedure

(1) Immerse slides in solution C for 5 min.
(2) Rinse in equal parts of acetone and benzene—two changes of 1 min each.
(3) Rinse in solution D for 2 min.
(4) Transfer to solution B for 1–2 min.
(5) Rinse in 100% ethanol—two changes of 1 min each.
(6) Rinse in 95% ethanol—two changes of 1 min each.
(7) Rinse in 80% ethanol for 1 min.
(8) Bleach in solution E for 3–5 min.
(9) Rinse in 80% ethanol for 1 min.
(10) Rinse in distilled water before staining.

Besides removing the reduced osmium, other methods employed to circumvent the difficulties encountered after OsO_4 fixation are: (1) the use of heat to facilitate the penetration of the stains, (2) the use of basic stains at high pH and (3) the use of metal stains (Berkowitz *et al.*, 1968).

Some evidence is available suggesting that polyester resins are more permeable to acid stains, while epoxy resins are more permeable to basic stains. It has also been suggested that polyester sections usually show a better differentiation and a higher image contrast than do epoxy resins (Bryant and Watson, 1967). De Martino *et al.* (1968) also indicated that Vestopal preserves the original shape of lipid droplets better than does Epon or Araldite, independently of fixation and dehydration. Epon and Araldite tend to induce excessive shrinkage of lipid droplets. However, it should be noted that both polyester and epoxy resins are lipid-solvents. In general, being an aliphatic resin, Epon sections are more permeable to staining solutions than are sections of Araldite, which is aromatic in nature. In this connection, it is admitted that the available evidence is inadequate to reach a definite conclusion.

Stained sections can be viewed with the light microscope without mounting them with a coverslip. If mounting is needed, a neutral medium is preferred, but stain fading is a problem. Among the commonly used mounting media, stain fading is maximum in Canada balsam and least in immersion oil. A medium amount of fading occurs in synthetic media (Permount, Harleco, Gurr Depex and Lipshaw). Heat, light and xylene catalyse the fading.

SELECTED STAINING METHODS FOR SEMITHIN SECTIONS

Numerous stains and staining combinations have been applied to semithin sections of resin-embedded tissue for light microscopy. The type of tissue and embedding medium, the method of fixation and the objective of the study apparently determine the staining method to be used. Rapid staining of epoxy sections for survey purposes can be obtained by 1% azure II in 1% borax, 1% toluidine blue and 1% azure blue in 1% borax, or 1% basic fuchsin in 50% acetone. A few selected methods of polychromatic staining are given below. For a detailed review of staining methods, the reader is referred to Hayat (1975) and Lewis and Knight (1977).

Azure B for Plant Tissues (Hoefert, 1968)

Procedure

(1) Cut 0.5 μm sections.
(2) Transfer the section to a drop of 10% acetone on a clean slide and heat slightly to expand the section and evaporate the acetone.
(3) Flood the section with 0.2% azure B solution in 1% sodium bicarbonate at pH 9, place the slide on a hot-plate at 50 °C for 2–5 min and then rinse rapidly in tap-water.
(4) Air-dry and mount the coverslip in Epon; do not use heat for polymerisation.

Results

Blue: nucleoli and primary walls.
Light blue: secondary walls.
Grey: cytoplasm.
Blue-grey: nuclei.
Blue-green: chloroplasts.

Basic Fuchsin and Methylene Blue (Sato and Shamoto, 1973)

Staining solution

Sodium phosphate (monobasic)	0.5 g
Basic fuchsin	0.25 g
Methylene blue	0.2 g
Boric acid solution (0.5%)	15.0 ml
Distilled water	70.0 ml
Sodium hydroxide (0.72%, pH 6.8)	10.0 ml

Although a pH range of 6.8–8.0 yields best results, precipitation may occur in the staining solution at pH 8.0 or above. Methylene blue staining is excessive at pH above 8.0. Although the staining solution can be used for several months, it should be shaken before use.

Procedure

(1) Cut sections 0.5–2 μm thick and heat-fix them on a glass slide.
(2) Place 1 ml of the staining solution on the slide with the attached sections, and heat for 4–5 s at 45–50 °C.
(3) Rinse in running tap-water and then allow to dry at room temperature.
(4) Cover sections with a drop of immersion oil and coverslip, and seal with the Epon mixture.
(5) When staining is insufficient, the procedure can be repeated.

Results

Red: mitochondria, myelin, and lipid droplets.
Pink: erythrocytes in blood vessels, glomeruli and tubules in kidney, smooth muscle cells, axoplasm and chondroblasts.
Brilliant pink: collagen.
Reddish purple: elastic lamina and zymogen granules.
Bluish purple: nuclei.
Blue: collagen, elastic lamina and connective tissue.

Methylene blue–Azure II–Basic Fuchsin

Solution A

Methylene blue	0.13 g
Azure II	0.02 g
Glycerol	10 ml
Methanol	10 ml
Phosphate buffer (pH 7)	30 ml
Distilled water	50 ml

Solution B

Basic fuchsin	0.1 g
Ethanol (50%)	10 ml
Distilled water	90 ml

Immerse slides in solution A for 20–30 min at 65 °C, rinse in water, immerse in solution B for 1–3 min and rinse in water.

Results

Dark blue: nucleus and goblet cells.
Blue: striated muscle and endothelial cells.
Light blue: microvilli, smooth muscle, cytoplasm and lipid droplets.
Blue green: RBC and fat.
Red: glycogen, collagen and mucous granules.
Pink: connective tissue cells and fibres, collagen and cartilage.
Violet: elastin, mucous granules and leucocytes.

Haematoxylin–Malachite Green and Basic Fuchsin (Berkowitz *et al.*, 1968)

Staining solutions

Iron chloride haematoxylin (Lillie, 1965)

Distilled water	75 ml
HCl (concentrated)	0.5 ml
Ferric chloride crystals ($FeCl_3 \cdot 6H_2O$)	0.62 g
Ferrous sulphate crystals ($FeSO_4 \cdot 7H_2O$)	1.12 g

After these salts are dissolved, add 25 ml freshly prepared 1% alcoholic solution of haematoxylin.

Malachite green

Ethanol (30%)	100 ml
Azure B	0.4 g
Malachite green	1.0 g
Aniline	1 ml
Phenol crystal	1 g

The stain does not dissolve completely; refrigerate for 3 days and use supernatant. Do not filter the solution.

Basic fuchsin

Ethanol (50%)	5 ml
Basic fuchsin	2 g
Distilled water	45 ml

Procedure

(1) Cut sections 0.5–2.0 μm thick and heat-fix them to a clean glass slide.
(2) Immerse the slide in 2% NaOH in absolute alcohol for 10 min to dissolve the resin.
(3) Wash the slide in four changes of absolute ethanol for 4 min each and then place it in phosphate buffer (pH 7.0) for 5 min.
(4) Rinse the slide in three changes of distilled water and then place in buffer at pH 4 for 5 min.

(5) Wash for 5 min in tap-water and stain in iron haematoxylin for 20 min (staining should be controlled microscopically).
(6) Rinse rapidly in tap-water and examine (chromatin and mitochondria must be light-grey and the cytoplasm clear; if overstained, destain with 4% ferric alum).
(7) Stain in malachite green for 2 min.
(8) Rinse rapidly in tap-water.
(9) Immerse in basic fuchsin on a hot-plate at ~37 °C for 20–120 s and rinse rapidly in tap-water.
(10) Air-dry and mount in Araldite (press coverslip to obtain the thinnest possible layer of plastics and polymerise at room temperature).

Note: The duration of staining is dependent upon section thickness.

Results

Bright blue-green: myelin, lipid droplets, nucleoli and oligodendrocytes.
Purplish pink: nuclei and astrocytes.

Haematoxylin and Phloxine B (Shires *et al.*, 1969)

Procedure

(1) Cut sections 0.5–2.0 μm thick.
(2) Transfer the sections to a 1 in × 3 in glass slide freshly coated with a thin layer of albumin adhesive.
(3) Dry the sections by placing the slide on a hot-plate at 50–60 °C (drying must be complete) and then immerse it in xylene for 1 h.
(4) Transfer the slide to a mixture (1:1) of xylene and absolute ethanol for 2 min and repeat.
(5) Transfer the slide to 95% ethanol for 2 min and then to three changes of distilled water for 10 min (be sure that the alcohol is completely removed).
(6) Immerse OsO_4-fixed tissues in 0.1% aqueous $KMnO_4$ for ~2 min, followed by bleaching with 1.0% aqueous oxalic acid until brownish residues are removed.
(7) Wash with three changes of distilled water for ~7 min and then incubate in haematoxylin (Harris' or Ehrlich's) at 60 °C for 12–24 h.
(8) Wash thoroughly in six changes of distilled water for ~15 min (slightly alkaline water—pH 9.0—is preferred for removing all of the acetic acid of the staining solution and developing tissue basophilia).
(9) Counterstain with 0.5% aqueous phloxine B for ~10 min.
(10) Remove excess stain with filter paper and allow the sections to air-dry.
(11) Wash in two changes of absolute ethanol until unbound stain is removed.
(12) Clear with a mixture (1:1) of xylene and absolute ethanol for 2 min, followed by three changes of xylene for 1 min each.
(13) Apply a coverslip with synthetic resin.

Results

Pale blue to deep blue or black: chromatin, nucleoli, karyolymph, basophilic cytoplasm, mitochondria, plasma and nuclear membranes, anisotropic myofibrils, mast cell granules and elastic membranes of blood vessels.
Pale pink to dark red: collagen fibrils, reticulum, goblet cell mucins, hyaline cartilage matrix, stereocilia of epididymal epithelial cells, cytoplasm of some cell types and erythrocytes.
Varying shades of green: fat droplets and perichondrocyte matrix.
Uncoloured: Golgi complex.

Methyl Green and Methyl Violet (Sievers, 1971)

Staining solution

Methyl green	1 g
Borax	1 g
Distilled water	100 ml

Methyl violet is the usual contaminant of commercial methyl green. Only a freshly prepared solution should be used. It should be noted that methyl green fades within a few days after staining.

Procedure

(1) Cut sections 0.3–2.0 μm thick and stain for 15–20 min at room temperature.
(2) Rinse briefly in distilled water.
(3) Dry the sections on a clean glass slide on a hot-plate at 70 °C.
(4) Mount coverslip with a synthetic resin.

Results

Dark green: nucleoli and chromatin.
Light green: cytoplasm.
Blue-violet: collagenous and elastic connective tissue.

Toluidine Blue and Acid Fuchsin

Procedure

(1) Cut sections 1 μm thick and heat-fix them on a clean glass slide.

(2) Stain with 0.5% toluidine blue in phosphate buffer (pH 7.0–7.4) for 5–8 min at 85 °C.
(3) Wash in tap-water and oxidise with 2% periodic acid for 0.5–2.0 min at room temperature.
(4) Wash and stain with 0.5–1.0% acid fuchsin for 5–8 min at 85 °C.
(5) Wash, dry and mount.

Results
- Dark blue: nuclei.
- Bright blue: muscle.
- Red: collagen.
- Deep red: elastica.

5 Negative Staining

INTRODUCTION

Negative staining is a simple and rapid method of studying the morphology and structure of particulate specimens such as viruses, cell components (e.g. ribosomes), cell fragments (e.g. membranes) and isolated macromolecules (e.g. protein). Negative staining allows determination of shapes at the molecular level using high-resolution electron microscopy. Negatively stained specimens often show well-preserved order. The reason is that, in addition to providing contrast, negative stains preserve macromolecular structure as well as minimise irradiation damage to the specimen.

The specimen is embedded in a negative stain—a metal such as phosphotungstic acid (PTA). On drying, the electron-dense metal atoms envelope the specimen. The difference between the specimen and the surrounding heavy metal atoms with respect to density produces the necessary contrast. The specimen appears light, surrounded by a dark background of dried stain. The electron beam passes through the low electron density of the specimen, but not through the metallic background. The specimen substructure is revealed by the penetration by the stain into its holes and crevices. In other words, the structure is inferred from the distribution which the specimen imposes upon the stain. The clarity of specimen detail depends on the degree to which the stain remains amorphous as it dries, as well as on the thickness of the dried negative stain envelope.

The negative staining method is based on the principle that there is no reaction between the stain and the specimen. This absence of reaction is accomplished by using the stain at a pH at which the attraction between proteins and stain is negligible. In positive staining, on the other hand, the atoms of heavy metals are attached to specific sites on the stained molecule. Some of the commonly used metal salts (e.g. uranyl acetate) are applied for both negative and positive staining. Images of exclusively negatively stained specimens are difficult to achieve, because positive staining is always a small part of these images. The details of contrast mechanism responsible for positive staining are presented in Chapter 4 and elsewhere (Hayat, 1975).

Many factors affect the appearance of negatively stained specimens. The shape and size of specimens are influenced by both the mode of negative stain application and the stain itself. Pre-fixation with an aldehyde allows internal components to be revealed. The pH and concentration of the stain, the concentration of the specimen, temperature, duration of staining, the nature of the support film and the length of time it takes for the specimen to dry on the support film influence the final image of the specimen.

Rigorous precautions should be taken in handling infectious materials. All potentially infectious specimens must be inactivated. Ultraviolet irradiation kills most viruses. It is desirable to keep a container of disinfectant (e.g. hypochlorite solution containing an anionic detergent) on the workbench to dispose of contaminated materials such as filter paper.

NEGATIVE STAINS

An ideal negative stain should show minimal interaction with the specimen; high solubility in water; high density, melting and boiling points; and resistance towards granulation under the electron beam. In view of these and other requirements, only a few substances qualify as negative stains. The choice of a stain is determined by the objective of the study and the type of specimen under examination. The results obtained with one negative stain should be checked with those yielded by a second stain, for each stain reveals somewhat different aspects of specimen morphology. In fact, a negative stain may give excellent results with one specimen, but totally unsuccessful results with another type of specimen. This becomes a serious limitation when a specimen is negatively stained for the first time. The solution is to use more than one type of stain. Uranyl acetate and PTA are the most commonly used negative stains. These negative stains are discussed below; other negative stains are discussed elsewhere (Hayat, 1986b; Hayat and Miller, 1989).

POTASSIUM PHOSPHOTUNGSTATE

Potassium phosphotungstate ($24WO_3 \cdot 2H_3PO_4 \cdot 48H_2O$) was the first stain specifically used as a negative stain (Hall, 1955). The water content of this compound varies appreciably. It has an anhydrous weight density of ~4, which is three times greater than that of the specimen (density is ~1). It is available as white or slightly yellowish-green, slightly efflorescent crystals or crystalline powder, and is soluble in ~0.5 part water; it is also soluble in alcohol. The size of PTA polyanions (~1.5 nm) restricts their penetration into intramolecular cavities of smaller dimensions. Under the electron beam, PTA migrates less than uranyl acetate.

Phosphotungstate is prepared by titrating phosphotungstic acid to a pH near nautrality with 0.1 M KOH. At neutral pH or slightly above, there is little interaction between the stain and protein components, since this pH is above the isoelectric point of the protein, where PTA has a net negative charge. This stain can be used at a pH range of 6.7–8.0 and at concentrations of 0.5–3.0%. It is stable for long periods if care is taken to prevent a fall in pH by dissolution of CO_2.

The stain has the advantage of tolerating high concentrations of non-volatile salts (e.g. phosphate buffer) without adverse effects. Staining and other effects of PTA on specimens (e.g. bacteriophages) are predictable. In other words, the results of PTA staining are consistent. The protein moiety of bacteriophages is well preserved and easily detected after staining. However, PTA yields uneven staining of bacteriophages; the stain accumulates more around the head than around the tail. Phosphotungstate shows relatively low negative contrast compared with uranyl acetate and poor adhesion of bacteriophages to the support film (Ackermann, personal communication). This stain may cause disruption of phage heads. It lacks a stabilising effect on membranes and mitochondria, and may damage them. Destructive effects of PTA on unfixed erythrocyte membranes have been reported (Reynolds, 1973; Harris *et al.*, 1974). Unlike uranyl acetate, PTA does not have any fixation effect at the pH at which it is usually used, and may favour dissociation of quaternary protein structure into smaller units. However, PTA may show some fixation effect at a low pH (3.5–5.0). Specimens stained with PTA are not stable for any length of time unless stored under vacuum. Phosphotungstate is recommended for negative staining of viruses, but should always be used in conjunction with another negative stain.

URANYL ACETATE

Uranyl acetate ($C_4H_6O_6U$) was introduced as a negative stain by Van Bruggen *et al.* (1960). The density of uranyl acetate is 2.87 g cm^{-3} and the diameter of its ions is 0.4–0.5 nm. It is available as a dihydrate, yellow, crystalline powder, and is incompletely soluble in 10 parts of water, owing to the presence of basic salt. Its solutions can be prepared in concentrations ranging from 0.2% to 2.0%. This compound takes about 20–30 min to dissolve in distilled water. The staining solution can be stored for ~3 weeks in a dark-brown bottle at 4 °C; it can be stored for longer periods if oxygen is excluded—that is, by completely filling the bottle. However, the exclusion of oxygen is not essential to obtain good results. Uranyl acetate can be used at a pH range of 2.0–5.0 (usual pH is 3.7–4.5). Thus, it is used below the isoelectric point of the protein (pH 4.5), where the net charge is positive. Uranyl salts are unstable at pH values higher than 6.0. Owing to its radioactivity, uranyl salts should be handled with utmost care (Hayat, 1986b). It is the most widely used negative stain for specimens other than viruses.

Uranyl acetate also acts as a fixative at the pH at which it is used for negative staining. Uranyl acetate stabilises lipids and thus minimises the deleterious effect of drying on some virus particles and cell organelles. Ribosomes associated with microsomes are preserved and negatively stained much better with uranyl acetate than with PTA (Mrena, 1980).

Uranyl acetate spreads easily over a wide range of specimen particle concentrations. In its presence, adhesion of particles to the support films is good, and relatively high negative contrast is produced. Uranyl acetate gives higher contrast than that yielded by PTA. Background granularity is minimal but increases when exposed to the electron beam. Uranyl salts produce an under-focused refractive grain of higher contrast than that produced by other stains. Both the acid pH and the uranyl ions (which bind to carboxyl and phosphate groups) contribute to surface staining by uranyl acetate. Effects of various negative stains on cell wall morphology of Gram-negative bacteria indicate that in 2% uranyl acetate (pH 4.2) the cell surface is mostly smooth (Fig. 5.1A); there is little stain penetration into the cell wall (Wells and Horne, 1983). On the other hand, the cell surface of such bacteria appears highly convoluted with 3% ammonium molybdate (pH 6.5) or 2% methylene tungstate (pH 6.5); these two stains penetrate the surface grooves (Fig. 5.1B). However, uranyl acetate can penetrate the surface membranes after 3–5 min, and details of the internal organisation of a cell organelle or virus particle can be detected. For example, oxysomes on the inner membranes of mitochondria can be easily seen.

Certain bacteriophage parts are seen better in uranyl acetate than in PTA. Pentagonal heads, indicative of an icosahedral shape, are more easily seen in uranyl acetate than in PTA (Ackermann *et al.*, 1974). Specimens prepared with uranyl acetate are stable and can be stored for a long time; these characteristics are important for teaching and demonstration.

The unpredictability of staining is one of the major disadvantages of uranyl acetate. In the case of bacteriophages, it may give three kinds of staining on the

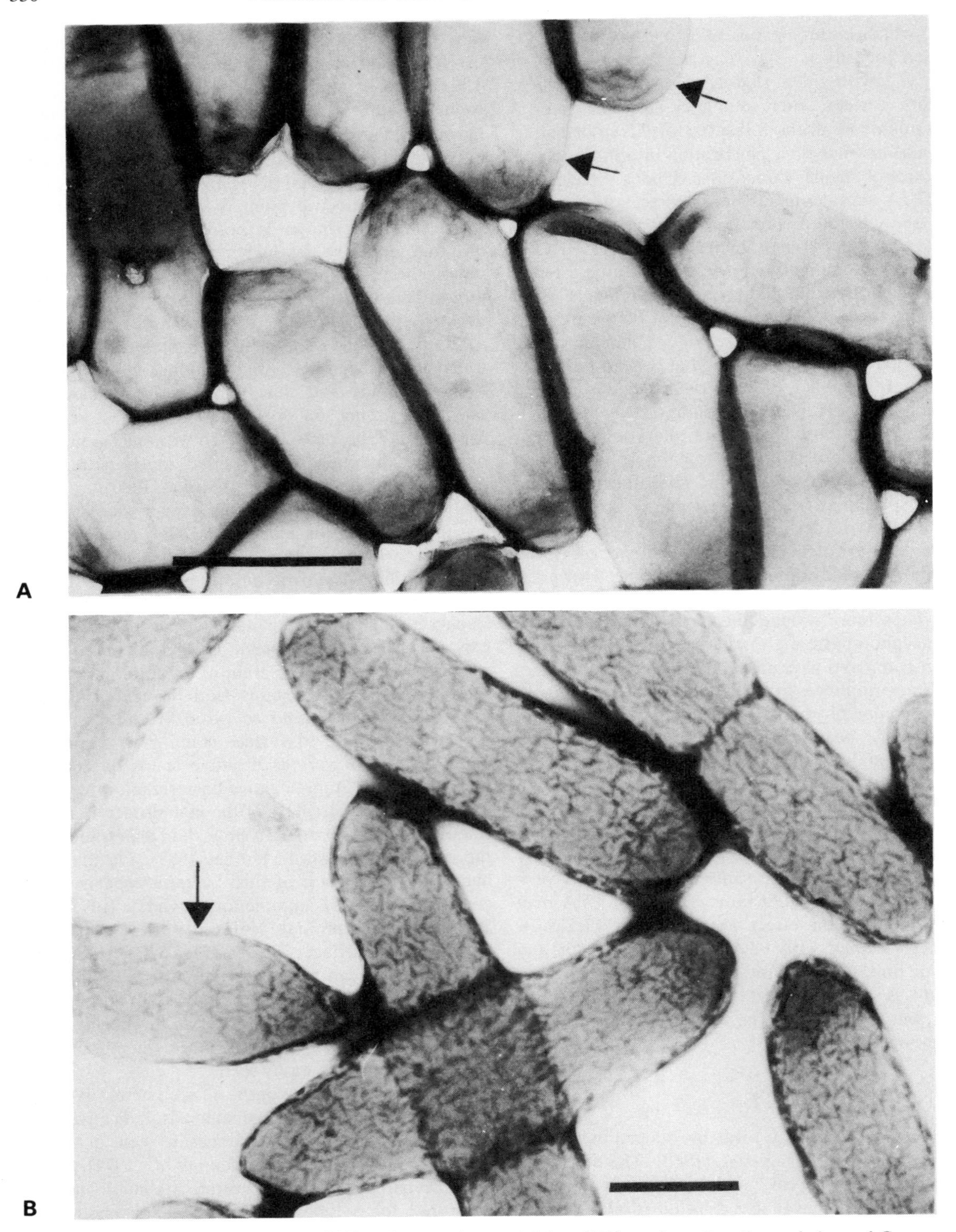

Figure 5.1 Effects of uranyl acetate (2%) and ammonium molybdate (3%) on the cell wall morphology of Gram-negative bacteria. A: Very little penetration by uranyl acetate into the cell surface is clear. B: Ammonium molybdate has penetrated into the surface of the bacteria. From Wells and Horne (1983)

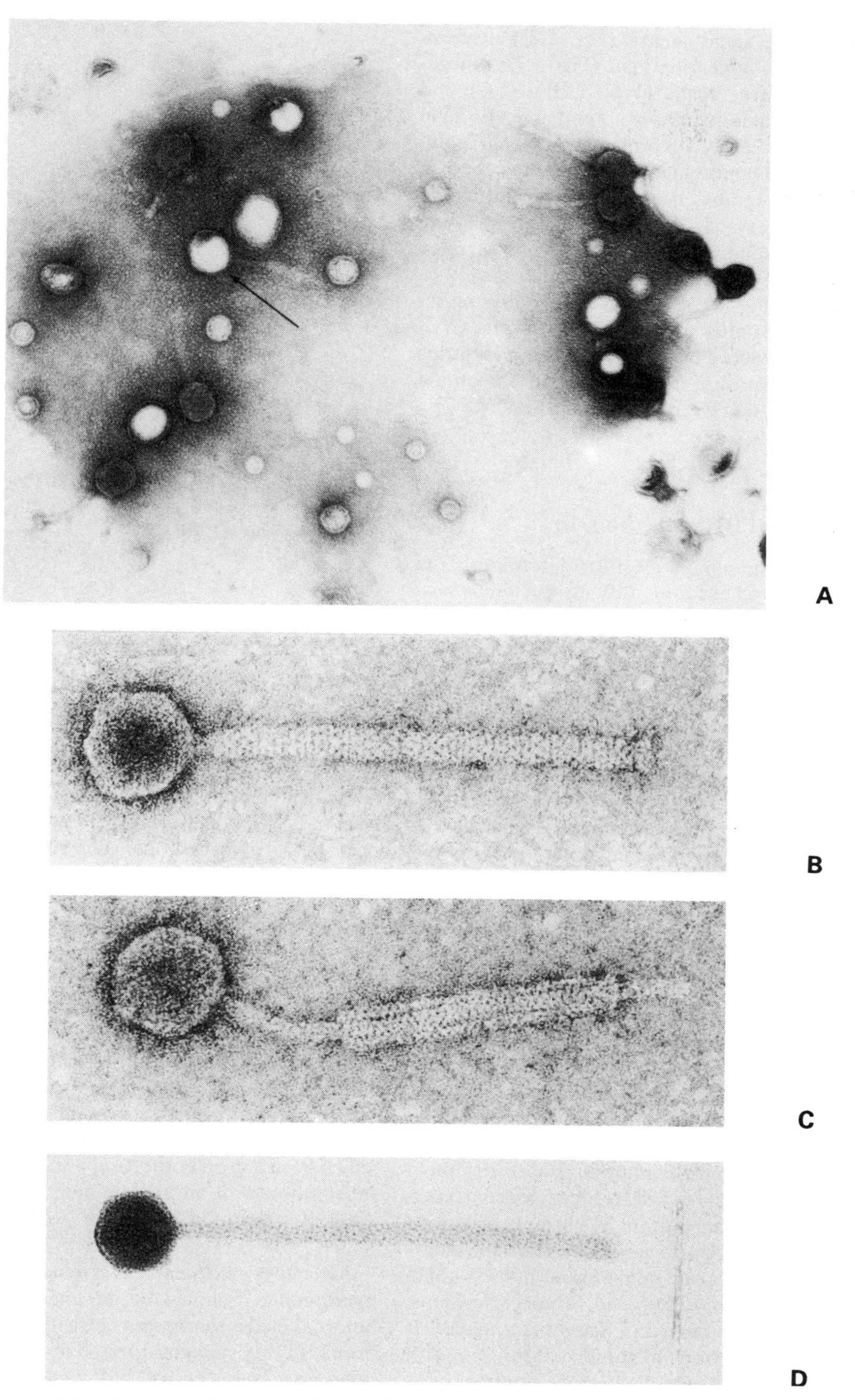

Figure 5.2 The unpredictability of negative staining by uranyl acetate is shown by bacteriophages. Three kinds of staining on the same grid are shown. See text for details. Courtesy W.-C. Ackermann

same grid and the same square (Fig. 5.2); negative staining with heads appearing light (Fig. 5.2A), negative staining with 'grey heads' (Figs. 5.2B and C), and positive staining that is invariably accompanied by head shrinkage (Fig. 5.2D) (Ackermann, personal communication). Purely proteinaceous structures tend to be swollen and sometimes hard to see. The formation of uranyl acetate crystals of various dimensions and shapes (e.g. flat solid patches, blades with serrated edges, and long or short needles) is another disadvantage of this stain (Fig. 5.3). Uranyl acetate crystallises in intramolecular cavities upon complete drying. Crystal formation is favoured by elevation of the pH during fixation and/or exposure to irradiation. Sometimes crystals can be avoided by preparing another grid.

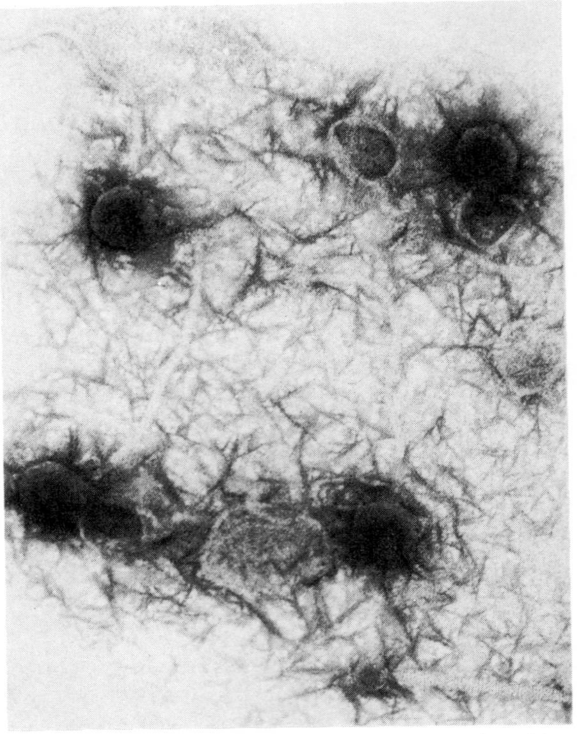

Figure 5.3 Crystallisation of uranyl acetate on the grid on exposure to the electron beam. Crystal needles are scattered among the bacteriophages. Courtesy W.-C. Ackermann

GENERAL METHODS

BASIC CONSIDERATIONS

Because negative staining is used for determining the basic structure as well as for identifying specimens such as viruses, the following remarks apply primarily to such specimens. Most virus particles range in diameter from 20 nm to 500 nm and can be easily studied with negative staining. The optimal concentration of virions needed in a specimen for electron microscopy depends on both the type of specimen and the type of virion. The ratio between virus and contaminants is important in determining the optimal virus concentration. A concentration of 10^5 virions/ml in urine is satisfactory, for this specimen has few contaminants, whereas the same concentration is unsatisfactory for sputum, which is rich in contaminants. A concentration of 10^5 small cubic virions/ml may be unsatisfactory, unless immunoelectron microscopy is used (Almeida, 1980). This method produces antigen–antibody aggregates that are larger than unaggregated virions. For the majority of viruses, a concentration of 10^6 virions/ml of starting viral material is required if the specimens are to be easily visualised with negative staining.

Both viable and non-viable virus particles in a sample are seen in the electron microscope. A specimen containing 10^6 virions/ml may contain as few as 10^3 infectious virions/ml (Almeida, 1980). Since concentration of almost all components present in a sample occurs on the grid during the drying phase, any low-molecular-weight proteins and organic salts present in the sample should be minimised or eliminated. These extraneous materials tend to obscure the specimen image. If specimens are taken from a liquid culture, the grid should be briefly rinsed with distilled water after specimen adsorption and before staining, in order to remove most of the extraneous materials. For dilution of a concentrated sample, distilled water should be used. It lyses cellular structures but usually leaves viruses undamaged. If needed, concentration of the virus by centrifugation should be carried out at the lowest speed (15 000 *g* for 1 h). After the final pellet has been obtained, the last drops of fluid in the tube should be removed; otherwise, when the pellet is resuspended in distilled water, the low-molecular-weight material present will contaminate the specimen.

For general purposes, the pieces of equipment and supplies needed are as follows: fine forceps, Formvar–carbon-coated 400-mesh copper grids, glass slides, Pasteur pipettes, torn pieces of filter paper, a washing bottle containing distilled water, 4% solution of PTA (pH 6.0) in a drop bottle, and a container of hypochlorite solution with an anionic detergent. For acid-labile specimens the pH of the stain is raised to 8.0 with 1 N KOH.

After the specimen grid is placed into the electron microscope, it should be left there for some minutes before the electron beam is turned on, because the grid needs to be vacuum-dried before being irradiated; otherwise, wet stain will boil under electron bombardment and impair the structural details. It is desirable to first stabilise the grid by irradiation from the over-focused condenser system for some minutes before it is focused (Lickfeld, 1976). This practice prevents sagging (due to heating during focusing) of the square of the grid that is directly under the electron beam. Focusing should be performed at the lowest possible beam current giving adequate screen brightness. An image intensifier is very helpful.

ONE-STEP (SIMULTANEOUS) METHOD

The one-step method is the simplest and most rapid means of negative staining, and it is especially suitable for specimens that are unstable at low ionic strength. The choice of this method depends on the stability of the specimen in the stain environment. Certain unfixed specimens, when mixed with the stain, tend to be damaged.

A 1:1 mixture of specimen and stain is used. As a guide, 1 volume of virus suspension (10^6–10^{10}/ml) is mixed with 1–2 volumes of 2% PTA (pH 7.5). If needed, an appropriate volume of a wetting agent (e.g. bacitracin in a final concentration of 50 μg/ml) can be added to this mixture. A carbon-coated (6–10 nm) 400-mesh copper grid is picked up by the edge with self-clamping forceps and placed on a flat surface with the support film side up. With the use of a fine Pasteur pipette (or a 5 μl Eppendorf pipette), a small drop of the mixture is placed onto the grid to form a 'bead' extending nearly to the edge of the grid. The pipette is held vertically. After about 10–90 s, the droplet is touched from one side with the torn edge of a filter paper to remove excess liquid, leaving a thin monolayer, which is allowed to dry at room temperature; alternatively, the grid can be vacuum-dried. The amount of stain left on the grid is empirically determined for each type of specimen. The residual film rapidly dries and is ready for examination in the electron microscope. The grid should be examined as soon as possible. When uranyl acetate is used, the grid can be stored without adverse effects.

The negative stain is deposited very rapidly over both the exposed support film and the surfaces of the specimen, and the stain penetrates any existing cavities or crevices so that they can be observed. If the specimen is allowed to remain in contact with the staining solution for longer periods, binding of the stain with the specimen can occur, resulting in positive staining. A lapse of as little as 90 s can cause positive staining. Positive staining is usually complete after 2 min.

The 'drop' method described has a limitation in that small particles (e.g. capsomeres) and large particles (e.g. viruses) adsorb to the film at different rates. Small particles adsorb rapidly because they exhibit faster Brownian motion. Erroneous conclusions may thus be drawn regarding the relative proportions of particles of different sizes in a sample. This difficulty can be circumvented by using a nebuliser to deposit the mixture—an approach termed the spray method. However, because of the danger of infectivity, the spray method is not recommended.

NEGATIVE STAIN–CARBON METHOD (Horne and Pasquali-Ronchetti, 1974)

The negative stain–carbon method allows the preparation of viruses and other small particulate specimens from very highly concentrated or crystalline suspensions. It provides ordered paracrystalline monomolecular arrays of protein molecules of viral or non-viral origin. Thus, image analysis by optical diffractometry, filtering and image reconstruction is facilitated, and a greater quantitative understanding of the molecular structure is achieved. The method provides a means for obtaining a very small specimen thickness (1.5–1.0 nm) for high-resolution viewing.

Suspensions containing virus material (e.g. brome mosaic virus or cowpea chlorotic mosaic virus) are added to an equal volume of 3% ammonium molybdate (pH 5.2) (or 4% methylene tungstate) and then mixed mechanically in a Whirmixer for 10 s (3% ammonium molybdate might etch the virus particles at high pH). The final concentration of the virus in the negative stain ranges from 0.4 to 8.0 mg/ml, depending on the concentration of the virus in the original suspension. A small volume of the negative stain and virus mixture (0.2 ml) is spread carefully on the surface of a precut and freshly cleaved mica sheet having one end pointed and the other flat (Figs. 5.4 and 5.5). The pointed end assists in the initial release of the final specimens at the liquid–air interface (as explained later). The excess liquid is carefully removed with a pointed filter paper, leaving a

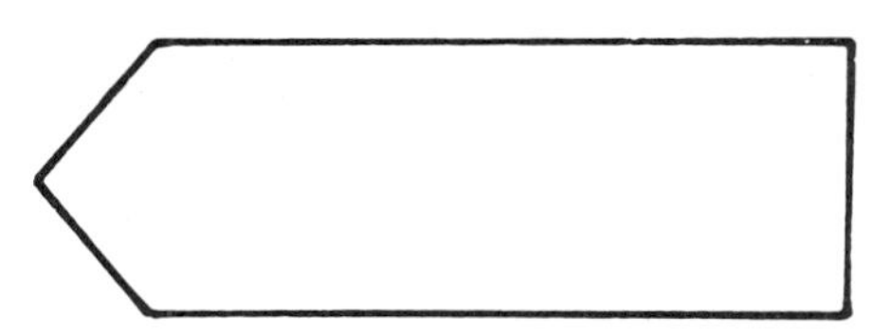

Figure 5.4 Diagram showing the shape of a cut mica piece before cleaving with a razor-blade. The diagram on the top is side view. From Horne and Pasquali-Ronchetti (1974)

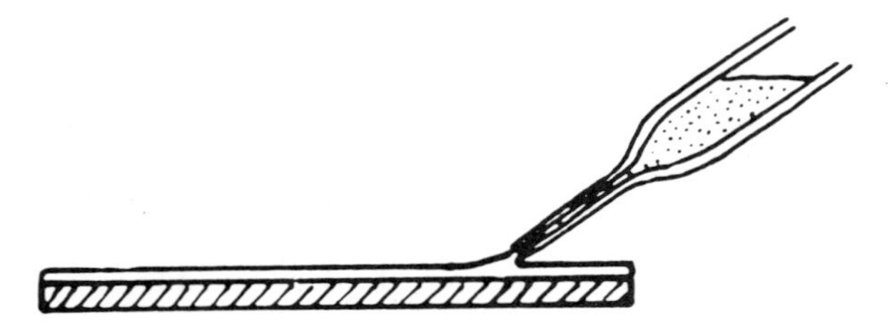

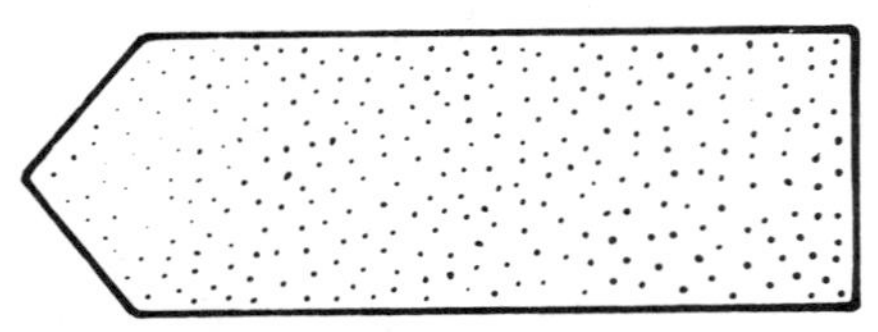

Figure 5.5 The mixture of the first negative stain and virus suspension is spread with a fine pipette on the freshly cleaved mica surface. From Horne and Pasquali-Ronchetti (1974)

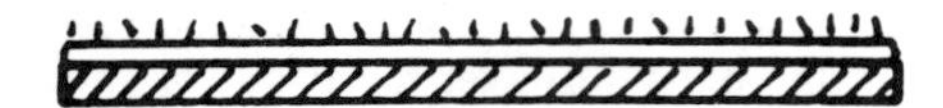

Figure 5.6 Carbon is evaporated onto the air-dried stain and virus. After removal from the evaporator, the carbon surface is carefully marked with four black spots. From Horne and Pasquali-Ronchetti (1974)

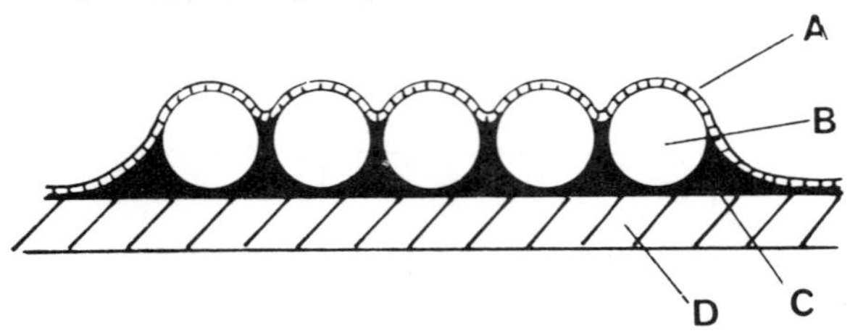

Figure 5.7 Schematic drawing of the mica piece (D), negatively stained virus (B and C), and thin carbon layer (A) before being floated onto the surface of the second negative stain. From Horne and Pasquali-Ronchetti (1974)

thin liquid film at the mica surface, and then allowed to dry at room temperature. After drying, the mica sheet is placed in the evaporator and coated with a very thin layer of carbon (Figs. 5.6 and 5.7). The carbon film is marked by placing four black spots (Fig. 5.6) with a soft felt pen at suitable positions in order to locate the film at a later stage, because the film deposited at the correct thickness is normally invisible to the naked eye. If the film is easily visible after being released from the mica, it is too thick for high-resolution work.

The carbon–virus film is separated from the mica surface according to the standard procedure used for carbon films, except that the film is floated on a second negative stain (e.g. 1% uranyl acetate at pH 4.0), instead of distilled water, in a Petri dish (Fig. 5.8). Grids covered with perforated films are placed underneath the floating specimen film and then raised with a pair of forceps to collect the virus and carbon film.

The success of this method depends on the purity and stability of the viruses in suspension. When virus suspensions stored at 4 °C for 24 h or longer are processed by this method, a considerable reduction is seen in the areas formed by the regular arrays (Fig. 5.9B) (Wells *et al.*, 1981). The addition of polyethylene glycol (PEG) (mol. wt. 6000) to stored virus suspensions results in the formation of extensive continuous areas of crystalline arrays (Fig. 5.9C). About 4 μl of 10% PEG in distilled water is added to 4 μl of virus suspension, and then 100 μl of 3% ammonium molybdate adjusted to pH 5.6 is added. The addition of PEG facilitates the preservation of virus structure.

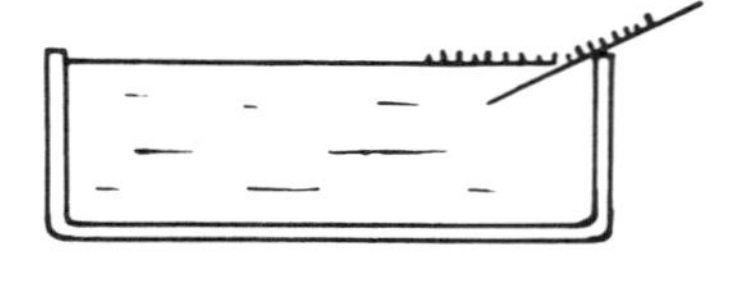

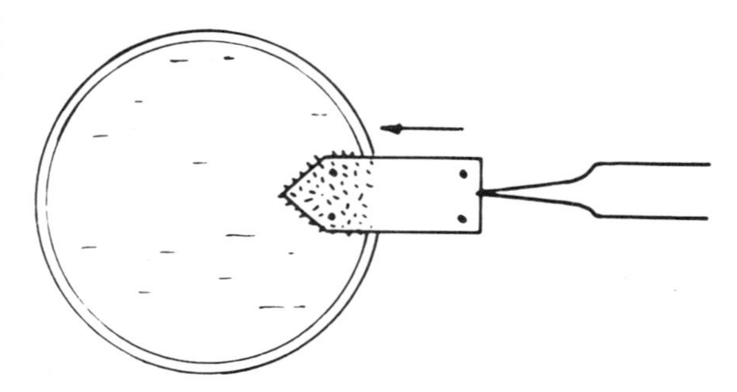

Figure 5.8 The mica piece is slowly immersed into the Petri dish containing the second stain, and the specimen film areas are identified by the four black spots. From Horne and Pasquali-Ronchetti (1974)

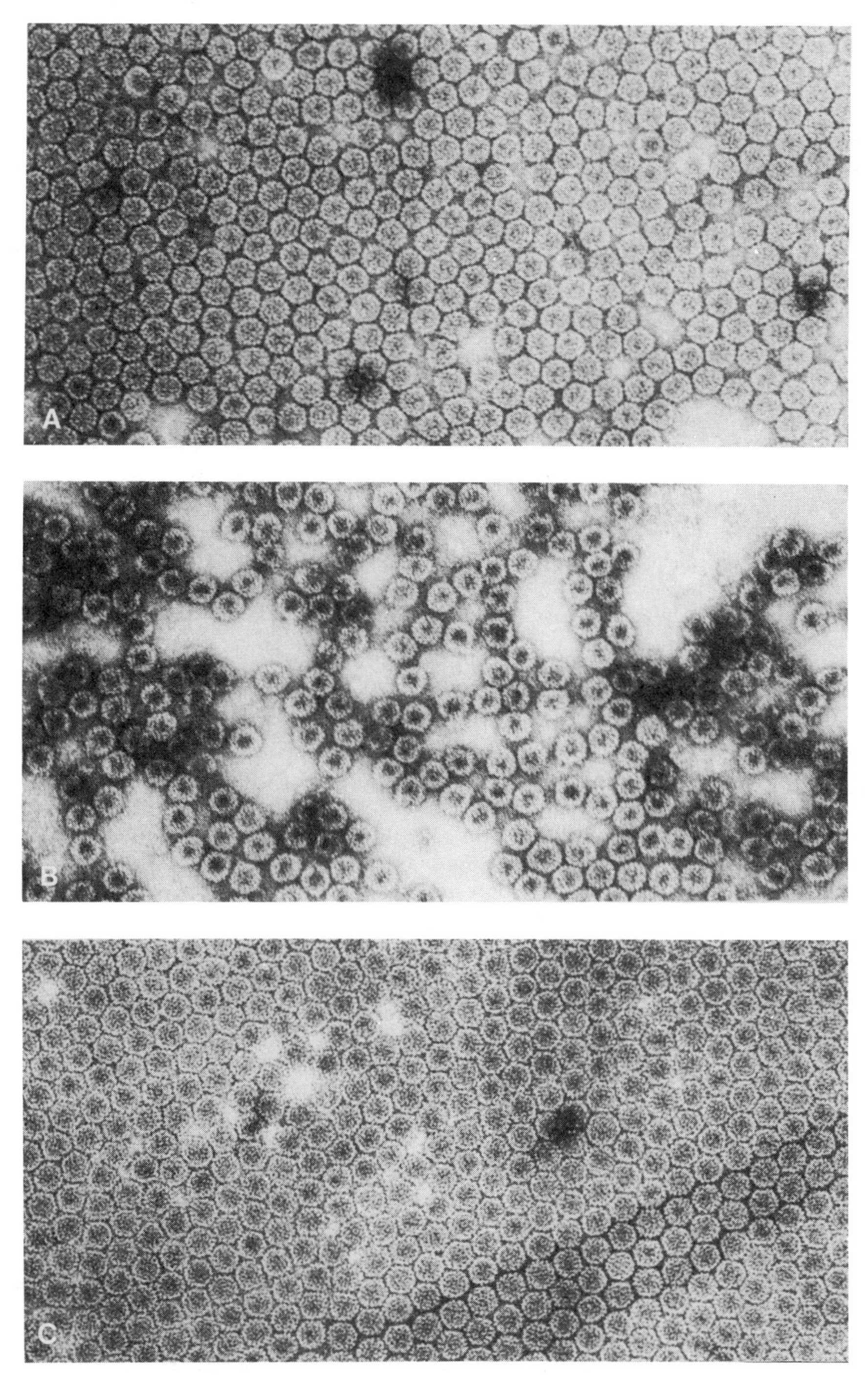

Figure 5.9 Broad bean mottle virus negatively stained with 3% ammonium molybdate (pH 5.6). A: Freshly prepared sample mixed with the stain, showing non-crystalline aggregates at the mica surface. B: The sample was stored for 4 days at 4 °C before staining. The virus particles are randomly dispersed with an increase in the amorphous background material, which is attributed to virus capsid components from dissociated particles. C: A two-dimensional crystalline array of the virus is formed from the same stored suspension as in B, but with PEG added to the mixture of virus and negative stain. From Wells *et al.* (1981)

As is true of every method, this method has certain limitations. Air-drying during the first negative staining may cause virus particles to collapse. Rehydration during the second negative staining (uranyl acetate) may be damaging to specimens such as large viruses (adenoviruses). Furthermore, dry specimens are penetrated by the stain, which may obscure the surface details. This method has been most successful for isometric and filamentous plant viruses; it has not yet produced satisfactory images of enveloped animal viruses.

TWO-STEP (SEQUENTIAL) METHOD

The two-step method involves the deposition of specimens onto a carbon-coated grid, followed by negative staining. This approach allows a longer duration of adsorption (1–30 min), which is required if either the particle concentration is very low or diffusion (Brownian motion) is reduced because of the presence of gradient components such as glycerol and sucrose (Nermut, 1982a). Another advantage of this approach is that the specimens can be suspended in buffers or gradient components and washed after adsorption to the support film. After adsorption, specimens can be treated on the grids with various reagents such as enzymes, detergents or fixative solutions before negative staining.

In this method, a carbon-coated, 400-mesh grid is floated, with the support film side facing down, on a few drops of the specimen suspension on a watch glass. Alternatively, a drop of the specimen suspension can be deposited on a carbon-coated grid that has been placed, with the support film side facing up, on a flat surface. A needle-gridholder can be used to handle several grids simultaneously. After enough time has elapsed to permit specimen particles to adsorb to the support film, the grid is washed with the aid of a Pasteur pipette. Another method is to transfer the grid onto a large drop of distilled water; to avoid breaking thin carbon film, the grid is transferred with a wire loop. Adequate rinsing is necessary to prevent possible precipitation caused by a reaction between sucrose or phosphate (present in the sample) and a negative stain such as uranyl acetate. Water is replaced by an appropriate negative stain, which is drained off after about 10 s. Longer staining (3–5 min) is needed if penetration of the stain into the specimen is required. At least two grids should be treated with one type of stain and two grids with another type of stain.

Very dilute specimens require special methods. The necessary concentrations of very dilute specimens can be obtained on Formvar–carbon-coated grids either by allowing prolonged adsorption time (about 20 min) or by using the agar- or paper-filtration technique. For agar filtration, a drop of the virus suspension is placed on an agar block and allowed to soak until a thin film remains. The grid is placed on top of the film for a few seconds; it is then removed, rinsed and stained.

According to one modification, a carbon-coated grid is deposited for 10 min on a drop of virus suspension placed on a dental wax plate; this duration is determined by the virus concentration. Washing with distilled water and negative staining are carried out underneath the grid, so that the virus suspension drop is drained off into a capillary pipette, and distilled water is simultaneously added with another pipette. In this approach the liquid is replaced underneath the grid without exposing it to surface tension forces during transfer from one drop to another. Subsequently, the distilled water is replaced by the negative stain (e.g. 1% uranyl acetate, pH 4.4). The grid is picked up with forceps and is dried by touching it with a piece of filter paper for about 2 s. If the film is strongly charged, the specimen particles will cling to the surface and withstand considerable washing. However, all washings should be done carefully.

Staining after Fixation

A carbon-coated grid is placed on a drop of virus suspension. After the necessary adsorption time, the grid is transferred onto a drop of 1% glutaraldehyde in phosphate-buffered saline (pH 7.2) for 5–10 min. The grid is then placed on 1–2 drops of distilled water and finally on a drop of negative stain such as 3% ammonium molybdate (pH 6.5). In every step the grid is placed with the support film side facing down. For quick fixation before negative staining of certain specimens such as bacteria, the grid carrying the adsorbed specimens is held for several seconds in the neck of a bottle containing OsO_4 solution.

The following simple device used in the author's laboratory allows manifold treatments of several tabbed grids simultaneously. This method is especially useful when each of the treatments is of very short duration. Moreover, capillary problems encountered when the grid is floated on a drop of the fluid are avoided. A piece of an appropriate length of double adhesive tape is attached to a microscope glass slide, so that the longitudinal edge of the tape is exactly over the edge of the slide (Fig. 5.10). The carbon-coated tabbed grids (coated side up) are carefully attached to the edge of the tape-slide, so that the tab rests on the tape while the grid is hanging free in the air; as many as 13 grids can be attached simultaneously. With the use of a Pasteur pipette, a drop of the specimen suspension is

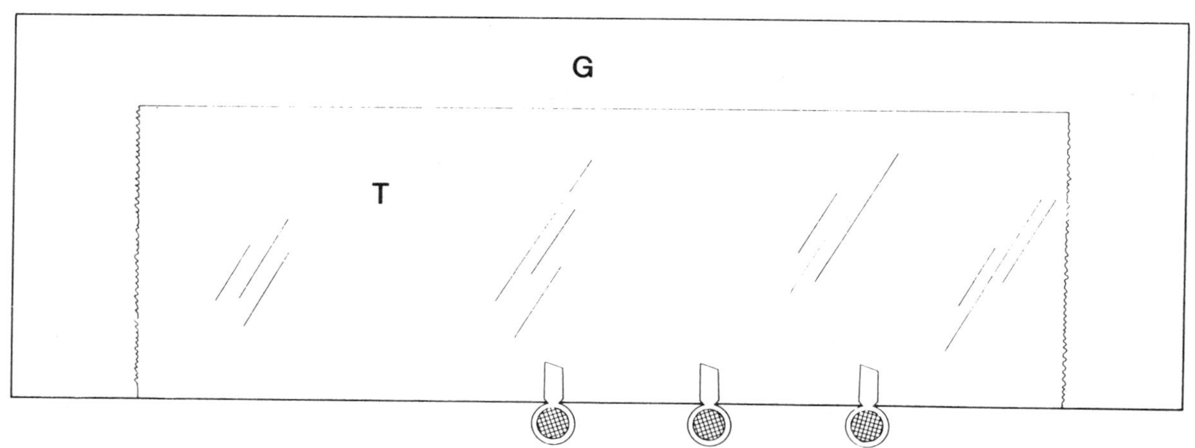

Figure 5.10 Diagram of the device that is easy to prepare and facilitates negative staining of as many as 13 grids at a time. The tab of the grid should be attached to the adhesive tape very carefully. The slide–tape–grid set-up should be placed on a support about 26 cm above the workbench prior to its use. G, glass slide; T, double adhesive tape

placed on the grid and allowed to stand for long enough (about 2 min) for it to be adsorbed on the grid surface. Excess fluid is removed by touching the rim of the grid with a piece of filter paper, so that a thin film of the fluid remains on the grid. A drop of the staining solution is placed on the grid with another Pasteur pipette. After a very short period of time (20–40 s), the excess stain is removed by touching the rim of the grid with a filter paper. The removal of the stain should be complete. If needed, the grid with the adsorbed specimens can be rinsed with a drop of distilled water before negative staining.

SINGLE- OR DOUBLE-CARBON-LAYER METHOD

Another variation of the two-step method was developed by Valentine *et al.* (1968) to visualise antibodies. This is referred to as the single-carbon-layer technique. Because thin, hydrophilic carbon films are used, this technique is suitable for high-resolution electron microscopy of smaller specimens such as ribosomes and parts of purified virus (capsomeres). It involves partial flotation of a small carbon film from a piece of freshly cleaved mica onto the specimen suspension. The carbon film is withdrawn by lifting the mica from the suspension, and the attached layer of carbon traps some of the specimen particles between it and the mica. This mica is inserted into a solution of the negative stain so that the carbon film is completely floated off the mica substrate.

In the double-carbon-layer technique (Lake, 1979), the first step is performed as in the single-layer technique. In the second step, however, the carbon layer with adsorbed specimen particles is rapidly floated off the mica and allowed to touch the wall of the uranyl acetate container, so that the particles become sandwiched in a shell of negative stain between two carbon films (Fig. 5.11C, p. 339).

In both variations a 400-mesh grid, coated with an ahesive (its preparation is described in the next paragraph), is placed on top of the carbon film(s), which is lifted from the staining solution by placing a piece of absorbent paper (for example, newsprint with very dark printing) on top of the grid. The grid is lifted from the staining solution after the solution has begun to wet the paper (Fig. 5.11D).

Adhesive-covered grids are prepared by dissolving the adhesive tape from a 3 mm strip of Scotch brand transparent tape (not magic mending tape) in 5 ml of chloroform (Lake, 1979). This solution is poured over grids arranged on the bottom of a Petri dish, and the chloroform is allowed to evaporate. After completely drying, the grids are picked up with finely tipped forceps. In this technique the specimens are fully embedded in stain, giving a two-sided image. This technique is particularly useful for studying ribosomes, but it has not been used systematically in virus research.

ONE-SIDE NEGATIVE STAINING METHOD

On drying, most specimens are surrounded by the negative stain from below as well as from above. Important structural information about certain specimens can be obtained by staining either from above or from below only. For example, by staining only the lower half of icosahedral viruses, very useful information on their geometry can be obtained. Negative staining from below can be accomplished by drying a film of 2% PTA on carbon-coated Formvar films under an infrared lamp. A dense suspension of specimen is transferred to these dry grids and then drained off immediately. Alternatively, virus particles can be adsorbed to Formvar-coated grids, which are then floated on a negative stain (3–5%) for 30–60 s (Nermut, 1982a); an increase in staining time would lead to staining also from above. After drying, the grids are coated with carbon to strengthen the Formvar film.

Negative staining from above can be obtained by first adsorbing the specimen on carbon-coated grids and then staining with, for example, PTA (pH 6.0) for 10 s (Müller and Peters, 1963). This approach causes the specimens to collapse on support films during air-drying, so that the stain has very little access to the near-side face of the specimen. Alternatively, specimens such as Orfvirus can be fixed on carbon-coated grids with formaldehyde vapours and then sprayed with PTA (Nagington *et al.*, 1964). Duration of spraying is a critical factor in achieving negative staining only from above. The pitfall is the presence of artefacts caused by air-drying from water; specimens such as enveloped viruses are especially prone to drying damage.

PAPER-FILTRATION METHOD (Nermut, 1972, 1982a)

In the paper-filtration method, a collodion (0.2%) film floating on water is picked up with a clean Whatman No. 1 filter paper and semidried at room temperature. Squares of about 3 × 3 cm are cut off and floated on water to determine whether the film is intact. The square is placed on the lower part of a Millipore filter holder connected to a water-pump. A drop of virus suspension is placed in the middle of the square (and spread over a larger area if required), and the water-pump is turned on. After the drop has disappeared and a wet film is left, the filtration is stopped and the square is removed from the filter-holder. The area with the material is cut off (usually as 2–4 small pieces) and immediately floated on negative stain such as 2% PTA (pH 6.8). After about 40 s, each piece of film is picked up on a grid and dried on filter paper. A thin layer of carbon is then evaporated onto the grids. The staining occurs from below, so there is a chance that a one-sided image can be obtained.

PSEUDOREPLICA METHOD

The pseudoreplica method is probably the best method among other negative staining methods used for the detection of viruses such as herpesviruses, arboviruses, haemorrhagic fever virus and rotaviruses. This method has a high sensitivity for virus detection and is more sensitive than the ultracentrifugation method and the enzyme-linked immunosorbent assay. The pseudoreplica method has the following advantages over some other rapid methods (McCombs *et al.*, 1980). As the fluid is absorbed on the agarose or agar surface, any virus present is concentrated on that surface. The agarose diffusion reduces the salt content of the specimen, preventing salt from crystallising and confusing interpretation. This method requires very small quantities of specimen (0.025 ml) and takes less than 30 min to complete. Ultracentrifugation is not needed, because virus concentration is obtained by fluid dialysis into and evaporation from the agarose. Also, partial purification of viruses occurs through diffusion of particles up to 15 nm in diameter, macromolecules and interfering salts (Doane and Anderson, 1977), resulting in clearer background during examination in the electron microscope.

One drop of virus suspension containing more than 10^4 particles/ml is placed on an agarose block (1.5 × 1.5 × 0.6 cm, solidified from 1.5% agarose solution), spread with a glass rod and allowed to semidry (Kellenberger and Arber, 1957; Palmer *et al.*, 1975). (Agarose is preferred to agar, since some viruses tend to stick more strongly to the latter, resulting in diminished virus recovery on Formvar film.) The preparation is covered with two drops of 0.1% Formvar or collodion and placed on edge to drain and dry the film. All four edges are trimmed with a razor-blade, and the Formvar film is floated onto the surface of the stain (e.g. ammonium molybdate, pH 6.5). The grids are then placed on the film and removed with a low-absorption filter paper. After excess stain has been drained off, the grids are air-dried and given a light carbon coating. A modification of this method is explained in Fig. 5.12 (p. 340).

The above procedure requires a large volume of stain, which may retain virions, preventing its use for other specimens. An alternative procedure is as follows. A drop of the viral suspension is transferred on an agar block (1 cm^2) placed on the edge of a glass slide. After the liquid has been absorbed into the agar in 50 min, a drop of 0.5% Formvar from a Pasteur pipette is placed onto the agar and excess is drained onto filter paper. Within 40 s the Formvar dries, and with a new, cleaned razor-blade the block is trimmed so that its four sides are smooth. The Formvar film is floated off the

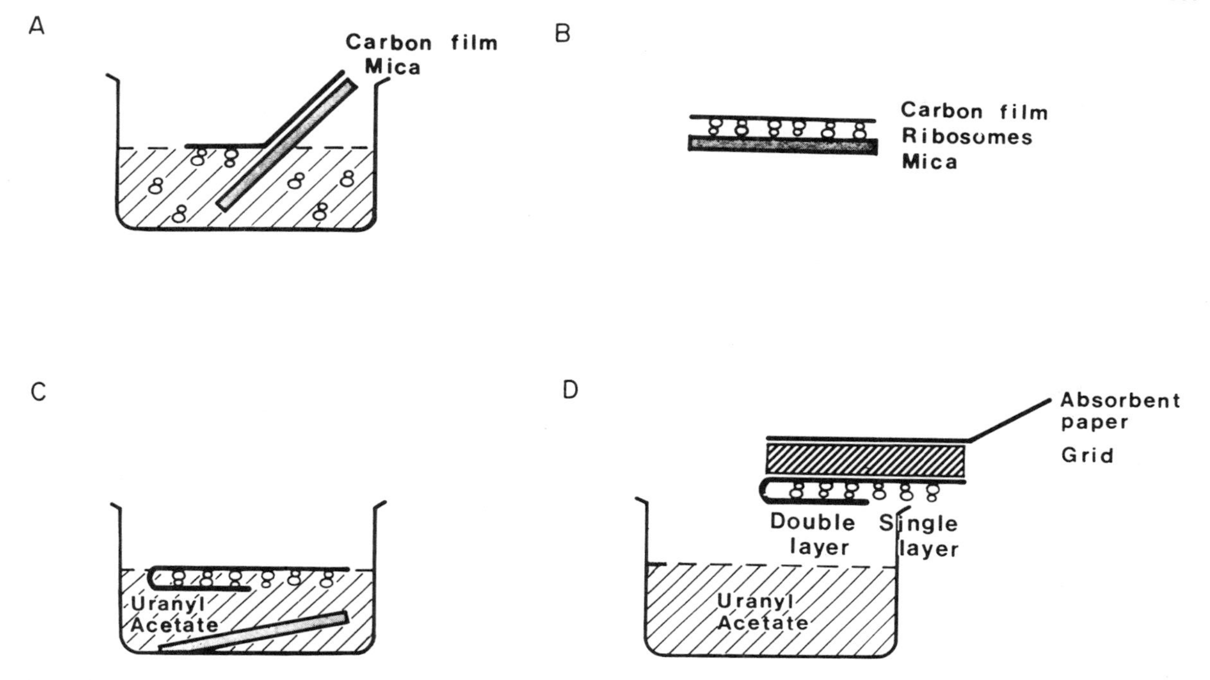

Figure 5.11 A diagram of the technique for preparing double-carbon-layer negatively stained specimen grids: A, adsorbing the ribosomes to the carbon film; B, the carbon film after withdrawing the mica; C, floating the carbon film off the mica and onto the surface of the negative stain; D, lifting the grid from the negative stain, showing a double-carbon-layer grid. From Lake (1979)

agar onto the water surface. A grid(s) is placed on the floating film, underneath which are the virions. A small piece (1–2 cm^2) of filter paper is placed over the grid and membrane floating on the water surface. When the paper becomes wet, it is lifted out by the edge with forceps, carrying the grid. The virions now are on top of the film, and are negatively stained.

AGAR-FILTRATION METHOD (Kellenberger and Bitterli, 1976)

The conventional method for preparing negatively stained particles suffers from selectivity—i.e. different species of particles in a mixture adsorb with unequal efficiencies to the support film. This selectivity might alter the proportion of observed particles by factors of up to 10^3, as compared with the initial proportion in the mixture (Dubochet and Kellenberger, 1972). Although the spray droplet method does not show any selectivity, particles in the size range of viruses have to be in a concentration of about $2–4 \times 10^{11}$ particles/ml, and they must be suspended in negative stain dissolved either in distilled water or buffers of low concentrations of volatile salts. The agar filtration method requires a concentration of about $2–4 \times 10^{10}$ particles/ml in any physiological medium. This method consists of filtering a particle suspension through a collodion film into an underlying, slightly dehydrated 1.0–1.5% agar in physiological medium. The film acts as a filter, which is then used as the support film.

Agar plates are prepared by adding 15 g of agar to 1 litre of a 1% solution of tryptone in distilled water. After being sterilised in an autoclave, the agar is poured into glass Petri dishes (10 cm in diameter) while its temperature is still about 50 °C. Several L-shaped stainless steel carriers are positioned into the dishes before pouring. The agar layer should be 5–8 mm thick above the carriers. Utmost care should be taken to avoid contamination from various sources such as fingerprints. The solidified and cooled agar plates are dehydrated 15–20% in a ventilated oven at 37–45 °C. These plates are immediately cleaned with distilled petrol ether to remove contaminants from the agar surface and the sides of the Petri dish. After a few minutes have been allowed for the ether to evaporate, 0.5–0.8% collodion in amyl acetate is poured over the surface of the agar. The plate is immediately turned upside down and deposited in an inclined position (Fig. 5.13, p. 341) on a table covered with wet filter paper. Within 3–4 h all the amyl acetate has evaporated. The plates can be stored at room temperature or in the refrigerator after being sealed in polyethylene bags or aluminium foil.

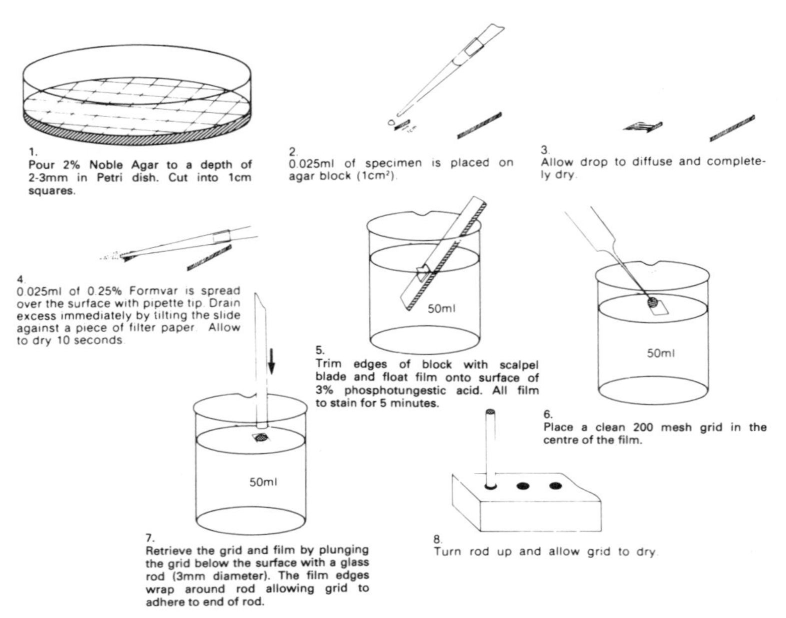

Figure 5.12 Details of the pseudoreplica method. From McCombs *et al.* (1980)

The collodion–agar plates are prepared for filtration by cutting agar blocks over the L-carrier with a razor-blade (Fig. 5.14A). The remaining agar is removed except for these blocks. A drop (10 μl) of the particle suspension is deposited on each of the three blocks (Fig. 5.14B) and spread with a glass spreader made out of a Pasteur pipette (Fig. 5.14C). To achieve uniform spreading, the plate is moved back and forth on a flat table while the spreader is held stationary. The spreader should not touch the collodion film. The Petri dish is closed to avoid drying by evaporation.

Within 10–30 min, filtration is completed on 80–100% of the surface which is visible in reflected light. Wet parts on the block will not be used later. The block is exposed to formaldehyde vapours for 10 min. Alternatively, the block can be exposed to OsO_4 vapours for 10 min for the observation of nucleoids of Gram-negative bacteria such as *E. coli*. These types of bacteria are relatively small. The L-carrier is seized with a forceps and obliquely introduced into distilled water. The film floats off on the surface of the water (Fig. 5.14D), and is picked up onto the grids from below with a 'fishing device' (Fig. 5.14E). This device is placed on a filter paper, so that the film becomes stretched over the grids. While still soaked underneath, each grid is transferred to fresh filter paper to remove the liquid below the film (Fig. 5.14F). The grid is now ready for observation. Floating the film on water is not recommended for counting particles. For this purpose, the film is floated onto 8–10% solution of PTA. The procedure for particle counting is described by Hayat and Miller (1989).

FREEZE-DRY NEGATIVE STAINING

Freeze-dry negative staining combines good preservation and high-resolution surface images of enveloped viruses (Nermut, 1977a, 1982a). Freeze-drying avoids formation of drying artefacts due to surface tension forces, and the spreading technique prevents orientation artefacts. About 0.5 ml of a virus suspension of high concentration is mixed with 0.01 ml of 0.1% cytochrome *c* in 0.1 M ammonium acetate (pH 7.0). This mixture is spread on the surface of distilled water in a plastics Petri dish (4 cm in diameter) and, if necessary, slightly compressed. The film is touched with a glow-discharged, carbon-coated grid. The grid with an adsorbed monolayer of virus particles is washed on 3 drops of 0.1 M ammonium acetate or distilled water for 2 min or longer. It is transferred onto a drop of negative stain (e.g. 3% ammonium molybdate, pH 6.5) for 10–30 s. The excess stain is drained off over a Dewar bottle containing liquid nitrogen, and the grid is immediately immersed in the liquid nitrogen. With hydrophilic support films and a high concentration of particles on the grid, the drain time is 2–3 s. After 10 s the grid is quickly transferred onto a precooled (−150 °C with Freon) specimen stage of a freeze-etch unit (Balzer). The chamber is evacuated and the specimen stage temperature brought to −80 °C. Simultaneously, the knife arm is cooled down to −150 °C and then moved over the specimen stage for 20–30 min. The knife arm cooling is switched off and the specimen stage warmed up to 30 °C. It is preferable to heat the knife arm before opening the chamber. The dried

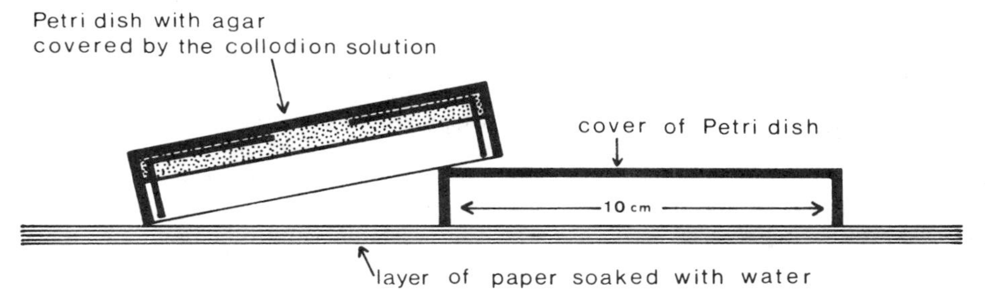

Figure 5.13 Disposition of the Petri dish over a layer of very wet paper during evaporation of the amyl acetate. From Kellenberger and Bitterli (1976)

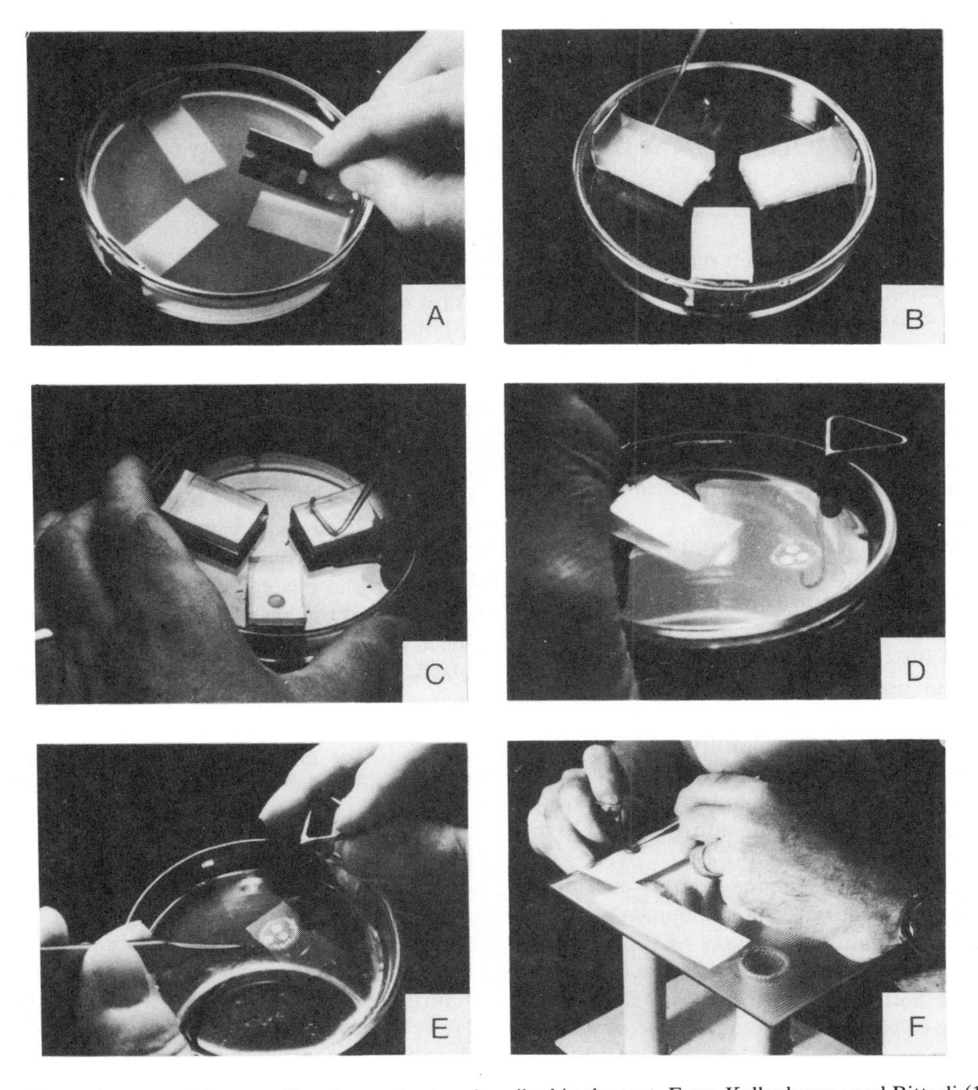

Figure 5.14 The main steps of the agar filtration method as described in the text. From Kellenberger and Bitterli (1976)

negative stain, in the form of a white powder on the grid, is removed by gentle blowing from a Pasteur pipette, so that only a fine layer of it remains attached to the virus particles. Too much or too little powder remaining on the virus will result in poor images. The grid should be observed immediately or stored in a vacuum.

The described method is more difficult than air-dry negative staining, and only about 50% of the grids processed yield good results. The percentage of useful grids increases if the following conditions are strictly observed (Nermut, 1982a). When the spreading technique is used, a reasonably dense monolayer of virus particles must be formed on hydrophilic grids; the stain drop should be drained off without haste, so that only a fine film remains on the grid; drying must be completed in 20–30 min; and the vacuum chamber is opened only when the specimen stage has been warmed up to room temperature.

GENERAL METHODS FOR VIRUSES

Human Viruses

Direct electron microscopy based on negative staining is the most rapid method for the examination and identification of virus particles. Treatment and prognosis of cases such as smallpox and herpes encephalitis depend on rapid and correct viral diagnosis. This is possible with direct electron microscopy because virus particles have characteristic morphologies—that is, shape, size and ultrastructure. However, within a group individual virions are difficult to differentiate on the basis of appearance alone.

Direct electron microscopy can be accomplished within minutes or hours. It can visualise any virus type present, and previous knowledge of the type of virus is not necessary. Furthermore, it can be used for viruses that are difficult or impossible to culture, either because they do not grow in routine cell cultures or because they have lost their infectivity during transportation (Kjeldsberg, 1980). Morphology of a virus particle is not damaged by the loss of its infectivity. Faecal viruses of hepatitis A and various diarrhoeal conditions are essentially non-cultivatable. Orfvirus grows with difficulty.

Other methods (e.g. immunoelectron microscopy, fluorescence procedure, radioimmunoassay and enzyme-linked immunosorbent assay—ELISA) used for diagnostic virology are sensitive and specific, but require specific antibody. These methods demonstrate only one type of virus at a time, and some of them are time-consuming. Although thin sectioning combined with positive staining is useful for the demonstration of intracellular virus particles in biopsy and autopsy tissues and cell cultures, it is a lengthy procedure. The primary objective of this method is to demonstrate the precise localisation of virus particles. In spite of the rigorous treatments involved in this method, virus particles are fairly well-preserved. This method yields less information on the ultrastructure of virus particles. The immunoelectron microscopy method is discussed on p. 346. The following general protocols are recommended for the demonstration of viruses in a wide variety of human materials.

ALLANTOIC FLUID

Simple centrifugation of the allantoic fluid at 15 000 *g* for 1 h followed by negative staining is recommended.

BIOPSY OR AUTOPSY TISSUES

Biopsy or autopsy tissues are cut into small pieces (1 mm^3) and frozen and thawed for a few minutes on a cold metal planchet on dry ice to rupture the cells and release the virus particles; a few drops of distilled water are added and the suspension is negatively stained.

BLISTER FLUIDS

Blister or vesicular fluids (pustular fluids) often contain a large amount of morphologically recognisable virus particles, and can be examined without prior centrifugation. The blister fluid is aspirated with a needle and placed directly on a coated grid for negative staining. If the blister contains only a very small amount of fluid, a drop of sterile water or saline is injected into the blister and then aspirated. Thus, a larger volume of fluid than the blister itself could yield can be obtained.

Alternatively, fluid can be collected directly on the grid from the blister by placing a coated grid on the underside of the blister. This can also be accomplished by a simple device which can be fabricated from a plastic-shafted throat swab (Bozzola, 1987). Viruses that may be present include herpes and pox. The presence of herpes simplex virus type I in the human lip blister is not uncommon. Varicella zoster virus has been found in the vesicle fluid of cancer patients.

BRAIN TISSUE

A 10% suspension of brain tissue is made in distilled water with a glass Teflon homogeniser; a few strokes suffice. Homogenates of infected brain tissue are usually sufficiently rich in virus to render concentration by centrifugation before negative staining unnecessary. However, in some cases virus-infected cells may be distributed unevenly in the brain tissue. In such cases brain tissue (biopsy or autopsy) is ground in a tissue homogeniser, and a small amount of an appropriate buffer is added, followed by centrifugation at a low speed (~1500 *g*) for 1–2 min. If the supernatant appears cloudy, the speed and/or duration of centrifugation can be increased, but should not exceed 2200 *g* for 10 min. The pellet is discarded, and a drop of the supernatant is negatively stained. Alternatively, brain tissue can be processed for thin sectioning by standard or rapid procedures. Viruses that may be present in the brain tissue include herpes, mumps, rabies, picorna, HIV, adenovirus, and paramyxovirus.

BREAST MILK

Breast milk can be concentrated by centrifugation for negative staining of virus particles, but it is poor in these particles. Particles resembling type B oncoviruses have been observed in milk (Moore *et al.*, 1969). Type D oncovirus-like particles have also been demonstrated in breast milk (Chopra *et al.*, 1973).

CELL CULTURES

Cell cultures inoculated with viruses usually show typical cytopathic effects after the lapse of several days. Early confirmation of the growth of virus can be accomplished by negative staining. Examinations of inoculated cell cultures by negative staining can differentiate myxoviruses from paramyxoviruses, adenoviruses from herpesviruses, and vaccinia virus (an orthopoxvirus) from enterovirus (Field, 1982). Thin sectioning can also be used to detect virus particles in cell cultures.

Cells inoculated with virus are scraped off the culture tube into the growth medium, and this cell suspension is centrifuged at 50 000 *g* for 90 min. The pellet is transferred to a cold metal planchet on dry ice for a few minutes and then thawed to rupture the cells and release virus particles. A drop of distilled water is added to the suspension before negative staining. Cells can also be ruptured by sonication.

CEREBROSPINAL FLUID

Since insufficient quantities of viruses are found in the cerebrospinal fluid (CSF), their detection is not easy. However, they can be detected by ultracentrifuging relatively large volumes (2–5 ml) of fluid. An airfuge is recommended for a small amount of fluid. Immunoelectron microscopy with pooled serum may be used as an adjunct procedure. Viruses that may be present include enterovirus, herpes, mumps and rubella. Viral antigens have been demonstrated in the CSF by immunofluorescence in the case of viral encephalitis or meningitis infection (Taber *et al.*, 1973).

EYE BIOPSY

Processing of eye biopsy sample is carried out as for brain. Viruses that may be present include herpes, cytomegalovirus (CMV), rubella and adenovirus.

FAECES

Direct electron microscopy of negatively stained human faeces is a routine diagnostic method. Faecal virology based on negative staining is playing a key role in the diagnosis of diseases such as diarrhoea. Large numbers of virions are usually excreted in the faeces during the course of illness. In retrovirus infections, as many as 10^{10} particles per gram of faeces may be excreted. Although faecal specimens can be examined by direct negative staining without prior centrifugation if virus content is very high, usually centrifugation precedes staining.

The primary objective of processing faecal specimens before negative staining is to remove salts, low-molecular-weight substances and water, and to concentrate virus particles. This objective can be accomplished by centrifugation. Small virus particles are more readily detected by centrifugation than by direct examination or the pseudoreplica method. Faecal specimens are suspended in phosphate-buffered saline (10–20%) and centrifuged at 3000 *g* for 30 min to remove bacteria and cell debris. The pellet is discarded and the supernatant is recentrifuged in a few drops of distilled water before negative staining. By this procedure the virus suspension is concentrated and partly purified and freed from high salt content. High-speed ultracentrifugation is not recommended, for it may eliminate virus particles, especially if they are aggregated.

Alternatively, lyphogel (a selectively absorbent polyacrylamide hydrogel) has been used to concentrate virus particles in the faeces without affecting their morphology (Whitby and Rodgers, 1980). Another alternative method involves concentration of enteroviruses from faecal specimens by adsorption onto a polyelectrolyte and then by elution into a smaller volume for negative staining (Chaudhary and Westwood, 1972). This method has proven successful for the detection of 10^6 poliovirus particles/ml. In the case of non-bacterial gastroenteritis, rotavirus is the most common virus, followed by adenovirus. Other viruses that may be present include enterovirus, hepatitis A, astrovirus, calcivirus, enteric coronaviruses, parvovirus and Norwalk agent. A detailed account of immunonegative staining techniques of viruses has been given by Kjeldsberg (1986).

HARD TISSUES

Hard tissues are cut into small fragments and placed in a glass or porcelain mortar with a little silver sand.

After a small amount of distilled water has been added, the specimens are ground with the pestle until the tissues are further fragmented. More water is added with continued grinding until a relatively smooth suspension is obtained. The suspension is clarified by centrifugation for 10 min in a bench centrifuge. The pellet is discarded and the supernatant is centrifuged at 15 000 *g* for 1 h.

LIVER OR KIDNEY BIOPSY

Processing of liver or kidney biopsy is carried out as for brain tissue. Herpes, CMV and adenovirus may be present. In kidney, antibody–antigen complexes in glomerulonephritis may resemble myxovirus nucleocapsids (Miller, 1986).

NODULES

A nodule or a piece of it can be processed as for brain tissue. *Molluscum contagiosum*, a poxvirus, has been identified in the finger nodule.

RESPIRATORY SECRETIONS

Since nasopharyngeal secretions contain large quantities of morphologically identifiable virus particles, they can be examined directly on the coated grid without prior concentration. If the specimen is thick with mucus, a small amount of water (10%) is mixed well with the sample and ultrasonicated if necessary (Miller, 1986). The sample is centrifuged at ~1500 *g* for 1–3 min. The pellet is discarded and the supernatant is examined. Alternatively, samples can be examined after ultracentrifugation or airfuging.

Coronaviruses, orthomyxoviruses (influenza) and paramyxoviruses (mumps) may be found in the nasopharyngeal secretions, although, in general, coronavirus is responsible for respiratory infection. Norwalk agent particles can be detected in a sample of vomit concentrated 100-fold by centrifugation before negative staining. In immunocompromised patients, almost any viral agent can be present in the lungs (Miller, 1986).

SERUM

Although serum is not very suitable for direct detection of viruses with the electron microscope, the following procedure can be tried. Serum (0.2–1.0 ml) is diluted with an equal volume of distilled water and then centrifuged at 15 000 *g* for 1 h. The supernatant is discarded and the pellet, which may not be clearly visible, is brought to its original volume in distilled water. This is centrifuged, the supernatant is discarded and the pellet is negatively stained. Immunoelectron microscopy based on negative staining is the best method of detecting viral antigens in the serum. Viruses that may be present include rubella, hepatitis A and B, Ebola, parvovirus and HIV.

SKIN LESIONS

Specimens from skin lesions are rich in virus particles and can be directly examined with the electron microscope after negative staining without prior centrifugation; however, older lesions contain fewer virions. Dried smears of lesion scrapings can be rehydrated in a minimal quantity of distilled water. Crude extracts of solid tissue in distilled water can also be used for direct electron microscopy. Poxvirus and herpesvirus are easily detectable in vesicle fluid and crusts from patients suffering from smallpox or chickenpox. Skin scabs are homogenised in a Dounce homogeniser in a small volume of distilled water and the homogenate is negatively stained without prior centrifugation. Clinical specimens collected on cotton swabs or spatulas may be resuspended in distilled water and negatively stained without prior treatment. Human monkeypox virus (Breman *et al.*, 1980) and varicella zoster virus (Macrae *et al.*, 1969) have been detected with electron microscopy.

SKIN TUMOURS

Orfvirus, the aetiological agent of ectyma contagiosum, can be easily detected in skin tumour samples without centrifugation. Poxvirus has been found in benign human tumours.

Fractionated cell extracts of human tumours were used by Mak *et al.* (1974) for studying retrovirus in human tumours by negative staining. In general, thin section methods have been used for the detection of virus in tumours.

SKIN WARTS

Some skin warts contain only a small number of virus particles, and genital warts usually have a low viral content. The best method of extracting virus from biopsies of genital warts is by light grinding to disrupt only the surface layers instead of complete homogenisation of the tissue. Studies of the wart virus lesions of epidermodysplasia verruciformis show that negative staining is somewhat more sensitive than thin sections in detecting virions in early malignant lesions where viral content is very low (Yabe and Sadakane, 1975). Negative staining has been used for detecting papillomavirus in lesions of focal epithelial hyperplasia of the oral mucosa (Goodfellow and Calvert, 1979). Papovavirus can be detected in common warts and genital warts with negative staining (Oriel and Almeida, 1970).

SPUTUM

It is difficult to completely remove associated mucus from sputum specimens. Sputum is diluted 1:4 with PBS and a homogeneous suspension is obtained in a homogeniser. The suspension is clarified by centrifugation for 10 min in a bench centrifuge. The pellet is discarded and the supernatant is centrifuged at 15 000 *g* for 1 h to obtain a pellet for negative staining.

TEARS

Processing of tears is the same as for CSF. Rapid discovery of the virus obviates the use of antibiotics. Viruses that may be present include adeno-, herpes-, echo-, entero- and HIV. The presence of adenovirus in the tear drop of a patient with eye infection is not uncommon.

TISSUE SCRAPINGS

Tissue scrapings (e.g. conjunctive cells) are diluted with sufficient distilled water to disrupt any cells present, and then are stained.

URINE

Direct examination of urine with the electron microscope does not usually result in easy detection of virus particles. According to a general method, cloudy or contaminated urine (as large a volume as possible) is clarified by centrifugation at 3000 *g* for 15 min in a bench centrifuge to remove bacteria and exfoliated cells. The pellet is discarded, and the supernatant is recentrifuged at 15 000–48 000 *g* for 1–2 h to pellet the virus. The pellet is used for negative staining. Large numbers of cytomegalovirus can be found by centrifuging a large volume (30–50 ml) of urine and negative staining of the concentrated virus; the presence of cytomegalovirus in urine of congenitally infected infants is not uncommon. Papovavirus can be detected in urine from renal-transplanted patients, and mumps virus can be seen in urine of patients with parotitis or orchitis. Papillomavirus particles have been demonstrated in urine from pregnant women (Lecatsas and Boes, 1979). Herpesviruses are easy to identify in urine, owing to the presence of the characteristic envelope and one or more layers constituting the nucleocapsid.

Plant Viruses

RAPID PROCEDURES

The following rapid procedures are recommended for preparing plant viruses.

(1) A 2 mm^2 piece of leaf or any other plant tissue is ground up in 0.1 ml of PTA on a clean glass slide, using a glass rod with a blunt tip. One small drop of 0.05% bacitracin may be added to aid virus release from the tissue. A grid is floated, with the support film side down, on a droplet of this suspension for 30 s. Alternatively, the suspension droplet can be picked up in a capillary tube and transferred onto a carbon-coated grid clamped in forceps. After 30 s most of the droplet is drawn off by using a piece of filter paper between the forceps blade until the grid appears dry.

(2) Plant tissue (2 mm^3) is placed in 2 drops of methylamine tungstate and 1 drop of water on a glass slide. The tissue is squashed in the stain and the droplet of suspension is drained to the edge of the slide, leaving the coarse debris behind. This droplet is transferred to a carbon-coated 400-mesh grid.

(3) Plant tissue is ground, and a grid (film side facing down) is floated on a drop of this suspension for 30 s–15 min. The grid is rinsed with 10–20 droplets of a negative stain. This treatment removes debris and usually leaves distinct virus particles on the grid. To avoid the production of precipitates, it is recommended that uranyl salts not be ground up with the tissue.

(4) A fragment of cigarette tobacco is ground up in 0.01 ml of PTA on a glass slide. A droplet of this suspension is transferred to a coated grid by using one of the preceding or other means. This simple procedure gives a good preparation of tobacco mosaic virus.

VIRUSES IN CRUDE EXTRACT

A small drop of 2% PTA (pH 6.5) is placed on a carbon-coated 400-mesh grid (Hitchborn and Hills, 1965). A piece of epidermis, peeled from the undersurface of a virus-infected leaf, is laid, with torn surface downward, onto the stain for a few seconds; it is gently drawn over the surface of the stain and is then removed. Excess stain is removed from the grid by momentarily touching its edge with filter paper. The grid is examined immediately. The cell contents deposited on the grids are mainly those from broken spongy mesophyll cells, most of the epidermal cells remaining intact. The pH of the stain can be varied from 5.5 to 6.9 without any apparent effect.

An alternative method involves the use of sap, extruded from macerated infected leaf tissue. The sap is clarified by centrifugation at a low speed either directly or after being frozen for 2 h at −10 °C and then thawed. The supernatant is diluted 1:10 with distilled water; undiluted sap is too dense to be useful. Specimens are prepared as before. The tissue-strip method is preferred to the sap method.

The following method is useful for extracting plant viruses from small amounts of infected tissue (Duncan and Roberts, 1981). A watch-glass serves as a mortar and a matched glass rod as the pestle for 100–500 mg of tissue. For smaller amounts of tissue (0.1–7.0 mg), a 0.5 ml microanalysis tube acts as the mortar and a tapered-to-match glass rod as the pestle. About 300 mg of tissue and 25 mg of 600-mesh washed carborundum powder (acting as an abrasive) are placed in the watch-glass. After the tissue has been ground for 1–2 min, cells and organelles are disrupted and the virus particles are released. Only enough buffer or fixative solution is added initially to produce a fine paste. The extract (about 1.5 ml) is transferred to a tube for centrifugation for 5–15 min at 8000 *g* in a microangle centrifuge. The supernatant is discarded; the pellet can be further fixed and/or washed with a buffer. A drop of virus suspension is stained as usual.

IMMUNOELECTRON MICROSCOPY

Immunoelectron microscopy (IEM) in conjunction with negative staining is used to visualise the interaction between an antibody and components on the surface of a macromolecular system. In other words, IEM uses antibodies (IgG) as specific markers for the localisation of components on the surface of biological complexes such as viruses and ribosomes. This approach combines the advantages of high specificity of serological reactions with extreme accuracy of electron microscopic imaging. The approach allows simultaneous identification of morphology and immunology of macromolecular systems. Antisera are raised against a particular component of the complex, and then the whole complex is incubated with antibodies; the binding site of the antibody delineates the position of the component under study. The resolution obtained is ~2 nm. The methods given later are relevant to viral studies; for IEM of ribosomes, the reader is referred to Stöffler and Stöffler-Meilicke (1984).

Classical IEM for viral diagnosis involves the visualisation of clumps of viruses specifically formed with homologous antibodies. Such clumps are formed as a result of the antigen–antibody reaction when viral suspension is mixed with specific antiserum. Although with regard to sensitivity classical IEM is superior to direct electron microscopy, the former is time-consuming because of the need for overnight incubation and high-speed centrifugation to sediment the antibody–virus complexes. The reactants must be in optimal proportions, as otherwise aggregates will not be formed. These limitations have been overcome by employing immunosorbent electron microscopy (ISEM). Immunosorbent electron microscopy is considerably more sensitive and reliable than direct electron microscopy and classical IEM.

Further refinements of the visualisation of the antigen–antibody reactions is achieved by labelling of one of the reactants, directly or indirectly, with an electron-opaque marker such as colloidal gold. This approach is useful for the identification of viruses in both plant and animal specimens. Furthermore, detection of viral antigens of unrecognisable structures such as fragments of disrupted virus particles or incompletely synthesised virions is possible. Human antisera can be used as primary antibody and gold–IgG conjugates as secondary antibody for direct labelling of viruses.

Immunoelectron microscopy methods have certain disadvantages. Since virus particles may aggregate spontaneously, it is difficult to distinguish between spontaneous clumping and immune aggregation. The ultrastructure of virions may be obscured by antibody molecules, coating with serum proteins may result in increased background noise, and contrast imparted by the negative stain may decrease. These limitations, however, can be minimised by using highly diluted antisera. Another minor disadvantage is that at present only one type of virus is detected at a time, depending on the specificity of the antiserum. Major variations of IEM are discussed below.

CLASSICAL IMMUNOELECTRON MICROSCOPY

The classical immunoelectron microscopy method involves mixing the virus with the antibody in suitable proportions and incubating the mixture. The incubation is usually carried out overnight, although much shorter times have been used either at 37 °C or at room temperature. A complex is formed as a result of bridging the virion by the antibody molecules; this complex is centrifuged to obtain a pellet, which is resuspended in a small amount of distilled water and mixed with a negative stain before placing it on a Formvar–carbon-coated grid. The result is an aggregation of the viruses in a sample that may not show any virus in the original preparation, owing to low concentration. When an excess of antibody is present, viral aggregation is prevented and the virus is surrounded with a halo of antibody molecules.

The advantages of classical immunoelectron microscopy are: (1) both the virus (or antigenic site) and the antibody (IgG, IgM or Fab) can be visualised; (2) antibodies can trap virions from a poor preparation in recognisable aggregates; and (3) the centrifugation following aggregation can concentrate and partly purify the immune complex. The limitiations are: (1) incubation and centrifugation take a long time; (2) centrifugation may cause non-specific clumping; (3) the initial virus preparation must be suspended in a suitable medium; and (4) the virus and antibody preparations must be relatively clean, as otherwise centrifugation will also pellet impurities.

IMMUNOSORBENT ELECTRON MICROSCOPY

Derrick (1973) introduced a method, serologically specific electron microscopy, in which electron microscope grids are coated with antibodies for specific trapping of viruses. This method subsequently came to be known as solid-phase immunoelectron microscopy (SPIEM) because one of the reactants (antibody) is first adsorbed to a 'solid phase' (the grid). Coating of the grid with virus followed by treatment with antibody is not recommended, for it may cause some aggregation due to specific migration on the grid (Almeida *et al.*, 1980). This method has been renamed immunosorbent electron microscopy (ISEM) by Roberts *et al.* (1982), and this terminology is followed in this volume; other terminologies have been suggested (Katz and Straussman, 1984). Many variations of the ISEM method are in use, and they are presented below.

ANTIGEN-COATED GRID METHOD

The practical application of the antigen-coated grid method was introduced by Milne and Luisoni (1975). In this technique, virus is adsorbed onto grids and rinsed, and then the antiserum is added, decorating or covering the antigenic sites by the antibody molecules. The decoration of virus particles with a halo of antibody around them is the best evidence for a specific immune reaction. Since virus particles are not free to move, clumping does not occur. This method is applicable to virus-infected tissues, fractions taken from sucrose or caesium chloride gradients, or purified virions. It is best suited for high-resolution work which requires clear visualisation of the antibody attachment site.

Procedure (Milne and Luisini, 1975)

In the first step a relatively large number of virions are trapped by the grid, and in the second step the identity of these virions is confirmed. The grid is incubated for 5 min in the antiserum diluted, for example, to 1:10–1:100. After a rinse with 20 drops of phosphate buffer, the virus preparation is transferred onto the grid for 15 min. The grid is rinsed in 20 drops of buffer and drained, and then a drop of antiserum is placed onto the grid for 15 min. The grid is rinsed in 20 drops of buffer, followed by 30 drops of distilled water, and then negatively stained.

PROTEIN A-COATED GRID METHOD

Immunosorbent electron microscopy has a limitation in that it requires the use of highly diluted sera, which causes problems when only low-titre antisera are available. This problem can be circumvented by using protein A in conjunction with the ISEM method (Shukla and Gough, 1979). This modification is called PA–CGT. Since protein A has affinity for the Fc part of the IgG molecules, this protein can selectively adsorb these molecules from the serum, leaving their antibody reactive sites (Fab) free. The result is a considerable increase in the binding of antibody molecules to the grid. Protein A functions as an antibody-'capturing' molecule. This, in turn, is the reason for increased virus (antigen) trapping efficiency in the presence of protein A. Thus, antisera of much lower dilutions (e.g. 1:40 for tobacco mosaic virus and 1:80 for Erysimum latent virus) (Shukla and Gough, 1983) than are optimal for classical ISEM can be used. Protein A allows the use of poor antisera that are ineffective with classical ISEM.

The increase in sensitivity of virus detection due to the intermediate layer of protein A is most apparent when low dilutions of antiserum (10^{-1}–10^{-2}) and high virus concentrations are employed (Nicolaieff *et al.*, 1982). The PA–CGT is ~25-fold more sensitive than the ISEM method. However, the advantage of protein A in detecting low concentrations of virus in the presence of a large number of antibodies on the grid is less than significant.

Protein A permits the use of antisera at high concentrations, which is an advantage when low-titred antisera are used. Compared with the ISEM method, the success of the PA–CGT is less dependent on antiserum concentration. The latter is also useful for the characterisation and measurement of cross-reactions (Svensson and von Bonsdorff, 1982).

Another advantage of protein A is that it allows the detection of distant heterologous relationships among virions with relatively short trapping times (Lesemann, 1982). Additional relationships can be detected by increasing the trapping time from 15 min to 2 h (Roberts, 1986). Such relationships are not detectable by classical ISEM.

Many parameters of specimen processing affect trapping efficiency and specificity. These include temperature and duration of virus incubation; concentrations of virus, antiserum and protein A; duration of treatments with protein A and antiserum; buffer rinse; negative staining solutions; and grids. Unless expertly prepared locally, commercially available carbon-coated 400-mesh copper grids give superior results (Katz and Straussman, 1984). Treatment with protein A solution in PBS (pH 7.2) at a concentration of 10–100 μg/ml for 5–20 min is recommended. The antiserum can be diluted 1:100–1:500 in PBS. Treatment with the antiserum lasts for 15–20 min at room temperature. Less debris is found on the grids at room temperature than at 37 °C, but the trapping efficiency is slightly higher at 37 °C (Katz and Straussman, 1984).

The minimal amount of virions (Sindbus virus from infected tissue culture fields) that can be detected is $\sim 3 \times 10^7$ virions/ml (Katz and Straussman, 1984). This sensitivity is ~10-fold and 4-fold higher than those obtained with direct electron microscopy and classical IEM, respectively. The specificity of the PA-CGT is achieved either by prolonged rinsing or by adding BSA to the rinsing buffer and virus solution. This technique can be completed in less than 3 h when grids have been precoated with protein A and antiserum; such precoated grids can be stored for weeks before use. Grids usually are treated with protein A at 10–100 $\mu g/ml^{-1}$ before they are floated on an antiserum. The technique is recommended as a simple and rapid diagnostic approach for any plant, animal or insect virus, provided that optimal processing conditions have been determined for each new specimen.

SERUM-IN-AGAR-DIFFUSION METHOD

The serum-in-agar diffusion method does not require a purified or concentrated viral preparation and can be performed on a crude sample obtained from the primary isolation culture. It further differs from classical IEM in not requiring an overnight incubation. Treatment of grids with protein A seems to result in higher sensitivity for virus detection than that obtained with the original serum-in-agar method. The reason may be that in the original method adsorption of viruses on the grid is non-specific and may be inhibited by the presence of serum proteins or contaminants in the viral preparation. On the other hand, in the modified method viruses are adsorbed with homologous antibody first, followed by their specific binding to the grids through protein A. In this case, contaminants do not inhibit the binding of virus to protein A.

Procedure

Formvar–carbon-coated 400-mesh copper grids are made hydrophilic by glow discharge. The grids are floated on drops of protein A at a concentration of 10 μg/ml in PBS (pH 7.4) for 20 min to trap virus particles. They are rinsed three times in drops of PBS and used without drying. The serum-in-agar is prepared by mixing 0.1 ml of the antiserum with 5 ml of 1.5% Noble agar in a water-bath at 55 °C. This mixture is pipetted into multicup disposable microtitre plates, each cup being three-quarters filled and solidified at room temperature. About 30 μl of the viral preparation is layered on agar containing serum in each cup. The grids coated with protein A are floated (coated side down) on the specimen in the cups and incubated for 1 h at 37 °C. About 0.1 ml of distilled water is added to each cup to float the grids from the agar surface. Each grid is taken out of the cup, rinsed three times by floating on drops of distilled water and negatively stained with 2% PTA. The results of this method are shown in Fig. 5.15 (opposite). For additional details, see Anderson and Doane (1973) and Furui (1986).

PROTEIN-COATED BACTERIA TECHNIQUE

The protein-coated bacteria technique (PA-CBT) was introduced by Katz *et al.* (1980). In this technique protein A-containing *Staphylococcus aureus* cells are coated with a layer of viral antibodies by incubating them with specific viral antiserum. Such bacteria are able to search out the corresponding virus particles from a suspension. Virions attached at the periphery of these cells are easily visible in the electron microscope. It has been demonstrated that a suspension containing 10^7 antibody-coated bacteria traps the total number of plant viruses present in 1 ml of a 500 ng/ml virus preparation, and that ~100 virions (30 nm diameter) are visible at the periphery of each cell (Nicolaieff *et al.*, 1982). This technique is relatively difficult to carry out, and viruses are visualised only at the periphery of the bacterial cell. A greater fraction of the attached virus particles may be seen by disrupting the cells. The protein A-coated grid method yields more reproducible results than those obtained with the PA–CBT.

Procedure

Staphylococcus aureus cells are sensitised with viral antibodies by mixing 1 ml of a bacterial suspension containing 10^8 cells with 1 ml of viral antiserum diluted 1:10 000. After incubation with stirring for 30–60 min at 37 °C, the cells are centrifuged and resuspended in 0.1 ml of PBS (pH 7.0). One millilitre of this suspension of antibody-coated cells is mixed with 1 ml of a purified virus preparation and incubated for 1 h at

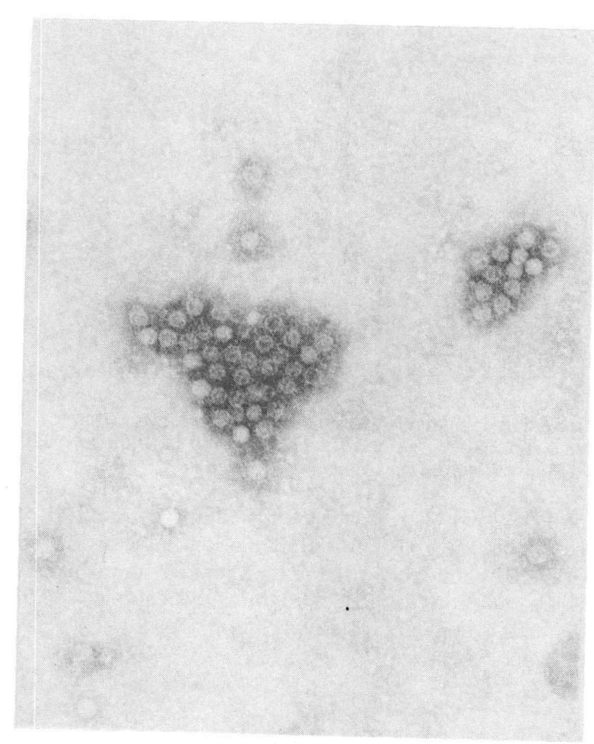

Figure 5.15 Poliovirus particle trapped on a grid coated with 10 μg/ml of protein A by the protein A–serum-in-agar diffusion method. The particles show aggregation and decoration. 88 000×. Courtesy S. Furui

37 °C. The cells are centrifuged and resuspended in 0.1 ml of PBS. Drops of this suspension (50 μl) are placed on a sheet of Parafilm, and Formvar–carbon-coated grids are floated for 5 min on the drops. The grids are rinsed on drops of PBS, and are negatively stained with 1% PTA (pH 7.0).

Alternatively, after incubation of the bacteria with the virus preparation, the cells are not centrifuged, but are allowed to settle by gravitation for 4½ h on the surface of grids. The grids are rinsed with PBS before negative staining.

VIRUS IMMUNE COMPLEX ELECTRON MICROSCOPY

Staphylococcus aureus cells have been used in the virus immune complex electron microscopy (VICEM) method introduced by Lee *et al.* (1983). This method is useful for the detection of virus particles within immune complexes in clinical specimens obtained from patients with viral infections. Since most clinical specimens contain only small quantities of virions, this method 'extracts' and concentrates virus particles from such samples. Immune complexes are precipitated with 5% polyethylene glycol 6000, adsorbed to protein A of *S. aureus* and eluted (with brief sonication) in 2 M propionic acid. The elute contains morphologically intact viruses free from background serum proteins.

Procedure (Lee *et al.*, 1983)
About 1 ml of serum containing immune complexes is mixed with 0.2 ml of 0.65 M EDTA buffer (pH 7.4) at 4 °C, Then 3.8 ml of 6.6% prechilled PEG 6000 is added and the tubes are incubated overnight at 4 °C. The precipitate is pelleted by centrifugation at 2000 *g* for 15 min, washed with 5% cold PEG solution and resuspended in 1 ml of PBS at room temperature.

Heat-killed, formalin-fixed *S. aureus* cells are prepared according to the method of Kessler (1975) and stored at −70 °C (203 K) at 10% in PBS. Before use, this cell preparation is treated for 30 min with PBS containing 2% gelatin and 0.05% Tween 20. It is washed three times and finally resuspended in the same buffer to give a 10% suspension. The immune complex is incubated with gentle shaking in 50 μl of 10% *S. aureus* suspension for 30 min at 20 °C. The suspension is centrifuged at 1000 *g* for 15 min, and the pellet is washed with 0.85% saline (pH 7.4) and resuspended in 0.1 ml of 2 M propionic acid. This is sonicated (20 W) for 10 s and centrifuged at 1000 *g* for 15 min, and the supernatant is negatively stained.

IMMUNOGOLD STAINING METHOD

Correct identification of small spherical viruses in clinical specimens and cell structures by direct negative staining is often difficult. This difficulty is also encountered in detecting spherical virions (e.g. luteoviruses) in plant specimens; such virions usually are present in low concentrations. Different virus types of similar size and without a distinct morphology are difficult to distinguish by conventional negative staining. Moreover, the identification of small, round viruses from non-viral material is problematical. These problems can be overcome by decorating the virus particles by the immunogold staining method.

Immunogold staining is more sensitive than classical IEM and consistently more sensitive than direct electron microscopy. Visualisation of the antigen–antibody reactions is achieved by labelling one of the reactants, directly or indirectly, with colloidal gold. This electron-opaque label is easily detectable in the electron microscope. Commerically available gold conjugates are usually used in dilutions of 1:2–1:10, and 10 μl of dilution is sufficient for each specimen (Kjeldsberg, 1986, 1989).

Gold labelling in suspension requires ~2 h incubation with antiserum and gold complex, whereas the 'on-grid' method requires only 10–15 min. The latter demands a relatively high concentration of virus particles, so that very short incubation durations with antiserum and gold complex can be used to avoid non-specific background staining. The success of this method depends on the optimisation of processing parameters such as the duration of incubation, and antiserum and gold conjugate dilutions. Excellent preparations of both colloidal gold sols and colloidal gold protein complexes are commercially available (ProbeTech Inc., Buckingham, Pa.).

ON-GRID METHOD

Rapid detection and identification of plant-, animal- and insect-infecting viruses can be accomplished with the on-grid method. Detection of plant viruses, including elongated (Fig. 5.16) and spherical ones (e.g. luteoviruses), in leaf-dip preparations can be accomplished in 10 min by specific secondary labelling with gold–IgG complexes (Lin, 1984). Since these complexes are electron-opaque, they enhance the visibility of decorating antibodies. This method is also termed gold-labelled antibody decoration (GLAD) (Pares and Whitecross, 1982).

The on-grid method allows detection of viral protein in both negatively stained preparations and thin sections. Individual virions can be decorated with several gold particles. The use of high antiserum dilutions (1:500) does not diminish specific labelling, while it decreases background staining. The sensitivity of this method can be increased and the background staining decreased by precoating the grid with virus-specific antiserum from an animal species different from that used for production of the primary antibody (Kjeldsberg, 1986). The duration of exposure of the grid to the virus preparation is increased when virus concentration is low, to obtain sufficient virus trapping on the grid. The size of the colloidal particles selected is determined in part by the size of the virus and the magnification used.

Procedure

A Formvar–carbon-coated grid is floated on a drop of virus preparation for ~5 min, and then it is drained by touching with a filter paper strip. The grid is floated for 1–3 min on a drop of virus-specific rabbit antiserum diluted 1:500 in 0.1 M PBS, washed six times on drops

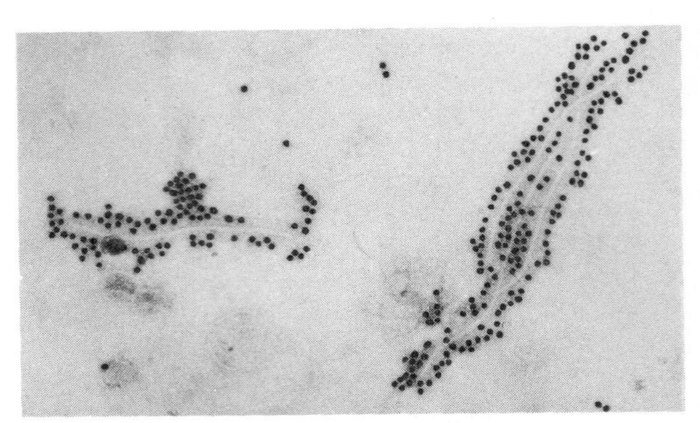

Figure 5.16 Immunoelectron microscopy for identification of wheat streak mosaic virus in the leaf-dip preparation by using secondary labelling with goat anti-rabbit gold–IgG complexes. A direct association of virus particles and labelled gold–IgG complexes is present. Courtesy N.-S. Lin

of PBS, and then floated for 1–3 min on a drop of gold-labelled goat antirabbit IgG complexes. It is drained by touching with a filter paper strip, washed six times on drops of PBS and then twice on drops of distilled water, and negatively stained with 1–2% PTA or uranyl acetate.

SUSPENSION METHOD

The suspension method is used for detecting viral antigens in fluid specimens. This method is recommended for detecting small quantities of virus, since low background staining allows long durations of incubation, which are necessary to form aggregates when the concentration of the reactants is low (Kjeldsberg, 1986). In this method viral preparation, antiserum and protein A–gold complex are mixed in suspension, and the colloidal gold-labelled immune-aggregates are dried on a grid by agar diffusion.

Equal volumes (25 μl) of the viral preparation and antiserum dilution are mixed in a microtitre plate well and incubated for ~45 min at 37 °C. A 25 μl volume of protein A–gold complex is added to the virus antibody suspension and the mixture incubated for another 45 min at 37 °C. About 50 μl of this mixture is transferred to a microtitre well, containing a Formvar–carbon-coated grid placed on top of 1% agar and allowed to diffuse into the agar for ~30 min. The grid is rinsed with two drops of PBS followed by one drop of distilled water, and then negatively stained.

GRID-CELL-CULTURE TECHNIQUE

The grid-cell-culture technique (GCCT) was introduced by Hyatt *et al.* (1987). It has the advantage of analysing and identifying viruses without purification or concentration. This technique avoids the problem of non-specific cellular debris. The GCCT allows the identification of extracellular and intracellular viruses as well as intracellular virus-specific structures. It also reveals extracellular budding of enveloped viruses. This is possible because virus-infected cells grown on grids are negatively stained. Unlike other techniques in immunoelectron microscopy–negative staining, viruses are already adsorbed to the grid. The GCCT can also be used in conjunction with protein A–gold, and is useful for identifying viruses by morphology and specific antigen–antibody reactions.

Procedure

Gold grids (200 mesh) are coated with Formvar–carbon and irradiated with UV (360 nm) for 10 min. They are rinsed with 0.1% Nonidet P40 in sterile PBS (pH 7.4) or culture medium to render them hydrophilic and then rinsed three times in PBS. Rinsing PBS and culture medium contain 100 units/ml penicillin and 100 μg/ml streptomycin. Cells are grown on these grids, which are then transferred to sterile Petri dishes and infected with virus. Grids are fixed with 2.5% glutaraldehyde in 0.1 M cacodylate buffer (pH 7.2) for 20 min, rinsed with distilled water and negatively stained with 2% PTA (pH 6.5) for 10–30 s.

Cytoskeletons are prepared by treating the grids for 2 min in each of the following solutions:

(1) PSMK containing 10 mM PIPES buffer (pH 6.8), 300 mM sucrose, 2.5 mM $MgCl_2$ and 100 mM KCl.
(2) PSMK containing 0.1% glutaraldehyde and 0.5% Nonidet P40.
(3) PSMK containing 1% Nonidet P40.

Grids are washed in PSMK for 2 min and post-fixed with 2.5% glutaraldehyde in PSMK for 20 min, rinsed in distilled water and post-fixed with 1% aqueous OsO_4 for 10 min. After dehydration and critical-point drying, cytoskeletons can be either viewed directly or coated with carbon or carbon–platinum. Virus-specific tubules can be seen within infected cells.

6 Support Films

INTRODUCTION

Thin sections of epoxy and polyester resins are sufficiently strong to withstand the electron bombardment and, as such, do not require any support film. Therefore, in order to gain contrast and resolution uncoated grids can and should be used for sections of these embedding media, provided that 300–400-mesh/in grids are employed. Since it is not easy to obtain artefact-free support films with a high transparency to electrons, films should be used only when their presence is absolutely necessary. The adherence of sections to the unsupported grids is not a problem when the sections are mounted on the matt (dull) surface of a grid rinsed in acetone prior to use.

On the other hand, in certain cases the use of a support film is required. Unless a section covers the entire grid hole and is free from holes, it may drift and break up under the electron beam, making photography almost impossible. Grids having larger openings than 200 mesh/in usually require a support film, irrespective of the type of embedding medium used. Also, the examination of sections of water-miscible embedding media such as glycol methacrylate or Lowicryl K4M usually requires the use of support films on the grids. Fragile and very thin sections are unstable under the electron beam without a support film. The study of serial sections commonly requires grids having one large hole or slit, and these must have a film for specimen support. Particulate specimens such as bacteria, viruses and cell fragments must also be mounted on a support film for examination. Negative staining requires grids coated with a support film. Support films can make available relatively large areas for observation, and thus a rarely occurring object can be located much more easily. Furthermore, holey films are used for astigmatism correction and for replica purposes for the study of surfaces.

The mass thickness of the support film influences contrast and its mechanical stability influences image clarity. Thus, the two most important criteria used for the selection of a material for preparing support films are: (1) high transparency (minimum intrinsic structure) to electrons and (2) adequate strength to withstand electron bombardment. Other criteria include a high signal-to-noise ratio (the noise arising from its structure should be small compared with the signal' from the specimen), clean surface, absence of surface irregularities and ease of preparation. High transparency is necessary in order to minimise the scattering of electrons, since the penetration power of electrons in the 25–100 kV range is already rather low. Strength is important, because exceptionally strong materials can be made very thin so that they may be highly transparent to electrons. Strength or stability of a film is also important in order to minimise the drift and shrinkage due to the heating effect of the electron beam. An increase in film thickness usually results in decreased transparency and increased strength. It is apparent, therefore, that the film thickness influences both the transparency and the strength of the film. Thus, an ideal material should be able to form thin films of high transparency and yet of adequate strength. If the film is too thick, it becomes very hot and unstable.

Many materials which form strong films are strong electron scatterers and thus give films of low transparency. The materials having lower density, on the other hand, form films that are not always adequately strong. Apparently, not many materials are suitable to be used as support films. However, films of a few materials such as carbon, graphite oxide and certain plastics have proved satisfactory. Since all materials scatter electrons, the best that can be done is to select a material having relatively low atomic number and density.

As stated earlier, the thickness of the support film has an important effect on the contrast and resolution of the image. When the thickness or mass density of the film is comparable to that of the specimen under investigation, a film can attenuate the intensity of important structural details in the image. What, then, is the useful thickness? The practical limit to the film thickness is determined, apparently, by its mechanical strength. In the case of organic materials such as nitrocellulose (collodion or Parlodion) or polyvinyl formal (Formvar), this limit is ~10–20 nm. On the other hand, very thin, stable films of the order of 1–2 nm can be prepared from carbon.

Because support films must be very thin, they are self-supporting over only limited areas, and thus they are most stable and useful when mounted on grids. Relatively thick films are needed for grids having large openings (50–200 mesh), to avoid splitting, while very thin films can be mounted on grids containing very small openings (>300 mesh). Films can be mounted successfully on openings as large as 6 mm in diameter.

MATERIALS FOR SUPPORT FILMS

Support films are commonly prepared from collodion (nitrocellulose), butvar, Formvar, polystyrene, carbon or graphite oxide, and each has certain advantages. The physicochemical properties of these materials have been discussed by Baumeister and Hahn (1978). Other materials which have been used as support films include glass (van Itterbeek *et al.*, 1952), silicon (Polivoda and Vinetskii, 1959), mica (Heinemann, 1970), vermiculite (Baumeister and Hahn, 1978), and metals such as aluminium (Müller and Koller, 1972), beryllium (Vollenweider *et al.*, 1973), boron (Dorignac *et al.*, 1979) and tantalum (Peters, 1986). The usefulness of most of these supports has been reviewed by Baumeister and Hahn (1978).

PLASTIC FILMS

Collodion or Formvar is probably the most common material used to prepare a clean relatively smooth support film for routine electron microscopy. Both collodion and Formvar films are somewhat hydrophilic (contact angle→0 °); aluminium and silicon monoxide films are strongly hydrophilic when freshly made, whereas carbon films are hydrophobic. Collodion is soluble in ethyl or amyl acetate and acetone, and collodion films are rather easy to prepare. Particulate droplets placed on this film spread sufficiently to be examined as unaggregated particles. The wetting properties of collodion can be improved further by serum albumin. Since collodion has comparatively little mechanical strength, films tend to tear as a result of electrostatic charge built up by the electron beam, especially when films are unshadowed and are 15 nm or less in thickness. This is the reason why films of ~30 nm thickness are usually employed. Films of this plastic cast on glass are much stronger than those spread on water.

The strength of collodion films can be greatly improved by pre-irradiating the grid with a low-intensity electron beam at a low magnification (~1000×) (Johansen, personal communication). The radiation damage (mass loss) which occurs cross-links the carbonaceous 'backbone' of the collodion structure to a higher degree of mechanical strength.

Formvars are prepared by causing a polyvinyl alcohol to react with formaldehyde. Formvar is soluble in ethylene dichloride, chloroform or dioxane. This film is prepared by casting Formvar on glass or water. The films of this plastic are stronger than those of collodion, because of their greater irradiation resistance. Formvar films show 10–100-fold better stability and 60% less mass loss during irradiation, compared with collodion films. However, these films are more difficult to prepare. The possible difficulties include the failure of the film to detach from the glass slide, uneven thickness and the formation of small holes. Furthermore, since Formvar is less hydrophilic than collodion, particles in a particulate droplet aggregate readily. The stock solutions are less stable than those of collodion.

Polystyrene is another plastic which, owing to its inherent radiation resistance, has been used as a support film (Baumeister and Hahn, 1975). This plastic has high electron transparency because of its low density (1.05 g/cm^3). The film can be made hydrophilic by short UV treatment. It has been claimed that no additional stabilisation by carbon is necessary, even when working at high beam intensities. Polystyrene is soluble in benzene, diethyl ether and methyl isobutyl ketone. The disadvantage of this film is its brittleness and poor adhesion to grids. Other plastic films used are butvar and Pioloform (trade name for a polyvinyl formal film).

All plastic films are prone to decomposition by the electron beam. Under the electron beam, the mass loss has been estimated to be 65–85% for collodion, 15–40% for Formvar (Reimer, 1965) and 5% for polystyrene. The shrinkage of plastic-supported sections may be as much as 10% in one dimension (Watson, 1955). Even if the sections are stable, they may be distorted, owing to unstable support film. The net result is reduction in contrast and shrinkage of the sections, causing distortion of the fine structure. The persistent tendency of plastic films to drift when first irradiated can be minimised by reducing the section thickness and by initial stabilising with the electron beam at low intensities. The inherent weakness of plastic films can be overcome by stabilising them with a thin layer of carbon.

ESTIMATION OF PLASTIC FILM THICKNESS

The thickness of a plastic film can be estimated by viewing it while it is floating on water under the reflected light. Films showing any interference colour, preferably over a dark background, are too thick for routine work. A bluish colour indicates a thickness of 75–100 nm. 'Colourless' films are 15–30 nm in thickness, and are too fragile to be used for routine work. Films showing a pale grey colour (30–50 nm) are suitable for routine work.

The film thickness can be estimated more accurately by calibrating, by interferometric methods, by an apparatus devised by Revell *et al.* (1955). The apparatus is calibrated for a given plastic, solvent and solution concentration. It should be noted that a plastic film is of irregular thickness. Reimer (1967) has presented a table which can be used to derive the thickness of plastic cast films. Baumeister and Hahn (1978) have given a calibration curve for cast polystyrene films withdrawn from the polymer solution at a constant speed by means of a motor-driven apparatus.

PREPARATION OF PLASTIC FILMS

Collodion or Formvar films can be prepared by casting them either on the glass or on the water surface, although films cast on glass are considered to be somewhat stronger. Collodion films are usually prepared from solutions ranging in concentration from 0.5% to 2%, while Formvar films are made from solutions that range in concentration from 0.1% to 0.5%. Other films used are prepared from 0.25% solution of cellulose acetobutyrate (Triafol) in ethyl acetate (Reichelt *et al.*, 1977) and from 1–0.2% solution of Pioloform in chloroform (Umar, 1982). In general, the thickness of the film desired determines the concentration of the solution to be used; solutions of higher concentration apparently yield thicker films. Excessively low initial concentrations of plastics result in thin films that are difficult to strip from the glass slide.

The use of stored grids with plastic films is not recommended, because storage may result in loosened films. Therefore, if possible, only freshly coated grids should be used. If necessary, the polymer solutions can be stored in bottles equipped with ground-in joints to prevent excessive evaporation of the solvent.

Methods for collecting plastic films of different types are given later. The following general method is recommended for collecting large numbers of coated grids; this approach is especially useful for collecting slot and large-mesh hexagonal grids for serial sectioning studies (Trett and Crouch, 1984). Another advantage of this method is that it provides a means of safe handling of the coated grids for carbon coating. Cleaned grids are placed on the plastic film floating on the surface of water. A strip of Parafilm (wax film) (80 × 25 mm) is placed over a glass slide, which is held (Parafilm side downward) above one end of the film at 45 ° to the surface of the water (Fig. 6.1). The slide is gently pushed beneath the water at the same angle. The film will adhere to the Parafilm and, along with the grids, is drawn beneath the surface. Once the film and grids are completely submerged, the slide is rotated until the Parafilm, support film and grids are uppermost, and

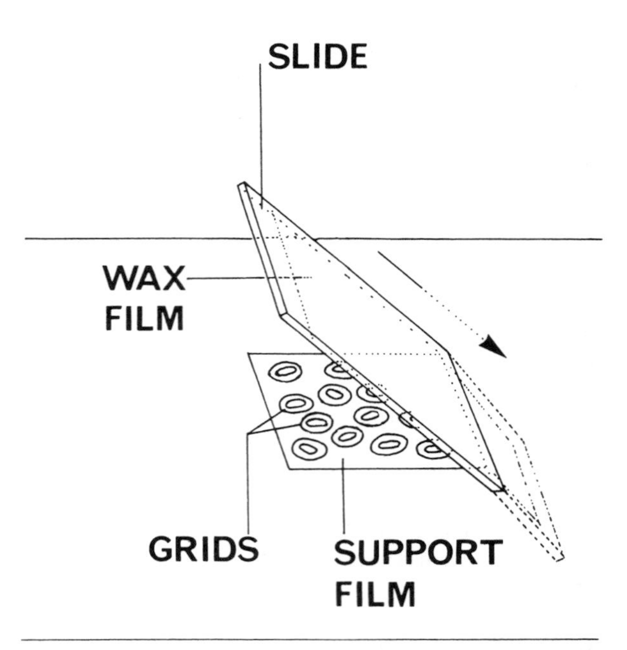

Figure 6.1 General method for collecting a large number of grids coated with a plastic support film. The diagram shows the position of the glass slide above one end of the plastic film at 45° to the surface of the water. The arrow indicates the direction of the movement of the slide. From Trett and Crouch (1984)

then slowly withdrawn, allowing the surface tension to remove any surplus water.

Another method for preparing Formvar films cast on glass and using Domino Rack has been presented by Moran and Rowley (1987). This method produces thin support films particularly useful for slot grids to mount serial thick or thin section. Although the method is relatively lengthy, it produces hole-free films. This method is recommended when other methods given here do not produce satisfactory results.

It is not easy to obtain clean, streak-free and hole-free Formvar (or collodion) films. One of the major factors that control the production of hole-free films is the moisture content of the atmosphere in which the slides dry after being dipped into the Formvar solution. Although the area in which the Formvar-coated slides are dried can be made locally dry by using a dehumidifier or a heat lamp, these procedures are not always successful in producing a controlled environment. Moreover, dust from the air tends to contaminate the films. The danger of inhaling noxious fumes of ethylene dichloride persists.

These problems can be circumvented by using a controlled-environment chamber (Keene, 1984). The materials needed are: (1) a 'glove bag' (inflated size is 42 × 42 × 26 cm) (Cat. 17-17, Instruments for Research

& Industry, Cheltenham, PA); (2) a regulated source of dry, pure nitrogen gas; (3) a 0.35–0.5% (w/v) solution of Formvar in ethylene dichloride in a *clean* bottle; (4) new glass slides cleaned in 90% ethanol and wiped dry with lint-free paper; (5) two covered Petri dishes containing enough phosphorus pentoxide (P_2O_5) powder to form a thin layer; (6) filter papers; and (7) a flexible hose (outer diameter 13 mm).

The bag is formed in such a way as to allow the user to place the necessary materials inside through a large sealable opening, and to have access to these materials via gloves protruding into the chamber. The bag is then purged with dry nitrogen gas via a flexible hose inserted into a smaller opening in the bag. The purge is carried out by inflating the bag with dry nitrogen gas and then collapsing it; this procedure is repeated several times until the humid air in the bag has been replaced with the gas. Residual moisture can be removed by uncovering the dishes containing the desiccant P_2O_5. This desiccant is a strong skin irritant. The slides are coated with the Formvar and then dried inside the chamber. Floating the films from the slides onto the water surface and coating the grids are carried out outside the chamber, according to the standard procedures described later. The fumes can be exhausted by opening the bag in a fume-hood.

Formvar Film Cast on Glass

(1) Prepare a 0.3–5.0% solution of Formvar in ethylene dichloride (solvent) at room temperature and transfer it to a Coplin jar: the jar is filled to a depth of 7–10 cm (Fig. 6.2A–E). It is essential that the solvent be dry. Clean thoroughly a new standard glass slide with soap and water to remove the surfactant. It can be bathed overnight in aqua regia, rinsed with water followed by weak solution of ammonia (to neutralise any residual acid), and finally with distilled water.

(2) Dry the slide and dip it about two-thirds of the way into the Formvar solution (Fig. 6.2A); after 2–3 s remove it from the jar. Keep the jar covered when not in use, as otherwise water in the polymer solution will result in holes in the film. Drain the slide onto a filter paper (Fig. 6 2B), and air-dry vertically for 1–10 min in dust-free air. Carry out this procedure by propping the slide up at an angle against the object, so that only the uncoated end of the slide touches the object. The drying should be accomplished under cover. Alternatively, dry the slide in a desiccator.

(3) Fill a large trough to the brim with distilled water and, immediately before use, clean the water surface by sweeping it with a sheet of lens paper.

(4) Scrape four edges of the slide with a razor-blade (Fig. 6.2C); a few lines scribed across the slide afford additional routes for capillary permeation of water. The separation of the film from the slide occurs when water penetrates the space between the film and the glass surface by capillary force. During scraping, the contamination of the film surface with glass and plastic particles should be avoided. After scraping, the slide can be placed in a freezer for ~6 min to facilitate the separation of the film from the slide. Alternatively, some workers prefer to breathe onto the slide before lowering it into the water; this treatment may be sufficient for the water to penetrate.

(5) Hold the slide at its uncoated end and lower *slowly* into the water at an angle of ~45° from the horizontal (Fig. 6.2D). To release the film from the slide, tease the water surface tension for a few seconds with the slide until the film begins to be released. If the film is incompletely released, it should be teased away from the slide corner with a needle. Under an appropriate reflective lighting and dark background, film release can be observed. The film surface appears shiny white, while the water surface appears non-reflective. Submerge the slide gradually, leaving the film from its upper side floating on the water. If stripping of the film from the glass slide is a problem, a piece of LKB glass strip used for the preparation of glass knives is recommended. Easy separation of the film from the glass slide can also be accomplished by keeping the scraped slide overnight in a moist chamber prior to lowering it into water. A simple method is to support the slide on glass rods and place wet filter paper under the slide in a Petri dish, which is then covered. Under reflected light, the thinnest films appear grey, while gold films are too thick. The concentration of Formvar can be varied to obtain the desired thickness.

(6) With forceps, quickly but carefully place 200-mesh grids (previously rinsed in acetone), matte side down, on good areas of the floating film (Fig. 6.2E). Avoid wrinkled and dusty portions of the film. Exerting light pressure on each grid by *gently* tapping it with the forceps ensures good adherence to the film. If the adherence of the film to the grids is a problem, the grids can be dipped (dull side down) into 0.5% Formvar solution and then blotted by touching the shiny side on a filter paper.

(7) Place a piece of lens paper or Parafilm over the floating film-grids and lift off the water surface.

(8) Place the lens paper, with grids upward, in a covered Petri dish or a desiccator, where they can remain until needed.

(9) Pick up each grid with the forceps as the film tears around the grid, leaving the film on the grid intact. If the film tears on the grid, the film around the grid should be perforated with a needle before picking up the grid. If needed, these grids can be coated with a

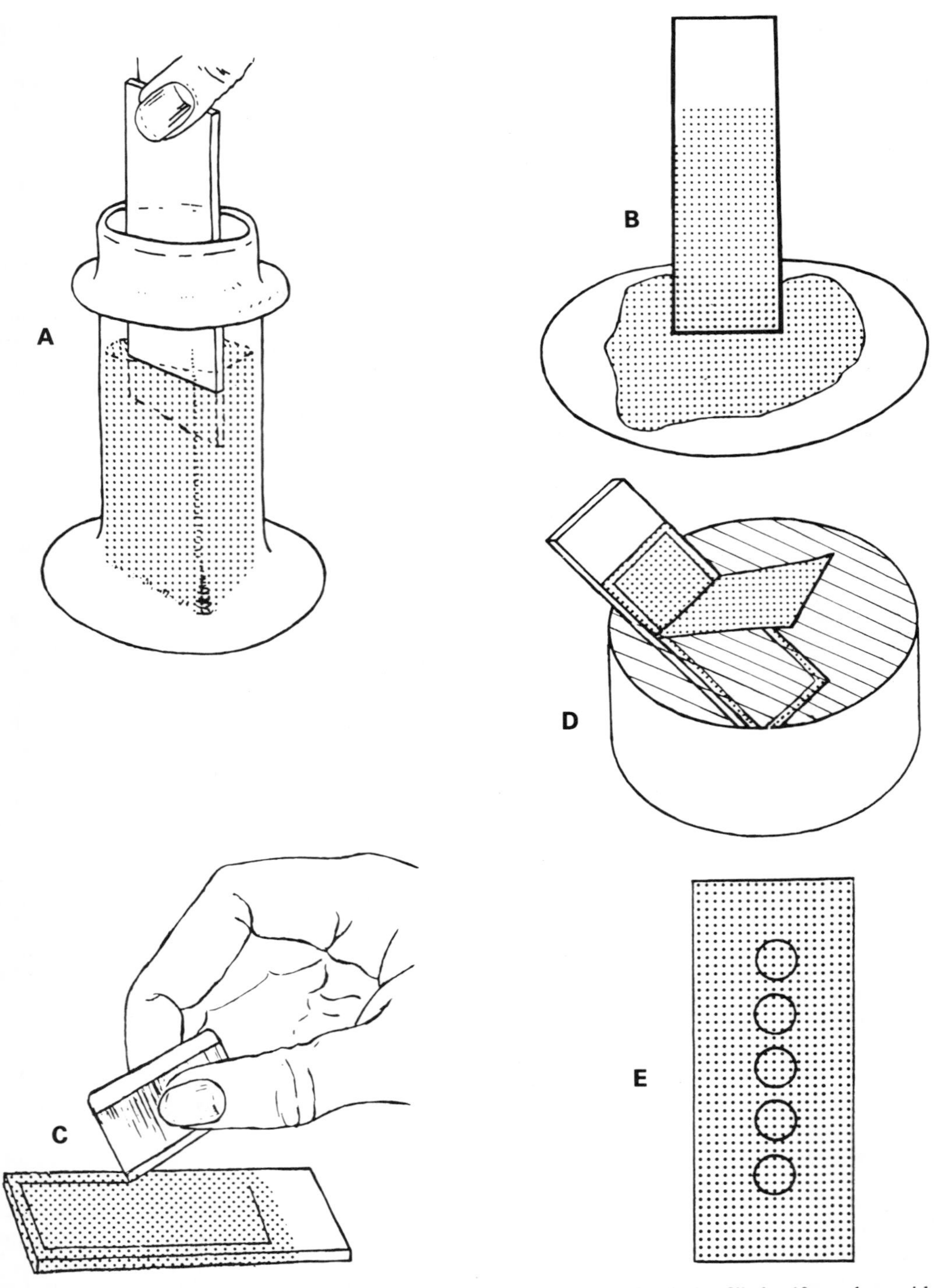

Figure 6.2 Preparation and mounting of plastic support films on grids. A: A Coplin jar filled ~10 cm deep with Formvar solution and a glass slide in the process of being dipped into it. B: The glass slide is drained onto a piece of filter paper for air-drying in a dust-free place. C: The four edges of the film deposited and dried on the glass slide are scraped with a razor-blade. D: The glass slide with the scraped film is lowered at an angle of 30–40° into the water surface in a suitable container. E: Five grids, with appropriate distance between them, are placed on the floating film

thin layer of carbon to increase the stability of the plastic film.

If difficulty is encountered in coaxing the film off the glass slide in this method, the slide can be pretreated with Victawet, a commercially available wetting agent (Edwards *et al.*, 1984) (E. F. Fullam, Inc.). Slides are dipped in 100% ethanol and then wiped dry. A tungsten wire basket is secured in a vacuum evaporator, and with fine forceps a small piece of Victawet (1 mm^3) is placed in the basket and then gently forced towards the bottom, or conical end. Three slides are placed directly under and 5 cm from the tungsten wire basket. The bell-jar is evacuated and the current is *slowly* increased across the tungsten wire. When heated properly, the Victawet melts and evaporates over the slides. About 30 s is needed to completely evaporate the Victawet. These slides are now ready for use.

Formvar Film Cast on Water

Formvar films are difficult to strip from clean glass slides. If the glass surface is not clean, the film may be contaminated. Since the glass surface itself is structurally uneven, the films are highly surface-structured. Ultrathin Formvar films can be made at an air–water interface by the drop method (Davison and Colquhoun, 1985). These films show improved flatness and stability, and their surface is smoother than that of collodion and glass-stripped Formvar films. The films prepared by the drop method are comparable in quality to pure carbon films made of cleaned mica; both of these films are useful for shadowing and negative staining preparations. The quality of these Formvar films does not seem to be affected by humidity variations or even water in the Formvar solution. Formvar concentrations of 0.25–0.4% in ethylene dichloride are recommended. Lower concentrations (~0.1%) produce nets consisting of many tiny holes, which, after carbon stabilisation, can be used for high-resolution studies.

(1) Place one drop (~15 μl) of Formvar solution from a Pasteur pipette from a height of ~2 mm onto the surface of scrupulously clean, double-distilled water. The film formed is almost invisible, but occupies a circular area ~8 cm in diameter.

(2) Place the grids (shining side down) on the central region of the film, at which time the film will become visible because of the presence of wrinkles around the grid edges.

(3) Touch the surface of the film grids with a piece of Parafilm and lift it. The film will adhere to the Parafilm, sandwiching the grids in between.

(4) Place the grids on filter paper in a Petri dish and cover it.

(5) After obtaining one film, discard the water in the container and replace it with fresh water for making the next film.

If relatively high concentrations of Formvar (2%) need to be used, films can be made by placing a very small drop of the Formvar solution onto the surface of a mixture containing 30% sucrose, 1% propylene glycol and 0.1% acetic acid (Day, 1984). Owing to its surface tension, pure water is not used for floating the film in this method. The drop of the Formvar solution formed at the tip of a Pasteur pipette is too large. Therefore, a fine tip must be drawn out with heat before this pipette can be used. Grids are placed on the floating film (avoiding the central area), and the grids along with the film are picked up on a piece of Parafilm. Grids are washed in distilled water to remove traces of the mixture.

Formvar Film Cast on Mica

The drop method described earlier cannot be used for making thicker Formvar films (20–30 nm). Mica instead of glass is an excellent substrate for making such films. The cleaved mica surface is exceedingly clean and smooth and the films strip easily from it. These films are very clean and uniform in thickness, and are less susceptible to methanol degradation than are Formvar films made on glass (Davison and Colquhoun, 1985). The latter characteristic is useful for post-staining the sections with heavy metal salts dissolved in methanol. Only freshly cleaved surface of high-quality mica (no dark inclusion spots) should be used. To make the Formvar film, the cleaved mica surface is submerged into 0.25–0.5% Formvar solution in ethylene dichloride, and then allowed to drain in a solvent vapour atmosphere. The film strips easily without sticking.

Collodion Film Cast on Glass

(1) Prepare a stock solution of 0.5% collodion in amyl acetate.

(2) Prepare and mount the film on grids as described for Formvar film cast on glass (p. 355). Alternatively, place carefully a piece of wire gauze (~1 in square) on the bottom of the water-trough and then place individually the desired number of grids on the wire gauze. Lift the wire gauze out of the water carrying the grids and the film, or lower the water level in the trough sufficiently so that the film is deposited over the grids.

Collodion Film Cast on Water

(1) Completely fill a trough ~20 cm in diameter with distilled water.

(2) Place a piece (~3 cm^2) of wire gauze (~100 mesh/in) on the bottom of the dish and place a number of grids some distance apart from one another on the gauze.

(3) With a pipette allow one or two drops of 2% solution of collodion in amyl acetate to fall on the water surface at the centre of the dish. These drops should quickly spread, owing to surface tension, into a thin film.

(4) After the solvent is evaporated, remove the solid film left behind with a needle to clean the water surface.

(5) Form a second film in the same way as described above.

(6) Lift the wire gauze carrying the grids out of the water in such a way that the floating film is deposited smoothly over the grids.

(7) Allow the grids to dry prior to use.

Butvar Film Cast on Water

Butvar B-98 (Monsanto Co.) is a polyvinyl butyral resin, containing 20% polyvinyl alcohol. It was introduced as a support film by Oliver (1973). Handley and Olsen (1979) have presented a simple procedure for preparing Butvar support films. The resulting films are very hydrophilic and can be used for negative staining of hydrophobic specimens such as isolated lipoproteins (Handley *et al.*, 1981). These films are thought to be more hydrophilic and elastic and mechanically stronger than Formvar and collodion films. Butvar films adhere tenaciously to the grid surface, and are excellent for single-slot grids for serial sections. These films can be placed in and taken out of high-vacuum chambers many times without popping (Milroy, 1985).

Support films are prepared from a stock solution of 0.25% Butvar in chloroform. This solution is heated to 50 °C with continuous stirring to ensure complete solubility of the resin and cooled to 40–42 °C just before film casting. A hot-plate must be used and the temperature not allowed to go above 50 °C, as chloroform boils at ~55 °C. Grids are rendered more adhesive by pretreatment with a 0.15% solution of Butvar that has been heat-treated. This is accomplished by transferring the grids on a piece of filter paper and then placing one drop of this solution on each grid. The grids are allowed to dry.

A glass slide is washed with 70% alcohol, dried and then dipped into the Butvar solution. It is drained in a vertical position. The upper part of the slide contains the thinnest film (40 nm). The slide is air-dried and scored with a razor-blade, and then the film is released by flotation on the water surface. The pretreated grids are placed on the thin part of the film. The film along with the grids is retrieved from the water with Parafilm (5 × 10 mm). This is carried out by gently touching one edge of the Butvar film with the Parafilm, lowering the latter into complete contact with the former, and then lifting it out. The coated grids are separated by scoring around their periphery. Alternatively, Butvar concentration can be reduced to 0.2% to facilitate separation of coated grids. Coated grids can be stored for up to 1 year when left on the Parafilm without any deterioration. The Butvar solution can be stored in a brown bottle at 22 °C for several months.

The following recommendations are made for Butvar films: silver films (~60 nm) for serial and conventional electron microscopy; grey films for additional support for electron microscopic autoradiography; and thick, gold films for high-voltage electron microscopy (Milroy, 1985).

TRANSFER OF PLASTIC FILMS ONTO GRIDS

In addition to the methods for transferring the floating films directly onto the grids described above, methods employed to deposit film onto individual grids include a special ring-tool (Bradley, 1965), a wire loop (Gay and Anderson, 1954) and a plastic ring.

In the ring-tool method, using a razor-blade the plastic film on the glass slide is scratched into squares of 0.25 in, which are floated on the water surface. Grids are placed on top of metal pegs with the same diameter as that of the grids. The floating squares of film are lifted on a ring-tool and placed onto the grids.

In the wire loop method a plastic film is transferred to a small wire loop with a small handle. The loop is then inserted into the trough fluid at an angle of ~45° and aligned under the floating ribbon. As the loop is raised, the ribbon is picked up on the film. The film with the ribbon is then transferred to a grid with a slit opening. This is accomplished by placing the grid on top of a Lucite rod ($\frac{1}{8}$ in diameter) mounted on the condenser of a compound microscope and attaching the loop to the microscope stage by means of a Lucite circular disc. The disc serves to support the loop. By manipulating the disc, the ribbon is optically superimposed over a slit in the grid. By raising the condenser, the Lucite rod with the grid is passed through the loop and the plastic film with the ribbon is attached to the grid. A modification of this method was employed by Sjöstrand (1967).

In the plastic ring method, plastic rings ($\frac{5}{8}$in diameter and $\frac{1}{16}$ in thick) are placed on a wire screen which rests on a support ($3\frac{1}{4}$ in diameter and 2 in high) cut from the middle of a plastic bottle. The bottle is notched at the top and bottom to facilitate the free flow of water. This assembly is placed in a container (8 in diameter and 3 in high) filled with distilled water to cover the rings. Four drops of a 2% collodion solution in amyl acetate are released on the water surface. A uniformly thin film is formed, provided that care has been taken to remove water bubbles and debris from the water surface and to protect the surface from air currents. After the film is formed, the water is siphoned from the container until the film comes in contact with the rings. The screen is removed and placed on a paper towel to drain and dry in a dust-free area.

PLASTIC COATING OF SINGLE-HOLE OR SLOTTED GRIDS

It is not easy to direct and deposit precisely the floating plastic film on the grids that have been placed under the film. To avoid unnecessary stress to the plastic film during its deposition on relatively thick, single-hole or slotted grids, they can be positioned over a perforated plate (grid coating plate No. 66-2, Ted Pella Inc.) before lowering the floating film on them. To allow easy removal of the grids, a small groove (1 mm long and 0.1 mm deep) is drilled on one side of each hole of one-third of the plate (Fig. 6.3) (Larramendi, 1987). One-third of the plate (2.5 × 7.6 cm) is about the size of the film that is floated off a glass slide. The same area of the plate is coated with Pelco Synapstick glue, which is placed in a Petri dish. The plate is supported by two glass rods to prevent it from adhering to the dish. The dish–plate is allowed to dry for 20 min at 22 °C, baked in an oven at 150 °C for 10 min and then coated before use.

The grids are gently placed (without pressing) over the holes on the sticky surface of the plate. The glue holds the grids firmly in place. The plastic film is floated off the glass slide on the surface of the water. Holding it by the non-treated side, the plate-grid is introduced into the water and then brought up, catching the film on the surface. The grids, with the film on top, are dried on the plate at 22 °C in a dust-free place. The grids are left on the plate until use. The plate can be cleaned with xylene and recoated with glue.

VERMICULITE FILMS

Vermiculite films are considered to have an advantage over carbon films that the latter substrate can introduce

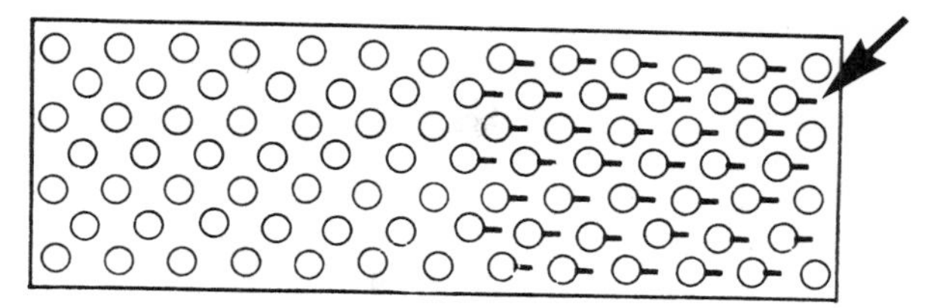

Figure 6.3 Schematic drawing of a perforated Pelco plate to which grooves (arrow) have been drilled for grid pick up. From Larramendi (1987)

confusion in the apparent structure of a specimen if the electron micrograph is inadvertently or intentionally taken under defocused conditions. Vermiculite films, on the other hand, show less sensitivity of texture to focal conditions. Like mica and graphite, vermiculite produces single-crystal films.

Vermiculite is a native hydrated magnesian alumino-silicate. It is similar to the micas in that it occurs as plate-shaped crystals consisting of superimposed silicate layers between which are double sheets of water. The negative charge of the silicate layers is balanced by various interlayer cations (Baumeister and Hahn, 1974). The high ion exchange capacity of vermiculite renders the selective adsorption phenomenon for biomolecular systems possible by producing anion or cation surface vacancies.

Exfoliation of vermiculite can be carried out either chemically (Manley *et al.*, 1971) or by heat treatment (Baumeister and Hahn, 1974).

Chemical exfoliation involves introduction of organic ions (butylammonium) into the interlayer position. Single crystals of vermiculite containing such organic interlayers swell macroscopically in water. Swelling is anisotropic and only occurs in a direction perpendicular to the plane of the silicate layers. When swollen crystals are subjected to shearing forces (ultrasonication), the weak interlayer forces give way and the individual layers disperse. This approach is time-consuming.

Thermic exfoliation is a superior approach. Owing to the explosive release of water upon sudden heating, the vermiculite crystals expand normally to the basal cleavage. Vermiculites with high water content (>12%) are used. Dehydration of vermiculite is complete at 900 °C.

A third (and the best) approach to producing extremely thin crystals involves exfoliation with hydrogen peroxide (Baumeister and Hahn, 1976, 1978). The decomposition of hydrogen peroxide between the silicate layers, which may be catalysed by the inorganic interlayer cations, and the evolution of oxygen yield satisfactory separations. Single-crystal, very thin films (2–3 nm) can be produced with this approach.

(1) Place 10 g of vermiculite in excess hydrogen

peroxide (30%) for 24–48 h at 4 °C.

(2) Collect vermiculite flakes on a filter paper and transfer to an oven heated to 100 °C; this treatment causes the crystals to expand immediately. The bulk volume may increase by a factor of 30.

(3) Resuspend these exfoliated crystals in double-distilled water, and treat with an ultrasonic disintegrator (20 kHz) for several minutes to separate the layers completely.

(4) Allow the suspension to settle for 15 min, and discard incompletely exfoliated vermiculite.

(5) Centrifuge the supernatant at ~10 000 *g* for 20–30 min.

(6) Discard the sediment and dilute the supernatant to a final concentration of 40–60 μg vermiculite/ml.

(7) Place a drop of this suspension on a hydrophilised perforated support film on a grid for ~10 min and then remove the excess suspension.

CARBON FILMS

Carbon films are uniformly amorphous and highly transparent to electrons, because carbon has a relatively low atomic weight. Owing to their strongly interconnected three-dimensional network structure, carbon films have remarkable mechanical stability, even when as thin as 1–2 nm. They are characterised by good electric conductivity, which prevents accumulation of charges. Another advantage of carbon films is that they routinely can be made to cover relatively large areas in a uniform manner. Because of the above-mentioned advantages, carbon films are most frequently used for high-resolution work. The structure of carbon films is influenced by evaporation conditions (vacuum, rate of evaporation, direct or indirect evaporation and target temperature), heat treatment, electron irradiation and ageing.

Carbon films have the disadvantage of being extremely hydrophobic, and thus cause particles in the water suspension of particulate specimens to aggregate. The primary reason for hydrophobicity is the presence of contaminants such as adsorbed hydrocarbons. Also, these films show defocus-dependent granularity in phase contrast. The preparation of carbon films is relatively laborious, for a number of operations are involved, in each of which there is some risk of contamination. In fact, it has been indicated that certain specimens supported by carbon are contaminated faster under the electron beam than those mounted on plastics films (Kölbel, 1974). Nevertheless, these minor disadvantages are offset by the greater mechanical strength and stability under the electron beam. It should be noted that since pure carbon sublimes above 3500 °C without preceding liquefaction, carbon films are exceedingly thermostable. Furthermore, carbon films can be rendered hydrophilic by storing them in a refrigerator, by placing them in a bell-jar and producing a discharge with a Tesla coil, or by adding a small amount of ethanol to the solution. The aggregation of particles can be reduced by employing the high-pressure spray method instead of drop application.

Carbon films are preferred to plastics films because the former are stronger and much more stable under electron bombardment, and show a much finer structure. In fact, carbon films thicker than 10 nm are considered to be virtually unbreakable under the electron bombardment. Very thin films of the order of 2 nm are commonly employed for high-resolution electron microscopy. Williams and Glaeser (1972) have reported a method for producing carbon films as thin as 0.4–0.16 nm, which can withstand spray-drop deposition of biomolecules and are stable under electron bombardment. Lamvik *et al.* (1987) have introduced an oil-free vacuum system for the production of thin, contaminant-free, hydrophilic carbon films.

PREPARATION OF CARBON FILMS

Carbon can be evaporated *in vacuo* to form a uniform amorphous film. Usually carbon films are deposited on the grids prior to collecting the sections, although sections can be collected on uncoated grids, stained and then stabilised by evaporating a carbon film. For general purposes, carbon films are made by evaporating the carbon as a thin layer from a carbon arc. To accomplish this, a vacuum evaporator is required. The vacuum evaporator consists essentially of a glass bell-jar evacuated by a diffusion pump. Inside it are two carbon rods (each 4–7 mm in diameter), one of which is fixed and the other of which is held lightly against it with the tips in contact. The pressure to accomplish contact is maintained by means of springs or by gravity. Either both tips of the two carbon rods are pointed or only one tip is pointed and the other is squared. The advantage of the latter arrangement is that, although the contact area remains small, the amount of misalignment which can be tolerated is increased. After sharpening, the approximate final diameter of one tip is 0.25 mm and that of the other tip is 1.5 mm.

To make carbon films with reproducible thickness, either one or both of the rods should be cylindrical; a diameter of ~1 mm is suitable. As opposed to a conical carbon rod, the cylindrical one can evaporate with a constant current and thereby reduce the photon energy exposed to the substrate. The latter factor is particularly important when carbon is evaporated onto plastics supports (Johansen, personal communication).

Evaporation of carbon is carried out at a pressure of 10^{-2}–10^{-5} Torr when an alternating current of ~50 A at 20 V is passed through the electrodes. It is recommended that the carbon be deposited as soon as the required pressure is achieved. However, carbon rods should be allowed to glow to red-hot before actual evaporation, in order to achieve effective electrical current between the ground ends of the carbon rods and to release gas pockets within the graphite rods and, hence, reproducible results. The heat produced in the region of contact between the two points is sufficient to cause rapid evaporation of the carbon. The target is placed 10–15 cm from the source, and the exposure should not last longer than a few seconds.

The thickness of the carbon film deposited is controlled by the rate of evaporation of carbon, which, in turn, is controlled primarily by the initial compression in the spring. In order to detect the subjective thickness of the film deposited, a clean piece of white porcelain, or similar surface, with a drop of vacuum oil on it is placed near the target. It is estimated that when the area of the porcelain without the oil shows a light-brown colour (only the non-coated surface of the porcelain is darkened when carbon is evaporated) compared with the white colour of the oil-covered area the film thickness is 5 nm, which is a satisfactory thickness for most studies. For more accurate thickness estimation methods, the reader is referred to Baumeister and Hahn (1978) and Johansen *et al.* (1980). Grey carbon films are claimed to be more stable than those of brown colour, and are obtained in a better vacuum (10^{-5} mbar) and after degassing the carbon rods, prior to evaporation, by preheating with a current of 20–25 A.

Whether or not rapid evaporation of carbon has an advantage over a slow evaporation is not clear. Indirect evaporation of carbon yields smoother and less grainy films than those produced by direct evaporation. The former method prevents larger carbon clusters from reaching the target. In an indirect evaporation unit, a stop is positioned between the evaporation source and the target, so that there is no direct sight from source to target. A versatile unit for both direct and indirect evaporation has been designed by Johansen (1974). Kölbel (1976) has given the conditions under which homogeneous carbon films can be produced.

Carbon Film Prepared by Evaporating Carbon Filament

The thickness of carbon films produced by conventional methods (carbon rod evaporation) is difficult to control accurately, for intense photon radiation prevents accurate measurements with quartz monitors during film deposition. The accuracy in thickness measurement is especially critical for thin films (2 nm). Furthermore, photon radiation emitted during evaporation and carbon scattering in a low vacuum may generate crystallites (~1 nm) and impart a resolvable structure to the films. These problems can be diminished by evaporating a carbon filament under high vacuum (10^{-5} Torr) (Peters, 1984). According to Vesely and Woodisse (1982), evaporation can also be accomplished at a low vacuum (10^{-2} Torr) by burning a carbon filament with a high current. This method is very rapid and efficient, and the films produced seem to be less influenced by the defocusing-dependent granulation visible at high magnifications (>50 000 ×) (Namork, 1985).

To obtain reproducible results, once optimal conditions have been determined, voltage, length of carbon filament, number of strands in the filament, distance from the source and vacuum should be the same. Reproducible carbon deposition is obtained without spark generation with filaments of 6–15 mm in length. A constant voltage supply is required, since voltage variations in the power line can account for variation of up to 20% in film thickness (Peters, 1984). The films produced are ~2.3 nm thick, which corresponds to a very light, hardly visible grey tone on filter paper. To prevent the bell-jar from becoming completely blackened during carbon filament evaporation, an inexpensive shield can be constructed (Smith, 1984). The shield is positioned by resting it on the clamps holding the carbon filaments. Carbon filaments should be handled with care as they are very brittle. They are commercially available (Balzers).

Substrates for Carbon Evaporation

Substrates used for preparing carbon films include glass, mica, plastics films, perforated plastics films, organic glass and glycerol. The films prepared on glass and mica are floated off onto a liquid surface, while in the other cases the substrate is dissolved after film formation. Very thin, continuous carbon films can be deposited onto the thicker carbon films containing numerous holes (Dubochet *et al.*, 1971). The thicker film is deposited on a collodion substrate which is dissolved away. The advantages and disadvantages of the various substrates have been summarised by Baumeister and Hahn (1978). Methods for the deposition of carbon films on various substrates are given below.

Carbon Film Deposited Directly on Grids

Very thin films of carbon can be deposited directly on

sections which have been mounted on grids. In order to facilitate the adhesion of carbon film to copper grids, they are treated with an adhesive solution (see below) prior to mounting the sections. A number of grids with mounted sections are arranged on a glass slide which is then placed in the vacuum evaporator. A very thin carbon film is sufficient to give necessary support to the sections. The advantage of this method is its convenience; however, it is not recommended for quality work.

Carbon Film Cast on Glycerin

(1) Prepare a 50% glycerin solution in glass-distilled water.

(2) Filter the solution through a 0.2 μm disposable filter to remove small particulate material.

(3) Transfer the solution to a Coplin jar.

(4) Clean a microscope glass slide with Kimwipes tissue paper and then pass it through a Bunsen burner flame.

(5) Dip the slide into the glycerin solution and allow it to stand on its short edge on bibulous paper to drain. Room temperature causes the glycerin to drain rapidly from the slide, leaving a thin, uniform coating on the slide surface. In humid atmospheres, glycerin solutions of lower concentrations (35%) can be used.

(6) Place the slide on the baseplate of a vacuum evaporator (equipped with a glow discharge unit) on a sheet of white paper.

(7) Pump down the evaporator to 0.1 Torr with the mechanical pump.

(8) Glow-discharge the slide at maximum current for 1 min.

(9) Immediately after this treatment, pump down the evaporator rapidly to $5\text{–}7 \times 10^{-3}$ Torr with a liquid nitrogen-cooled oil diffusion pump.

(10) Evaporate carbon slowly until a pale beige colour appears on the paper surrounding the slide.

(11) Score the slide with a razor-blade into squares, each of which is slightly larger than the diameter of a grid.

(12) Float off the film from the slide onto a clean water surface.

(13) Rinse copper grids (300–400 mesh) in acetone, air-dry and place them (shiny side up) on bibulous paper.

(14) Pick up the drop of water containing a square of film from beneath the water surface, using a wire loop attached to a wooden cotton-tipped applicator. The grid should be held stationary by a straight wire attached to a wooden stick while the water drop film is lowered onto the grid.

(15) Place the bibulous paper along with the coated grids in a refrigerator to dry the films slowly.

Coated grids can be stored in a closed container for over a month without apparent deterioration. The grid can be checked in the TEM for strength of film under the electron beam and uniformity of grain.

Carbon Film Prepared on Glass

Moderately thin films can be prepared by depositing a layer of carbon on a thoroughly cleaned glass slide and then floating it on a clean water surface. At times it is difficult to strip carbon films from the glass surface. If needed, glass slides can be cleaned with a detergent, although slides as received from the manufacturers are already coated with a detergent layer. This is the reason why films usually float away from glass slides used without additional cleaning.

The best method to facilitate the release of carbon films was developed by Münch (1964). After depositing the film, it is scratched into small squares a little larger than the size of the grid. A mixture of 0.5–1% hydrofluoric acid and 10% acetone in water is prepared in a flat plastics trough of an appropriate size. After removing the carbon layer from the edges of the slide by scraping with a razor-blade, the slide (with the carbon film on top) is immersed at a shallow angle to the liquid level in the trough. As the acetone reduces the surface tension and the hydrofluoric acid solution creeps between glass surface and carbon layer and dissolves the top surface of the glass, the carbon layer floats off. The floating carbon layer is picked up on grids from below.

Procedure

(1) With a detergent, clean and polish a glass microscope slide.

(2) Deposit a carbon film of 10 nm thickness on the slide in a vacuum evaporator as described on p. 360.

(3) Float the carbon film onto a clean water surface by immersing the whole slide at a shallow angle.

(4) Submerge the individual grid by holding in forceps and then bring it up through the floating film and dry on a filter paper. If needed, the adherence of the film to the grids can be facilitated by rinsing the grids in concentrated HCl for 2 min; this treatment results in the slight etching of the grids. Alternatively, the grids can be dipped in a 1% solution of polybutene in xylene, in an extremely dilute solution of rubber cement in carbon tetrachloride or in a solution made by dissolving cellulose adhesive tape in chloroform.

Carbon Film Prepared on Mica

Ultrathin (<3 nm) carbon films required for high-resolution electron microscopy should be prepared onto exceptionally smooth surfaces such as mica. This is also a versatile method for thicker (~10 nm) films. The removal of carbon films from the mica surface is relatively easy. However, too high a pre-glowing temperature removes the crystal water from the freshly cleaved mica surface and makes carbon removal difficult. Freshly cleaned mica sheets are kept for several hours under dust-protective conditions in a humidity box prior to evaporation. The thin water layer formed on the hydrophilic mica surface facilitates the ingress of water between the substrate and carbon film.

Procedure 1

(1) Evaporate graphite pencils in an evaporator at 2×10^{-6} Torr onto freshly cleaved mica sheet cut to 12×15 mm in size.

(2) Determine the amount of carbon deposited by placing a piece of white paper in close proximity to the mica.

(3) Remove the carbon film by slowly immersing the mica sheet with carbon side up into distilled water at a 45° angle.

(4) Grip one grid at a time with a pincer forceps and lock the forceps.

(5) Immerse the grid under the floating carbon film and then lift it.

(6) Rotate the forceps through 180° and remove the droplet of water now located over the grid with a piece of filter paper.

Procedure 2

A grid-coating trough (Smith, 1981) (Ladd Research Industries) (Fig. 6.4) can be used for coating the grids with carbon film evaporated on a mica or glass slide. This device circumvents the problem of manoeuvring the film over the grids that have been placed below the water surface in a container.

(1) Fill the trough to the top with *clean*, distilled water.

(2) Draw ~5 ml and 1 ml of this water into the large and small syringe, respectively. These amounts should be determined empirically as one becomes familiar with the trough. The water level should be adjusted with the syringe throughout the procedure. The large syringe is used for major adjustments in the water level and the smaller syringe for finer adjustments.

(3) Position the stainless steel tray so that the hole is on the window side of the trough.

(4) Cover the tray with a piece of filter paper cut to roughly the same size as the tray.

(5) Adjust the water level with syringes so that the surface bulges slightly above the top of the trough.

(6) Sink the grids below the water surface and onto the filter paper.

(7) Adjust the position of the tray so that its leading edge is 2 mm from the rails on the front wall.

(8) If needed, clean the water surface with a Pasteur pipette attached to a vacuum line or rubber bulb.

(9) If desired, the water level can be lowered by drawing on the syringe until it is just below the grids.

(10) Slide a mica or glass slide, upon which the substrate film has been deposited, smoothly down the rails into the water until it is stopped by the retaining pin at the bottom of the rails. This procedure results in the thin film being partially floated off the slide to cover the grids beneath. It is still attached to the slide, however, and is therefore held in its correct position and does not float away.

(11) Lower the water level carefully by using the syringes so that the film comes to rest on the grids; lowering the level even further allows the filter paper to drain.

(12) Remove the tray from the trough and air-dry the filter paper and grids. Care should be taken to avoid contamination of the substrated grids while they are drying. Depending upon the condition of the water, it can either be pumped back into the trough from the syringes for reuse or be discarded.

Carbon Film Using Plastic Substrate

Strong adhesion of carbon films to the copper grid can be accomplished by depositing them onto a plastic film, which subsequently is dissolved. The adhesion is due to plastic remnants on the grid bars. Collodion is preferred to other plastics. The disadvantages are that the carbon layer replicates the imperfect plastic surface and it is difficult to remove completely the plastic substrate.

Procedure

(1) Coat the grid with a thin film of collodion as described earlier.

(2) Arrange the grids on a clean glass slide (with coated side up) and then place the slide in a vacuum evaporator.

(3) Allow the grids to be coated with a carbon film 3–10 nm thick.

(4) Place the carbonised collodion-coated grids on a piece of wire gauze and then immerse the gauze in acetone at an angle of 45° or more for a few minutes to dissolve the plastics (for Formvar-coated grids, use chloroform instead of acetone). The duration of exposure to the solvent is crucial because the objective is to

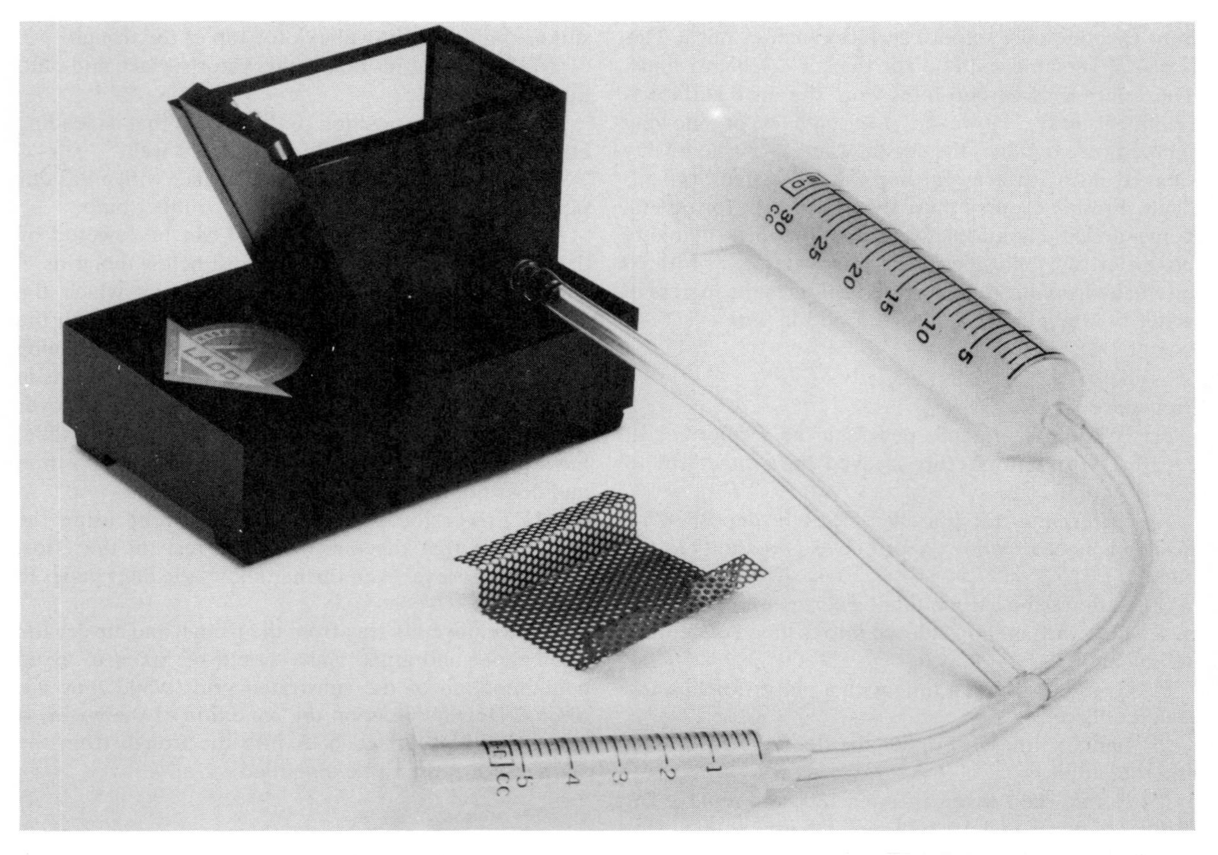

Figure 6.4 Water trough used to facilitate the coating of grids with carbon and/or plastic. The device consists of a trough made of black opaque Plexiglas (with the exception of one side wall, which is made of clear Plexiglas to allow the progress of the slide to be seen inside the trough), stainless steel tray, large and small syringes, T-coupling and tubing. From Smith (1981)

dissolve the plastics from the grid holes but leave the plastics between the grid bars and the carbon film.

A modified method was introduced by Stolinski and Gross (1969). It has the following advantages: (1) ~79% of the total grid area constitutes carbon film and is available for observation; (2) the film is extremely thin (~1.5 nm); (3) contamination is minimum, since the carbon or grid does not come in contact with water; and (4) as many as nine grids can be prepared at one time. The film is useful for supporting thin sections as well as whole specimens such as bacteria, viruses and cell fragments.

Procedure

(1) Fill a glass trough (20 cm diameter and 6 cm deep) with distilled water to a depth of 4.5 cm and place on a sheet of white paper.

(2) Prepare the solution for the substrate film by adding 3 parts of Belco cellulose to 1 part of amyl acetate by volume.

(3) Release several drops of this solution onto the water surface and allow to dry. The thick film thus produced will pick up the dust and other debris floating on the water surface.

(4) Remove the film with a glass rod by means of a twisting motion.

(5) Release one drop of cellulose solution onto the water surface from a height of 4 cm and allow to dry. The film should have an area of 3–4 cm with an even silver-to-gold interference colour. The colours can be observed by placing a tungsten bulb adjacent to the trough.

(6) Bend Athene-type grids (3.05 mm diameter) of hexagonal pattern by holding one part of the rim with fine forceps and pressing the opposite side onto a hard surface (shiny surface concave). This procedure facilitates handling of the grids at a later stage.

(7) Place as many as nine grids onto the floating

film; the distance between the grids should not be less than 6 mm.

(8) Push the film with grids to one side and position a lifting tool underneath the film so that the grids are aligned over the hole in the tool. The tool is made of stainless steel which has outer diameter of 7 cm and a punched hole 2.5 cm in diameter. The top end with the hole has a rounded edge, and the handle is 12 cm long.

(9) Raise the tool slowly to lift the film with grids out of the water. Before the tool is lifted away completely, gather the surplus film surrounding the tool by touching it with a glass rod.

(10) Upturn the tool with the film and grids and place it on a clean glass slide.

(11) After puncturing the film around the inner rim, lift the tool away so that the grids covered by the film are left on the slide.

(12) Dry the film by placing the slide briefly on a warm plate.

(13) Lift the grids individually and place them on another glass slide in a Petri dish.

(14) Place the grids in a vacuum evaporator and coat them with carbon; a very faint grey colour shown by a piece of white porcelain placed in the evaporator indicates a film thickness of ~1.5 nm.

(15) Place the carbon-coated grids on a 50-mesh stainless steel gauze (5 cm^2) and immerse under pure amyl acetate in a Petri dish for 4–12 h in order to dissolve the cellulose substrate.

(16) Remove the gauze with grids from the solvent and place on a filter paper for drying in a dust-free area.

Carbonised Plastic Film

As stated earlier, the simplest way of stabilising plastics films against drift and damage under the electron beam is to coat them with carbon. Shrinkage of plastic films is also minimised with this coating. The carbon layer is assumed to protect both the plastics film and the specimen, for irradiation in the electron microscope may cause cross-linking of the carbon layer with the specimen on the one side and with the plastic film on the other. However, various modifications of the plastics and carbon films and of the specimen caused by irradiation are too complex to be explained here. Very thin composite films of plastic and carbon having desirable features of both substrates can be prepared. It should be noted that carbon has a higher scattering power than collodion, because the former has a higher density (~2) than the latter (~1.4). It is apparent, therefore, that carbon films prepared for stabilising plastics films should be extremely thin. The ideal carbon film is shiny and shows no interference colour. One of the limitations is that the combined films are usually thicker than a pure carbon film, and thus unsuitable for high-resolution electron microscopy. For high-resolution work, pure carbon films or self-supporting specimens may have to be used. Plain collodion or Formvar films are no longer used for quality work.

Procedure

(1) Coat grids with extremely thin collodion or Formvar film.

(2) Arrange dry, plastics-coated grids on a clean glass slide (with coated side up) and then place the slide in a vacuum evaporator.

(3) Allow the grids to be deposited with ~5-nm-thick carbon film.

(4) Store the grids in covered Petri dishes.

As stated earlier, specimens supported by carbon or carbon-coated plastic films are also contaminated in the electron microscope. This problem can be minimised by partially coating the plastics films with carbon. The non-coated portion of the plastic film seems to be sufficiently stable to allow high-resolution electron microscopy. Partial coating of a plastic film is achieved by superposition of two grids during carbon evaporation (Kölbel, 1974). The underlying grid bears the plastics film to be strengthened and the upper grid (without plastic film) serves as a mask.

Carbon Film Supported by Perforated Plastic Substrate

High-resolution electron microscopy requires extremely thin carbon films. Since such films lack sufficient mechanical stability, they need to be supported by fine-meshed plastic nets. The following method produces perforated collodion films characterised by holes of diameters ranging from 1.75 μm to 7.9 μm; the average diameter is 4.5 μm (Fig. 6.5) (Lünsdorf and Spiess, 1986). Pseudo-holes are not formed.

Procedure (Lünsdorf and Spiess, 1986)

Flasks and other containers used for preparing the solutions should be cleaned with an acidic bath and then with deionised water. Collodion resin (Polaron Equipment, Watford, UK) is washed with ethanol and dried for 5 min under an electric light bulb (60 W, distance 200 mm). Exactly 0.1 g of collodion is completely dissolved overnight with stirring in 75 ml of ethyl acetate in a brown glass-stoppered flask. Exactly 125 μl of 10% aqueous solution of Brij 58 detergent (Serva, Heidelberg) and 0.92 g of 87% glycerol are

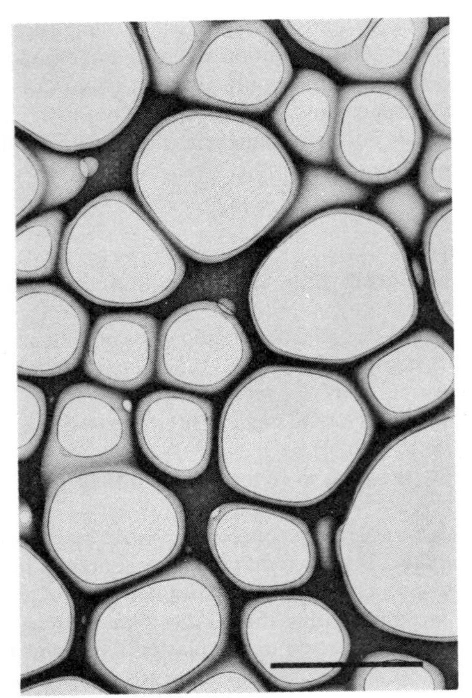

Figure 6.5 Electron microscopic view of the pure perforated support film, spanning the open area of a grid. Scale bar: 4 μm. Courtesy H. Lünsdorf

added to the collodion solution. To obtain a homogeneous dispersion of the ingredients, the flask is shaken thoroughly by hand for 2–3 min; vigorous shaking will result in holes of a small size. A stirring rod should not be used.

The turbid solution is poured into a glass beaker. A glass slide cleaned with acetone is dipped into the solution for 10 s and, after slow withdrawal, it is dried for 10 min, standing upright in a beaker. The perforated collodion film is floated off from the slide on double-distilled water in a glass dish (diameter 140–200 mm; height 40–70 mm). To separate the film, the slide is dipped slowly into the water at an angle of 30–40°. Grids (200–300-mesh) are placed, with dull side facing down, on the floating film, and then removed along with the perforated film, using filter paper. The grids are placed on filter paper in a Petri dish and dried at 22 °C.

Film-coated grids are baked in an oven at 180 °C for 10 min. This treatment causes the areas covered by a thin collodion film to perforate, producing a perforated film. The grids are covered with a thick layer of carbon and exposed to an intense plasma glow for 10–20 s to increase adhesiveness of the collodion film. These grids can be used to pick up thin carbon films to which negatively stained specimens have been adsorbed. After blotting with filter paper, the air-dried grid-specimen is viewed in the electron microscope. The mixture of collodion, Brij 58 and glycerol can be stored in a brown glass bottle (tightly stoppered) for at least 6 months.

GRAPHITE FILMS

Background noise arising from amorphous carbon films is a problem in high-resolution electron microscopy. At a resolution of ~0.3 nm, even thin carbon films (2 nm) show structural details that confuse the specimen image. Theoretically, sufficiently thin single-crystal films of light atoms should produce noiseless films. Crystals are, in principle, free from the background noise that is present in the carbon films. Single crystals of high-purity graphite yield low-noise support films, and these films of low scattering power are ideal for high-resolution electron microscopy. Useful areas of the order of 20 μm^2 can be obtained by ultrasonic exfoliation and centrifugation of high-purity powders of natural graphite (Johansen, 1975).

Several methods are available for preparing graphite films ~2 nm in thickness. Single-crystal films have been prepared by exfoliation of graphite (Johansen, 1975), by alkali splitting of graphite oxide and subsequent reduction with hydrazine hydrate (Dobelle and Beer, 1968), by splitting of crystals of graphite (Iijima, 1977) or by gluing a graphite crystal to a grid and then cleaving the crystal with adhesive tape (Hines, 1975). The method described below has been introduced by Formanek (1979) for preparing hydrophilic, crystalline films of graphite, which are especially useful for high-resolution electron microscopy and electron diffraction. One disadvantage of this method is the amount of time required to prepare graphite oxide suspension.

Although conventional graphite films are hydrophobic, hydrophilic films can be obtained from graphite oxide with $KMnO_4$ in concentrated sulphuric acid. The resulting graphite oxide still has the hexagonal lattice structure of graphite, but oxygen atoms linked to its surface and intercalated between its layers make it hydrophilic. The major difference between the hexagonal layers of graphite and graphite oxide is the enlarged interlayer spacing of ~0.62 nm in the latter, compared with 0.34 nm in the former. This separation of the carbon layers is caused by the intercalated oxygen atoms. Because thin plates of graphite oxide have a diameter of 2–5 nm, they are spread over the holes of a holey carbon film.

PREPARATION OF GRAPHITE OXIDE (Formanek, 1979)

A suspension of graphite oxide is prepared by stirring 2 g of thin crystalline plates of graphite and 2 g of sodium nitrate with 100 ml of concentrated sulphuric acid in a 500 ml Erlenmeyer flask placed in an ice-bath. Over a 2 h period, 12 g of $KMnO_4$ is *slowly* added to the suspension, and the temperature is not allowed to exceed 20 °C; the mixture is explosive at higher temperatures. The ice-bath is removed and the suspension stirred for ~12 h at room temperature. Then 100 ml of distilled water is added slowly and dropwise, causing violent effervescence and an increase in temperature to 98 °C. After cooling, 40 ml of 3% hydrogen peroxide is added to reduce residual permanganate and manganese dioxide to colourless and soluble manganese sulphate.

The suspended small graphite oxide discs give the suspension a golden-yellow colour. After dilution with 1 litre of distilled water, the suspension is centrifuged. The residue is washed and centrifuged five times in 1 litre of distilled water, and then suspended in 1 litre of distilled water and poured into a 1 litre measuring cylinder. After ~2 h sedimentation, the upper 500 ml of the supernatant is decanted and centrifuged. The pellet is dissolved in 100 ml of distilled water and dialysed twice against 10 litres of 0.01 N EDTA (pH 8.0) and five times against 10 litres of distilled water. The net yield is 100 ml of a suspension containing ~1.5 mg of graphite oxide/ml.

QUARTZ FILMS

Heat-resistant, electron-transparent quartz support films (40–60 nm) supported by a silicon grid can be produced by micromachining of silicon (Stenberg *et al.*, 1987). Hydrophilic surface properties of these films allow uniform spreading of specimens and heavy metal salts used for negative staining. The surface properties (including surface energy) of the film can be varied from hydrophilic to hydrophobic with silanes coupled to the surface. The preparation of these films is described below.

The grids are fabricated on a silicon wafer (250 μm thick) (Wacker Chemitronic). The wafer is oxidised at 1000 °C for 3 h in dry oxygen, producing an oxide layer of 110 nm on both sides. Rectangular holes (420 × 420 μm) in the oxide on the reverse side of the wafer are made by a standard photolithographic process, followed by etching in hydrofluoric acid. This step is followed by etching the silicon with a solution consisting of ethylenediamine, pyrocatechol and water at 115 °C. Since silicon dioxide is resistant to this treatment, the etching penetrates the whole wafer and stops at the front oxide. The silicon etch is anisotropic, etching towards different crystallographic directions at different speeds.

The etching results in front film areas of 70 × 70 μm, and the oxide thickness is 65 nm. Films of different oxide thicknesses can be prepared using a final etch in dilute hydrofluoric acid (1 part concentrated acid (47%): 50 parts water). The grid films are treated with dichromate sulphuric acid (or concentrated nitric acid) for 10 min and rinsed in distilled water. This treatment removes any organic material and makes the surface hydrophilic.

PERFORATED FILMS

For certain studies, such as particulate specimens, the mesh size (50–400) of commercial specimen grids or the hole diameters of drilled diaphragms (20–750 μm) are too large. Perforated or holey films (microgrids) with holes of various diameters (0.05–100 μm) (Fig. 6.6) are helpful in such studies. Perforated films with large open areas are also called micronets. In perforated films image contrast is higher in the regions of holes than in the areas with film. Since all supporting films reduce image contrast, perforated films provide an increased image contrast as well as support for thin sections. Also, films with small circular holes are useful for testing the symmetry of the image and for correcting astigmatism in the lens.

Perforated collodion, Formvar and carbonised plastics films are prepared by using three basic techniques: (1) the development and incorporation of local faults in the film; (2) the localised destruction of the film by physical or chemical treatment; and (3) the replication of perforated templates (e.g. etched eutectics or filters) (Baumeister and Seredynski, 1976). The numerous modifications of the three basic techniques reflect the difficulties in obtaining reproducible results. The difficulties include the presence of pseudo-holes and unsuitable average hole diameter. Ideally, the selected technique should allow a control on the size and number of perforations as well as the geometry of the hole.

The initial hydrophobicity of the glass slide plays a part in determining the hole pattern of the plastics film. An insufficiently hydrophobic slide encourages production of very small, distantly spaced holes, and even pseudo-holes. An excessively hydrophobic slide may cause very large or numerous holes, resulting in a weak network of holes. Production of pseudo-holes can be minimised by using plastics solutions of low concentration.

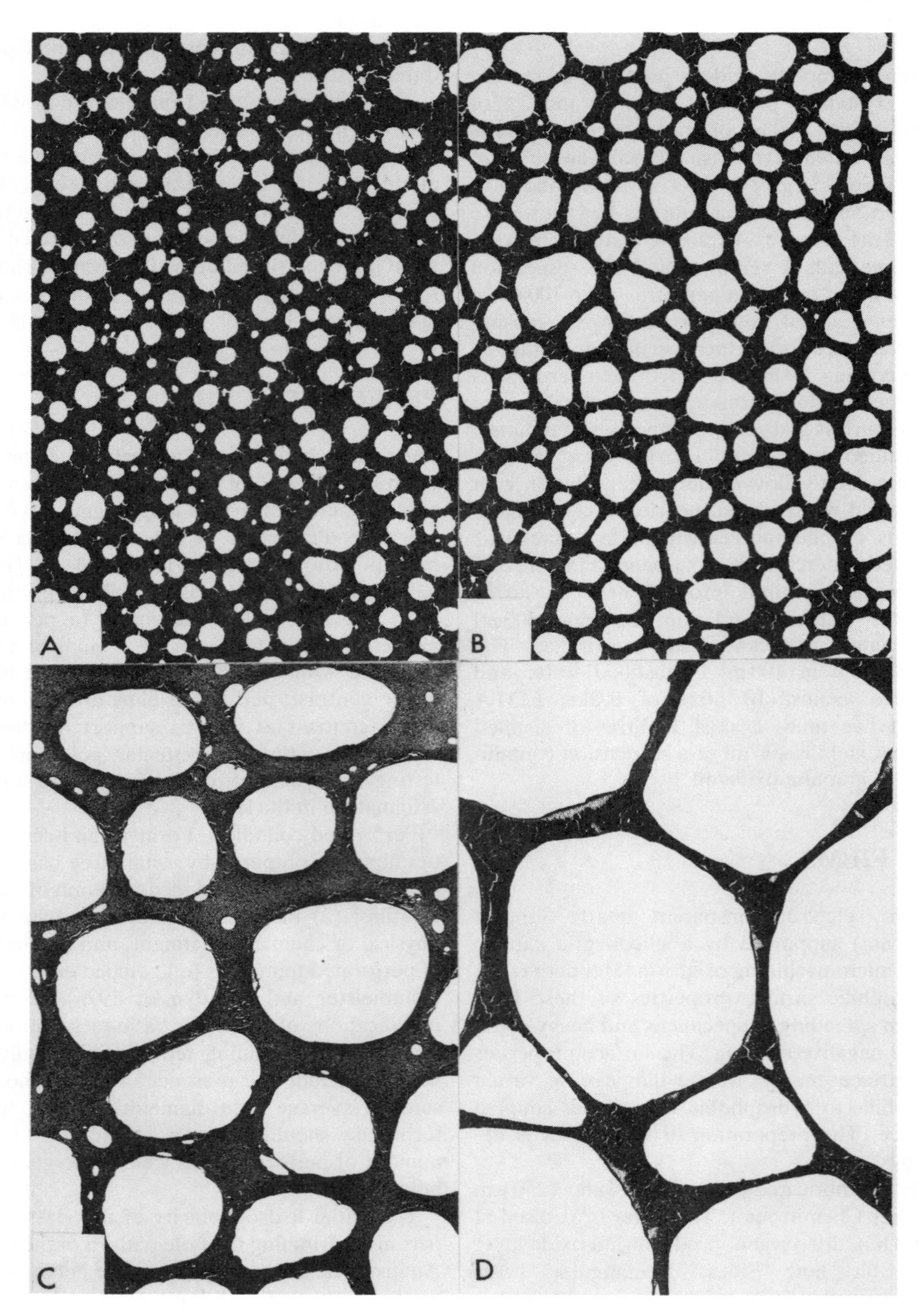

Figure 6.6 Gold-coated perforated films prepared by the glycerol method (10 000×). The average hole size increases with the concentration of glycerol in the Formvar–chloroform/glycerol emulsion. A: Glycerol content 0.1 ml/90 ml Formvar solution. B: Glycerol content 0.3 ml. C: Glycerol content 1 ml. D: Glycerol content 3 ml. From Baumeister and Hahn (1978)

It is desirable that one select a film with an appropriate perforation size, depending upon the objective of the study. The perforation size is also determined by the magnification at which the observation is to be made. For example, perforations of 0.2–0.5 μm in size are ideal when the grids are to be examined at a direct magnification of 100 000×, for at this magnification the area of the viewing field is comparable to the area of a single perforation.

Methods most commonly used to prepare plastics perforated films essentially involve the addition of a substance which is not miscible with the solvent used to prepare the plastics solution. This treatment results in an interference in the uniform drying of the film. For example, to prepare perforated Formvar films, water is introduced into the Formvar–ethylene dichloride solution. Since water and ethylene dichloride do not mix, during the last stages of solvent and water evaporation, a large number of air bubbles are evolved, which eventually produce perforations in the Formvar film.

Perforations in the plastics film can also be obtained by blowing a stream of warm air saturated with water into a glass vessel containing the coated slide. Alternatively, minute water droplets can be inserted into plastics films by gently breathing upon a freshly spread plastics film. Since the holes produced by such a treatment may be pseudo-holes, they are converted into real holes by lightly heating the coated slide on a Bunsen burner. The perforations obtained have a mean diameter of 1 μm. The number and size of the holes depend upon the duration and intensity of breathing. For an improved 'breathing' method, the reader is referred to Drahos and Delong (1960).

Another method for obtaining films with perforations is essentially based on cooling the hydrophobic glass surface to near the freezing temperature and forming by condensation minute water droplets on it. The size and number of perforations formed are determined primarily by the temperature of the glass surface, the surrounding environment's humidity and temperature, and the concentration of plastics solution. By this method perforations ranging in size from 0.01 μm to 10 μm can be obtained. The total open area in each grid is estimated to be 40%, 65% and 75%, with perforation sizes of 0.2–0.5 μm, 1–2 μm and 3–6 μm, respectively.

Although the above-mentioned procedures produce excellent results, it is preferable to add water directly to the plastics solution to form an emulsion (simultaneous method). The disadvantage of the former procedures is the difficulty in controlling a rather critical timing for the insertion of water droplets into the drying plastics film. The simultaneous method and other procedures for preparing perforated support films are given below. For a detailed review and preparation of support films, the reader should consult Baumeister and Hahn (1978).

PERFORATED FORMVAR FILM

Various methods are used to prepare perforated Formvar films. The following three methods (Harris, 1962; Elsner, 1971; Baumeister and Seredynski, 1976) are reliable, and perforations ranging from 0.05 μm to 25 μm in diameter can be obtained (Fig. 6.6).

Method 1

(1) Prepare a 0.25% solution of Formvar in ethylene dichloride.

(2) Add glycerol in the ratio of 1:32 and shake well to form an emulsion. This ratio should produce holes of ~7 μm in diameter.

(3) Dip a clean microscope slide into the emulsion and then by tilting allow it to drain and dry for ~10 min.

(4) Expose the slide to a jet of steam for 1 min.

(5) Float the film on water surface and mount on grids as described earlier.

Note: If desired, these films can be coated with carbon. The desired size and number of holes in the film can be obtained by changing the proportion of glycerol (see Table 6.1)

Table 6.1 Relationship of hole size to glycerol content for perforated Formvar films

Formvar solution (parts)	*Glycerol (parts)*	*Maximum hole diameter* (μm)
8	1	25
16	1	14
32	1	7
120	1	4

Method 2

By using the method given below, small, round perforations with smooth edges can be produced consistently. This method employs solvent treatment for smoothening the edges and enlarging the perforations formed in high humidity. Perforation size and edge characteristic can be controlled easily, and useful perforations as small as 0.15 μm can be obtained.

(1) Prepare a 0.2–0.4% solution of Formvar in ethylene dichloride.

(2) Dip a clean glass slide into a Coplin jar containing the Formvar solution, and by tilting allow it to drain

and dry for 1 min. Room temperature should be ~23 °C with 60–70% relative humidity; lower humidity will result in fewer perforations.

(3) Float the film on water surface and mount on grids as described earlier.

(4) Immerse the grids in chloroform–absolute methanol mixture (1:9) in the compartments of a depression plate. In order to reduce the solvent evaporation, the plate is placed on a blotted paper saturated with the solvent mixture in a covered Petri dish. Increasing the chloroform : methanol ratio results in increased dissolution of the mixture.

(5) Allow the grids to remain in the solvent mixture for 15–16 min at 22 °C.

(6) Wash the grids by dipping in 95% ethanol, 50% ethanol and distilled water (30 dips in each).

(7) Place the grids on a filter paper for drying at room temperature.

Note: If desired, these films can be coated with carbon by placing the grids along the edge of double-stick tape on a glass slide. Perforated carbon films can be obtained by removing Formvar from carbon-coated grids by washing with chloroform.

Method 3

(1) Prepare a solution by adding 0.18 g of Formvar to 90 ml of chloroform in an Erlenmeyer flask. The quantity of glycerol added ranges from 0.01 ml to 5.0 ml (the hole size increases with increasing concentrations of glycerol: Fig. 6.6). The solution is stirred with a magnetic stirrer until the polymer is completely dissolved. The solution in the flask is sonicated in the water-filled tank of an ultrasonic generator for 30 min.

(2) Clean a glass slide in a detergent solution, rinse with distilled water and rub dry with a clean cloth. The rubbing should be done parallel to the long axis of the slide.

(3) Immerse the slide into the emulsion for 5–10 s and then slowly withdraw (0.5 in/s) and dry under dust-free conditions.

(4) Place the slide for ~5 s in steam from a water-bath and allow to dry.

(5) To aid complete perforation, immerse the slide into a beaker containing acetone for 10–20 s.

(6) After drying, the film is floated off on the water surface and transferred onto the grids.

(7) To achieve good thermal and electrical conductivity and stability, the film can be coated with a thin layer (~25 nm) of carbon or gold.

SUPPORT FILM WITH LARGE HOLES (MICRONET) (Pease, 1975)

(1) Hydrophobise the glass slide by soaking it overnight in a saturated solution of ferric stearate in benzene. This solution is prepared by dissolving 1 part of ferric stearate in 100 parts of benzene, which is then brought to a boil. Any ferric soap can be substituted.

(2) Remove most of the original solution by dipping the slide in benzene, followed by vigorously rubbing it with thin paper wipes saturated with successive changes of benzene. All visible traces of the film should be removed. The slide remains hydrophobic.

(3) Dip the pretreated slide in 0.4–0.6% solution of collodion in amyl acetate, and dry by holding it in a vertical position.

(4) Condense water on the surface by exposing the slide to furiously boiling water. Exposure to the steam should be continued until the milkiness disappears, indicative of complete evaporation of amyl acetate. Thin regions with large holes are found near the top of the film, while thicker regions with smaller holes are located near the bottom. The drainage pattern is responsible for the inhomogeneous nature of the film.

(5) Bake the grid with the dried film at 170–180 °C to open up pseudo-holes. The micronets can be stabilised by coating with a thick layer of carbon or gold.

PERFORATED CARBON FILM

In the case of frozen, hydrated specimens, a small density difference between the organic specimen and water produces a relatively weak phase contrast signal in the image. This problem is compounded by the use of even thin, continuous carbon films, because carbon inelastically scatters electrons, contributing to background noise. Thus, maximisation of signal-to-noise ratio is crucial for the production of satisfactory images of frozen, hydrated specimens in the cryoelectron microscope. The problem of background noise can be significantly minimised by using carbonised perforated plastics films. The following method for preparing such films (10 nm thick) with 5–10 μm holes has been introduced by Murray and Ward (1987).

(1) Uniformly smear a glass slide with a finger. Do not wipe the slide too clean, but dust can be removed by wiping the slide with lens tissue.

(2) Immerse the slide into a 0.5% solution of collodion in amyl acetate, and remove excess collodion by allowing the slide to stand at 45° on filter paper saturated with water in a closed vessel for 10 min to evaporate the amyl acetate.

(3) Water droplets are distributed over the slide in the vessel through a Tygon tubing. The tube has been perforated with holes of 1 mm in diameter, and is positioned 6–8 cm from the surface of the slide. The size and distribution of holes in the dried film can be checked in the light microscope under phase contrast. The higher the temperature of the water droplets, the larger the holes which are more closely spaced.

(4) Immerse the slide with perforated collodion film in a mixture of amyl acetate, methanol and water (1:150:10) for 1 min. This treatment allows the amyl acetate to etch the plastics remaining in any pseudo-holes.

(5) Strip the film from the slide onto a clean water surface, and place 300-mesh grids at an appropriate distance from one another on the film.

(6) Lift the grids from the water surface, and after drying them at room temperature, coat them with a carbon film (10 nm).

(7) The collodion film under the carbon can be removed by placing the grids on a wire mesh over amyl acetate in a closed Petri dish for several hours.

Another method for preparing perforated carbon films was introduced by Koreeda (1980). Holes as small as ~5 nm in diameter can be produced with this method.

(1) Deposit Pt–Pd film ~0.5–1.0 nm thick on a cleaved surface of mica, followed by carbon film, ~10–40 nm thick.

(2) Strip the Pt–Pd–carbon film from the mica onto the water surface.

(3) Place the grids on the film and retrieve them, so that the carbon is positioned between the Pt–Pd and the grid.

(4) Transfer the grids in a glow-discharge tube housed in a bell-jar of a vacuum evaporator for ion-etching.

(5) Evacuate the bell-jar at 0.2 Torr and etch the film with current densities of 0.25–0.05 mA/cm^2 for up to 10 s.

(6) Evacuate the bell-jar at ~10^{-6} Torr.

(7) Fill the bell-jar with dried air (using three desiccating agents—silica gel, calcium chloride and phosphorus pentoxide) up to 20 Torr.

(8) Again evacuate the jar at 0.2 Torr and etch the film with 0.05 mA/cm^2 for 10–15 min.

The first ion-etching produces holes only in the Pt–Pd film, while the second ion-etching causes preferential etching of the carbon. Both the size and number of holes are controlled by: (1) the quantity of charge during the glow discharge, (2) the thickness of the Pt–Pd layer, and (3) the partial pressure of water vapour resulting from the desorption of adsorbed water in the bell-jar. The rate of the first ion-etching is greater than that of the second one. Apparently, excited water acts as the most effective ion species for the sputtering reaction.

PERFORATED COLLODION–CARBON–GRAPHITE OXIDE FILM (Formanek, 1979)

In this procedure a perforated collodion film is prepared first, which is covered by a thin carbon layer. The collodion is removed and then thin single-crystalline plates of graphite oxide are spread over the holes.

Calcium rhodanide (1 g/100 ml) is dissolved in a solution of 0.25% collodion in *n*-butyl acetate. A glass slide is dipped into this solution and dried. The collodion film is floated on a water surface, and copper grids are placed on the film, picked up with filter paper and either air-dried or dried at a certain distance from a 1000 W lamp. Calcium rhodanide is used for producing perforated collodion films. The size of the holes depends on the relative humidity of the air (Fig. 6.7) and on the mode of drying. Holes of ~0.1–8.0 μm are obtained if the film is normally dried in air of a relative humidity between 39% and 65%. The higher the relative humidity, the larger the average diameter of the holes (Fig. 6.7C). At a relative humidity of 43% and a distance of 30–60 cm from a 1000 W lamp, holes with diameters of ~0.1–3 μm are obtained. The larger the distance between the drying film and the lamp, the larger the holes.

A thin layer of carbon is deposited on the grids coated with perforated collodion film. The collodion is removed by placing the grids twice on several layers of filter paper soaked with acetone in a Petri dish. A drop of the graphite oxide suspension (see p. 367) is placed on each grid and immediately pipetted off. The graphite oxide film has a hexagonal symmetry vith periodicities of 0.21–0.12 nm.

WETTABILITY OF SUPPORT FILMS

A serious problem with the negative staining method is the unequal spreading of the specimens on support films. This problem is inevitable, because support films are normally hydrophobic and specimens are usually in aqueous suspensions. Although particle suspensions of sufficiently low surface tension can be spread adequately, most biological preparations fail to wet the support films uniformly. Wetting of the support films is a very complex phenomenon and is difficult to assess quan-

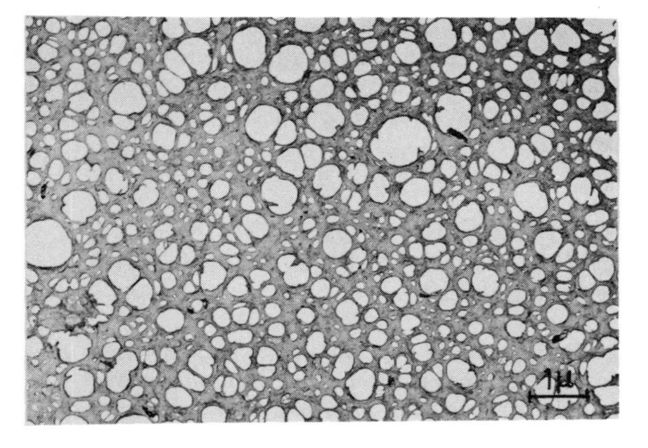

A

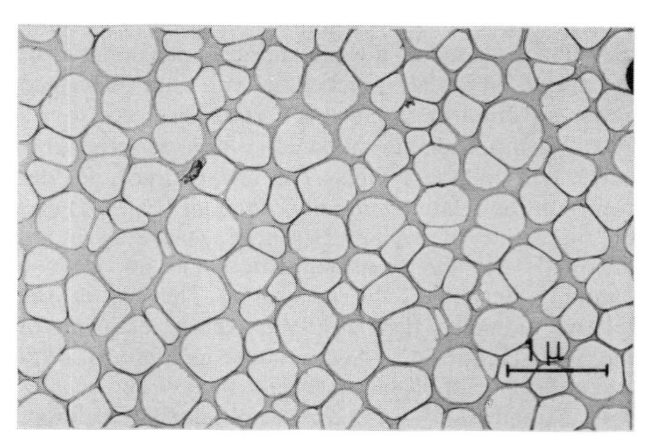

B

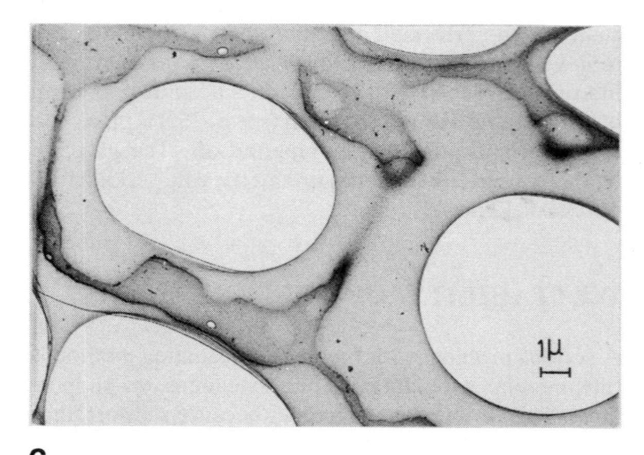

C

Figure 6.7 Perforated carbon films from a solution of 0.25% collodion and 1% calcium rhodanide in *n*-butyl acetate at different relative humidities. A: Relative humidity 39% (25 °C). B: Relative humidity 47% (22 °C). C: Relative humidity 67% (29 °C). From Formanek (1979)

titatively. Even traces of contamination and ageing may drastically alter the properties of a given support film. Freshly evaporated carbon films are hydrophobic, but ageing makes them hydrophilic. On the other hand, aluminium oxide films behave in the opposite way.

Hydrophobic films usually result from the coating process. If films are carbon-coated before use or are made of carbon alone, their hydrophobic state is influenced by the operating conditions and cleanliness of the coating unit. Both an ultrahigh-vacuum system and an absolutely clean vacuum unit result in the production of smooth, contaminant-free, thin and hydrophilic carbon films. Although the vacuum conditions during evaporation are known to affect the degree of hydrophobicity, no information is available about exactly what occurs during evaporation. An extremely clean unit with a baffled oil-diffusion pump produces better, stronger and less hydrophobic films than a less clean unit. Recently Lamvik *et al.* (1987) have introduced an oil-free vacuum system for the production of hydrophilic carbon films. This system can be evacuated to the low milliTorr range while completely avoiding all vapours and contaminants.

Some biological specimens contain soluble protein or surface-active impurities, both of which act as a surfactant. In the absence of such an agent, two main approaches are available. One approach is to modify the surface of the support film by irradiation (ion bombardment, UV irradiation). This modification can also be accomplished by exposing the film to a high-voltage glow discharge passed between metal electrodes in a continuously evacuated vessel. This treatment renders the support film readily wettable even by preparations having a high surface tension. Moreover, the treatment reduces the ability of the film to 'hold' virus particles as a result of the interaction between the virus particles and the surface charge of the film.

Glow-discharge treatment in a reduced atmosphere of air makes films hydrophilic, negatively charged and uniformly thinner. These films remain hydrophilic for periods ranging from several minutes to a few hours. Hydrophilicity of the carbon films obtained by the glow discharge procedure has not been surpassed by any other method discussed here. Such carbon films are ideal for both negative staining and adsorption freeze-drying of molecular and supramolecular suspensions. Positively charged specimens such as ferritin and cytochrome can be easily examined. The carbon surface is sufficiently charged to facilitate uniform spreading of a thin specimen film. The specimen film dries down evenly over the entire surface, even when the bulk of the liquid is removed by touching the grids with a piece of filter paper. However, subtle variables in glow discharge conditions may introduce considerable inconsistency in the hydrophilic character of these films, which is relevant in the study of molecules.

On the other hand, films treated with glow discharge in organic vapours (e.g. pentylamine) become hydrophobic, positively charged and uniformly thicker. Pentylamine (amylamine) is easily soluble in water, but on polymerisation in a plasma discharge it becomes insoluble. The effect of this polymerisation persists for at least several hours. Caution should be used, because pentylamine is a strong irritant.

This method involves essentially the deposition of basic radicals on carbon films in a reducing atmosphere produced with a glow discharge in organic vapours. Carbon films are at least partially covered with charged groups when they are ionically bombarded. Positively charged carbon films strongly adsorb negatively charged specimens such as nucleic acids, proteins such as RNA polymerase, repressors, and restriction endonucleases bound to DNA or glutaraldehyde-fixed proteins. Glow discharge does not produce a positively charged hydrophilic film. This difficulty can be circumvented by neutralising the negative charges of the hydrophilic surface by treating it with a solution of magnesium acetate before adding the specimen solution to be adsorbed (Portmann and Koller, 1976). A brief descrıption of the glow-discharge procedure follows.

Most coating units are equipped for routine glow discharge. The best glow discharge is achieved when an alternating voltage is applied to a continuous stream of vapour. An electric field of about 100 V/cm is required. Grids carrying support films are placed in a bell-jar, in which the pressure is reduced to either 10^{-1} Torr (for glow discharge in air) or 10^{-2} Torr (for glow discharge in organic vapours). The efficiency of the glow discharge for making surfaces more wettable is higher at lower pressures. To produce a glow discharge in organic vapours without stopping the action of the pump, the organic liquid is admitted at a very slow rate until the pressure is stable at a reading of about 10^{-1} Torr on the Pirani gauge.

The glow discharge is obtained by applying a tension of about 500 V to the electrodes for 10 s. During glow discharge, the space between the electrodes should be filled with a uniform glow. An excessive discharge will burn away the film, thus making it so fragile that it can withstand neither the use of macrodrops deposited with a pipette or transferred with a wire loop nor the virus–stain mixture sprayed with a glass Vaponefrin nebuliser (Vaponefrin Co., Edison, NJ). At this stage the high-tension supply is turned off, the valve connecting the bell-jar to the pump is closed and air is immediately admitted to the system. After glow discharge, the grids should be used as soon as possible to prevent atmospheric contamination of their surfaces; for example, for DNA adsorption they should be used within 1 h. Special care must be taken to exclude the possibility of injury by electricity or implosion of the bell-jar. A detailed discussion of mounting macromolecules on support films and the treatment of these films by glow discharge is presented by Dubochet *et al.* (1982).

A simple, inexpensive glow-discharge unit can be constructed from materials available in a laboratory (Aebi and Pollard, 1987). Glow discharge can be accomplished in 2–3 min in this unit. This duration is considerably shorter than that required in conventional vacuum evaporators. It is sufficiently inexpensive for each investigator to have his/her own unit, especially when the laboratory is not equipped with a vacuum evaporator. This unit can be used to render carbon surfaces either negatively charged (in a reduced atmosphere of air) or positively charged (in a reduced atmosphere of pentylamine). An intense purple-to-violet glow discharge between the two parallel electrodes is used to render films negatively charged, while an intense violet-to-blue discharge is used to render films positively charged. The construction and operation of this unit are explained below.

Construction (Figs. 6.8 and 6.9)

The top electrode (G) consists of one of the round 100 mm diameter aluminium plates, which is press-fitted onto the 11 mm diameter aluminium rod (J). The rod is fed through an 11 mm hole drilled in the top centre of the polycarbonate upper section of the desiccator (A1). The rod (J) is stabilised by a Plexiglas washer (N) with internal and external diameters of 11 mm and 25 mm, respectively, which is pushed into the neck of the cover. This feed-through assembly is fastened to the cover and vacuum-sealed with epoxy glue.

To allow adjustment of the distance between the two parallel electrodes, the aluminium plate constituting the bottom electrode (H) is mounted on a threaded brass rod (K) which screws onto a nut (L) of a proper size, mounted around a central hole on the 140 mm diameter Plexiglas ground-plate (I). A hole of ~5 mm in diameter is drilled through the ground-plate to assure equalisation of the pressure above and below the ground-plate when in place inside the evacuated desiccator.

The needle-type valve (C) is glued into a 10 mm hole in the polypropylene lower section of the desiccator (A2). The air inlet nozzle (A3) in the polycarbonate upper section of the desiccator is drilled out to an inner diameter of 8 mm, and the micrometric capillary valve

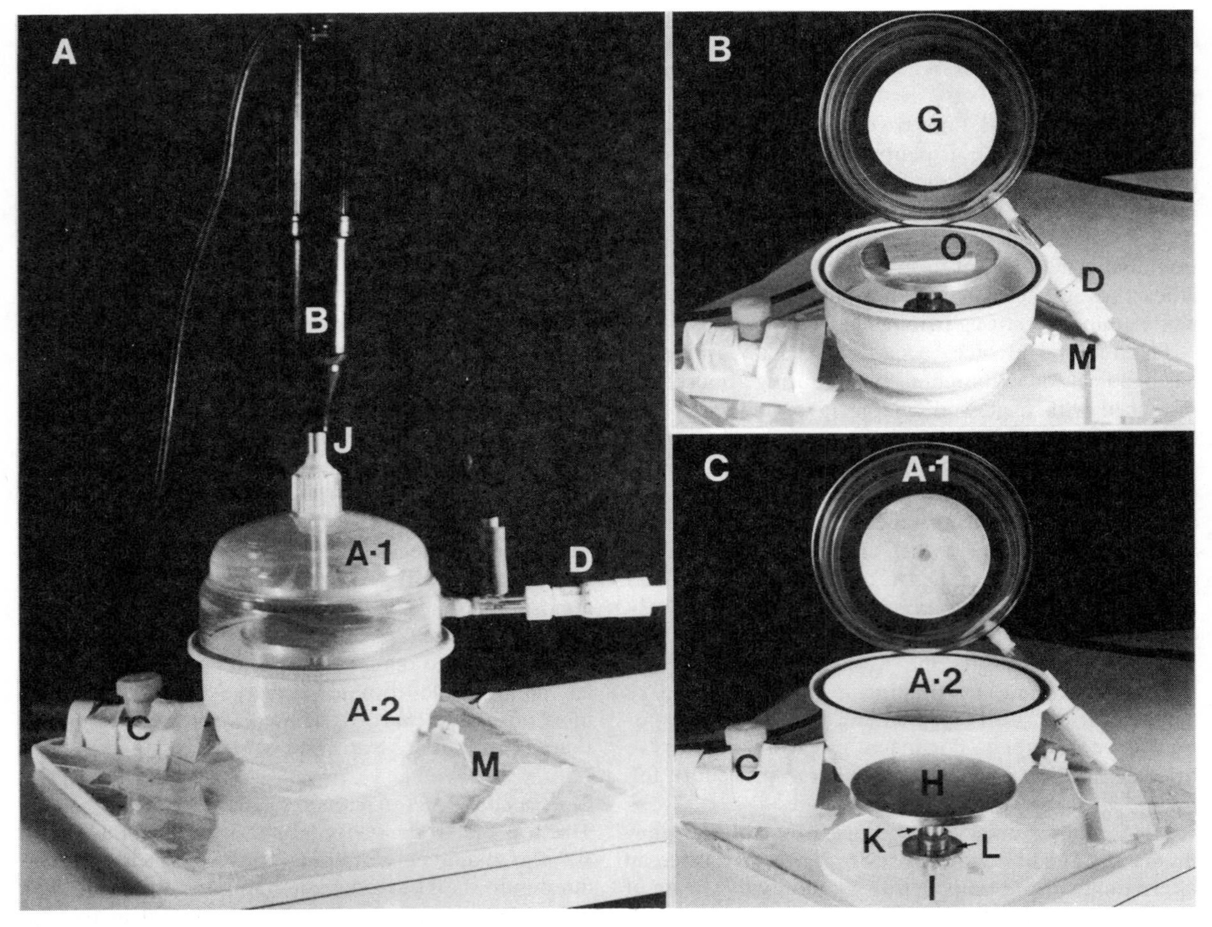

Figure 6.8 Photographs of the glow discharge unit; A, B and C show various views. A-1, Upper section of desiccator; B, high-frequency vacuum tester (Tesla coil); C, needle valve connecting chamber to vacuum line; D, micrometric capillary valve; G, upper electrode plate; H, lower, grounded electrode plate; I, Plexiglas ground plate; J, rod connecting upper electrode to the Tesla coil; K, thread, adjustable support for lower electrode; L, nut; M, ground wire; N, Plexiglas washer supporting electrode rod; O, grid carrier. From Aebi and Pollard (1987)

(D) is glued into the nozzle. To ground the bottom electrode (H), a copper wire (M) is fed through the lower section of the desiccator (i.e. by heating the wire over a flame before pushing it through the polypropylene wall) and vacuum-sealed with epoxy glue. The copper wire is connected to the bottom electrode by soldering it to the nut (L) fastened to the ground-plate. Good grounding is important for proper glow discharge.

It is necessary to remove the electrode tip supplied with the high-frequency vacuum tester (B) and connect the tester to the top of the 11 mm diameter aluminium rod (J). A simple direct-drive vacuum pump can be used to evacuate the chamber. To minimise back-streaming of oil vapour from the pump, a molecular sieve type of foreline trap should be placed into the vacuum line between the pump and the vacuum chamber.

Operation

The chamber is at atmospheric pressure, the needle-type valve is closed and the pump is off. Carbon-coated grids or glass coverslips are placed on a rectangular (40×60 mm), 5 mm thick aluminium grid carrier (O), scribed with one or more grooves to facilitate picking up the grids. While the upper section of the vacuum chamber (A1) is lifted up, the grid carrier is placed onto the bottom electrode (H). The upper section is put back in place and the micrometer capillary valve (D) is closed.

The pump is turned, and after ~5 s (to allow evacuation of the line between the pump and the vacuum chamber), the needle valve (C) is opened to evacuate the chamber. After the pressure has reached 10^{-1}–10^{-2} Torr (which is after 1–2 min with the mechanical pump used here), the vacuum tester is turned on with the

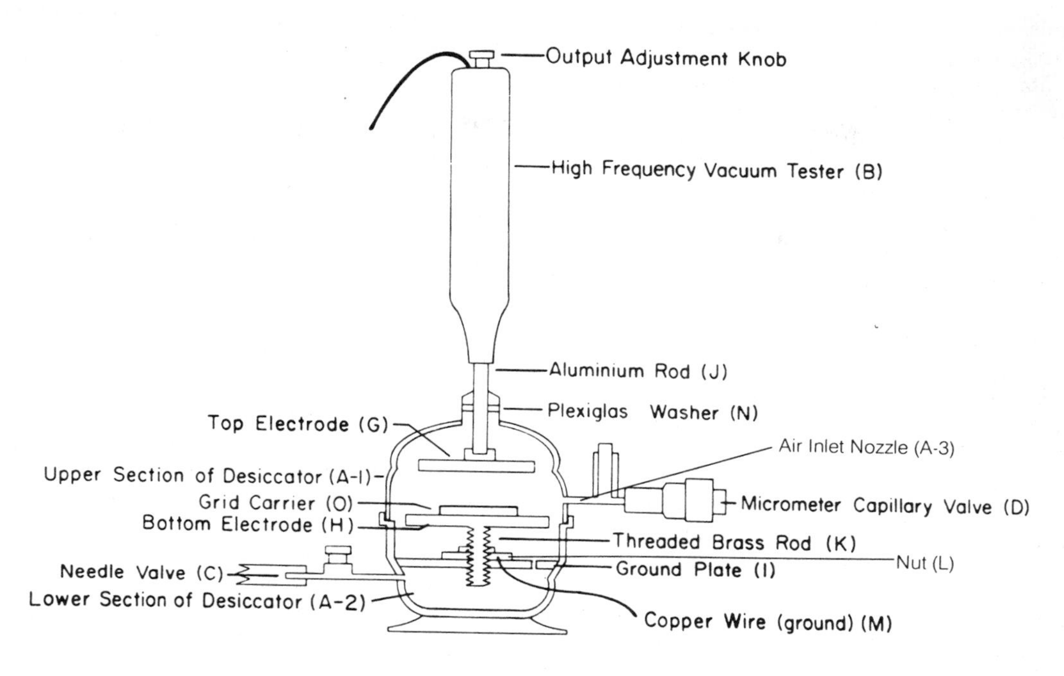

Figure 6.9 Drawing of a central vertical section through the glow discharge unit. Same symbols as Fig. 6.8. From Aebi and Pollard (1987)

output set with the adjustment knob to generate an intense purple-to-violet, ozone-generating glow discharge between the two parallel electrodes. The duration of the discharge for carbon films on grids is 5–15 s.

After glow discharge is completed, the needle valve is closed, and air is admitted slowly (to prevent the grids from being blown away) to the chamber through the micrometric capillary valve (D). The grid carrier is removed, the rotary pump is switched off and the needle valve is opened to ventilate the line between the pump and the vacuum chamber; this is done to prevent oil from backstreaming into the vacuum line. The whole cycle of operation takes 2–3 min.

Carbon films can be positively charged with this apparatus by connecting the air inlet nozzle of the micrometric capillary valve to a bottle of pentylamine and adjusting its vapour pressure in the vacuum chamber with the capillary valve such that an intense violet-to-blue glow discharge between the two parallel electrodes is obtained. Pentylamine slowly attacks the polycarbonate from which the upper section of the desiccator is made. If glow discharge in pentylamine is going to be routinely used, the chamber should be constructed from a glass desiccator.

Alternatively, carbon films can be made hydrophilic by exposure to strong UV radiation from a xenon arc for 40 min. This should be done just before using the support film. Such films facilitate uniform spreading of many types of specimens. Ultraviolet irradiation has the apparent advantage of not adding any material to the support film or the specimen. It is probable that some type of ionisation process produces the change, since the irradiation is accompanied by a strong smell of ozone. Although standard graphite films are hydrophobic, they can be made hydrophilic by linking oxygen atoms to the surface of the graphite oxide plates (Formanek, 1979); the preparation of graphite oxide is described on p. 367.

The second approach to modifying the surface of the support film is the addition of a surface-active compound to the specimen suspension, thus lowering its surface tension and facilitating spreading. Such compounds include sucrose, glycerol and bovine serum albumin (BSA). Although the use of BSA (0.005–0.05%) is preferred by some, it may increase the background granularity because of the large size of its molecule. The polypeptide antibiotic bacitracin (PB) is superior to BSA as a wetting agent. It is also more effective than BSA for lowering the surface tension, and its molecule is too small to be visualised by negative staining at a moderate magnification. Furthermore, PB allows specimen deposition by a pipette, loop or spray on the support film. Solutions of PB are stable for 2–3 weeks at 4 °C (Gregory and Pirie, 1973). As a guide, 1% PTA containing 50 μg/ml of PB is satisfactory. Bacitracin allows some extraction of virus particles from tissue when such tissue and the stain (with PB) are ground up simultaneously. At higher magnifications, PTA–PB or uranyl acetate–PB may give background noise.

Another approach is the attachment of cells to a positively charged surface via electrostatic interactions. Glass or plastics coverslips, freshly cleaved mica or carbon films can be coated with polylysine (Mazia *et al.*, 1974), as well as the cationic dyes ruthenium red or alcian blue (Sommer, 1977). Carbon-coated plastics films or plain carbon films on grids are floated on a drop of 0.25% polylysine (mol. wt. 4000) or 1% aqueous alcian blue for about 5 min. The grids are transferred onto a large drop of distilled water and then placed on clean filter paper (coated face down). Cell suspension is applied to the grids by quickly spreading a drop over the surface (Nermut, 1982b). Care should be taken to avoid the formation of air bubbles.

The adsorptive properties of the carbon can also be improved by treating it with the quaternary ammonium salt benzyldimethyl alkylammonium chloride (BAC) (mol. wt. ~350) (Vollenweider *et al.*, 1975). This ionic detergent possesses positively charged groups, and forms a monofilm. It can be used in concentrations as low as 10^{-5} g/ml for a few hours. Carbon films thus treated should be dried without rinsing. In certain studies, the use of BAC is not desirable. Low concentrations of BAC may cause the separation of DNA from histones in glutaraldehyde-fixed chromatin. The glow-discharge or alcian blue method is preferred for attaching glutaraldehyde-fixed chromatin to carbon films. It is better to modify the charges on the film than on the specimen. Of all the methods discussed above, glow discharge is the best approach, provided that the instrument is available.

Adsorption Properties of Support Films

Adsorption properties of support films are correlated with their surface charge. The presence of competing ions in the buffer also plays a part in the adsorption phenomenon. As stated earlier, the charge on the surface of the support film should be opposite to that of the specimen. Carbon and aluminium films (15 nm thick) have about the same surface charge densities (C/cm^2)—i.e. -3.1^{-8} and $+7.1^{-8}$, respectively (Sogo *et al.*, 1975). The surface potential of carbon and aluminium is −0.1 V and +0.5 V, respectively. Aluminium–beryllium films carry considerably higher positive potential at the surface (+1.2 V).

Since DNA behaves as an anion in a buffer of low ionic strength, it binds preferentially to positively charged films such as aluminium–beryllium (Sogo *et al.*, 1975). On the other hand, DNA in the presence of intercalating dyes (e.g. ethidium bromide) acts as a cation, and thus binds to negatively charged films such as carbon.

The appearance of certain macromolecules such as DNA depends on the type of support films used. Well-extended, uncoiled and unaggregated DNA molecules are obtained only when it is adsorbed to carbon, collodion or mica film in the presence of ethidium bromide (Sogo *et al.*, 1975). This dye also stiffens the DNA molecule, minimising its distortion.

7 Low-temperature Methods

INTRODUCTION: CRYOFIXATION

The goal of fixation is to preserve the cell structure as it was in the living state. The most common method used to achieve this goal is chemical fixation. An attempt is made through this method to preserve the size, shape, interactions, location and orientation of cell structures to be observed. Chemical fixation has long been providing extremely useful information concerning biological structure and function, and it will continue to do so. An overwhelming amount of information derived from analysis of chemically fixed specimens has been confirmed by studying rapidly frozen specimens, and this trend is expected to continue. However, though many features in chemically fixed and rapidly frozen specimens resemble each other, there are some important differences (Howard and Aist, 1979; Menco, 1984, 1986).

Chemical fixation does have certain limitations. Because of a normal delay in the diffusion of the fixative, it is thought that fixation only fixes the response of the cell to the fixative and not the living state. Many dynamic, electrophysiological processes occur in milliseconds, and are thus too fast to be captured by slow-interacting fixatives. Additionally, fixation with aldehydes involves oxygen consumption (Johnson and Rash, 1980) and hydrenium ion production. The implication is that fixation and asphyxiation occur simultaneously as the tissue hardens, thereby depriving the tissue of oxygen before fixation is completed. The effect of anaerobiosis existing during chemical fixation is typically seen on the cristal configuration of mitochondria, which has been discussed in detail by Hayat (1981).

Aldehyde-induced artefacts include shrinkage or swelling of organelles, membrane fusion (Hasty and Hay, 1977; Chandler, 1979), fusion of pinocytotic vesicles at the cell surface (Bretscher and Whytock, 1977), alteration in the fracture plane of membranes, rearrangement of intramembranous particles (Arancia *et al.*, 1980), a change in the appearance of membrane junctions and introduction of mesosomes in certain bacteria. Whether an aldehyde treatment increases or decreases intramembranous particles is still enigmatic. According to Menco (1984), the number of particles is lower in rapidly frozen than in chemically fixed specimens.

Aldehydes trigger elevated release of neurotransmitters (Hubbard and Laskowski, 1972). Fixatives tend to destroy the cytoplasmic pleomorphic canalicular system in plant cells (Mersey and McCully, 1978). Diffusion and extraction of various cell constituents, including ions, during chemical fixation is well established. Tissues pre-fixed with an aldehyde, rapidly frozen and freeze-substituted show relatively large and uniform extracellular spaces. Similar tissues conventionally dehydrated show collapse of such spaces. In other words, conventional dehydration is an important factor in causing shrinkage, while freeze-substitution is a gentle process. Artefacts introduced by conventional preparatory procedures have also been discussed in this volume and elsewhere (Hayat, 1981).

A better alternative to chemical fixation is cryofixation (physical fixation), a physical procedure used to preserve the distribution and structure of all components in a biological system. The principal aim and advantage of cryofixation is the near-instantaneous arrest of cellular metabolism (physiological processes), which is achievable with ultrarapid freezing. Ultrarapid freezing arrests intracellular movements in milliseconds; this is not possible with chemical fixatives, as their action is very slow. With cryofixation an attempt is made to capture cell morphology in its living state through stabilising and retaining soluble cell constituents by eliminating specimen contact with fixing and dehydration fluid. The major objectives of cryofixation include the preservation of native three-dimensional structure, solidification of soft tissues for sectioning or fracturing, prevention or reduction of the displacement of water-soluble substances, and maintenance of the viability of certain specimens.

The achievement of true vitrification (amorphous ice) is impossible with most biological specimens, because once the water in the superficial cell layers has become ice, the deeper layers are unable to dissipate heat fast enough. This is due to the low thermal conductivity of water and ice. The result is the formation of relatively large ice crystals in the core of the bulk specimen. Even the most thorough methods of rapid freezing, cold metal block freezing and cryogen jet freezing, can freeze without observable ice crystal damage only 600 μm surface layers; even these layers may have been damaged during trimming of the tissue block. Cryoultramicrotomy is much easier with a vitreous or near-vitreous sample. Nevertheless, ice crystal damage visible at the electron microscope level is avoidable through near-vitrification by ultrarapid freezing, resulting in the preservation of cell structures in the microcrystalline matrix. If the cooling rate is sufficiently rapid, achievement of a near-vitreous state with ice microcrystalline sizes less than 10^{-8} m is possible. The phenomenon of vitrification has also been discussed on p. 381.

Visible ice crystal damage can be easily avoided in monolayers of cells, cell fractions and specimens in the form of droplets or a thin layer. In the case of tissue blocks and other thick specimens, the best one can do is to reduce the size of ice crystals down to a dimension that will not interfere with ultrastructural details. This can be accomplished by using either ultrarapid cooling methods or, if one does not mind certain artefacts, cryoprotectants with or without pre-fixation with aldehydes, when deeper layers in the bulk specimen need to be studied. Artefacts introduced by fixatives and cryprotectants are discussed elsewhere in this volume.

It is pointed out that our understanding of alterations occurring in specimens in response to cryofixation is still rudimentary. The phenomenon of transition of liquid to solid water in a hydrated tissue is infinitely complex and difficult to control. Meaningful interpretation of electron micrographs of specimens that are cryofixed requires not only technical skill, but also an understanding of physicochemical processes that occur in a specimen during cooling and subsequent treatments. This understanding requires the knowledge of physics of ice, including homogeneous and heterogeneous nucleation, vitrification and devitrification, crystallisation and recrystallisation, thermal gradients arising during cooling, the dissipation of latent heat, the chemistry of aqueous solutions of electrolytes and non-electrolytes, and the physiology of the cell and the organism (Franks, 1977, 1985). The response of cells to different temperature regimens needs to be known.

Cryofixation can be followed by freeze-fracturing, freeze-etching, freeze-drying, freeze-substitution, cryoultramicrotomy and cryomicroscopy. Freeze-fracturing has become the most widely used approach and its contributions to the understanding of membrane molecular structure are well known. Considerable progress has been made in developing other approaches that follow cryofixation. Cryoultramicrotomy used in conjunction with cryofixation is the ideal way to ensure retention of soluble cell constituents, when cytochemistry is not applied; cellular materials tend to diffuse out during cytochemical procedures. However, cryoultramicrotomy is a demanding procedure, and so it should be used only for critical problems such as the localisation of antigenic sites and ionic species. Like other methods involving cryofixation, this method is still in its infancy in terms of understanding the physical principles involved.

The advantages of cryofixation can be further enhanced by reducing electron-beam-induced damage through cryomicroscopy of frozen-hydrated specimens. However, only small specimens such as lipid vesicles, viruses (Adrian *et al.*, 1984), bacteria (Lepault and Pitt, 1984), microtubules (Mandelokow *et al.*, 1986), actin filaments (Trinick *et al.*, 1986) and ribosomal subunits (Wagenknecht *et al.*, 1988) have been examined in the frozen-hydrated state, and even these specimens are plagued with very low contrast. The interpretation of the images of frozen-hydrated specimens is difficult, because the mechanisms by which contrast is generated are incompletely understood. Cryomicroscopy is neither possible nor necessary for routine studies. The success of the above-mentioned approaches depends primarily on the quality of cryofixation. Most of these methods are discussed later.

ADVANTAGES OF CRYOFIXATION

Theoretical and empirical advantages of cryofixation are as follows.

(1) Although not all aspects of the process of cryofixation are known, it is more amenable to rational interpretation than chemical fixation. The latter involves an almost infinite number of biochemical interactions.

(2) Artefacts produced by cryofixation are relatively easily interpretable, because only physical processes are involved.

(3) Little chemical alteration of cell constituents occurs by cryofixation. This results in minimum loss of enzyme activity and antigenicity. Extensive denaturation of proteins caused by chemical fixation is minimal in cryofixed specimens.

(4) Loss of cell constituents, common in chemical fixation, is absent during cryofixation. Cryofixation preserves electron-opaque materials in the cisternae of rough endoplasmic reticulum and within the Golgi cisternae and vesicles (Terracio *et al.*, 1981). These materials are difficult to preserve with chemical fixation. Cryofixation and freeze-substitution can result in dense mitochondrial matrix. Cryofixation facilitates the recognition of presynaptic vesicles attached to the cell membrane in the electric organ (Phillips and Boyne, 1984). This attachment is not recognisable with chemical fixation. Cryofixation followed by freeze-substitution can reveal lysosomal membrane structures not seen

with conventional methods (Nott *et al.*, 1985). Filaments within the β-cell secretory granules in pancreatic islets can be viewed with cryofixation, but not with chemical fixation (Dudek *et al.*, 1984).

(5) Cryofixation can arrest intracellular movements in milliseconds. This is important in the study of dynamic processes such as membrane fusion, exocytosis, endocytosis and ciliary beat, and theoretically allows the maintenance of all solute components in their thermodynamically unfavourable random distributions. It allows *in situ* localisation of diffusible cell components. However, capture of the earliest stages of molecular rearrangement that lead to and constitute membrane fusion may not be possible with freezing methods, since these events occur at the speed of molecular translation, which also limits the rate at which heat can be extracted from the system. Because freezing and fusion are occurring at about the same rate, fusion events probably cannot be captured with fidelity.

CRYOPROTECTANTS

Most commonly specimens are protected against ice crystal damage by the usage of cryoprotectants. Such a pretreatment is essential when cooling rates are not high enough to prevent ice crystal damage. It is very difficult to avoid extensive ice crystal formation and consequent structural damage when freezing tissue specimens larger than the critical size. Freezing of non-cryoprotected specimens by simple immersion in cold liquefied gases is difficult because the contact with a warm specimen heats the cold liquid, which warms and becomes less efficient. The coolant becomes too warm to freeze adequately, and tissue water crystallises and destroys the ultrastructure.

Cryoprotectants are substances that facilitate freezing of specimens with the formation of small ice crystals and acceptable preservation of cellular morphology. The most important protective influence of cryoprotectants is, apparently, through increasing the viscosity of the medium. They bind to or substitute for water and thereby reduce the amount of water available for freezing. In other words, cryoprotectants decrease the chemical potential of the water and thus retard the growth of ice crystals from the pool of nuclei present. The other influence exerted by cryoprotectants is through increasing the number of ice crystal nuclei. The more numerous the nucleation sites, the smaller the size of ice crystals. The quality of freezing is improved when the specimen fluid is split into small droplets. Nucleation is the creation of centres in the solution to which water molecules migrate to form a crystal. Normally the rate of crystal growth is greater than the rate of nucleation. An increase in solute concentration is accompanied by an increase in the rate of nucleation. Cryoprotectants tend to decrease the rates of nucleation and crystal growth. Another factor that affects the size of ice crystals is the rate at which water molecules diffuse to the crystal surface. Because the rate of diffusion depends on temperature, the lower the nucleation temperature, the slower the rate of crystal growth.

The use of cryoprotectants becomes necessary when the ice crystal growth needs to be controlled throughout a tissue block, as, for example, for cryosectioning (Skaer, 1982). The deeper layers of a tissue block, even when ultrarapidly frozen, show cellular distortion caused by ice crystals of relatively large size. Even the high-pressure freezing method does not prevent the formation of sizeable ice crystals in cell layers located deeper than 0.4–0.6 mm from the surface.

Either low-molecular-weight penetrating cryoprotectants (e.g. glycerol) or high-molecular-weight polymeric, non-penetrating compounds (e.g. sucrose) can be used to suppress ice crystal formation. These reagents are employed usually at concentrations of 10–30%, depending on the specimen under study, and can be used with or without prior chemical fixation. Specimens with very low water content and high surface area-to-volume ratio can be frozen with minimal damage by ice crystals in the absence of a cryoprotectant. Cryoprotectants, such as glycerol, that easily penetrate the membranes, should not be used when X-ray microanalysis needs to be carried out, because they cause intracellular ionic dislocation. On the other hand, non-penetrant hydrophilic polymers have been used for this type of study.

PENETRATING CRYOPROTECTANTS

Penetrating cryoprotectants produce the environment for a reduction of cell water content at temperatures low enough to reduce the damaging effect of the concentrated solutes on the cells. They permeate cell membranes and directly prevent the formation of large-size ice crystals. The most commonly used low-molecular-weight penetrating cryoprotectants are glycerol and dimethyl sulphoxide (Me_2SO or DMSO). Other penetrating cryoprotectants include methyl and ethyl alcohols and ethylene glycol. Glycerol is used quite commonly for freeze-fracture investigations. It has been well studied as a cryoprotectant by Luyet and Kroener (1966). Since glycerol penetrates cells very slowly at 0 °C (273 K) but rapidly at 20 °C (293 K), it may be regarded either as a penetrating or as a non-penetrating agent. Full protection of cells occurs only when glycerol penetrates the cells.

Glycerol penetrates animal tissues rapidly, but generally does not penetrate plant tissues. For example, glycerol does not penetrate *Nitella flexilis* cells, but DMSO does (Taylor, 1988). In this case, although glycerol does not act as an intracellular cryoprotectant, it may perform this function by osmotically drawing water out of cells. Plant tissues can be pretreated with glutaraldehyde, rendering their membranes leaky and thus facilitating the penetration by glycerol. Glycerol confers increased conformational stability to native globular proteins. This cryoprotectant has been shown to change lipid membrane structures (Bearer *et al.*, 1982) and to prohibit water sublimation from a frozen, fractured surface (Bearer and Orci, 1986). Glycerol reduces deformation of macromolecules during freeze-fracturing (Sleytr and Robards, 1977).

Both glycerol and DMSO enhance phase separation, which may produce artefacts such as aggregation of cellular components. Glycerol causes a rearrangement of microtubules, which is proportional to the concentration of the cryoprotectant and duration of exposure to it (Indi *et al.*, 1986). This adverse effect is more pronounced on free microtubules than on those showing intertubule linkages. As stated above, glycerol is known to change membrane permeability, allowing cellular contents to leak. This results in a change in cell pH, which, in turn, may cause aggregation of microtubules. Treatment with glycerol is thought to cause a reduction in the number of microvilli (Murphy *et al.*, 1982). Other artefacts produced by glycerol include clustering of intramembrane particles, and blebbing and vesiculation of internal membranes. The cause of these artefacts seems to be that glycerol perturbs hydrophobic interactions (Franks and Reid, 1973). Some of these artefacts can be minimised by utilising pre-fixation with glutaraldehyde. However, this and other aldehydes tend to produce their own artefacts such as redistribution of intramembranous particles (Willison and Brown, 1979). Chemical fixation also makes membranes permeable to intracellular ionic species. Other artefacts produced by glycerol have been reviewed by Skaer (1982) and Gilkey and Staehelin (1986). In spite of the above limitations, glycerol is simple to use and is likely to remain popular as a cryoprotectant. Like most other cryoprotectants, glycerol is used in rather high concentratons (>20%).

Dimethyl sulphoxide as a cryoprotectant has been thoroughly studied by Rasmussen and MacKenzie (1968). It penetrates more rapidly than glycerol. This reagent has a high vapour pressure that permits freeze-drying or deep-etching (Stolinski and Breathnach, 1975). It is toxic to cells at the concentrations necessary to preserve morphology, and produces a range of artefacts (McIntyre *et al.*, 1974). This reagent tends to disrupt membrane integrity rather easily, resulting in solute leakage from the cell. For these reasons it is not recommended for preserving cell ultrastructure.

NON-PENETRATING CRYOPROTECTANTS

When used in conjunction with very low cooling rates, non-penetrating cryoprotectants osmotically remove water from the interior of cells primarily during the initial phases of freezing at temperatures between −10 °C and −20 °C (263 and 253 K), as these additives become concentrated in the extracellular spaces. Commonly used non-penetrating cryoprotectants are polyvinylpyrrolidone (PVP), hydroxyethyl starch (HES) and dextran. They are used usually at concentrations of 15–25%, depending on the type of specimen under study. These compounds are hydrophilic polymers of high molecular weight, and do not penetrate cells easily, although they may be taken up pinocytotically (Barnard, 1980). It is thought that these compounds retard extracellular ice nucleation by raising the viscosity of the extracellular medium (MacKenzie, 1977), allowing supercooling of the cytoplasm. It has been indicated that the formation of extracellular ice crystals somehow seeds the growth of intracellular ice (Mazur, 1984).

It is thought that non-penetrating cryoprotectants generally do not interfere significantly and irreversibly with cell function. It has been indicated that, over a short duration, 25% solutions of both PVP and HES are less toxic to cells and tissues compared with the same concentration of glycerol (Echlin *et al.*, 1977). The relatively low toxicity of PVP and HES makes them more suitable as cryoprotectants for morphological as well as analytical studies, compared with glycerol and DMSO.

However, both PVP and HES can cause cell shrinkage through colligative or non-colligative osmotic pressure effects (Skaer, 1982). It has been shown that PVP exerts osmotic effects on cells, resulting in dramatic changes of the cell volume unless its concentration is adjusted to isotonicity (Halgunset *et al.*, 1984). Therefore, the use of PVP requires scrupulous adjustment of osmolarity and pH to avoid adverse side-effects. Dextran has the advantage of having a relatively low toxicity. According to Franks *et al.* (1977), it should be used at a concentration of >40% to ensure near-vitrification. However, such a high concentration of dextran is expected to cause osmotic perturbations. Dextran and other high-molecular-weight cryoprotectants provide an extracellular matrix for electrolytes (Gupta *et al.*, 1976), reduce mechanical trauma in certain tissues (Barnard, 1980), facilitate sectioning and decrease the risk of accidental dehydration during specimen handling prior to freezing.

Prefreezing infusion of the tissue blocks (0.5–1.0 mm^3) with 1.6–2.3 M sucrose (pH 7.0–7.4) seems to prevent ice crystal damage and imparts plasticity, facilitating cryosectioning at −90 to −120 °C (183 to 153 K) (Tokuyasu, 1986). Sucrose appears to provide a satisfactory supporting matrix for cryosectioning.

Mixtures of cryoprotectants can be used. The advantage is that the total concentration of the solute is increased in the mixture without increasing excessively the concentration of any one cryoprotectant. A mixture of bovine serum and sucrose has been used for cryoprotecting yeast suspensions (Roomans and Sevéus, 1976), while 10% glycerol in 20% sucrose has been employed by Willison and Rowe (1980).

VITRIFICATION

Vitrification is a process which allows solidification of the solution into amorphous ice (glass) without any formation of ice crystals. Vitrification avoids any phase separation, and the vitrified state is supposed to be a faithful representation of the liquid state. Cryoelectron microscopy of vitrified specimens allows their study in their native aqueous environment. Images of simple, very small, vitrified specimens appear to be free from any artefact (Lepault, 1985). The vitrification temperature of water—i.e. the temperature above which vitrified water would transform into the crystalline state—is about −143 °C (130 K) (Yannas, 1968). Complete vitrification of pure water or aqueous solutions occurs at cooling rates in excess of $\sim 3 \times 10^6$ K/s (Bald, 1986).

Vitrification guarantees that the entire water content in the suspension is conserved. The requirements for vitrification are stringent. Ultrarapid freezing must be used, which demands that the thickness of the specimen should not exceed 1 μm (Dubochet and McDowall, 1983). After freezing, vitrification demands that the specimen temperature be kept below −143 °C (130 K) to avoid devitrification.

Water has been vitrified from the liquid state without ice crystals (Brüggeler and Mayer, 1980). Vitrified water was first viewed in the electron microscope by Dubochet *et al.* (1982). This vitrification was accomplished by the ultrarapid freeze method including cryogen jet and cold metal block freezing procedures. Vitrified virus and T4 polyhead suspensions have been viewed by cryoelectron microscopy (Lepault *et al.*, 1983; Lepault, 1985). Flattening of these specimens is avoided by embedding them in an ice layer. Freezing of specimens in vitreous ice maintains them in an environment close to that in which they normally function and so minimises artefacts from dehydration and staining while preserving structure to high resolution (Stewart and Vigers, 1986).

Up till now the specimens that have been truly vitrified are suspensions (thin films). Whether true vitrification can be accomplished in tissues is uncertain. On the basis of theoretical models, Bald (1986) has indicated that at normal pressures no single constant critical cooling rate value exists that will limit ice crystal size to an acceptable level for good-quality microscopy during cryofixation. Perhaps near-vitrification can be achieved in the outer layers of a tissue block.

At present, tissue specimens can be vitrified if the physical properties of the specimen water are altered. This is accomplished by applying cryoprotectants which lower the point of freezing, elevate the recrystallisation temperature and decrease the critical rate of freezing necessary for vitrification (Moor, 1973). About 10–15% sucrose can be added to vitrify specimens of sectionable size. The absence of ice crystals can be verified by the smoothness of the aqueous background in the replica of the frozen specimen (Costello *et al.*, 1984; Menco, 1986). However, cryoprotectants have adverse side-effects. An excellent, up-to-date review including the methodology of cryoelectron microscopy of vitrified specimens has been presented by Dubochet *et al.* (1988).

LIQUID CRYOGENS

Cryoprotected specimens are usually frozen in nitrogen slush, Freon 22 or liquid propane. At present, with more emphasis on rapid freezing of non-cryoprotected specimens, there is ample research effort on better ways of freezing. An effective liquid cryogen must have high thermal conductivity and heat capacity, a low melting point at atmospheric pressure, a high specific heat, and a wide temperature range between its freezing and boiling points. Liquid propane cooled by liquid nitrogen is a very good liquid cryogen for general use. It comes very close to fulfilling the above criteria. It has a low melting point (−189.6 °C; 84 K), high boiling point (−42 °C; 231 K), and good thermal conductivity and specific heat. When propane is supercooled at −194 °C (79 K), it cools very fast. Also, it is readily available and is relatively inexpensive. It has been suggested that a mixture of propane and isopentane is superior to propane (Jehl *et al.*, 1981; Ward and

Murray, 1987). The details of this superiority are given on p. 383.

Ethane produces faster cooling rates than those provided by propane or propane–isopentane with the plunge-freeze method (Silvester *et al.*, 1982; Bald, 1984; Ryan *et al.*, 1987). Ethane, when supercooled, cools even faster, and gives the best measured cooling rates. When ethane and propane are supercooled, they easily solidify; to avoid this, these coolants should be employed at ~5 °C above their freezing points. Liquid nitrogen is a more efficient coolant than either ethane or propane, provided that vapour formation is suppressed (Bald, 1984). However, in practical terms, propane cools faster than boiling liquid nitrogen.

Liquid nitrogen is not recommended for routine studies. When specimens are dipped into liquid nitrogen, they cool relatively slowly because of the formation of an insulating layer of gas around the specimen. Nitrogen slush provides faster cooling than does liquid nitrogen at its boiling point (−196 °C; 77 K). This is especially true when nitrogen slush is used at elevated pressures. Although liquid helium has a very low temperature (<−271 °C; 2.2 K) and very high heat conduction capacity (8.03 mW m^{-1} K^{-1} at 4 K) (Fernández-Morán, 1960), it is not recommended because, like nitrogen, a film of gaseous helium insulates the specimen from the coolant and thus retards the cooling rate. Helium, moreover, is tedious to use, must be maintained under vacuum, is required in large quantities, and is very expensive. However, liquid helium can be used as a coolant for metal blocks. The choice of a cryogen among the commonly used liquid cryogens is thought to be of less importance than the velocity of the cryogen past the specimen. Physical criteria for efficient cooling in liquid coolants are discussed by Robards and Crosby (1983).

It is cautioned that propane is highly flammable; it is combustible at concentrations as low as 2% (20 000 ppm) in air. Pure propane gas is odourless and is heavier than air. However, at the temperature at which it is usually used (about −188 °C; 85 K), propane's vapour pressure is exceedingly low and is different from its flash-point (−104 °C; 169 K). Thus, hazards are much less when it is actually being used. Liquid helium is the most volatile liquid known.

RATE OF COOLING

In cryofixation the most important factor is the rate at which heat can be extracted from the region of interest in the tissue specimen. Typically, specimens should be frozen at a cooling rate of or exceeding 10^4 °C/s, until liquid nitrogen temperature (−196 °C; 77 K) has been achieved. At this temperature, the molecular structure of cells is stable. One of the major problems encountered during freezing of the tissue is not the different thermal gradients and thermal conductivities of the cryogen, but the poor thermal conductivity of the tissue, which reduces the rate of freezing and results in ice crystal damage. If the rate of heat release during the phase change equals the rate of heat removal by conduction before the phase transformation is complete, the specimen will warm, followed by rapid growth of ice crystals (Stephenson, 1956). If, on the other hand, the rate of heat removal by conduction through the specimen exceeds the maximum rate at which heat is released by crystallisation, the cooling will be maintained and either very small ice crystals or complete vitrification of the specimen will occur.

The rate of cooling is in part dependent upon the nature, size and geometry of the specimen, the depth of the region of interest in the specimen, the initial temperature of the specimen, the presence or absence of high concentrations of salt or natural organic cryoprotectants in the cytoplasm, the rate of specimen entry into the cryogen and the method of cryofixation. The cooling rate is also limited by convection from the cryogen to the specimen surface (Zasadzinski, 1988). As a result of this limitation, the cooling rate is proportional to the ratio between specimen surface area and volume, independent of the thermal conductivity of the specimen.

Theoretically, the cooling rate increases in the following order of specimen geometry: flat (cooled from one side), flat (cooled from both sides), cylindrical and spherical. However, very thin specimens of equivalent volume cool even more rapidly. Ultrarapid freezing of very small (1–2 μm) specimens such as tissue culture cells can be easily achieved throughout its volume. Similar freezing of the entire volume of a thin planar specimen (up to 40 μm in thickness) can be obtained by cooling it from both sides. The closer the region of interest to the cold surface, the higher the rate of cooling.

As stated earlier, a definite correlation exists between heat transfer and the cooling rate. The cooling rate is inversely proportional to the thermal density of the specimen. The order of heat transfer is ethane > propane > Freon 22 > Freon 12. Heat transfer is more or less proportional to the cryogen velocity. For commercial jet freeze devices and plunge freezing, the cryogen velocities can approach 100 m/s and >0.5 m/s, respectively. Thus, the relative cooling rate of jet frozen specimens 100 μm thick is comparable to that of plunge-frozen specimens 1 μm thick that show vitreous ice (Adrian *et al.*, 1984). Significant improvements in the cooling rate can be achieved by increasing the velocity of the cryogen. This can be obtained by jet freezing or by using higher immersion velocities.

For routine studies, specimens should be projected into the liquid cryogen at a moderate speed (1–5 m/s). Specimens can be vibrated at ~50 Hz as they are plunged into the liquid cryogen, or the cryogen can be kept in constant motion with sonication to increase the rate of cooling. The purpose is to maximise the exposure of the specimen surface to new, cold cryogen (Robards and Sleytr, 1985).

METHODS OF FREEZING

CONVENTIONAL FREEZE PROCEDURE

According to its strictest definition, cryofixation is a superior alternative to chemical fixation. However, for conventional cryofixation, tissue specimens are briefly pretreated with aldehydes to avoid structural artefacts that otherwise will be introduced by the cryoprotectants (antifreeze agents). If needed, soft specimens of small size and plant tissues can be encapsulated in an inert substance (e.g. agar, bovine serum albumin, PVP and dextran) that confers mechanical strength on the specimens; these substances can be used at a concentration of ~5%.

The next step is impregnation of the specimens with a cryoprotectant to prevent the formation of large ice crystals during freezing. Both the freezing-point depression and the reduced nucleation rate can be achieved with cryoprotectants such as glycerol. Specimens are frozen by plunging them into a liquefied organic gas such as propane at a temperature of from −160 °C to −190 °C (113 to 83 K), resulting in a freezing rate of ~100 °C (173 K)/s. Although this rate of freezing can be increased by using helium, tissue specimens are still not vitrified. The thickness of the specimen should not exceed 0.5 mm. Conventional cryofixation is adequate only when the results obtained have been confirmed by ultrarapid freezing of unfixed, uncryoprotected specimens.

PLUNGE-FREEZE METHOD FOR CONVENTIONAL AND ULTRARAPID FREEZING

Biological specimens were being frozen in cryogens for transmission electron microscope studies as early as the 1950s (Glick and Malmstrom, 1952). The plunge-freeze method (quench-freeze method) involves dipping the bulk specimen into a liquid cryogen. This method is the simplest, least expensive and most widely used procedure, whether ultrarapid freezing is needed or not. The tissue block is dipped into the cryogen by hand or by plunging the tissue with the aid of spring-loaded or gravity-driven devices to increase the rate of travel of the specimen through the cryogen. Frozen specimens can be processed for cryosectioning, freeze-substitution, freeze-drying or freeze-fracturing.

Cryogens such as liquid propane, a mixture of propane and isopentane, liquid nitrogen, solid–liquid nitrogen slush, ethane and Freons have been used for plunge freezing. The cryogens are cooled by liquid nitrogen, and are maintained above their freezing points. One of the best coolants is contaminated liquid propane used at about −180 °C to −190 °C (93 to 83 K). The useful heat sink capacity of propane is higher than that of any other commonly used liquid cryogen. If the specimen is contained within a metal holder during plunging into a highly subcooled liquid, nitrogen will be the most efficient fluid (Bald, 1985). Bald (1984) indicates that liquid nitrogen maintained near its melting point (−210 °C; 63.1 K) at a pressure in excess of the critical value of 33.5 atm produces the fastest cooling. It is concluded that although subcooled liquid nitrogen at the pressure indicated above produces the fastest cooling theoretically, for practical purposes contaminated propane is the best coolant.

However, it has been suggested that some other coolants are better than propane for plunge freezing. Jehl *et al.* (1981) recommend the addition of isopentane to pure propane to depress its freezing point below the temperature of liquid nitrogen. Ward and Murray (1987) recommend the addition of 0–4% of isopentane to natural propane (at least 96% pure) instead of pure propane to avoid freezing at −196 °C (77 K). This mixture has been used as a cryogen for plunge-freezing and vitrifying aqueous suspensions of biological specimens (e.g. fragments of cilia and flagella) (Ward and Murray, 1987). According to K. P. Rayan (personal communication), ethane is better than propane for plunge-freezing when the former is used at −185 °C (88 K).

Vapour formation is presumed to occur during plunging of metal specimens at atmospheric pressure due to their high conductivity, thus retarding their rate of cooling (Bald, 1984). With tissue specimens, this does not seem to occur. High plunge velocity results in improved cooling rates, because the vapour barrier is swept away at such velocities. On the other hand, in metal specimens, cooling appears to be enhanced and to occur over a constant distance as plunge velocity is increased, perhaps because vapour formation is suppressed (Ryan *et al.*, 1987). It is apparent that for maximum transfer of heat from the biological specimens in the shortest possible time, the entrance and travelling velocity of the specimen into the cryogen should be high. Plunging should be performed as rapidly as possible through a sufficient depth of cryogen. Ideally, specimens should continue to be plunged throughout the freezing process, so that cryogen renewal all around the specimen is continuous. The best results are obtained when cooling is completed during plunge motion (Ryan and Purse, 1985a). It seems that cooling is accomplished by forced convection, which is the most efficient process of heat transfer and is dependent on travelling velocity.

As stated above, the small size of the tissue specimen is of utmost importance in achieving rapid freezing. Cooling rate at a given depth from the specimen surface decreases with increasing thickness of specimens (Bald, 1983). It is thought that instantaneous heat flux and cooling rate are dependent less on the plunge velocity in the range of 0.75–1.75 m/s (Walford, 1969) than on the physical size of the specimen and the depth or time of plunge at a given velocity (Bald, 1985). However, with short plunge times, the average cooling rate for the specimen between 0 °C and −100 °C (173 K) undergoes significant increase with increasing plunge velocity (Robards and Crosby, 1983).

Procedures

The simplest and an inexpensive conventional freezing procedure is rapid manual immersion of the specimen into nitrogen slush. Thin specimens (<5 μm) are recommended. The device for this procedure can be constructed in many laboratories. The equipment needed includes a small vacuum chamber, a vacuum line already fitted in the laboratory (or a small rotary pump) and an insulated container made of Styrofoam (polystyrene) for liquid nitrogen. Immediately after evacuation has started, grey nitrogen slush appears over the whole surface of the liquid nitrogen. When the surface ice turns sugar-white, evacuation is interrupted by admitting some air. The frozen nitrogen surface is immediately cleared, becomes transparent and sinks. This freezing–thawing cycle may be repeated three times, so that the nitrogen should become frozen throughout its entire depth (Sevéus, 1978). Finally, within 1–2 min after the removal of the vacuum, the solid nitrogen begins to become liquid at a pressure of 1 atm. During this very brief slush period, a specimen is immersed into it, resulting in a rapid heat transfer due to the reduction in the formation of insulating nitrogen gas at the specimen surface at temperatures below the boiling point of nitrogen (−196 °C) (Schooley, 1985). The temperature of the nitrogen should be kept low for several minutes if a number of specimens need to be frozen in sequence. If the tissue is frozen inadvertently in grey nitrogen slush, a sound of boiling will be heard. However, the absence of the sound of boiling does not preclude the possibility that vapour barrier formation has occurred.

The above procedure does not result in ultrarapid freezing. More rapid specimen freezing is accomplished by mechanically propelling the specimen into the cryogen. The velocity of specimen entry into and travelling through the cryogen should be high in order to achieve quick heat transfer. Although several authors have described home-built plunge-freezers (e.g. Völker *et al.*, 1984; Ryan and Purse, 1985b), versatile but expensive devices are commercially available (e.g. Reichert KF80).

Improved plunge-freezing of small tissue blocks (<0.3 mm^3) mounted on aluminium foil attached to a modified Reichert FC4 pin has been reported by Ryan and Purse (1984). The edge of this pin is bevelled and a strip of foil is attached with epoxy glue to form a streamlined arch over the front of the pin (Fig. 7.1). The arch shape of the foil allows efficient cryogen flow over the specimen as well as separating it from the thermal mass of the metal support. The tissue–foil–pin assembly is plunged into a 7:3 mixture of propane and isopentane (−190 °C; 83 K) at ~1 ms/s. Ryan and Purse (1984) claim that up to a 200 μm region from the leading tissue block face is well frozen. The improved freezing is thought to occur as a result of the separation of the specimen from the thermal mass of the metallic support pin. In other words, cooling is relatively slow in

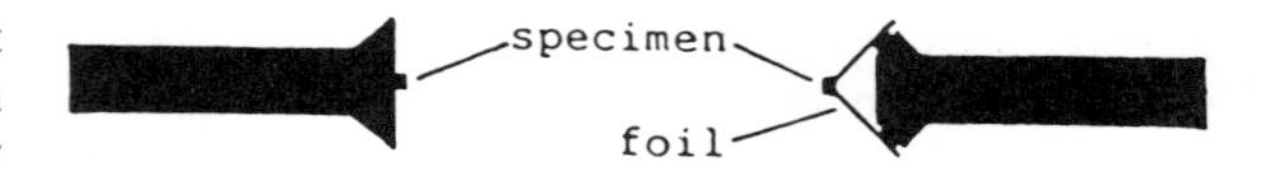

Figure 7.1 Left: a standard, low-mass aluminium support used in the Reichert FC4 cryoultramicrotome system with cryoprotected specimens for routine immunocytochemical studies. Right: modified support with a bridge of aluminium foil attached by epoxy adhesive. The specimen is placed on the bridge, where it is in a streamlined position and removed from contact with the thermal mass of the pin. Fresh tissue specimens adhere well for cryosectioning; pre-fixed tissue blocks need a thin layer of cryoadhesive, such as Tissue-Tek. Narrow bridges may exhibit chatter during cryosectioning; this can be overcome by filling the space under the bridge with *n*-heptane at −90 °C (183 K). The modified supports are particularly useful in cryosubstitution studies. Specimens on the modified supports are sensitive to passing through a cold gas layer above the coolant. Other low-mass supports can be made by soldering a wire frame to the pin and attaching a foil sheet using epoxy, similar to a tennis racket with an angled head. Another design features a wire V-frame attached to the pin; long thin specimens are attached between the ends of the arms. Scale bar: 1 mm. From Ryan and Purse (1984)

specimens in contact with solid metal supports; streamlining and coolant flow are also important. Fast freezing is possible when the cold gas layer above the cryogen is reduced or eliminated by raising the cryogen unit for plunging.

In small specimens (0.25 mm diameter) mounted on foil supports cooling is related to plunge velocity; faster plunge velocities result in fast cooling rates. However, in relatively large specimens (1 mm diameter) cooling occurs mainly after they come to rest (Ryan and Purse, 1985a). In specimens of 0.5 mm diameter cooling is related to plunge velocity at low immersion speeds, but then it decreases with increasing plunge velocity. Further developments in the use of aluminium foil as specimen support are awaited.

COLD METAL BLOCK FREEZE METHOD

In 1932 Gersch touched the sliced tissue with a copper plate which had been cooled in liquid air, and then the plate–slice was transferred back into liquid air. Subsequently, freezing of specimens against a cold metal surface was attempted by Simpson (1941) and Eränko (1954), and further developed for electron microscopy by Van Harreveld and Crowell (1964). The cold metal block freeze method ('slam-freeze method') involves bringing rapidly exposed, thin and flat specimens into contact with a polished metal surface (copper or silver) which has been uniformly cooled to −190 °C (83 K) with liquid nitrogen or to −254 °C (19 K) with liquid helium (synthetic sapphire has better thermal properties than those of copper, and can be used instead of copper: Meisner and Hagins, 1978). The melting points of liquid helium and liquid nitrogen are −271.4 °C (1.75 K) and −210 °C (63 K), respectively.

Thermal characteristics of metals differ variously with temperature. Silver functions less efficiently at −269 °C (4 K) than at −257.4 °C (15.6 K) (Bald, 1983). Little difference is found between silver and copper at −269 °C (4 K), while at −196 °C (77 K) copper is better than silver. At −254 °C (19 K) both copper and silver are superconductors of heat. The thermal conductivity of copper is 500 times greater than that of glass. Only ultrapure copper (>99.99%) has the high thermal conductivity necessary for maximum freezing rates. Copper containing 0.056% iron impurity has a thermal conductivity of only ~13% of that shown by ultrapure copper (at −235 °C; 20 K) (White and Woods, 1955). The transfer of heat by copper (and other metals) improves significantly after lowering of the temperature. After a brief use, copper blocks develop microscratches, which can be eliminated by repolishing. The copper surface should be meticulously cleaned and polished. The details of handling and polishing of copper plates are given by Chiovetti *et al.* (1987).

The metal surface is protected from condensation of frost and atmospheric oxygen by passing a stream of nitrogen boil-off gas over its surface. This method provides cooling rates of from 25 000 °C/s to 50 000 °C/s up to a depth of 10–20 μm from the freezing face of a fresh, untreated tissue block. The well-frozen region contains ice crystals ~10 nm in size; ice crystals increase in size exponentially with depth. With helium, freezing occurs in 2 ms or less up to a depth of 10 μm (Heuser *et al.*, 1979). Although helium provides the fastest possible cooling rates, for routine studies liquid nitrogen is preferred, owing to the high cost of and difficulty in obtaining and working with the former.

Although the cold metal block freezing method is generally used for freezing tissues, other types of specimens, such as cell suspensions, have also been frozen (Wagner and Andrews, 1985). This method is uniquely suited to the study of the precise timing of physiological events happening in the millisecond range (Heuser *et al.*, 1979; Gilkey and Staehelin, 1986). Another advantage of this method is that the well-frozen layer of the specimen is physically flat and parallel to the specimen support (Metz, 1981). The freezing starts on the freezing front along the area of contact with the copper block and proceeds inwardly. Higher cooling rates are obtained with this method compared with those achieved by the plunge freezing method. In general, the depth of the ice crystal-free zone obtained with the cold metal block freeze method is greater than that given by the plunge freeze method.

As with all other cooling methods, the cold metal block method has certain limitations. The removal of heat is only from one side of the specimen. Since the hydrated, deformable tissue specimen is brought very rapidly into close contact with the surface of the cold metal block, some degree of distortion of the specimen surface is likely (Robards and Sleytr, 1985). In order to achieve satisfactory thermal contact for ultrarapid freezing, the specimen must be flat and oriented parallel to the metal surface. With irregularly shaped specimens, sufficient thermal contact for rapid freezing is obtained by the plunge-freeze method. The malleability of copper can result in an uneven surface for the attachment of cells. Polishing and etching of copper is time-consuming, and the polishing grit can get embedded in the copper and thus attached to cells. Although the possibility that copper may react with cells has been mentioned (Edwards *et al.*, 1986), the specimen is frozen long before any significant chemical interactions can occur between the specimen and the copper block.

The use of the term 'slamming' for the cold metal

block freeze method is inappropriate, for the tissue is not slammed (smashed) against a cold metal block. Although the tissue is brought in contact with the metal surface very rapidly, it is gently apposed and not slammed. In fact, one of the critical engineering features in the Gentleman Jim quick-freezing device is the ability to rapidly but gently appose the tissue against the copper.

Procedures

Both simple and complex devices and procedures are available for cold metal block freezing. According to the simplest procedure, the copper block is cooled in liquid nitrogen, from which, immediately before freezing, it is lifted slightly above the surface (Dempsey and Bullivant, 1976). Immediately the specimen is picked up with forceps and pushed against the copper block (−196 °C; 77 K), good freezing being obtained up to a distance of ~12 μm from the block specimen surface. A drawback of this procedure is that the block is probably covered with a thin layer of liquid nitrogen, which will dramatically slow the freezing process.

Alternatively, the specimen, mounted on a spring-loaded support on the underside of a vertically falling rod, is brought into rapid contact with the polished surface of a copper block cooled from below. Heath (1984) has introduced an apparatus for freezing with minimal manipulation against a liquid helium-cooled metal block. Liquid helium is expensive and unavailable in many laboratories. A simple and inexpensive cold metal device using a dry liquid nitrogen-cooled copper surface has been introduced by Allison *et al.* (1987). It employs a 'slamming' plunger to provide rapid and uniform contact between the tissue and the metal block. The tissue is brought in contact with a dry copper surface with minimal rebound off the block. A dry copper surface produces a high thermal conductivity. Rebound is undesirable, for it temporarily interrupts contact between the tissue and the metal block during the critical initial cooling phase. Heart tissue cryofixed with this device is shown in Fig. 7.2. The above-mentioned devices can be constructed from available materials in most laboratories. These devices are far less expensive than commercial equivalents.

In a variation of the cold metal block freeze process the polished metal surface moves towards the specimen. Copper blocks are attached to the jaws of a pair of pliers which are used to clamp-freeze the tissue (Hagler *et al.*, 1984). Alternatively, a pair of pliers uses blocks of frozen Freon 22 to achieve a gentle clamping action (Somlyo *et al.*, 1985). Both variants yield good results when used skilfully. Devices for controlled and reproducible, bounce-free contact of the specimen with the frozen metal block are commercially available (Reichert, Buffalo, NY, and Buckinghamshire, England; Ted Pella, Tustin, CA; Polaron, Watford, England; R.M.C., Inc., Tucson, AZ).

CRYOGEN JET FREEZE METHOD

The cryogen jet freeze method was introduced by Moor *et al.* (1976) and subsequently elaborated upon by Müller *et al.* (1980). This method involves spraying liquid propane, cooled by liquid nitrogen to −180 °C to −190 °C (~93 K), onto the specimen at high speed. The cooling rate increases with increasing speed of the propane jet. This method takes advantage of increased heat convection when a jet of liquid cryogen is shot onto the surface of a bulk specimen. The high-velocity propane jet (~100 m/s) (Gilkey and Staehelin, 1986) provides a cooling rate of about 30 000 °C/s (Costello *et al.*, 1982). This heat exchange is ~2–30 times faster than that provided by the plunge-freeze method. The cooling rate near the specimen surface is four times greater for jet spraying compared with plunge-freezing (Bald, 1985). However, as the specimen thickness increases, the superiority of jet spraying over the plunging process disappears. This decline is not applicable to spherical or block specimens. Overall, the jet freeze method is superior to plunge-freezing.

Jet freezing allows near-vitrification of any specimen that can be prepared as a thin layer. The use of a sandwich preparation is recommended with this method in order to prevent possible loss of or damage to the specimen by the force of the propane jet. Because heat is withdrawn simultaneously and rapidly from both sides of the specimen, the depth of the frozen region in the specimen is as great as or greater than that achieved with any other method at atmospheric pressure (Gilkey and Staehelin, 1986).

Theoretically, the entire volume of a specimen up to 40 μm thick can be rapidly frozen with propane-jet freezing. In practical terms, propane-jet freezing shows satisfactory freezing to a depth of ~15 μm. However, it has been claimed that tissue samples up to 40 μm thick can be ultrarapidly frozen by a double-propane-jet freezer (MF7200; R.M.C., Inc., Tucson, AZ). In the MF7200 unit, propane is delivered at −190 °C (83 K), the specimen is frozen between gold or copper holders only 50 μm thick, and the specimen is held in such a way that it cannot be dislodged by high-velocity jets. Haggis (1986) has obtained satisfactory ultrarapid freezing by modifying the double-propane-jet freezer, QFO 101 (Balzers). Thin copper plates (<100 μm) used in the sandwich increase the rate of cooling. Frozen sandwiches can be used for ultrarapid freezing of aqueous suspensions and emulsions, tissue culture

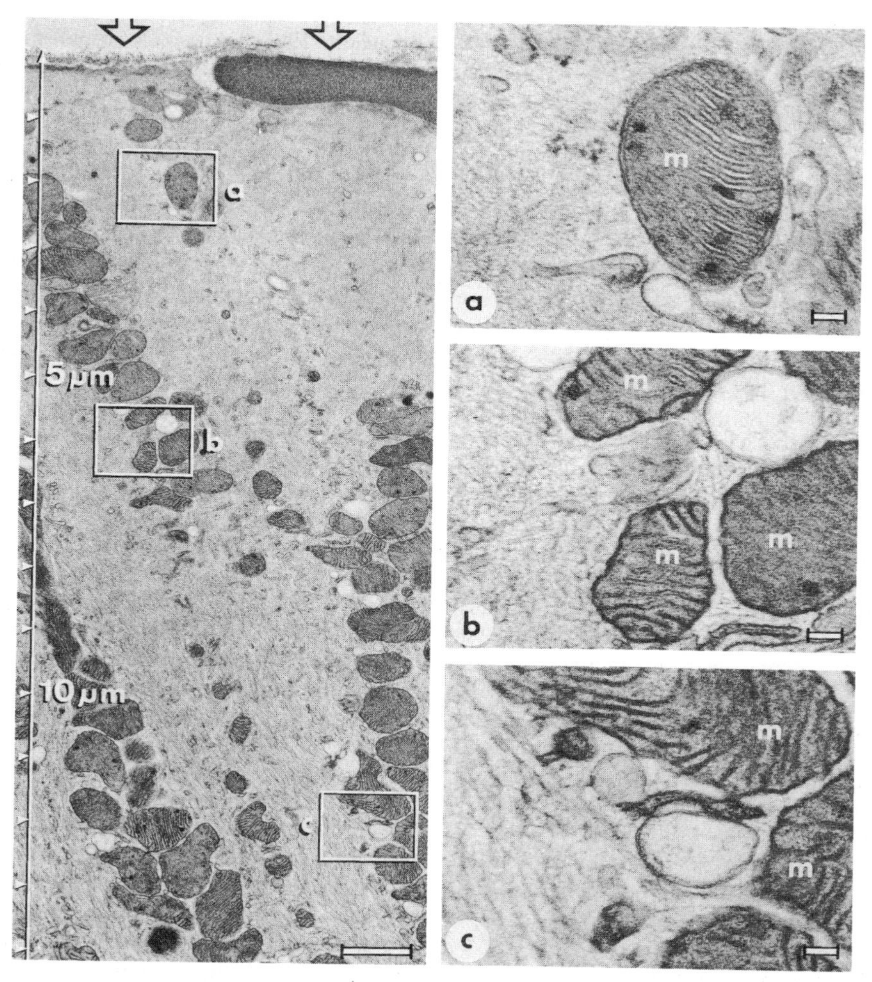

Figure 7.2 Electron micrographs showing the gradient of good to adequate ultrastructural preservation that results when a tissue untreated with cryoprotective agents is correctly prepared for electron microscopy by cryofixation. In this example, a mouse heart was quickly excised, cryofixed without chemical treatment by contacting the tissue against the surface of a dry copper block cooled to the temperature of liquid nitrogen, freeze-substituted at $-80\,°C$ (193 K) in 4% OsO_4 in acetone, and embedded in Epon prior to thin sectioning. From the point of contact with the copper block (open arrows) to a depth of 10 µm into the tissue the ultrastructure is well preserved by cryofixation. Between the myofibrils, which are sectioned obliquely and therefore do not show typical cross-striations, lie large mitochondria (m) with well-defined, closely packed cristae. Selected fields (a–c) taken at higher magnification from the indicated areas at left show a definite gradient of freezing. At a depth of 2–3 µm into the tissue (a) ice crystals are absent, while at 6 µm (b) ice crystals of 20 nm first appear in the myofibrils but not in the mitochondria. Although the mitochondria remain ice-free to a depth of 12 µm (c), the myofilaments are disrupted by ice crystals 30–40 nm in diameter. Stained with uranyl acetate and lead citrate. Bar on left is 1 µm; bars on (a)–(c) are 100 nm. From Allison *et al.* (1987)

cells, suspensions of a wide variety of animal and plant cells, and tissues.

The limitations of the cryogen jet freeze method are: (1) as in all other rapid freeze methods, freezing is not always uniform throughout the specimen; (2) damage to exposed surface structures of the specimen by the mechanical impact forces is possible; (3) after freeze-fracturing, the replica may be difficult to separate from the metal carrier; (4) the frozen specimen does not easily adhere to the smooth surface of the metal carrier unless the latter has been sufficiently roughened; and (5) possible toxic effect of copper or anoxia on cell viability, especially when assemblage of the sandwich is prolonged. Although jet freezing is the best general-purpose method, the cold metal block method is ideal for rapid freezing of tissues.

Procedure

Many laboratories construct their own versions of the apparatus for processing specimens with the jet freeze method. The procedure given below is based on the work of Gilkey and Staehelin (1986). A hollow metal cylinder is cooled to near liquid nitrogen temperature (about −190 °C; 83 K) while being flushed with dry nitrogen to avoid the formation of frost and condensation of liquid oxygen in the cylinder. Gaseous propane is admitted into the cylinder sufficiently slowly to allow it to condense and fill the cylinder with liquid propane. The propane is allowed to cool to near liquid nitrogen temperature, but should not be permitted to freeze.

The specimen, along with a spacer, is placed between two specially shaped copper or gold plates (~50 μm thick) of very low mass to form a sandwich; the spacer prevents the specimen from being pressed and damaged. The spacer should be as thin as feasible; 300-mesh electron microscope grids of copper (~15 μm thick) or gold (~12 μm thick) can be used as spacers. Slot grids of copper (~70 μm thick) or gold (~35 μm thick) can be used for larger cells. The sandwich is clamped in a holder and placed between two jets, through which liquid propane under 10 atm is blown from opposite directions onto both sides of the specimen to freeze it. The frozen specimen is separated from the holder and transferred to liquid nitrogen.

SPRAY-FREEZE METHOD

The spray-freeze method was introduced by Williams (1954) and refined by Bachmann and Schmitt (1971). This method is based on the supposition that freezing in a microcrystalline state is achieved when an efficient cryogen acts upon specimens of very small size (droplets). The method achieves a high surface-to-volume ratio by splitting the specimen into very small droplets (10–50 μm diameter). Spray-freezing also takes advantage of increased heat convection by vigorously injecting the spray droplets into a cryogen fluid. Besides these two factors which increase cooling rates, the very low heat content of specimen droplets encourages higher cooling rates. Also, the reduced nucleation within very small spray droplets is an advantage of this method. The cooling rates achieved are fast enough to prevent observable ice crystal damage. This method provides higher rates of cooling than those obtained with any other method. The rate of cooling of the droplets may be higher than 100 000 °C/s (Mayer and Brüggler, 1982). Aqueous suspensions of cells and subcellular fractions are ideally suited for this method. The quality of ultrastructural preservation achieved with this method is satisfactory.

The limitations of the spray-freezing method are: (1) the specimens that can be used are limited to very small unicellular organisms (e.g. bacteria), cell fragments and other particulates not larger than ~10 μm in diameter; (2) spray gun, frozen specimen droplets and adhesive are difficult to handle; (3) the surface area of the replicas useful for viewing is limited; (4) smearing of fracture faces with butylbenzene, used to collect the frozen droplets, by the microtome knife is not uncommon; and (5) cells may be damaged by the shearing forces when the specimens are forced through the air-brush. This method should be used only for special studies.

Procedure

An aqueous suspension is sprayed in the form of fine droplets (10–20 μm in diameter) by a commercial air-brush (atomiser) at high velocity into a cooled vessel containing liquid propane near its freezing point (−190 °C; 83 K); the atomiser can be used at pressures between 0.5 and 1.5 atm. The final ratio between specimen and suspending medium is usually 1:1 (Robards and Sleytr, 1985). The distance of the spray gun tip from the cryogen should be ~50–100 mm. Using a sample of 0.2–0.3 ml, the spraying is done at intervals of ~1 s to slow down the warming of the cryogen. The droplets should be propelled into liquid propane in a liquid state and not, as could happen as a result of low velocity, frozen within the gaseous phase of the cryogen. Both the minute size of the specimen droplets and the speed at which they move through the cryogen maximise the rate of heat transfer, resulting in an extremely high rate of cooling.

The frozen specimens are recovered by transferring the vessel to a temperature-controlled block kept at −85 °C (188 K) with liquid nitrogen boil-off in a glove-box and evaporating the propane under vacuum. A fine powder of frozen specimens is found on the bottom and sides of the vessel. An inert organic medium such as butylbenzene is added at the same temperature to make a paste with the specimens. Butylbenzene is used because it has a low freezing point (−95 °C; 178 K) but does not freeze-substitute the frozen specimens during the brief period for which it is allowed to remain liquid (Gilkey and Staehelin, 1986). Small aliquots of the paste are transferred with a cold platinum loop onto precooled specimen carriers, which are dropped into liquid nitrogen to solidify the butylbenzene. The specimens are now ready for freeze-facturing. The details of the construction of an inexpensive apparatus for spray-freezing have been presented by Bachmann and Schmitt (1971) and Lang *et al.* (1976). The method's low cost makes it suitable for use in laboratories that are unable to purchase expensive, commercial apparatus.

HIGH-PRESSURE FREEZE METHOD

The high-pressure freeze method was introduced by Moor and Riehle (1968). It takes advantage of the suppression of ice nucleation and growth in water at high pressures. Under hydrostatic pressure of 2100 bar, water in the specimens does not begin to crystallise until the temperature reaches −90 °C (183 K), at which the rate of ice crystal formation is significantly reduced. At 2045 atm, the melting point of pure water is lowered from 0 °C to −22 °C (251 K). The critical ultrarapid freezing rate of water in specimens is lowered to −100 °C (173 K)/s (Riehle and Höchli, 1973). The cumulative effect of the above changes is significant lowering of the rate of ice crystal nucleation and growth; the changes incurred by pressure freezing exert a cryoprotective effect equivalent to pretreatment with ~20% glycerol.

Although high-pressure freezing does not accomplish freezing as fast as other methods of rapid freezing, it deserves special interest because of its potential to freeze near vitrification relatively thick specimens. Theoretically, specimens up to 0.6 mm thick can be immobilised in <10 ms, producing ice crystals <10 nm in diameter. This thickness is 20–30 times greater than that available with plunge-freezing, cold metal block freezing or cryogen jet freezing. Although both cryogen jet and cold metal block freezing can stabilise cytoplasm uniformly in <10 ms, the maximum thickness of well-frozen cytoplasm is usually <20 μm with cold metal block freezing and 40 μm with cryogen jet freezing. Since most plant cells have diameters of 50–100 μm, these two methods are restricted in their usefulness for the study of plant cells. On the other hand, high-pressure freezing can preserve cells within 50–100 μm from the specimen surface (Craig and Staehelin, 1988). Most plant cell components, including endoplasmic reticulum, Golgi apparatus, nuclear membrane and pores, and microtubules, are well preserved. Transient membrane events such as those associated with vesicle fusion and vesicle blebbing also show good preservation.

Although a pressure of 2000 bar is lethal to cells and tissues, specimens are exposed to this pressure for less than 20 ms before freezing. A disadvantage of the high-pressure method is its adverse effect on sensitive structures. In the case of some plant cells, the high-pressure artefacts include wrinkling of the plasma membrane. This artefact consists of long tears or folds in the plasma membrane; these folds may be up to 100 nm wide and up to several micrometres long in the freeze-fractured images (Craig and Staehelin, 1988). The collapse of intercellular spaces under high pressure before freezing is thought to cause mechanical stresses in surrounding cells which result in such artefacts. Some large-scale, but slight deformation of whole plant cells may also occur (Craig and Staehelin, 1988). Menco (1986) has also reviewed the method's limitations.

In the correct implementation of this method, a jet of isopropanol followed by liquid nitrogen is spread onto a sandwich preparation of the specimen. The isopropanol serves to pressurise the specimen chamber before the specimen is cooled by the liquid nitrogen. Liquid nitrogen can be used as the cryogen because the pressure achieved is far above the critical pressure of nitrogen (35 atm), so no vapour barrier is formed. The pressure is maintained for 0.5 s, long enough to complete the freezing of the thick samples usable with the method.

A variation of the high-pressure freeze method involves the application of an appropriate amount of pressure during freezing, so that the cryogen (liquid nitrogen) is prevented from forming excessive vapour and so the *Leidenfrost* phenomenon does not occur (Bald and Robards, 1978). Significant depression of T_m and T_h of the specimen does not occur under these conditions. An expensive prototype apparatus (HPM 010) for freezing under high pressure is commercially available (Balzers, Hudson, NH, USA; Berkhamsted, England; Furstentum, Liechtenstein).

POPSICLE-FREEZE METHOD

The popsicle-freeze method involves clamping of tissue between slabs of melting cryogen, which assume the shape of the tissue during freezing (Somlyo *et al.*, 1985). This method allows rapid freezing of organs *in situ* with a spring-loaded, hand-held clamping device which holds two apposing melting Freon 22 popsicles. Since the Freon is at the melting point while coming in contact with the tissue, the former moulds itself exactly to the unevenness of the tissue surface.

A small incision is made in the abdomen of the anaesthetised animal and the Freon clamper is activated to snap-freeze the tissue. The frozen specimen, along with the clamp, is plunged into liquid nitrogen. The specimen is removed by opening the jaws of the clamper under liquid nitrogen, and then allowing the specimen to float free by placing the heads of the clamper with the tissue in a large beaker of melting Freon.

The advantage of this method is that certain types of tissues can be frozen in a near-physiological condition at −164 °C (109 K). This method has only been employed for preserving elemental composition at a high spatial resolution. The preservation of tissue morphology is adequate, showing signs of rather extensive ice crystals.

PUNCH-FREEZE METHOD

The punch-freeze method is similar to the popsicle-freeze method in that the specimen is punched out of the living organ by means of a syringe chilled with liquid propane or liquid nitrogen (Von Zglinicki *et al.*, 1986).

SPECIMEN PREPARATION BY SANDWICH FREEZING

The sandwich freeze method was introduced by Gulik-Krzywicki and Costello in 1978, and since then many variations and refinements of this method have been developed. Sandwiched specimens can be cooled by the plunge-freeze method, the propane jet freeze method, or the high-pressure freeze method. This method exploits a high surface-to-volume ratio by producing a thin specimen layer (~10 μm thick) between metallic plates. Near-vitrification of thin slices of tissue or cellular suspensions (<20 μm) can be achieved. The specimen is sandwiched between two thin metal plates. Compared with other methods, sandwich freezing allows a better control of ambient conditions such as ion concentration and surface tension forces. The housing of the thin specimen within a well-defined closed space permits such a control, which is not afforded to the specimens on a free surface, as in the cold metal block method. Another advantage of sandwich freezing is that very small volumes of the specimen are needed. Several sandwiches can be prepared from a specimen volume of 1 μl, which is a fraction of the amount required for spray freezing. The processing of radioactive or infectious specimens can be safely accomplished with sandwich freezing.

The ideal specimen carrier should be very thin, possessing a high strength: weight ratio. These two criteria are fulfilled by titanium (Handley *et al.*, 1981b). The tensile strength of titanium (860 MPa) is much greater than that of the more commonly used copper (211 MPa). Hence, thinner carriers can be made from titanium, thereby increasing the surface area: volume ratio; such a ratio will increase the cooling rate (Zasadzinski, 1988). Titanium carriers as thin as 4 μm have given excellent results (Handley *et al.*, 1981b); copper carriers cannot be made as thin. Commercial copper carriers usually used are 100 μm thick. Toxic effects of copper on certain cell populations cannot be disregarded.

The sandwich method was combined with the propane jet process by Müller *et al.* (1980), which led to the development of a synchronous double-jet cryofixation device (Balzers). In this device, two copper plates sandwich the specimen, and the sandwich is fixed in a holder, which is placed between two nozzles through which liquid propane (−180 °C; 93 K) is shot from opposite directions onto the copper plates. This approach takes advantage of increased heat convection by shooting a jet of liquid cryogen onto the surface of a sandwiched specimen. Compared with cooling from one side only, cooling from opposite directions results in a fourfold increase in the ultrarapidly frozen layer of the specimen (Müller *et al.*, 1980). Hippe (1984) has introduced a unit for double-jet cryofixation. Using stainless steel ring tube, a double jet of liquid propane cooled to −190 °C (83 K) with dry nitrogen (6 atm) is directed on both sides of the copper–copper-sandwiched specimen. The use of spacer rings is not necessary, except for mechanically sensitive specimens such as protoplasts.

It is thought that the double-jet method has the disadvantage that two jets may not hit the two copper surfaces of the sandwich precisely at the same time (Plattner and Knoll, 1983). The one-side cryogen jet sandwich method does not have this disadvantage, but gives only one-sided freezing. A simple, inexpensive, one-side propane jet procedure was developed by Pscheid *et al.* (1981). In this procedure the sandwich consists of one copper sheet and a thermoinsulator sheet for the back of the sandwich. The propane is shot from a pressure vial onto the copper plate. The cooling rate obtained with this procedure is ~18 000 K/s (Knoll *et al.*, 1982). Details of this procedure are outlined below.

Sandwich Freezing of Monolayer Cell Cultures (Pscheid *et al.*, 1981)

The following method has the advantage of processing the cells as they were grown on their substrate. Small discs of the supporting plastics substrate (Thermanox, Miles Labs.) are cut with the undisturbed cells grown on them. The cells are sandwiched by placing a thin copper or gold plate over the disc, with a spacer ring in between. The spacer helps to avoid cell injury during sandwiching. Specimen grids with the central portion removed can be used as spacers; they provide a thickness ranging from 12 μm to 30 μm. This sandwich now has the metal plate on top and the lower side is thermally insulated by the plastics substrate.

A jet of liquid propane (−190 °C; 83 K), obtained by condensing propane gas with liquid nitrogen, is shot onto the metal holder. The propane is cooled by pouring it into a simple brass container that stands on its brass support rod within a liquid nitrogen-filled Dewar container. To freeze the metal holder, the propane container is briefly pressurised with dry nitrogen by inserting a nitrogen pressure pistol into the Teflon connecting piece (contained in the lid of the propane container) and briefly operating it. The sandwich is brought into the propane jet by using insulated tweezers. Care should be taken that the copper side faces the jet. Excess propane should be removed either by a jerky movement of the sandwich or by passing it through a cold dry nitrogen stream. In this step the specimen temperature does not exceed −150 °C (123 K). The specimens are plunged into liquid nitrogen for storage.

Freezing Tissues *in situ*

Intracellular concentrations and distribution of ions tend to change as a result of ischaemic and traumatic effects on the organ under study. Even short-lasting ischaemia (~30 s) prior to cryofixation can cause ionic shifts (Von Zglinicki *et al.*, 1986). Similarly, stoppage of blood flow for more than ~10 s results in notable ionic shifts between cells and extracellular space in organs such as heart and liver. It is known that even under a brief anaerobic condition intracellular concentrations of ions such as K^+, Na^+ and Cl^- undergo major changes. Absence of aerobic glycolysis creates low membrane potential, high intracellular NaCl and low K^+ concentrations. Damage to membranes during tissue excision prior to cryofixation causes elemental redistribution. The above-mentioned alterations are unacceptable in X-ray microanalysis.

In order to minimise both ischaemic and traumatic injury, cryoballistic devices have been introduced (Monroe *et al.*, 1968; Chang *et al.*, 1980). By using these devices the tissue specimen under study is not separated from the circulation system before onset of cryofixation. Recently a simple hand-held device for simultaneous excision and freezing was introduced by Von Zglinicki *et al.* (1986). This device prevents ionic shifts caused by traumatic effects, even in the outermost cells of the tissue block. The device can excise tissues from organs within the circulatory system while the heart of the animal is still beating. As a result, the shift in the intracellular concentrations of ions is small. On the other hand, because the speed of cryofixation obtained with this device is slower than that achieved with other advanced freezing methods, the quality of ultrastructural preservation is less than ideal. The pistol device (Chang *et al.*, 1980) has the advantage of giving a reproducible excision speed.

The main part of the device introduced by Von Zglinicki *et al.* (1986) is a long, stainless steel tube (1 mm inner diameter and 2.1 mm outer diameter) having a sharp circular tip and a polished inner wall. The length of the tube is adjusted according to the depth of the tissue to be studied. The device is chilled with liquid propane at the initial temperature of −196 °C (77 K) and plunged into the organ, leather gloves being worn. A cylindrical specimen block (1 mm in diameter and 0.5–1.5 mm in height) is punched out. The needle remains within the organ for $< \sim 2$ s, and its temperature remains below −153 °C (120 K). The frozen tissue is cryosectioned for X-ray microanalysis.

A high-velocity impact device for obtaining multiple, contiguous myocardial biopsies was introduced by Hearse *et al.* (1981). The device is first explosively projected against the tissue to be collected at 10^5 mm/s (100 m/s) and then rapidly cooled to −135 °C (138 K) in a cold halocarbon. Although this device allows minimal time between excision and freezing, it is less efficient than that introduced by Von Zglinicki *et al.* (1986).

STORAGE OF FROZEN SPECIMENS

Frozen specimens can be stored indefinitely in liquid nitrogen. The requirement is that the temperature of the specimen does not rise above the recrystallisation temperature of the specimen (−70 °C for plant tissues and −50 °C for animal tissues). Storage systems are commercially available, but are rather expensive. Polyethylene scintillation vials are not recommended for storage. An inexpensive, simple storage device for small specimens (~5 mm in diameter) using liquid nitrogen has been introduced by Rigler and Patton (1984). Another advantage of this device is that its length allows the specimen to be continuously immersed while the cap is removed from the unit.

This device can be constructed from commonly available laboratory materials in less than 20 min. It can fit into any Thermos-type Dewar and is reusable. The device consists of any number of disposable polyethylene pipettes joined together by two supports cut from polyethylene micromoulds (Polysciences, Inc., Warrington, PA) (Fig. 7.3). The conical bottom is cut from each mould, and a pipette is inserted into the modified mould. The pipette tip is removed and re-

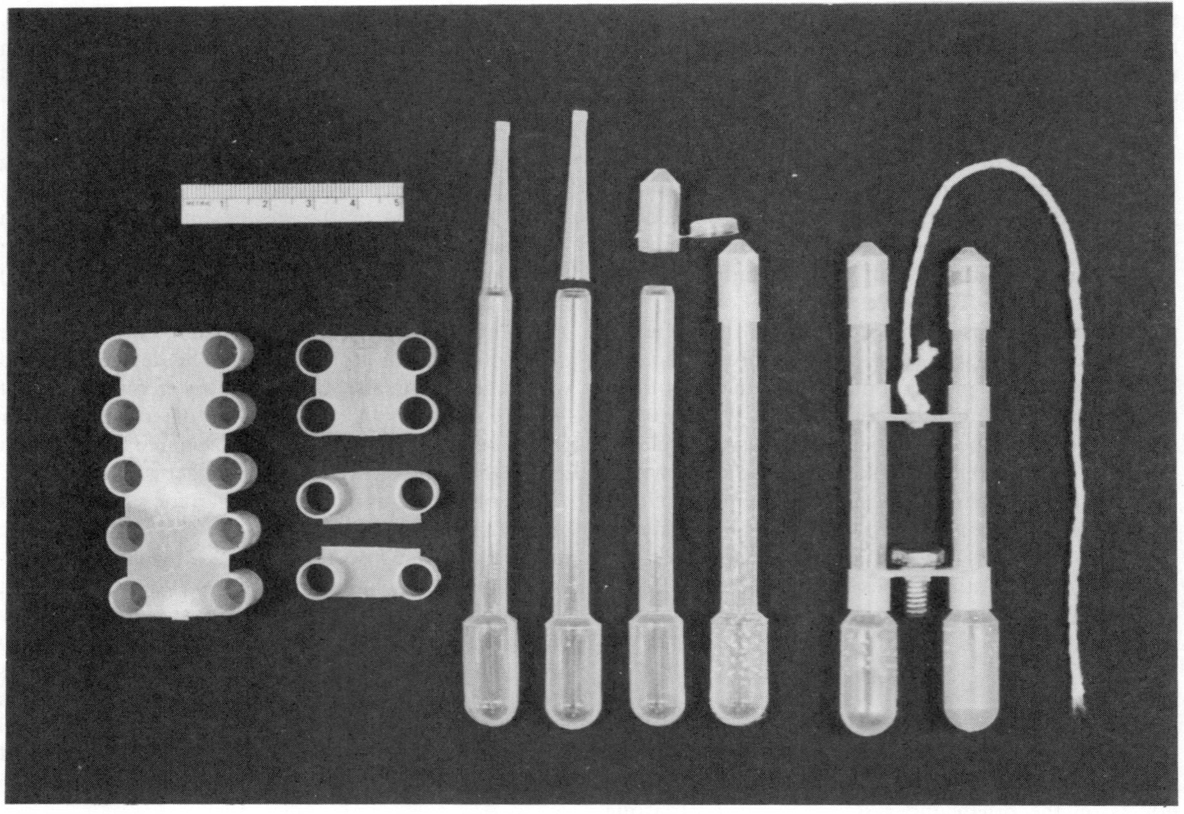

Figure 7.3 The basic components for construction of a cryogenic storage device. From left to right: Micro-mould, supports, polyethylene pipettes (units) and a completed device. From Rigler and Patton (1984)

placed with a BEEM capsule whose cap has been discarded (Fig. 7.3). The entire length of the pipette is perforated with a 12-gauge needle, except the lower 1–1.5 cm of the bulb. The device is weighted at the specimen end with a metallic object such as a nut. A string or wire is tied to the upper end of the device (Fig. 7.3) to retrieve it from storage.

The device is immersed in liquid nitrogen in a Styrofoam container (minimum diameter 12 cm and depth 6 cm). The capsule is removed and the device is tilted nearly horizontally under liquid nitrogen with the help of the string. Frozen specimens are transferred with precooled forceps into the device, and the pipette is capped with the capsule. While still submerged in liquid nitrogen, the device is tilted upward slightly and then sharply tapped, so that the specimen slides down into the bulb. If needed, the specimen can be carefully pushed down with a wooden applicator stick before capping the pipette. Specimens are retrieved by removing the cap and rapidly inverting the device over a catch-pan which has been immersed in liquid nitrogen. Catch-pans may be made from folded aluminium foil. If the specimens adhere to each other after prolonged storage, they can be separated by sharply tapping the bulb with a spatula.

Williamson (1984) has also introduced a simple device for the storage of six small specimens in liquid nitrogen. Specimens are individually placed in six blind holes (8 mm diameter, 7 mm deep) drilled in an aluminium disc (32 mm diameter, 10 mm deep). Further details of this device are given by Williamson (1984).

FREEZE-DRYING

Freeze-drying involves the dehydration of a frozen tissue block through the sublimination of ice. Unlike the freeze-substitution process, organic solvents are not used during freeze-drying. Thus, the frozen tissue is not exposed to the adverse effects of solvents. Water vapour leaving the specimen is immediately trapped on the surface of a nearby condenser, provided that the mean-free path for water molecules is comparable with or greater than the specimen–condenser spacing (Ingram and Ingram, 1983). The pressure in the apparatus determines the mean free path. The freeze-drying is thus a vacuum distillation process rather than a simple vacuum drying.

Freeze-drying of small tissue pieces is completed usually in about 3 days. Relatively large tissue blocks require longer periods at low temperatures for complete freeze-drying. In this case the elevation of temperature to ambient should be gradual, to ensure complete drying of the core of the specimen. Relatively large specimens after abrupt warming may contain undried portions which can cause rehydration of the entire specimen, including the peripheral regions.

Since the frozen tissue is warmed to increase vapour pressure and facilitate sublimation, this increased temperature tends to cause undesirable effects such as damage to the ultrastructure due to ice recrystallisation. The production of ice crystal reticulation may occur during freeze-drying when it is not carefully controlled, even though ice crystal formation has been held to a minimum during rapid freezing (Dudek and Boyne, 1986). According to these workers, for low- and intermediate-resolution studies, 10–100 nm and 1–10 nm ice crystal reticulations, respectively, are acceptable.

Another undesirable effect of freeze-drying is intracellular, whole-cell and entire tissue shrinkage. Ultrarapidly cryofixed specimens are thermodynamically unstable at low temperatures. Water tends to stabilise by forming ice crystals, causing partitioning in the specimen (Lyon *et al.*, 1985). Pure ice may form outside the cell, resulting in dehydration and shrinkage of cell organelles (Plattner and Bachman, 1982). Approximately 7% shrinkage of the entire cell during freeze-drying has been reported (Boyde and Franc, 1981). As much as 20% shrinkage of relatively large tissue blocks may accompany freeze-drying (Boyde *et al.*, 1977).

The presence of salts, sucrose or Tris buffer during rapid freezing followed by freeze-drying may result in the production of artefactual filament-like and trabeculum-like structures (Miller *et al.*, 1983). Extensive washing of the specimens in distilled water seems to be necessary to remove such materials before freezing, but carries the risk of osmotic and physiological damage to the specimen. Displacement of elements and other diffusible cell constituents will occur because these substances in the aqueous phase tend to migrate towards the nearest surface(s) upon drying (Roos and Barnard, 1985). Nevertheless, freeze-drying has been widely used for studying electrolyte distribution, on the assumption that any redistribution of elements is below the X-ray spatial resolution of the electron probe. This method is used extensively for scanning electron microscopy.

Procedure

The rapidly frozen tissue specimen is transferred onto the precooled specimen-holder housed in a glass cylinder (condenser), which has been cooled in liquid nitrogen. The specimen is warmed to −123 °C (150 K) under vacuum (10^{-3} Torr). Sufficient pressure should be used to prevent rehydration of the specimen. The specimen is allowed to dry under vacuum at −110 °C (163 K) for 24 h and then at an incremental temperature rise of 10 °C/h until the temperature reaches −55 °C. The apparatus is left to reach room temperature in ~24 h. The tissue on a specimen-holder is placed on a piece of copper gauze above a small beaker containing 0.1 g of OsO_4 crystals under a fume-hood. The beaker and gauze are placed inside a large vessel, containing a molecular sieve (type 5A, calcium aluminosilicate) to prevent rehydration, and closed to form an airtight seal (Lyon *et al.*, 1985). After about 12–16 h, the tissue is embedded in a resin for thin sectioning. Well-preserved area is restricted to ~15 μm in thickness in the periphery of the tissue specimen. Variable quality of preservation of the smooth muscle ultrastructure with freeze-drying is shown in Fig. 7.4. This procedure applies only to a cold-metal-block-frozen specimen.

Alternatively, both OsO_4 vapour fixation and resin infiltration can be accomplished under vacuum within the freeze-drying apparatus. In some studies, formaldehyde vapours can be used instead of OsO_4 vapours. It is difficult to suggest a ready-made freeze-drying apparatus based on the quality of electron micrographs published in the literature, because many laboratories construct their own apparatus. Several apparatuses are commercially available, including the Coulter-Terracio unit (Ted Pella, Burlington, VT). Chiovetti *et al.* (1987) have introduced a freeze-drying apparatus (Figs. 7.5 on p. 395 and 7.6 on p. 396) that is a modification of the design presented by Coulter and Terracio (1977). The details of the design and operation of this apparatus are given by Chiovetti *et al.* (1987), and the results obtained with the apparatus are shown in Fig. 7.4.

Linner *et al.* (1986) have introduced an ultra-high-vacuum molecular distillation drying apparatus to remove amorphous-phase tissue water without apparent rehydration. The rapidly frozen specimen is transferred under liquid nitrogen to a vacuum chamber (10^{-8} mbar) where the temperature is equilibrated at −192 °C (86 K). The specimen is incrementally warmed (1 °C/h) and the tissue water gradually removed. The specimen is resin-embedded in the vacuum with or

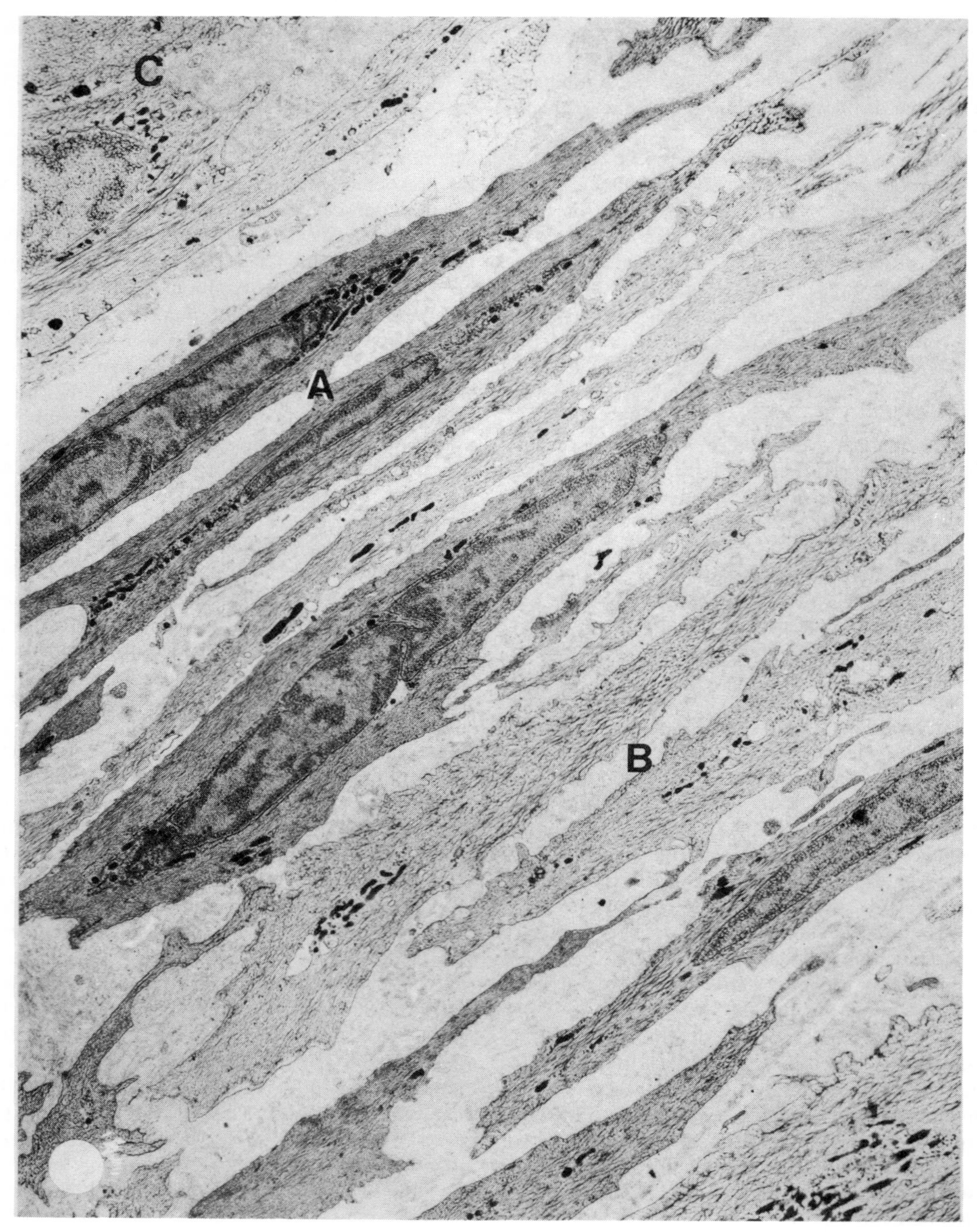

Figure 7.4 A low-magnification view of smooth muscle cells in the rabbit renal artery. This tissue was freeze-dried, osmicated and embedded in Spurr's low-viscosity resin. Well-frozen cells (with more dense cytoplasm) are interspersed among cells that show varying degrees of freezing damage (damaged cells have less dense cytoplasm). The cells at A exhibit typical well-frozen morphology. Nuclei are similar to classically fixed and embedded tissue, and the cytoplasm is homogeneous and moderately electron-dense. The cells at B and C show progressive freezing damage. In B there is some aggregation of contractile filaments, with open spaces between the filaments. The area at C exhibits obvious freezing damage. Notice the loss of contrast in both the nucleus and the cytoplasm, the aggregation of contractile filaments and ice reticulation in the nucleus. 7500×. From Chiovetti *et al.* (1987)

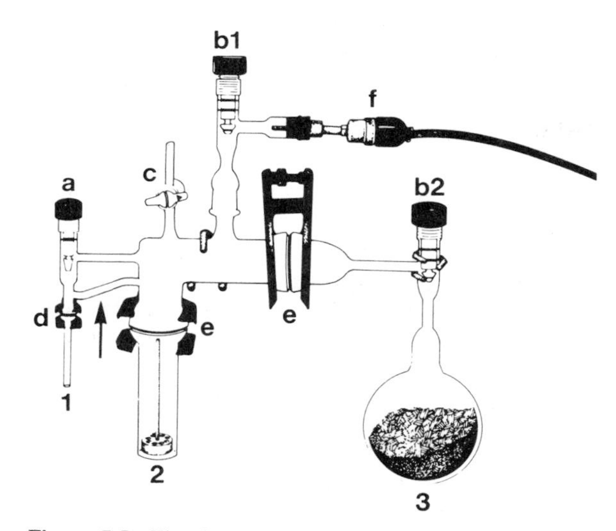

Figure 7.5 The freeze-drying apparatus consists of three chambers connected by a manifold. The small chamber (1) holds crystals of osmium tetroxide for vapour fixation. The aluminium specimen-holder is shown enclosed in the glass specimen chamber (2), and the molecular sieve chamber (3) is on the right. The other components and sources are as follows. Teflon stopcocks: (a) 0–5 mm, Rotaflo TF6/18; (bl and b2) 0–10 mm, Rotaflo TF6/24 (source: Ace Glass Incorporated, Vineland, N.J.); (c) hollow-bore ground-glass vacuum stopcock (source: Eck and Krebs Scientific Laboratory Glass Apparatus, Inc., Long Island City, N.Y.). O-ring joints: (d) size 12/5; (e) size 65/40; (source: Kontes Co., Vineland, N.J.). Pinch clamps: clamp for (d): no. 12A; clamp for (e): no. 65 (source: Thomas Scientific, Swedesboro, N.J.). (f) Vacuum gauge, DV-6 (source: Teledyne Hastings-Raydist, Hampton, Va). *Note:* The arrow points to a solid piece of glass between the main body of the dryer and valve A, which stabilises the valve. From Chiovetti *et al.* (1987)

without vapour-phase osmication. Contrary to published claims, specimens processed with this method are not vitrified.

FREEZE-SUBSTITUTION

Freeze-substitution involves the replacement of ice in the frozen tissue with an organic solvent at a higher temperature than that at which the specimen was frozen. The solvent, in turn, is replaced by a resin. Solvents that have been used include acetone, methanol, ethanol, heptane and diethyl ether; acetone is the most commonly used solvent. Freeze-substitution is normally carried out at −80 to −95 °C (193 to 178 K). An organic solvent is employed which dissolves ice at these temperatures. Compared with acetone, methanol is better capable of dissolving ice at −80 to −95 °C (193 to 178 K); the latter dissolves ice in a few hours. Methanol also tolerates a higher water content than does acetone.

The extraction and translocation of solutes in the frozen and freeze-substituted tissue are thought to be less than those involved in the conventionally fixed and dehydrated tissues at room temperature. In the rapidly frozen and properly freeze-substituted tissue, movement of organelles is less likely. Fixation occurs at a low temperature during the substitution process, resulting in less extraction. Similarly, the preservation of ultrastructure is improved by the slow and continuous dehydration, and the loss of elements from the specimen is reduced at low temperatures. The preservation of cell constituents is further improved by embedding at a low temperature (−35 to −60 °C) (238 to 213 K) in an embedding medium such as Lowicryl. However, when the organic solvent front passes through the tissue and displaces water even at a low temperature, the degree to which extraction and translocation of diffusible substances occur is uncertain. It remains to be seen whether the original subcellular element distribution is preserved quantitatively even at low temperatures.

The effects of initial freezing and freeze-substitution on the preservation of cell constituents are complementary to each other. It means that poor freeze-substitution results in unsatisfactory preservation of cellular morphology, irrespective of the superior initial freezing. Temperature rise and contamination by frost condensation before, during and after transfer of frozen specimens into the substitution medium should be avoided (Völker *et al.*, 1984). The danger of recrystallisation of frozen specimens is imminent when they are exposed to air during transfer into the substitution medium. While transferring the specimens from the cryogen to the freeze-substitution medium, even a few seconds in air may be sufficient to raise the temperature to a value at which devitrification and ice crystal growth may take place (Franks, 1977). If frost condenses within the vessel of freezing, which may also serve for freeze-substitution, the capacity of the substitution medium to resorb water will be minimised. By using the same test-tube for freezing and freeze-substitution, the dangers of access of warm air and humidity can be avoided (Zalokar, 1966).

Usually OsO_4 is the fixative used in the freeze-substitution step. The presence of OsO_4 in the organic solvent results in darkening of the tissue at about −20 °C (253 K), indicating the reaction and preservation of unsaturated lipids. Since only a superficial 10–15 μm region of the tissue block is well preserved, penetration by OsO_4 is not a problem. If needed, stabilisation of proteins during freeze-substitution can be accomplished by adding an aldehyde (glutaraldehyde or acrolein) to the organic solvent. Uranyl acetate

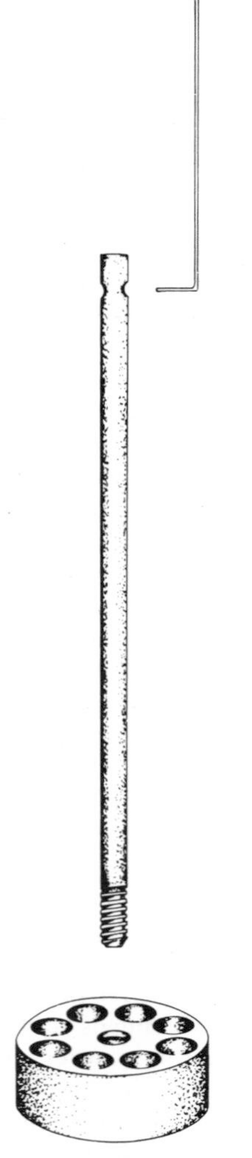

Figure 7.6 The specimen-holder consists of an aluminium disc 45 mm in diameter and 14 mm thick. Eight specimen wells (9 mm diameter, 8 mm deep) are machined in the disc. The handle, 6 mm in diameter and 140 mm in length, has a threaded portion at the bottom that allows the unit to be disassembled for cleaning. A small-gauge wire fits into a hole in the top of the handle to facilitate transfer of the specimen holder into the glass specimen container. From Chiovetti *et al.* (1987)

can also be added to further stabilise the cell structure. While uranyl ions may react at very low temperatures, OsO_4 reacts only slowly with unsaturated lipids at −30 °C (243 K). Glutaraldehyde will cross-link proteins at −50 °C (223 K), but is much more efficient at −30 °C (243 K) and above (Humbel and Müller, 1986). Like other cryo-observations (e.g. freeze-drying, freeze-fracturing and freeze-etching), the freeze-substitution method can be used not only for studying cell morphology, but also for localising enzymatic activity (Saito and Takizawa, 1986) and antigenicity (Hisano *et al.*, 1984). The use of OsO_4 adversely affects the preservation of antigenicity.

Procedures

A scintillation vial containing 1% OsO_4 in 100% acetone is chilled to −80 °C (193 K) with dry ice. Frozen tissue specimens are immediately transferred into this vial, which is then partially submerged in liquid nitrogen until the solution at the sides of the vial is frozen. The vial is capped (which is optional) and transferred to a Styrofoam box filled with dry ice and 100% ethanol. The tissue is freeze-substituted at −80 °C (193 K) for 2 days to replace ice with acetone, allowed to gradually warm to −20 °C (253 K) in the freezer for 1 h and then to 4 °C (269 K) for 1 h, and finally to room temperature. The vial is uncapped to release the gas that dissolves in the solvent during the freeze-substitution. If the tissue is not blackened from the osmication, the substitution solution is replaced with 100% ethanol for 1 h to facilitate secondary blackening. If needed, the tissue is exposed to 10% uranyl acetate in 100% ethanol for 1 h at room temperature.

The tissue is infiltrated and embedded in a resin. Infiltration with monomeric resins at temperatures above 0 °C (273 K) may result in the extraction of cell constituents. Further specimen damage may occur during polymerisation unless it is accomplished at low temperatures. Thin sections can be post-stained with salts of uranium and lead (Fig. 7.7).

Alternatively, simultaneous fixation, staining and freeze-substitution can be carried out in a mixture of 3% glutaraldehyde, 1% OsO_4 and 0.5% uranyl acetate in methanol at −95 °C (178 K). After an appropriate time, this mixture is replaced with 100% acetone. If needed, ruthenium red can be added to the above substitution solution. The choice of the ingredients of the substitution solution depends on the specimen type and objective of the study. Figure 7.7 shows the typical image of freeze-substituted muscle.

As stated above, freeze-substitution used in combination with low-temperature embedding is a superior approach to minimise loss of soluble cell constituents. However, the extraction of soluble pools of carbohydrates, proteins and lipids still remains a problem in this low-temperature processsing. A lipid loss of ~5% and 25% in acetone and methanol, respectively, has been demonstrated during freeze-substitution at −90 °C (183 K) and low-temperature embedding in Lowicryl at −70 °C (203 K) (Weibull *et al.*, 1984). Lipid extraction by Lowicryl embedding was negligible. This lipid loss is much less than that observed in conventionally processed specimens. Relatively low lipid loss in acetone needs to be noted.

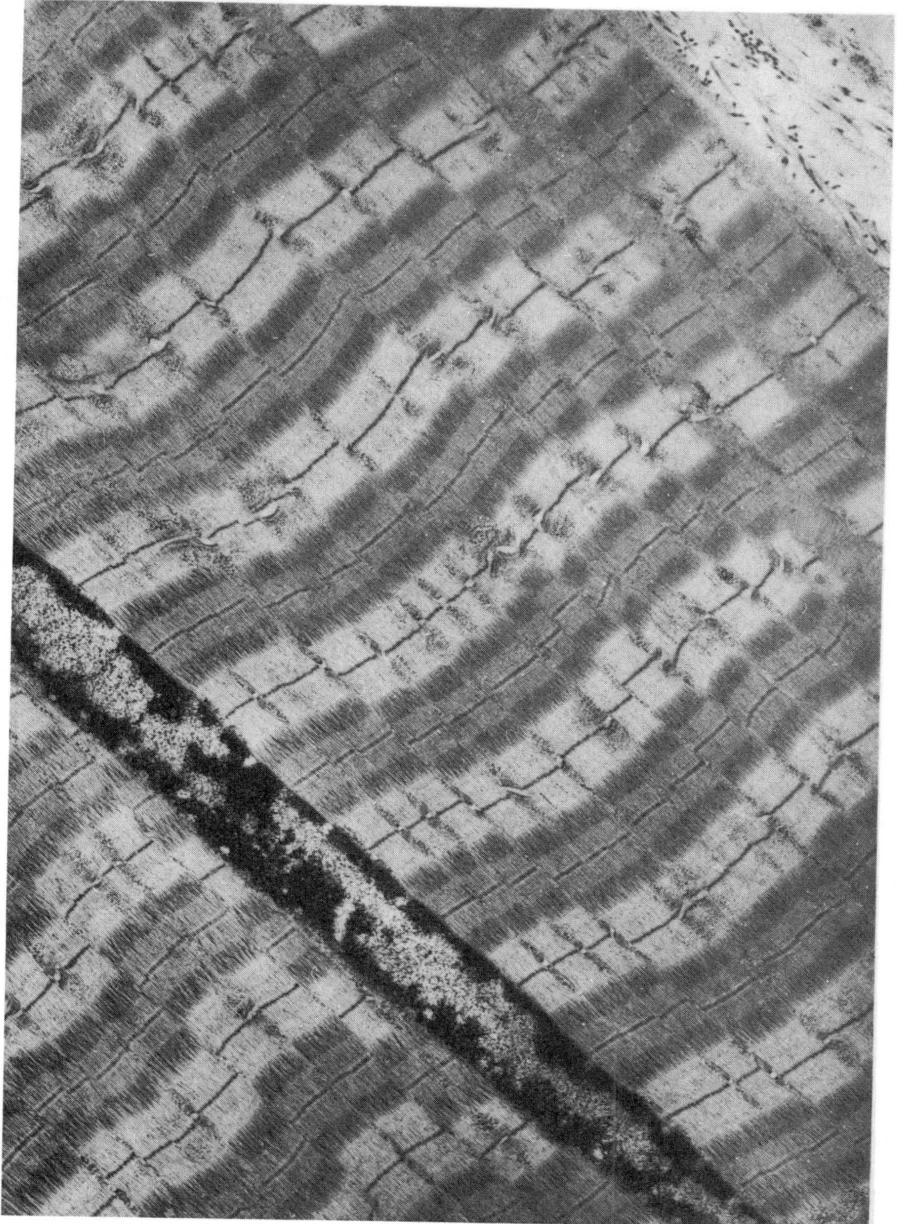

Figure 7.7 Freeze-substituted frog sartorius muscle. The tissue was cryofixed by the cold metal (copper) block freeze method, using liquid nitrogen. Freeze substitution was carried out with 3% OsO_4 in acetone for 2 days at −80 °C (193 K). Temperature was increased to 0 °C (273 K) at a rate of 4 °C/h. The tissue was treated with pure acetone for 1 h, and then the temperature was increased to room temperature in the Reichert–Jung freeze-substitution apparatus. Embedding was done in Spurr's resin. The thin section was cut exactly perpendicular to the contact plane of the specimen with the metal block, and post-stained with salts of uranium and lead. The lessening quality of ultrastructural preservation is clearly visible from the right upper corner to the left lower corner. 10 500×. Courtesy Ludwig Edelmann

When water-miscible Lowicryl embedding media are used, specimens are lightly fixed with aldehydes in the absence of OsO_4 treatment. Without OsO_4, contrast reversal of cellular membranes is observed in some cases. Staining of membranes can be enhanced by using tannic acid. Frozen specimen sandwiches are transferred into the substitution fluid at −90 °C (183 K), gently separated and substituted at −90 °C (183 K), −60 °C

(213 K) and −35 °C (238 K) for 24 h in each step. The low-temperature infiltration and embedding procedure is described elsewhere in this volume.

FREEZE-FRACTURING

The freeze-fracture method was introduced by Steere (1957) and Moor *et al.* (1961). Freeze-fracturing involves fracturing of a frozen specimen, followed by the preparation of a metal–carbon replica of the fracture surface in a frozen state. The fractured surface of the actual specimen is replicated with a heavy metal and then a supporting carbon film is deposited on top of the replica film. The most important contribution of the freeze-fracture method is to the understanding of membrane structure, although protoplasmic components are also studied with this method.

Freeze-fracturing has also been used in conjunction with labelling with ferritin and colloidal gold to relate to underlying cellular structures (Pinto da Silva, 1986). Freeze-fracturing can also be used in combination with autoradiography (Nobiling, 1986) and negative staining (Nermut, 1977a,b). The use of these techniques is limited.

Membranes split during freeze-fracturing along their central hydrophobic plane, exposing intramembranous surfaces. When monolayers of cells attached to a positively charged flat support are freeze-fractured, the fracture plane runs preferentially through the membrane attached to the support, thereby providing pure preparations of outer membrane leaflets (Nermut and Williams, 1977). The fracture plane often follows the contours of membranes and leaves bumps or depressions where it passes around vesicles and other cell organelles. In the replica, smooth areas represent the face of the lipid monolayer, while particles represent protein or non-bilayer lipid conformations present in the interior of the membrane. The replica provides *en face* view of large areas of intramembrane structures that cannot be seen in thin sections. The cross-fractured portions of freeze-fractured specimens are relatively smooth as compared with the areas of freeze-etched (real surfaces exposed) specimens. The fracturing process provides more accurate insight into the molecular architecture of membranes than any other ultrastructural method.

Freeze-fracturing can be carried out without chemical fixation or after fixation with glutaraldehyde. If chemical fixation is needed, highly vascularised organs and tissues should be fixed by vascular perfusion to obtain rapid diffusion of the fixative. Fixation with glutaraldehyde prior to cryofixation stabilises the membrane structure by introducing covalent bonds and minimises the artefactual aggregation of protein particles induced by glycerol when used as a cryoprotectant. Structural deformation occurring during fracturing can also be minimised by chemical fixation, or alternatively by infiltration with glycerol or by lowering the fracturing temperature. However, glutaraldehyde may induce significant changes in the fracturing of membranes.

Specimens of low water content (e.g. seeds, yeast, fungal spores and bacteria) can be cryofixed without chemical fixation and cryoprotection (Robards and Sleytr, 1985). Specimens of high water content usually require cryoprotection in order to obtain ultrastructural preservation of satisfactory quality, unless ultrarapid freezing methods are used. Specimen cryoprotection has been discussed elsewhere in this volume.

Procedure (modified from Müller and Pscheid, 1981)
The frozen specimen should be rapidly transferred onto the cold-table (−150 °C; 123 K) of the freeze-fracturing apparatus. The vacuum chamber is immediately evacuated to avoid excessive frost on the specimens. The temperature is raised to −100 °C (173 K) until the visible frost has disappeared and the vacuum pressure reached to $\sim 4 \times 10^{-6}$ mbar; this takes ~30 min. The knife arm is cooled with liquid nitrogen, and the vacuum falls below 2×10^{-6} mbar. The table temperature is adjusted between −100 °C (173 K) and −150 °C (123 K). After ~5–15 min, the fracturing can be started. Lower temperatures and better vacuums are used in modern devices.

With successive fracturing cycles, the desired specimen plane is exposed. While fracturing, the tissue specimen can be positioned in such a way that the best-frozen surface layer is parallel to the fracturing knife, so that large-size replicas can be obtained. On the other hand, to avoid losing the best-frozen layer of the tissue, it is placed perpendicular to the knife. Usually multisample holders are available so that more than one specimen can be fractured at the same time (Balzer's multisample specimen holders). The last knife cycle should not be completed, but stopped as soon as its edge has passed the specimen. To avoid mechanical damage to the fractured surface, the knife is lifted ~0.1 mm. In this position the cold knife remains to protect the fractured surface from contaminating water vapour. The cold knife acts as a cryopump. For fracturing, the samples are immediately replicated.

The gun for platinum–carbon is heated and the freeze-fractured surface is unilaterally shadowed with 1–2 nm layer at an angle of 45° and further stabilised with a 10 nm layer of carbon. Alternatively, it can be rotary shadowed with platinum–carbon from an angle of 15–25 ° and stabilised by a carbon evaporation at

75°. With rotary shadowing, the rotation speed should be adjusted so that the rate of deposition is not greater than 0.05 nm/s. A rotary speed of ~120 rev/min is usually maintained. The thickness of the film can be controlled with a quartz crystal monitor.

To prevent excessive frost on the cooled parts in the vacuum chamber, they should be brought to room temperature before opening the chamber. Replicas are released in deionised water, cleaned with 40% chromic acid, rinsed in distilled water and retrieved on bare or Formvar-coated copper grids. Alternatively, replicas may be floated onto and cleaned on household bleach and then rinsed on distilled water. Replica transfer is carried out with a platinum loop of a diameter of 0.1 mm. Figure 7.8 shows freeze-fractured microsomal membranes and the complementarity of particles and pits.

CLEANING OF REPLICAS

Since problems arise during cleaning the replicas, a brief discussion on how to avoid them is in order. Two major problems encountered are: (1) fragmentation of the original replica and (2) rolling up of the fragile film in bleach. Another problem is that replicas of certain tissues do not float off onto the surface of the cleaning solution after they have been immersed. Cleaning of the submerged replicas adhering to the tissue block with household bleach usually is ineffective. Concentrated solutions of chromic acid may turn opaque during digestion of the tissue, making recovery of the replica difficult. After treatment with chromic acid, replicas may appear spotted with small, chrome-containing precipitates.

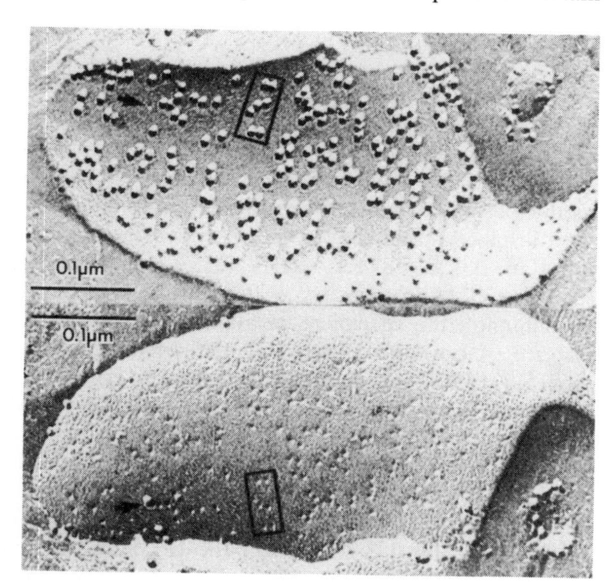

Figure 7.8 Freeze-fracture complementary (stereo) pair(s) of purified microsomal membranes containing Na, K-ATPase and showing complementarity of particles and pits. Purified microsomal membranes from porcine kidney outer medulla were ultrarapidly frozen in liquid propane and fractured in a Cryofract 250 at ~5 K and 10^{-10} Torr. Fractured faces were shadowed immediately with 1.0 nm Pt/C at an angle of 20° and followed by 10 nm carbon at 90° angle. 194 000×. From Ting-Beall *et al.* (1986)

De Mazière *et al.* (1985) have introduced an effective method for complete cleaning of replicas without breaking them; this method is given below. Methanol is solidified in small plastics containers (e.g. caps of PVC bottles) by floating them on the surface of liquid nitrogen. Replicated specimen support discs are placed on solid methanol with cold tweezers. After 24 h at room temperature, the tissue with the intact replica on top is removed from the specimen carrier and immersed in a mixture (1:1) of household bleach (containing 10–12% sodium hypochlorite) and 10% aqueous ethanol. This mixture is obtained by placing two drops of the ethanol and one drop of the bleach in the well of a porcelain spot plate. After 10–60 min, the replica is released fron the largely dissolved tissue. After 1–3 h, the replica is picked up before the thin film curls. If the replica is rolled up, it can be stretched with a hair or an extremely thin glass rod.

The replica is soaked in full-strength bleach for 3–5 h. To digest the last organic debris, the replica is transferred overnight to a 50% saturated sodium hydroxide solution at room temperature. Transfer of the replica from one solution to another is done very carefully, using a glass bead fused to the end of a Pasteur pipette. To ensure rinsing of both sides of the replica, it is submerged in three baths of double-distilled water at 20 °C (293 K). Immersion of the replica is accomplished by firmly plunging in the glass bead. In the third bath, the replica is floated and picked up on a Formvar-coated copper grid.

FREEZE-ETCHING

The freeze-etching method was introduced by Steere (1957). The process of etching is the controlled sublimation of water vapour, resulting in lowering the water level in the specimen. Sublimation means transition from solid state to gas state, bypassing the fluid state. Actually, freeze-etching is a slight modification of freeze-fracturing, in that the sample is left at a temperature at which some of the ice is allowed to sublime away (etching) to expose macromolecules or membrane surfaces just below the fracture plane before replication. Extracellular details such as membrane surfaces and intracellular details (e.g. cell cytoskeleton) can be studied with this method. Thus, real membrane surfaces and intra- and extracellular structures, which would otherwise be masked from view, can be visualised.

When sublimation is less than ~100 nm, it is called normal etching. When sublimation is up to a few micrometres, it is termed deep etching. At a pressure of 10^{-7} mbar, water begins to sublime near −115 °C (158 K) at a rate of 0.1 nm/s, and increases exponentially with warming (e.g. near −105 °C the rate is 10 nm/s). Increased rate of water sublimation results in increased forces at the etching surface, which may distort cellular details. The water should be removed at a slow rate (1–10 nm/s) and the duration of etching is adjusted to achieve the desired depth of etching.

Procedure

As in the case of freeze-fracturing, specimens are frozen with any means of cryofixation, without cryoprotection. The frozen specimens are fractured with a knife, followed by sublimation of water under vacuum as the temperature is raised. The parts of the cell exposed by fracturing and etching process are shadowed with platinum; organic material is dissolved away and the replica is floated off for viewing in the electron microscope. Normal etching is carried out at about −100 °C (173 K) in a vacuum better than 10^{-6} Torr for about 1–10 min before shadowing and replication. To avoid contamination of the fractured surface and to improve the local vacuum, the specimen should be protected during the time of etching by a cold shroud, such as the liquid nitrogen-cooled knife in the Balzer's or Cressington freeze-etching unit (Aggerback and Gulik-Krzywicki, 1986). The specimens are cooled to −150 °C (123 K) for shadowing and carbon deposition, and otherwise treated as for freeze-fracturing. Figure 7.9 shows a quick-frozen, deep-etched mitochondrion. The mitochondrion shows connections between its outer and inner membranes as well as between crystal membranes.

EMBEDDING AT LOW TEMPERATURE

Frozen tissue specimens can be either freeze-dried or freeze-substituted in conjunction with low-temperature embedding. Embedding media compatible under varying conditions of water content, solvent polarity and low-temperature environment are available. The future trend is to embed unfixed and unstained specimens in water-miscible resins.

FREEZE-DRYING AND EMBEDDING

Frozen specimens are dehydrated without using organic solvents (acetone), since the frozen aqueous phase is removed by sublimation. By using a low-viscosity embedding medium, freeze-dried specimens are directly infiltrated without diluting the resin with an organic solvent. Moreover, specimens are infiltrated and embedded at a low temperature in one of the Lowicryl resins. Chemical fixation is optional, for this method allows freezing of fixed or unfixed specimens.

Tissue specimens are frozen by one of the cooling methods described elsewhere in this volume, and freeze-dried in a freeze-dryer (Figs. 7.5, 7.6) introduced by McGuffee *et al.* (1981). This dryer is based on a liquid nitrogen-cooled, 5 Å molecular sieve that functions as a cryosorption pump. The dryer can be pumped down to 1.5×10^{-5} Torr in ~4–20 min with a mechanical pump. The details of freeze-drying are discussed by Chiovetti *et al.* (1987). Specimens can be infiltrated overnight at −20 °C (253 K) with Lowicryl K4M; at this temperature this embedding medium is relatively non-viscous. Infiltration at lower temperatures (−50 to −70 °C; 223 to 203 K) can be accomplished with Lowicryl K11M and HM23.

FREEZE-SUBSTITUTION AND EMBEDDING

Frozen specimens are freeze-substituted in the substitution medium consisting of 2% OsO_4, 0.5% uranyl acetate and 3% glutaraldehyde in pure anhydrous methanol (Hippe and Hermanns, 1986); acetone can be used in place of methanol. The specimens are substituted at −90 °C (183 K), −60 °C (213 K) and −35 °C (238 K) for 24 h in each step. After rinsing three times in methanol, the specimens are infiltrated with a 1:1 mixture of Lowicryl HM20 and methanol for 6 h at −60 °C (213 K), and then in 2:1 mixture overnight at the same temperature. The last 12 h infiltration is carried out at −50 °C (223 K). Polymerisation is accomplished under two UV lamps (15 W, 360 nm UV fluorescent light) at a distance of 40 cm at −35 °C (238 K) for 24 h. Before sectioning, specimen blocks are hardened for 2–3 days under the UV light conditions at room temperature. Thin sections can also be stained with tannic acid.

LIMITATIONS OF LOW-TEMPERATURE METHODS

Artefacts can arise in each of the preparatory steps

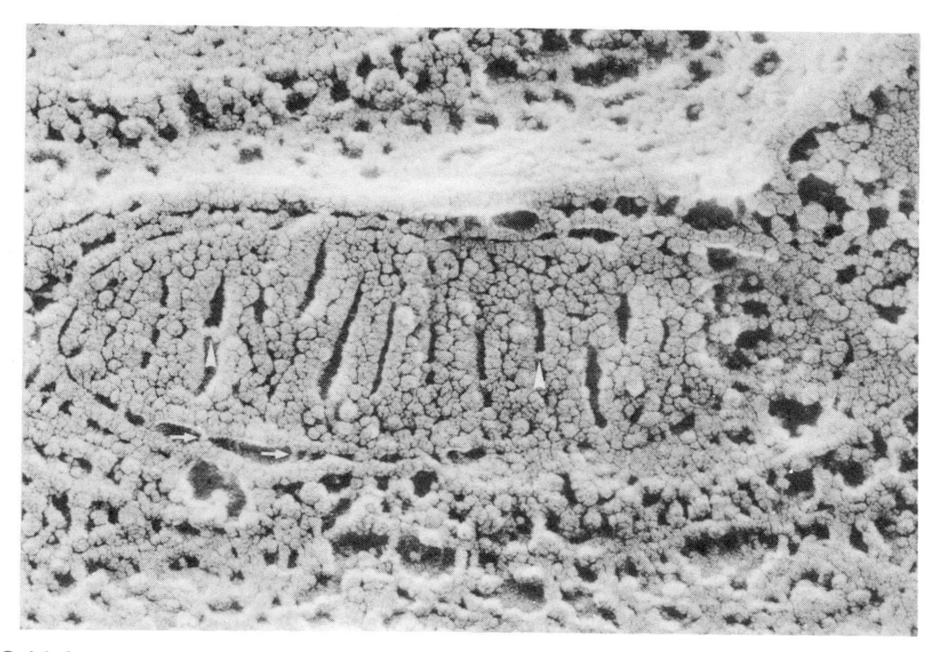

Figure 7.9 Quick-frozen and deep-etched image of a mitochondrion in rat anterior pituitary cell. The tissue was frozen by 'slamming' it against a pure copper block cooled with liquid helium. The frozen tissue was freeze-fractured within 20 μm from its surface at −120 °C (153 K) and deeply-etched for 10 min at −90 °C (183 K) at 1×10^{-6} Torr. It was rotary shadowed with platinum–carbon at an angle of 30° and shadowed with carbon at an angle of 90°. The shadowed specimen was immersed in 30% chromic acid to completely dissolve the tissue; the replica was rinsed carefully with distilled water and then picked up on a 300-mesh copper grid. Outer and inner membranes of the mitochondrion show connections (arrows), crossing the intermembranous space. The membranes of cristae are also connected by filaments (arrowheads), crossing the intercristal space. Many filaments terminate at the outer membrane of the mitochondrion. 120 000×. From Senda and Fujita (1987)

from the collection of the specimens to the viewing of replicas, whole mounts or sections in the electron microscope. In other words, artefacts can arise prior to and/or during freezing of specimens. Artefacts can also be formed during freeze-drying, freeze-substitution, freeze-fracturing (-etching) and cryoultramicrotomy. Artefacts introduced by cryoprotectants are listed below and elsewhere in this volume.

It is not easy to recognise artefacts produced in various steps of low-temperature processing. Fracturing artefacts are especially difficult to recognise. One has to learn to identify such artefacts. One way to achieve this learning is by preparing freeze-fracture replicas of homogeneous specimens such as glycerol, in which any structure seen is related to the fracturing process. Also, one should use the information about similar specimens obtained with other established methods. Most of the major artefacts are listed below.

(1) Since cells respond to trauma within milliseconds and collection of specimens may take from several seconds to several minutes prior to freezing, harmful conditions, including anoxia, may set in, resulting in artefacts. Such artefacts are common when tissues are excised prior to freezing, unless their collection and freezing occur simultaneously. Excision of the unfixed tissue from the body as well as dissection of partially fixed tissue can cause physical trauma, resulting in stretch artefacts (regions devoid of freeze fracture particles) in membranes.

(2) Cryoprotectants such as glycerol can cause swelling of mitochondria, vesiculation of the membranes of rough endoplasmic reticulum, aggregation of intramembrane particles into clusters and the formation of blisters at the plasma membrane (McIntyre *et al.*, 1974). These artefacts can be avoided by fixation with glutaraldehyde, although internal membrane systems will still vesiculate if infiltration by the cryoprotectant is too rapid. If penetrating cryoprotectants are used, they will alter membrane permeability, affecting dislocation of soluble ions and molecules.

(3) The most common artefacts present in unfixed, uncryoprotected, frozen specimens within the temperature range from 0 °C (273 K) to −40 °C (233 K) are ice crystals large enough to be seen in the electron microscope. Formation of extra- and intracellular ice crystals causes direct as well as indirect stresses on the specimen. The indirect stress results in dehydration and

increased solute concentrations in extra- and intracellular fluids. Solidification of buffering solutes will change the pH. Dehydration can cause cell shrinkage, which, in turn, will bring about general deformation. Intracellular freezing of the water can destroy the cytoplasmic membrane systems (Akert *et al.*, 1981).

The nucleus is the most susceptible intracellular organelle to ice crystal formation. Ice crystal damage appears usually as small holes or clear spaces in the cytoplasm or in the organelle matrix in freeze-dried or freeze-substituted specimens (Gilkey and Staehelin, 1986). Such spaces may be outlined by fibrillar material, which represents precipitates of cellular materials. The fibrillar material in some cases resembles a microtrabecular system (Miller *et al.*, 1983). Ice crystallisation may also cause mechanical rupturing of cell membranes, which, in turn, may lead to an irreversible loss of compartmentalisation (Robards and Sleytr, 1985). Another effect of ice crystals is collapse of microtubules and distortion of the cytoskeletal system. Bulk specimens may show ice crystal puncture artefacts when ice crystals grow through the specimen surface. Extracellular crystallisation may cause dehydration and cell shrinkage.

Slow freezing rates can bring about a variety of ultrastructural changes in the membranes. As stated above, intracellular ice causes mechanical stress, which, in turn, sandwiches the membrane between intra- and extracellular ice. Slow freezing may cause a lateral displacement of membrane-integrated proteins by lipid phase separation (intramembranous segregation). The net result is aggregation of freeze-fracture particles in the form of patches. This aggregation, in turn, may be accompanied by the appearance of particle-free patches in the membrane. Another result of slow freezing is loss of preferential path of freeze-fracturing along the membrane interior at the damaged regions of the membrane. Mild ice crystal damage in freeze-fractured specimens is seen as light lumpiness or unevenness of membrane contours (Gilkey and Staehelin, 1986) and granularity of the aqueous phase. Some of the artefacts mentioned above can be avoided or minimised by increasing the cooling rates.

(4) Cooling destabilises the native structure in context with thermodynamics. Freezing causes denaturation of globular proteins and conformational changes of other polymers; it may also cause a decrease in rotational mobility of lipids. The stress owing to temperature reduction (cold shock injury) affects the membrane lipids and enzymes. The change of water from a liquid to a solid state is accompanied by an increase in volume of ~9%, which can cause damage to inappropriately prepared specimens. Large-size tissue specimens may show thermal stress cracking.

(5) High-velocity plunging (e.g. >2 m/s) causes an overall compression in prefixed tissue blocks, sometimes up to ~50% compression, and deformation in fresh tissue specimens (K. P. Ryan, personal communication). High-velocity undamped contact of the tissue with cold metal block may cause similar deformation.

(6) At a very high freeze-drying temperature (−35 °C, 238 K), eutectic melting may occur, resulting in the creation of surface tension forces between the vapour phase and the fluid eutectic. This can cause collapse of cellular structures. Recrystallisation may occur above −50 °C (223 K) (animal system) or −70 °C (203 K) (plant system).

(7) The energy released during fracturing may cause plastic deformation, including aggregation or stretching of protein particles in some membranes. Structural deformation during fracturing is always expected. Some lipid collapse may occur in the membranes during fracturing and replication. Lack of complementarity of opposite fracture faces of a membrane in single replicas is common, although procedures for obtaining complementary replicas are available (Fetter and Costello, 1986; Ting-Beall *et al.*, 1986; Schinz, 1987) (Fig. 7.8). A large knife advance during fracturing can result in deep fracture steps. Without stereo-imaging the replica, density of intramembranous particles is underestimated, especially that of the extracellular fracture face (E-face) (Kordylewski *et al.*, 1986). Generally, more particles appear on the protoplasmic fracture face (P-face) than on the E-face, which in part is dependent on how membrane particles segregate during freeze-fracturing.

Small fragments of tissue may be transferred from the knife onto the fractured surface. They may be etched and replicated and appear as flakes and tiny chips on the membrane faces (Akert *et al.*, 1981). Water vapour subliming from the chips will condense on fracture faces and form particle-like artefacts.

If the fracture method is carried out less than skilfully, condensation of contaminants from the vacuum can increase the protein particle size or even produce structures that resemble protein particles (Staehelin and Bertaud, 1971). The most critical time when condensation artefacts are introduced is the period between fracturing and replication. Water or hydrocarbon vapour condensation is the primary source of contamination, resulting in artefacts such as warts or flat plaques. A specimen temperature lower than −100 °C (163 K) or a vacuum pressure higher than 10^{-6} Torr can produce condensation during fracturing. The condensation of water vapour on the specimen can be avoided by maintaining an appropriate specimen temperature in relation to the attainable vacuum pressure. Immediate coating of the fractured surface is also helpful.

The replica is produced from a fracture face of a frozen specimen under physical conditions that are different from those existing *in vivo*. The replica is difficult to probe by standard cytochemical methods because of its inertness. However, labelling of freeze-fractured plasma and intracellular membranes with ferritin or colloidal gold can be accomplished (Pinto da Silva, 1984; Kan and Pinto da Silva, 1989).

Insufficient shadowing results in poor contrast and loss of fine structural details. Excessive shadowing masks ultrastructural details. Too steep a shadowing angle causes the evaporated film to completely hide small structural details. Low-angle shadowing may blur the fracture lines across membrane leaflets (Akert *et al.*, 1981). Suboptimal ratio of platinum and carbon during shadowing may cause some of the problems listed above. Excessive carbon may give rise to black specks on the fracture surface. If the replica is insufficiently cleaned, specimen residues may appear as black spots on the electron micrograph. Excessive electron beam intensity may recrystallise the platinum, resulting in the appearance of small black grains on the electron micrograph.

(8) Cryoultramicrotomy has many attendant artefacts, including cutting-induced devitrification and possible melting, chatter, section deformation and crevasses, as well as knife marks (Chang *et al.*, 1983) (see p. 405).

Production of artefacts is significantly reduced by using specimens of the smallest possible size. The smaller the specimen, the faster it will cool. Tissue blocks of ~0.2 mm^3 or slices of ~0.1 mm thickness cut with a Vibratome should be prepared as quickly as possible. Freeze-drying, freeze-substitution and freeze-fracturing (-etching) provide information on different aspects of the specimen. Therefore, they must be chosen carefully and their results interpreted accordingly. There is no perfect method for all uses. It should be borne in mind that every method has shortcomings. Any one method should not be used in isolation, especially by an inexperienced worker. A specimen should be subjected to as many methods as possible, including the conventional method. Detailed discussions of artefacts produced during freeze-fracturing (-etching) and replica preparation have been presented by Rash and Hudson (1979), Willison and Rowe (1980), Akert *et al.* (1981), Robards and Sleytr (1985), Menco (1986) and Fujikawa (1988).

HAZARDS

Low-temperature methodology encounters five main hazards: (1) direct contact with extremely cold liquid, gas or solid material; (2) inhalation of cryogenic gases; (3) high inflammability of cryogenic fluids when they are in the gaseous state; (4) explosion of liquefied gases when vaporising under uncontrolled conditions within a sealed container; and (5) toxicity of cryogenic gases. Liquid oxygen (boiling point: −183 °C, 90 K) will condense on objects cooled to liquid nitrogen temperature (−196 °C, 77 K) or liquid propane temperature (−189 °C, 84 K). Oxygen is a powerful oxidant, and can cause many substances to self-ignite. Condensed oxygen will increase the hazard involved in using propane, and the liquid oxygen, being near its boiling point and on the surface of the propane, will greatly slow the initial cooling of the specimen due to vapour barrier formation. Condensation of liquid oxygen on the surface of cryogens such as propane can be avoided by covering (but not sealing) the container when not freezing a specimen.

One should be cognisant of extremes of pressure in laboratory equipment such as freeze-fracture apparatus and gas cylinders. Apparatus must be mechanically stable; accidental toppling could be disastrous. Protective clothing, non-porous gloves and eye goggles must be worn while handling cryogens. Cryogens should always be handled in such a way that they cannot be captured near body tissue. Various devices should be handled with a cloth rather than gloves. One should not wear high boots, for they can capture liquid nitrogen. Insulated forceps should be used to handle specimens under a liquid cryogen. Liquid gas must always be used in a fume-hood. Ample ventilation will prevent the build-up of high concentrations of gas. A flammable gas sensor (Draeger Safety, Chesham, England) should be kept in the room where propane is in use. A detailed discussion of hazards encountered in the low-temperature laboratory has been presented by Robards and Sleytr (1985), and safe use of propane and other liquefied gases has been discussed by Ryan and Liddicoat (1987).

CRYOULTRAMICROTOMY

Chemical fixation, dehydration and embedding cause destruction of biological activity, and extraction and/or dislocation of many soluble components, and limit to a great extent chemical and immunological reactions. Cryoultramicrotomy was introduced by Bernhard and Leduc (1967) as an electron microscopy procedure to avoid or minimise these adverse effects. Ice serves as the embedding medium. Cryoultramicrotomy was introduced primarily to examine biopsy, surgical and other specimens without delay, especially for diagnostic purposes. The usefulness of frozen sections in demonstrating enzyme activity and in autoradiography has been established.

The main objective of cryoultramicrotomy is to provide thin sections of a specimen for transmission electron microscopy. This method implies that chemical composition of the specimen remains almost unchanged from that in the living state. Another implication is that the ultrastructure preservation is very close to that in the natural state. The reason for this optimism is that the method allows observation of a specimen that has not been subjected to chemical fixation, embedding and staining. Cryoultramicrotomy is almost indispensable in a meaningful application of X-ray microanalysis, for most electrolytes are either lost or displaced during conventional processing. Cryoultramicrotomy retains these substances in physiological amounts in specimens. The distribution of water can also be studied in frozen hydrated sections. A detailed procedure for obtaining thin cryosections for X-ray microanalysis has been presented by Zierold (1986).

The chemical composition, especially antigenicity, of biomolecules can be maintained with cryofixation and cryoultramicrotomy. This method enables post-sectioning labelling, providing accessibility to all antigens located both in the cytoplasm and on the cell surface. Penetration of labelled antibodies into cryosections is not a problem. It is also possible to localise cellular sites of antibody synthesis following stimulation by an antigen. The method allows the localisation of receptor sites to hormones and subcellular sites of action of drugs. Cryoultramicrotomy, once mastered, is simpler and faster than freeze-drying and freeze-substitution, although the latter procedures can be used in conjunction with the former to elucidate structure–function correlation.

Several cryoultramicrotomes are commercially available (CryoNova, LKB Instruments; freezing attachment for Reichart-Jung Cryoultramicrotome; frozen, thin sectioner attachment for MT6000 Sorvall ultramicrotome, Du Pont Co.). A cryoultramicrotome with a new feedback-advance system has been introduced by Wolf (1987). It is a software-controlled machine that cuts thin cryosections of a precise thickness with good reproducibility. All steering of the machine is carried out by an AIM microcomputer. This machine is not commercially available, although it has a patent. With these instruments, it is now possible to obtain thin cryosections of a wide range of specimens, including plant and animal tissues, single cells and suspensions of cells.

Although basic principles governing cryoultramicrotomy are now clearly established and its possibilities and limitations are well understood, it still has some serious unsolved problems that are discussed later; a few examples will suffice here. The quality of ultrastructural preservation is poor. However, in certain studies, chemical information is more important than the high quality of ultrastructural preservation. As long as chemical information can be localised in identifiable cell components, the procedure is acceptable. Poor contrast and radiation damage commonly seen in frozen, hydrated thin sections can be significantly diminished by using freeze-dried cryosections. Elimination of water prevents radiation damage. However, removal of water will affect molecular configuration and ionic distribution. Freeze-dried sections can be further stabilised by osmium vapour (Frederik and Busing, 1981). This treatment will introduce its own artefacts at least at molecular level.

Cryoultramicrotomy can also be used in combination with immunocytochemistry, especially with colloidal gold as the electron-opaque marker (Tokuyasu, 1986; van Bergen en Henegouwen, 1989). To improve the quality of ultrastructural preservation and lessen cellular damage, specimens can be lightly fixed with glutaraldehyde or osmium vapour prior to freezing. Various techniques are available to obtain thin cryosections of unfixed, lightly fixed or fixed specimens. These sections can be obtained from either embedded or unembedded specimens. Freezing renders the tissue sufficiently rigid to be mounted in a cryoultramicrotome without using an embedding medium.

Introduction of cold stages in the electron microscope to allow viewing the frozen hydrated thin sections is providing exciting new information. A desirable revolutionary innovation would be the development of a procedure permitting rapid freezing and cutting thin cryosections of unfixed, hydrated specimens simultaneously.

SECTIONING

The quality of thin cryosections is ultimately related to the treatments the specimen has undergone before sectioning. For example, the more rapid the cryofixation, the better is the quality of cryosections. The presence of ice crystals in the specimen adversely affects the quality of thin cryosections. Specimens having large ice crystals are excessively brittle and do not produce satisfactory cryosections. While rapid cryofixation requires a very small tissue block ($<1\ mm^3$), a specimen of this small size is difficult to mount on the cryoultramicrotome holder. The solution is to carry out cryofixation of the tissue block attached to the holder such as an aluminium pin (Zierold, 1987). Alternatively, small, frozen tissue blocks can be attached to the holder with a cryoglue such as heptane (melting point: −90 °C, 183 K) or toluene (melting point: −95 °C; 178 K).

The specimen and knife temperature is set about −120 °C (153 K). Glass knives with a scoring angle of 45° and a clearance angle of 4° are recommended. Only best glass knives that have been shadowed with tungsten (Roberts, 1975) should be used. Such knives last longer and allow smoother cutting. The sharpest part of the cutting edge is near the check marks rather than at the farthest end from these marks (Fig. 3.9). Diamond knives with a clearance angle of 4° can be used. A cutting speed of 1 mm/s is satisfactory to produce sections 120 nm thick. The cutting face of the frozen tissue block should be small enough to yield cryosections with a width less than 200 μm. Thin cryosections appear as transparent curling sheets on a glass knife. Cryosections usually do not produce ribbons.

Agreement on whether and to what extent local melting of cryosections occurs during cutting is lacking. According to Frederik and Busing (1981), cutting-induced melting is absent at −180 °C (193 K). It is thought that there is no difference in principle between the cutting process in cryoultramicrotomy and the fracturing process in the preparation of freeze-fracturing replicas (Frederik *et al.*, 1982). Even if melting of a superficial zone of the cryosection were to occur, its effect may not show up in a transmission image, since such an image constitutes the majority of the section. The majority of the section does not contain melting artefacts. On the other hand, Chang *et al.* (1983) report that a grainy surface of some cryosections represents a cutting-induced melting of a surface followed by freezing.

SECTION TRANSFER

A number of methods are available for transferring the cryosections onto the grid. A vacuum pipette can be used to suck the sections out horizontally from the knife edge. Alternatively, cryosections can be transferred with a hollow plastics straw (5 mm in diameter) plugged with cotton wool, to the pointed end of which is glued a white dog hair (Barnard, 1982). A single hair probe precooled with liquid nitrogen can also be used to transfer the cryosections. Another technique involves an antiroll plate that consists of a piece of a glass coverslip attached close to and parallel to the glass cutting edge. The simplest and most effective approach is to transfer the cryosection with the aid of a precooled single-hair probe onto a precooled grid that has been coated with either Butvar (or Formvar) or carbon, or both, and has been placed on the knife face ~1 mm from the cutting edge. A shelf on which to rest the grid can be made by placing a piece of electrical tape on the knife face ~5 mm from the cutting edge.

Adhesion of dry cryosections to the grid can be facilitated by placing them between two precooled coated grids which are pressed together by a cold, polished metal rod (Zierold, 1987). The grids are carefully unfolded, and the grid with adhering sections is used. This grid can be stored in a cold nitrogen gas atmosphere. If static electricity is a problem during sectioning, it can be neutralised by an electrical discharge device that generates both negative and positive ions above the sections. The static build-up on the grids can be dissipated by lightly coating them with carbon, if they already have not been coated with carbon.

To avoid ultrastructural damage and elemental displacement due to rehydration, cryosections should be transferred fron the cryoultramicrotome to the electron microscope with a cryotransfer stage (Gatan Model 626) at a temperature of −130 °C (143 K) (Hagler and Buja, 1986). The problem of water vapour deposition during cryo-observation in the electron microscope can be prevented by an additional protection with solidified coolant such as ethane or Freon 22 (Frederik and Busing, 1986). The evaporating coolant in the high vacuum of the electron microscope will carry away the water films.

LIMITATIONS OF CRYOULTRAMICROTOMY

Cryofixation prior to cryoultramicrotomy allows the study of only superficial layers of a tissue block. The section thickness obtained with cryoultramicrotomy is ~100 nm, which is a constraint in achieving high-resolution images. Since immunolabelling of cryosections requires thawing, the specimens are subjected to chemical fixation and cryoprotection. For a number of enzyme cytochemical reactions, the presence of a certain minimal amount of enzyme activity is necessary. Thin cryosections usually do not contain enough enzyme activity to react with the substrate and form an easily detectable reaction product. Frozen-hydrated sections have the disadvantage of low contrast and excessive mass loss due to radiation in the electron microscope. Specimen shrinkage and redistribution of diffusible substances in freeze-dried cryosections should be kept in mind. Section deformation during cryosectioning is neither well understood nor uncommon. Thin cryosections cut with a glass knife often show knife marks perpendicular to the cutting edge. These sections may also show deformation lines with periodicities ranging from 150 nm to 300 nm parallel to the cutting edge, resulting in a change in the shape of the cells and their components. Typical deformation of vitrified specimens is compression along the cutting direction and a network of crevasses (Chang *et al.*, 1983). Cryoultramicrotomy has played only a minor role in electron microscope autoradiography. It is apparent that great expectations for the use of cryoultramicrotomy are only partly fulfilled. Cryoultramicrotomy is a tedious, time-consuming and relatively expensive method, and so should be used only when other methods have proved unsatisfactory. This topic has also been discussed in detail in the volumes authored and edited by Robard and Sleytr (1985) and Steinbrecht and Zierold (1987), respectively.

References

Abad, A. (1988). A study of section wrinkling on single-hole, coated grids using TEM and SEM. *J. Electron Microsc. Tech.*, **8**: 217

Abramczuk, J. (1972). Effect of formalin fixation on the dry mass of isolated rat liver nuclei. *Histochemie*, **29**: 207

Abrunhosa, R. (1972). Microperfusion fixation of embryos for ultrastructural studies. *J. Ultrastruct. Res.*, **41**: 176

Acetarin, J.-D., Carlemalm, E., Kellenberger, E. and Villiger, W. (1987). Correlation of some mechanical properties of embedding resins with their behaviour in microtomy. *J. Electron Microsc. Tech.*, **6**: 63

Ackerman, G. A. (1964). Histochemical differentiation during neutrophil development and maturation. *Ann. N.Y. Acad. Sci.*, **113**: 537

Ackerman, G. A. and Freeman, W. H. (1979). Membrane differentiation of developing haemic cells of the bone marrow demonstrated by changes in concanavalin A surface labeling. *J. Histochem. Cytochem.*, **27**: 1413

Ackermann, H.-W., Jolicoeur, P. and Berthiamme, L. (1974). Advantages et inconvénients de l'acétate d'uranyle en virologie comparée: étude de quatre bactériophages caudés. *Can. J. Microbiol.*, **20**: 1093

Adachi, K., Adachi, M., Katoh, M. and Fukami, A. (1968). On a measuring method of the film thickness of biological ultrathin sections. *J. Electron Microsc.*, **17**: 280

Adams, C. W. M., Abdulla, Y. H. and Bayliss, O. B. (1967). Osmium tetroxide as a histochemical and histological reagent. *Histochemie*, **9**: 68

Adams, C. W. M. and Bayliss, O. B. (1968). Reappraisal of osmium tetroxide and OTAN histochemical reactions. *Histochemie*, **16**: 162

Adamson, I. Y. R. and Bowden, D. H. (1970). The surface complexes of the lung. A cytochemical partition of phospholipid surfactant and mucopolysaccharides. *Am. J. Path.*, **61**: 359.

Adomian, G. E., Laks, M. M. and Billingham, M. E. (1977). Contraction bands in human hearts: pathology or artifact? *Proc. 35th Ann. EMSA Meet.*, p. 578. Claitor's Pub. Division, Baton Rouge, La.

Adrian, M., Dubochet, J., Lepault, J. and McDowell, A. W. (1984). Cryoelectron microscopy of viruses. *Nature, Lond.*, **308**: 32

Aebi, U. and Pollard, T. D. (1987). A glow discharge unit to render electron microscope grids and other surfaces hydrophilic. *J. Electron Microsc. Tech.*, **7**: 29

Aggerback, L. P. and Gulik-Krzywick, , T. (1986). Studies of lipoproteins by freeze-fracture and etching electron microscopy. In: *Methods in Enzymology*, Vol. 128 (J. P. Segrest and J. J. Albers, Eds.), pp. 457–471. Academic Press, Orlando

Ainsworth, S. K., Ito, S. and Karnovsky, M. J. (1972). Alkaline bismuth reagent for high resolution ultrastructural demonstration of periodate-reactive sites. *J. Histochem. Cytochem.*, **20**: 995

Ainsworth, S. K., and Karnovsky, M. J. (1972). An ultrastructural staining method for enhancing the size and electron opacity of ferritin in thin section. *J. Histochem. Cytochem.*, **20**: 225

Åkerström, B. and Björck, L. (1986). A physiochemical study of protein G, a molecule with unique immunoglobulin G-binding properties. *J. Biol. Chem.*, **261**: 10240

Akert, K., Sandri, C. and Moor, H. (1981). Freeze-etching in neuroanatomy. In: *Techniques in Neuroanatomical Research* (Ch. Heym and W.-G. Forsmann, Eds.), pp. 41–54. Springer-Verlag, Berlin

Akey, C. W. and Edelstein, S. J. (1983). Equivalence of the projected structure of thin catalase crystals preserved for electron microscopy by negative stain, glucose or embedding in the presence of tannic acid. *J. Mol. Biol.*, **163**: 575

Akey, C. W., Szalay, M. and Edelstein, S. J. (1984). Trigonal catalase crystals: a new molecular packing assignment obtained from sections preserved with tannic acid. *Ultramicroscopy*, **13**: 103

Albersheim, P. and Killias, U. (1963a). Histochemical localization at the electron microscope level. *Am. J. Bot.*, **50**: 732

Albersheim, P. and Killias, U. (1963b). The use of bismuth as an electron stain for nucleic acids. *J. Cell. Biol.*, **17**: 93

Albin, T. B. (1962). Handling and toxicology. In: *Acrolein* (C. W. Smith, Ed.), p. 234. Wiley, New York

Alderson, T. (1964). Crosslinking of fibrous protein by formaldehyde. *Nature, Lond.*, **187**: 485

Aldrich, H. C., Beimborn, D. B. and Schonheit, P. (1987). Creation of artifactual internal membranes during fixation of *Methanobacterium thermoautotrophicum*. *Can. J. Microbiol.*, **33**: 844

Allison, D. P., Daw, C. S. and Rorvik, M. C. (1987). The construction and operation of a simple inexpensive slam freezing device for electron microscopy. *J. Microsc.*, **147**: 103

Allman, K. (1971). Elektronenmikroskopische Untersuchungen uber die Eisen III-Bindong in glutaraldehydfixierter Insektenflugmuskulatur. *Cytobiologie*, **3**: 282

Almeida, J. D. (1980). Practical aspects of diagnostic electron microscopy. *Yale J. Biol. Med.*, **53**: 5

Almeida, J. D., Stannard, L. M. and Shersby, A. S. M. (1980). A new phenomenon (SMOG) associated with solid phase immune electron microscopy. *J. Virol. Meth.*, **1**: 325

Altman, L. G., Schneider, B. G. and Papermaster, D. S. (1984). Rapid embedding of tissues in Lowicryl K4M for immunoelectron microscopy. *J. Histochem. Cytochem.*, **32**: 1217

Amsterdam, A. and Schramm, M. (1966). Rapid release of the zymogen granule protein by osmium tetroxide and its retention during fixation by glutaraldehyde. *J. Cell. Biol.*, **29**: 199

Anderson, N. and Doane, F. W. (1973). Specific identification of enteroviruses by immunoelectron microscopy using serum-in-agar diffusion method. *Can. J. Microbiol.*, **19**: 585

Anderson, P. J. (1967). Purification and quantitation of glutaraldehyde and its effects on several enzyme activities in skeletal muscle. *J. Histochem. Cytochem.*, **15**: 652

Anderson, R. G. W. and Brenner, R. M. (1971). Accurate placement of ultrathin sections on grids: control by sol-gel phases of a gelatin flotation fluid. *Stain Technol.*, **46**: 1

Anderson, T. F. (1951). Techniques for the preservation of three-dimensional structure in preparing specimens for the electron microscope. *Trans. N.Y. Acad. Sci.*, **13**: 130

Anderson, W. A. (1972). Methods for electron microscopic localization of glycogen. In: *Techniques of Biochemical and Biophysical Morphology*, Vol. 1 (D. Glock and R. M. Rosenbaum, Eds.), pp. 1–23. Wiley-Interscience, New York

Anderson, W. A. and André, J. (1968). The extraction of some cell components with pronase and pepsin from thin sections of tissue embedded in an Epon–Araldite mixture. *J. Microscopie*, **7**: 343

Angermüller, S. and Fahimi, H. D. (1982). Imidazole-buffered osmium tetroxide: an excellent stain for visualization of lipids in transmission electron microscopy. *Histochem. J.*, **14**: 823

Anthony, A., Colurso, G. J., Bocan, T. M. A. and Doebler, J. A. (1984). Interferometric analysis of intrasection thickness variability associated with cryostat microtomy. *Histochem. J.*, **16**: 61

Anwar, R. A. and Oda, G. (1966). The biosynthesis of desmosine and isodesmosine. *J. Biol. Chem.*, **241**: 4638

Arancia, G., Rosati Valente, F. and Trovalusi Crateri, P. (1980). Effects of glutaraldehyde and glycerol on freeze-fractured *Escherichia coli*. *J. Microsc.*, **118**: 161

Arborgh, B., Bell, P., Brunk, U. and Collins, V. P. (1976). The osmotic effect of glutaraldehyde. A transmission electron microscopy, scanning electron microscopy and cytochemical study. *J. Ultrastruct. Res.*, **56**: 339

Arbuthnott, E. R. (1974). Routine collection of flat area sections for electron microscopy as applied to a detailed study of axon dimensions. *J. Microsc.*, **101**: 219

Argagnon, J. and Enjalbert, L. (1964). Technique d'inclusion pour la microscopie électronique utilisant un polyester: Le Rhodester 1108 CPSL. *J. Microscopie*, **3**: 339

Armbruster, B. L., Carlemalm, E., Chiovetti, R., Garavito, R. M., Hobot, J. A., Kellenberger, E. and Villiger, W. (1982). Specimen preparation for electron microscopy using low temperature embedding resins. *J. Microsc.*, **126**: 77

Arnold, J. D., Berger, A. E. and Allison, O. L. (1971). Some problems of fixation of selected biological samples for SEM examination. *Proc. 4th. Ann. SEM Symp.* (O. Johari, Ed.), p. 249. IITRI, Chicago

Arstila, A. V., Hirsimäki, P. and Trump, B. F. (1974). Studies on the subcellular pathophysiology of sublethal chronic injury. *Beitr. Path.*, **152**: 211

Ashford, A. E., Allaway, W. G., Gubler, F., Lennon, A. and Sleegers, J. (1986). Temperature control in Lowicryl K4M and glycol methacrylate during polymerization: is there a low-temperature embedding method. *J. Microsc.*, **144**: 107

Ashton, F. T. and Schultz, J. (1971). The three-dimensional fine structure of chromosomes in a prophase *Drosophila* nucleus. *Chromosoma*, **35**: 383

Ashworth, C. T., Leonard, J. S., Eigenbrodt, E. H. and Wrightsman, F. J. (1966). Hepatic intracellular osmiophilic droplets: effect of lipid solvents during tissue preparation. *J. Cell Biol.*, **31**: 301

Auer, G. (1972). Cytochemical properties of nuclear chromatin as demonstrated by the colloidal iron binding technique. *Exp. Cell Res.*, **75**: 237

Baba, N. Nakamura, S.-I., Kino, I. and Kanaya, K. (1986). Three dimensional reconstruction from serial section images by computer graphics. *J. Electron Microsc. Tech.*, **3**: 401

Bachhuber, K., Böhme, H., Westphal, C. and Frösch, D. (1987). Ultrastructure and histochemistry of blue-green algae freeze-substituted at 190 K by Nanoplast MUV 116. *J. Microsc.*, **147**: 323

Bacchuber, K. and Frösch, D. (1983). Melamine resins, a new class of water-soluble embedding media for electron microscopy. *J. Microsc.*, **130**: 1

Bachmann, L. and Schmitt, W. W. (1971). Improved cryofixation applicable to freeze-etching. *Proc. Natl Acad. Sci. U.S.A.*, **68**: 2149

Bachofen, M., Weibel, E. R. and Roos, B. (1975). Postmortem fixation of human lungs for electron microscopy. *Am. Rev. Resp. Dis.*, **111**: 247

Bahr, G. F. (1954). Osmium tetroxide and ruthenium tetroxide and their reactions with biologically important substances. *Exp. Cell Res.*, **7**: 457

Bahr, G. F. (1955). Continued studies about the fixation with osmium tetroxide. Electron stains IV. *Exp. Cell Res.*, **9**: 277

Bahr, G. F., Bloom, G. and Friberg, U. (1957). Volume changes of tissues in physiological fluids during fixation in osmium tetroxide or formaldehyde and during subsequent treatment. *Exp. Cell Res.*, **12**: 342

Bahr, G. F. and Zeitler, E. (1965). The determination of the dry mass in populations of isolated particles. *Lab. Invest.*, **14**: 955

Bain, J. M. and Gove, D. W. (1971). Rapid preparation of plant tissues for electron microscopy. *J. Microsc.*, **93**: 159

Baker, J. R. and McCrae, J. M. (1966). The fine structure resulting from fixation by formaldehyde: the effects of concentration, duration, and temperature. *J. Roy. Microsc. Soc.*, **58**: 391

Bald, W. B. (1983). Optimizing the cooling block for the quick-freeze method. *J. Microsc.*, **131**: 11

Bald, W. B. (1984). The relative efficiency of cryogenic fluids used in the rapid quench cooling of biological samples. *J. Microsc.*, **134**: 261

Bald, W. B. (1985). The relative merits of various cooling methods. *J. Microsc.*, **140**: 17

Bald, W. B. (1986). On crystal size and cooling rate. *J. Microsc.*, **143**: 89

Bald, W. B. and Robards, A. W. (1978). A device for the rapid freezing of biological specimens under precisely controlled and reproducible conditions. *J. Microsc.*, **112**: 3

Baldwin, K. M. (1973). A study of electrical uncoupling using ruthenium red. *J. Cell Biol.*, **59**: 15a

Balyuzi, H. H. M. and Burge, R. E. (1970). Structure in embedding media for electron microscopy. *Nature, Lond.*, **227**: 489

Banerjee, T. K. and Yamada, K. (1984). Histochemical analysis of urea-unmasked glycosaminoglycans in the skin of rat and mouse. *Histochem. J.*, **16**: 1325

Barajas, L. (1970). The ultrastructure of the juxtaglomerular apparatus as disclosed by three-dimensional reconstructions from serial sections *J. Ultrastruct. Res.*, **33**: 116

Barnard, T. (1980). Ultrastructural effects of the high molecular weight cryoprotectants Dextran and polyvinyl pyrrolidone on liver and brown adipose tissue *in vitro*. *J. Microsc.*, **120**: 93

Barnard, T. (1982). Thin frozen-dried cryosections and biological x-ray microanalysis. *J. Microsc.*, **126**: 317

Barnes, B. G. and Chambers, T. C. (1961). A simple and rapid method for mounting serial sections for electron microscopy. *J. Biophys. Biochem. Cytol.*, **9**: 724

Barnicot, N. A. (1967). An electron microscopic study of newt mitotic chromosomes by negative staining. *J. Cell Biol.*, **32**: 585

Barrett, J. M., Heidger, P. M. and Kennedy, S. W. (1975). Chelated bismuth as a stain in electron microscopy. *J. Histochem. Cytochem.*, **23**: 780

Barrnett, R. J., Perney, D. P. and Hagström, P. E. (1964). Additional new aldehyde fixatives for histochemistry and electron microscopy. *J. Histochem. Cytochem.*, **12**: 36

Bartl, P. (1962). Freeze-substitution method using a water-miscible embedding medium. In: *Proceedings of the 5th International Congress on Electron Microscopy*, Vol. 2, p. 4. Academic Press, New York

Bartl, P. (1964). A simple device for rapid polymerization of electron microscopy specimens by UV-light under controlled conditions. *J. Microscopie*, **3**: 573

Bartlett, P. A., Bauer, B. and Singer, S. J. (1978). Synthesis of water-soluble undecagold cluster compounds of potential importance in electron microscopic and other studies of biological systems. *J. Am. Chem. Soc.*, **100**: 5085

Baschong, W., Baschong-Prescianotto, C., Wurtz, M., Carlemalm, E., Kellenberger, C. and Kellenberger, E. (1984). Preservation of protein structures for electron microscopy by fixation with aldehyde and/or OsO_4. *Eur. J. Cell Biol.*, **35**: 21

Baschong, W., Lucocq, J. M. and Roth, J. (1985). 'Thiocyanate gold': small (2–3 nm) colloidal gold for affinity cytochemical labelling in electron microscopy. *Histochemistry*, **83**: 409

Bastacky, J. and Hayes, T. L. (1985). Safety in the scanning electron microscope laboratory. *Scanning*, **7**: 255

Bastholm, L., Scopsi, L. and Nielsen, M. H. (1986). Silver-enhanced immunogold staining of semithin and ultrathin cryosections. *J. Electron Microsc. Tech.*, **4**: 175

Battaglia, E. and Maggini, F. (1968). Use of osmium tetroxide for staining nucleolus in squash technique. *Caryologia*, **21**: 287

Bauman, D. and Mendell, J. R. (1974). Method of reembedding tissue for electron microscopy. *Stain Technol.*, **49**: 118

Baumeister, W. and Hahn, M. H. (1974). Suppression of lattice periods in vermiculite single crystal specimen supports for high resolution electron microscopy. *J. Microsc.*, **101**: 111

Baumeister, W. and Hahn, M. H. (1975). Radiation resistant plastic specimen supports. *Naturwissenschaften*, **62**: 527

Baumeister, W. and Hahn, M. (1976). An improved method for the preparation of single crystal specimen supports. *Micron*, **7**: 247

Baumeister, W. and Hahn, M. (1978). Specimen supports. In: *Principles and Techniques of Electron Microscopy: Biological Applications*, Vol. 8 (M. A. Hayat, Ed.). Van Nostrand Reinhold, New York

Baumeister, W. and Seredynski, J. (1976). Preparation of perforated films with predeterminable hole size distributions. *Micron*, **7**: 49

Baur, P. S. and Stacey, T. R. (1977). The use of PIPES buffer in the fixation of mammalian and marine tissues for electron microscopy. *J. Microsc.*, **109**: 315

Baur, P. S. and Walkinshaw, C. H. (1974). Fine structure of tannin accumulations in callus cultures of *Pinus elliotii*. *Can. J. Bot.*, **52**: 615

Bearer, E. L., Duzgunes, N., Friend, D. S. and Papahadjopoulos, D. (1982). Fusion of phospholipid vesicles arrested by quick-freezing. The question of lipidic particles as intermediates in membrane fusion. *Biochim. Biophys. Acta*, **693**: 93

Bearer, E. L. and Orci, L. (1986). A simple method for quick-freezing. *J. Electron Microsc. Tech.*, **3**: 233

Beckmann, H.-J. and Dierichs, R. (1982). Lipid extracting properties of 2,2-dimethoxypropane as revealed by electron microscopy and thin layer chromatography. *Histochemistry*, **76**: 407

Bedi, K. S. (1987). A simple method of measuring the thickness of semithin and ultrathin sections. *J. Microsc.*, **148**: 107

Beer, M., Stern, S., Carmalt, D. and Mohlenrich, K. H. (1966). Determination of base sequence in nucleic acids with the electron microscope. V. The thymine specific reactions of osmium tetroxide with deoxyribonucleic acid and its components. *Biochemistry*, **5**: 2283

Beer, M., Wiggins, J. W., Alexander, R., Schettino, R., Stoeckert, C. and Piez, K. (1979). Electron microscopy of selectively stained collagen. *Proc. 37th Ann. EMSA Meet.*, pp. 28–29. Claitor's Pub. Division, Baton Rouge, La.

Beeseley, J. E., Beckford, U. and Chanter, S. M. (1984). Double immunogold staining method for the simultaneous localization of two tumour associated epithelial cell antigens in breast tissue. *Proc. 8th Eur. Cong. Electron Microsc.*, Vol. 3: 1599–1600

Behnke, O., Ammitzbøll, T., Jessen, H., Klokker, M., Nilausen, K., Tranum-Jensen, J. and Olsson, L. (1986). Non-specific binding of protein-stabilized gold sols as a source of error in immunocytochemistry. *Eur. J. Cell Biol.*, **41**: 326

Behnke, O. and Rostgaard, J. (1964). Your 'third hand' in mounting serial sections on grids for electron microscopy. *Stain Technol.*, **39**: 205

Behrman, E. J. (1984). The chemistry of osmium tetroxide fixation. *Sci. Biol. Spec. Prep.*, 1

Bella, A. and Kim, Y. S. (1973). Iron binding of gastric mucins. *Biochem. Biophys. Acta*, **304**: 5800

Bendayan, M. (1981). Electron microscopical localization of nucleic acids by means of nuclease–gold complexes. *Histochem. J.*, **13**: 699

Bendayan, M. (1982). Double immunocytochemical labelling applying the protein A–gold technique. *J. Histochem. Cytochem.*, **30**: 81

Bendayan, M. (1984a). Protein A–gold electron microscopic immunocytochemistry: methods, applications and limitations. *J. Electron Microsc. Tech.*, **1**: 243

Bendayan, M. (1984b). Enzyme–gold electron microscopic cytochemistry: a new affinity approach for the ultrastructural localization of macromolecules. *J. Electron Microsc. Tech.*, **1**: 349

Bendayan, M. (1986). Facts and artifacts in colloidal gold postembedding cytochemistry. *Proc. 44th Ann. Meet. EMSA.*, p. 44. San Francisco Press, San Francisco

Bendayan, M. (1987). Introduction of the protein G–gold complex for high resolution immunocytochemistry. *J. Electron Microsc. Tech.*, **6**: 7

Bendayan, M., Nanci, A. and Kan, F. W. K. (1987). Effect of tissue processing on colloidal gold cytochemistry *J. Histochem. Cytochem.*, **35**: 983

Bendayan, M. (1989a). Protein A–gold and protein G–gold post-embedding immunoelectron microscopy. In: *Colloidal Gold: Principles, Methods, and Applications*, Vol. 1 (M. A. Hayat, Ed.). Academic Press, San Diego and London

Bendayan, M. (1989b). Preparation and application of enzyme–gold complex. In: *Colloidal Gold: Principles, Methods, and Applications*, Vol. 2 (M. A. Hayat, Ed.). Academic Press, San Diego and London

Benedeczky, I. and Smith, A. D. (1972). Ruthenium red staining of the hamster adrenal medulla. *Histochemie*, **32**: 213

Benedetti, E. L. and Bertolini, B. (1963). The use of phosphotungstic acid (PTA) as a stain for the plasma membranes. *J. Roy. Microsc. Soc.*, **81**: 219

Benedetti, E. L. and Emmelot, P. (1967). Studies on plasma membranes. IV. The ultrastructural localization and content of sialic acid in plasma membranes isolated from rat liver and hepatoma. *J. Cell Sci.*, **2**: 499

Benhamou, N. (1989). Preparation and application of lectin–gold complexes. In: *Colloidal Gold: Principles, Methods, and Applications*, Vol. 1 (M. A. Hayat, Ed.). Academic Press, San Diego and London

Bennett, H. S., Wyrick, A. D., Lee, S. W. and McNeil, J. H. (1976). Science and art in preparing tissues embedded in plastic for light microscopy, with special reference to glycol methacrylate, glass knives and simple stains. *Stain Technol.*, **51**: 71

Bennett, P. M. (1974). Decrease in section thickness on exposure to the electron beam: The use of tilted sections in estimating the amount of shrinkage. *J. Cell Sci.*, **15**: 693

Berger, J. M. and Bencosme, S. A. (1971). Fine structural cytochemistry of granules in atrial cardiocytes. *J. Moll. Cell Card.*, **3**: 111

Bergh Weerman, M. A. v. d. and Dingemans, K. P. (1984). Rapid deparaffinization for electron microscopy. *Ultrastruct. Pathol.*, **7**: 55

Berkowitz, L. R., Fiorello, O., Kruger, L. and Maxwell, D. S. (1968). Selective staining of nervous tissue for light microscopy following preparations for electron microscopy. *J. Histochem. Cytochem.*, **16**: 808

Bernfield, M. R. and Banerjee, S. D. (1972). Acid mucopolysaccharide (glycosaminoglycan) at the epithelial–mesenchymal interface of mouse embryo salivary glands. *J. Cell Biol.*, **52**: 664

Bernhard, W. (1968). Une méthode de coloration regressive à l'usage de la microscopie électronique. *C. R. Acad. Sci. (D) Paris*, **267**: 2170

Bernhard, W. (1969). A new staining procedure for electron microscopical cytology. *J. Ultrastruct. Res.*, **27**: 250

Bernhard, W. and Leduc, E. H. (1967). Ultrathin frozen sections. I. methods and ultrastructural preservation. *J. Cell Biol.*, **34**: 757

Berriman, J., Bryan, R. K., Freeman, R. and Leonard, K. R. (1984). Methods for specimen thickness determination in electron microscopy. *Ultramicroscopy*, **13**: 351

Berthold, C.-H., Rydmark, M. and Corneliuson, O. (1982). Estimation of sectioning compression and thickness of ultrathin sections through Vestopal W-embedded cat spinal roots. *J. Ultrastruct. Res.*, **80**: 42

Bertram, J. F., Sampson, P. D. and Bolender, R. P. (1986). Influence of tissue composition on the final volume of rat liver blocks prepared for electron microscopy. *J. Electron Microsc. Tech.*, **4**: 303

Bhagwat, A. G. and Wong, P. (1972). Effect of pH in direct OsO_4 fixation on glycogen staining as shown by electron microscopy. *Stain Technol.*, **47**: 39

Biddlecomb, W. H., Ballard, K. J. and Elder, H. Y. (1971). Adaptation of the standard Cambridge Huxley ultramicrotome knife clamp to accept triangular machine-made glass knives. *J. Microsc.*, **93**: 163

Bienz, K., Egger, D. and Pasamontes, L. (1986). Electron microscopic immunocytochemistry: silver enhancement of colloidal gold marker allows double labelling with the same primary antibody. *J. Histochem. Cytochem.*, **34**: 1337

Bird, M. M. (1984). Regions of putative acetylcholine receptors at synaptic contacts between neurons maintained in culture and subsequently fixed in solutions containing tannic acid. *Cell Tiss. Res.*, **235**: 85

Birrell, G. B., Habliston, D. L., Hedberg, K. K. and Griffith, O. H. (1986). Silver-enhanced colloidal gold as a cell surface marker for photoelectron microscopy. *J. Histochem. Cytochem.*, **34**: 339

Björck, L. and Kronvall, G. (1984). Purification and some properties of streptococcal protein G, a novel IgG-binding reagent. *J. Immunol.*, **133**: 969

Björck. I., Petersson, B. A. and Sjöquist, J. (1972). Some physiochemical properties of protein A from *Staphylococcus aureus*. *Eur. J. Biochem.*, **29**: 579

Björkman, N. and Hellström, B. (1965). Lead–ammonium acetate, a staining method for electron microscopy free of contamination by carbonate. *Stain Technol.*, **40**: 169

Black, J. T. (1971a). Glass knife geometry as seen by the

SEM. *Proc. 29th Ann. EMSA Meet.*, p. 454. Claitor's Pub. Division, Baton Rouge, La.

Black, J. T. (1971b). Ultramicrotomy of embedding plastics. *Appl. Polymer Symp.*, **16**: 105

Black, J. T. and Boldosser, W. G. (1971). Surface structure in microtomed sections caused by glass and diamond knives. In: *Proc. 29th Ann. EMSA Meet*, p. 456. Claitor's Pub. Division, Baton Rouge, La.

Black, M. M. and Ansley, H. R. (1966). Histone specificity revealed by ammoniacal silver staining. *J. Histochem. Cytochem.*, **14**: 177

Black, M. M. and Ansley, H. R. (1969). Selective staining of histones with ammoniacal silver. In: *Analytical Methods of Protein Chemistry (Including Polypeptides)* (P. Alexander and H. P. Lundgren, Eds.), pp. 1–22. Permagon Press, New York

Blanquet, P. R. and Loiez, A. (1974). Colloidal iron used at pHs lower than 1 as electron stain for surface proteins. *J. Histochem. Cytochem.*, **22**: 368

Blass, J., Verriest, C., Leau, A. and Weiss, M. (1976). Monomeric glutaraldehyde as an effective crosslinking reagent for proteins. *J. Leather Chemists Assoc.*, **71**: 121

Bloom, F. E. and Aghjanian, G. K. (1968). Fine structural and cytochemical analysis of the staining of synaptic junctions with phosphotungstic acid. *J. Ultrastruct. Res.*, **22**: 361

Bobak, M. and Herich, R. (1969). The use of $AgNO_3$ and glutaraldehyde in electron microscope studies for fixation of chromosomal matrix. *Mikroskopie*, **24**: 270

Bodian, D. (1970). An electron microscopic characterization of classes of synaptic vesicles by means of controlled aldehyde fixation. *J. Cell Biol.*, **44**: 115

Bohman, S. O. (1974). The ultrastructure of the rat renal medulla as observed after improved fixation methods. *J. Ultrastruct. Res.*, **47**: 329

Bohnam, S. O. and Maunsbach, A. B. (1970). Effect on tissue fine structure of variations in colloid osmotic pressure of glutaraldehyde fixatives. *J. Ultrastruct. Res.*, **30**: 195

Bonnard, C., Papermaster, D. S. and Kraehenbuhl, J.-P. (1984). The streptavidin–biotin bridge technique: application in light and electron microscope immunocytochemistry. In: *Immunolabelling For Electron Microscopy* (J. M. Polak and I. M. Varndell, Eds.), pp. 95–111. Elsevier, Amsterdam

Bontoux, J., Dauplan, A. and Margignan, R. (1969). Stabilisation des colloides mineraux. Effect de la lasse moleculaire du stabilisant, des dimensions micellaires, et essais d'evaulation de la couche protectrice. *J. Chim. Phys. Physio-Chim. Biol.*, **66**: 1259

Boothroyd, B. (1964). The problem of demineralization in thin sections of fully calcified bone. *J. Cell Biol.*, **20**: 165

Bosman, F. T. and Go, P. M. N. Y. H. (1981). Polyethylene glycol embedded tissue sections for immunoelectron microscopy. *Histochemistry*, **73**: 195

Boucher, R. M. G. (1972). Advances in sterilization techniques. *Am. J. Hosp. Pharm.*, **29**: 661

Boucher, R. M. G. (1974). Potentiated acid 1,5-pentanediol solution—a new chemical sterilizing and disinfecting agent. *Am. J. Pharm.*, **31**: 546

Boucher, R. M. G., Last, A. J. and Smith, D. K. (1973). Biocidal mechanisms of saturated dialdehydes. *Proc. West Pharmacol.*, **16**: 282

Bowes, D., Bullock, G. R. and Winsey, N. J. P. (1970). A method for fixing rabbit and rat hind limb skeletal muscle by perfusion. *Proc. 7th Ann. Int. Cong. Electron Microsc.*, **1**: 397

Bowser, S. S. and Rieder, C. L. (1986). Section thickness terminology: a source of confusion in microscopy. *J. Microsc.*, **143**: 319

Boyde, A., Bailey, E., Jones, S. J. and Tamarin, A. (1977). Dimensional changes during specimen preparation for scanning electron microscopy. *Scanning Electron Microsc.*, **1**: 507

Boyde, A. and Franc, F. (1981). Freeze-drying shrinkage of glutaraldehyde-fixed liver. *J. Microsc.*, **122**: 75

Boyles, J., Anderson, L. and Hutcherson, P. (1985). A new fixative for the preservation of actin filaments: fixation of pure actin filament pellets. *J. Histochem. Cytochem.*, **33**: 1116

Bozzola, J. J. (1987). Clinical sampling device for rapid viral diagnosis by transmission electron microscopy. *J. Electron Microsc. Tech.*, **5**: 243

Brack, C. (1973). Use of uranyl formate staining for the electron microscopic visualization of DNA–protein complexes. *Experientia*, **29**: 768

Bradbury, S and Meek, G. A. (1960). A study of potassium permanganate 'fixation' for electron microscopy. *Quart. J. Microsc. Sci.*, **101**: 241

Bradbury, S. and Stoward, P. J. (1967). The specific cytochemical demonstration in the electron microscope of periodate-reactive mucosubstances and polysaccharides containing *vic*-glycol groups. *Histochemie*, **11**: 71

Bradley, A. J. and Illingworth, J. W. (1936). The crystal structure of $H_3PW_{12}O_{40}{\cdot}29H_2O$. *Proc. Roy. Soc. Lond. A*, **157**: 113

Bradley, D. E. (1965). The preparation of specimen support films. In: *Techniques for Electron Microscopy* (D. H. Kay and V. E. Cosslett, Eds.), p. 58. F. A. Davis, Philadelphia

Breathnach, A. S. and Goodwin, D. (1965). Electron microscopy of guinea pig epidermis stained by the osmium–iodide technique. *J. Anat.*, **100**: 159

Breman, J. G., Ruti, K., Steniowski, M. V., Zanotto, E., Gromyko, A. I. and Arita, I. (1980). *World Health Organization Bulletin*, **58**: 165

Bretscher, M. S. and Whytock, S. (1977). Membrane-associated vesicles in fibroblasts. *J. Ultrastruct. Res.*, **61**: 215

Bretschneider, A., Burns, W. and Morrison, A. (1981). 'Pop-off' technic. The ultrastructure of paraffin-embedded sections. *Am. J. Clin. Path.*, **76**: 450

Briarty, L. G. (1986). Quantitative morphological analysis of botanical micrographs. In: *Botanical Microscopy* (A. W. Robards, Ed.), pp. 91–127. Oxford Scientific Publications, Oxford

Brightman, M. W. and Reese, T. S. (1969). Junctions between intimately apposed cell membranes in the vertebrate brain. *J. Cell Biol.*, **40**: 648

Brodie, D. A. (1982). Bismuth chemistry and the glutaraldehyde insensitive staining reaction. *Tiss. Cell*, **14**: 39

Brodie, D. A., Huie, P., Locke, M. and Ottensmeyer, F. P. (1982a). The correlation between bismuth and uranyl staining and phosphorus content of intracellular structures as determined by electron spectroscopic imaging. *Tiss. Cell*, **14**: 621

Brodie, D. A., Locke, M. and Ottensmeyer, F. P. (1982b). High resolution microanalysis for phosphorus in Golgi complex beads of insect fat body tissue by electron spectroscopic imaging. *Tiss. Cell*, **14**: 1

Brody, I. (1971). An electron microscopic study of non-fixed and non-dehydrated normal human stratum corneum. *Z. Zellforsch.*, **118**: 97

Brooks, B. R. and Klamerth, O. L. (1968). Interaction of DNA with bifunctional aldehydes. *Eur. J. Biochem.*, **5**: 178

Brooks, J. C. and Carmichael, S. W. (1987). Ultrastructural demonstration of exocytosis in intact and saponin-permeabilized cultured bovine chromaffin cells. *Am. J. Anat.*, **178**: 85

Brown, G. L. and Locke, M. (1978). Nucleoprotein localization by bismuth staining. *Tiss. Cell*, **10**: 365

Brown, J. N. (1983a). A fluid exchange apparatus for the processing of biological specimens for electron microscopy. *Microsc. Acta*, **87**: 329

Brown, J. N. (1983b). An improved apparatus for staining large numbers of electron microscope ultrathin sections simultaneously. *Micron*, **14**: 69

Brown, R. M. Jr. (1969). Burnishing uncoated grids before mounting ultrathin section: A means of assuring adhesion. *Stain Technol.*, **44**: 158

Brown, R. M. Jr. and Arnott, H. J. (1971). A photographic method for producing true three-dimensional electron micrographs. *Protoplasma*, **72**: 105

Brüggler, P. and Mayer, E. (1980). Complete vitrification in pure liquid water and dilute aqueous solutions. *Nature, Lond.*, **288**: 569

Brunings, E. A. and Priester, W. de (1971). Effect of mode of fixation on formation of extrusions in the midgut epithelium of Calliphora. *Cytobiologie*, **4**: 487

Bryant, V. and Watson, J. H. L. (1967). A comparison of light microscopy staining methods applied to a polyester and three epoxy resins. *Henry Ford Hosp. Med. Bull.*, **15**: 65

Bucek, M. T. and Arnott, H. J. (1972). A method for setting the clearance angle. In: *Proc. 30th Ann. EMSA Meet.*, p. 388. Claitor's Pub. Division, Baton Rouge, La.

Buchanan, G. M. (1982). Comparison of 2,2-dimethoxypropane and ethanol dehydration: a morphometric evaluation. *Proc. 40th Ann. Meet. EMSA*, pp. 360–361

Buckley, I. K. (1973a). Studies in fixation for electron microscopy using cultured cells. *Lab. Invest.*, **29**: 398

Buckley, I. K. (1973b). The lyosomes of cultured chick embryo cells. A correlated light and electron microscopic study. *Lab. Invest.*, **29**: 411

Buma, P. (1989). Study of exocytosis with colloidal gold and other methods. In: *Colloidal Gold: Principles, Methods, and Applications*, Vol. 2 (M. A. Hayat, Ed.). Academic Press, San Diego and London

Buma, P., Roubos, E. W. and Buijs, R. M. (1984). Ultrastructural demonstration of exocytosis of neural, neuroendocrine and endocrine secretions with an *in vitro* tannic acid (TARI-) method. *Histochemistry*, **80**: 247

Buma, P. and Nieuwenhuys, R. (1987). Ultrastructural demonstration of oxytocin and vasopressin release sites in the neural lobe and median eminence of the rat by tannic acid and immunogold methods. *Neurosci. Lett.*, **74**: 151

Bundgaard, M. (1984). Choice of section thickness in serial-section electron microscopy. *J. Ultrastruct. Res.*, **88**: 287

Burdett, I. D. J. and Rogers, H. J. (1970). Modifications of the appearance of mesosomes in sections of *Bacillus licheniformis* according to the fixation procedures. *J. Ultrastruct. Res.*, **30**: 354

Burke, C. N. and Geiselman, C. W. (1971). Exact anhydride epoxy percentage for electron microscopy embedding (Epon). *J. Ultrastruct. Res.*, **36**: 119

Burns, J. (1970). Preparation of thin epoxy resin sections from thick sections of paraffin-embedded material. *J. Clin. Pathol.*, **23**: 643

Burr, F. A. (1972). Modifications of the silver methenamine test for the ultrastructural demonstration of cystine-containing proteins. *J. Histochem. Cytochem.*, **20**: 296

Burr, F. A. (1973). Staining of a protein polysaccharide model with silver methenamine. *J. Histochem. Cytochem.*, **21**: 386

Burr, F. A. and Evert, R. F. (1972). A cytochemical study of the wound-healing protein in *Bryopsis hypnoides*. *Cytobios*, **6**: 199

Burton, K. (1967). Oxidation of pyrimidine nucleosides and nucleotides by osmium tetroxide. *Biochem. J.*, **104**: 686

Bushman, R. J. and Taylor, A. B. (1968). Extraction of absorbed lipid (linoleic acid-1-^{14}C) from rat intestinal epithelium during processing for electron microscopy. *J. Cell Biol.*, **38**: 252

Butler, J. K. (1974). A precision hand trimmer for electron microscope tissue blocks. *Stain Technol.*, **49**: 129

Butler, J. K. (1980). The use of Ralph–Bennett glass knives for specimen block trimming. *Stain Technol.*, **55**: 323

Buys, C. H. C. M. and Osinaga, J. (1980). Abundance of protein-bound sulfhydryl and disulfide at chromosomal nucleolus organizing regions. *Chromosoma*, **77**: 1

Caldwell, J. B. and Milligan, B. (1972). The sites of reaction of wool with formaldehyde. *Text. Res. J.*, **42**: 122

Cammermeyer, J. (1978). Is the solitary dark neuron a manifestation of postmortem trauma to the brain inadequately fixed by perfusion? *Histochemistry*, **56**: 97

Capco, D. G., Krochmalnic, G. and Penman, S. (1984). A new method of preparing embedment-free sections for transmission electron microscopy: applications to the cytoskeletal framework and other three-dimensional networks. *J. Cell Biol.*, **98**: 1878

Capco, D. G. and McGaughey, R. W. (1986). Cytoskeletal reorganization during early mammalian development: analysis using embedment-free section. *Dev. Biol.*, **115**: 446

Cardamone, J. J. Jr. (1982). A simple and inexpensive multiple grid staining device. *EMSA Bull.*, **12**: 78

Carlemalm, E., Armbruster, B., Chiovetti, B., Hobot, J., Garavito, M., Acetarin, J.-D., Villiger, W. and Kellenberger, E. (1982b). Limitations and developments of the embedding technique. *Ultramicroscopy*, **9**: 419

Carlemalm, E., Garavito, R. M. and Villiger, W. (1982a).

Resin development for electron microscopy and an analysis of embedding at low temperature. *J. Microsc.*, **126**: 123

Carlemelm, E. and Kellenberger, E. (1982). The reproducible observation of unstained embedded cellular material in thin sections: visualization of an integral membrane protein by a new mode of imaging for STEM. *EMBO J.*, **1**: 63

Carlemalm, E., Villiger, W., Hobot, J. A., Acetarin, J.-D. and Kellenberger, E. (1985). Low temperature embedding with Lowicryl resins: two new formulations and some applications. *J. Microsc.*, **140**: 55

Carlemalm E., Villiger, W. and Kellenberger, E. (1986). New resins for low temperature embedding that are designed for immunocytochemistry and for observing unstained biological material. *Proc. 11th Int. Cong. Electron Microsc.*, pp. 2167–2168. Kyoto, Japan

Carling, D. E., White, J. A. and Brown, M. F. (1977). The influence of fixation procedure on the ultrastructure of the host-endophyte interface of vesicular-arbuscular mycorrhizae. *Can. J. Bot.*, **55**: 48

Carson, F., Lynn, J. A. and Martin, J. H. (1972). Ultrastructural effect of various buffers, osmolality, and temperature on paraformaldehyde fixation of the formed elements of blood and bone marrow. *Tex. Rep. Biol. Med.*, **30**: 125

Carson, J. H., Martin, J. H. and Lynn, J. A. (1973). Formalin fixation for electron microscopy: a reevaluation. *Am. J. Clin. Pathol.*, **59**: 365

Carstensen, E. L., Aldridge, W. G., Child, S. Z., Sullivan, P. and Brown, H. H. (1971). Stability of cells fixed with glutaraldehyde and acrolein. *J. Cell Biol.*, **50**: 529

Casley-Smith, J. R. (1967). Some observations on the electron microscopy of lipids. *J. Roy. Microsc. Soc.*, **81**: 235

Casley-Smith, J. R. and Crocker, K. W. J. (1975). Estimation of section thickness, etc. by quantitative electron microscopy. *J. Microsc.*, **103**: 351

Casley-Smith, J. R. and Day, A. J. (1966). The uptake of lipid and lipoprotein by macrophages *in vitro*: an electron microscopical study. *Quart. J. Exp. Physiol.*, **51**: 1

Casteels, R., van Breeman, C. and Wuytack, F. (1972). Effect of metabolic depletion of the membrane permeability of smooth muscle cells and its modification by La^{3+}. *Nature, New Biol.*, **239**: 249

Cataldo, A. M. and Broadwell, R. D. (1986). Cytochemical identification of cerebral glycogen and glucose-6-phosphatase activity under normal and experimental conditions. I. Neurons and glia. *J. Electron Microsc. Tech.*, **3**: 413

Cattini, P. A. and Davies, H. G. (1983). Kinetics of lead citrate staining of thin sections for electron microscopy. *Stain Technol.*, **58**: 29

Cattini, P. A. and Davies, H. G. (1984). Observations on the kinetics of uranyl acetate and phosphotungstic acid staining of chromatin in thin sections for electron microscopy. *Stain Technol.*, **59**: 291

Causton, B. E. (1980). The molecular structure of resins and its effects on the epoxy embedding resins. *Proc. Roy. Microsc. Soc.*, **15**: 185

Causton, B. E. (1981). Resins: toxicity, hazards, and safe handling. *Proc. Roy. Microsc. Soc.*, **16**: 265

Chambers, R. W., Bowling, M. C. and Grimley, P. M. (1968). Glutaraldehyde fixation in routine histopathology. *Arch. Path.*, **85**: 18

Champy, C. (1913). Granules et stubstances réduisant l'iodure d'osmium. *J. Anat. Physiol. Norm. Pathol. (Paris)*, **49**: 323

Chandler, D. E. (1979). Quick freezing avoids specimen preparation artifacts in membrane fusion studies. In: *Freeze-Fracture: Methods, Artifacts, and Interpretations* (J. E. Rash and C. S. Hudson, Eds.), pp. 81–87. Raven Press, New York

Chandler, D. E. (1984). Comparison of quick frozen and chemically fixed sea urchin eggs: structural evidence that cortical granule exocytosis is preceded by a local increase in membrane mobility. *J. Cell Sci.*, **72**: 23

Chandler, J. A. and Battersby, S. (1976). X-ray microanalysis of zinc and calcium in ultrathin sections of human sperm cells using the pyroantimonate techniques. *J. Histochem. Cytochem.*, **24**: 740

Chang, C.-H., Beer, M. and Marzilli, L. G. (1977). Osmium-labelled polynucleotides. The reaction of osmium tetroxide with deoxyribonuleic acid and synthetic polynucleotides in the presence of tertiary nitrogen donor ligands. *Biochemistry*, **16**: 33

Chang, J.-J., McDowall, A. W., Lepault, J., Freeman, R., Walter, C. A. and Dubochet, J. (1983). Freezing, sectioning and observation artefacts of frozen hydrated sections for electron microscopy. *J. Microsc.*, **132**: 109

Chang, S. H., Megner, W. J., Pendergrass, R. E., Bulger, R. E., Berezesky, I. K. and Trump, B. F. (1980). A rapid method of cryofixation of tissues *in situ* for ultracryomicrotomy. *J. Histochem. Cytochem.*, **28**: 47

Chang, W. W. L. and Bencosme, S. A. (1968). Selective staining of secretory granules of adrenal medullary cells by silver methenamine; a light and electron microscope study. *Can. J. Physiol. Pharmacol.*, **46**: 745

Chaplin, A. J. (1985). Tannic acid in histology: a historical perspective. *Stain Technol.*, **60**: 219

Chaplin, G. L. (1972). Facilitating diamond knife ultramicrotomy by the use of reflecting metallic foil. *Stain Technol.*, **47**: 47

Chapman, J. A. and Hulmes, D. J. S. (1984). Electron microscopy of the collagen fibril. In: *Ultrastructure of the Connective Tissue Matrix* (A. Ruggeri and P. M. Motta, Eds.), pp. 1–33. Martinus Nijhoff, Boston

Chappard, D., Alexandre, C., Camps, M., Monthéard, J. P. and Riffat, G. (1983). Embedding iliac bone biopsies at low temperature using glycol and methyl methacrylates. *Stain Technol.*, **58**: 299

Chappard, D., Alexandre, C., Palle, S., Monthéard, J. P. and Riffat, G. (1986). Improved stability of a purified glycol methacrylate preparation: comments. *Stain Technol.*, **61**: 185

Chappard, D., Palle, S., Alexandre, C., Vico, L. and Riffat, G. (1987). Bone embedding in pure methyl methacrylate at low temperature preserves enzyme activities. *Acta Histochem.*, **81**: 183

Chaudhary, R. K. and Westwood, J. C. N. (1972). Use of polyelectrolytes and electron microscopy for detection of viruses from stool. *Appl. Microbiol.*, **24**: 270

Chedid, A. and Nair, V. (1972). Diurnal rhythm in endoplasmic reticulum of rat liver: electron microscopic study. *Science, N.Y.*, **175**: 179

Cheung, D. T. and Nimni, M. E. (1982a). Mechanism of cross-linking of proteins by glutaraldehyde. I. Reaction with model compounds. *Conn. Tiss. Res.*, **10**: 187

Cheung, D. T. and Nimni, M. E. (1982b). Mechanism of crosslinking of proteins by glutaraldehyde. II. Reaction with monomeric and polymeric collagen. *Conn. Tiss. Res.*, **10**: 201

Cheung, D. T., Perelman, N., Ko, E. C. and Nimni, M. E. (1985). Mechanism of crosslinking of proteins by glutaraldehyde. III. Reaction with collagen in tissues. *Conn. Tiss. Res.*, **13**: 109

Chien, K., van de Velde, R. L. and Heusser, R. C. (1982). A one-step method for reembedding paraffin embedded specimens for electron microscopy. *Proc. 40th Ann. EMSA Meet.*, pp. 356. Claitor's Pub. Division, Baton Rouge, La.

Childs, G. V., Unabia, G. and Ellison, D. (1986). Immunocytochemical studies of pituitary hormones with PAP, ABC, and immunogold techniques: evolution of technology to best fit the antigen. *Am. J. Anat.*, **175**: 307

Chiovetto, R., McGuffee, L. J., Little, S. A., Wheeler-Clark, E. and Brass-Dale, J. (1987). Combined quick freezing, freeze-drying, and embedding tissue at low temperature and in low viscosity resins. *J. Electron Microsc. Tech.*, **5**: 1

Chopra, H., Ebert, P., Woodside, N., Kvedar, J., Albert, S. and Brennan, M. (1973). Electron microscopic detection of simian-type virus particles in human milk. *Nature, New Biol.*, **243**: 159

Clark, A. W. (1976). Changes in the structure of neuromuscular junctions caused by variations in osmotic pressure. *J. Cell Biol.*, **69**: 521

Clark, M. A. and Ackermann, G. A. (1971). Osmium–zinc iodide reactivity in human blood and bone marrow cells. *Anat. Rec.*, **170**: 81

Clark, T. S. and Rochlani, S. P. (1970). Retrieval of paraffin embedded material for ultrastructure study. *Proc. 28th Ann. EMSA Meet.*, p. 232

Clawson, C. C. and Good, R. A. (1971). Micropapillae: a surface specialization of human leukocytes. *J. Cell Biol.*, **48**: 207

Cleton, M. I., Roelofs, J. M., Blok-van Hoek, C. J. G. and De Bruijn, W. C. (1986). Integrated image and X-ray microanalysis of hepatic lysosomes in a patient with idiopathic hemosiderosis before and after treatment by phlebotomy. *Scanning Electron Microsc.*, **3**: 999

Cliff, W. J. (1971). The ultrastructure of aortic elastica as revealed by prolonged treatment with OsO_4. *Exp. Mol. Pathol.*, **15**: 220

Coalson, J. (1983). A simple method of lung perfusion fixation. *Anat. Rec.*, **205**: 233

Coetzee, J. and van der Merwe, C. F. (1985a). Penetration rate of glutaraldehyde in various buffers into plant tissue and gelatin gels. *J. Microsc.*, **137**: 129

Coetzee, J. and van der Merwe, C. F. (1985b). Effect of glutaraldehyde on the osmolarity of the buffer vehicle. *J. Microsc.*, **138**: 99

Coetzee, J. and van der Merwe, C. F. (1987). Some characteristics of the buffer vehicle in glutaraldehyde-based fixatives. *J. Microsc.*, **146**: 143

Cohen, H. and Kretzer, F. (1982). Imaging of outer segment periodicities in unstained cryoultramicrotomy sections of the frog retina. *J. Microsc.*, **128**: 287

Coimbra, A. and Lopes-Vaz, A. (1971). The presence of lipid droplets and the absence of stable sudanophilia in osmium-fixed human leukocytes. *J. Histochem. Cytochem.*, **19**: 551

Cole, M. B. (1968). A simple apparatus for ultraviolet polymerization of soluble embedding media in electron microscopy. *J. Microscopie*, **7**: 441

Cole, M. B. (1982). Glycol methacrylate embedding of bone and cartilage for light microscopic staining. *J. Microsc.*, **127**: 139

Cole, M. B. (1984). Methods and results of testing 'low acid' glycol methacrylate (GMA) for light microscopic cytochemistry. *J. Histochem. Cytochem.*, **32**: 555

Cole, M. B. and Sykes, S. M. (1974). Glycol methacrylate in light microscopy: a routine method for embedding and sectioning animal tissues. *Stain Technol.*, **49**: 387

Cole, M. D., Wiggins, J. W. and Beer, M. (1977). Molecular microscopy of labelled polynucleotides: stability of osmium atoms. *J. Mol. Biol.*, **117**: 387

Cole, T. S. and Shierenberg, E. (1986). Laser microbeam-induced fixation for electron microscopy: visualization of transient developmental features in nematode embryos. *Experientia*, **42**: 1046

Collin, R. J., Griffith, W. P., Phillips, F. L. and Skapski, A. C. (1974). Staining and fixation of unsaturated membrane lipids by osmium tetroxide: crystal structure of a model osmium(VI) di-ester. *Biochem. Biophys. Acta*, **354**: 152

Collins, C. J. and Guild, W. R. (1969). Irreversible effects of formaldehyde on DNA. *Biochem. Biophys. Acta*, **157**: 107

Colquhoun, W. R. (1980). Bevelled sections. *J. Electron Microsc. Tech.*, **29**: 218

Colquhoun, W. R. and Rieder, C. L. (1980). Contrast enhancement based on rapid dehydration in the presence of phosphate buffer. *J. Ultrastruct. Res.*, **73**: 1

Colvin, J. R. and Leppard, G. G. (1971). The non-uniform distribution of protein in plant cell walls. *J. Microsc.*, **11**: 285

Connelly, P. S. (1977). Reembedding of tissue culture cells for comparative and quantitative electron microscopy. *Proc. 35th Ann. EMSA Meet.*, p. 552. Claitor's Pub. Division, Baton Rouge, La.

Constantin, L. L., Franzini-Armstrong, C. and Podolsky, R. J. (1965). Localization of calcium-accumulating structures in striated muscle fibers. *Science, N.Y.*, **147**: 158

Cope, G. H. (1968). Low temperature embedding in water miscible methacrylates after treatment with antifreezes. *J. Roy. Microsc. Soc.*, **88**: 259

Cope, G. H. and Williams, M. A. (1968). Quantitative studies on neutral lipid preservation in electron microscopy. *J. Roy. Microsc. Soc.*, **88**: 259

Cosslett, A. (1960). The effect of the electron beam on thin sections. *Proc. Eur. Reg. Conf. Electron Microsc.*, **2**: 678. Delft, Netherlands

Costello, M. J., Felter, R. and Corless, J. M. (1984). Optimum conditions for the plunge freezing of sandwiched samples. In: *Science of Biological Specimen Preparation for Microscopy and Microanalysis* (J.-P. Revel, R. Barnard and G. H. Haggis, Eds.), pp. 105–115. SEM Inc., AMF O'Hare, Ill.

Costello, M. J., Felter, R. and Höchli, M. (1982). Simple procedures for evaluating the cryofixation of biological samples. *J. Microsc.*, **125**: 125

Cottell, D. C. and Hooper, A. C. B. (1985). Some limitations of the use of tannic acid as a marker of damaged skeletal muscle fibers. *J. Microsc.*, **139**: 331

Coulter, H. D. and Terracio, L. (1977). Preparation of biological tissues for electron microscopy by freeze-drying. *Anat. Rec.*, **187**: 477

Coupland, R. E. (1965). *The Natural History of the Chromaffine Tissue*. Longmans, London

Couve, E. (1986). Controlled mounting of serial sections for electron microscopy. *J. Electron Microsc. Tech.*, **3**: 453

Cox, G. C. (1968). Ultrathin sectioning artifacts resembling plasmodesmata. *J. Microsc.*, **89**: 225

Cox, G. C. and Juniper, B. (1973). Electron microscopy of cellulose in entire tissue. *J. Microsc.*, **97**: 343

Craig, S. and Staehelin, L. A. (1988). High pressure freezing of intact plant tissue: evaluation and characterization of novel features of the endoplasmic reticulum and associated membrane systems. *Eur. J. Cell Biol.*, **46**: 80

Criegee, R., Marchand, B. and Wannowius, H. (1942). Zur Kenntnis der organischen Osmium-Verbindungen. *Ann. Chem.*, **550**: 99

Curran, P. F. (1972). Effect of silver ion on permeability properties of frog skin. *Biochim. Biophys. Acta*, **288**: 90

Daddow, L. Y. M. (1986). An abbreviated method of the double lead stain technique. *J. Submicrosc. Cytol.*, **18**: 221

Dae, M. W., Heymann, M. A. and Jones, A. L. (1982). A new technique for perfusion fixation and contrast enhancement of fetal lamb myocardium for electron microscopy. *J. Microsc.*, **127**: 301

Daems, W. Th. and Persijn, J. P. (1963). Sections staining with heavy metals of osmium-fixed mouse liver tissue. *J. Roy. Microsc. Soc.*, **81**: 199

Diamon, T., Mizuhira, V. and Uchida, K. (1978). Ultrastructural localization of calcium around the membrane of the surface connected system in the human platelet. *Histochemistry*, **55**: 271

Dallam, R. D. (1957). Determination of protein and lipid lost during osmic acid fixation of tissues and cellular particulates. *J. Histochem. Cytochem.*, **5**: 178

Dalton, A. J. and Zeigel, R. F. (1960). A simple method of staining thin sections of biological material with lead hydroxide for electron microscopy. *J. Biophys. Biochem. Cytol.*, **1**: 409

Daniel, F. B. and Behrman, E. J. (1975). Reactions of osmium ligand complexes with nucleosides. *J. Am. Chem. Soc.*, **97**: 7352

Daniel, F. B. and Behrman, E. J. (1976). Osmium(VI) complexes of the 3′,5′-dinucleoside monophosphates, ApU and UpA. *Biochemistry*, **15**: 565

Dankelman, W. and Daemen, J. M. H. (1976). Gas chromatographic and nuclear magnetic resonance determination of linear formaldehyde oligomers in formalin. *Anal. Chem.*, **48**: 401

Danscher, G. (1981). Localization of gold in biological tissue: a photochemical method for light and electron microscopy. *Histochemistry*, **71**: 81

Danscher, G. and Nörgaard, J. O. R. (1983). Light microscopical visualization of colloidal gold on resin-embedded tissue. *J. Histochem. Cytochem.*, **31**: 1394

Darley, J. J. (1972). A suction device for safe and fast handling of electron microscope grids. *Stain Technol.*, **47**: 167

Darley, J. J. and Ezoe, H. (1976). Potential hazards of uranium and its compounds in electron microscopy: a brief review. *J. Microsc.*, **106**: 85

Darrah, H. K., Hedley Whyte, J. and Hedley Whyte, E. T. (1971). Radioautography of cholesterol in lung. An assessment of different tissue processing techniques. *J. Cell Biol.*, **49**: 345

D'Arrigo, J. S. (1975). Axonal surface charges: evidence for phosphate structure. *J. Membrane Biol.*, **22**: 255

Dauwalder, M. and Whaley, W. G. (1973). Staining of cells of *Zea mays* root apices with the osmium-zinc iodide and osmium impregnation techniques. *J. Ultrastruct. Res.*, **45**: 279

Davies, H. G. and Spencer, M. (1962). The variation in the structure of erythrocyte nuclei with fixation. *J. Cell Biol.*, **14**: 445

Davis, J. M. and Himwich, W. A. (1971). The amino acid, norepinephrine, and serotonin content of rat brain fixed with glutaraldehyde. *Brain Res.*, **33**: 568

Davison, E. and Colquhoun, W. (1985). Ultrathin Formvar support films for transmission electron microscopy. *J. Electron Microsc. Tech.*, **2**: 35

Davison, E. A. and Rieder, C. L. (1985). The use of glass fibres as an alternative to hair tools. *J. Electron Microsc. Tech.*, **2**: 395

Day, T. (1984). Formvar films. *Proc. Roy. Microsc. Soc.*, **19**: 77

Deamer, D. W. and Crofts, A. (1967). Action of Triton X-100 on chloroplast membranes. *J. Cell Biol.*, **33**: 395

Debray, H., Pierce-Cretel, A., Spik, G. and Montreuil, J. (1983). Affinity of ten insolubilized lectins towards various glycopeptides with the *N*-glycosamine linkage and related oligosaccharides. In: *Lectins*, Vol. 3 (H. Popper, L. Bianchi, F. Gudat and W. Reulter, Eds.), pp. 335–350. Walter de Gruyter, Berlin

de Brouckère, L. and Casimir, J. (1948). Préparation d'hydrosols d'or homéodisperses très stables. *Bull. Soc. Chim. Belg.*, **57**: 517

De Bruijn, W. C. (1968). A modified OsO_4-(double) fixation procedure which selectively contrasts glycogen. *Proc. 4th Eur. Reg. Conf. Electron Microsc.*, p. 65

De Bruijn, W. C. (1973). Glycogen, its chemistry and morphologic appearance in the electron microscope. I. A modified OsO_4 fixative which selectively contrasts glycogen. *J. Ultrastruct. Res.*, **42**: 29

De Bruijn, W. C. and den Breejen, P. (1974). Selective glycogen contrast by hexavalent osmium oxide compounds. *Histochem. J.*, **6**: 61

De Bruijn, W. C. and den Breejen, P. (1976). Glycogen, its chemistry and morphological appearance in the electron microscope. III. Identification of the tissue ligands involved in the glycogen contrast staining reaction with the osmium-(VI)–iron(II) complex. *Histochem. J.*, **8**: 121

De Bruijn, W. C. and van Buitenen, J. M. H. (1980). X-ray microanalysis of aldehyde-fixed glycogen contrast-stained by $Os^{VI}{\cdot}Fe^{II}$ and $Os^{VI}{\cdot}Ru^{IV}$ complexes. *J. Histochem. Cytochem.*, **28**: 1242

De Bruijn, W. C. and van Buitenen, J. M. H. (1981). X-ray microanalysis of non-aldehyde-fixed glycogen contrast-stained by $Os^{VIII}O_4$, $Os^{III}{\cdot}Fe^{III}$, or $Os^{VI}{\cdot}Fe^{II}$ complex *in vitro*. *Histochem J.*, **13**: 125

De Bruijn, W. C. and Cleton, M. I. (1985). Application of Chelex standard beads in integrated morphometrical and X-ray microanalysis. *Scanning Electron Microsc.*, **II**: 715

De Bruijn, W. C., Memelink, A. A. and Riemersma, J. C. (1984). Cellular membrane contrast differentiation with osmium triazole and tetrazole complexes. *Histochem J.*, **16**: 37

Deer, B. C. (1983). A loop for the transfer of thin tissue sections. *J. Microsc.*, **130**: 115

Deetz, J. S. (1980). Ph.D. Dissertation. The Ohio State University, p. 75. Univ. Microfilm International No. 80-22256, Ann Arbor, Mich.

Deetz, J. S. and Behrman, E. J. (1981). Reaction of osmium reagents with amino acids and proteins. *Int. J. Peptide Protein Res.*, **17**: 495

de Groot, J. C. M. J., Veldman, J. E. and Huizing, E. H. (1987). An improved fixation method for guinea pig cochlear tissues. *Acta Otolaryngol.*, **104**: 234

DeLamater, E. D., Johnson, E., Schoen, T. and Whitaker, C. (1971). The use of styrenes as embedding media for electron microscopy. *Proc. 29th Ann. Meet. EMSA*, p. 488. Claitor's Pub. Division, Baton Rouge, La.

Dellmann, H.-D. and Pearson, C. B. (1977). Better epoxy resin embedding for electron microscopy at low relative humidity. *Stain Technol.*, **52**: 5

DeMason, D. A. and Stillman, J. I. (1986). Identification of phosphate granules occurring in seedling tissue of palm species (*Phoenix dactylifera* and *Washingtonia filifera*). *Planta*, **167**: 321

De Mazière, A. M. G. L., Aertgeerts, P. and Scheuermann, D. W. (1985). A modified cleansing procedure to obtain large freeze-fracture replicas. *J. Microsc.*, **137**: 185

De Mey, J. (1983). The preparation of immunoglobulin gold conjugates (IGS reagents) and their use as markers for light and electron microscopy. In: *Immunohistochemistry* (A. C. Cuello, Ed.), pp. 347–372. Wiley, New York

De Mey, J. (1984). Colloidal gold as marker and tracer in light and electron microscopy. *EMSA Bull.*, **14**: 54

De Mey, J. (1985). Gold probe particle size: why are small probes 'better' probes? *Proc. Roy. Microsc. Soc.*, **20**: IM3

De Mey, J., Moeremans, M., Geuens, G., Nuydens, R. and de Brabander, M. (1981). High resolution light and electron microscopic localization of tubulin with the IGS (immunogold staining) method. *Cell. Biol. Int. Rep.*, **5**: 889

Dempsey, G. P. and Bullivant, S. (1976). Copper block method for freezing non-cryoprotected tissue to produce ice crystal free regions for electron microscopy. I. Evaluation using freeze-substitution. *J. Microsc.*, **106**: 251

Dennison, M. E. (1971). Zinc iodide–osmium tetroxide reaction in the spinal cord of goldfish. *Brain Res.*, **27**: 357

De Noyer, E., Van Grieken, R., Adams, F. and Natusch, N. (1982). Laser microprobe mass analysis. *Anal. Chem.*, **54**: A26

Derenzini, M., Farabegoli, F. and Marinozzi, V. (1986). An improved periodic acid–thiosemicarbazide–osmium technique to reveal glycoconjugates at the molecular level in situ. *J. Histochem. Cytochem.*, **34**: 1161

Derksen, J. W. M. and Willart, E. (1976). Cytochemical studies on RNP complexes produced by puff 2-48BC in *Drosophilia hydei*. Uranyl acetate and phosphotungstic acid staining. *Chromosoma*, **55**: 57

Dermer, G. B. (1970). The fixation of pulmonary surfactant for electron microscopy. *J. Ultrastruct. Res.*, **31**: 229

Dermer, G. B. (1973). Specificity of phosphotungstic acid used as a section stain to visualize surface coats of cells. *J. Ultrastruct. Res.*, **45**: 183

Derrick, K. S. (1973). Quantitative assay for plant viruses using serologically specific electron microscopy. *Virology*, **56**: 652

DeVore, E. C., Holt, S. L., Asplung, R. O. and Catalano, A. W. (1971). Preparation and properties of Fe^{3+}–amino acid complexes. I. A novel complex of Fe^{3+} and glycine. *Arch. Biochim. Biophys.*, **146**: 658

Dhir, S. P. and Boatman, E. S. (1972). Location of polysaccharide on *Chlamydia psittaci* by silver methenamine staining and electron microscopy. *J. Bact.*, **111**: 267

Diers, L. and Schieren, M. T. (1972). Der Einfluss einiger elektronenmikroskopischer Fixierungs—und Einbettungsmittel auf die Chloroplasten—und Zellgrösse bei *Elodea*, *Protoplasma*, **74**: 321

Di Giamberardino, L., Koller, T. and Beer, M. (1969). Electron microscopic study of the base sequence in nucleic acids. IX. Absence of fragmentation and of crosslinking during reaction with osmium tetroxide and cyanide. *Biochim. Biophys Acta*, **182**: 523

Dijk, F., Oosterbaan, J. A. and Hulstaert, C. E. (1985). A rapid method for obtaining monomeric glutaraldehyde. *Histochemistry*, **83**: 573

Doane, F. W. and Anderson, N. (1977). Electron and immunoelectron microscopic procedures for diagnosis of viral infections. In: *Comparative Diagnosis of Viral Diseases*, Vol. 2 (E. Kurstsk and C. Kurstsk, Eds.), p. 505. Academic Press, New York

Dobelle, W. H. and Beer, M. (1968). Chemically cleaved graphite support films for electron microchemistry. *J. Cell Biol.*, **39**: 733

Doerr-Schott, J. (1989). Colloidal gold for multiple staining. In: *Colloidal Gold: Principles, Methods, and Applications*, Vol. 1 (M. A. Hayat, Ed.). Academic Press, San Diego and London

Doerr-Schott, J. and Lichte, C. M. (1986). A triple ultrastructural immunogold staining method: application to the simultaneous demonstration of three hypophyseal hormones. *J. Histochem. Cytochem.*, **34**: 1101

Doggenweiler, C. F. and Zambrano, F. (1981). Extraction of

phospholipids from aldehyde-fixed membranes. *Arch. Biol. Med. Exp.*, **14**: 343

Dorignac, D., Maclachlan, M. E. C. and Jouffrey, B. (1979). Low-noise boron supports for high resolution electron microscopy. *Ultramicroscopy*, **4**: 85

Douglas, W. H. J., Redding, R. A. and Stein, M. (1973). A method for improving preservation of the lamellar substructure of osmiophilic inclusion bodies present in rat type II alveolar pneumonocytes. *J. Cell Biol.*, **59**: 84a

Dowell, W. C. T. (1959). Unobstructed mounting of serial sections. *J. Ultrastruct. Res.*, **2**: 388

Drahos, V. and Delong, A. (1960). A simple method for obtaining perforated supporting membranes for electron microscopy. *Nature, Lond.*, **186**: 104

Dreher, K. D., Schulman, J. H., Anderson, O. R. and Roels, O. A. (1967). The stability and structure of mixed lipid monolayers and bilayers. I. Properties of lipid and lipoprotein monolayers on OsO_4 solutions and the role of cholesterol, retinal, and tocopherol in stabilizing lecithin monolayers. *J. Ultrastruct. Res.*, **19**: 586

Dubochet, J., Ducommun, M., Zollinger, M. and Kellenberger, E. (1971). A new preparation method for dark field electron microscopy of biomacromolecules. *J. Ultrastruct. Res.*, **35**: 147

Dubochet, J. and Kellenberger, E. (1972). Selective adsorption of particles to the supporting film and its consequences on particle counts in electron microscopy. *Microsc. Acta*, **72**: 119

Dubochet, J., Lepault, J., Freeman, R., Berriman, J. A. and Homo, J.-C. (1982). Electron microscopy of frozen water and aqueous solutions. *J. Microsc.*, **128**: 219

Dubochet, J., Adrian, M., Chang, J.-J., Homo, J.-C., Lepault, J., McDowall, A. W. and Schultz, P. (1988). Cryoelectron microscopy of vitrified specimens. *Quart. Rev. Biophys.*, **21**: 129

Dubochet, J. and McDowall, A. W. (1983). Frozen hydrated sections. In: *Science of Biological Specimen Preparation* (J.-P. Revel *et al.*, Eds.), pp. 147–152. SEM Inc., AMF O'Hare, Ill.

Dudek, R. W. and Boyne, A. F. (1986). An excursion through the ultrastructural world of quick-frozen pancreatic islets. *Am. J. Anat.*, **175**: 217

Dudek, R. W., Boyne, A. F. and Charles, T. M. (1984). Novel secretory granule morphology in physically fixed pancreatic islets. *J. Histochem. Cytochem.*, **32**: 929

Duncan, G. H. and Roberts, I. M. (1981). Extraction of virus particles from small amounts of material for electron microscope serology. *Micron*, **12**: 171

Dunn, R. F. (1972). Graphic three-dimensional representations from serial sections. *J. Ultrastruct. Res.*, **33**: 116

Echlin, P., Skaer, H. le B., Gardiner, B. O. C., Franks, F. and Asquith, M. H. (1977). Polymeric cryoprotectants in the preservation of biological ultrastructure. II. Physiological effects. *J. Microsc.*, **110**: 239

Eddy, A. A. and Johns, P. (1965). Preliminary observations on the proteins of the erythrocyte membranes with special reference to their cytological behavior. *Soc. Chem. Ind. Monograph*, **19**: 24

Edie, J. W. and Karlsson, U. L. (1972). A routine method for object thickness determination in the transmission electron microscope. *J. Microscopie*, **13**: 13

Edwards, C. A., Walker, G. K. and Avery, J. K. (1984). A technique for achieving consistent release of Formvar film from clean glass slides. *J. Electron Microsc. Tech.*, **1**: 203

Eisenstein, R., Sorgentn, N. and Kuettner, K. E. (1971). Organization of extracellular matrix in epiphyseal growth plate. *Am. J. Path.*, **65**: 515

Elbers, P. F. (1966). Ion permeability of the egg *Limnaea stagnalis* L. On fixation for electron microscopy. *Biochim. Biophys. Acta*, **112**: 318

Elbers, P. F., Veruergaert, P. H. J. T. and Demel, R. (1965). Tricomplex fixation for phospholipids. *J. Cell Biol.*, **24**: 23

Elgjo, R. F. (1976). Platelets, endothelial cells, and macrophages in the spleen. An ultrastructural study on perfusion-fixed organs. *Am. J. Anat.*, **145**: 101

Elias, P. M., Park, H. D., Patterson, A. E., Lutzner, M. A. and Wetzel, B. K. (1972). Osmium tetroxide–zinc iodide staining of Golgi elements and surface coats of hydras. *J. Ultrastruct. Res.*, **40**: 87

Elleder, M. and Lojda, Z. (1968a). Remarks on the detection of osmium derivatives in tissue sections. *Histochemie*, **13**: 276

Elleder, M. and Lojda, Z. (1986b). Remarks on the 'OTAN' reaction. *Histochemie*, **14**: 47

Ellis, E. A. (1986). Araldites, low viscosity epoxy resins and mixed resin embedding: formulations and uses. *Bull. EMSA*, **16**: 53

Ellis, E. A. and Anthony, D. W. (1979). A method for removing precipitate from ultrathin sections resulting from glutaraldehyde–osmium tetroxide fixation. *Stain Technol.*, **54**: 282

Ellis, L. F., Van Frank, R. M. and Kleinschmidt, W. J. (1969). Uranyl acetate and rotary shadowing to increase the contrast of small aggregated virus particles. In: *Proc. 27th Ann. EMSA Meet.*, p. 424. Claitor's Pub. Division, Baton Rouge, La.

Elsner, P. R. (1971). A simple, reliable method for preparing perforated Formvar films. *Proc. 29th Ann. EMSA Meet.*, p. 460. Claitor's Pub. Division, Baton Rouge, La.

Emerman, M. and Behrman, E. J. (1982). Cleavage and cross-linking of proteins with osmium(VII) reagents. *J. Histochem. Cytochem.*, **30**: 395

Epstein, M. A. (1959). Observations on the fine structure of Type 5 adenovirus. *J. Biophys. Biochem. Cytol.*, **6**: 523

Eränko, O. (1954). Quenching of tissue for freeze-drying. *Acta Anat.* **22**: 331

Erdos, G. W. (1987). Simplified flat embedding with Lowicryl K4M and LR White. *J. Electron Microsc. Tech.*, **5**: 111

Ericsson, J. L. E., Saladino, A. J. and Trump, B. F. (1965). Electron microscopic observations on the influence of different fixatives on the appearance of cellular ultrastructure. *Z. Zellforsch. Mikroskop. Anat.*, **66**: 161

Erlandson, R. A. (1964). A new Maraglas, DER 732, embedment for electron microscopy. *J. Cell Biol.*, **22**: 704

Eskelinen, S. and Saukko, P. (1982). The use of slowly increasing glutaraldehyde concentrations preserves the shape of erythrocytes under the influence of an osmotic pressure gradient or detergents. *J. Ultrastruct. Res.*, **81**: 403

Essner, E. (1974). Hemoproteins. In: *Electron Microscopy of Enzymes: Principles and Methods*, Vol. 2 (M. A. Hayat, Ed.), p. 1. Van Nostrand Reinhold, New York

Estes, L. W. and Apicella, J. V. (1969). A rapid embedding technique for electron microscopy. *Lab. Invest.*, **20**: 159

Etherton, J. E. and Botham, C. M. (1970). Factors affecting lead capture methods for the fine localization of rat lung acid phosphatase. *Histochem J.*, **2**: 507

Evans, L. V. and Holligan, M. S. (1972). Correlated light and electron microscopic studies on brown algae. II. Physical production in Dictyota. *New Phytol.*, **71**: 1173

Eyring, E. J. and Ofengand, J. (1967). Reaction of formaldehyde with heterocyclic imino nitrogen or purine and pyrimidine nucleosides. *Biochemistry*, **6**: 2500

Fahimi, H. D. and Drochmans, P. (1965). Essais de standardisation de la fixation au glutaraldehyde. I. Purification et determination de la concentration du glutaraldehyde. *J. Microscopie*, **4**: 725

Fahmy, A. (1967). An extemporaneous lead citrate for electron microscopy. *Proc. 25th Ann. EMSA Meet.*, p. 148. Claitor's Pub. Division, Baton Rouge, La.

Fahrenbach, W. H. (1984). Continuous serial thin sectioning for electron microscopy. *J. Electron Microsc. Tech.*, **1**: 387

Fakan, S. and Hernandez-Verdun, D. (1986). The nucleolus and the nucleolar organizer regions. *Biol. Cell*, **56**: 189

Falk, H. (1969). Rough thykaloids: polysomes attached to chloroplast membranes. *J. Cell Biol.*, **42**: 582

Fanning, J. C. and Cleary, E. G. (1985). Identification of glycoproteins associated with elastin-associated microfibrils. *J. Histochem. Cytochem.*, **33**: 287

Faraday, M. (1857). Experimental relations of gold (and other metals) to light. *Phil. Trans. Roy. Soc.*, **147**: 145

Faulk, W. P. and Taylor, G. M. (1971). An immunocolloid method for the electron microscope. *Immunocytochemistry*, **8**: 1081

Favard, P. and Carasso, N. (1973). The preparation and observation of thick biological sections in the high voltage electron microscope. *J. Microsc.*, **97**: 59

Favard, P., Ovtracht, L. and Carasso, N. (1971). Observation de specimens biologiques en microscopie électonique à haute tension. I. Coupes épaisses. *J. Microscopie*, **12**: 301

Feldman, I., Rich, K. E. and Agarwal, R. P. (1970). Uranyl ion–adenin nucleotide complexes. In: *Effects of Metals on Cells, Subcellular Elements and Macromolecules* (J. Maniloff, J. R. Coleman, and M. W. Miller, Eds.). Charles C. Thomas, Springfield, Ill.

Fernández-Gómez, M. E., Stockert, J. C., López-Sáez, J. F. and Giménez-Martín, G. (1969). Staining plant cell nucleoli with $AgNO_3$ after formalin–hydroquinone fixation. *Stain Technol.*, **44**: 48

Fernández-Morán, H. (1953). A diamond knife for ultrathin sectioning. *Exptl Cell Res.*, **5**: 255

Fernández-Morán, H. (1960). Low temperature preparation techniques for electron microscopy of biological specimens based on rapid freezing with liquid halium II. *Ann. N.Y. Acad. Sci.*, **85**: 689

Fernández-Morán, H. (1986). Megavolt and cryoelectron microscopy of diamond knife edges. *Ultramicroscopy*, **20**: 317

Fetter, R. D. and Costello, M. J. (1986). A procedure for obtaining complementary replicas of ultrarapidly frozen sandwiched samples. *J. Microsc.*, **141**: 277

Field, A. M. (1982). Diagnostic virology using electron microscopic techniques. In: *Adv. Virus Res.*, **27**: 1–69

Fieser, L. F. and Fieser, M. (1956). *Organic Chemistry*, p. 207. Heath, Boston, Mass.

Filshie, B. K. and Rodgers, G. E. (1961). The fine structure of α-keratin. *J. Mol. Biol.*, **3**: 784

Fineran, B. A. (1971). A device for holding LKB-ultratome chunks during preliminary block trimming. *J. Microsc.*, **94**: 83

Finlay-Jones, J.-M. and Papadimitriou, J. M. (1972). Demonstration of pulmonary surfactant by tracheal injection of tricomplex salt mixture: electron microscopy. *Stain Technol.*, **47**: 59

Fleischer, S., Fleischer, B. and Stoeckenius, W. (1967). Fine structure of lipid depleted mitochondria. *J. Cell Biol.*, **32**: 193

Forbes, M. S. (1986). Dog hairs as section manipulators. *EMSA Bull.*, **16**: 67

Forbes, M. S., Hawkey, L. A., Jirge, S. K. and Sperelakis, N. (1985). The sarcoplasmic reticulum of mouse heart: its divisions, configurations, and distribution. *J. Ultrastruct. Res.*, **93**: 1

Forbes, M. S., Plantholt, B. A. and Sperekalis, N. (1977). Cytochemical staining procedures selective for sarcotubular systems of muscle: applications and modifications. *J. Ultrastruct. Res.*, **60**: 306

Forbes, M. S. and Sperelakis, N. (1979). Ruthenium red staining of skeletal and cardiac muscles. *Cell Tiss. Res.*, **200**: 367

Ford, D. J. (1978). The reaction of glutaraldehyde with protein. Ph.D. thesis, University of Cincinnati. Univ. Microfilm International, Ann Arbor, Mich.

Formanek, H. (1979). Preparation of hydrophilic, single crystalline specimen supports of graphite oxide. *Ultramicroscopy*, **4**: 227

Forsgreen, A. and Sjöquist, J. (1966). Protein A from *S. aureus*: pseudoimmune reaction with human γ-globulins. *J. Immunol.*, **97**: 822

Forssmann, W. G., Siegrist, G., Orci, L., Girardier, L., Picket, R. and Rouiller, C. (1967). Fixation par perfusion pour le microscope électronique essai de generalisation. *J. Microscopie*, **6**: 279

Fox, C. H., Johnson, F. B., Whiting, J. and Roller, P. P. (1985). Formaldehyde fixation. *J. Histochem. Cytochem.*, **33**: 845

Frank, J. S., Beydler, S. and Mottino, G. (1987). Membrane structure in ultrarapidly frozen, unrepeated, freeze-fractured myocardium. *Circ. Res.*, **61**: 141

Franke, W. W., Krien, S. and Brown, J. R. M. (1969). Simultaneous glutaraldehyde–osmium tetroxide fixation with postosmification. *Histochemie*, **19**: 162

Franklin, R. M. and Martin, M.-T. (1980). Staining and histochemistry of undecalcified bone embedded in a water-miscible plastic. *Stain Technol.*, **55**: 313

Franklin, R. M., Martin, M.-T. and Longato, R. (1981). A simple procedure for extracting methacrylic acid from water-miscible methacrylates. *Stain Technol.*, **56**: 283

Franks, F. (1977). Biological freezeing and cryofixation. *J. Microsc.*, **111**: 3

Franks, F. (1985). *Biophysics and Biochemistry at Low Temperatures.* Cambridge University Press, Cambridge

Franks, F., Asquith, M. H., Hammond, C. C., Skaer, H. le B. and Echlin, P. (1977). Polymeric cryoprotectants in the preservation of biological ultrastructure. I. Low temperature states of aqueous solutions of hydrophilic polymers. *J. Microsc.*, **110**: 223

Franks, F. and Reid, D. S. (1973). Thermodynamic properties. In *Water—A Comprehensive Treatise*, Vol. 2 (F. Franks, Ed.), p. 232. Plenum Press, New York

Frasca, J. M. and Parks, V. R. (1965). A routine technique for double-staining ultrathin sections using uranyl and lead salts. *J. Cell Biol.*, **25**: 157

Fraser, T. W. (1976). The anti-static pistol as an aid to ultrathin sectioning. *J. Microsc.*, **106**: 97

Frater, R. (1981). Neutralization of acid in glycol methacrylate and the use of cyclohexanol as a plasticizer. *Stain Technol.*, **56**: 99

Frauenfelder, H., Petsko, G. A. and Tsernoglou, D. (1979). Temperature-dependent X-ray diffraction as a probe of protein structural dynamics. *Nature, Lod.*, **280**: 558

Frederick, P. M. and Busing, W. M. (1981). Strong evidence against section thawing whilst cutting on the cryo-ultratome. *J. Microsc.*, **122**: 217

Frederick, P. M. and Busing, W. M. (1986). Cryo-transfer revised. *J. Microsc.*, **144**: 215

Frederick, P. M., Busing, W. M. and Persson, A. (1982). Concerning the nature of the cryosectioning process. *J. Microsc.*, **125**: 167

Fregerslev, S., Blackstad, T. W., Fredens, K. and Holm, M. J. (1971). Golgi potassium dichromate–silver nitrate impregnation. Nature of the precipitate studied by X-ray powder diffraction methods. *Histochemie*, **25**: 63

Frens, G. (1973). Controlled nucleation for the regulation of particle size in monodisperse gold solutions. *Nature, Lond.*, **241**: 20

Friedman, M. M. (1985). A simple Plexiglas holder that secures epoxy specimen blocks for precise trimming in a dissecting microscope. *J. Electron Microsc. Tech.*, **2**: 281

Friend, D. S. and Brassil, G. E. (1970). Osmium staining of endoplasmic reticulum and mitochondria in the rat adrenal cortex. *J. Cell Biol.*, **46**: 252

Friend, D. S. and Murray, M. J. (1965). Osmium impregnation of the Golgi apparatus. *Am. J. Anat.*, **117**: 135

Frigero, N. A. and Nebel, B. R. (1962). The quantitative determination of osmium tetroxide in fixatives. *Stain Technol.*, **37**: 347

Frösch, D. and Westphal, C. (1985). Choosing the appropriate section thickness in the melamine embedding technique. *J. Microsc.*, **137**: 177

Frösch, D., Westphal, C. and Bachhuber, K. (1985). A determination of the thickness and surface relief in reembedded sections of an epoxy- and melamine-resin containing ferritin as size standard. *Ultramicroscopy*, **17**: 141

Frösch, D., Westphal, C. and Böhme, H. (1987). Improved preservation of glycogen in unfixed cyanobacteria embedded at −82 °C in Nanoplast. *J. Histochem. Cytochem.*, **35**: 119

Fryer, P. R., Wells, C. and Ratcliffe, A. (1983). Technical difficulties overcome in the use of Lowicryl K4MEM embedding resin. *Histochemistry*, **77**: 141

Fujikawa, S. (1983). Tannic acid improves the visualization of the human erythrocyte membrane skeleton by freeze-etching. *J. Ultrastruct. Res.*, **84**: 289

Fujikawa, S. (1988). Artificial biological membrane ultrastructural changes caused by freezing. *Electron Microsc. Rev.*, **1**: 113

Fujita, K., Yamada, K., Sato, T., Takagi, I. and Sakashita, T. (1977). A rapid method for embedding biological tissues in a low viscosity epoxy resin, Quetol 651, for electron microscopy. *J. Electron Microsc.*, **26**: 165

Fukami, A. and Adachi, K. (1964). On an adhering method of thin film specimens to specimen grids. *J. Electron Microsc.*, **13**: 26

Fulcher, R. G. and McCully, M. E. (1971). Histological studies on the genus *Fucus*. V. An autoradiographic and electron microscopic study of the early stages of regeneration. *Can. J. Bot.*, **49**: 161

Fullagar, K. (1966). The role of the LKB Knifemaker in ultramicrotomy. *Sci. Tools*, **13**: 39

Furness, J. B., Costa, M. and Blessing, W. W. (1977). Simultaneous fixation and production of catecholamine fluorescence in central nervous tissue by perfusion with aldehydes. *Histochem. J.*, **9**: 745

Furness, J. B., Heath, J. W. and Costa, M. (1978). Aqueous aldehyde (Faglu) methods for the fluorescence histochemical localization of catecholamines and for ultrastructural studies of central nervous tissue. *Histochemistry*, **57**: 285

Furui, S. (1986). Use of protein A in the serum-in-agar diffusion method in immune electron microscopy for detection of virus particles in cell culture. *Microbiol. Immunol.*, **30**: 1023

Fuscaldo, K. E. and Jones, H. H. (1959). A method for the reconstruction of three-dimensional models from electron micrographs of serial sections. *J. Ultrastruct. Res.*, **3**: 1

Futaesaku, Y., Mizuhira, V. and Nakamura, H. (1972). A new fixation method using tannic acid for electron microscopy and some observation of biological specimens. *Proc. Int. Cong. Histochem. Cytochem.*, **4**: 155

Galas, A. M. R., Hursthouse, M. B., Behrman, E. J., Midden, W. R., Green, G., and Griffith, W. P. (1981). The X-ray crystal structures of the oxo-osmium complexes, $OsO_2(OH)_2$ phen and OsO_2py_4. *Trans. Met. Chem.*, **6**: 194

Galey, F. R. and Nilsson, S. E. G. (1966). A new method for transferring sections from the liquid surface of the trough through staining solutions to the supporting film of a grid. *J. Ultrastruct. Res.*, **14**: 405

Garcia-Patrone, M. and Algranati, I. D. (1976). Artifacts induced by glutaraldeyde fixation of ribosomal particle. *Mol. Biol. Rep.*, **2**: 507

Gardner, H. A. and Silver, M. D. (1971). The chemical basis of the argentaffin reaction applied to glutaraldehyde fixed tissue. *Histochemie*, **28**: 95

Garner, G. E. and Steever, R. G. E. (1970). Treatment of diamond knives with Aerosol OT for uniform wetting of their edges in ultramicrotomy. *Stain Technol.*, **45**: 186

Garrett, J. R. (1965). Electron microscopial evaluation of the osmic acid-sodium iodide nerve staining technique on salivary tissue. *Acta. Anat.*, **62**: 325

Garrett, J. R., Davies, K. J. and Parsons, P. A. (1972). Consumers' guide to glutaraldehyde. *Proc. Roy. Microsc. Soc.*, **7**: 116

Gas, N., Inchauspe, G., Azum, M. C. and Stevens, B. (1984). Bismuth staining of a nucleolar protein. *Exp. Cell Res.*, **151**: 447

Gasic, G., Berwick, L. and Sorrentino, M. (1968). Positive and negative colloidal iron as cell surface electron stains. *Lab. Invest.*, **18**: 63

Gay, H. and Anderson, T. F. (1954). Serial sections for electron microscopy. *Science, N.Y.*, **120**: 1071

Geoghegan, W. D. and Ackerman, G. A. (1977). Adsorption of horseradish peroxidase, ovomucoid and anti-immunoglobulin to colloidal gold for the indirect detection of concanavalin A, wheat germ agglutinin and goat anti-human immunoglobulin G on cell surfaces at the electron microscopic level: a new method, theory, and application. *J. Histochem. Cytochem.*, **25**: 1187

Geoghegan, W. D., Scillian, J. J. and Ackerman, G. A. (1978). The detection of human B lymphocytes by both light and electron microscopy utilizing colloidal gold labelled anti-immunoglobulin. *Immunol. Commun.*, **7**: 1

Gerrits, P. O. (1987). *Fundamental Aspects of Tissue Processing When Applying Glycol Methacrylate.* Doctorate Dissertation, Rijksuniversiteit, Groningen

Gerrits, P. O. and van Leeuwen, M. B. M. (1985). Glycol methacrylate embedding in histotechnology: factors which influence the evolution of heat during polymerization at room temperature. *J. Microsc.*, **139**: 303

Gerrits, P. O. and van Leeuwen, M. B. M. (1987). Glycol methacrylate embedding in histotechnology: the hematoxylin–eosin stain as a method for assessing the stability of glycol methacrylate sections. *Stain Technol.*, **62**: 181

Gerrits, P. O., van Leeuwen, M. B. M., Boon, M. E. and Kok, L. P. (1987). Floating on a water bath and mounting glycol methacrylate and hydroxypropyl methacrylate sections influence final dimensions. *J. Microsc.*, **145**: 107

Gerrits, P. O. and Smid, L. (1983). A new, less toxic polymerization system for the embedding of soft tissues in glycol methacrylate and subsequent preparing of serial sections. *J. Microsc.*, **132**: 81

Gersch, I. (1932). The Altmann technique for fixation by drying while freezing. *Anat. Rec.*, **53**: 309

Geuze, H. J., Slot, J. W., van der Ley, P. A. and Scheffer, R. C. T. (1981). Use of colloidal gold particles in double-labelling immunoelectron microscopy of ultrathin frozen tissue sections. *J. Cell Biol.*, **89**: 653

Geyer, G., Helmke, U. and Christner, A. (1971). Ultrahistochemical demonstration of alcian blue stained mucosubstances by the sulfide–silver reaction. *Acta Histochem.*, **40**: 80

Geyer, G., Linss, W. and Richter, H. (1972a). A PAT–Protargol method. *Acta Histochem.*, **42**: 185

Geyer, G., Linss, W. and Schaaf, P. (1972b). The distribution pattern of anionic sites at the human erythrocyte surface as revealed by the colloidal iron method. *Acta Histochem.*, **2**: 112

Geyer, G., Linss, W. and Stibenz, D. (1973). Ultrahistochemical demonstration of positively charged groups at the surface of human red blood cells. *Acta Histochem.*, **46**: 244

Ghosh, B. K. (1971). Grooves in the plasmalemma of *Saccharomyces cerevisiae* seen in glancing sections of double aldehyde-fixed cells. *J. Cell Biol.*, **48**: 192

Giammara, B. L. (1981). The Grid-All: a multiple grid handling, staining, and storage device for use in electron microscopy. *Proc. 39th Ann. EMSA Meet.*, p. 552. Claitor's Pub. Division, Baton Rouge, La.

Gicquand, C., Turcotte, A. and St-Pierre, S. (1983). Peptides from *Amanita virosa*: viroidin and viroisin are more effective than phalloidin in protecting actin in vitro against osmic acid. *Eur. J. Cell Biol.*, **32**: 171

Gielink, A. J., Sauer, G. and Ringoet, A. (1966). Histoautoradiographic localization of calcium in oat plant tissue. *Stain Technol.*, **41**: 281

Gil, J. (1972). Effect of tricomplex fixation on lung tissue. *J. Ultrastruct. Res.*, **40**: 122

Gil, J. and Weibel, E. R. (1968). The role of buffers in lung fixation with glutaraldehyde and osmium tetroxide. *J. Ultrastruct. Res.*, **25**: 331

Gil, J. and Weibel, E. R. (1969/70). Improvements in demonstration of lining layer of lung alveoli by electron microscopy. *Respir. Physiol.*, **8**: 13

Gilëv, V. P. and Melnikova, E. J. (1968). Study of tissue ultrastructure using a method of cutting of unembedded specimens. In: *Proc. 4th Eur. Reg. Conf. Electron Microsc.*, p. 31. Tipografia Poliglotta Vaticana, Rome

Gilkey, J. C. and Staehelin, L. A. (1986). Advances in ultrarapid freezing for the preservation of cellular ultrastructure. *J. Electron Microsc. Tech.*, **3**: 177

Gillett, R. and Gull, K. (1972). Glutaraldehyde—its purity and stability. *Histochemie*, **30**: 162

Gillett, R., Jones, G. E. and Partridge, T. (1975). Distilled glutaraldehyde: its use in an improved fixation regime for cell suspensions. *J. Microsc.*, **105**: 325

Gillis, J. M. and Wibo, M. (1971). Accurate measurement of the thickness of ultrathin section by interference microscopy. *J. Cell Biol.*, **49**: 947

Ginsburg, H. and Wolosin, J. M. (1979). Effects of uranyl ions of lipid bilayer membranes. *Chem. Phys. Lipids*, **23**: 125

Glauert, A. M. (1974). Fixation, dehydration and embedding of biological specimens. In: *Practical Methods in Electron Microscopy* (A. M. Glauert, Ed.). North-Holland, Amsterdam

Glauert, A. M. (1987). Section thickness terminology: a source of confusion in microscopy. *J. Microsc.*, **145**: 241

Glauert, A. M. and Glauert, R. H. (1958). Araldite as an embedding medium for electron microscopy. *J. Biophys. Biochem. Cytol.*, **4**: 191

Glauert, A. M. and Mayo, C. R. (1973). The study of the three-dimensional structural relationships in connective tissue by high voltage electron microscopy. *J. Microsc.*, **97**: 83

Glick, D. and Malmstrom, B. G. (1952). Simple and efficient freeze-drying apparatus for preparation of embedded tissue. *Exp. Cell Res.*, **3**: 125

Glick, D. and Scott, J. E. (1970). Phosphotungstic acid not a stain for polysaccharide. *J. Histochem. Cytochem.*, **18**: 455

Godkin, S. E. and Keith, C. T. (1975). Ultramicrotome chucks for flat castings. *Stain Technol.*, **50**: 63

Goff, C. W. and Oster, M. O. (1974). Formation of 235-nanometer absorbing substance during glutaraldehyde fixation. *J. Histochem. Cytochem.*, **22**: 913

Goldstein, D. J. and Horobin, R. W. (1970). Rate and equilibrium factors in staining by alcian blue. *Proc. Roy. Microsc. Soc.*, **6**: 24

Goldstein, D. J. and Horobin, R. W. (1974). Rate factors in staining by alcian blue. *Histochem J.*, **6**: 157

Gomez-Dumm, C. L. and Echave Llanos, J. M. (1970). Variations in the Golgi complex of mouse somatotrops at different times in a 24-hour period. *Experientia*, **26**: 177

Good, N. E., Winget, G. D., Winter, W., Connolly, T. N., Izawa, S. and Singh, R. M. M. (1966). Hydrogen ion buffers for biological research. *Biochemistry*, **5**: 467

Goodenough, D. A. and Revel, J.-P. (1971). The permeability of isolated and *in situ* mouse hepatic gap junctions studied with enzymatic tracers. *J. Cell Biol.*, **50**: 81

Goodfellow, A. and Calvert, H. (1979). A case of focal epithelial hyperplasia of the oral mucosa from the UK. *Br. J. Dermatol.*, **101**: 341

Goodman, S. L., Hodges, G. M. and Livinston, D. C. (1980). A review of the colloidal gold marker system. *Scanning Electron Microsc.*, **2**: 133

Goodman, S. L., Hodges, G. M., Trejdosiewicz, L. K. and Livinston, D. C. (1979). Colloidal gold probes. A further evaluation. *Scanning Electron Microscopy*, **3**: 619.

Goodman, S. L., Hodges, G. M., Trejdosiewicz, L. K. and Livingston, D. C. (1981). Colloidal gold markers and probes for routine application in microscopy. *J. Microsc.*, **123**: 201

Goodpasture, C. and Bloom, S. E. (1975). Visualization of nucleolar organizer regions in mammalian chromosomes using silver staining. *Chromosoma*, **53**: 37

Gorbsky, G. and Borisy, G. (1985). Reversible embedment cytochemistry. *Proc. Natl Acad. Sci. USA*, **82**: 6889

Gorbsky, G. and Borisy, G. (1986). Reversible embedment cytochemistry (REC): a versatile method for the ultrastructural analysis and affinity labelling of tissue sections. *J. Histochem. Cytochem.*, **34**: 177

Gordon, C. N. and Kleinschmidt, A. K. (1968). High contrast staining of individual nucleic acid molecules. *Biochim. Biophys. Acta*, **155**: 305

Gordon, M. and Bensch, K. G. (1968). Cytochemical differentiation of the guinea sperm flagellum with phosphotungstic acid. *J. Ultrastruct. Res.*, **24**: 33

Gorycki, M. A. (1978). Methods for precisely trimming block faces for ultramicrotomy. *Stain Technol.*, **53**: 63

Gorycki, M. A. and Oberc, M. A. (1978). Cleaning diamond knives before or during sectioning. *Stain Technol.*, **53**: 51

Gorycki, M. A. and Sohm, E. K. (1979). Modifying a Porter-Blum MT-2 ultramicrotome knife holder to accept Ralph knives. *Stain Technol.*, **54**: 293

Goudswaard, J., Van der Donk, J. A., Noordzij, A., Van Dam, R. H. and Vaerman, J. P. (1978). Protein A reactivity of various mammalian immunoglobulins. *Scand. J. Immunol.*, **8**: 21

Gray, E. G. and Willis, R. A. (1968). Problems of electron microscopy of biological tissues. *J. Cell Sci.*, **3**: 309

Gregory, D. W. and Pirie, J. S. (1973). Wetting agents for biological electron microscopy. I. General considerations and negative staining. *J. Microsc.*, **99**: 261

Griffin, J. L. (1963). Motion picture analysis of fixation for electron microscopy. *Amoeba proteus. J. Cell Biol.*, **19**: 77A

Griffith, D. L. and Bondareff, J. F. (1972). Localization of thiamine pyrophosphatase in synaptic vesicles. *J. Cell Biol.*, **55**: 97A

Griffith, J. D. and Bonner, J. F. (1973). Chromatin-like aggregates of uranyl acetate. *Nature, New Biol.*, **244**: 80

Griffith, W. P. (1965). Osmium and its compounds. *Quart. Rev. Chem. Soc.*, **19**: 254

Griffiths, G., Brands, R., Burke, B., Louvard, D. and Warren, G. (1982). Viral membrane proteins acquire galactose trans Golgi cisternae during intracellular transport. *J. Cell Biol.*, **95**: 781

Griffiths, G., McDowall, A., Back, R. and Dubochet, J. (1984). On the preparation of cryosections for immunocytochemistry. *J. Ultrastruct. Res.*, **89**: 67

Griffiths, G., Simons, K., Warren, G. and Tokuyasu, K. T. (1983). Immunoelectron microscopy using thin, frozen sections: application to studies of the intracellular transport of Semiliki Forest virus spike glycoproteins. In: *Methods in Enzymology*, Vol. 96 (S. Fleischer and B. Fleischer, Eds.), pp. 466–495. Academic Press, New York

Grimelius, L. (1969). An electron microscopic study of silver stained adult human pancreatic islet cells, with reference to new silver nitrate procedure. *Acta Soc. Med. Upsala*, **74**: 28

Grimley, P. M. (1967). Precision ultramicrotomy of narrow specimen embedments for electron microscopy—theory and practice. *J. Roy. Microsc. Soc.*, **87**: 383

Gruber, H. E., Marshall, G. J., Kirchen, M. E., Kang, J. and Massry, S. G. (1985). Improvements in dehydration and cement line staining for methacrylate embedded human bone biopsies. *Stain Technol.*, **60**: 337

Grund, S., Eichberg, J. and Asmussen, F. (1982). A specific embedding resin (PVK) for fine cytological investigations in the photoemission electron microscope. *J. Ultrastruct. Res.*, **80**: 89

Gugleilmotti, V. (1976). Device for manual trimming of tissue blocks for ultramicrotomy. *Stain Technol.*, **51**: 135

Gulik-Krzywicki, T. and Costello, M. T. (1978). The use of low temperature X-ray diffraction to evaluate freezing methods used in freeze-fracture electron microscopy. *J. Microsc.*, **112**: 103

Gunning, B. E. S. and Hardham, A. R. (1977). Estimation of the average section thickness in ribbons of ultrathin sections. *J. Microsc.*, **109**: 337

Gupta, B. L., Hall, T. A., Maddrel, S. H. P. and Moreton, R. B. (1976). Distribution of ions in a fluid transporting epithelium determined by electron probe X-ray microanalysis. *Nature, Lond.*, **264**: 284

Gustafson, G. T. and Pihl, E. (1967). Staining of mast cell acid glycosamino-glycans in ultrathin sections by ruthenium red. *Nature, Lond.*, **216**: 697

Guth, L. and Watson, P. K. (1968). A correlated histochemical and quantitative study on cerebral glycogen after brain injury in the rat. *Exp. Neurol.*, **22**: 590

Haapala, O. K. and Nygrén, T. (1973). Localization of histones and their synthesis by ammoniacal silver reaction in meristematic root tip cells of *Allium cepa*. *Histochemie*, **34**: 257

Habeeb, A. F. S. A. and Hiramoto, R. (1968). Reaction of proteins with glutaraldehyde. *Arch. Biochem. Biophys.*, **126**: 16

Hackenbrock, C. R. (1966). Ultrastructural bases for metabolically linked mechanical activity in mitochondria. I. Reversible ultrastructural changes with change in metabolic steady state in isolated liver mitochondria. *J. Cell Biol.*, **30**: 269

Hacker, C. S., Springall, D. R., Van Noorden, S., Bishop, A. E., Grimelius, L. and Polak, J. M. (1985). The immunogold–silver staining method. A powerful tool in histopathology. *Virchows Arch.* (*Pathol. Anat.*) **406**: 449

Hacker, G. W. (1989). Silver-enhanced colloidal gold for light microscopy. In: *Colloidal Gold: Principles, Methods, and Applications*, Vol. 1 (M. A. Hayat, Ed.). Academic Press, San Diego and London

Haggis, G. H. (1986). Study of the conditions necessary for propane-jet freezing of fresh biological tissues without detectable ice formation. *J. Microsc.*, **143**: 275

Hagler, H. K., Burton, K. P., Willerson, J. T. and Buja, L. M. (1984). New techniques for the elemental analysis of the myocardium: application to the study of ischemia. *Ann. N.Y. Acad. Sci.*, **428**: 68

Hagler, H. K. and Buja, L. M. (1986). Effect of specimen preparation and section transfer techniques on the preservation of ultrastructure, lipids and elements in cryosections. *J. Microsc.*, **141**: 311

Hakanson, R., Owman, Ch., Sporrong, B. and Sundler, F. (1971). Electron microscopic classification of amine producing endocrine cells by selective staining of ultrathin sections. *Histochemie*, **27**: 226

Hake, T. (1965). Studies on the reaction of OsO_4 and $KMnO_4$ with amino acids, peptides, and proteins. *Lab. Invest.*, **14**: 470

Halgunset, K. E. Tvedt, Kopstad, G. and Haugen, O. A. (1984). Osmotic effects of polyvinyl pyrrolidone (PVP) used as a cryoprotectant for energy dispersive X-ray microanalysis (EDX). *J. Ultrastruct. Res.*, **88**: 289

Hall, C. E. (1955). Electron densitometry of stained virus particles. *J. Biophys. Biochem. Cytol.*, **1**: 1

Hall, C. E., Jakus, M. A. and Schmitt, F. O. (1945). The structure of certain muscle fibrils as revealed by the use of electron stains. *J. Appl. Phys.*, **16**: 459

Hampton, J. C. (1965). Effects of fixation on the morphology of Paneth cell granules. *Stain Technol.*, **40**: 283

Hanaichi, T., Sato, T., Iwamoto, T., Malavasi-Yamashiro, J., Hoshino, M. and Mizuno, N. (1986). A stable lead by modification of Sato's method. *J. Electron Microsc.*, **35**: 304

Handley, D. A., Alexander, J. T. and Chion, S. (1981b). The design and use of a simple device for rapid quench freezing of biological samples. *J. Microsc.*, **121**: 273

Handley, D. A., Arbeeny, C. M., Witte, L. D. and Eder, H. A. (1981b). Colloidal gold-low density lipoprotein conjugates as membrane receptor probes. *Proc. Natl Acad. Sci. USA*, **78**: 368

Handley, D. A. and Olsen, B. R. (1979). Butvar B-98 as a thin support film. *Ultramicroscopy*, **4**: 479

Hanker, J. S., Deb, C. and Seligman, A. M. (1966). Staining tissue for light and electron microscopy by bridging metals with multidentate ligands. *Science, N.Y.*, **152**: 1631

Hanker, J. S., Kasler, F., Bloom, M. G., Copeland, J. S. and Seligman, A. M. (1967). Coordination polymers of osmium: the nature of osmium black. *Science, N.Y.*, **156**: 1737

Hanker, J. S., Seaman, A. R., Weiss, L. P., Ueno, H., Bergmann, R. A. and Seligman, A. M. (1964). Osmiophilic reagents: new cytochemical principles for light and electron microscopy. *Science, N.Y.*, **146**: 1039

Hanker, J. S. (1975). Oxidoreductases. In: *Electron Microscopy of Enzymes: Principles and Methods*, Vol. 4 (M. A. Hayat, Ed.). Van Nostrand Reinhold, New York and London

Hanstede, J. G. and Gerrits, P. O. (1983). The effects of embedding in water-soluble plastics on the final dimensions of liver sections. *J. Microsc.*, **131**: 79

Hardy, P. M., Hughes, G. J. and Rydon, H. N. (1976a). Formation of quaternary pyridinium compounds by the action of glutaraldehyde on proteins. *J. Chem. Soc. Chem. Commun.*, 157

Hardy, P. M., Hughes, G. J. and Rydon, H. N. (1977). Identification of a 3-(2-piperidyl)pyridinium derivative ('anabilysine') as a crosslinking entity in a glutaraldehyde-treated protein. *J. Chem. Soc. Chem. Commun.*, 759

Hardy, P. M., Nicholls, A. C. and Rydon, H. N. (1969). The nature of glutaraldehyde in aqueous solution. *Chem. Comm.*, **D-1**: 565

Hardy, P. M., Nicholls, A. C. and Rydon, H. N. (1972). The hydration and polymerization of succinaldehyde, glutaraldehyde and adipaldehyde. *J. Chem. Soc.*, **15**: 2270

Hardy, P. M., Nicholls, A. C. and Rydon, H. N. (1976b). The nature of the cross-linking of proteins by glutaraldehyde. Part I. Interaction of glutaraldehyde with the amino groups of 6-aminohexanoic acid and of a *N*-acetyl-lysine. *J. Chem. Soc. Perkin*, **1**: 958

Hardy, P. M., Hughes, G. J. and Rydon, H. N. (1979). The nature of the cross-linking of proteins by glutaraldehyde. Part 2. The formation of quaternary pyridinium compounds by the action of glutaraldehyde on proteins and the identification of 3-(2-piperidyl)-pyridinium derivatives, anabilysine, as a cross-linking entity. *J. Chem. Soc. Perkin*, **1**: 2282

Harper, I. S. (1986). Glutaraldehyde-induced permeabilization of cell membranes. *J. Microsc.*, **141**: RP3

Harris, J. R., Price, M. R. and Willison, M. (1974). A comparative study on rat liver and hepatoma nuclear membranes. *J. Ultrastruct. Res.*, **48**: 17

Harris, P. (1962). Some structural and functional aspects of the mitotic apparatus in sea urchin embryos. *J. Cell Biol.*, **14**: 475

Hart, R. K., Kassner, T. F. and Maurin, J. K. (1970). The contamination of surfaces during high energy electron irradiation. *Phil. Mag.*, **21**: 453

Hartmann, R. (1984). A new embedding medium for cryosectioning eggs of high yolk and lipid content. *Eur. J. Cell Biol.*, **34**: 206

Harvey, D. M. R. and Kent, B. (1981). Sodium localization in *Suaeda maritima* leaf cells using zinc uranyl acetate precipitation. *J. Microsc.*, **121**: 179

Haselkorn, R. and Doty, P. (1961). The reaction of formaldehyde with polynucleotides. *J. Biol. Chem.*, **236**: 2738

Haskins, R. H. and Nesbitt, C. R. (1975). The use of an ionizing unit for improved ultrathin sectioning. *Can. J. Bot.*, **53**: 423

Hasty, D. L. and Hay, E. D. (1977). Freeze-fracture studies of the developing cell surface. I. The plasmalemma of the corneal fibroplast. *J. Cell Biol.*, **72**: 667

Hasty, D. L. and Hay, E. D. (1978). Freeze-fracture studies of the developing cell surface. II. Particle-free membrane blisters on glutaraldehyde-fixed corneal fibroplasts are artifacts. *J. Cell Biol.*, **78**: 756

Haudenschild, C., Baumgartner, H. R. and Studer, A. (1972). Significance of fixation procedure for preservation of arteries. *Experientia*, **29**: 828

Hauser, H., Howell, K., Dawson, R. M. C. and Bowyer, D. E. (1980). Rabbit small intestinal brush border membrane preparation and lipid composition. *Biochim. Biophys. Acta*, **602**: 567

Hausmann, K. (1977). Artifactual fusion of membranes during preparation of material for electron microscopy. *Naturwissenschaften*, **64**: 95

Haviernick, S., Lalague, E. D., Corvellec, M.-R., Pelletier, L. and Calamba, A. D. (1984). The use of Hank's pipes buffers in the preparation of human normal leukocytes for TEM observation. *J. Microsc.*, **135**: 83

Hayashi, M., Maramatsu, T. and Hara, I. (1972). Surface properties of synthetic phospholipids. *Biochim. Biophys. Acta*, **255**: 98

Hayat, M. A. (1969). Uranyl acetate as a stain and fixative for heart tissue. *Proc. 27th Ann. EMSA Meet.*, p. 412. Claitor's Pub. Division, Baton Rouge, La.

Hayat, M. A. (1973). Specimen preparation. In: *Electron Microscopy of Enzymes: Principles and Methods*, Vol. 1 (M. A. Hayat, Ed.). Van Nostrand Reinhold, New York and London

Hayat, M. A. (Ed.). (1973–1977). *Electron Microscopy of Enzymes: Principles and Methods*, Vols. 1–5. Van Nostrand Reinhold, New York

Hayat, M. A. (1975). *Positive Staining for Electron Microscopy*. Van Nostrand Reinhold, New York

Hayat, M. A. (1978). *Introduction to Biological Scanning Electron Microscopy*. University Park Press, Baltimore

Hayat, M. A. (1981). *Fixation For Electron Microscopy*. Academic Press, San Diego and London

Hayat, M. A. (1986a). Glutaraldehyde: role in electron microscopy. *Micron Microsc. Acta*, **17**: 115

Hayat, M. A. (1986b). *Basic Techniques for Transmission Electron Microscopy*. Academic Press, San Diego and London

Hayat, M. A. (Ed.) (1987). *Correlative Microscopy in Biology: Instrumentation and Methods*. Academic Press, Orlando and London

Hayat, M. A. (1989a). *Cytochemical Methods*. Wiley, New York

Hayat, M. A. (Ed.) (1989b). *Colloidal Gold: Principles, Methods, and Applications*. Academic Press, San Diego and London

Hayat, M. A. and Giaquinta, R. (1970). Vapor fixation prior to fixation by immersion for electron microscopy. *Proc. 7th Int. Cong. Electron Microsc.*, p. 391. Grenoble, Soc. Francaise Microsc. Electronique, Paris

Hayat, M. A. and Miller, S. A. (1989). *Negative Staining: Methods and Applications*. McGraw-Hill, New York

Hayat, M. A. and Zirkin, B. R. (1973). Critical point-drying method. In: *Principles and Techniques of Electron Microscopy: Biological Applications*, Vol. 3 (M. A. Hayat, Ed.). Van Nostrand Reinhold, New York and London.

Haydon, G. B. (1969a). An electron-optical lens effect as possible source of contrast in biological preparations. *J. Microsc.*, **90**: 1

Haydon, G. B. (1969b). Electron phase and amplitude images of stained biological thin sections. *J. Microsc.*, **89**: 73

Hazel, F. and Ayers, G. H. (1931). Migration studies with ferric oxide sols. II. Negative sols. *J. Phys. Chem.*, **35**: 3148

Hearse, D. J., Yellon, D. M., Chappell, D. A., Wyse, R. K. H. and Ball, G. R. (1981). A high velocity impact device for obtaining multiple, contiguous, myocardial biopsies. *J. Mol. Cell Cardiol.*, **13**: 197

Heath, I. B. (1984). A simple and inexpensive liquid helium cooled 'slam freezing' device. *J. Microsc.*, **135**: 75

Heath, I. B., Rethoret, K. and Moens, P. B. (1984). The ultrastructure of mitotic spindles from conventionally fixed and freeze-substituted nuclei of the fungus *Saprolegnia*. *Eur. J. Cell Biol.*, **35**: 284

Heinemann, K. (1970). A comment on mica as electron microscope specimen support film. In: *Proc. 28th Ann. EMSA Meet.*, p. 526. Claitor's Pub. Division, Baton Rouge, La.

Helander, H. F. (1962). On the preservation of gastric mucosa. *Proc. 5th Int. Cong. Electron Microsc. Philadelphia*, Vol. 2, pp. 1–7. Academic Press, New York

Helander, H. F. (1969). Surface topography of ultramicrotome sections. *J. Ultrastruct. Res.*, **29**: 373

Helander, K. G. (1983). Thickness variations within individual paraffin and glycol methacrylate sections. *J. Microsc.*, **132**: 223

Helander, K. G. (1984). The Ralph knife in practice. *J. Microsc.*, **135**: 139

Helander, K. G. (1987). Studies on the rate of dehydration of histological specimens. *J. Microsc.*, **145**: 351

Heller, W. and Pugh, T. L. (1960). 'Steric' stabilization of colloidal solutions by adsorption of flexible macromolecules. *J. Polymer Sci.*, **47**: 203

Hendy, R. (1971). Electron microscopy of lipofucsin pigment stained by the Schmorl and Fontana technique. *Histochemie*, **26**: 311

Henrikson, R. C. (1974). Lanthanum binding by mouse cecal mucosa. *Cell Tiss. Res.*, **148**: 309

Henrikson, R. C. and Stacey, B. D. (1971). The barrier to diffusion across ruminal epithelium: a study by electron microscopy using horseradish peroxidase, lanthanum, and ferritin. *J. Ultrastruct. Res.*, **34**: 72

Hepler, P. K. (1980). Membranes in the mitotic apparatus of barley cells. *J. Cell Biol.*, **86**: 490

Hepler, P. K. (1981). The structure of the endoplasmic reticulum revealed by osmium tetroxide–potassium ferricyanide staining. *Eur. J. Cell Biol.*, **26**: 102

Hernandez, W., Rambourg, A. and Leblond, C. P. (1968). Periodic acid–chromic acid–methenamine silver technique for glycoprotein detection in the electron microscope. *J. Histochem. Cytochem.*, **16**: 507

Hernández-Verdún, D., Hubert, J., Bourgeois, C. A. and Bouteille, M. (1980). Ultrastructural localization of Ag–NOR stained proteins in the nucleolus during the cell cycle and in other nucleolar structures. *Chromosoma*, **79**: 349

Heuser, J. E., Reese, T. S., Dennis, M. J., Jan, Y., Jan, L. and Evans, L. (1979). Synaptic vesicle exocytosis captures by quick freezing and correlated with quantal transmitter release. *J. Cell Biol.*, **81**: 275

Heywood, P., Hodge, L., Davis, F. and Simmons, T. (1977). A simple method for holding electron microscope grids during autoradiography of serial sections. *J. Microsc.*, **110**: 167

Hicks, D. and Molday, R. S. (1985). Localization of lectin receptors on bovine photoreceptor cells using dextran–gold markers. *Invest. Ophthalmol. Vis. Sci.*, **26**: 1002

Highton, P. J. and Beer, M. (1963). An electron microscope study of extended single polynucleotide chains. *J. Mol. Biol.*, **7**: 70

Hillier, J. and Gettner, M. E. (1950). Improved ultrathin sectioning of tissue for electron microscopy. *J. Appl. Phys.*, **21**: 889

Hillman, H. and Deutsch, K. (1978). Area changes in slices of rat brain during preparation for histology or electron microscopy. *J. Microsc.*, **114**: 77

Hills, G. S. and Plaskitt, A. (1968). A protein stain for the electron microscopy of small isometric plant virus particles. *J. Ultrastruct. Res.*, **25**: 323

Hinckley, C. C., Ostenburg, P. S. and Roth, W. J. (1982). Osmium carbohydrate polymers. *Polyhedron*, **1**: 335

Hines, R. L. (1975). Graphite crystal film preparation by cleavage. *J. Microsc.*, **104**: 257

Hinkley, R. E. Jr. (1973). Axonal microtubules and associated filaments stained by alcian blue. *J. Cell Biol.*, **13**: 753

Hippe, S. (1984). Rapid cryofixation by a simple propane-double-jet device adapted to a modified specimen table of the BIOTECH 2005. *Mikroskopie*, **41**: 289

Hippe, S. and Hermanns, M. (1986). Improved structural preservation in freeze-substituted sporidia of *Ustilago avenae*—a comparison with low-temperature embedding. *Protoplasma*, **135**: 19

Hirano, A., Zimmerman, H. M. and Levine, S. (1964). The fine structure of cerebral fluid accumulation. *Am. J. Path.*, **45**: 195

Hiraoka, J.-I. (1972). A holder for mass treatment of grids, adapted especially to electron staining and autoradiography. *Stain Technol.*, **47**: 297

Hirs, C. H. W. (1964). Enzyme structure and function with particular reference to bovine ribonuclease and chymotypsin. In: *Comprehensive Biochemistry*, Vol. 12 (M. Florkin and E. H. Stotz, Eds.), pp. 261–279. Elsevier, Amsterdam

Hirsch, J. G. and Fedorko, M. E. (1968). Ulstrastructure of human leukocytes after simultaneous fixation with glutaraldehyde and osmium tetroxide and postfixation in uranyl acetate. *J. Cell Biol.*, **38**: 615

Hisano, S., Adachi, T. and Daikoku, S. (1984). Immunolabelling of adenohypophysial cells with protein A–colloidal gold–antibody complex for electron microscopy. *J. Histochem. Cytochem.*, **32**: 705

Hitchborn, J. H. and Hills, G. J. (1965). The use of negative staining in the electron microscopic examination of plant viruses in crude extracts. *Virology*, **27**: 528

Hobot, J. A. (1989). Lowicryls and low temperature embedding for colloidal gold methods. In: *Colloidal Gold: Principles, Methods, and Applications*, Vol. 2 (M. A. Hayat, Ed.). Academic Press, San Diego and London

Hodges, G. M., Smolira, M. A. and Livingston, D. C. (1984). Scanning electron microscope immunocytochemistry in practice. In: *Immunolabeling for Electron Microscopy* (J. M. Polak and I. M. Varndell, Eds.), pp. 189–233. Elsevier, Amsterdam

Hodgson, C. P. and Cunningham, W. P. (1987). A synthetic corundum knife for ultrathin sectioning. *J. Microsc.*, **146**: 103

Hoefert, L. L. (1968). Polychromatic stains for thin sections of *Beta* embedded in epoxy resin. *Stain Technol.*, **43**: 145

Hoff, S. F. and MacInnis, A. J. (1983). Ultrastructural localization of phenothiazines and tetracycline: a new histochemical approach. *J. Histochem. Cytochem.*, **31**: 613

Hoffman, H. P. and Avers, C. J. (1973). Mitochondrion of yeast: Ultrastructural evidence for one giant branched organelle per cell. *Science, N.Y.*, **181**: 749

Hohl, H. R., Hamamoto, S. T. and Hemmes, D. E. (1968). Ultrastructure aspects of cell elongation, cellulose synthesis, and spore differentiation in *Acytoselium leptosomum*, a cellular slime mould. *J. Bot.*, **55**: 783

Hohman, W. R. (1974). Ultramicroincineration of thin-sectioned tissue. In: *Principles and Techniques of Electron Microscopy: Biological Applications*, Vol. 4 (M. A. Hayat, Ed.), pp. 129–158. Van Nostrand Reinhold, New York

Holec, B. and Ciampor, F. (1978). Accessory equipment for the transfer and orientation of ultrathin serial sections. *Sci. Tools*, **25**: 15

Holgate, C. S., Jackson, P., Cowen, Ph. N. and Bird, C. C. (1983). Immunogold–silver staining: new method of immunostaining with enhanced sensitivity. *J. Histochem. Cytochem.*, **31**: 938

Hoogeveen, J. T. (1970). Thermoconductometric investigation of phosphatidylcholine in aqueous tertiary butanol solutions in the absence and presence of metal ions. In: *Effects of Metals on Cells, Subcellular Elements, and Macromolecules* (J. Maniloff, J. R. Coleman and M. W. Miller, Eds.), p. 207. Charles C. Thomas, Springfield, Ill.

Hopkinson, H. E., Battersby, S. and Anderson, T. J. (1987). The nature of breast dense core granules: uranaffin reactivity. *Histopathology*, **11**: 1149

Hopwood, D. (1967a). Some aspects of fixation with glutaraldehyde: A biochemical and histochemical comparison of the effects of formaldehyde and glutaraldehyde fixation on various enzymes and glycogen with a note on penetration of glutaraldehyde into liver. *J. Anat.*, **101**: 83

Hopwood, D. (1967b). The behaviour of various glutaraldehydes on Sephadex G-10 and some implications for fixation. *Histochemie*, **11**: 289

Hopwood, D. (1969a). Fixatives and fixation: a review. *Histochem J.*, **1**: 323

Hopwood, D. (1969b). Fixation of proteins by osmium tetroxide, potassium dichromate and potassium permanganate. *Histochemie*, **18**: 250

Hopwood, D. (1970). The reactions between formaldehyde, glutaraldehyde, and osmium tetroxide, and their fixation effects on bovine serum albumin and on tissue blocks. *Histochemie*, **24**: 50

Hopwood, D. (1972). Theoretical and practical aspects of glutaraldehyde fixation. A review. *Histochem J.*, **4**: 267

Hopwood, D. (1975). The reactions of glutaraldehyde with nucleic acids. *Histochem J.*, **7**: 267

Hopwood, D., Allen, C. R. and McCabe, M. (1970). The reactions between glutaraldehyde and various proteins. An investigation of their kinetics. *Histochem J.*, **2**: 137

Hori, I. (1986). Surface specializations of planarian gastrodermal cells as revealed by staining with ruthenium red. *J. Morphol.*, **188**: 69

Horisberger, M. (1978). Agglutination of erythrocytes using lectin-labelled spacers. *Experientia*, **34**: 721

Horisberger, M. (1981). Colloidal gold: a cytochemical marker for light and fluorescent microscopy and for transmission and scanning electron microscopy. *Scanning Electron Microsc.*, **2**: 9

Horisberger, M. (1984). Lectin cytochemistry. In: *Immunolabelling for Electron Microscopy* (J. M. Polak and I. M. Varndell, Eds.), pp. 249–258. Elsevier, Amsterdam

Horisberger, M. (1985). The gold method as applied to lectin cytochemistry in transmission and scanning electron microscopy. In: *Techniques in Immunocytochemistry*, Vol. 3 (G. R. Bullock and P. Petrusz, Eds.), pp. 155–178. Academic Press, London

Horisberger, M. (1989). Colloidal gold for scanning electron microscopy. In: *Colloidal Gold: Principles, Methods, and Applications*, Vol. 1 (M. A. Hayat, Ed.). Academic Press, San Diego and London

Horisberger, M. and Rosset, J. (1977). Colloidal gold, a useful marker for transmission and scanning electron microscopy. *J. Histochem. Cytochem.*, **25**: 295

Horisberger, M. and Clerc, M.-F. (1985). Labeling of colloidal gold with protein A: a quantitative study. *Histochemistry*, **82**: 219

Horisberger, M. and Vauthey, M. (1984). Labeling of colloidal gold with protein: A quantitative study using β-lactoglobulin. *Histochemistry*, **80**: 13

Horisberger, M. and Vonlanthen, M. (1979). Location of mannan and chitin on thin sections of budding yeasts with gold markers. *Arch. Microbiol.*, **115**: 1

Horne, R. W. and Pasquali-Ronchetti, I. (1974). A negative staining–carbon film technique for studying viruses in the electron microscope. I. Preparation procedures for examining icosahederal and filamentous viruses. *J. Ultrastruct. Res.*, **47**: 361

Horobin, R. W. and Goldstein, D. J. (1972). Impurities and staining characteristics of alcian blue samples. *Histochem. J.*, **4**: 391

Horobin, R. W. and Tomlinson, A. (1976). The influence of the embedding medium when staining for electron microscopy: the penetration of stains into plastic sections. *J. Microsc.*, **108**: 69

Hott, M. and Marie, P. J. (1987). Glycol methacrylate as an embedding medium for bone. *Stain Technol.*, **62**: 51

Howard, R. J. and Aist, J. R. (1979). Hyphal tip cell ultrastructure of the fungus *Fusarium*: improved preservation by freeze substitution. *J. Ultrastruct. Res.*, **66**: 224

Howard, V. and Eins, S. (1984). Software solutions to problems in stereology. *Acta Stereol.*, **3**: 139

Howell, S. L., Young, D. A. and Lacy, P. E. (1969). Isolation and properties of secretory granules from rat islets of Langerhans. III. Studies on the stability of the isolated beta granule. *J. Cell Biol.*, **41**: 167

Hubbard, J. E. and Laskowski, M. B. (1972). Spontaneous transmitter release and ACh sensitivity during glutaraldehyde fixation of rat diaphragm. *Life Sci.*, **11**: 781

Hubbell, H. R. (1985). Silver staining as an indicator of active ribosomal genes. *Stain Technol.*, **60**: 285

Hulstaert, C. E. and Blaauw, E. H. (1986). Cellular contrast depending on buffer and period of rinsing after glutaraldehyde fixation. *Ultramicroscopy*, **19**: 97

Humbel, B. and Müller, M. (1986). Freeze-substitution and low temperature embedding. In: *The Science of Biological Specimen Preparation for Microscopy and Microanalysis* (J.-P. Revel, T. Barnard and G. H. Haggis, Eds.), pp. 175–183

Humphreys, W. J. (1977). Health and safety hazards. *SEM*, Vol. 1: 537

Huxley, H. E. (1957). The double array of filaments in cross-striated muscle. *J. Biophys. Biochem. Cytol.*, **3**: 631

Huxley, H. E. (1959). Some aspects of staining of tissue for sectioning. *J. Roy. Microsc. Soc.*, **78**: 30

Huxley, H. E. and Zubay, G. (1961). Preferential staining of nucleic acid containing structures for electron microscopy. *J. Biophys. Biochem. Cytol.*, **11**: 273

Hwang, Y.-C. (1970). A modification for orientation by the use of silicone rubber moulds for reembedding tissue in epoxy resins. *J. Electron Microsc.*, **19**: 189

Hyatt, A. D., Eaton, B. T. and Lunt, R. (1987). The grid–cell–culture technique: the direct examination of virus-infected cells and progeny viruses. *J. Microsc.*, **145**: 97

Hyde, J. M. and Peters, D. (1970). The influence of pH and osmolality on fixation of the fowlpox virus core. *J. Cell Biol.*, **46**: 179

Idelman, S. (1964). Modification de la technique de Luft en vue de la conservation des lipides en microscopie electronique. *J. Microscopie*, **3**: 715

Idelman, S. (1965). Conservation des lipides par les techniques utilieses en microscopie electronique. *Histochemie*, **5**: 18

Iijima, S. (1977). Thin graphite support films for high resolution electron microscopy. *Micron*, **8**: 41

Imoto, M. and Choe, S. (1955). Vinyl polymerization. V. Decomposition of *sym*-substituted benzoyl peroxides in the presence of dimethylaniline. *J. Polymer. Sci.*, **15**: 485

Indi, S., Wakley, G. and Stebbings, H. (1986). Effects of glycerol and freezing on the appearance and arrangement of microtubules in three different systems: a freeze-substitution study. *Tiss. Cell*, **18**: 331

Ingram, F. D. and Ingram, M. J. (1983). Influence of freeze-drying and plastic embedding on electrolyte distributions. In: *Science of Biological Specimen Preparation* (J.-P. Revel, T. Barnard and G. H. Haggis, Eds.), pp. 167–174. SEM Inc., AMF O'Hare, Ill.

Isler, H. (1974). Simple mechanical device for orienting tissue blocks on the ultramicrotome. *J. Microsc.*, **102**: 225

Ito, S. (1961). The endoplasmic reticulum of gastric parietal cells. *J. Biophys. Biochem. Cytol.*, **11**: 333

Ito, S. and Karnovsky, M. J. (1968). Formaldehyde–glutaraldehyde fixatives containing trinitro compounds. *J. Cell Biol.*, **39**: 168a

Ivanov, V. I., Michenkova, L. E. and Timofeev, V. P. (1967). The nature of Cu^{++} and Ag^{+} complexes with DNA. *J. Mol. Biol.*, **1**: 682

Izard, C. and Libermann, C. (1978). Acrolein. *Mut. Res.*, **47**: 115

James, T. H. (1977). *The Theory of the Photographic Process*. Macmillan, New York

Jehl, B., Bauer, R. Dörge, A and Rick, R. (1981). The use of propane/isopentane mixtures for rapid freezing of biological specimens. *J. Microsc.*, **123**: 307

Jensen, O. A., Prause, J. U. and Laursen, H. (1981). Shrinkage in preparatory steps for SEM. *Albrecht Graefes Arch. Klin. Ophthalmol.*, **215**: 233

Jensen, R. H. and Davidson, N. (1966). Spectrophotometric, potentiometric and density gradient ultracentrifugation studies of the binding of silver ion by DNA. *Biopolymers*, **4**: 17

Jésior, J.-C. (1982). The grid sectioning technique: a study of catalase platelets. *EMBO J.*, **1**: 1423

Jésior, J.-C. (1985). How to avoid compression: a model study of latex sphere grid sections. *J. Ultrastruct. Res.*, **90**: 135

Jésior, J.-C. (1986). How to avoid compression. II. The influence of sectioning conditions. *J. Ultrastruct. Mol. Struct. Res.*, **95**: 210

Jessen, H. (1973). Electron cytochemical demonstration of sulfhydryl groups in keratohyalin granules and in the peripheral envelope of cornified cells. *Histochemie*, **33**: 15

Jewell, G. G. and Saxton, C. A. (1970). The ultrastructural demonstration of compounds containing 1,2-glycol groups in plant cell walls. *Histochem J.*, **2**: 17

Johannessen, J. V. (1973). Rapid processing of kidney biopsies for electron microscopy. *Kid. Intern.*, **3**: 46

Johannessen, J. V. (1977). Use of paraffin material for electron microscopy. *Pathol. Ann.*, **12**: 189

Johansen, B. V. (1974). Bright field electron microscopy of biological specimens. II. Preparation of ultrathin carbon support films. *Micron*, **5**: 209

Johansen, B. V. (1975). Bright field electron microscopy of biological specimens. IV. Ultrasonic exfoliated graphite as 'low-noise' support films. *Micron*, **6**: 165

Johansen, B. V. (1978). Negative staining. In: *Electron Microscopy in Human Medicine*, Vol. 1 (J. V. Johannessen, Ed.), pp. 84–98. McGraw-Hill, New York

Johansen, B. V., Namork, E., Oygarden, K. and Holtet, T. (1980). Film thickness measurements with a simple vibrating quartz monitor. *J. Ultrastruct. Res.*, **73**: 97

Johansson, O. (1983). The Vibratome–Ralph knife combination: a useful tool for immunohistochemistry. *Histochem J.*, **15**: 265

Johnson, E. M. and Capowski, J. J. (1984). Principles of reconstruction and three-dimensional display of serial sections using a computer. In: *The Microcomputer in Cell and Neurobiology Research* (R. R. Mize, Ed.). Elsevier, New York

Johnson, P. C. (1976). A rapidly setting glue for resectioning and remounting epoxy embedded tissue. *Stain Technol.*, **51**: 275

Johnson, T. J. A. (1985). Aldehyde fixatives: quantification of acid-producing reactions. *J. Electron Microsc. Tech.*, **2**: 129

Johnson, T. J. A. (1986). Glutaraldehyde fixation chemistry. In: *Science of Biological Specimen Preparation*, pp. 51–62. SEM Inc., AMF O'Hare, Ill.

Johnson, T. J. A. and Rash, J. E. (1980). Fixation reaction consumes O_2 and is inhibited by tissue anoxia. *J. Cell Biol.*, **87**: 231a

Johnson, W. H., Latta, H. and Osvaldo, L. (1973). Variation in glomerular ultrastructure in rat kidneys fixed by perfusion. *J. Ultrastruct. Res.*, **45**: 149

Joly, M. (1965). *A Physico-Chemical Approach to the Denaturation of Proteins*. Academic Press, London

Jones, D. (1969a). The reaction of formaldehyde with unsaturated fatty acids during histological fixation. *Histochem. J.*, **1**: 459

Jones, D. (1969b). Acrolein as a histological fixative. *J. Microsc.*, **90**: 75

Jones, D. G. (1973). Some factors affecting the PTA staining of synaptic junctions. A preliminary comparison of PTA stained junctions in the various regions of the CNS. *Z. Zellforsch.*, **143**: 301

Jones, G., Gallant, P. and Butler, W. H. (1977). Improved techniques in light and electron microscopy. *J. Path.*, **121**: 141

Jones, G. J. (1974). Polymerization of glutaraldehyde at fixative pH. *J. Histochem. Cytochem.*, **22**: 911

Jones, R. and Reid, L. (1973a). The effect of pH on alcian blue staining of epithelial acid glycoproteins. I. Sialomucins and sulphomucins (singly or in simple combinations). *Histochem J.*, **5**: 9

Jones, R. and Reid, L. (1973b). The effect of pH on alcian blue staining of epithelial acid glycoproteins. II. Human bronchial submucosal gland. *Histochem J.*, **5**: 9

Jones, R. T. and Trump, B. F. (1975). Cellular and subcellular effects of ischaemia on the pancreatic acinar cell. *In vitro* studies of rat tissue. *Virchows Arch. B Cell Path.*, **19**: 325

Joó, F., Halász, N. and Párducz, Á. (1973). Studies on the fine structural localization of zinc iodide–osmium reaction in the brain. I. Some characteristics of localization in the perikarya of identified neurons. *J. Neurocytol.*, **2**: 393

Jordan, E. G. and Saunders, A. M. (1976). The presentation of three-dimensional reconstructions from serial sections. *J. Microsc.*, **107**: 205

Jost, P. C., Brooks, U. J. and Griffith, O. H. (1973). Fluidity of phospholipid bilayers and membranes after exposure to osmium tetroxide and glutaraldehyde, *J. Mol. Biol.*, **76**: 313

Jost, P. C. and Griffith, O. H. (1973). The molecular reorganization of lipid bilayers by osmium tetroxide. A spin-label study of orientation and restricted *y*-axis anistropic motion in model membrane systems. *Arch. Biochem. Biophys.*, **159**: 70

Kádár, A., Gardner, D. L. and Bush, V. (1973). Glycosaminoglycans in developing chick-embryo aorta revealed by ruthenium red: an electron microscope study. *J. Pathol.*, **108**: 275

Kageyama, M., Takagi, M., Parmley, R. T., Toda, M., Hirayama, H. and Toda, Y. (1985). Ultrastructural visualization of elastic fibers, with a tannate–metal salt method. *Histochem. J.*, **17**: 93

Kaibara, M. and Kikkawa, Y. (1971). Osmiophilia of the saturated phospholipid dipalmitoyl lecithin, and its relationship to the alveolar lining layer of mammalian lung. *Am. J. Anat.*, **132**: 61

Kalina, M. and Pease, D. C. (1977a). The preservation of ultrastructure in saturated phosphatidyl cholines by tannic acid in model systems and type II pneumocytes. *J. Cell Biol.*, **74**: 726

Kalina, M. and Pease, D. C. (1977b). The probable role of phosphatidyl cholines in the tannic acid enhancement of cytomembrane electron contrast. *J. Cell Biol.*, **74**: 742

Kalt, M. R. and Tandler, B. (1971). A study of fixation of early amphibian embryos for electron microscopy. *J. Ultrastruct. Res.*, **36**: 633

Kalucheva, J. V., Winarova, K. R. and Toshkoff, D. (1966). On the technique of methacrylic embedding of samples for ultramicrotome sectioning. *Mikroskopie*, **21**: 316

Kan, F. W. K. and Pinto da Silva, P. (1989). Colloidal gold label fracture cytochemistry. In: *Colloidal Gold: Principles, Methods, and Applications*, Vol. 2 (M. A. Hayat, Ed.). Academic Press, San Diego and London

Karnovsky, M. J. (1961). Simple methods for 'staining with lead' at high pH in electron microscopy. *J. Biophys. Biochem. Cytol.*, **11**: 729

Karnovsky, M. J. (1965). A formaldehyde–glutaraldehyde fixative of high osmolarity for use in electron microscopy. *J. Cell Biol.*, **27**: 137A

Karnovsky, M. J. (1971). Use of ferrocyanide-reduced osmium tetroxide in electron microscopy. *Proc. 14th Ann. Meet. Am. Soc. Cell Biol.*, p. 146

Kashiwa, H. K. and Thiersch, N. J. (1984). Evaluation of potassium pyroantimonate/sucrose/glutaraldehyde concentration and incubation time as essential variables for localizing calcium bound to organic compounds in epiphyseal chondrocytes. *J. Histochem. Cytochem.*, **32**: 105

Kasten, F. H. (1960). The chemistry of Schiff's reagent. *Int. Rev. Cytol.*, **10**: 1

Katz, D., Straussman, Y., Shahar, A. and Kohn, A. (1980). Solid-phase immune electron microscopy (SPIEM) for rapid viral diagnosis. *J. Immunol. Meth.*, **38**: 171

Katz, D. and Straussman, Y. (1984). Evaluation of immunosorbent electron microscopic techniques for detection of Sindbis virus. *J. Virol. Meth.*, **8**: 243

Kaufmann, R. (1980). Recent LAMMA studies of physiological calcium distributions in retina tissues. *SEM*, Vol. 2: 641–646. SEM Inc., AMF O'Hare, Chicago

Kawakami, H. and Hirano, H. (1986). Rearrangement of the open-canalicular system of the human blood platelet after incorporation of surface-bound ligands. *Cell Tiss. Res.*, **245**: 465

Kawana, E., Akert, K. and Sandri, C. (1969). Zinc iodide–osmium tetroxide impregnation of nerve terminals in the spinal cord. *Brain Res.*, **16**: 325

Keene, D. R. (1984). A method for producing dust-, streak-, and hole-free Formvar films in laboratories having high atmospheric humidity. *Stain Technol.*, **59**: 56

Kellenberger, E. and Arber, W. (1957). Electron microscopical studies of phage multiplication I. A method for quantitative analysis of particle suspensions. *Virology*, **3**: 245

Kellenberger, E. and Bitterli, D. (1976). Preparation and counts of particles in electron microscopy: application of negative stain in the agar filtration method. *Microsc. Acta*, **78**: 131

Kellenberger, E., Carlemalm, E., Villiger, W., Roth, J. and Garavito, R. M. (1980). Low denaturation embedding for electron microscopy of thin sections. *Chemische Werke Lowi GmbH* (Beuthener Strasse 2, Postfach 1660, D-8264 Waldkraiburg, F.R.G.), pp. 1–59

Kellenberger, E., Ryter, A. and Sechaud, J. (1958). Electron microscope study of DNA-containing plasms. II. Vegetative and mature phage DNA as compared with normal bacterial nucleoids in different physiological states. *J. Biophys. Biochem. Cytol.*, **4**: 671

Kessler, S. W. (1975). Rapid isolation of antigens from cells with a *Staphyloccoccus* protein A–antibody adsorbent: parameters of the interaction of antibody–antigen complexes with protein A. *J. Immunol.*, **115**: 1617

Kikkawa, Y. and Kaibara, M. (1972). The distribution of osmiophilic lamellae within the alveolar and bronchiolar walls of the mammalian lungs as revealed by osmium–ethanol treatment. *Am. J. Anat.*, **134**: 203

Kimble, J. E., Sorensen, S. C. and Møllgaard, K. (1973). Cell junctions in the subcommissural organ of the rabbit as revealed by the use of ruthenium red. *Z. Zellforsch.*, **137**: 375

King, D. G., Kammlade, N. and Murphy, J. (1982). A simple device to help reembed thick plastic sections. *Stain Technol.*, **57**: 307

Kirkeby, S. and Moe, D. (1986). Studies on the actions of glutaraldehyde, formaldehyde, and mixtures of glutaraldehyde and formaldehyde on tissue proteins. *Acta Histochem.*, **79**: 115

Kjeldsberg, E. (1980). 40–43 nm virus-like particles associated with gastoenteritis. *Ultrastruct. Path.*, **1**: 457

Kjeldsberg, E. (1986). Immunonegative stain techniques for electron microscopic detection of viruses in human faeces. *Ultrastruct. Path.*, **10**: 553

Kjeldsberg, E. (1989). Immunogold labeling of viruses in suspension. In: *Colloidal Gold: Principles, Methods, and Applications*, Vol. 1 (M. A. Hayat, Ed.). Academic Press, San Diego and London

Klein, R. L., Yen, S.-S. and Thureson-Klein, A. (1972). Critique on the K–pyroantimonate method for semi-quantitative estimation of cations in conjunction with electron microscopy. *J. Histochem. Cytochem.*, **20**: 65

Kleinschmidt, V. A. and Zahn, R. K. (1959). Über Desoxyribonucleinsäure-Molekeln in Protein Mischfilmen. *Z. Naturforsch.*, **14b**: 770

Knobler, R. L., Stempak, J. G. and Laurencin, M. (1978). Preparation and analysis of serial sections in electron microscopy. In: *Principles and Techniques of Electron Microscopy: Biological Applications*, Vol. 2 (M. A. Hayat, Ed.). Van Nostrand Reinhold, New York

Knoll, G., Oebel, G. and Plattner, H. (1982). Sandwich propane jet cryofixation methods: comparison of cooling rates obtained with a commercial double-jet device and a simple one-side jet procedure. *Proc. 10th Int. Cong. Electron Microsc.*, Vol. 3, p. 181. Hamburg, Germany

Kobayasi, T. (1983). A simple staining device for grids. *Proc. Roy. Microsc. Soc.*, **18**: 220

Kobs, S. F. and Behrman, E. J. (1987). Reactions of osmium tetroxide with imidazoles. *Inorg. Chim. Acta*, **138**: 113

Koelle, G. B., Davis, R., Smyrl, E. G. and Fine, A. V. (1974). Refinement of the bis-(thioacetoxy)aurate(I) method for the electron microscopic localization of acetylcholinesterase and nonspecific cholinesterase. *J. Histochem. Cytochem.*, **22**: 252

Kolb-Bachofen, V. (1977). Electron microscopic localization of acid phosphatase in *Tetrahymena pyriformis*: the influence of activities of lysosomal enzymes on fixation and structural preservation. *Cytobiologie*, **15**: 135

Kölbel, H. K. (1970). Der Wert des Methacrylat unter heute gebräuchlichen Einbettungsmitteln für die Elektronenmikroskopie. *Mikroskopie*, **26**: 251

Kölbel, H. K. (1974). Partially carbon coated support films—qualities and application. *Mikroskopie*, **30**: 208

Kölbel, H. K. (1976). Kohle-Trägerfilme für die hochauflösende Elektronenmikroskopie—Verbesserung von Eigenschaften und Herstellungstechnik. *Mikroskopie*, **32**: 1

Koller, T. (1965). Mounting of ultrathin sections with the aid of an electrostatic field. *J. Cell Biol.*, **27**: 441

Komnick, H. (1962). Elektronenmikroskopische Lokalisation von Na^+ und Cl^- in Zellen und Geweben. *Protoplasma*, **55**: 414

Kondo, H. (1984). Polyethylene glycol (PEG) embedding and subsequent deembedding as a method for the structural and immunocytochemical examination of biological specimens by electron microscopy. *J. Electron Microsc. Tech.*, **1**: 227

Kondo, H. and Ushiki, T. (1985). Polyethylene glycol (PEG) embedding and subsequent deembedding as a method for the correlation of light microscopy, scanning electron microscopy and transmission electron microscopy. *J. Electron Microsc. Tech.*, **2**: 457

König, N. and Loos, H. V. (1980). Two useful techniques in three-dimensional electron microscopy: quarter-micron serial sectioning and stereomicroscopy. *J. Neurosci. Meth.*, **2**: 79

Konwiński, M., Abramczuk, J., Barańska, W. and Szymkowiak, W. (1974). Size changes of mouse ova during preparation for morphometric studies in the electron microscope. *Histochemistry*, **42**: 315

Kordylewski, L., Karrison, T. and Page, E. (1986). Developmental changes in P-face and E-face particle densities of *Xenopus* cardiac muscle plasma membrane. *Tiss. Cell*, **18**: 793

Koreeda, A. (1980). A new holey film formed by an ion-etching method. *J. Electron Microsc.*, **29**: 61

Korn, A. H., Feairheller, S. H. and Filachione, E. M. (1972). Glutaraldehyde: nature of the reagent. *J. Mol. Biol.*, **65**: 525

Korn, E. D. and Weisman, R. A. (1966). I. Loss of lipids during preparation of amoeba for electron microscopy. *Biochim. Biophys. Acta*, **116**: 309

Kraft, L. M., Joyce, K. and D'Amelio, E. D. (1983). Removal of histological sections from glass for electron microscopy: use of Quetol 651 resin and heat. *Stain Technol.*, **58**: 41

Krames, B. and Page, E. (1968). Effects of electron microscopic fixatives on cell membranes on the perfused rat heart. *Biochim. Biophys. Acta*, **150**: 24

Krstić, R. (1972). Die Einwirkung von Kälte auf mit Zinkjodid-Osmium tetroxyd reagierende synaptische Bläschen in den Nervenendigungen im Corpus pineale der ratte. *Z. Anat. Entwickl. Gesch.*, **135**: 301

Kühn, K., Grassmann, W. and Hofmann, U. (1957). Über die Bindung des Phosphorwolfransäure im Kollagen. *Naturwissenschaften*, **44**: 538

Kuo, J., Husca, G. L. and Lucas, L. N. D. (1981). Forming and removing stain precipitates on ultrathin sections. *Stain Technol.*, **56**: 199

Kuran, H. and Olszewska, M. J. (1974). Effect of some buffers on the ultrastructure, dry mass content and radioactivity of nuclei of *Haemanthus katharinae. Folia Histo-Cytochemica*, **12**: 173

Kuran, H. and Olszewska, M. J. (1977). Effects of some buffers on the ultrastructure, dry mass content, and radioactivity of nuclei of *Haemanthus katharinae. Microsc. Acta*, **79**: 69

Kushida, H. (1960). A new polyester embedding method for ultrathin sectioning. *J. Electron Microsc.*, **9**: 113

Kushida, H. (1961a). A styrene–methacrylate resin embedding method for ultrathin sectioning. *J. Electron Microsc.*, **10**: 16

Kushida, H. (1961b). A new embedding method for ultrathin sectioning using a methacrylate resin with three-dimensional polymer structure. *J. Electron Microsc.*, **10**: 194

Kushida, H. (1962a). A study of cellular swelling and shrinkage during fixation, dehydration, and embedding in various standard media. *Proc. 5th Int. Cong. Electron Microsc.*, Vol. 2. p. 10. Academic Press, New York

Kushida, H. (1962b). On ultraviolet polymerization of styrene resins in embedding for electron microscopy. *J. Electron Microsc.*, **11**: 128

Kushida, H. (1962c). Uranyl nitrate as a catalyst for ultraviolet polymerization in embedding. *J. Electron Microsc.*, **11**: 253

Kushida, H. (1964). Improved methods for embedding with Durcupan. *J. Electron Microsc.*, **13**: 139

Kushida, H. (1966a). Staining of the thin section with lead acetate. *J. Electron Microsc.*, **15**: 93

Kushida, H. (1966b). Block staining with lead acetate. *J. Electron Microsc.*, **15**: 90

Kushida, H. (1967). A new embedding method employing DER 736 and Epon 812. *J. Electron Microsc.*, **16**: 278

Kushida, H. (1969). A new rotary shaker for fixation, dehydration, and embedding. *J. Electron Microsc.*, **18**: 137

Kushida, H., Kushida, T. and Iijima, H. (1985). An improved method for both light and electron microscopy of identical sites in semi-thin tissue sections under 200 kV transmission electron microscope. *J. Electron Microsc.*, **34**: 438

Kushida, H., Kushida, T., Iijima, H. and Aida, S. (1986). An improved method for embedding with Quetol 651 and ERL 4206 for stereoscopic observation of thick sections under 400 kV transmission electron microscope. *Proc. 11th Int. Cong. Electron Microsc.*, p. 2177. Kyoto

Kuthy, E. and Csapó, Z. (1976). Peculiar artefacts after fixation with glutaraldehyde and osmium tetroxide. *J. Microsc.*, **107**: 177

Lachapelle, M. and Lafontaine, J. G. (1987). Observations on the ultrastructural preservation of the nucleus in the *Myxomycete polycephalum* as observed in resinless sections. *J. Electron Microsc. Tech.*, **5**: 227

Lackie, P. M., Hennessy, R. J., Hacker, G. W. and Polak, J. M. (1985). Investigation of immunogold–silver staining by electron microscopy. *Histochemistry*, **83**: 545

Ladman, A. J. (1973). Immersion fixation of dog retina: factors affecting heterochromatin stainability of photoreceptor nuclei. *Anat. Rec.*, **175**: 365

Laiho, K. U., Shelburne, J. D. and Trump, B. F. (1971). Observations on cell volume, ultrastructure, mitochondrial conformation and vital-dye uptake in Ehrlich ascites tumor cells. *Am. J. Path.*, **65**: 203

Lake, J. A. (1979). Practical aspects of immune electron microscopy. In: *Enzyme Structure*, Part H (C. H. W. Hirs and S. N. Timasheff, Eds.), *Methods in Enzymology*, Vol. 61, pp. 250–257. Academic Press, New York

Lamvik, M. K., Voreades, D., Norton, P., LeFurgey, A. and Ingram, P. (1987). A vacuum system for the production of clean carbon films. *J. Electron Microsc. Tech.*, **5**: 153

Lane, N. J. (1972). Fine structure of a Lepidopteran nervous system and its accessibility to peroxidase and lanthanum. *Z. Zellforsch.*, **131**: 205

Lane, N. J. and Treherne, J. E. (1970). Lanthanum staining of neurotubules in axons from cockroach ganglia. *J. Cell Sci.*, **7**: 217

Lang, R. D. A., Crosby, P. and Robards, A. W. (1976). An inexpensive spray-freezing unit for preparing specimens for freeze-etching. *J. Microsc.*, **108**: 101

Lange, R. H. (1976). Tilting experiments in the electron microscope. In: *Principles and Techniques of Electron Microscopy: Biological Applications*, Vol. 6 (M. A. Hayat, Ed.). Van Nostrand Reinhold, New York

Langenberg, W. G. (1979). Chilling of tissue before glutaraldehyde fixation preserves fragile inclusions of several plant viruses. *J. Ultrastruct. Res.*, **66**: 120

Langenberg, W. G. (1982). Silicone additive facilitates epoxy plastic sectioning. *Stain Technol.*, **57**: 79

Langenberg, W. G. and Sharpee, R. L. (1978). Chromic acid–formaldehyde fixation of nucleic acids of bacteriophage ϕ6 and infectious bovine rhinotracheitis virus. *J. Gen. Virol.*, **39**: 377

Langford, L. A. and Coggeshall, R. E. (1980). The use of potassium ferricyanide in neural fixation. *Anat. Rec.*, **197**: 297

Langone, J. J. (1982). Protein A of *Staphylococcus aureus* and related immunoglobulin receptors produced by streptococci and pneumococci. *Adv. Immunol.*, **32**: 157

Larramendi, P. C. H. (1987). Simplified and reliable procedure for Formvar coating single hole or slotted grids. *J. Electron Microsc. Tech.*, **5**: 383

Larsson, L. (1975). Effects of different fixatives on ultrastructure of the developing proximal tubule in the rat kidney. *J. Ultrastruct. Res.*, **51**: 140

Latta, H. (1962). The plasma membrane of glomerular epithelium. *J. Ultrastruct. Res.*, **6**: 407

Latta, H. and Hartmann, J. F. (1950). Use of a glass edge in thin sectioning for electron microscopy. *Proc. Soc. Exp. Biol. Med.*, **74**: 436

Lawn, A. M. (1960). The use of potassium permanganate as an electron-dense stain for sections of tissues embedded in epoxy resin. *J. Biophys. Biochem. Cytol.*, **7**: 197

Lawton, J. R. and Harris, P. J. (1978). Fixation of senescing plant tissues: sclerenchymatous fibre cells from the flowering stem of grass. *J. Microsc.*, **122**: 307

Lecatasas, G. and Boes, E. (1979). Papillomavirus in pregnancy urine. *Lancet*, September 8: 533

Ledingham, J. M. and Simpson, F. O. (1972). The use of *p*-phenylene in the block to enhance osmium staining for electron microscopy. *Stain Technol.*, **47**: 239

Leduc, E. H. and Holt, S. J. (1965). Hydroxypropyl methacrylate, a new water-miscible embedding medium for electron microscopy. *J. Cell Biol.*, **26**: 137

Lee, D.-H. (1967). Identification of argyrophilic cells in pancreatic islets by light and electron microscopy in osmium-fixed, plastic-embedded sections. *Z. Zellforsch.*, **77**: 1

Lee, F. K., Nahmias, A. J., Nahmias, D. G. E. and McDougal, J. S. (1983). Demonstration of virus particles within immune complexes by electron microscopy. *J. Virol. Meth.*, **8**: 167

Lee, R. M. K. W., McKenzie, R., Kobayashi, K., Garfield, R. E., Forrest, J. B. and Daniel, E. E. (1982). Effects of glutaraldehyde fixative osmolarities on smooth muscle cell volume, and osmotic reactivity of the cells after fixation. *J. Microsc.*, **125**: 77

Lehner, T., Nunn, R. E. and Pearse, A. G. E. (1966). Electron microscopy of paraffin embedded material in amyloidosis. *J. Path. Bact.*, **91**: 297

Leknes, I. L. (1985). An improved method for transferring semithin epoxy sections from the microtome knife to microscope slides. *Stain Technol.*, **60**: 58

Lenard, J. and Singer, S. J. (1968). Alterations of the conformation of proteins in red blood cell membranes and in solution by fixatives used in electron microscopy. *J. Cell Biol.*, **37**: 117

Lepault, J. (1985). Cryoelectron microscopy of helical particles TMV and T4 polyheads. *J. Microsc.*, **140**: 73

Lepault, J., Booy, F. P. and Dubochet, J. (1983). Electron microscopy of frozen biological suspensions. *J. Microsc.*, **129**: 89

Lepaul, J. and Pitt, T. (1984). Projected structure of unstained, frozen-hydrated *Bacillus brevis*. *EMBO J.*, **3**: 101

Leppard, G. G. and Colvin, J. R. (1972). Electron opaque

fibrils and granules in and between the cell walls of higher plants. *J. Cell Biol.*, **53**: 695

Leppard, G. G., Colvin, J. R., Rose, D. and Martin, S. M. (1971). Lignofibrils on the external cell wall surface of cultured plant cells. *J. Cell Biol.*, **50**: 63

Lesemann, D.-E. (1982). Conditions for the use of Protein A in combination with the Derrick method. *Acta Hort.*, **127**: 159

Leskes, A., Siekevitz, P. and Palade, G. E. (1971). Differentiation of endoplasmic reticulum in hepatocytes. I. Glucose-6-phosphatase distribution *in situ*. *J. Cell Biol.*, **49**: 264

Lever, J. D. (1960). A method of staining sectioned tissues with lead for electron microscopy. *Nature, Lond.*, **186**: 810

Lewin, S. (1966). Reaction of salmon sperm deoxyribonucleic acid with formaldehyde. *Arch. Biochem. Biophys.*, **113**: 584

Lewis, P. R. (1983). Fixatives: hazards and safe handling. *Proc. Roy. Microsc. Soc.*, **18**: 164

Lewis, P. R. and Knight, D. P. (1977). Staining methods for sectioned material. In: *Practical Methods in Electron Microscopy* (A. M. Glauert, Ed.). North-Holland, New York

Lickfeld, K. G. (1976). Transmission electron microscopy of bacteria. *Meth. Microbiol.*, **9**: 127

Lickfield, K. G. (1985). Ein Beitrag zur Frage welche Kräfte und Faktoren Dünnstscheiden bewirken. *J. Ultrastruct. Res.*, **93**: 101

Lilley, D. M. J. and Palaček, E. (1984). The supercoil-stabilized cruciform of Co|E| is hyper-reactive to osmium tetroxide. *EMBO J.*, **3**: 1187

Lillie, R. D. (1965). *Histopathological Technic and Practical Histochemistry*. McGraw-Hill, New York

Lim, B. S. and Solomon, J.-D. (1975). Electron microscopic study of freshly fixed and time-delayed fixed tissue. *Proc. 33rd Ann. EMSA Meet.* Claitor's Pub. Division, Baton Rouge, La.

Lin, N. S. (1984). Gold–IgG complexes improve the detection and identification of viruses in leaf dip preparations. *J. Virol. Meth.*, **8**: 181

Linner, J. G., Livesey, S. A., Harrison, D. S. and Steiner, A. L. (1986). A new technique for removal of amorphous phase tissue water without ice crystal damage: a preparative method for ultrastructural analysis and immunoelectron microscopy. *J. Histochem. Cytochem.*, **34**: 1123

Linss, W., Geyer, G., Richter, H. and Helmke, U. (1971). Quantitative untersuchung der sauren Bindungsorte an menschlichen Erythrocytenschatten nach Trypsinvorbehandlung. *Z. Mikrosk. Anat. Forsch.*, **84**: 497

Linss, R., Geyer, G., Richter, H. and Helmke, U. (1972). Quantitative Untersuchung über die Verteilung saurer Bindungsorte an menschlichen Erythrocytenschatten. *Acta Biol. Germ.*, **28**: 7

Lisbeth, M. K., Karen, J. and Elisa, D. D. (1983). Removal of histological sections from glass for electron microscopy: use of Quetol 651 resin and heat. *Stain Technol.*, **58**: 41

Litman, R. B. and Barrnett, R. J. (1972). The mechanism of the fixation of tissue components by osmium tetroxide via hydrogen bonding. *J. Ultrastruct. Res.*, **38**: 63

Liu, C.-C. (1987). A simplified technique for low temperature methyl methacrylate embedding. *Stain Technol.*, **62**: 155

Lo, H. K., Malinin, T. I. and Malinin, G. I. (1987). A modified periodic acid–thiocarbohydrazide–silver proteinate staining sequence for enhanced contrast and resolution of glycogen depositions by transmission electron microscopy. *J. Histochem. Cytochem.*, **35**: 393

Locke, M. and Huie, P. (1976a). Nucleoprotein localization by bismuth staining. *Microsc. Soc. Can.*, **3**: 96

Locke, M. and Huie, P. (1976b). The beads in the Golgi complex/endoplasmic reticulum region. *J. Cell Biol.*, **70**: 384

Locke, M. and Huie, P. (1977). Bismuth staining for light and electron microscopy. *Tiss. Cell*, **9**: 347

Locke, M. and Huie, P. (1980). The nucleolus during epidermal development in an insect. *Tiss. Cell*, **12**: 175

Locke, M., Krishnan, N. and McMahon, J. T. (1971). A routine method for obtaining high contrast without staining sections. *J. Cell Biol.*, **50**: 540

Lockwood, W. R. (1964). A reliable and easily sectioned epoxy embedding medium. *Anat. Rec.*, **150**: 129

Lombardi, L., Prenna, G., Okoliesanyi, L. and Gautier, A. (1971). Electron staining with uranyl acetate: Possible role of free amino groups. *J. Histochem. Cytochem.*, **19**: 161

Loomis, T. A. (1979). Formaldehyde toxicity. *Arch. Pathol. Lab. Med.*, **103**: 321

Louw, J., Wolfe-Coote, S. A. and Day, R. S. (1986). Observations on the effects on tissue dimensions of altering osmium tetroxide molarity with sucrose or sodium chloride. *J. Electron Microsc. Tech.*, **4**: 65

Low, F. N. and Clevenger, M. R. (1962). Polyester–methacrylate embedments for electron microscopy. *J. Cell Biol.*, **12**: 615

Lowry, O. H., Passonneau, J. V., Hasselberger, F. X. and Schulz, D. W. (1964). Effect of ischemia on known substrates and co-factors of the glycolytic pathway in brain. *J. Biol. Chem.*, **239**: 18

Lózsa, A. (1974). Uranyl acetate as an excellent fixative for lipoproteins after electrophoresis on agarose gel. *Clin. Chim. Acta*, **53**: 43

Lucocq, J. M. and Baschong, W. (1986). Preparation of protein–colloidal gold complexes in presence of commonly used buffers. *Eur. J. Cell Biol.*, **42**, 332

Luft, J. H. (1956). Permanganate—a new fixative for electron microscopy. *J. Biophys. Biochem. Cytol.*, **2**: 799

Luft, J. H. (1959). The use of acrolein as a fixative for light and electron microscopy. *Anat. Rec.*, **133**: 305

Luft, J. H. (1964). Electron microscopy of cell extraneous coats as revealed by ruthenium red staining. *J. Cell Biol.*, **23**: 54A

Luft, J. H. (1971a). Ruthenium red and violet. II. Fine structural localization in animal tissues. *Anat. Rec.*, **171**: 369

Luft, J. H. (1971b). Ruthenium red and violet. I. Chemistry, purification, methods of use for electron microscopy, and mechanisms of action. *Anat. Rec.*, **171**: 347

Luft, J. H. (1973). Embedding media—old and new. In: *Advanced Techniques in Biological Electron Microscopy* (J. K. Koehler, Ed.). Springer-Verlag, New York

Luft, J. H. and Wood, R. L. (1963). The extraction of tissue protein during and after fixation with OsO_4 in various buffer systems. *J. Cell Biol.*, **19**: 46A

Lumb, W. V. (1963). *Small Animal Anesthesia.* Lea and Febiger, Philadelphia

Lünsdorf, H. and Spiess, E. (1986). A rapid method of preparing perforated supporting foils for the thin carbon films used in high resolution transmission electron microscopy. *J. Microsc.*, **144**: 211

Luther, P. K., Lawrence, M. C. and Crowther, R. A. (1988). A method for monitoring the collapse of plastic sections as a function of electron dose. *Ultramicroscopy*, **24**: 7

Luyet, B. J. and Kroener, C. (1966). The temperature of the 'glass transition' in aqueous solutions of glycerol and ethylene glycol. *Biodynamica*, **10**: 33

Luzardo-Baptista, M. (1972). Correlation between molecular structure and glycogen ultrastructure. *Ann. Histochim.*, **17**: 141

Lyon, R., Appleton, J., Swindin, K. J., Abbot, J. J. and Chesters, J. (1985). An inexpensive device for freeze drying and plastic embedding tissues at low temperatures. *J. Microsc.*, **140**: 81

McAuliffe, W. G. (1983). A note on the purification of alcian blue. *Stain Technol.*, **58**: 374

McCarthy, D. A., Pell, B. K., Holburn, C. M., Moore, S. R., Perry, J. D., Goddard, D. H. and Kirk, A. P. (1985). A tannic acid based preparation procedure which enables leucocytes to be examined subsequently by either SEM or TEM. *J. Microsc.*, **137**: 57

Mack, J. P. (1964). A holder for trimming tissue blocks for electron microscopy. *Stain Technol.*, **39**: 177

McClintock, J. and Locke, M. (1982). Lead staining in the Golgi complex. *Tiss. Cell*, **14**: 541

McCombs, W. B., McCoy, C. E. and Holton, O. D. (1980). Electron microscopy for rapid viral diagnosis. *Tex. Soc. Electron Microsc. J.*, **11**: 9

McCutchen, C. W. and Tice, L. W. (1973). The chatterbox: A device for hearing chatter as sections are cut. In: *Proc. 31st Ann. EMSA Meet.*, p. 360. Claitor's Pub. Division, Baton Rouge, La.

McDonald, K. (1984). Osmium ferricyanide fixation improves microfilament preservation and membrane visualization in a variety of animal cell types. *J. Ultrastruct. Res.*, **86**: 107

McDowell, E. M. and Trump, B. F. (1976). Histological fixatives suitable for diagnostic light and electron microscopy. *Arch. Path. Lab. Med.*, **100**: 405

McGuffee, L. J., Hurwitz, L., Little, S. A. and Skipper, B. E. (1981). A ^{45}Ca autoradiographic and stereological study of freeze-dried smooth muscle of the guinea-pig vas deferens. *J. Cell Biol.*, **90**: 201

McIntyre, J. A., Gilula, N. B. and Karnovsky, M. J. (1974). Cryoprotectant induced redistribution of intramembranous particles in mouse lymphocytes. *J. Cell Biol.*, **60**: 192

MacKenzie, A. P. (1977). Non-equilibrium freezing behaviour of aqueous systems. *Phil. Trans. Roy. Soc. Lond. B*, **278**: 167

McKinney, R. V. (1969). Facilitation of sealing metal troughs to glass knives by use of an alcohol hand torch and dental baseplate wax. *Stain Technol.*, **44**: 44

Maclaren, J. A. and Sweetman, B. J. (1966). The preparation of reduced and S-alkylated wool keratins using tri-*n*-butylphosphine. *Austr. J. Biol. Sci.*, **19**: 2355

McLay, A. L. C., Anderson, J. D. and McMeekin, W. (1987). Microwave polymerization of epoxy resin: rapid processing technique in ultrastructural pathology. *J. Clin. Path.*, **40**: 350

McMillan, P. N. and Luftig, R. B. (1973). Preservation of erythrocyte ultrastructure achieved by various fixatives. *Proc. Natl Acad. Sci. USA*, **70**: 3060

McNary, W. F. Jr. and El-Bermani, A.-W. (1970). Differentiating type I and type II alveolar cells in rat lung by OsO_4–NaI staining. *Stain Technol.*, **45**: 215

McNelly, N. A. and Hinds, J. W. (1975). Rescuing poorly embedded tissue for electron microscopy: A new and simple technique of re-embedding. *Stain Technol.*, **50**: 209

McPherson, A. (1976). The analysis of biological structure with X-ray diffraction techniques. In: *Principles and Techniques of Electron Microscopy: Biological Applications*, Vol. 6 (M. A. Hayat, Ed.). Van Nostrand Reinhold, New York and London

Macrae, A. D., Field, A. M., MacDonald, J. R., Meurisse, E. V. and Porter, A. A. (1969). Laboratory differential diagnosis of vesicular skin rashes. *Lancet*, **2**: 313

MacRae, E. K. M. and Meetz, G. D. (1970). Electron microscopy of the ammoniacal silver reaction for histones in the erythopoietic cells of the chick. *J. Cell Biol.*, **45**: 235

Maillet, M. (1963). Le réactif au tétroxyde d'osmium-iodure du zinc. *Z. Mikr. Anat. Forsch.*, **70**: 397

Maillet, M. (1968). Étude critique des fixation au tetroxyde d'osmium-iodure. *Bull. Ass. Anat., Paris*, **1**: 233

Mak, T. W., Manaster, J., Howatson, A. F., McCulloch, E. A. and Till, J. E. (1974). Particles with characteristics of leukoviruses in cultures of marrow cells from leukemic patients in remission and relapse. *Proc. Natl Acad. Sci. USA*, **71**: 4336

Mallinger, R., Geleff, S. and Böck, P. (1986). Histochemistry of glycosaminoglycans in cartilage ground substance: alcian blue staining and lectin-binding affinities in semithin Epon sections. *Histochemistry*, **85**: 121

Mandelokow, E.-M., Rapp, R. and Mandelokow, E. (1986). Microtubule structure studied by quick freezing: cryoelectron microscopy and freeze-fracture. *J. Microsc.*, **141**: 361

Manley, J. H., Williams, D. L. and Okinaka, R. (1971). Vermiculite lamellae as substrate for transmission electron microscopy. *J. Microsc.*, **93**: 73

Mar, H., Tsukuda, T., Gown, A. M., Wight, T. N. and Baskin, D. G. (1987). Correlative light and electron microscopic immunocytochemistry on the same section with colloidal gold. *J. Histochem. Cytochem.*, **35**: 419

Mar, H. and Wight, T. N. (1989). Correlative light and electron microscopic immunocytochemistry on preembedded resin sections with colloidal gold. In: *Colloidal Gold: Principles, Methods, and Applications*, Vol. 2 (M. A. Hayat, Ed.). Academic Press, San Diego and London

Marchese-Ragona, S. P. and Johnson, S. P. S. (1982). A simple method for the progressive infiltration of resin into a dehydrated biological sample. *Proc. Roy. Microsc. Soc.*, **17**: 311

Marchetti, A., Bistocchi, M. and Tognetti, A. R. (1987). Silver enhancement of protein A–gold probes on resin-embedded ultrathin sections. An electron microscopic localization of mouse mammary tumor virus (MMTV) antigens., *Histochemistry*, **86**: 371

Marenus, K. D. (1985). Antikeratin antibody staining on ultrathin sections of epidermal cells prepared by low-denaturation embedding. *J. Ultrastruct. Res.*, **91**: 92

Marikovsky, Y. and Danon, D. (1969). Electron microscope analysis of young and old red blood cells stained with colloidal iron for surface charge evaluation. *J. Cell Biol.*, **43**: 1

Marinozzi, V. (1961). Silver impregnation of ultrathin sections for electron microscopy. *J. Biophys. Biochem. Cytol.*, **9**: 121

Marinozzi, V. (1963a). Techniques nouvelles de coloration de tissus inclus dans des matières plastiques pour l'etude au microscope optique à haute résolution. *Z. Wiss. Mikrosk.*, **65**: 220

Marinozzi, V. (1963b). The role of fixation in electron staining. *J. Roy. Microsc. Soc.*, **81**: 141

Marinozzi, V. (1964). Cytochimie ultrastructurale du nucléole-RNA et proteines intranucléolaires. *J. Ultrastruct. Res.*, **10**: 433

Marinozzi, V. (1968). Phosphotungstic acid (PTA) as a stain for polysaccharides and glycoproteins in electron microscopy. *Proc. 4th Europ. Reg. Conf. Electron Microsc. Rome*, Vol. 2, p. 55. Tipografia Poliglotta Vaticana, Rome

Marshall, A. T. (1980). Frozen-hydrated bulk specimens. In: *X-ray Microanalysis in Biology* (M. A. Hayat, Ed.). University Park Press, Baltimore

Marshall, J. and Sheen, F. (1970). A modified knife holder for the Cambridge Huxley ultramicrotome. *J. Microsc.*, **92**: 155

Martin, C. J., Lam, D. P. and Marini, M. A. (1975). Reaction of formaldehyde with the histidine residues of proteins. *Bioorg. Chem.*, **4**: 22

Martino, C. de, Natali, P. G., Bruni, C. B. and Accinni, L. (1968). Influence of plastic embedding media on staining and morphology of lipid bodies. *Histochemie*, **16**: 350

Marzilli, L. G., Hanson, B. E., Kistenmacher, T. J., Epps, L. A. and Stewart, R. C. (1976). Isomerism about a dioxobridge. Spectrosopic and chemical studies on oxo-osmium(VI) esters. *Inorg. Chem.*, **15**: 1661

Mascorro, J. A. and Kirby, G. S. (1986). Physical characteristics of 'old' Epon 812 and various Epon-like replacements. *Proc. 44th Ann. Meet. Electron Microsc. Soc. Am.*, pp. 222–223

Mascorro, J. A., Ladd, M. W. and Yates, R. D. (1976). Rapid infiltration of biological tissues utilizing *n*-hexenyl succinic anhydride (HXSA)/vinylcyclohexene dioxide (VCD), an ultra-low viscosity embedding medium. *Proc. 34th Ann. Meet. Electron Microsc. Soc. Am.*, p. 346

Maser, M. D., Powell, T. E. III and Philpott, C. W. (1967). Relationships among pH, osmolality, and concentration of fixative solutions. *Stain Technol.*, **42**: 175

Massie, H. R., Samis, H. V. and Baird, M. B. (1972). The effects of the buffer HEPES on the division potential of WI-38 cells. *In Vitro*, **7**: 191

Mathieu, O., Claassen, H. and Weibel, E. R. (1978). Differential effect of glutaraldehyde and buffer osmolarity on cell dimensions: a study on lung tissue. *J. Ultrastruct. Res.*, **63**: 20

Matukafs, V. J., Panner, B. J. and Orbison, J. L. (1967). Studies on ultrastructural identification and distribution of protein–polysaccharide in cartilage matrix. *J. Cell Biol.*, **32**: 365

Maul, G. G., Maul, H. M., Slogma, J. E., Lieberman, M. W., Stein, G. S., Hsu, B. Y. L. and Borun, T. W. (1972). Time sequence of nuclear pore formation in phytohaemagglutinin-stimulated lymphocytes and in HeLa cells during cell cycle. *J. Cell Biol.*, **55**: 433

Maupin, P. and Pollard, T. D. (1983). Improved preservation and staining of HeLa cell actin filaments, clothrin-coated membranes, and other cytoplasmic structures by tannic acid–glutaraldehyde–saponin fixation. *J. Cell Biol.*, **96**: 51

Maupin-Szamier, P. and Pollard, T. D. (1978). Actin filament destruction by osmium tetroxide. *J. Cell Biol.*, **77**: 837

Maxwell, M. H. (1975). Application of Durcupan processing to avian tissues with particular reference to the demonstration of lipid. *Micron*, **6**: 53

Mayer, E. and Brüggeller, P. (1982). Vitrification of pure liquid water by high pressure jet freezing. *Nature, Lond.*, **298**: 715

Mayer, P. (1896). Über Schleimfarburg. *Mitth. Zool. Stat. Neapel*, **12**: 303

Mayo, M. A. and Cocking, E. C. (1969). Detection of pinocytic activity using selective staining with phosphotungstic acid. *Protoplasma*, **68**: 231

Mazhul, L. A., Dobrov, E. N. and Tikhonenko, T. I. (1978). Tight binding of RNA to protein in particles of rod-like virus under the action of formaldehyde. *Biokhimiya*, **43**: 138

Mazia, D., Sale, W. S. and Schatten, G. (1974). Polylysine as an adhesive for electron microscopy. *J. Cell Biol.*, **63**: 212a

Mazur, P. (1984). Freezing of living cells: mechanisms and implications. *Am. J. Physiol.*, **247**: C125

Medina, F.-J., Solanilla, E. L., Sánchez-Pina, M. A., Férnandez-Gómez, M. E. and Risueño, M. C. (1986). Cytological approach to the nucleolar functions detected by silver staining. *Chromosoma*, **94**: 259

Meek, K. M., Scott, J. E. and Nave, C. (1985). An X-ray diffraction analysis of rat tail tendons treated with cupromeronic blue. *J. Microsc.*, **139**: 205

Meek, K. M. and Weiss, J. B. (1979). Differential fixation of poly(L-arginine) and poly(L-lysine) by tannic acid and its application to the fixation of collagen in electron microscopy. *Biochim. Biophys. Acta*, **587**: 112

Meisner, J. and Hagins, W. A. (1978). Fast freezing of thin tissues by thermal conduction into sapphire crystals at 77 Kelvin. *Biophys. J.*, **21**: 149a

Melcher, A. H. and Chan, J. (1978). The relationship between section thickness and the ultrastructural visualization of collagen fibrils: importance in studies on resorption of collagen. *Arch. Oral Biol.*, **23**: 231

Meller, J. W. (1927). *Comprehensive Treatise On Organic and Theoretical Chemistry*. Longmans, Green, New York

Menco, B. Ph. M. (1984). Ciliated and microvillous structures of rat olfactory and nasal respiratory epithelia. A study using ultrarapid cryofixation followed by freeze-substitution or freeze-etching. *Cell. Tiss. Res.*, **235**: 225

Menco, B. Ph. M. (1986). A survey of ultra-rapid cryofixation methods with particular emphasis on applications to freeze-fracturing, freeze-etching, and freeze-substitution. *J. Electron Microsc. Tech.*, **4**: 177

Mentré, P. and Halpern, S. (1988). Localization of cations by pyroantimonate. II. Electron probe microanalysis of calcium and sodium in skeletal muscle of mouse. *J. Histochem. Cytochem.*, **36**: 55

Mersey, B. and McCully, M. E. (1978). Monitoring the course of fixation in plant cells. *J. Microsc.*, **116**: 49

Meszler, R. M. and Gennaro, J. F. (1969). In: *Biology of the Reptile*, Vol. 2 (C. Gans, Ed.), p. 305. Academic Press, London

Metz, J. (1981). Quick-freezing methods in neuroanatomy. In: *Techniques in Neuroanatomical Research* (Ch. Heym and W.-G. Forssmann, Eds.), pp. 267–276. Springer-Verlag, Berlin

Miller, K. R., Prescott, C. S., Jacobs, T. L. and Lassignol, N. L. (1983). Artifacts associated with quick-freezing and freeze-drying. *J. Ultrastruct. Res.*, **82**: 123

Miller, S. E. (1986). Detection and identification of viruses by electron microscopy. *J. Electron Microsc. Tech.*, **4**: 265

Millonig, G. (1961). A modified procedure for lead staining of thin sections. *J. Biophys. Biochem. Cytol.*, **11**: 736

Millonig, G. (1964). Study on the factors which influence preservation of fine structure. In: *Symp. on Electron Microscopy* (P. Buffa, Ed.), p. 347. Consiglio Nazionale delle Ricerche, Roma

Millonig, G. and Marinozzi, V. (1968). Fixation and embedding in electron microscopy. In: *Advances In Optical And Electron Microscopy*, Vol. 2 (R. Barer, and V. E. Cosslett, Eds.), p. 251. Academic Press, New York

Milne, R. G. and Lusoni, E. (1975). Rapid high-resolution immune electron microscopy of plant viruses. *Virology*, **68**: 270

Milroy, A. M. (1985). A method for eliminating unnecessary serial thin sectioning for electron microscopy. *J. Electron Microsc. Tech.*, **2**: 399

Minick, O. T. (1963). Low temperature storage of epoxy embedding resins. *Stain Technol.*, **38**: 131

Minio, F., Lombardi, L. and Gautier, A. (1966). Mise en évidence et ultrastructure du glycogène hépatique influence des techniques de préparation. *J. Ultrastruct. Res.*, **16**: 339

Minnasian, H. and Huang, S. (1979). Effect of sodium azide on the ultrastructural preservation of tissues. *J. Microsc.*, **117**: 243

Mishima, Y. (1964). Electron microscopic cytochemistry of melanosomes and mitochondria. *J. Histochem. Cytochem.*, **12**: 784

Mitchell, C. D. (1969). Preservation of the lipids of the human erythrocyte stroma during fixation and dehydration for electron microscopy. *J. Cell Biol.*, **40**: 869

Mizuhira, V. (1973). Demonstration of the elemental distribution in biological tissues by means of the electron microscope and electron probe X-ray microanalyzer. *Acta Histochem. Cytochem.*, **6**: 44

Mizuhira, V. and Futaesaku, Y. (1972). New fixation for biological membranes using tannic acids. *Acta Histochem. Cytochem.*, **5**: 233

Mizuhira, V., Shiina, S., Miake, K., Ishida, M., Nakamura, H., Yotsumoto, H. and Namae, T. (1971). Comparative examination between the chemical and physical methods to the demonstration of the ionic localization in the tissues. I. In the case of potassium antimonate method. *Proc. 29th Ann. EMSA Meet.*, pp. 408–409.

Mohr, W. P. and Cocking, E. C. (1968). A method of preparing highly vacuolated, senescent, or damaged plant tissue for ultrastructural study. *J. Ultrastruct. Res.*, **21**: 171

Molin, S.-O., Nygren, H. and Dolonius, L. (1978). A new method for the study of glutaraldehyde-induced crosslinking properties in proteins with special reference to the reaction with amino groups. *J. Histochem. Cytochem.*, **26**: 412

Mollenhauer, H. H. (1964). Plastic embedding mixtures for use in electron microscopy. *Stain Technol.*, **39**: 111

Mollenhauer, H. H. (1976). Improved specimen lighting in ultramicrotomy by painting reflective surfaces on specimen blocks. *J. Microsc.*, **107**: 203

Mollenhauer, H. H. (1978). Improved technique for pipetting solutions during tissue processing for electron microscopy. *Stain Technol.*, **113**: 215

Mollenhauer, H. H. (1986a). Surfactants as resin modifiers and their effect on sectioning. *J. Electron Microsc. Tech.*, **3**: 217

Mollenhauer, H. H. (1986b). Stain contamination and embedding in electron microscopy. *Proc. 44th Ann. Meet. EMSA*, p. 50. San Francisco Press, San Francisco

Mollenhauer, H. H. (1986c). A simple charge neutralization device for ultrathin sectioning. *J. Electron Microsc. Tech.*, **4**: 173

Mollenhauer, H. H. (1987). Comparison of surface roughness of sections cut by diamond, sapphire, and glass knives. *J. Electron Microsc. Tech.*, **6**: 81

Mollenhauer, H. H. and Morré, D. J. (1978). Contamination of thin sections, cause and elimination. *Proc. 9th Int. Cong. Electron Microsc.*, Vol. II, p. 78

Monneron, A. and Bernhard, W. (1969). Fine structural organization of the interphase nucleus in some mammalian cells. *J. Ultrastruct. Res.*, **27**: 266

Monroe, R. G., Gamble, W. J., La Farge, C. G., Gamboa, R., Morgan, C. L., Rosenthal, A. and Bullivant, S. (1968). Myocardial ultrastructure in systole and diastole using ballistic cryofixation. *J. Ultrastruct. Res.*, **22**: 22

Monsan, P., Puzo, G. and Mazarguil, H. (1975). Étude du mécanisme d'établissement des liaisons glutaraldéhyde-protéines. *Biochimie*, **57**: 1281

Moor, H. (1973). Cryotechnology for the structural analysis of biological material. In: *Freeze-etching Technique and Applications* (E. L. Benedetti, and P. Favard, Eds.), pp. 11–19. Soc. Francaise de Microscopie Electronique, Paris

Moor, H., Mühlethaler, K., Waldner, H. and Frey-Wyssling, A. (1961). A new freezing ultramicrotome. *J. Biophys. Biochem. Cytol.*, **10**: 1

Moor, H. and Riehle, U. (1968). Snap-freezing under high pressure: a new fixation technique for freeze-etching. *Proc. 4th Eur. Reg. Conf. Electron Microsc.*, Vol. 2, p. 33. Rome, Italy

Moore, D. H., Sarkar, N. H., Kelly, C. E., Pillsbury, N. and Charney, J. (1969). Type B particles in human milk. *Tex. Rep. Biol. Med.*, **27**: 1027

Moran, D. T. and Rowley, J. C. (1987). Biological specimen

preparation for correlative light and electron microscopy. In: *Correlative Microscopy in Biology: Instrumentation and Methods* (M. A. Hayat, Ed.), pp. 1–22. Academic Press, New York

Morel, F. M. M., Baker, R. F. and Wayland, H. (1971). Quantitation of human red blood cell fixation by glutaraldehyde. *J. Cell Biol.*, **48**: 91

Moreno, F. J., Hernandez-Verdun, D., Masson, C. and Bouteille, M. (1985). Silver staining of the nucleolar organizer regions (NORs) on Lowicryl and cryoultrathin sections. *J. Histochem. Cytochem.*, **33**: 389

Moretz, R. C., Akers, C. K. and Parsons, D. F. (1969). Use of small angle X-ray diffraction to investigate disordering of membranes during preparation for electron microscopy. II. Aldehydes. *Biochim. Biophys. Acta*, **193**: 12

Morgan, A. J. (1980). Preparation of specimens: changes in chemical integrity. In: *X-ray Microanalysis in Biology* (M. A. Hayat, Ed.). University Park Press, Baltimore

Mortensen, N. J. McC. and Morris, J. F. (1977). The effect of fixation conditions on the ultrastructural appearance of gastrin cell granules in the rat gastric pyloric antrum. *Cell Tiss. Res.*, **176**: 251

Mowry, R. W. (1958). Improved procedure for the staining of acid polysaccharides by Müller's colloidal (hydrous) ferric oxide and its combination with the Feulgen and the periodic acid–Schiff reactions. *Lab. Invest.*, **7**: 566

Mowry, R. W. (1959). Effect of periodic acid used prior to chromic acid on the staining of polysaccharides by Gomori's methenamine silver. *J. Histochem. Cytochem.*, **7**: 288

Mrena, E. (1980). A modification of negative staining for the study of isolated microsomes. *Philips Electron Opt. Bull. EM*, **114**: 6

Mueller, W. C. and Beckman, C. H. (1974). Ultrastructure of the phenol-storing cells in the roots of banana. *Physiol. Plant Path.*, **4**: 187

Mueller, W. C. and Beckman, C. H. (1976). Ultrastructure and development of phenolic-storing cells in cotton roots. *Can. J. Bot.*, **54**: 2074

Mueller, W. C. and Rodehorst, E. (1977). The effect of some alkaloids on the ultrastructure of phenolic-containing cells in the endodermis of cotton roots. *Proc. 35th Ann. Meet. Electron Microsc. Soc. Am.*, p. 544.

Mühlphordt, H. (1982). The preparation of colloidal gold particles using tannic acid as an additional reducing agent. *Experientia*, **38**: 1127

Müller, G. and Peters, D. (1963). Substrukturen des Vaccinevirus, dargestellt durch Negativkontrastierung. *Arch. Gesamte Virusforsch.*, **13**: 435

Müller, L. L. and Jacks, T. J. (1975). Rapid chemical dehydration of samples for electron microscopic examinations. *J. Histochem. Cytochem.*, **23**: 107

Müller, M. and Koller, T. (1972). Preparation of aluminium oxide films for high resolution electron microscopy. *Optik*, **35**: 287

Müller, M., Meister, N. and Moor, H. (1980). Freezing in a propane jet and its application in freeze-fracturing. *Mikroskopie*, **36**: 129

Müller, W. and Pscheid, P. (1981). An improved freeze-fracturing procedure preventing contamination artefacts at fracturing temperatures below 163 K (−100 °C) in an unmodified Balzers unit. *Mikroskopie*, **39**: 143

Mumaw, V. R. and Munger, B. L. (1969). Uranyl acetate-oxalate, an *en bloc* stain as well as a fixative for lipids associated with mitochondria. *Anat. Rec.*, **169**: 383

Mumaw, V. R. and Munger, B. L. (1971). Uranyl acetate as a fixative—from pH 2.0 to 8.0. *Proc. 29th Ann. EMSA Meet.*, p. 490. Claitor's Pub. Division, Baton Rouge, La.

Münch, G. (1964). Simplified preparation method for carbon replicas and carbon films for specimen support in electron microscopy. *Rev. Sci. Instrum.*, **35**: 524

Muñoz-Guerra, S. and Subirana, J. A. (1982). Crosslinked polyvinyl alcohol: a water soluble polymer as embedding medium for electron microscopy. *Mikroskopie*, **39**: 346

Munton, T. J. and Russell, A. D. (1970). Aspects of the action of glutaraldehyde on *Escherichia coli*. *J. Appl. Bact.*, **33**: 410

Murakami, Y. (1976). Puncture perfusion of small tissue pieces for scanning electron microscopy. *Arch. Histol. Jap.*, **39**: 99

Murphy, C. R., Swift, J. G., Mukherjee, T. M. and Rogers, A. W. (1982). Loss of microvilli caused by glycerol treatment of uterine epithelial cells. *Acta Anat.*, **144**: 361

Murray, J. M. and Ward, R. (1987). Preparation of holey carbon films suitable for cryoelectron microscopy. *J. Electron Microsc. Tech.*, **5**: 285

Nagele, R. G., Roisen, F. J. and Lee, H. (1983). A method for studying the three-dimensional organization of cytoskeletal elements of cells: improvements in the polyethylene glycol technique. *J. Microsc.*, **129**: 179

Nagington, J., Newton, A. A. and Horne, R. W. (1964). The structure of Orf virus. *Virology*, **23**: 261

Namork, E. (1985). Preparation of carbon films from evaporated fibers and rods at fore-vacuum pressure. *J. Electron Microsc. Tech.*, **2**: 45

Napolitano, L. M., LeBaron, F. and Scaletti, J. (1967). Preservation of myelin lamellar structure in the absence of lipid. *J. Cell Biol.*, **34**: 817

Napolitano, L. M., Scaletti, J. and LeBaron, F. (1968). Further observations on the fine structure of myelin. *J. Cell Biol.*, **39**: 98a

Neaves, W. B. (1973). Permeability of Sertoli cell tight junctions to lanthanum after ligation of ductus deferens and ductuli efferentes. *J. Cell Biol.*, **59**: 559

Nehls, R. and Schaffner, G. (1976). Specific negative staining of proteins *in situ* with iron tannin. *Cytobiologie*, **13**: 285

Neiss, W. F. (1984). Electron staining of the cell surface coat by osmium–low ferrocyanide. *Histochemistry*, **80**: 231

Neiss, W. F. (1988). Enhancement of the periodic acid–Schiff (PAS) and periodic acid–thiocarbohydrazide–silver proteinate (PA–TCH–SP) reaction in LR White sections. *Histochemistry*, **88**: 603

Nelson, B. K. and Flaxman, B. A. (1973). Use of high intensity illumination to aid alignment of knife edge and block face for ultramicrotomy. *Stain Technol.*, **48**: 13

Nemetschek, Th., Riedl, H. and Jonak, R. (1979). Topochemistry of the binding of phosphotungstic acid to collagen. *J. Mol. Biol.*, **133**: 67

Nermut, M. V. (1972). Negative staining of viruses. *J. Microsc. (Oxford)*, **96**: 351

Nermut, M. V. (1977a). Freeze-drying for electron microscopy. In: *Principles and Techniques of Electron Microscopy: Biological Applications*, Vol. 7. (M. A. Hayat, Ed.). Van Nostrand Reinhold, New York and London

Nermut, M. V. (1977b). Negative staining in freeze-drying and freeze-fracturing. *Micron*, **8**: 211

Nermut, M. V. (1982a). The 'cell monolayer technique' in membrane research. *Eur. J. Cell Biol.*, **28**: 160

Nermut, M. V. (1982b). Advanced methods in electron microscopy of viruses. In: *New Developments in Practical Virology* (C. R. Howard, Ed.), pp. 1–58. Alan. R. Liss, New York

Nermut, M. V. and Ward, B. J. (1974). Effect of fixatives on the fracture plane in red blood cells. *J. Microsc.*, **102**: 29

Nermut, M. V. and Williams, L. D. (1977). Freeze-fracturing of monolayers (capillary layers) of cells, membranes and viruses. Some technical considerations. *J. Microsc.*, **110**: 121

Nevalainen, T. J. and Anttinen, J. (1977). Ultrastructural and functional changes in pancreatic acinar cells during autolysis. *Virchows Arch. B. Cell Path.*, **24**: 197

Newcomb, W., Murrin, S. F. and Wood, S. M. (1984). A simpler and more economical method for storing glass knives. *Stain Technol.*, **58**: 376

Newman, G. R. (1987). Use and abuse of LR White. *Histochem. J.*, **19**: 118

Newman, G. R. (1989). LR White embedding medium for colloidal gold methods. In: *Colloidal Gold: Principles, Methods, and Applications*, Vol. 2 (M. A. Hayat, Ed.). Academic Press. San Diego and London

Newman, G. R. and Hobot, J. A. (1989). Role of tissue processing in colloidal gold methods. In: *Colloidal Gold: Principles, Methods, and Applications*, Vol. 2 (M. A. Hayat, Ed.). Academic Press, San Diego and London

Newman, G. R., Jasani, B. and Williams, E. D. (1983). A simple post-embedding system for the rapid demonstration of tissue antigens under the electron microscope. *Histochem. J.*, **15**: 543

Nichols, D. B., Cheng, H. and LeBond, C. P. (1974). Variability of the shape and argentaffinity of the granules in the enteroendocrine cells of the mouse duodenum. *J. Histochem. Cytochem.*, **22**: 929

Nicolaieff, A., Katz, D. and Van Regenmortel, M. H. V. (1982). Comparison of two methods of virus detection by immunosorbent electron microscopy (ISEM) using protein A. *J. Virol. Meth.*, **4**: 155

Nicolson, G. L. (1973). Anionic sites of human erythrocyte membranes. I. Effects of trypsin, phospholipase C, and pH on the topography of bound positively charged colloidal particles. *J. Cell Biol.*, **57**: 373

Nicolson, G. L. and Painter, R. G. (1973). Anionic sites of human erythrocyte membranes. II. Antispectrin-induced transmembrane aggregation of the binding sites for positively charged colloidal particles. *J. Cell Biol.*, **59**: 395

Niebauer, G., Krawczyk, W. S., Kidd, R. L. and Wilgram, G. F. (1969). Osmium zinc iodide reactive sites in the epidermal Langerhans cell. *J. Cell Biol.*, **43**: 80

Nielson, A. J. and Griffith, W. P. (1979). Reactions of osmium tetroxide with protein side chains and unsaturated lipids. *J. Chem. Soc. Dalton*, 1084–1088

Niklowitz, W. J. (1971). The interaction of hippocampal granular cells with OsO_4-zinc iodide stain. *Acta Anat.*, **80**: 114

Niklowitz, W. J. (1972). The interaction of hippocampal tissue of the rabbit with OsO_4 -zinc iodide stain after treatment with 3-acetylpyridine, reserpine, and iproniazid. *Acta Anat.*, **81**: 570

Nilsson, J. R. (1974). Effects of DMSO on vacuole formation, contractile vacuole function, and nuclear division in *Tetrahymena pyriformis* GL. *J. Cell Sci.*, **16**: 39

Nir, I. and Hall, M. O. (1974). The ultrastructure of lipid-depleted rod photoreceptor membranes. *J. Cell Biol.*, **63**: 587

Nobiling, R. (1986). Freeze-fracture autoradiography—a method for the localization of diffusible substances at electron microscopic level. *Proc. 11th Int. Cong. Electron Microsc.*, Vol. 3, p. 1969. Kyoto, Japan.

Nordmann, J. J. (1977). Ultrastructural appearance of neurosecretory granules in the sinus gland of the crab after different fixation procedures. *Cell Tiss. Res.*, **185**: 557

Norenburg, J. L. and Barrett, J. M. (1987). Steedman's polyester wax embedment and deembedment for combined light and scanning electron microscopy. *J. Electron Microsc. Tech.*, **6**: 35

Normann, T. C. (1964). Staining thin sections with lead hydroxide without contamination. *Stain Technol.*, **39**: 50

Norton, T. N., Gelfand, M. and Brotz, M. (1962). Studies in the histochemistry of plasmalogens. I. The effect of formalin and acrolein on the plasmalogens of adrenal and brain. *J. Histochem. Cytochem.*, **10**: 375

Nott, J. A., Moore, M. N., Mavin, L. J. and Ryan, K. P. (1985). The fine structure of lysosomal membranes and endoplasmic reticulum in the digestive cells of *Mytilus edulis* exposed to anthracene and phenanthrene. *Mar. Envir. Res.*, **17**: 226

Nurden, A. T., Horisberger, M., Savariau, E. and Caen, J. P. (1980). Visualization of lectin binding sites on the surface of human platelets using lectins adsorbed to gold granules. *Experientia*, **36**: 1215

Nyhlén, L. (1975). A modified method for preparing Teflon-tipped probes for manipulation of ultrathin sections. *Stain Technol.*, **50**: 365

O'Brien, T. P., Kuo, J., McCully, M. E. and Zee, S.-Y. (1973). Coagulant and non-coagulant fixation of plant cells. *Aust. J. Biol. Sci.*, **26**: 1231

Ochs, R., Lischwe, M., O'Leary, P. and Busch, H. (1983). Localization of nucleolar phosphoproteins B23 and C23 during mitosis. *Exp. Cell Res.*, **146**: 139

Ockleford, C. D. (1975). Redundancy of washing in the preparation of biological specimens for transmission electron microscopy. *J. Microsc.*, **105**: 193

Ockleford, C. D. and Tucker, J. B. (1973). Growth, breakdown, repair and rapid contraction of microtubular axopodia in the heliozoan *Actinophrys sol*. *J. Ultrastruct. Res.*, **44**: 396

Ogura, H. and Oda, T. (1973). A method for the recovery of

inadequately epoxy resin-embedded tissues for electron microscopy. *J. Electron Microsc.*, **22**: 365

Ohnishi, A., Offord, K. and Dyck, P. J. (1974). Studies to improve fixation of human nerves. I. Effect of duration of glutaraldehyde fixation on peripheral nerve morphometry. *J. Neurol. Sci.*, **23**: 223

Ohno, S. (1980). Morphometry for determination of size distribution of peroxisomes in thick sections by high voltage electron microscopy. I. Studies on section thickness. *J. Electron Microsc.*, **29**: 230

Oliveira, L., Burns, A., Bisalputra, T. and Yang, K.-C. (1983). The use of an ultra-low viscosity medium (VCD/HXSA) in the rapid embedding of plant cells for electron microscopy. *J. Microsc.*, **132**: 195

Oliver, R. M. (1973). Negative stain electron microscopy of protein molecules. In: *Methods In Enzymology* (C. H. W. Hirs and S. N. Timasheff, Eds.). Academic Press, New York

Ongun, A., Thomson, W. W. and Mudd, J. B. (1968). Lipid fixation during preparation of chloroplasts for electron microscopy. *J. Lipid Res.*, **9**: 416

Oriel, J. D. and Almeida, J. D. (1970). Demonstration of virus particles in human genital warts. *Br. J. Vener. Dis.*, **46**: 37

Osborne, M. P. and Thornhill, R. A. (1974). The zinc iodide–osmium reactive sites in the sensory epithelia of the frog labyrinth. *J. Neurocytol.*, **3**: 459

Ostendorf, M. L., Niedorf, H. K. and Blümcke, S. (1971). Elecktronenmikroskopische Untersuchungen an menschlichen Leukozyten nach Osmium-Zink Imprägnation. *Z. Zellforsch.*, **121**: 358

Owman, C. and Rudeberg, C. (1970). Light, fluorescent, and electron microscopical studies on the pineal organ of pike, *Esox lucius L.*, with special regard to 5-hydroxytryptamine. *Z. Zellforsch. Anat.*, **107**: 522

Paavola, L. G. (1977). The corpus luteum of the guinea-pig. Fine structure at the time of maximum progesterone secretion and during regression. *Am. J. Anat.*, **150**: 565

Page, S. G. and Huxley, H. E. (1963). Filament lengths in striated muscle. *J. Cell Biol.*, **19**: 369

Palay, S. L. and Chan-Palay, V. (1973). High voltage electron microscopy of the central nervous system in Golgi preparations. *J. Microsc.*, **97**: 41

Palladini, G., Lauro, G. and Basile, A. (1970). Observations sur la spécificité de la coloration aux acides phosphotungstique et phosphomolybdique. *Histochemie*, **24**: 315

Palmer, E. L., Martin, M. L. and Gary, G. W. (1975). The ultrastructure of disrupted herpes virus nucleocapsids. *Virology*, **65**: 260

Pares, R. D. and Whitecross, M. I. (1982). Gold-labelled antibody decoration (GLAD) in the diagnosis of plant viruses by immunoelectron microscopy. *J. Immunol. Meth.*, **51**: 23

Park, P., Ohno, T., Kato-Kikuchi, H. and Miki, H. (1987). Alkaline bismuth stain as a tracer for Golgi vesicles of plant cells. *Stain Technol.*, **62**: 253

Park, R. B., Kelly, J., Drury, S. and Sauer, K. (1966). The Hill reaction of chloroplasts isolated from glutaraldehyde-fixed spinach leaves. *Proc. Natl Acad. Sci. USA*, **55**: 1056

Parker, M. L. (1984). The use of hydrogen peroxide to accelerate staining of Spurr-embedded sections. *Ann. Bot.*, **53**: 121

Parsons, D., Belloto, D. J. Schultz, W. W. and Buja, M. (1984). Toward routine cryoultramicrotomy. *EMSA Bull.*, **14**: 49

Paula-Barbosa, M. (1975). The duration of aldehyde fixation as a 'flattening factor' of synaptic vesicles. *Cell Tiss. Res.*, **164**: 63

Paula-Barbosa, M., Sobrinho-Simões, M. A. and Gray, E. G. (1977). The effects of different methods of fixation on central nervous system synaptic pinocytotic vesicles. *Cell Tiss. Res.*, **178**: 323

Payne, C. M., Nagle, R. B. and Borduin, V. (1984). Methods in laboratory investigation: an ultrastructural cytochemical stain specific for neuroendocrine neoplasms. *Lab. Invest.*, **51**: 350

Peachey, L. D. (1958). Thin sections. I. A study of section thickness and physical distortion produced during microtomy. *J. Biophys. Biochem. Cytol.*, **4**: 233

Peachey, L. D., Fotino, M. and Porter, K. R. (1974). Biological applications of high voltage microscopy. *Proc. 3rd Int. Cong. High Volt. Electron Microsc.*, pp. 405–513. Oxford, England

Peachey, L. D. and Franzini-Armstrong, C. (1977). Three-dimensional visualization of the T-system of frog muscle using high voltage electron microscopy and a lanthanum stain. *Proc. 35th Ann. EMSA Meet.*, p. 570. Claitor's Pub. Division, Baton Rouge, La.

Pearse, A. D. and Marks, R. (1976). Further studies on section thickness measurement. *Histochem. J.*, **8**: 383

Pease, D. C. (1966a). The preservation of unfixed cytological detail by dehydration with 'inert' agents. *J. Ultrastruct. Res.*, **14**: 356

Pease, D. C. (1966b). Polysaccharides associated with the exterior surface of epithelial cells—kidney, intestine, brain. *Anat. Rec.*, **154**: 400

Pease, D. C. (1970). Phosphotungstic acid as a specific electron stain for complex carbohydrates. *J. Histochem. Cytochem.*, **18**: 455

Pease, D. C. (1975). Micronets for electron microscopy. *Micron*, **6**: 85

Pease, D. C. (1980). Ultrathin sectioning of fixed but unembedded tissue. In: *Proc. 38th Ann. Meet. EMSA*, p. 650. Claitor's Pub. Division, Baton Rouge, La.

Pease, D. C. (1982). Unembedded, aldehyde-fixed tissue, sectioned for transmission electron microscopy. *J. Ultrastruct. Res.*, **79**: 250

Pease, D. C. and Bouteille, M. (1971). The tridimensional ultrastructure of native collagenous fibrils, cytochemical evidence for a carbohydrate matrix. *J. Ultrastruct. Res.*, **35**: 339

Pedler, C. and Tilly, R. (1966). A new method of serial reconstruction from electron micrographs. *J. R. Microsc. Soc.*, **86**: 189

Pellegrino de Iraldi, A. (1974). ZIO staining of monoaminergic granulated vesicles. *Brain Res.*, **66**: 227

Pellegrino de Iraldi, A. and Suburo, A. M. (1970). Electron staining of synpatic vesicles using Champy–Maillet technique. *J. Microsc.*, **91**: 99

Pépin, R. (1977). Epoxy resin embedding in gelatin capsules: Improved orientation of thin specimens. *Cytologia*, **42**: 37

Peracchia, C. and Mittler, B. S. (1972). Fixation by means of glutaraldehyde–hydrogen peroxide reaction products. *J. Cell Biol.*, **53**: 234

Peracchia, C. and Robertson, J. D. (1971). Increase in osmiophilia of axonal membranes of crayfish as a result of electrical stimulation, asphyxia, or treatment with reducing agents. *J. Cell Biol.*, **51**: 223

Perera, F. and Petito, C. (1982). Formaldehyde: a question of cancer policy. *Science, N.Y.*, **216**: 1285

Perlin, M. and Hallum, J. V. (1971). Effect of acid pH on macromolecular synthesis in L cells. *J. Cell Biol.*, **49**: 66

Peters, A., Hinds, P. L. and Vaughn, J. E. (1971). Extent of stain penetration in sections prepared for electron microscopy. *J. Ultrastruct. Res.*, **36**: 37

Peters, K. and Richards, F. M. (1977). Chemical crosslinking: reagents and problems in studies of membrane structure. *Ann. Rev. Biochem.*, **46**: 523

Peters, K.-R. (1980). Improved handling of structural fragile cell-biological specimens during electron microscopic preparation by the exchange method. *J. Microsc.*, **118**: 429

Peters, K.-R. (1984). Precise and reproducible deposition of thin and ultrathin carbon films by flash evaporation of carbon yarn in high vacuum. *J. Microsc.*, **133**: 17

Peters, K.-R. (1986). Rationale for the application of thin, continuous metal films in high magnification electron microscopy. *J. Microsc.*, **142**: 25

Peters, T. and Ashley, C. A. (1967). An artifact in autoradiography due to binding of free amino acids to tissues by fixatives. *J. Cell Biol.*, **33**: 53

Petris, S. de (1965). Ultrastructure of the cell wall of *Escherichia coli*. *J. Ultrastruct. Res.*, **12**: 247

Pexieder, T. (1977). The role of buffer osmolarity in fixation for SEM and TEM. *Experientia*, **32**: 806

Pfeiffer, S. W. (1982). Use of hydrogen peroxide to accelerate staining of ultrathin Spurr sections. *Stain Technol.*, **57**: 137

Pfenninger, K. H. (1971). The cytochemistry of synaptic densities. I. An analysis of the bismuth iodide impregnation method. *J. Ultrastruct. Res.*, **34**: 103

Pfenninger, K., Sandri, C., Abert, K. and Eugster, C. H. (1969). Contribution to the problem of structural organization of the presynaptic area. *Brain Res.*, **12**: 10

Phillips, R. (1962). Comment on localized, short spaced periodic variation in contrast on methacrylate sections. *J. Roy. Microsc. Soc.*, **81**: 41

Phillips, T. E. and Boyne, A. F. (1984). Liquid nitrogen-based quick freezing: experiences with bounce-free delivery of cholinergic nerve terminals to a metal surface. *J. Electron Microsc. Tech.*, **1**: 9

Pickett-Heaps, J. D. (1967). Preliminary attempts at ultrastructural polysaccharide localization in root tip cells. *J. Histochem. Cytochem.*, **15**: 442

Pihakaski, K. and Suoranta, U.-M. (1985). Effects of different epoxies on avoiding wrinkles in thin sections of botanical specimens. *J. Electron Microsc. Tech.*, **2**: 7

Pinching, A. J. and Brooke, R. N. L. (1973). Electron microscopy of single cells in the olfactory bulb using Golgi impregnation. *J. Neurocytol.*, **2**: 157

Pinto da Silva, P. (1984). Freeze-fracture cytochemistry. In: *Immunolabelling for Electron Microscopy* (J. M. Polak and I. M. Varndell, Eds.), pp. 179–188. Elsevier, Amsterdam

Pinto da Silva, P. (1986). Molecular cytochemistry of freeze-fractured cells: freeze-etching; fracture-label; label-fracture. *Proc. 11th Int. Cong. Electron Microsc.*, Vol. 3, p. 1991. Kyoto, Japan

Pisam, M., Caroff, A. and Rambourg, A. (1987). Two types of chloride cells in the gill epithelium of a freshwater-adapted euryhaline fish: *Lebistes reticulatus*; their modifications during adaptation to saltwater. *Am. J. Anat.*, **179**: 40

Pizzolato, P. and Lillie, R. D. (1973). Mayer's tannic acid–ferric chloride stains for mucins. *J. Histochem. Cytochem.*, **21**: 56

Plattner, H. and Bachmann, L. (1982). Cryofixation. A tool in biological ultrastructural research. *Int. Rev. Cytol.*, **79**: 237

Plattner, H. and Knoll, G. (1983). Cryofixation of biological materials for electron microscopy by the methods of spray-, sandwich-, cryogen jet- and sandwich-cryogen-jet freezing: a comparison of techniques. In: *Science of Biological Specimen Preparation* (J.-P. Revel *et al.*, Eds.), pp. 139–146. SEM Inc., AMF O'Hare, Ill.

Podolsky, R. J., Hall, T. and Hatchett, S. L. (1970). Identification of oxalate precipitates in striated muscle fibers. *J. Cell Biol.*, **44**: 699

Polivoda, A. I. and Vinetskii, Y. P. (1959). A method of preparing quartz films for electron microscopy in studies of the fine structure of erythrocytes. *Biofizika*, **4**: 599

Pollard, T. D. and Korn, E. D. (1973). Electron microscopic identification of actin associated with isolated amoeba membranes. *J. Biol. Chem.*, **248**: 448

Ponzo, M., Rozzi, Z., Picardi, R. and De Filippo, M. G. (1973). Cytochemical aspects of the zymogen in rat pancreas: electron microscopic study. *Ann. Histochem.*, **18**: 124

Poolsawat, S. S. (1973). The hardness of epoxy embedding compounds for ultrathin sectioning. *Proc. 31st Ann. Meet. EMSA*, p. 364.

Porter, K. R. and Anderson, K. L. (1982). The structure of the cytoplasmic matrix preserved by freeze-drying and freeze-substitution. *Eur. J. Cell Biol.*, **29**: 83

Porter, K. R. and Blum, J. (1953). A study in microtomy for electron microscopy. *Anat. Rec.*, **117**: 685

Portmann, R. and Koller, Th. (1976). The divalent cation method for protein-free spreading of nucleic acid molecules. *Proc. 6th Eur. Cong. Electron Microsc.*, Vol. 2, p. 546. Jerusalem, Israel

Poste, G. and Papahadjopoulos, D. (1976). Lipid vesicles as carriers for introducing materials into cultured cells: influence of vesicle lipid composition on mechanism(s) of vesicle incorporation into cells. *Proc. Natl Acad. Sci. USA*, **73**: 1603

Powell, K. A. (1987). Application of a thin carbon coating improves the selected surface technique for double immunostaining at the EM level. *Proc. Roy. Microsc. Soc.*, **22**: 157

Pratt, S. A. and Napolitano, I. (1969). Osmium binding to the surface coat of intestinal microvilli in the cat under various conditions. *Anat. Rec.*, **165**: 197

Prentø, P. (1985). A reliable and simple method for oriented epoxy embedding of tissue sections and strips using horizontal polyethylene BEEM capsules. *Stain Technol.*, **60**: 120

Pscheid, P., Schudt, C. and Plattner, H. (1981). Cryofixation of monolayer cell cultures for freeze-fracturing without chemical pretreatments. *J. Microsc.*, **121**: 149

Purdy-Ramos, S. I. (1987). Dry diamond knife approach using the LKB ultramicrotome III or V. *EMSA Bull.*, **17**: 94

Puvion-Dutilleul, F., Pichard, E., Laithier, M. and Leduc, E. H. (1987). Effect of dehydrating agents on DNA organization in herpes viruses. *J. Histochem. Cytochem.*, **35**: 635

Pyper, A. S. (1970). A glass plier kit for making glass knives for electron microscopy. *Lab. Pract.*, **19**: 491

Quintarelli, G., Bellocci, M. and Zito, R. (1973). Structural features of insoluble elastin. *Histochemie*, **37**: 49

Quintarelli, G., Cifonelli, J. A. and Zito, R. (1971b). On phosphotungstic acid staining. II. *J. Histochem. Cytochem.*, **19**: 648

Quintarelli, G. and Dellovo, M. C. (1966). Age changes in the localization and distribution of glycosaminoglycans in human hyaline cartilage. *Histochemie*, **7**: 141

Quintarelli, G., Zito, R. and Cifonelli, J. A. (1971a). On phosphotungstic acid staining. I. *J. Histochem. Cytochem.*, **19**: 641

Rambourg, A. (1967). An improved silver methenamine technique for the detection of periodic acid-reactive complex carbohydrates with the electron microscope. *J. Histochem. Cytochem.*, **15**: 409

Rambourg, A. and Leblond, C. P. (1967). Electron microscope observations on the carbohydrate-rich cell coat present at the surface of cells in the rat. *J. Cell Biol.*, **32**: 27

Rambourg, A. (1974). Staining of intracellular glycoproteins. In: *Electron Microscopy and Cytochemistry* (E. Wisse, W. Th. Daems, I. Molenaar and P. van Duijn, Eds.), pp. 245–253. Elsevier, New York

Rambourg, A., Hernandez, W. and Leblond, C. P. (1969). Detection of complex carbohydrates in the Golgi apparatus of rat cells. *J. Cell Biol.*, **40**: 395

Rampley, D. N. (1967). Embedding media for electron microscopy. *Lab. Pract.*, **16**: 591

Rampley, R. D. and Morris, A. (1972). A rapid method for polyester embedding. *Proc. 5th Eur. Reg. Conf. Electron Microsc. Manchester*, p. 224. Institute of Physics, London

Rash, J. E. and Hudson, C. S. (1979). *Freeze Fracture: Methods, Artifacts, And Interpretations.* Raven Press, New York

Rasmussen, D. H. and MacKenzie, A. P. (1968). Phase diagram for the system water/dimethyl sulfoxide. *Nature, Lond.*, **220**: 1315

Rasmussen, K. E. and Albrechtsen, J. (1974). Glutaraldehyde. The influence of pH, temperature, and buffering on the polymerization rate. *Histochemistry*, **38**: 19

Redmond, B. L. and Bob, C. (1984). A fixation/dehydration/infiltration apparatus that minimizes human exposure to harmful chemicals. *J. Electron Microsc. Tech.*, **1**: 97

Reedy, M. K. (1968). Ultrastructure of insect flight muscle. I. Screw sense and structural grouping in the rigor cross-bridge lattice. *J. Mol. Biol.*, **31**: 155

Reichelt, R., König, T. and Wangermann, G. (1977). Preparation of microgrids as specimen supports for high resolution electron microscopy. *Micron*, **8**: 29

Reichle, R. (1972). A Teflon tipped probe for easy manipulation of ultrathin sections in the knife trough. *Stain Technol.*, **47**: 171

Reid, N. (1974). Ultramicrotomy. In: *Practical Methods in Electron Microscopy*, Vol. 3 (A. M. Glauert, Ed.). North-Holland, Amsterdam

Reimann, B. (1961). Zur Verwendbarkeit von Rutheniumrot als elektronenmikroskopische kontrastierungsmittle. *Mikroskopie*, **16**: 224

Reimer, L. (1959). Quantitative Untersuchung zur Massenabnahme von Einbettungsmitten (Methacrylat, Vestopal und Araldite) unter Electionenbeschuss. *Z. Naturforsch.*, **146**: 566

Reimer, L. (1965). Irradiation changes in organic and inorganic objects. *Lab. Invest.*, **14**: 1082

Reimer, L. (1967). *Elektronenmikroskopische Untersuchungs—und Praparationsmethoden.* 2. Aufl. Springer-Verlag, Berlin

Reis, K. J., Ayoub, E. M. and Boyle, M. D. P. (1984). Streptococcal Fc receptors. II. Comparison of the reactivity of a receptor from a group C streptococcus with staphylococcal protein A. *J. Immunol.*, **132**: 3098

Reith, A., Kraemer, M. and Vassy, J. (1984). The influence of mode of fixation, type of fixative and vehicles on the same rat liver: a morphometric/stereologic study by light and electron microscopy. *Scan. Electron Microsc.*, **2**: 645

Resch, J., Tunkel, D., Stoeckert, C. and Beer, M. (1980). Osmium-labelled polysaccharides for atomic microscopy. *J. Mol. Biol.*, **138**: 673

Revell, R. S. M., Agar, A. W. and Lee, A. M. (1955). The preparation of uniform plastic films. *Br. J. Appl. Phys.*, **6**: 23

Reymond, O. L. (1986). The diamond knife semi: a substitute for glass or conventional diamond knives in the ultramicrotomy of thin and semithin sections. *Bas. Appl. Histochem.*, **30**: 487

Reynolds, E. S. (1963). The use of lead citrate at high pH as an electron opaque stain in electron microscopy. *J. Cell Biol.*, **17**: 208

Reynolds, J. A. (1973). Red cell membranes: fact and fancy. *Fed. Proc.*, **32**: 2034

Richards, F. M. and Knowles, J. R. (1968). Glutaraldehyde as a cross-linking agent. *J. Mol. Biol.*, **37**: 231

Richards, G. J. and Da Prada, M. (1977). Uranaffin reaction: a new cytochemical technique for the localization of adenine nucleotides in organelles storing biogenic amines. *J. Histochem. Cytochem.*, **25**: 1322

Richards, R. R. (1979). Fluorocarbon coating as a means of improving paraffin sectioning with Ralph glass knives. *Stain Technol.*, **54**: 292

Richardson, R. L., Hinton, D. M. and Campion, D. R. (1983). An improved method for storing and using stains in electron microscopy. *J. Electron Microsc.*, **32**: 216

Richardson, W. D. and Davies, H. G. (1980). Quantitative observations on the kinetics and mechanisms of binding of electron stains to thin sections through hen erythrocytes. *J. Cell Sci.*, **46**: 253

Ridgway, R. L. and Chestnut, M. H. (1984). Processing small tissue specimens in acrylic resins for ultramicrotomy: improved handling and orientation. *J. Electron Microsc. Tech.*, **1**: 205

Riede, U. N., Lobingen, A. Grünholz, P., Steimer, R. and Sandritter, W. (1976). Einfluss einer einstündigen Autolyse auf die quantitative Zytoarchitektur der Rattenleberzelle (Eine ultrastrukturellmorphometrische). *Beitr. Path.*, **157**: 391

Rieder, C. L. (1981). Thick and thin serial sectioning for the three-dimensional reconstruction of biological ultrastructure. In: *Methods in Cell Biology*, Vol. 22 (J. Turner, Ed.), pp. 215–249. Academic Press, New York

Rieder, C. L. and Bowser, S. S. (1983). Factors which influence light microscopic visualization of biological material in sections prepared for electron microscopy. *J. Microsc.*, **132**: 71

Riehle, U. and Höchli, M. (1973). The theory and technique of high pressure freezing. In: *Freeze-Etching: Techniques and Applications* (E. L. Benedetti and P. Favard, Eds.), pp. 31–61. Sociéte Francais de Microscopie Electronique, Paris

Riemersma, J. C. (1963). Osmium tetroxide fixation of lipids: nature of the reaction product. *J. Histochem. Cytochem.*, **11**: 436

Riemersma, J. C. (1970). Chemical effects of fixation in biological systems. In: *Some Biological Techniques in Electron Microscopy* (D. F. Parsons, Ed.). Academic Press, New York

Riemersma, J. C., Alsbach, E. J. J. and De Bruijn, W. C. (1984). Chemical aspects of glycogen contrast staining by potassium osmate. *Histochem. J.*, **16**: 123

Rigler, M. W. and Patton, J. S. (1984). A simple inexpensive cryogenic storage device for microscopy specimens. *J. Microsc.*, **134**: 335

Rinehart, J. F. and Abul-Haj, S. K. (1951). An improved method for histologic demonstration of acid mucopolysaccharides in tissues. *A.M.A. Arch. Pathol.*, **52**: 189

Ringo, D. L., Brennan, E. F. and Cota-Robles, E. H. (1982). Epoxy resins are mutagenic: implications for electron microscopists. *J. Ultrastruct. Res.*, **80**: 280

Ringo, D. L., Read, D. B. and Cota-Robles, E. H. (1984). Glove materials for handling epoxy resins. *J. Electron Microsc. Tech.*, **1**: 417

Risueño, C., Fernández-Gomez, E. and Giménez-Martín, G. (1973). Nucleoli under the electron microscope by silver impregnation. *Mikroskopie*, **29**: 292

Riva, A. (1974). A simple and rapid staining method for enhancing the contrast of tissues previously treated with uranyl acetate. *J. Microscopie*, **19**: 105

Rivlin, P. K. and Raymond, P. A. (1987). Use of osmium tetroxide–potassium ferricyanide in reconstructing cells from serial ultrathin sections. *J. Neurosci. Meth.*, **20**: 23

Robards, A. W. (1968). On the ultrastructure of differentiating secondary xylem in willow. *Protoplasma*, **65**: 449

Robards, A. W. and Crosby, P. (1983). Optimization of plunge freezing: linear relationship between cooling rate and entry velocity into liquid propane. *Cryo-Letters*, **4**: 23

Robards, A. W. and Sleytr, U. B. (1985). *Low Temperature Methods in Biological Electron Microscopy*. Elsevier, Amsterdam

Roberts, I. M. (1970). Reduction of compression artifacts in ultrathin sections by the application of heat. *J. Microsc.*, **92**: 57

Roberts, I. M. (1975). Tungsten coating—a method of improving glass microtome knives for cutting ultrathin sections. *J. Microsc.*, **103**: 113

Roberts, I. M. (1986). Immunoelectron microscopy of extracts of virus-infected plants. In: *Electron Microscopy of Proteins*, Vol. 5 (J. R. Harris and R. W. Horne, Eds.), p. 293. Academic Press, London

Roberts, I. M., Milne, R. G. and van Regenmortel, M. H. V. (1982). Suggested terminology for virus/antibody interactions observed by electron microscopy. *Intervirology*, **18**: 147

Roberts, W. J. (1935). A new procedure for the detection of gold in animal tissues. *Proc. Roy. Acad. Sci., Amsterdam*, **38**: 540

Robertson, E. A. and Schultz, R. L. (1970). The impurities in commercial glutaraldehyde and their effect on the fixation of brain. *J. Ultrastruct. Res.*, **30**: 275

Robertson, W. M., Storey, B. and Griffiths, B. S. (1984). An interference technique for measuring the thickness of semithin and thick sections. *J. Microsc.*, **133**: 121

Rogers, G. E. and Filshie, B. K. (1962). Electron staining and fine structure of keratins. In: *Proc. 5th Int. Cong. Electron Microsc. Philadelphia*, Vol. 2, p. 1. Academic Press, New York

Roholl, P. J. M., Leene, W., Kapsenberg, M. L. and Vos, J. G. (1981). The use of tannic acid fixation for the electron microscope visualization of fluorochrome-labelled antibodies attached to cell surface antigens. *J. Immunol. Meth.*, **42**: 285

Roland, J.-C. (1978). General preparation and staining of thin section. In: *Electron Microscopy and Cytochemistry of Plant Cells* (J. L. Hall, Ed.). Elsevier/North-Holland, New York

Romano, E. L. and Romano, M. (1977). Staphlococcal protein A bound to colloidal gold: a useful reagent to label antigen–antibody sites in electron microscopy. *Immunochemistry*, **14**: 711

Roomans, G. M. and Sevéus, L. A. (1976). Subcellular localization of diffusible ions in the yeast *Saccharomyces cerevisciae*: quantitative microprobe analysis of thin freeze-dried sections. *J. Cell Sci.*, **21**: 119

Roos, A. and Baron, W. F. (1981). Intracellular pH. *Physiol. Rev.*, **61**: 296

Roos, N. and Barnard, T. (1985). A comparison of subcellular element concentrations in frozen-dried, plastic-embedded, dry-cut sections and frozen-dried cryosections. *Ultramicroscopy*, **17**: 335

Roozemond, R. C. (1969). The effect of fixation with formaldehyde and glutaraldehyde on the composition of phospholipids extractable from rat hypothalamus. *J. Histochem. Cytochem.*, **17**: 482

Rosa, J. J. and Sigler, P. B. (1974). The site of covalent attachment in the crystalline osmium–$tRNA^{fMET}$ isomorphous derivative. *Biochemistry*, **13**: 5102

Rosenberg, M., Bartl, P. and Lesko, J. (1960). Water-soluble methacrylate as an embedding medium for the preparations of ultrathin sections. *J. Ultrastruct. Res.*, **4**: 298

Rosenthal, A. S., Moses, H. L., Beaver, D. L. and Schuffman, S. S. (1966). Lead ion phosphatase histochemistry. I. Nonenzymatic hydrolysis of nucleoside phosphates by lead ion. *J. Histochem. Cytochem.*, **14**: 698

Rossi, G. L. (1975). Simple apparatus for perfusion fixation for electron microscopy. *Experientia*, **31**: 998

Rossi, G. L., Lugenbuhl, H. and Probst, D. (1970). A method for ultrastructural study of lesions found in conventional histological sections. *Virchows Arch. Abt. A. Pathl. Anat.*, **350**: 216

Rostgaard, J. and Buchmann, B. (1974). Problems in ultraviolet polymerization of embedding media for electron microscopy. *J. Microsc.*, **102**: 187

Rostgaard, J. and Tranum Jensen, J. (1980). A procedure for minimizing cellular shrinkage in electron microscope preparation: a quantitative study on frog gall bladder. *J. Microsc.*, **119**: 213

Roth, J. (1982). The preparation of protein A–gold complexes with 3 nm and 15 nm gold particles and their use in labelling multiple antigens on ultrathin sections. *Histochem. J.*, **14**: 791

Roth, J., Bendayan, M., Carlemalm, E., Villiger, W. and Garavito, M. (1981). Enhancement of structural preservation and immunocytochemical staining in low temperature embedded pancreatic tissue. *J. Histochem. Cytochem.*, **29**: 663

Roth, J. and Binder, M. (1978). Colloidal gold, ferritin and peroxidase as markers for electron microscopic double labelling lectin techniques. *J. Histochem. Cytochem.*, **26**: 163

Roth, L. E., Jenkins, R. A., Johnson, C. W. and Robinson, R. W. (1963). Additional stabilizing conditions for electron microscopy of the mitotic apparatus of giant amoebae. *J. Cell Biol.*, **19**: 62A

Roth, W. J. and Hinckley, C. C. (1981). Synthesis and characterization of osmyl–amino acid complexes. *Inorg. Chem.*, **20**: 2023

Rothstein, A. (1970). A reappraisal of the action of uranyl ions on cell membranes. In: *Effects of Metals on Cells, Subcellular Elements, and Macromolecules.* (J. Maniloff, J. R. Coleman and M. W. Miller, Eds.). Charles C. Thomas, Springfield, Ill.

Rothstein, A. and Meier, R. (1951). The relationship of the cell surface to metabolism. VI. The chemical nature of uranium-complexing groups of the cell surface. *J. Cell Comp. Physiol.*, **38**: 245

Roubos, E. W. and van der Wal-Divendal, R. M. (1980). Ultrastructural analysis of peptide-hormone release by exocytosis. *Cell Tiss. Res.*, **207**: 267

Rowden, G. (1968). A simple system to alleviate handling damage to electron microscopy grids. *J. Roy. Microsc. Soc.*, **88**: 595

Rowden, G. and Lewis, M. G. (1974). Experience with a three-hour electron microscopy biopsy service. *J. Clin. Pathol.*, **27**: 505

Rowley, G. and Moran, D. T. (1975). A simple procedure for mounting wrinkle-free sections on Formvar-coated slot grids. *Ultramicroscopy*, **1**: 151

Rowley, J. R. (1971). Resolution of channels in the exine by translocation of colloidal iron. *Proc. 29th Ann. EMSA Meet.*, p. 352. Claitor's Pub. Division, Baton Rouge, La.

Rubbo, S. D., Gardner, J. F. and Webb, R. L. (1967). Biocidal activities of glutaraldehyde and related compounds. *J. Appl. Bact.*, **30**: 78

Ruddell, C. L. (1971). Embedding media for 1–2 micron sectioning. 3. Hydroxyethyl methacrylate–benzoyl peroxide activated with pyridine. *Stain Technol.*, **46**: 77

Ruddell, C. L. (1983). Initiating polymerization of glycol methacrylate with cyclic diketo carbon acids. *Stain Technol.*, **58**: 329

Russo, E., Giancotti, V., Cosimi, S. and Crane-Robinson, C. (1981). Identification of an heterologous histone complex using reversible crosslinking. *Int. J. Biol. Macromol.*, **3**: 367

Ryan, K. P. and Liddicoat, M. I. (1987). Safety considerations regarding the use of propane and other liquefied gases as coolants for rapid freezing purposes. *J. Microsc.*, **147**: 337

Ryan, K. P. and Purse, D. H. (1984). Rapid freezing: specimen supports and cold gas layers. *J. Microsc.*, **136**: RP5.

Ryan, K. P. and Purse, D. H. (1985a). Plunge-cooling of tissue blocks: determinants of cooling rates. *J. Microsc.*, **140**: 47

Ryan, K. P. and Purse, D. H. (1985b). A simple plunge-cooling device for preparing biological specimens for cryotechniques. *Mikroskopie*, **42**: 247

Ryan, K. P., Purse, D. H., Robinson, S. G. and Wood, J. W. (1987). The relative efficiency of cryogens used for plunge-cooling biological specimens. *J. Microsc.*, **145**: 89

Rybicka, K. (1981). Simultaneous demonstration of glycogen and protein in glycosomes of cardiac tissue. *J. Histochem. Cytochem.*, **29**: 4

Ryter, A. and Kellenberger, E. (1958). Etude au microscope électronique de plasma contenant de l'acide desoxyribonucleique. I. Les nucléotides des bactéries en croissance active. *Z. Naturforsch.*, **13b**: 597

Sabatini, D. D., Bensch, K. and Barrnett, R. J. (1962). New means of fixation for electron microscopy and histochemistry. *Anat. Rec.*, **142**: 274

Sabatini, D. D., Bensch, K. and Barrnett, R. (1963). Cytochemistry and electron microscopy. The preservation of cellular structure and enzymatic activity by aldehyde fixation. *J. Cell Biol.*, **17**: 19

Sabatini, D. D., Miller, F. and Barrnett, R. J. (1964). Aldehyde fixation for morphological and enzyme histochemical studies with the electron microscope. *J. Histochem. Cytochem.*, **12**: 57

Saetersdal, T. S., Myklebust, R., Justesen, N. P. B. and Olsen, W. C. (1974). Ultrastructural localization of calcium in the pigeon papillary muscle as demonstrated by cytochemical studies and X-ray microanalysis. *Cell Tiss. Res.*, **155**: 57

Safer, D., Hainfeld, J., Wall, J. S. and Reardon, J. E. (1982). Biospecific labeling with undecagold: visualization of the biotin-binding site on avidin. *Science, N.Y.*, **218**: 290

Saffitz, J. E., Gross, R. W., Williamson, J. R. and Sobel,

B. E. (1981). Autoradiography of phosphatidyl choline. *J. Histochem. Cytochem.*, **29**: 371

Saito, T. and Takizawa, T. (1986). Rapid freeze-substitution enzyme cytochemistry. *Proc. XIth Int. Cong. Electron Microsc.*, Kyoto, Vol. 3, p. 1995.

Sakai, T. (1980). Relation between thickness and interference colours of biological ultrathin section. *J. Electron Microsc.*, **29**: 369

Salema, R. and Brandão, I. (1973). The use of PIPES buffer in the fixation of plant cells for electron microscopy. *J. Submicrosc. Cytol.*, **5**: 79

Sampson, H. W., Dill, R. E., Matthews, J. L. and Martin, J. H. (1970). An ultrastructural investigation of calcium-dependent granules in the rat neurophil. *Brain Res.*, **22**: 157

Sanan, D. A., van der Merwe, E. L. and Loehner, A. (1985). Comparison of the effects of immersion and perfusion fixation on the ultrastructure of mitochondria from severely ischaemic myocardium. *South Afric. J. Sci.*, **81**: 564

Sannes, P. L., Katsuyama, T. and Spicer, S. S. (1978). Tannic acid–metal salt sequences for light and electron microscopic localization of complex carbohydrates. *J. Histochem. Cytochem.*, **26**: 55

Sannes, P. L., Spicer, S. S. and Katsuyama, T. (1979). Ultrastructural localization of sulfated complex carbohydrates with a modified iron diamine procedure. *J. Histochem. Cytochem.*, **27**: 1108

Sato, T. (1967). A modified method for lead staining of thin sections. *J. Electron Microsc.*, **16**: 133

Sato, T. and Shamoto, M. (1973). A simple rapid polychrome stain for epoxy-embedded tissue. *Stain Technol.*, **48**: 223

Sawers, J. R. (1972). Optimizing sectioning with diamond knives. In: *Proc. 30th Ann. EMSA Meet.*, p. 386. Claitor's Pub. Division, Baton Rouge, La.

Saxena, O. G. (1967). New titrimetric microdetermination of osmium. *Microchem. J.*, **12**: 609

Scala, C., Govoni, E., Cenacchi, G. and Cantelli, G. B. (1977). Improvements in embedding media for electron microscopy: A silicone-epoxy copolymer. *J. Submicrosc. Cytol.*, **9**: 97

Schierenberg, E., Cole, T., Carlson, C. and Sidio, W. (1986). Computer-aided three-dimensional reconstruction of nematode embryos from EM serial sections. *Exp. Cell. Res.*, **166**: 247

Schiff, R. I. and Gennaro, J. F. (1979). The role of the buffer in the fixation of biological specimens for transmission and scanning electron microscopy. *Scanning*, **2**: 135

Schinz, R. (1987). Complementary replica of freeze-fractured human lymphocyte nuclei. *Experientia*, **43**: 423

Schliwa, M. (1977). Influence of calcium on intermicrotubule bridges within heliozoan axonemes. *J. Submicrosc. Cytol.*, **9**: 221

Schneider, H. and Sasaki, P. J. (1976). A blade guide for hand trimming resin blocks for ultramicrotomy. *Stain Technol.*, **51**: 238

Schnepf, E. (1972). Structural modifications in the plasmalemma of *Aphelidium*-infected *Scenedesmus* cells. *Protoplasma*, **75**: 155

Schnepf, E., Hausmann, K. and Herth, W. (1982). The osmium tetroxide–potassium ferrocyanide (OsFeCN) staining technique for electron microscopy: a critical evaluation using ciliates, algae, mosses, and higher plants. *Histochemistry*, **76**: 261

Schoenwolf, G. C. (1982). A simple method for fabricating inexpensive storage boxes for 6.35 mm-thick glass knives. *Stain Technol.*, **57**: 185

Schofield, B. H., Williams, B. R. and Doty, S. B. (1975). Alcian blue staining of cartilage for electron microscopy: application of the critical electrolyte concentration principle. *Histochem. J.*, **7**: 139

Schooley, C. (1985). A poor person's guide to cryotechnique: part 1. *EMSA Bull.*, **15**: 98

Schreil, W. H. (1964). Studies on the fixation of artificial and bacterial DNA plasms for the electron microscopy of thin sections. *J. Cell. Biol.*, **22**: 1

Schröder, M. (1980). Osmium tetroxide *cis* hydroxylation of unsaturated substances. *Chem. Rev.*, **80**: 187

Schultz, R. L. and Case, N. M. (1970). A modified aldehyde perfusion technique for preventing certain artifacts in electron microscopy of the central nervous system. *J. Microsc.*, **92**: 69

Schultz, R. L. and Wagner, D. O. (1986). Membrane alterations in cerebral cortex when using PIPES buffer. *J. Neurocytol.*, **15**: 461

Schulz, A. (1977). A reliable method of preparing undecalcified human bone biopsies for electron microscopic investigation. *Microsc. Acta*, **80**: 7

Schwab, M. E. and Thoenen, H. (1978). Selective binding, uptake and retrograde transport of tetanus toxin by nerve terminals in the rat iris. *J. Cell Biol.*, **77**: 1

Schwarzacher, H. G. and Wachtler, F. (1983). Nucleolus organizer regions and nucleoli. *Hum. Genet.*, **63**: 89

Scopsi, L. (1989). Silver-enhanced colloidal gold method. In: *Colloidal Gold: Principles, Methods, and Applications*, Vol. 1 (M. A. Hayat, Ed.). Academic Press, San Diego and London

Scopsi, L. and Larsson, L.-I. (1985). Increased sensitivity in immunocytochemistry. Effects of double application of antibodies and of silver intensification on immunogold and peroxidase–antiperoxidase staining techniques. *Histochemistry*, **82**: 321

Scopsi, L. and Larsson, L.-I. (1986). Colloidal gold probes in immunocytochemistry. An optimization of their application in light microscopy by use of silver intensification procedures. *Med. Biol.*, **64**: 139

Scopsi, L., Larsson, L.-I., Bastholm, L. and Nielsen, M. H. (1986). Silver-enhanced colloidal gold probes as markers for scanning electron microscopy. *Histochemistry*, **86**: 35

Scott, J. E. (1971). Phosphotungstate: a 'universal' (non-specific) precipitant for polar polymers in acid solution. *J. Histochem. Cytochem.*, **19**: 689

Scott, J. E. (1972a). The histochemistry of alcian blue. Note on the presence and removal of boric acid as the major diluent in alcian blue 8GX. *Histochemistry*, **29**: 129

Scott, J. E. (1972b). Histochemistry of alcian blue. III. The molecular biological basis of staining by alcian blue 8GX and analogous phthalocyanins. *Histochemie*, **32**: 191

Scott, J. E. (1972c). Histochemistry of alcian blue. II. The structure of alcian blue 8GX. *Histochemie*, **30**: 215

Scott, J. E. (1973a). Affinity, competition and specific interactions in the biochemistry and histochemistry of polyelectrolytes. *Trans. Biochem. Soc.*, **1**: 787

Scott, J. E. (1973b). Phosphotungstic acid 'Schiff-reactive' but not a 'glycol reagent'. *J. Histochem. Cytochem.*, **21**: 1084

Scott, J. E. and Dorling, J. (1965). Differential staining of acid glycosaminoglycans (mucopolysaccharides) by alcian blue in salt solutions. *Histochemie*, **5**: 221

Scott, J. E., Dorling, J. and Stockwell, R. A. (1968). Reversal of protein blocking of basophilia in salt solutions: implications on the localization of polyanions using alcian blue. *J. Histochem. Cytochem.*, **16**: 383

Scott, J. E. and Glick, D. (1971). The invalidity of 'phosphotungstic acid as a specific electron stain for complex carbohydrates'. *J. Histochem. Cytochem.*, **19**: 63

Scott, J. E. and Stockwell, R. A. (1967). On the use and abuse of the critical electrolyte concentration approach to the localization of tissue polyanions. *J. Histochem. Cytochem.*, **15**: 111

Scott, J. E. and Willett, I. (1966). Binding of cationic dyes to nucleic acid and other biological polyanions. *Nature, Lond.*, **209**: 985

Scott, J. R. and Thurston, E. L. (1975). Ultramicrotome specimen chuck for improved viewing in block trimming. *Stain Technol.*, **50**: 290

Sealock, R. (1980). Identification of regions of high acetylcholine receptor density in tannic acid fixed postsynaptic membranes from electric tissue. *Brain Res.*, **199**: 267

Seigel, N. J., Spargo, B. H. and Kashgarian, M. (1973). An evaluation of routine electron microscopy in the examination of renal biopsies. *Nephron*, **10**: 209

Seliger, W. G. (1968). The production of large, epoxy-embedded, 50-μ sections by precision sawing; a preliminary to survey for ultrathin sectioning. *Stain Technol.*, **43**: 269

Seligman, A. M., Hanker, J. S., Wasserkrug, H., DiMochowski, H. and Katzoff, L. (1965). Histochemical demonstration of some oxidized macromolecules with thiocarbohydrazide (TCH) or thiosemicarbazide (TSC) and osmium tetroxide. *J. Histochem. Cytochem.*, **13**: 629

Seligman, A. M., Wasserkrug, H. L. and Hanker, J. S. (1966). A new staining method (OTO) for enhancing contrast of lipid-containing membranes and droplets in osmium tetroxide fixed tissue with osmiophilic thiocarbohydrazide (TCH). *J. Cell Biol.*, **30**: 424

Senda, T. and Fujita, H. (1987). Ultrastructural aspects of quick-freezing, deep-etching replica images of the cytoskeletal system in anterior pituitary secretory cells of rats and mice. *Arch. Histol. Jap.*, **50**: 49

Sevéus, L. (1978). Preparation of biological material for X-ray microanalysis of diffusible elements. *J. Microsc.*, **112**: 269

Sewell, B. T., Bouloukos, C. and von Holt, C. (1984). Formaldehyde and glutaraldehyde in the fixation of chromatin for electron microscopy. *J. Microsc.*, **136**: 103

Shah, D. O. (1969). Interaction of uranyl ions with phospholipid and cholesterol monolayer. *J. Colloid Interf. Sci.*, **29**: 210

Shah, D. O. (1970). The effect of potassium permanganate on lecithin and cholesterol monolayer. *Biochim. Biophys. Acta*, **211**: 358

Shalla, T. A., Carroll, T. W. and Dezoeten, G. A. (1964). Penetration of stain in ultrathin sections of tobacco mosaic virus. *Stain Technol.*, **39**: 257

Shannon, W. A. (1974). A simplified agar–polyethylene disc method for the Sorvall TC-2 tissue sectioner. *Stain Technol.*, **49**: 109

Shearer, T. P. and Hunsicker, L. G. (1980). A rapid method for embedding tissues for electron microscopy using 1,4-dioxane and polybed 812. *J. Histochem. Cytochem.*, **28**: 465

Shelton, E. and Mowczko, W. E. (1977). Membrane blebs: a fixation artifact. *J. Cell Biol.*, **75**: 206a

Sheridan, W. F. and Barrnett, R. J. (1967). Cytochemical studies of chromosomal ultrastructure. *J. Cell Biol.*, **35**: 125a

Shigenaka, Y., Watanabe, K. and Kaneda, M. (1973). Effects of glutaraldehyde and osmium tetroxide on hypotrichous ciliates, and determination of the most satisfactory fixation methods for electron microscopy. *J. Protozool.*, **20**: 414

Shinagawa, Y. and Uchida, Y. (1961). On the specimen damage of spinal cord due to polymerization of embedding media. *J. Electron Microsc.*, **10**: 86

Shinagawa, Y., Yahara, S. and Uchida, Y. (1962). Polymerization of epoxy resins in the cold for electron microscopy. *J. Electron Microsc.*, **11**: 133

Shinji, Y., Shinji, E. and Mizuhira, V. (1974). The simple staining technique for 1,2-glycol groups by alkaline bismuth solution, unlike demonstration of periodate reactive sites. *Proc. 8th Inter. Cong. Electron Microsc.*, Vol. 2, p. 136. Canberra.

Shinji, Y., Shinji, E. and Mizuhira, V. (1975). A new electron microscopic histo-cytochemical staining method: demonstration of glycogen particles. *Acta Histochem. Cytochem.*, **8**: 139

Shires, T. K., Johnson, M. and Richter, K. M. (1969). Hematoxylin staining of tissues embedded in epoxy resins. *Stain Technol.*, **44**: 21

Shore, I. and Moss, J. (1988). The use of Formvar films on both sides of a section to facilitate the selected surface technique for double immunostaining at the electron microscope level. *Histochem. J.*, **20**: 183

Shukla, D. D. and Gough, K. H. (1979). The use of protein A, from *Staphylococcus aureus*, in immune electron microscopy for detecting plant virus particles. *J. Gen. Virol.*, **45**: 533

Shukla, D. D. and Gough, K. H. (1983). Characteristics of the protein A-immunosorbent electron microscopic technique for detecting plant virus particles. *Acta Phytopath. Hung.*, **18**: 173

Sievers, J. (1971). Basic two-dye stains for epoxy-embedded 0.3–1 μ sections. *Stain Technol.*, **46**: 195

Silva, M. T. (1967). Electron microscopic study on the effect of the oxidation of ultrathin sections of *Bacillus cerus* and *Bacillus megaterium*. *J. Ultrastruct. Res.*, **18**: 345

Silva, M. T. (1971). Changes induced in the ultrastructure of the cytoplasmic and intracytoplasmic membranes of several gram-positive bacteria by variations in OsO_4 fixation. *J. Microsc.*, **93**: 227

Silva, M. T., Guerra, F. C. and Magãlhaes, M. M. (1968). The fixative action of uranyl acetate in electron microscopy. *Experientia*, **24**: 1974

Silva, M. T., Mota, J. M. S., Melo, J. V. C. and Guerra, F. C. (1971). Uranyl salts as fixatives for electron microscopy. Study of the membrane ultrastructure and phospholipid loss in Bacilli. *Biochim. Biophys. Acta*, **233**: 513

Silva, M. T. and Sousa, J. C. F. (1973). Ultrastructure of the cell wall and cytoplasmic membrane of gram-negative bacteria with different fixation techniques. *J. Bact.*, **113**: 953

Silva, M. T., Sousa, J. C. F., Polónia, J. J., Macedo, M. A. E. and Parente, A. M. (1976). Bacterial mesosomes: Real structures or artifacts. *Biochim. Biophys. Acta*, **443**: 92

Silverman, L. and Glick, D. (1969). The reactivity and staining of tissue proteins with phosphotungstic acid. *J. Cell. Biol.*, **40**: 761

Silverman, L., Schreiner, B. and Glick, D. (1969). Measurement of thickness within sections by quantitative electron microscopy. *J. Cell Biol.*, **40**: 768

Silvester, N. R. and Burge, R. E. (1959). A quantitative estimation of the uptake of two new electron stains by the cytoplasmic membrane of ram sperm. *J. Biophys. Biochem. Cytol.*, **6**: 179

Silvester, N. R., Marchese-Ragona, S. and Johnstone, D. N. (1982). The relative efficiency of various fluids in the rapid freezing of protozoa. *J. Microsc.*, **128**: 175

Simionescu, N. and Simionescu, M. (1976a). Galloylglucoses of low molecular weight as mordants in electron microscopy. I. Procedure and evidence for mordanting effect. *J. Cell Biol.*, **70**: 622

Simionescu, N. and Simionescu, M. (1976b). Galloylglucoses of low molecular weight as mordants in electron microscopy. II. The moiety and functional groups possibly involved in the mordanting effect. *J. Cell Biol.*, **70**: 622

Simon, G. T., Thomas, J. A., Chorneyko, K. A. and Carlemalm, E. (1987). Rapid embedding in Lowicryl K4M for immunoelectron microscopic studies. *J. Electron Microsc. Tech.*, **6**: 317

Simpson, W. L. (1941). An experimental analysis of the Altman technique of freezing drying. *Anat. Rec.*, **80**: 173

Simson, J. A. V. (1977). The influence of fixation on the carbohydrate cytochemistry of rat salivary gland secretory granules. *Histochem. J.*, **9**: 645

Simson, J. A. V. and Spicer, S. S. (1975). Selective subcellular localization of cations with variants of the potassium (pyro) antimonate technique. *J. Histochem. Cytochem.*, **23**: 575

Singer, M. N., Krishnan, N. and Fyfe, D. A. (1972). Penetration of ruthenium red into peripheral nerve fibers. *Anat. Rec.*, **173**: 375

Singley, C. T. and Solursh, M. (1979). The use of tannic acid for the ultrastructural visualization of hyaluronic acid. *Histochemistry*, **65**: 93

Sitte, H. (1984). Process of ultrathin sectioning. In: *Science of Biological Specimen Preparation for Microscopy and Microanalysis*, pp. 97–104. SEM Inc., Chicago

Sjödahl, J. (1977). Structural studies in 4 repetitive Fe-binding regions in protein A from *Staphylococcus aureus*. *Eur. J. Biochem.*, **78**: 471

Sjöhalm, J., Bjerkén, A. and Sjöquist, J. (1973). Protein A from *Staphylococcus aureus*. XIV. The effect of nitration of protein A with tetranitromethane and subsequent reduction. *J. Immunol.*, **110**: 1562

Sjöholm, I. (1975). Protein A from *Staphylococcus aureus*. Spectropolarimetric and spectrophotometric studies. *Eur. J. Biochem.*, **51**: 55

Sjöstrand, F. S. (1963). A new ultrastructural element of the membranes in mitochondria and of some cytoplasmic membranes. *J. Ultrastruct. Res.*, **9**: 340

Sjöstrand, F. S. (1967). *Electron Microscopy of Cells and Tissues*, Vol. 1. Academic Press, New York

Sjöstrand, F. S. (1969). In: *Physical Techniques in Biological Research*, Vol. 3, Part C (A. W. Pollister, Ed.), pp. 169–200. Academic Press, New York and London

Sjöstrand, F. S. and Barajas, L. (1968). Effect of modifications on conformation of protein molecules on structure of mitochondrial membranes. *J. Ultrastruct. Res.*, **25**: 121

Skaer, H. (1982). Chemical cryoprotection for structural studies. *J. Microsc.*, **125**: 137

Skaer, R. J. (1981). A new *en bloc* stain for cell membranes: tannic methylamine tungstate. *J. Microsc.*, **123**: 111

Skaer, R. J. and Whytock, S. (1976). The fixation of nuclei and chromosomes. *J. Cell Sci.*, **20**: 221

Skaer, R. J. and Whytock, S. (1977). Chromatin-like artifacts from nuclear sap. *J. Cell Sci.*, **26**: 301

Sleytr, U. B. and Robards, A. W. (1977). Plastic deformation during freeze-cleavage: a review. *J. Microsc.*, **110**: 1

Slot, J. W. and Geuze, H. J. (1981). Sizing of protein A–colloidal gold probes for immunoelectron microscopy. *J. Cell Biol.*, **90**: 533

Slot, J. W. and Geuze, H. J. (1982). Ultracryotomy of polyacrylamide embedded tissue for immunoelectron microscopy. *Biol. Cell*, **44**: 325

Slot, J. W. and Geuze, H. J. (1984). Gold markers for single and double immunolabeling of ultrathin cryosections. In: *Immunolabelling for Electron Microscopy* (J. M. Polak and I. M. Varndell, Eds.), pp. 129–142. Elsevier, Amsterdam.

Slot, J. W. and Geuze, H. J. (1985). A new method of preparing gold probes for multiple-labelling cytochemistry. *Eur. J. Cell Biol.*, **38**: 87

Small, J. V. (1984). Polyvinylalcohol, a water soluble resin suitable for electron microscope immunocytochemistry. *Proc. 8th Eur. Conf. Electron Microsc.*, Vol. 3, p. 1799

Smith, A. R. and Wren, M. J. (1983). A Ralph knife holder assembly for use with the Sorvall Porter-Blum MT-1 ultramicrotome. *Stain Technol.*, **58**: 235

Smith, J. M. (1984). Imaging cytoskeletal networks in thick sections. *J. Electron Microsc.*, **33**: 378

Smith, J. T., Funckes, A. J., Barak, A. J. and Thomas, L. E. (1957). Cellular lipoproteins. I. The insoluble lipoprotein of whole liver cells. *Exp. Cell Res.*, **13**: 96

Smith, J. W. and Stuart, R. J. (1971). Silver staining of ribosomal proteins. *J. Cell Sci.*, **9**: 253

Smith, K. and Luther, P. D. (1976). Problems associated with particulate contamination of cell cultures with relation to ultrathin sectioning techniques. *J. Microsc.*, **108**: 317

Smith, M. (1984). Carbon coating. *Proc. Roy. Microsc. Soc.*, **19**: 77

Smith, P. R. (1981). A trough designed to facilitate the

coating of electron microscope grids. *Philips Electron Optics Bull.*, **115**: 13

Smith, R. E. and Farquhar, M. G. (1966). Lysosome function in the regulation of the secretory process in the cells of the anterior pituitary gland. *J. Cell Biol.*, **31**: 319

Smithwick, E. B. (1985). Cautions, common sense, and rationale for the electron microscopy laboratory. *J. Electron Microsc. Tech.*, **2**: 193

Snodgress, A. B., Dorsey, C. H., Bailey, G. W. H. and Dickson, L. G. (1972). Conventional histopathologic staining methods compatible with Epon-embedded, osmicated tissue. *Lab. Invest.*, **26**: 329

Sogo, J. M., Portmann, R., Kaufmann, P. and Koller, Th. (1975). Adsorption of DNA molecules to different support films. *J. Microsc.*, **104**: 187

Soloff, B. L. (1973). Buffered potassium permanganate–uranyl acetate–lead citrate staining sequence for ultrathin sections. *Stain Technol.*, **48**: 159

Soma, L. R. (Ed.) (1971). *Textbook of Veterinary Anesthesia.* Williams and Wilkins, Baltimore

Somlyo, A. V., Bond, M., Silcox, J. C. and Somlyo, A. P. (1985). Direct measurements of intracellular elemental composition utilizing a new approach to freezing in vivo. *Proc. 43rd Ann. Meet. EMSA*, p. 10. San Francisco Press, San Francisco

Sommer, J. R. (1977). To cationize glass. *J. Cell Biol.*, **75**: 245a

Sotelo, J. R., Garcia, R. B. and Wettstein, R. (1973). Serial sectioning study of some meiotic stages in *Scaptericus borelli* (Grylloidea). *Chromosoma*, **42**: 307

Sottocasa, G. L. (1967). An electron transport system associated with the outer membrane of liver mitochondria. A biochemical and morphological study. *J. Cell Biol.*, **32**: 415

Spaur, R. C. and Moriarty, G. C. (1977). Improvements of glycol-methacrylate. I. Its use as an embedding medium for electron microscopy studies. *J. Histochem. Cytochem.*, **25**: 163

Spicer, S. S. (1965). Diamine methods for differentiating mucosubstances histochemically. *J. Histochem. Cytochem.*, **13**: 211

Sprumont, P. and Musy, J.-P. (1971). pH effect on electron microscopical contrast with iron salt solutions. *Histochemie*, **26**: 228

Spurr, A. R. (1969). A low-viscosity epoxy resin embedding medium for electron microscopy. *J. Ultrastruct. Res.*, **26**: 31

Staehelin, L. A. and Bertaud, W. S. (1971). Temperature and contamination dependent freeze-etch images of frozen water and glycerol solutions. *J. Ultrastruct. Res.*, **37**: 146

Stang, E. and Johansen, B. V. (1988). Improved glass knives for ultramicrotomy: a procedure for tungsten coating the knife edge. *J. Ultrastruct. Mol. Struct. Res.*, **98**: 328

Stastńa, J. and Trávník, P. (1971). Electron microscopic detection of PAS-positive substances with thiosemicarbazide. *Histochemie*, **27**: 63

Stathis, E. C. and Fabrikanos, A. (1958). Preparation of colloidal gold. *Chem. Ind. (London)*, **27**: 860

Steedman, H. F. (1950). Alcian blue 8GS: a new stain for mucin. *Quart. J. Microsc. Sci.*, **91**: 477

Steere, R. L. (1957). Electron microscopy of structural detail in frozen biological specimens. *J. Biophys. Biochem. Cytol.*, **3** : 45

Stein, O. and Stein, Y. (1967). Lipid synthesis, intracellular transport, storage, and secretion. I. Electron microscopic radioautographic study of liver after injection of tritiated palmitate or glycerol in fasted and ethanol-treated rats. *J. Cell Biol.*, **33**: 319

Steinbrecht, R. A. and Zierold, K. (Eds.) (1987). *Cryotechniques in Biological Electron Microscopy.* Springer-Verlag, New York

Stempak, J. G. and Ward, R. T. (1964). An improved staining method for electron microscopy. *J. Cell Biol.*, **22**: 697

Stenberg, M., Stemme, G. and Nygren, H. (1987). An improved negative staining technique using a thin quartz membrane as sample support. *Stain Technol.*, **62**: 231

Stephenson, J. L. (1956). Ice crystal growth during the rapid freezing of tissues. *J. Biophys. Biochem. Cytol.*, **2**: 45

Stephenson, J. L. M. and Hawes, C. R. (1986). Stereology and stereometry of endoplasmic reticulum during differentiation in the maize root cap. *Protoplasma*, **131**: 32

Sterling, C. (1970). Crystal-structure of ruthenium red and stereochemistry of its pectic stain. *Am. J. Bot.*, **57**: 172

Sternberger, L. A. (1979). *Immunocytochemistry*, 2nd edn. Wiley, New York

Sternberger, L. A., Donati, E. J., Cuculis, J. J. and Petrali, J. P. (1965). Indirect immunuranium technique for staining of embedded antigen in electron microscopy. *Exp. Mol. Path.*, **4**: 112

Stevens, B. J. and Swift, H. (1966). RNA transport from nucleus to cytoplasm in *Chironomus* salivary glands. *J. Cell Biol.*, **31**: 55

Stevens, C. L., Chay, T. R. and Loga, S. (1977). Rupture of base pairing in double-stranded poly(riboadenylic acid). Poly(riboadenylic acid) by formaldehyde: medium chain lengths. *Biochemistry*, **16**: 3727

Stevens, J. K., Davis, T. L., Friedman, N. and Sterling, P. (1980). A systematic approach to reconstructuring microcircuitry by electron microscopy of serial sections. *Brain Res. Rev.*, **2**: 265

Stewart, M. and Vigers, G. (1986). Electron microscopy of frozen-hydrated biological material. *Nature, Lond.*, **319**: 631

Stockert, J. C. (1977). Sodium tungstate as a stain in electron microscopy. *Biol. Cellul.*, **29**: 211

Stockert, J. C. and Colman, O. D. (1974). Observations on nucleolar staining with osmium tetroxide. *Experientia*, **30**: 751

Stockert, J. C. and Juarranz, A. (1980). Temperature-dependent staining reaction of chromatin by alcian blue. *Z. Naturforsch.*, **35c**: 1092

Stockinger, L. and Graf, J. (1965). Elektronenmikroskopische Analyse der Osmium-Zinkjodid Methode. *Mikroskopie*, **20**: 16

Stoeckenius, W. (1960). Osmium tetroxide fixation of lipids. In: *Proc. Europ. Conf. Electron Microsc.*, Vol. 2. Delft

Stoeckenius, W. and Mahr, S. C. (1965). Studies on the reaction of osmium tetroxide with lipids and related compounds. *Lab. Invest.*, **14**: 458

Stoeckert, C. J. Jr., Beer, M., Wiggins, J. W. and Wierman,

J. C. (1984). Histone positions within the nucleosome using platinum labeling and the scanning transmission electron microscope. *J. Mol. Biol.*, **177**: 483

Stöffler, G. and Stöffler-Meilicke, M. (1984). Immunoelectron microscopy of ribosomes. *Ann. Rev. Biophys. Bioeng.*, **13**: 303

Stolinski, C. and Breathnach, A. S. (1975). *Freeze-Fracture Replication of Biological Tissues*. Academic Press, London

Stolinski, C. and Gross, M. (1969). A method for making thin large surface area carbon supporting films for use in electron microscopy. *Micron*, **1**: 340

Stollar, D. and Grossman, L. (1962). The reaction of formaldehyde with denatured DNA: spectrophotometric immunologic and enzymic studies. *J. Mol. Biol.*, **4**: 31

Stratton, C. J. (1976). The investigation of vinylcyclohexane dioxide as a polar dehydrant for the improved retention of lipids in the Spurr embedment. *Tiss. Cell*, **8**: 729

Stratton, C. J., Erickson, T. B. and Wetzstein, H. Y. (1982). The lipid solubility of fixative, staining and embedding media, and the introduction of LX-112 and poly/bed-812 as dehydrants for epoxy resin embedment. *Tiss. Cell*, **14**: 13

Strauss, E. W. and Arabian, A. A. (1969). Fixation of long-chain fatty acids in segments of jejunum from golden hamster. *J. Cell Biol.*, **43**: 140a

Sturrock, R. R. (1984). Identification of mitotic cells in the central nervous system by electron microscopy of reembedded semithin sections. *J. Anat.*, **138**: 657

Subbaraman, L. R., Subbaraman, J. and Behrman, E. J. (1972). Studies on the formation and hydrolysis of osmate (VI) esters. *Inorg. Chem.*, **11**: 2621

Subbaraman, L. R., Subbaraman, J. and Behrman, E. J. (1973). The reaction of oxo-osmium(VI)-pyridine complexes with thymine glycols. *J. Org. Chem.*, **38**: 1499

Summers, R. G. and Rusanowski, P. C. (1973). A scanning electron microscopic evaluation of section and film mounting for transmission electron microscopy. *Stain Technol.*, **48**: 337

Sutton, J. S. (1968). Potassium permanganate staining of ultrathin sections for electron microscopy. *J. Ultrastruct. Res.*, **21**: 424

Svensson, L. and von Bonsdorff, C.-H. (1982). Solid-phase immune electron microscopy (SPIEM) by use of protein A and its application for characterization of selected adenovirus serotypes. *J. Med. Virol.*, **10**: 243

Swift, J. A. (1966). The electron histochemical demonstration of sulfhydryl and disulfide in electron microscopic sections, with particular reference to the presence of these chemical groups in the cell wall of the yeast *Pityrosporum ovale*. In: *Proc. 6th Int. Cong. Electron Microsc. Kyoto*, Vol. 2, p. 63

Swift, J. A. (1968). The electron histochemistry of cystine-containing proteins in the transverse sections of human hair. *J. Roy. Microsc. Soc.*, **88**: 449

Swift, J. A. (1969). The electron histochemical demonstration of cystine-containing proteins in the guinea pig hair follicle. *Histochemie*, **19**: 88

Swift, J. A. (1973). The electron cytochemical demonstration of cystine disulphide bonds using silver-methenamine reagent. *Histochemie*, **35**: 307

Swift, J. A. and Adams, B. J. (1966). Nucleic acid cytochemistry of mitochondria and chloroplasts. *J. Histochem. Cytochem.*, **14**: 744

Swinehart, P. A., Bentley, D. L. and Kardong, K. V. (1976). Scanning electron microscopic study of the effects of pressure on the luminal surface of the rabbit aorta. *Am. J. Anat.*, **145**: 137

Szczesny, T. M. (1978). Holder assembly for 'Ralph' type glass knives. *Stain Technol.*, **53**: 48

Szirmai, J. A. (1963). Quantitative approaches in the histochemistry of mucopolysaccharides. *J. Histochem. Cytochem.*, **11**: 24

Taber, L. H., Mirkovic, R. R., Adam, V., Ellis, S. S., Yow, M. D. and Melnick, J. L. (1973). Rapid diagnosis of enterovirus meningitis by immunofluorescent staining of CSF leukocytes. *Intervirology*, **1**: 127

Tahmisian, T. N. (1964). Using of the freezing point to adjust the tonicity of fixing solutions. *J. Ultrastruct. Res.*, **10**: 182

Takagi, I., Sato, T. and Yamada, K. (1979). A rapid method for embedding fractionated samples for electron microscopy. *J. Electron Microsc.*, **28**: 316

Takagi, M., Parmley, R. T., Denys, F. R. and Kageyama, M. (1983). Ultrastructural visualization of complex carbohydrates in epiphyseal cartilage with the tannic acid–metal salt methods. *J. Histochem. Cytochem.*, **31**: 783

Takagi, M., Parmley, R. T., Spicer, S. S., Denys, F. R. and Setser, M. E. (1982). Ultrastructural localization of acidic glycoconjugates with the low iron diamine method. *J. Histochem Cytochem.*, **30**: 471

Takata, K., Arii, T., Yamagishi, S. and Hirano, H. (1984). Use of colloidal gold and ruthenium red in stereohigh-voltage electron microscopic study of Con A-binding sited in mouse macrophages. *Histochemistry*, **81**: 441

Takeda, M. (1969). Virus identification in cytologic and histologic material by electron microscopy. *Acta Cytol.*, **13**: 206

Takeuchi, I. K. (1981). Differential staining of nucleoli and chromatin by sodium tungstate. *J. Electron Microsc.*, **30**: 150

Takeuchi, I. K. (1987). Electron microscopic study on the bismuth staining of nucleoli in growing mouse oocytes. *Acta Histochem. Cytochem.*, **20**: 295

Tanaka, Y., De Camilli, P. and Meldolesi, J. (1980). Membrane interactions between secretion granules and plasmalemma in three exocrine glands. *J. Cell Biol.*, **84**: 438

Tandler, B. and Walter, R. J. (1977). Epon–Maraglas embedment for electron microscopy. *Stain Technol.*, **52**: 238

Tas, J. (1977a). The alcian blue and combined alcian blue–safranin O staining of glycosaminoglycans studies in a model system and in mast cells. *Histochem. J.*, **9**: 205

Tas, J. (1977b). Polyacrylamide films as tools for investigating qualitative and quantitative aspects of the staining of glycosaminoglucans with basic dyes. *Histochem. J.*, **9**: 267

Tashima, T., Kawakami, U., Harada, M., Sakata, T., Satoh, N., Nakagawa, T. and Tanaka, H. (1987). Isolation and identification of new oligomers in aqueous solution of glutaraldehyde. *Chem. Pharm. Bull.*, **35**: 4169

Taylor, D. P. (1988). Direct measurement of the osmotic effects of buffers and fixatives in *Nitella flexilis*. *J. Microsc.*, **150**: 71

Telford, J. N. and Racker, E. (1973). A method for increased contrast of mitochondrial inner membrane spheres in thin sections of Epon–Araldite embedded tissue. *J. Cell Biol.*, **57**: 580

Temmink, J. H. M., Collard, J. G., Spits, H. and Roos, E. (1975). A comparative study of four cytochemical detection methods of concanavalin A binding sites on the cell membrane. *Exp. Cell Res.*, **92**: 307

Terracio, L., Bankston, P. W. and McAteer, J. A. (1981). Ultrastructural observations on tissues processed by a quick-freezing, rapid-drying method: comparison with conventional specimen preparation. *Cryobiology*, **18**: 55

Terzakis, J. A. (1968). Uranyl acetate, a stain and a fixative. *J. Ultrastruct. Res.*, **22**: 168

Thaete, L. G. (1979). Lead and uranium stain artifacts in electron microscopy: a technique for minimizing their occurrence. *J. Microsc.*, **115**: 195

Thiéry, G. and Rambourg, A. (1976). A new staining technique for studying thick sections in the electron microscope. *J. Microsc. Biol. Cell*, **26**: 103

Thiéry, J. P. (1967). Mise en évidence des polysaccharides sur coupes fines en microscopie électronique. *J. Microsc.*, **6**: 987

Thiéry, J. P. and Bader, J. P. (1966). ultrastructure des îlots de langerhans du pancréas humain normal et pathologique. *Ann. Endocrinol.*, **27**: 625

Thomas, R. S. (1962). Demonstration of structure-bound mineral constituents in thin-sectioned bacterial spores by ultramicro-incineration. *Proc. 5th Int. Cong. Electron Microsc., Philadelphia*, Vol. 2, RR-11. Academic Press, New York

Thompson, C. F. and Gottlieb, F. J. (1972). A technique for reconstructing surface and near surface structures of cylindrically shaped specimens. *Mikroskopie*, **28**: 125

Thompson, E. and Colvin, J. R. (1970). Electron cytochemical localization of cystine in plant cell walls. *J. Microsc.*, **91**: 87

Thorball, N. (1982). The electron microscopy of fluorescent dextrans (FITC-dextrant) in thin sections of tissues. *Experientia*, **38**: 876

Thornthwaite, J. T., Thomas, R. A., Leif, S. B., Yopp, T. A., Cameron, B. F. and Leif, R. C. (1978). The use of electronic cell volume analysis with the AMAC II to determine the optimum glutaraldehyde fixative concentration for nucleated mammalian cells. *Proc. SEM Symp.*, Vol. 2, p. 1123. SEM Inc., AMF O'Hare, Ill.

Thornton, V. F. and Howe, C. (1974). The effect of change of background color on the ultrastructure of the pars intermedia of the pituitary of the eel (*Anguilla anguilla*). *Cell Tiss. Res.*, **151**: 103

Thorpe, J. R. and Harvey D. M. R. (1979). Optimization and investigation of the use of 2,2-dimethoxypropane as a dehydration agent for plant tissues in transmission electron microscopy. *J. Ultrastruct. Res.*, **68**: 186

Thurston, E. L. (1978). Health and safety hazards in the SEM laboratory: update 1978. *Scanning Electron Microsc.*, **2**: 849

Thyberg, J., Hinek, A., Nilsson, J. and Friberg, U. (1979). Electron microscopic and cytochemical studies of rat aorta. Intracellular vesicles containing elastin-like and collagen-like material. *Histochem. J.*, **11**: 1

Timms, B. G. and Chandler, J. A. (1984). Endogeneous elements in the prostate. An X-ray microanalytical study of freeze-dried frozen sections and histochemical localization of zinc by potassium pyroantimonate. *Histochem. J.*, **16**: 733

Ting-Beall, H. P. (1980). Interactions of uranyl ions with lipid bilayer membranes. *J. Microsc.*, **118**: 221

Ting-Beall, H. P., Burgess, F. M. and Robertson, J. D. (1986). Particles and pits matched in native membranes. *J. Microsc.*, **142**: 311

Tinglu, G., Ghosh, A. and Ghosh, B. K. (1984). Subcellular localization of alkaline phosphatase in *Bacillus lichenformis* 749/C by immunoelectron microscopy with colloidal gold. *J. Bacteriol.*, **159**: 668

Tisher, C. C., Weaver, B. A. and Cirkene, W. J. (1972). X-ray microanalysis of pyroantimonate complex in rat kidney. *Am. J. Path.*, **69**: 255

Tokuyasu, K. T. (1978). A study of positive staining of ultrathin frozen sections. *J. Ultrastruct. Res.*, **63**: 287

Tokuyasu, K. T. (1980a). Immunocytochemistry on ultrathin frozen sections. *Histochem. J.*, **12**: 381

Tokuyasu, K. T. (1980b). Adsorption staining method for ultrathin frozen sections. *Proc. 38th Meet. Electron Microsc. Soc. Am.*, p. 760. Claitor's Pub. Division, Baton Rouge, La.

Tokuyasu, K. T. (1983a). Present state of immunocryoultramicrotomy. *J. Histochem. Cytochem.*, **31**: 164

Tokuyasu, K. T. (1983b). Visualization of longitudinally-oriented intermediate filaments in frozen sections of chicken cardiac muscle by a new staining method. *J. Cell Biol.*, **97**: 562

Tokuyasu, K. T. (1986). Application of cryoultramicrotomy to immunocytochemistry. *J. Microsc.*, **143**: 139

Tomsig, J. L. and Pellegrino de Iraldi, A. (1987). Effect of collidine (2,4,6-trimethylpyridine) on the osmiophilia and chromaffin reaction in the synaptic vesicles of rat pineal nerves. *Histochemistry*, **87**: 21

Tomsig, J. L. and Pellegrino de Iraldi, A. (1988). Effect of collidine (2,4,6-trimethylpyridine) on rat pineal gland and vas deferens nerves. *Histochemistry*, **89**: 301

Tooze, J. (1964). Measurement of some cellular changes during fixation of amphibian erythrocytes with osmium tetroxide solutions. *J. Cell Biol.*, **22**: 551

Torack, R. M. and LaValle, M. (1970). The specificity of the pyroantimonate technique to demonstrate sodium. *J. Histochem. Cytochem.*, **18**: 635

Tramezzani, J. H., Vidal, O. R., Cannata, M. A. and Chiocchio, S. R. (1966). The G.S.F. Technique for electron microscopy differentiation of genetic material. *Acta Physiol. Latin Am.*, **16**: 391

Tranzer, J.-P., da Prada, M. and Pletscher, A. (1972). Storage of 5-hydroxytryptamine in megakaryocytes. *J. Cell Biol.*, **52**: 191

Trelstad, R. L. (1969). The effect of pH on the stability of purified glutaraldehyde. *J. Histochem. Cytochem.*, **17**: 756

Trett, M. W. (1981). A question of static electricity? *Proc. Roy. Microsc. Soc.*, **16**: 41

Trett, M. W. and Crouch, J. A. (1984). The collection of

plastic films on specimen support grids for transmission electronmicroscopy. *Proc. Roy. Microsc. Soc.*, **19**: 250

Trigaux, G. A. (1960). New epoxides for plastics and coatings. *Mod. Plastics*, **38**: 147

Trinick, J., Cooper, J., Seymour, J. and Egelman, E. H. (1986). Cryoelectron electron microscopy and three-dimensional reconstruction of actin filaments. *J. Microsc.*, **141**: 349

Trump, B. F., Goldblatt, P. J. and Stowell, R. E. (1962). An electron microscope study of early cytoplasmic alterations in hepatic parenchymal cells of mouse liver during necrosis *in vitro*. *Lab. Invest.*, **11**: 986

Tsuchiya, A. and Ogawa, K. (1973). Ultracytochemistry of the periodic acid (PA)-phosphotungstic acid (PTA) reaction. *J. Electron Microsc.*, **22**: 290

Tuohy, M., McConchie, C., Knox, R. B., Szarski, L. and Arkin, A. (1987). Computer-assisted three-dimensional reconstruction technology in plant cell image analysis: applications of interactive computer graphics. *J. Microsc.*, **147**: 83

Tzaphlidou, M., Chapman, J. A. and Al-Samman, M. H. (1982b). A study of positive staining for electron microscopy using collagen as a model system. II. Staining by uranyl ions. *Micron*, **13**: 133

Tzaphlidou, M., Chapman, J. A. and Al-Samman, M. H. (1982b). A study of positive staining for electron microscopy using collagen as a model system. II. Staining by uranyl ions. *Micron*, **13**: 133

Ulmer, D. D. and Vallee, B. L. (1969). Trace substances in environmental health. II. In: *Proc. Univ. Missouri Ann. Conf. Trace Substances Health*, 2nd edn. D. D. Hemphill, 7, Columbia, University of Missouri

Umar, H. (1982). A new apparatus for coating grids with a thin layer of pioloform. *Mikroskopie*, **39**: 233

Valentine, R. C. (1958). Quantitative electron staining of virus particles. *J. Roy. Microsc. Soc.*, **78**: 26

Valentine, R. C., Shapiro, B. M. and Stadtman, E. R. (1968). Regulation of glutamine synthetase. XII. Electron microscopy of the enzyme from *Escherichia coli*. *Biochemistry*, **7**: 2143

Vallee, B. L. and Ulmer, D. D. (1972). Biochemical effects of mercury, cadmium and lead. *Ann. Rev. Biochem.*, **41**: 91

van Bergen en Henegouwen, P. and Leunissen, J. (1984). An improved method for producing gold particles of particular sizes. *Ultramicroscopy*, **14**: 407

van Bergen en Henegouwen, P. (1989). Methods for labeling cryosections with colloidal gold. In: *Colloidal Gold: Principles, Methods, and Applications*, Vol. 1 (M. A. Hayat, Ed.). Academic Press, San Diego and London

van Bergen en Henegouwen, P. M. P. and Leunissen, J. L. M. (1986). Controlled growth of colloidal gold particles and implications for labelling efficiency. *Histochemistry*, **85**: 81

van Bergen en Henegouwen, P. (1988). Methods for labeling cryosections with colloidal gold. In: *Colloidal Gold: Principles, Methods, and Applications*, Vol. 1 (M. A. Hayat, Ed.). Academic Press, San Diego and London

Van Bruggen, E. F. J., Wiebenger, E. H. and Gruber, M. (1960). Negative staining electron microscopy of proteins at pH values below their isoelectric points: its application to hemocyanin. *Biochim. Biophys. Acta*, **42**: 171

van den Pol, A. N. (1985). Silver intensified gold and peroxidase as simultaneous ultrastructural markers for pre- and postsynaptic neurotransmitters. *Science, N.Y.*, **228**: 332

van den Pol, A. N. (1986). Tyrosine hydroxylase immunoreactive neurons throughout the hypothalamus receive glutamate decarboxylase immunoreactive synpases: a double preembedding immunocytochemical study with particulate silver and HRP. *J. Neurosci.*, **6**: 877

van der Voort, H. TM., Valkenburg, J. A. C., van Spronsen, E. A., Woldringh, C. L. and Brakenhoff, G. J. (1987). Confocal microscopy in comparison with electron and conventional light microscopy. In: *Correlative Microscopy in Biology: Instrumentations and Methods* (M. A. Hayat, Ed.), pp. 59–81. Academic Press, Orlando and London

van Duijn, P. (1961). Acrolein–Schiff, a new staining method for proteins. *J. Histochem. Cytochem.*, **9**: 234

van Emburg, P. R. and De Bruijn, W. C. (1984). Enhanced cellular membrane contrast in a marine alga by osmium-azole complexes. *Proitoplasma*, **119**: 48

Van Harreveld, A. and Crowell, J. (1964). Electron microscopy after rapid freezing on a metal surface and substitution fixation. *Anat. Rec.*, **149**: 381

Van Harreveld, A. and Fifkova, E. (1972). Release of glutamate from the retina during glutaraldehyde fixation. *J. Neurochem.*, **19**: 237

Van Harreveld, A. and Khattab, F. I. (1968). Perfusion fixation with glutaraldehyde and postfixation with osmium tetroxide for electron microscopy. *J. Cell Sci.*, **3**: 579

Van Harreveld, A. and Khattab, F. I. (1969). Changes in extracellular space of the mouse cerebral cortex during hydroxyadipaldehyde fixation and osmium tetroxide post-fixation. *J. Cell Sci.*, **4**: 437

Van Iren, F., Ven Essen-Joolen, L., Van der Diun-Schouten, P., Van der Sluijs, P. B. and De Bruijn, W. C. (1979). Sodium and calcium localization in cells and tissues by precipitation with antimonate: a quantitative study. *Histochemistry*, **63**: 273

Van Iterson, W. (1984). *Inner Structures of Bacteria*. Van Nostrand Reinhold, New York

Van Itterbeek, A., de Greve, L., van Veelen, G. F. and Tuynman, C. A. F. (1952). Glass layers as supports for electron microscopy. *Nature, Lond.*, **170**: 795

Van Reempts, J. L., Borgers, M., De Nollin, S. R., Garrevoet, T. C. and Jacob, W. A. (1984). Identification of calcium in the retina by the combined use of ultrastructural cytochemistry and laser microprobe mass analysis. *J. Histochem. Cytochem.*, **32**: 788

Van Stevenick, J. and Booij, H. L. (1964). The role of polyphosphates in the transport mechanism of glucose in yeast cells. *J. Gen. Physiol.*, **48**: 43

van Winkle, J. L. (1962). Cited in *Acrolein* (C. W. Smith, Ed.). John Wiley, New York

Varndell, I. M. and Polak, J. M. (1984). Double immuno-staining procedures: techniques and applications. In: *Immunolabelling for Electron Microscopy* (J. M. Polak and I. M. Varndell, Eds.), pp. 155–177. Elsevier, Amsterdam

Vassallo, G., Capello, C. and Solica, E. (1971). Grimelius' silver stain for endocrine cell granules, as shown by electron microscopy. *Stain Technol.*, **46**: 7

Venable, J. H. and Coggeshall, R. (1965). A simplified lead citrate stain for use in electron microscopy. *J. Cell Biol.*, **25**: 407

Verwey, E. F. (1940). A type-specific protein derived from the *Staphylococcus*. *J. Exp. Med.*, **71**: 635

Verwey, E. J. W. and Overbeek, J. T. C. (1948). *Theory of the Stability of Lyophobic Colloids*. Elsevier, Amsterdam

Vesely, D. and Woodisse, S. (1982). Carbon coating with a carbon fibre filament. *Proc. Roy. Microsc. Soc.*, **17**: 137

Vidíc, B. (1973). Uptake of marker particles by *in vitro* ventilated and perfused rat lung. *Am. J. Anat.*, **138**: 521

Vio-Cigna, M., Pebusque, M. J. and Seite, R. (1982). Improvements in selective silver staining of nucleolar organizer regions in block tissues at the ultrastructural level. *Biol. Cell*, **44**: 329

Völker, W., Frick, B. and Robenek, H. (1985). A simple device for low temperature polymerization of Lowicryl K4M resin. *J. Microsc.*, **138**: 91

Völker, W., Meschede, H., Weil, J. and Thurm, U. (1984). Immersion-freezing and freeze-substitution of irregularly shaped specimens with the aid of a cryo-injector. *Mikroskopie*, **41**: 351

Vollenweider, H. J., Koller, T. and Kübler, O. (1973). Aluminium–beryllium alloy films as specimen supports for high resolution electron microscopy. *J. Microsc.*, **16**: 247

Vollenweider, H. J., Sogo, J. M. and Koller, T. (1975). A routine method for protein-free spreading of double- and single-stranded nucleic acid molecules. *Proc. Natl Acad. Sci. USA*, **72**: 83

von Hippel, P. H. and Wong, K. Y. (1971). Dynamic aspects of native DNA structure: kinetics of the formaldehyde reaction with calf thymus DNA. *J. Mol. Biol.*, **61**: 587

Von Zglinicki, T., Rimmler, M. and Purz, H. J. (1986). Fast cryofixation technique for X-ray microanalysis. *J. Microsc.*, **141**: 79

Vrensen, G. and De Groot, D. (1974a). Phosphotungstic acid staining and the quantitative stereology of synapses. In: *Electron Microscopy and Cytochemistry* (E. Wisse, W. Th. Daems, I. Molenaar and P. van Duijn, Eds.). American Elseier, New York

Vrensen, G. and De Groot, D. (1974b). Osmium–zinc iodide staining and the quantitative study of central synapses. *Brain Res.*, **74**: 131

Vuillet, J. (1987). Correlated light and electron microscopic method for the visualization of the same in vitro cell using radioautography and serial sectioning. In: *Correlative Microscopy in Biology: Instrumentations and Methods* (M. A. Hayat, Ed.), pp. 165–172. Academic Press, Orlando and London

Vye, M. V. (1971). A comparative evaluation of four methods for staining of glycogen in thin sections. *Lab. Invest.*, **24**: 452

Vye, M. V. and Fischman, D. A. (1970). The morphological alteration of particulate glycogen by *en bloc* staining with uranyl acetate. *J. Ultrastruct. Res.*, **33**: 278

Wagenknecht, T., Grassucci, R. and Frank, J. (1988). Electron microscopy and computer image-averaging of ice-embedded large ribosomal subunits from *Escherichia coli*. *J. Mol. Biol.*, **199**: 137

Wagner, R. C. (1976). Tannic acid as a mordant for heavy metal stains. *Proc. 34th Ann. EMSA Meet.*, p. 316. Claitor's Pub. Division, Baton Rouge, La.

Wagner, R. C. and Andrews, S. B. (1985). Ultrastructure of the vesicular system in rapidly frozen capillary endothelium of the rete mirabile. *J. Ultrastruct. Res.*, **90**: 172

Walford, F. J. (1969). Transient heat transfer from a hot nickel sphere moving through water. *Int. J. Heat Mass Transfer*, **12**: 1621

Walker, J. F. (1964). *Formaldehyde*, 3rd edn. Reinhold, New York

Wall, J. S., Hainfeld, J. F., Bartlett, P. A. and Singer, S. J. (1982). Observation of an undecagold cluster compound in the scanning electron microscope. *Ultramicroscopy*, **8**: 397

Wallstrom, A. C. and Iseri, O. A. (1972). Ultrasonic cleaning of diamond knives. *J. Ultrastruct. Res.*, **41**: 561

Walter, F. (1961). Ultramikrotomie. I. Das Ultramikrotom nach Fernández-Morán. *Leitz. Mitt. Wiss. Tech.*, **1**: 236

Walton, J. (1979). Lead aspartate, an *en bloc* contrast stain particularly useful for ultrastructural enzymology. *J. Histochem. Cytochem.*, **27**: 1337

Walz, B. (1982). Ca^{2+}-sequestering smooth endoplasmic reticulum in an invertebrate photoreceptor. I. Intracellular topography as revealed by OsFeCN staining and in situ Ca accumulation. *J. Cell Biol.*, **93**: 839

Warchol, J. B., Brelinska, R. and Herbert, D. C. (1982). Analysis of colloidal gold methods for labeling proteins. *Histochemistry*, 76: 567

Ward, R. T. (1958). Prevention of polymerization damage in methacrylate embedding media. *J. Histochem. Cytochem.*, **6**: 398

Ward, R. T. (1972). A section lifter designed for attachment to an ultramicrotome. *Stain Technol.*, **47**: 257

Ward, R. T. (1977). Some observations on glass-knife making. *Stain Technol.*, **52**: 305

Ward, R. T. and Murray, J. M. (1987). Natural propane cryogen for frozen-hydrated biological specimens. *J. Electron Microsc. Tech.*, **5**: 275

Wassef, M. (1979). A cytochemical study of interchromatin granules. *J. Ultrastruct. Res.*, **69**: 121

Watson, M. L. (1955). The use of carbon films to support tissue sections for electron microscopy. *J. Biophys. Biochem. Cytol.*, **1**: 183

Watson, M. L. and Aldridge, W. G. (1961). Methods for use of indium as an electron stain for nucleic acids. *J. Biophys. Biochem. Cytol.*, **11**: 257

Weakley, B. S. (1979). A variant of the pyroantimonate technique suitable for localization of calcium in ovarian tissue. *J. Histochem. Cytochem.*, **27**: 1017

Weakley, B. S. (1981). *A Beginner's Handbook in Biological Transmission Electron Microscopy*. Churchill Livingston, London

Webster, B. R. and del Cerro, M. (1981). A simple and sensitive vibration monitor for micrography and general laboratory use. *Microsc. Acta*, **84**: 179

Weed, H. G., Krochmalnic, G. and Penman, S. (1985). Poliovirus metabolism and the cytoskeletal framework: detergent extraction and resinless section electron microscopy. *J. Virol.*, **56**: 549

Wegner, K. W. (1971). Easy and accurate collection of thin

serial sections by means of a grid support. *Mikroskopie*, **27**: 289

Weibel, E. R. (1979). *Stereological Methods: Practical Methods for Biological Morphometry*. Academic Press, London

Weibel, E. R. and Bolender, R. P. (1973). Stereological techniques for electron microscopic morphometry. In: *Principles and Techniques of Electron Microscopy: Biological Applications*, Vol. 3. Van Nostrand Reinhold, New York

Weibel, E. R. and Knight, B. W. (1964). A morphometric study on the thickness of the pulmonary air–blood barrier. *J. Cell Biol.*, **21**: 367

Weibull, C. (1986). Temperature rise in Lowicryl resins during polymerization by ultraviolet light. *J. Ultrastruct. Mol. Struct. Res.*, **97**: 207

Weibull, C., Christiansson, A. and Carlemalm, E. (1983). Extraction of membrane lipids during fixation, dehydration and embedding of *Acholeplasma laidlawii*-cells for electron microscopy. *J. Microsc.*, **129**: 201

Weibull, C., Villiger, W. and Carlemalm, E. (1984). Extraction of lipids during freeze-substitution of *Acholeplasma laidlawii*-cells for electron microscopy. *J. Microsc.*, **134**: 213

Weinshelbaum, E. I. and Pittman, J. M. (1972). Application of silver stains to gastric endocrine cells in plastic sections. *Histochemie*, **29**: 134

Wells, B. (1974). A convenient technique for the collection of ultrathin serial sections. *Micron*, **5**: 79

Wells, B. and Horne, R. W. (1983). The ultrastructure of *Pseudomonas avenae*. I. Paracrystalline surface layer and extracellular material. *Micron*, **14**: 11

Wells, B., Horne, R. W. and Shaw, P. J. (1981). The formation of two-dimensional arrays of isometric plant viruses in the presence of polyethylene glycol. *Micron*, **12**: 37

Werner, G., Morgenstern, E. and Neumann, K. (1974). Improved contrast in dry ultrathin frozen sections. *Histochem. J.*, **6**: 111

Westfall, J. A. (1961). Obtaining flat serial sections for electron microscopy. *Stain Technol.*, **36**: 36

Westfall, J. A. and Healy, D. L. (1962). A water control device for mounting serial ultrathin sections. *Stain Technol.*, **37**: 118

Wetmur, J. G. and Davidson, N. (1966). Properties of DNA of bacteriophage N1, a DNA with reversible circularity. *Biochem. Biophys. Res. Commun.*, **25**: 684

Wetzel, M. G., Wetzel, B. K. and Spicer, S. S. (1966). Ultrastructural localization of acid mucosubstances in the mouse colon with iron-containing stains. *J. Cell Biol.*, **30**: 299

Weybull, C. (1970). Estimation of the thickness of thin sections prepared for electron microscopy. Philips Analytical Equipment, EM45

Whitby, H. J. and Rodgers, F. G. (1980). Detection of virus particles by electron microscopy with polyacrylamide hydrogel. *J. Clin. Path.*, **33**: 484

White, D. L., Andrews, S. B., Faller, J. W. and Barrnett, R. J. (1976). The chemical nature of osmium tetroxide fixation and staining of membranes by X-ray photoelectron spectroscopy. *Biochim. Biophys. Acta*, **436**: 577

White, D. L., Mazurkiewicz, J. E. and Barrnett, R. J. (1979). A chemical mechanism for tissue staining by osmium tetroxide–ferrocyanide mixtures. *J. Histochem Cytochem.*, **27**: 1084

White, G. K. and Woods, S. B. (1955). Thermal and electrical conductivities of solids at low temperatures. *Can. J. Phys.*, **33**: 58

Whiteman, P. (1973a). The quantitative measurement of alcian blue–glycosaminoglycan complexes. *Biochem. J.*, **131**: 343

Whiteman, P. (1973b). The quantitative determination of glycosaminoglycans in urine with alcian blue 8GX. *Biochem. J.*, **131**: 351

Whiting, R. F. and Ottensmeyer, F. P. (1972). Heavy atoms in model compounds and nucleic acids imaged by dark field transmission electron microscopy. *J. Mol. Biol.*, **67**: 173

Wick, S. M. and Hepler, P. K. (1982). Selective localization of intracellular Ca^{2+} with potassium antimonate. *J. Histochem. Cytochem.*, **30**: 1190

Wigglesworth, V. B. (1964). The union of protein and nucleic acid in the living cell and its demonstration by osmium staining. *Quart. J. Microsc. Sci.*, **105**: 113

Wilander, E. and Lundqvist, M. (1987). Sequential immunocytochemical and silver staining of neuroendocrine cells in the same section. In: *Correlative Microscopy in Biology: Instrumentation and Methods* (M. A. Hayat, Ed.), pp. 157–163. Academic Press, Orlando and London

Wild, P., Bertoni, G., Schraner, E. M. and Beglinger, R. (1987). Influence of calcium and magnesium containing fixatives of the ultrastructure of parathyroids. *Micron Microsc. Acta*, **18**: 259

Williams, M. A., Kleinschmidt, J. A., Krohne, G. and Franke, W. W. (1982). Argyrophilic nuclear and nucleolar proteins of *Xenopus laevis* oocytes identified by gel electrophoresis. *Exp. Cell Res.*, **137**: 341

Williams, M. A. and Meek, G. A. (1966). Studies on thickness variation in ultrathin sections for electron microscopy. *J. Roy. Microsc. Soc.*, **85**: 337

Williams, R. C. (1954). *Biological Applications of Freezing and Drying to Electron Microscopy* (R. J. C. Harris, Ed.), pp. 303–328. Academic Press, New York

Williams, R. C. and Kallman, F. (1955). Interpretation of electron micrographs of single and serial sections. *J. Biophys. Biochem. Cytol.*, **1**: 301

Williams, R. C. and Glaeser, R. M. (1972). Ultrathin carbon support films for electron microscopy. *Science, N.Y.*, **175**: 1000

Williamson, F. A. (1984). A rapid-access system for the storage of small samples under liquid nitrogen. *J. Microsc.*, **134**: 125

Willison, J. H. M. and Brown, R. M. (1979). Pretreatment artifacts in plant cells. In: *Freeze-Fracture: Methods, Artifacts and Interpretation* (J. E. Rash and C. S. Hudson, Eds.), p. 51. Raven Press, New York

Willison, J. H. M. and Rowe, A. J. (1980). *Replica, Shadowing and Freeze-etching Techniques. Practical Methods in Electron Microscopy*, Vol. 8 (A. M. Galuert, Ed.). North-Holland, Amsterdam

Winborn, W. B. (1964). Epoxy embedments for electron microscopy. *Anat. Rec.*, **148**: 422

Winborn, W. B. and Seelig, L. L. (1970). Paraformaldehyde and *s*-collidine—a fixative for preserving large blocks of tissue for electron microscopy. *Tex. Rep. Biol. Med.*, **28**: 347

Winborn, W. B. and Seelig, L. L. (1974). Pattern of osmium deposition in the parietal cells of the stomach. *J. Cell Biol.*, **63**: 99

Wolf, B. (1987). New cryoultramicrotome with a feedback–advance system. *J. Electron Microsc. Tech.*, **7**: 185

Wolfe, S. L., Beer, M. and Zobel, C. R. (1962). The selective staining of nucleic acids in a model system and in tissue. *Proc. 5th Int. Cong. Electron Microsc. Philadelphia*, **2**: 0–6. Academic Press, New York

Wolman, M. (1955). Problems of fixation in cytology, histology and histochemistry. *Int. Rev. Cytol.*, **4**: 79

Wolman, M. (1957). Histochemical study of changes occurring during the degeneration of myelin. *J. Neurochem.*, **1**: 370

Wolosewick, J. J. (1980). The application of polyethylene glycol (PEG) to electron microscopy. *J. Cell Biol.*, **86**: 675

Wood, J. G. (1973). The effects of glutaraldehyde and osmium on the proteins and lipids of myelin and mitochondria. *Biochim. Biophys. Acta*, **329**: 118

Wood, R. L. and Luft, J. H. (1965). The influence of the buffer system on fixation with osmium tetroxide. *J. Ultrastruct. Res.*, **12**: 22

Wouters, C. H. and Ploem, J. S. (1987). Light and scanning electron microscopy in a combined instrument. In: *Correlative Microscopy in Biology: Instrumentations and Methods* (M. A. Hayat, Ed.), pp. 23–57. Academic Press, Orlando and London

Wrigglesworth, J. M. and Packer, L. (1969). pH-dependent conformational change in submitochondrial particles. *Arch. Biochem. Biophys.*, **133**: 194

Wright, M. J., Schröder, M. and Nielson, A. J. (1981). Tissue fixation and staining by osmium tetroxide: a possible role for alkaloids. *J. Histochem. Cytochem.*, **29**: 1347

Wroblewski, R. and Wroblewski, J. (1984). Freeze-drying and freeze-substitution combined with low temperature-embedding: preparation techniques for microprobe analysis of biological soft tissues. *Histochem. J.*, **81**: 464

Wyatt, J. H. (1970). Coating of electron microscope grids. *J. Electron Microsc.*, **19**: 283

Wyatt, J. H. (1971). Hydrophobic coating of forceps used for handling electron microscope grids. *Stain Technol.*, **46**: 213

Wyatt, J. H. (1972). An ultramicrotome knife trough for glass knives. *J. Electron Microsc.*, **21**: 89

Yabe, Y. and Sadakane, H. (1975). The virus of epidermodysplasia verruciformis: electron microscopic and fluorescent antibody studies. *J. Invest. Dermatol.*, **65**: 324

Yamada, E. and Ishikawa, H. (1972). High voltage electron microscopy for thick sections of biological materials combined with molecular tracers. *Proc. 30th Ann. Meet. Electron Microsc. Soc. Am.*, p. 480. Claitor's Pub. Division, Baton Rouge, La.

Yamamoto, I. and Maruyama, H. (1985). Usefulness of a sapphire knife in ultrathin sectioning of biological specimens. *J. Electron Microsc.*, **34**: 442

Yamamoto, N., Yamashita, S. and Yasuda, K. (1980). New embedding method for immunohistochemical studies using acrylamide gel. *Acta Histochem. Cytochem.*, **13**: 601

Yamane, T. and Davidson, N. (1962). On the complexing of deoxyribonucleic acid by silver (I). *Biochim. Biohys. Acta*, **55**: 609

Yang, G. C. H. and Shea, S. M. (1975). The precise measurement of the thickness of ultrathin sections by a resectioned section technique. *J. Microsc.*, **103**: 385

Yannas, I. (1968). Vitrification temperature of water. *Science, N.Y.*, **160**: 298

Yanoff, M. (1973). Formaldehyde-glutaraldehyde fixation. *Am. J. Opthalmol.*, **76**: 303

Yarom, R. and Meiri, U. (1973). Pyroantimonate precipitates in frog skeletal muscle. Changes produced by alterations in composition of bathing fluid. *J. Histochem. Cytochem.*, **21**: 146

Yarom, R., Peters, P. D. and Hall, T. A. (1974). Effect of glutaraldehyde and urea embedding on intracellular ionic elements: X-ray microanalysis of skeletal muscle and myocardium. *J. Ultrastruct. Res.*, **49**: 405

Yau, W. L., Or, S. B. and Ngai, H. K. (1985). A 'free-floating' technique for reprocessing paraffin sections for electron microscopy. *Med. Lab. Sci.*, **42**: 26

Yensen, J. (1968). Removal of epoxy resin from histological sections following halogenation. *Stain Technol.*, **43**: 344

Yeung, E. C. and Law, S. K. (1987). Serial sectioning techniques for a modified LKB historesin. *Stain Technol.*, **62**: 147

Yoshiyama, J. M., Goff, D. and Walton, J. (1980). Variations in sample preparation that affect contrast enhancement by lead aspartate. *Proc. 38th Ann. Meet. EMSA*, p. 654

Young, E. G. (1963). *Comprehensive Biochemistry*, Vol. 7, p. 25. Elsevier, New York

Yun, J. and Kenney, R. A. (1976). Preparation of cat kidney tissue for ultrastructural studies. *J. Electron Microsc.*, **25**: 11

Zacks, S. I. (1963). Mechanical stirrer for epoxy resin embedding media. *Stain Technol.*, **38**: 60

Zacks, S. I., Sheff, M. F. and Saito, A. (1973). Structure and staining characteristics of myofiber external lamina. *J. Histochem. Cytochem.*, **21**: 703

Zagon, I. S. (1970). *Carchesium polypinum*: cytostructure after protargol silver deposition. *Trans. Am. Microsc. Soc.*, **89**: 450

Zalokar, M. (1966). A simple freeze-substitution method for electron microscopy. *J. Ultrastruct. Res.*, **15**: 469

Zasadzinski, J. A. N. (1988). A new heat transfer model to predict cooling rates for rapid freezing fixation. *J. Microsc.*, **150**: 137

Zeitler, E. and Bahr, G. F. (1962). A photometric procedure for weight determination of submicroscopic particles quantitative electron microscopy. *J. Appl. Phys.*, **33**: 847

Zelechowska, M. G. and Potworowski, E. F. (1985). Improved adhesion of ultrathin sections to filmless grids. *J. Electron Microsc. Tech.*, **2**: 389

Zierold, K. (1986). Preparation of cryosections for biological microanalysis. In: *The Science of Biological Specimen Preparation for Microscopy and Microanalysis* (M. Müller, R. P. Becker, A. Boyd and J. J. Wolosewick, Eds.), pp. 119–127. SEM Inc., AMF O'Hare, Ill.

Zierold, K. (1987). Cryoultramicrotomy. In: *Cryotechniques in Biological Electron Microscopy* (R. A. Steinbrecht and K. Zierold, Eds.), pp. 132–148. Springer-Verlag, New York

Zimmerman, L. E., Font, R. L. and Ts'o, M. O. M. (1972). Application of electron microscopy to histopathologic diagnosis. *Trans. Am. Acad. Opthalmol. Otolaryngol.*, **76**: 101

Zobel, C. R. and Beer, M. (1961). Electron stains. I. Chemical studies on the interaction of DNA with uranyl salts. *J. Biophys. Biochem. Cytol.*, **10**: 335

Zobel, C. R. and Beer, M. (1965). The use of heavy metal salts as electron stains. *Int. Rev. Cytol.*, **18**: 363

Zsigmondy, R. (1905). *Zur Erkenntnis der Kolloide.* Jena

Subject Index